Digital Devices and Systems
WITH PLD APPLICATIONS

Digital Devices and Systems

WITH PLD APPLICATIONS

Michael Miller
DeVry Institute of Technology

Delmar Publishers

an International Thomson Publishing company I(T)P®

Albany • Bonn • Boston • Cincinnati • Detroit • London • Madrid
Melbourne • Mexico City • New York • Pacific Grove • Paris • San Francisco
Singapore • Tokyo • Toronto • Washington

NOTICE TO THE READER

Publisher does not warrant or guarantee any of the products described herein or perform any independent analysis in connection with any of the product information contained herein. Publisher does not assume, and expressly disclaims, any obligation to obtain and include information other than that provided to it by the manufacturer.

The reader is expressly warned to consider and adopt all safety precautions that might be indicated by the activities herein and to avoid all potential hazards. By following the instructions contained herein, the reader willingly assumes all risks in connection with such instructions.

The publisher makes no representation or warranties of any kind, including but not limited to, the warranties of fitness for particular purpose or merchantability, nor are any such representations implied with respect to the material set forth herein, and the publisher takes no responsibility with respect to such material. The publisher shall not be liable for any special, consequential, or exemplary damages resulting, in whole or part, from the readers' use of, or reliance upon, this material.

Production, Prepress, Printing and Binding by West International Publishing Company.

The ITP logo is a trademark under license.

Printed in the United States of America

PRODUCTION CREDITS
Production Management: Tobi Giannone, Michael Bass & Associates
Copyediting: Linda Purrington
Text Design: Joy Dickinson
Composition: BookMasters, Inc.
Cover Images: Courtesy of Tektronix, Inc.

British Library Cataloguing-in-Publication Data. A catalogue record for this book is available from the British Library.

Library of Congress Cataloging-in-Publication Data

Miller, Michael A.

Digital devices and systems with PLD applications / Michael A. Miller
p. cm.
Includes index.
ISBN 0-314-20151-3 (soft : alk. paper)
1. Digital electronics. I. Title.
TK7868.D5M54 1997
621.39'5—dc20 96-22291
CIP

For more information, contact:

Delmar Publishers
3 Columbus Circle, Box 15015
Albany, New York 12212-5015

International Thomson Publishing Europe
Berkshire House 168-173
High Holborn
London, WC1V 7AA
England

Thomas Nelson Australia
102 Dodds Street
South Melbourne, 3205
Victoria, Australia

Nelson Canada
1120 Birchmont Road
Scarborough, Ontario
Canada, M1K 5G4

International Thomson Editores
Campos Eliseos 385, Piso 7
Col Polanco
11560 Mexico D F Mexico

International Thomson Publishing GmbH
Konigswintarer Strasse 418
53227 Bonn
Germany

International Thomson Publishing Asia
221 Henderson Road
#05-10 Henderson Building
Singapore 0315

International Thomson Publishing Japan
Hirakawacho Kyowa Building, 3F
2-2-1 Hirakawacho
Chiyoda-ku, Tokyo 102
Japan

1 2 3 4 5 6 7 8 9 10 XXX 03 02 01 00 99 98 97

This text is dedicated to Judy for a lifetime of friendship and love.

Contents

Acknowledgment

I would like to acknowledge the fine efforts of those who took time to review the manuscript for this text and who have presented many worthwhile and useful suggestions, criticisms, and information. Thank you all.

Additionally, I would like to show my appreciation to the people at West Educational Publishing who gave me the opportunity and guidance required to complete this work. Specifically, Chris Conty and Liz Reidel who were the editorial team through most of the project's term; Michael Bass and Tobi Giannone who were responsible for production management and editorial services.

A. F. Adkins
Amarillo College (TX)

John Blankenship
DeVry-Atlanta (GA)

David Delker
Kansas State University (KS)

Thomas R. Durgin
Lamar University Institute of Technology (TX)

John A. Hamilton
Spartan School of Aeronautics (OR)

Ronald L. Harris
Weber State University (UT)

Stephen Harsany
Mt. San Antonio College (CA)

D. Bruce Hutchinson
Mohawk College (Ontario)

Paul D. Johnson
Grand Valley State University (MI)

Edward Kaufenberg
Blackhawk Technical College (WI)

Samuel I. Kraemar
Oklahoma State University (OK)

P. Erik Liimatta
Anne Arundel Community College (MD)

Robert E. Martin
Northern Virginia Community College (VA)

Alan Moltz
Naugatuck Valley Community Technical College (CT)

Karl Perusich
Purdue University at South Bend (IN)

Charles W. Pirtle
DeVry Institute of Technology (MO)

Dennis W. Suchecki
San Diego Mesa College (CA)

Charles Swain
Rochester Institute of Technology (NY)

Vic Quiros
DeVry Institute of Technology (MO)

Preface

Any text on introductory digital electronics must contain a lot of theory to lay the groundwork for the student who is new to digital circuitry, for understanding both basic and complex concepts. Topics such as logic functions and Boolean algebra must be presented. This text seeks to present this material in a friendly and thorough manner without causing malaise or becoming deeply mired in these topics.

Each chapter includes an opening list of key terms used throughout the chapter, a set of objectives that are met by discussion of the topics in the chapter, an overview of the fundamental concepts in the chapter, and where appropriate, a separate section on troubleshooting and design applications based on the concepts presented in the chapter. Practice problems are found at the end of each section of each chapter, with solutions at the end of the chapter. The topics covered are summarized, and a set of problems are presented at the close of each chapter. Solutions to odd-numbered problems are placed at the back of the text. The even-numbered problems are similar to the odd set, so that the student can practice solving problems related to a chapter's material and then be assigned even-numbered problems for homework.

The prerequisites for this text are minimal. A basic idea of voltage, current, resistance, and simple DC circuit operation is helpful. The student should know what multimeters and oscilloscope are and how to use them, in order to understand their application in later chapters. Most chapters in the text do not require that anything else be known ahead of time. However, a basic understanding of transistors and FETs is needed to fully understand the information in Chapter 5 concerning solid state memories, and in Chapter 9, which deals with logic families and device specifications. Knowledge of the operation of an op amp as a negative feedback amplifier and comparator is required for Chapter 7 on digital-to-analog and analog-to-digital conversions.

Chapter 1

This introductory chapter presents the terminology that provides the framework for studying digital electronics. The basic building blocks of logic circuits, the buffer, AND, OR, and inverter logic functions are discussed. Development of logic expressions and truth tables are included in this chapter.

Chapter 2

The second chapter covers the use of Boolean algebra, DeMorgan's theorem, and Karnaugh mapping in the development of logic expressions.

Chapter 3

Number systems—specifically, the decimal, binary, and hexadecimal systems—are presented in Chapter 3. Conversions from one system to another are explored, along with basic arithmetic operations. Logic circuits for performing addition and subtraction of binary numbers are developed and analyzed.

Chapter 4

NAND and NOR gates are used to build the next level of logic building block, the flip-flop. After the basic flip-flop types are shown and discussed, they are combined into counting circuits. State tables and diagrams are added as tools in analyzing synchronous circuits.

Chapter 5

Chapter 5 deals with logic circuits and devices that are used for storing binary data. Parallel-load and shift registers are explained. Decoders, encoders, multiplexers, and demultiplexers are included in this chapter and used in many of the register and memory circuits discussed in the chapter.

Different types of memory devices are presented that require basic knowledge of transistors and FETs by the student. Also helpful is an understanding of capacitor charge and discharge functions.

Chapter 6

This chapter is devoted completely to programmable logic devices (PLDs). It is my intent to present a concise picture of designing logic circuits using PLDs. To this end, the chapter has a detailed discussion of the use of CUPL software as a basis for creating and compiling PLD designs. PLDs are such an integral part of the digital field that it is important for today's technologist and technician to have a good grasp on how they are used and function. A student version of the CUPL compiler, called *CUPL Starter Kit,* is included with the text to allow students to create logic circuits using PLDs. In addition, files containing the PLD templates are included in the software on the disk included with the text.

Chapter 7

Conversion between analog and digital electronic circuits is the topic of Chapter 7. Basic types of resistor ladder digital-to-analog converters are analyzed and compared. Discussion of counter and noncounter-type analog-to-digital converters follow the digital-to-analog discussion. The benefits, specifications, and limitations of all circuits are shown.

Chapter 8

This chapter presents a generic computing system that I termed the BABE computer. This presentation brings many of the previously presented concepts into a system environment leading toward microprocessor-based computers made today. Specifically, the hardware design and instruction format is directed toward the Intel 80x86 line of microprocessors to lay a good introductory foundation for those microprocessors and computer systems that use them.

Chapter 9

This last chapter presents material on the specifications, advantages, and disadvantages of various logic families of devices. These primarily include TTL, ECL, NMOS, and CMOS devices. Data sheets detailing these devices are detailed to show how information about these devices is found and used.

Appendix A

Data sheets for all devices used in the text are provided in Appendix A.

Appendix B

This appendix is a reference for the CUPL Starter software package.

Appendix C

Answers to Odd–Numbered Problems.

Software Package

Disks are provided with the text which contain the CUPL Starter Compiler and copies of the PLD templates used throughout the text. There are DOS or Windows versions of the CUPL compiler included on the CUPL program disk.

Information about the 80x86 shareware assembler and debugger, A86/D86, including arrangements for school licensing, can be obtained from the program's author at

Eric Isaacson Software
416 East University Ave.
Bloomington, IN 47401-4739

His phone number is (812) 339-1811; Compuserve at GO ZIPKEY to any ! prompt. His Internet address is

71333.3154@compuserve.com

Digital Devices and Systems
WITH PLD APPLICATIONS

one

Digital and Binary Electronics and Computers

KEY TERMS

active
amplitude
analog
and
assert
base or radix
binary
buffer
central processing unit (CPU)
combinational logic
comparator
complement
computer
digital
digital logic
dynamic logic
exclusive OR or XOR
frequency
gate or logic gate
inactive
inverter or NOT
integrated circuit
logic 1 and logic 0
logic gate
logic probe
maxterms
microprocessor
minterms
NAND
negate
negative logic
NOR
NOT bar
operator
OR
parity generator
positive logic
product of sums (POS)
pulse
oscilloscope
signal flow
solid state
state
static combinational logic
sum of products (SOP)
truth table
weight

OBJECTIVES

After studying this chapter, the student will be able to

- Give a brief history of the development of digital circuits and devices.
- Explain how digital circuits and devices are applied in electronic equipment.
- Describe the similarities and differences between digital and analog electronic circuits.
- Define basic digital terms.
- Explain how logic circuits are described by truth tables.
- Describe the logic functions of the basic building blocks of all digital circuits, the AND, OR, and inverter gates.
- Explain how to troubleshoot and design fundamental digital circuits.

OVERVIEW

This chapter introduces the student to the broad concepts and applications of digital electronic systems. Some basic digital terms are defined and explained to lay the foundation for concepts that form the basis of digital circuits and devices used in everyday electronic equipment. Comparisons are made between digital and analog circuits and devices. We shall also see how digital and analog systems are integrated and interfaced from a functional standpoint. This introductory chapter gives the student a sense of the position digital electronics holds in today's world.

Our study of digital logic circuits begins with an exploration of the relationships among the digital logic functions of the buffer, inverter, and AND and OR logic gates. The use of truth tables to represent these logic functions and the devices associated with them is explained, and examples of their use are presented.

Logic devices are then combined to form new and different logic functions and circuits. Logic expressions that describe these devices and circuits are presented, along with the process of creating these expressions from applicable truth tables.

The format followed in this and future chapters is to first present theory and concepts so that you, the student, receive the foundation you need to understand the functions of the devices and circuits described. This theory section is followed by a troubleshooting section that details processes to uncover problems in circuits or systems, using the devices or circuit units discussed in the preceding theory section. The last section of the chapter describes the principles of designing digital circuits, using the ideas presented in the theory section. Examples and practice problems are presented throughout the chapter to help clarify concepts and to give you practice using the ideas presented. In the selected questions at the end of the chapter, two full sets of questions and problems are given. The odd-numbered set has answers at the back of the text so that you can try them and check your work. A similar set of even-numbered questions and problems that do not have answers are provided so that your instructors can assign them for you to complete.

1.1 Introduction

The field of digital electronics influences many aspects of our daily living. Besides personal computers, digital circuits appear in automobiles, microwave ovens, television sets, stereos, in stores in the form of the UPC (universal product code) scanners, in banks using automatic teller machines (ATM), and numerous other applications. Many wristwatches are digital, we have digital music on compact discs (CDs)—the list goes on and on. Behind each of these devices or services are similar basic digital electronic circuits. Before we can understand the full significance or impact of digital electronic circuits and devices on our lives, we must look at what preceded these circuits and devices.

In the early part of the twentieth century, electricity and electronic devices were in their infancy. Homes were lighted by light globes screwed into sockets that had wires connected to them. Electric current, supplied to the home, heated a filament element in these globes, causing it to flow and give off light. As time moved on, more uses of electricity evolved. There were machines that produced music recorded on disks and other machines that captured signals from radio transmitters and brought all manner of performers into our living rooms. Moving pictures with sound were projected onto large screens in theaters, and then one day actual pictures were beamed into our homes via television receivers, and these moving pictures were eventually colorized. Radar sets helped protect our warships during war, and radio transmitters and receivers allowed people to communicate over large distances.

Analog *branch of electronics dealing with circuits that operate with and/or produce electrical quantities, such as voltage, current, or power that regularly vary in value.*

Amplitude *measure of the quantity of an electrical characteristic, such as voltage or current, in a circuit.*

Frequency *the rate at which a cycle of an analog signal's amplitude varies within a specified time, usually 1 second.*

State *the current condition of an entity. As an example, the state of a light switch can be ON or OFF, UP or DOWN.*

Complement *the opposite or inverted binary state.*

All these devices involved the use of **analog** electronic circuits. They are driven by and produce electronic signals that continually vary in voltage, current, or power. A basic example of an analog signal is an electronic signal that correlates to sound. When we listen to music, we hear a constantly changing sound. The music varies in volume and pitch as the song dictates. These changes can be reproduced as a varying electronic voltage. Changes in the voltage amount, called **amplitude,** correspond to changes in the music's volume. The rate of amplitude change, called **frequency,** is similar to changes in music between high- and low-pitched notes. The higher the note, the faster the analog signal changes. Analog signals are susceptible to electronic noise that affects both the amplitude and frequency of the signals and can cause a certain amount of distortion. One familiar type of noise is the "static" heard on AM radio programs. Another is the "snow" that appears on television sets, particularly when there is a weather disturbance nearby.

About the middle of the twentieth century, the principles of Boolean logic were adapted to electronic circuits and digital electronics was born. Digital concepts and logic themselves were not new, but the application of these concepts to practical devices and circuits evolved during this time period. Digital electronics involves itself with circuits that operate in one of two conditions or **states.** These two states can represent many things that are, generally, opposites or **complements** of one another. For example, on and off are two opposing states. Others are

True and false
Black and white
Go and stop
Fast and slow
Top and bottom

In terms of electronics parameters, the two states can be voltage, current, or power values. The on state may be represented by +5 volts (+5 V) or a current flow of 20 milliamps (20 mA). The off state, being the opposite of the on state, could be 0 V (no +5 V) or 0 mA (no current flow). In contrast, analog electronic circuits use signals that constantly vary in amplitude, having different voltage, current, or power values at different times.

As technology advanced, digital electronic circuits evolved, resulting in devices that allowed circuits to be compressed into a smaller physical area. A greater number of circuits can thus be placed into a smaller area. The first such advance was the development of the transistor.

Before the advent of "solid state" devices such as transistors, electronic circuits such as radios, TVs, and audio amplifiers were designed around devices called vacuum tubes. These tubes were composed of several conducting elements enveloped by a glass tube as shown in Figure 1.1. The air was evacuated from these tubes so that its filament element would not burn up and disintegrate when current was applied to it. The purpose of these low-resistance filaments was to produce enough heat, when current ran through them, to free negatively charged electrons from a second element, called the cathode. These free electrons were attracted to a positively charged anode. The flow of these electrons from cathode to anode was controlled by a signal applied to a control grid located close to the cathode. The less negative the signal on the control grid, the more electrons flowed to the anode. Thus the vacuum tube could be used to amplify a small signal placed on the control grid by controlling a larger signal at the anode output of the tube. Vacuum tube sizes varied depending on the power requirements of the signals generated by them, but even the lowest power tubes were quite large and generated a considerable amount of heat. In the first half of the 1950s, a new device, the transistor, was developed to replace the triode vacuum tube that is pictured in Figure 1.1.

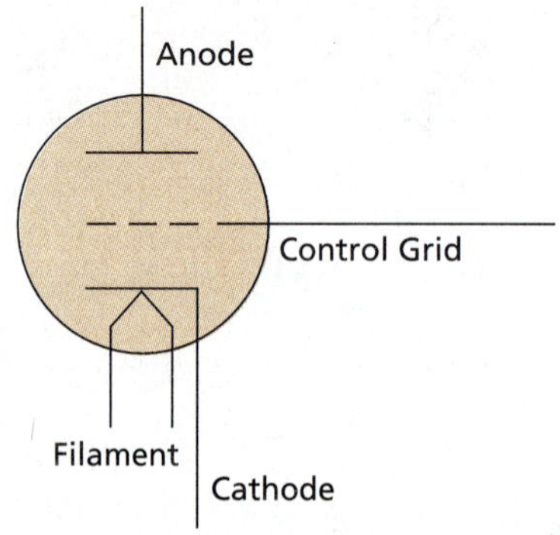

Figure 1.1 Vacuum Tube

Solid state *devices composed of semiconductor material such as transistors.*

An NPN bipolar transistor, shown in Figure 1.2, is a solid state semiconductor device that performs the same function as the triode vacuum tube, but in a smaller package and generating less heat. The term **solid state** means that the device is constructed of solid elements that are joined together as opposed to the open element design of vacuum tubes. These solid state elements were neither insulators nor conductors, thus neither preventing nor allowing current to flow easily. These elements are called *semiconductors* and are generally made from impure germanium or silicon. Current flow in a vacuum tube is controlled by applying a voltage to one of the vacuum tube elements, called the *control grid.* In a transistor, the same type of control is obtained by applying a current to the base lead of the transistor. The transistor, like the triode vacuum tube, can be used as an amplifier. A small current applied to the base input of the transistor controls a larger current flow from the transistor's emitter, out through its collector lead.

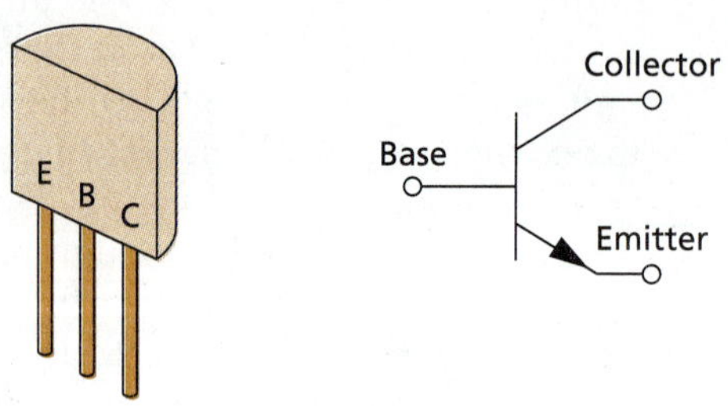

Figure 1.2 NPN Transistor

The transistor's reduced size and heat dissipation led the electronic industry toward miniaturization of electronic devices. Portable radios shrank from lunch box sizes down to pocket sizes. TVs became portable and no longer required a massive console to house the large vacuum tube circuits. This size reduction was very desirable in the electronics industry for the emerging digital electronics and computer fields, to allow minimization of many logic circuits.

Computers of the 1940s and 1950s were designed using vacuum tubes and electromechanical switches called relays shown in Figure 1.3. These computers usually occupied entire air-conditioned rooms and experienced regular breakdowns. Vacuum tube air leaks due to vacuum seal fractures would cause filaments to burn out. The mechanical contacts in the relays would get pitted or contaminated with dirt or dust or wear out from constant use. When transistors replaced both vacuum tubes and relays, computers began to quickly shrink in size. Because transistors had no mechanical elements, there were no parts that required regular maintenance or that wore out. This is not to say that transistors do not fail or that the overall computers did not require regular maintenance. However, the individual transistors did not require the extra care that relays or vacuum tubes required.

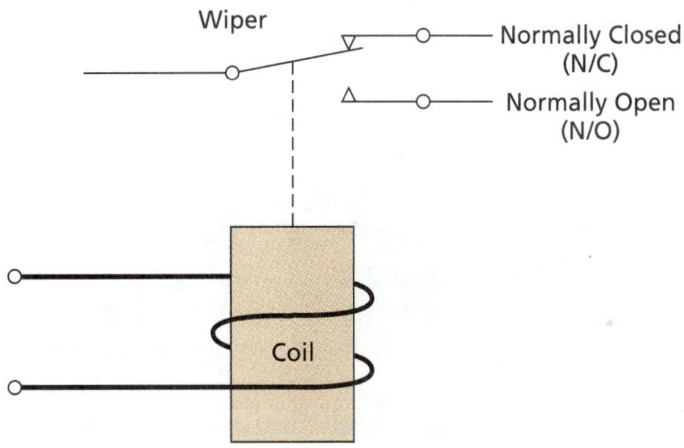

Figure 1.3 Relay

The next technological leap for the electronics industry came in the late 1960s with the development of the *field effect transistor (FET),* shown in Figure 1.4, and the **integrated circuit** (IC), shown in Figure 1.5. FETs are another

Integrated circuit *a digital device that contains a number of electronic logic circuits.*

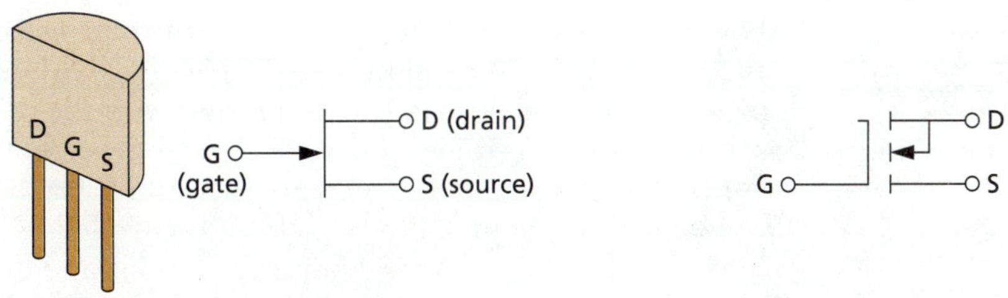

(a) Junction FET (JFET)

(b) Insulated Gate FET (IGFET or MOSFET)

Figure 1.4 Field Effect Transistors (FETs)

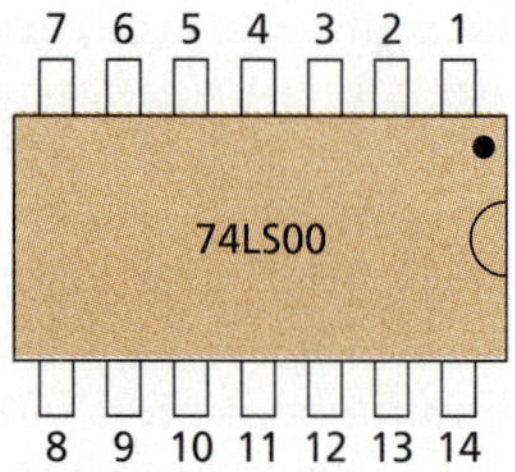

Figure 1.5 Integrated Circuit (IC

type of semiconductor device that performs the same function as vacuum tubes. Control signals are applied to a FET's gate input, which functions like the vacuum tube's control grid and transistor's base input, to control a larger signal between the FET's source and drain leads. Two different types of FET devices are used today, the *junction FET (JFET)* and the *insulated-gate metal oxide semiconductor FET (IGFET or MOSFET)*. The MOSFET pictured in Figure 1.4 is the device of choice in today's integrated circuits, because of its low power consumption and low heat dissipation, which leads to greater packing density (number of devices per package).

Manufacturing processes to integrate MOSFETs and other devices into a single unit led to the integrated circuit or IC package shown in Figure 1.5. These devices, which hold a group of electronic devices or circuits on a physical space called a *wafer* or *die*, pushed the process of miniaturization even further.

One cautionary note about FETs and the integrated circuits made by using them: Since MOSFET gates are isolated from the other elements of the device, they have a high input impedance, which is beneficial, because the input does not present much loading factor to outputs driving it. However, this high input impedance does make these devices susceptible to damage through static electricity charges when they are handled out of the system. Static electricity charges are not dissipated easily through this high impedance. Instead, they build up to a point where the device breaks down from the overvoltage of the charge. If such a device fails and needs to be replaced, special care must be taken in handling the new device to prevent it from being damaged by static charge buildup. Antistatic pads and ground straps have been used to reduce this danger. Safeguards are also built into the integrated devices themselves, to reduce the risk of static damage. Whenever you, the student, are handling MOSFET devices, handle them with care. These devices include NMOS, PMOS, CMOS, HMOS, HCMOS, and many other varieties of FET-based devices. If in doubt about the device you are using, check with your instructor before using it.

Microprocessor *central processing unit on a single IC chip.*

Central processing unit (CPU) *the part of the computer that directs the actions of the remaining parts and is responsible for executing mathematical and logical operations.*

Computer *electronic system that systematically executes a series of instructions to perform a given task.*

In the early 1970s, the process of placing more and more circuitry into a single IC achieved a major breakthrough with the development of the **microprocessor,** shown in Figure 1.6. The microprocessor is a **central processing unit (CPU)** in a single IC. CPUs are the central brain and controlling circuitry for a **computer.** Some of the functions controlled by a CPU include the movement of data between other computer sections and the execution of arithmetic and logic processes using some of these data. Placing all the CPU's functions into a single IC further reduced the overall size of computers. We have already seen where this reduction in size and power consumption has led. Lap-top or

Figure 1.6 Motorola 68060 Microprocessor. (Photo courtesy of Motorola)

notebook computers are made today that can easily be taken from one place to another and used along the way.

Section 1.1

Practice Problem

What precaution should you take when handling MOS devices? Why is this precaution necessary?

1.2 Binary Representation

Binary *a number system based on two distinct values, 1 and 0.*

Logic 1 *and* ***logic 0*** *represent the two opposing logic conditions of digital circuits and numbers.*

A **binary** system uses only two numbers: In contrast, the decimal system we all use in our daily lives has 10 (*decem*) numbers, 0 through 9. The binary system works well with digital circuits that have only two states instead of a continually variable analog state. The two numbers of the binary number system are 0 and 1. Opposite logic states can be used to represent the 0 and 1 numbers of the binary system. For instance, a logic 1 or high state is associated with the number 1 and is called a **logic one** or **1,** and the opposite state, logic 0 or low state, represents the number 0 and is called a **logic zero** or **0.** The main reason this works is that zero, meaning "nothing," is the opposite of 1 or "something."

An example of a binary system that is easily manipulated is a wall switch that controls a ceiling light. A person enters a room that is in a dark (or zero) state and flips on a light switch. The room turns from the dark zero state to the light, or logic 1, state. After that person is finished using the room, he or she leaves, turning off the light by setting the light switch to the off, or logic 0, position. This causes the room to return to a dark, logic zero state. This individual has just performed a digital binary operation. The switch used to control the light has only two positions—on and off. The state of the room controlled by the switch position has two states—light and dark. In addition to the binary characteristic of the control (the switch) and the result (light or dark), there is also a very specific relationship between that control and that result. That relationship can be shown using a diagram called a **truth table,** as in Figure 1.7.

Truth table *a diagram of columns and rows that illustrates the relationship between an entity's input and output states.*

Input SWITCH	Output ROOM
ON	LIGHT
OFF	DARK

Figure 1.7 Truth Table for Single Light Switch

1.3 Truth Tables

A truth table is a diagram illustrating the relationship between the input or control state of a binary digital system and the output or result state of using that input or control. The input side, usually on the left, lists all the possible states or combinations of states that the input(s) may have. In the case of our light switch, those states are ON and OFF. The right side of the truth table lists the results caused by each input combination: in this case, LIGHT and DARK. Since there are no other possibilities for our example, the truth table offers a complete description of how our light switch circuit works.

Suppose there is more than one input in the circuit, and you need to develop a truth table for this condition. For example, let's put together a truth table for a dual light switch arrangement that turns on the light if both switches are in the same position (UP or DOWN) and off if the switches are in opposite positions. There are two switches, called SWITCH A and SWITCH B, shown on the input side of the truth table in Figure 1.8. Given these two input controls, there are four possible combinations the switches can be placed in. Both can be UP, both can be DOWN, SWITCH A can be UP while SWITCH B is DOWN, and,

Inputs		Output
SWITCH A	SWITCH B	ROOM
DOWN	DOWN	LIGHT
DOWN	UP	DARK
UP	DOWN	DARK
UP	UP	LIGHT

Figure 1.8 Truth Table for Dual Light Switches

finally, SWITCH A can be DOWN while SWITCH B is UP. Begin the truth table by listing all these input combinations.

On the output side of the truth table, the resultant condition of the light for each input state is shown. When the switches are in the same direction, the light is on (LIGHT). When the switches are in opposite directions, the light is off (DARK). Thus the truth table in Figure 1.8 describes the complete operation of this light control. The light control illustrates one aspect of digital electronics, that is, controlling an event based on a given set of input conditions. The key to any use of digital electronics is based on the two state conditions of both inputs and outputs. The examples used already are ON/OFF, UP/DOWN, and LIGHT/DARK.

The previous paragraphs noted the use of a logic 1 and logic 0 to represent opposing states in a binary system. Two symbols, 1 and 0, are universally used to represent logic functions as listed in truth tables. The light switch truth table of Figure 1.7 is rewritten in Figure 1.9 using the symbols 1 and 0 to represent states of the switch input and light output. The UP position of the switches are represented by a 1 and the DOWN condition, a 0. Similarly, the LIGHT on state is a 1 and the DARK or off state is a 0. Notice that the use of 1's and 0's are made for both input and output states despite the physical actions that they represent. In the case of the switch inputs, 1 and 0 represent the positions UP and DOWN respectively, and for the output, the room conditions of LIGHT and DARK respectively.

Inputs		Output
SWITCH A	SWITCH B	ROOM
0	0	1
0	1	0
1	0	0
1	1	1

Figure 1.9 Truth Table Using 1's and 0's

Binary values 0 and 1 are also numerical quantities equal to the decimal value of 0 and 1 respectively. They are subject to the same mathematical rules and usage as the decimal number system we are more familiar with. The use of binary numbers is expanded in Chapter 3, in the discussion of digital arithmetic circuits.

EXAMPLE 1.1

Create a truth table, using logic 1 and logic 0 to represent input and output functions of a circuit that turns an ALARM on (logic 1) if SMOKE is detected (logic 1) or someone is BREAKING in to the home (logic 1). The alarm is OFF (logic 0) for all other combinations of inputs.

SMOKE	BREAK-IN	ALARM
0	0	
0	1	
1	0	
1	1	

(a) Inputs to Examples 1.1 Truth Table
0 = No SMOKE, No BREAK-IN
1 = SMOKE, BREAK-IN

SMOKE	BREAK-IN	ALARM
0	0	0
0	1	1
1	0	1
1	1	1

(b) Completed Truth Table for Example 1.1
0 = ALARM Off
1 = ALARM On

Figure **1.10** Example 1.1 Truth Table

SOLUTION Inputs SMOKE detected and BREAKIN are represented by logic 1. No SMOKE and no burglar BREAKIN are represented by a logic 0. The ALARM ON is a logic 1 and OFF a logic 0. On the input side of the truth table of Figure 1.10a, we show the four possible combinations of inputs.

0 0 = no SMOKE no BREAKIN
0 1 = no SMOKE BREAKIN
1 0 = SMOKE no BREAKIN
1 1 = SMOKE BREAKIN

The output in Figure 1.10b shows the results of applying the inputs, whenever there is SMOKE (logic 1) or someone is BREAKIN (logic 1), the ALARM is ON (logic 1). The only time the alarm is OFF (logic 0) is when there is no SMOKE (logic 0) and no one is BREAKIN (logic 0).

Section 1.3

Practice Problems

1. Create a truth table for a system that plays a tune when a question is answered correctly and sounds a buzzer when it is answered incorrectly. The input to the system is ANSWER, which has two states: RIGHT and WRONG. The output is SOUND, with two states: MUSIC and BUZZER.
2. Create a truth table illustrating a system that lights up a sign that says "Fasten seat belt" if the key is in the ignition and the belt is not fastened. The sign is off for any other combination of inputs. There are two inputs to the system: KEY with the states of IN and OUT (of the ignition) and BELT with the states of UNFASTENED and FASTENED. The output is SIGN and has the states LIT and OFF.
3. Convert the truth tables of Practice Problems 1 and 2 into tables using logic 1's and 0's for the states of each of the input and output variables.
4. Create a truth table using logic 1's and 0's to represent a system that SOUNDS (logic 1) a car's HORN based on these conditions: The DOOR is OPEN (logic 1) and the KEY is not in the lock, indicated by the key being OUT (logic 0) of the lock. The HORN does not sound (logic 0) if
   (a) the KEY is IN the lock (logic 1) or
   (b) the DOOR is NOT OPEN (logic 0).

1.4 Digital Logic

Digital logic *system that relates the output logic state of a digital circuit to the logic states of its inputs.*

Buffer *logic function where the output result is the same state as the input. This logic circuit is used when there is a requirement to change logic levels between circuits or to increase the drive capability of a gate's output.*

Inverter *or* ***NOT*** *logic function where the output result is the opposite logic state or complement of the input.*

AND *logic function where the result is logic 1 if all the input states are logic 1's and a logic 0 for all other input combinations.*

OR *logic function where the output state is a logic one if any input is a logic one.*

Gate *or* ***logic gate*** *a single logic circuit whose output is determined by the logic states of its inputs.*

Digital logic relates the output of digital devices to the states of their inputs. This logic is used to describe the operation of digital circuits and combinations of logic devices that make them work. Four basic logic building blocks are used to create most digital circuits. They are the **buffer, inverter,** and **AND** and **OR logic gates.** A gate is a logic circuit that has an output whose state is determined by the states of its inputs and the type of logic circuit the gate is. For example, the buffer is a logic gate with one input and one output. A buffer, whose logic symbol is shown in Figure 1.11, is defined as a logic gate whose output is the same state as its input. A truth table describing this relationship is included in Figure 1.11. On the input side, a single input can have only two states, 0 or 1. For the buffer gate, the logic state of the output matches the input logic state exactly. When the input is a 0, the output is a 0, and when the input is a 1, the output is a 1. On the truth table in Figure 1.11, the input of the buffer is labeled A and the output Y. Any symbols can be used for input and output, but letters are generally used, for convenience. Because the relationship between output and input is logically the same, an expression can be written that describes the relationship between the output (on the left side) and input (on the right). For the buffer gate, this expression is Y = A. Buffer logic defines a circuit function that has the same logic characteristics as a buffer gate. In other words, definitions for the digital gates can be applied to any logic situation that produces the same result as a defined logic function.

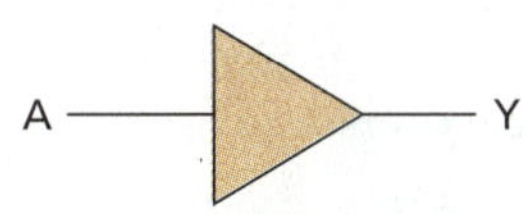

(a) Logic Symbol

A	Y
0	0
1	1

(b) Truth Table

Figure **1.11** Buffer Logic Gate

1.5 Application of Buffer Gates

Buffers are used when changes in electrical characteristics are required, but logic changes are not. For instance, if a higher current is required in a circuit a buffer with low input current, but high output current, may be needed. Today's microprocessors pack a lot of digital circuitry into a small package. To do so successfully, the heat dissipated by these microprocessors must be held to a minimum. Otherwise the device would consume itself in heat and "burn out." One way to reduce heat dissipation is to reduce the electrical power consumed by these devices. Lower power consumption results, in part, from keeping output driving currents low. However, the amount of circuitry in a computer external to the microprocessor keeps growing. Each time a card is added to a PC, additional functions are required and more drive current is needed. All personal computers use numerous buffer devices to change the low current of the microprocessor to the level of current required by the remaining computer circuits.

1.6 The Inverter Logic Gate and Function

The second building block, the inverter logic gate, is shown in Figure 1.12, along with its truth table and logic expression. An inverter is defined as a logic gate or logic function whose output is always in the opposite state from its input. The process of inverting a logic state is also known as *complementing* that logic state. Another definition of an inverter is a logic device or function whose output is the complement of its input. The symbol that represents inversion or the complement is a "bubble" or small circle appearing at an input or output to a gate or circuit. The inverter gate, itself, appears as a buffer with a bubble at either the input or output, but not at both. Labeling the input as pin A and the output as pin Y, as with the buffer, an expression can be written to describe the inverter that indicates the process of inversion by placing a bar over the inverted quantity. This bar is called a **NOT bar,** and the term NOT is often used to signify an inverting or complementing process. For the inverter gate, the logic expression is $Y = \overline{A}$, which is referred to as "Y equals NOT A."

NOT bar *bar over a logic variable or term indicating a logic inverted or "NOT" state.*

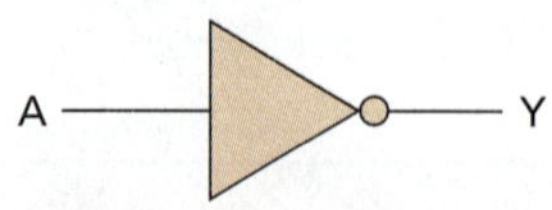

(a) Logic Symbol

A	Y
0	1
1	0

(b) Truth Table

$Y = \overline{A}$

(c) Logic Expression

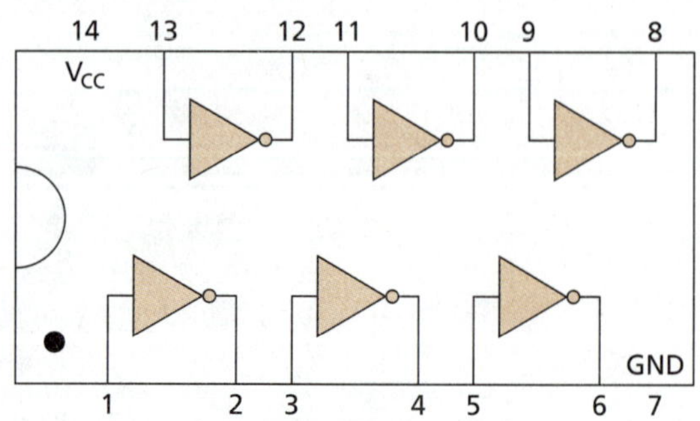

(d) 74LS04 Hex-Inverter IC

Figure **1.12** Inverter Logic Gate

Also shown in Figure 1.12 is the *dual in-line integrated circuit (IC)* pin diagram or the 74LS04 inverter gate included in the popular 74LSxx series of logic devices. Pin numbers of these devices generally begin with pin 1, indicated by a dot in one corner or by an indentation at one end, and are numbered consecutively in a counterclockwise direction. The pin in the lower right corner, when looking at a top view of the IC device (pin 7 for the 74LS04 inverter), is usually reserved for the ground. At the opposite corner (pin 14 for the 74LS04 inverter chip) is the power or V_{CC} pin. The 74LS04 is a six-section inverter chip—that is, it contains six inverter gates whose inputs and outputs are connected to the pins as shown. These gates are electrically and functionally independent of each other, having only a common positive source (V_{CC}) and ground connection.

Signal flow *circuit schematic showing desired output and input logic states for correct circuit operation.*

Active *the logic state used to turn on a circuit or function.*

The inverter gate can be shown with the inverting bubble on the input or the output lead as shown in Figure 1.13. Logically, they are the same gate, an inverter. Whether or not one places the inverting bubble at the input or output depends on a process called **signal flow**, which shows the logic symbol drawn to indicate the desired output level and the state of the input required to produce it. In the first inverter symbol, the desired output to place it in a usable or **active** state is a 0, indicated by the bubble at the output. In other words, the logic circuit driven by this gate needs a 0 logic level applied to its input to be driven on. The input state of the inverter required to produce this desired output is a logic 1, since there is no bubble at the input. We can say that when the input is active high or 1, the output is driven active low or 0.

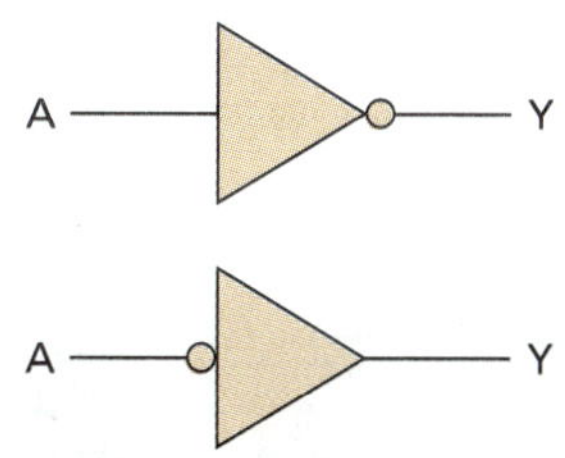

Figure 1.13 Inverter Gate Symbols

The second inverter shows an active low for the input, which generates an active high at the output. Once again, the process of inversion is the same, but this symbol illustrates a different signal flow requirement from the other gate. This gate is said to have an active high output, needing an active low signal at the input to put it into its active output state. Be aware that both inverter gates will invert their **inactive** as well as their active states. That is, the first gate will still produce a 1 if a 0 is applied to the input, but this would be the state that would turn off succeeding circuits, whereas the active state causes them to operate.

Inactive *Logic state that turns off a circuit or function.*

Assert *apply a signal or logic state that turns on a circuit or function.*

Negate *in terms of a logic signal, negate means to turn off a logic circuit or function.*

Two additional terms, **assert** and **negate**, are used to describe the application of logic signals to a gate or circuit. When an input is placed into its active state, the input is asserted or turned on. To negate an input is to turn it off or place it in its inactive state. Thus an inverter showing the bubble on the output can be said to have its output active low when its input is asserted high.

EXAMPLE 1.2

Describe the output of each of the logic gates in Figure 1.14.

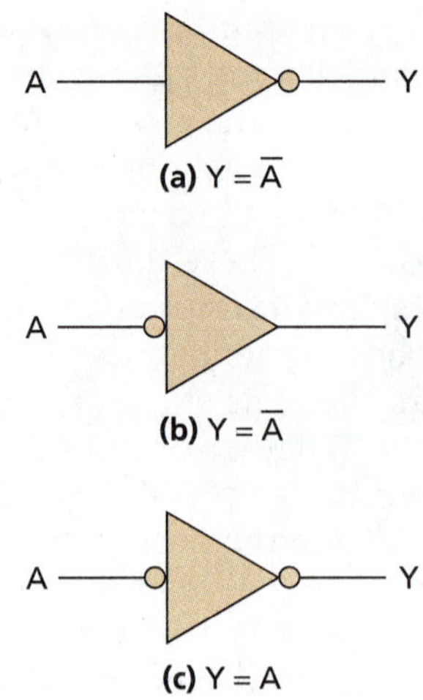

Figure 1.14 Signal Flow Logic Symbols

SOLUTION In the first gate, the input is buffered and then inverted, so the gate is an inverter. In the second, the input is first inverted and then buffered, which is another representation of the inverter gate. The third gate has a "double inversion." The input is inverted, buffered, and then inverted at the output so that the output matches the state of the input, making this gate a logic buffer.

EXAMPLE 1.3

Describe the buffer gates in Figures 1.11 and 1.14 with respect to active states and signal flow.

SOLUTION The input and output active states for the buffer in Figure 1.11 are both 1, while in Figure 1.14 they are 0. This means that the buffer symbol in Figure 1.11 would be used when the desired output signal is high; that is, the desired active output state is high and requires a high input signal to be asserted to achieve the active high output state. The buffer in Figure 1.14c is used when the desired output and required input active states are 0.

Section 1.6

Practice Problem

Draw two logic circuits using inverter gates to make a buffer gate. Draw one circuit using active high outputs, and one with active low outputs. *Hint:* Inverting a logic state twice yields the same output state as the input state. See Figure 1.14.

1.7 A Brief Introduction to Binary Numbers

We will deal with number systems and binary arithmetic later in the text when we discuss digital arithmetic circuits. However, there are some basic characteristics of binary numbers that can make life a bit easier for us right now in understanding truth table formats and logic expressions.

__Base__ or __radix__ defines the number system in use and has a value equal to the number of different values in that system, including the lowest value, zero. The decimal number system is a base 10 number system.

__Weight__ value of a digit based on its position in a number.

The binary system uses only two numbers, 0 and 1, so it has a **base** or **radix** number of 2. The decimal system uses 10 numerical symbols, 0 through 9, and is thus called a base 10 system. The full value of a decimal number is found by considering the value of the digits and their **weight.** A weight is the value of a digit based on its position in a number. In decimal, the number 364 has three digits, 3, 6, and 4. The weights of each digit increase as their position increases from right to left. The 4 is in the one's column, meaning that its weight is 1. The 6 is in the ten's columns, with a weight of 10, and the 3 is in the hundred's column, with a weight of 100. The total value of the number is found by adding the products of the digit values and their weights:

$$3 \times 100 + 6 \times 10 + 4 \times 1 = 300 + 60 + 4 = 364$$

The weight value is found by raising the base or radix value (10 in the case of decimal numbers, 2 for binary numbers) by the position number of the digit, starting with 0 for the rightmost digit. In decimal, the one's weight is $10^0 = 1$, ten's is $10^1 = 10$ and hundred's, $10^2 = 100$. With this brief explanation, we can extend this concept to assign some basic values to binary numbers.

Binary values are converted to decimal equivalent values by adding together the position weight of each binary digit that is a 1. This works because the position × weight product of a binary number is always zero or the weight value (weight × 1 = weight). First we find the weight of binary digits: $2^0 = 1$, $2^1 = 2$, $2^2 = 4$, $2^3 = 8$, and so on. Each position's weight is twice the previous position, that is, 1, 2, 4, 8, 16, 32, and so on. For binary numbers, the digit times weight product is "there" (× 1) or "not there" (× 0), so a binary number's equivalent decimal value can be found by adding the weights of binary digits that are a logic 1. Try this: Convert the binary number 1101 to a decimal number. The weights for logic 1 digits in the number are 8, 4, and 1. The lone 0 has a weight of 2, but $0 \times 2 = 0$, so the equivalent decimal value is the sum of the weights for each digit that is a 1: $8 + 4 + 1 = 13$. The binary number 1101 is equal to the decimal number 13.

There is one more thing you can do with the base 2 system before we discuss logic gates and truth tables further. You can find the number of input combinations that a truth table will have by raising the base 2 to an exponent equal to the number of inputs (n) used, or

$$\text{Number of input combinations} = 2^n$$

For instance, a two-input gate would have $2^2 = 4$ different input combinations applied to those two inputs. They would be 00, 01, 10, and 11. A three-input device and its corresponding truth table would show $2^3 = 8$ input combinations.

Section 1.7

Practice Problems

1. Convert the binary numbers 10011 and 11001 to decimal values.
2. How many possible input combinations are available in a four-input and a six-input gate?

1.8 AND Gates and Logic Function

The AND logic function is the first gate we will discuss that requires more than one input and gives us a more interesting logic function to work with. In Figure 1.15, the inputs of the AND gate are labeled A and B, and its output is Y. This illustrates a two-input AND gate, but keep in mind that an AND function is not limited to two inputs. Also, the use of the symbols A and B for input variables is arbitrary and could be anything, such as words, graphics characters, letters, or numbers. The common practice is to use letters for convenience and clarity.

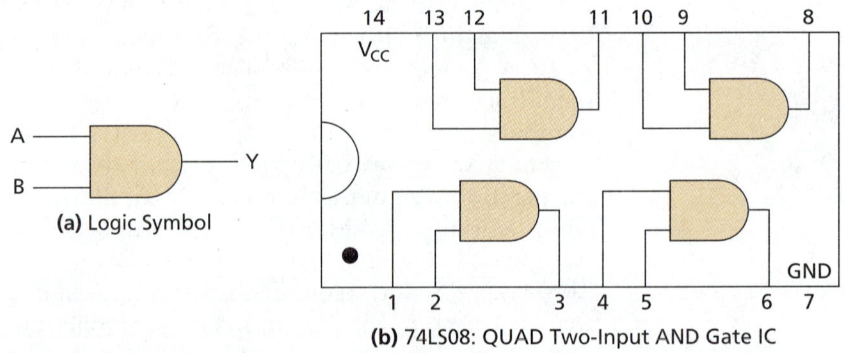

Figure 1.15 AND Logic Gates

The definition of the AND function is that the output will be a logic 1 when all inputs (A *AND* B in Figure 1.15) are a logic 1. A corollary of the AND function exists and is stated thus: If any input to an AND gate is a logic 0, then the output is forced into a logic 0 state.

This function of the logic 0 input forcing the output into a specific state explains why a logic device is called a gate. To illustrate, let's describe the function of a physical gate in front of a walk. When closed and locked, a physical gate prevents anyone from entering the walk. When open, the physical gate allows a person to pass onto the walk. An AND gate logically performs a similar function. When the gate is closed by placing a logic 0 on one input, the state of the other input is not "passed" onto the output, which is forced to a constant 0 state. When the gate input is raised to a logic 1 (the gate open), the state of the second input is "passed" to the output. This idea can be illustrated by examining the truth table for an AND gate.

On the AND truth table, the two inputs, labeled A and B, have 2^n or four different combinations of 0 and 1 states. Both can be placed into a logic 0 state, or both into a logic 1 state. The two remaining input combinations occur when one input is a logic 1 and the other a logic 0.

Positive logic *binary logic system in which a logic 1 state has a higher voltage value than a logic 0 state.*

Negative logic *binary logic system in which a logic 1 state has a lower voltage value than a logic 0 state.*

Another set of terms used when describing logic states is *high* for a logic 1 and *low* for a logic 0. This is a convention used when describing **positive logic,** because the logic 1 state is considered more positive, therefore above or higher, than the lower logic 0 state. However, there is absolutely no reason why the convention cannot be reversed, making the logic 0 the higher state. Systems that use this latter form of digital logic are using **negative logic.** Most of this text uses positive logic, except for a few cases where it will be clearly noted.

The standard convention for truth tables is to list inputs starting with both inputs low and then list the remaining input combinations in ascending binary order. Usually the rightmost input on the truth table is considered the least significant bit (LSB) when listing inputs in binary numerical order. As a result, the input side of the truth table for the two-input AND gate appears as

A	B	Y
0	0	
0	1	
1	0	
1	1	

The output of the two-input AND gate is driven high to logic 1 state only when both inputs are high, as in the last line of the truth table. As long as any input is low, the output is forced low, into a logic 0 state. This satisfies the definition and corollary for the AND function. The output is high when all inputs are high—in this case, when A *and* B are both high. When any input is low—A *or* B, for the two-input example—the output is forced low. The completed truth table appears as

A	B	Y
0	0	0
0	1	0
1	0	0
1	1	1

The logic symbol shown in Figure 1.15 is a symbol you will become very familiar with. It is the logic symbol for the AND function. When seen in a logic diagram, it says to the viewer that the output is a logic 1 when both inputs are logic 1 and is a logic 0 otherwise. The signal flow description of the AND gate says that the output of an AND gate is an active high when both input signals are asserted high.

Operators *symbol for a logic function such as '+' for the OR function.*

Several symbolic forms, called **operators,** are used to describe the AND function as a logic expression:

$$Y = A \cdot B$$

$$Y = (A)(B)$$

$$Y = AB$$

The AND operators, ·, parentheses, and omitted operator, used in these expressions are similar to multiplication operators used in mathematics. One reason for this association is a process called *Boolean algebra,* which includes identities and theorems that relate different logic expressions to each other. We explore these later, in Chapter 2.

The 74LS08, included in Figure 1.15, is a 74LSxx series quad two-input AND gate integrated circuit. The word *quad* refers to the number of gates in the chip package (four), each with two inputs.

We can apply the AND gate to solve some simple logic problems. For instance, let's show how an AND gate can be used to turn on a light when two doors and a window are closed and when the light is off when any of the doors or window are open. We will show the gate and develop its truth table and Boolean expression. This gate has three inputs (DOORA, DOORB, and WINDOW) and one output (LIGHT), but the principle is the same as the two-input AND gate. That is, the output is a logic 1 if all its inputs are logic 1 and logic 0 if any input is a logic 0. We shall select the closed condition of the doors and window as a logic 1 input state. A lighted light is the output logic 1 state. The opposite conditions are represented as logic 0's. After defining the logic states, construct a truth table to illustrate the input and the output states. The only time that the light is on (logic 1) is when all three items are closed (logic 1's). At all other times, the output is a 0. This fits the definition of the AND function:

DOORA	DOORB	WINDOW	LIGHT
0	0	0	0
0	0	1	0
0	1	0	0
0	1	1	0
1	0	0	0
1	0	1	0
1	1	0	0
1	1	1	1
0 = open	1 = closed	0 = off	1 = on

Notice on this table, since there are three inputs, that the total number of input combinations is 2^3 or 8. They are listed on the table in ascending binary order, following the same convention used with a two-input table.

The logic expression that describes this gate can be written in one of two ways:

$$LIGHT = DOORA \cdot DOORB \cdot WINDOW$$

$$LIGHT = (DOORA)(DOORB)(WINDOW)$$

The third AND expression form, using no operator between each input variable, does not work well, since the words used as input labels would run together and would not be easily distinguishable. The expression would come out as

LIGHT = DOORADOORBWINDOW

Last, the logic symbol is an AND gate with three inputs and one output, as shown in Figure 1.16.

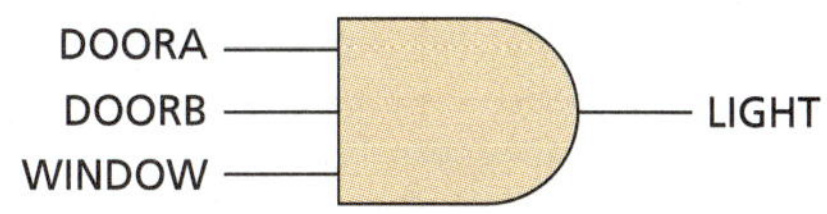

Figure 1.16 Three-Input AND Gate Application

Section 1.8

Practice Problems

The floppy disk circuit of a personal computer has a logic circuit that checks if the disk is correctly inserted before the PC begins to try to read the disk. The requirements are

1. The disk DOOR must be closed (logic 1).
2. The disk must be INserted (logic 1).
3. The disk must be right SIDE up (logic 1).

When these three criteria are met, a signal goes high, telling the computer that the disk drive is ready for use. Construct the truth table for the circuit that generates the READY signal and show its logic symbol and expression.

1.9 The OR Logic Gate

The OR logic function states that the output of the OR gate is a logic 1 or high if any input to the gate is a logic 1 or high. For example, given the two-input OR gate of Figure 1.17, if input A OR input B is a logic 1, then output Y is a logic 1. Output Y is a logic 0 only if both inputs are logic 0. The truth table describing the OR gate is included in Figure 1.17.

Compare the OR truth table to the truth table for the AND gate in Figure 1.15. Notice that the output states of the OR gate are totally different from the AND gate. They are definitely not all the opposite of the AND gate. As such, the OR gate and OR logic function are not opposite to or the complement of the AND function. The logic operator for the OR logic function is an arithmetic plus (+). The two-input OR gate of Figure 1.17 has an expression of Y = A + B. The 74LSxx series integrated circuit for a quad dual-input OR gate is the 74LS32, whose pin diagram is included in Figure 1.17.

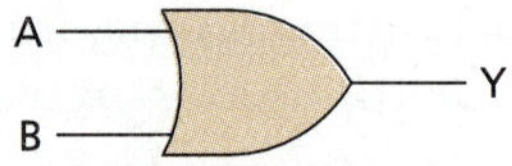

(a) Logic Symbol

A	B	Y
0	0	0
0	1	1
1	0	1
1	1	1

(b) Truth Table

Y = A + B

(c) Logic Expression

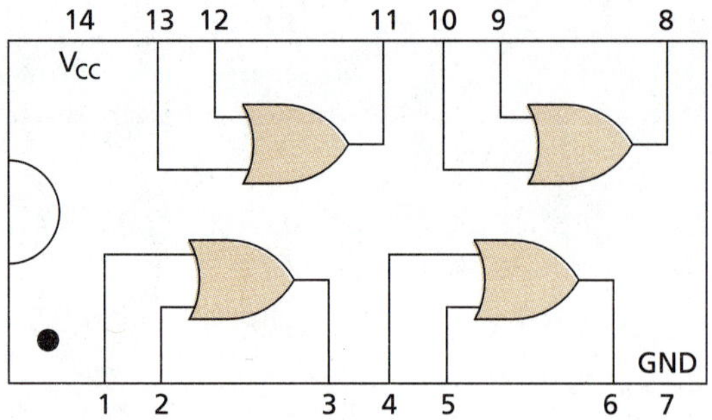

(d) 74LS32 QUAD Two-Input OR Gate IC

Figure 1.17 OR Logic Gate

As an example of the use of an OR logic function, let's create a truth table and a logic circuit to accomplish the following task: Turn on a light if door A or door B or window is closed. The light remains off when all three are open. The number of different possible input combinations is $2^3 = 8$, which are placed on a truth table in ascending binary order. The output state in this example is a high at all times except when all three items are open (DOORA, DOORB, and WINDOW = 000). The truth table describing this logic circuit is

DECIMAL	DOORA	DOORB	WINDOW	LIGHT
0	0	0	0	0
1	0	0	1	1
2	0	1	0	1
3	0	1	1	1
4	1	0	0	1
5	1	0	1	1
6	1	1	0	1
7	1	1	1	1

This table fits the description of a three-input OR gate in that the output is a logic 1 if any input is a logic 1 and is logic 0 only when all inputs are logic 0. Therefore, the logic circuit that satisfies the requirements of the problem is a three-input OR gate with a logic expression of LIGHT = DOORA + DOORB + WINDOW, as shown in Figure 1.18.

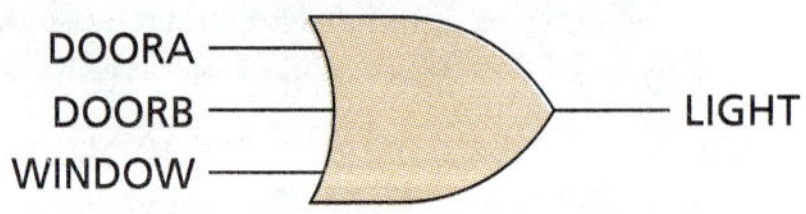

Figure 1.18 Three-Input OR Gate Application

Comparing the truth table in Section 1.7 for the three-input AND gate (which turned on the light only when both doors and window were closed) to this table, notice the difference in outputs for each set of input conditions. Inverting the outputs of one truth table does not create the output conditions of the other table. This once more verifies that the OR function is not simply the opposite of the AND function, but is an entirely different logic process.

Section 1.9

Practice Problems

1. Those annoying car buzzers that remind us to BUCKLE up or take the KEY out of the ignition, or that the OIL pressure is low, are included in every newer model car. Develop a truth table and gate for this type of BUZZER.
2. Using A to represent "buckle up," B to represent "key in ignition," and C to represent "low oil pressure," write the Boolean expression for the gate in Problem 1, whose output buzzer is represented by Y.

1.10 The NAND and NOR Logic Gates

Combinational logic *the use of different logic gates in a logic circuit to form a different logic function.*

NAND *not AND. Logic function where the output state is a logic zero if all inputs are logic 1 and a logic 1 state for all other input conditions.*

The buffer, inverter, AND, and OR gates make up the four basic logic blocks of most digital circuits. These functional blocks are used to make additional blocks, which are in turn used to make still larger ones until the final logic circuits are completed. This process causes logic circuits to be made from a combination of logic building blocks and is called **combinational logic.** Two gates, the NAND and NOR, are examples of the first level of this process.

The NOT AND or **NAND** gate of Figure 1.19 is made by connecting the output of an AND gate to the input of an inverter and combining the two gates into

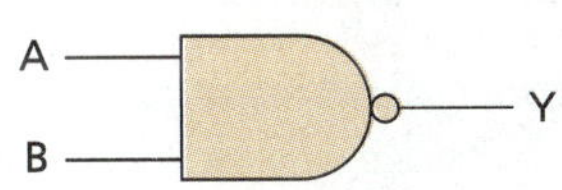

(a) Logic Symbol

A	B	Y
0	0	1
0	1	1
1	0	1
1	1	0

(b) Truth Table

$$Y = \overline{AB}$$

(c) Logic Expression

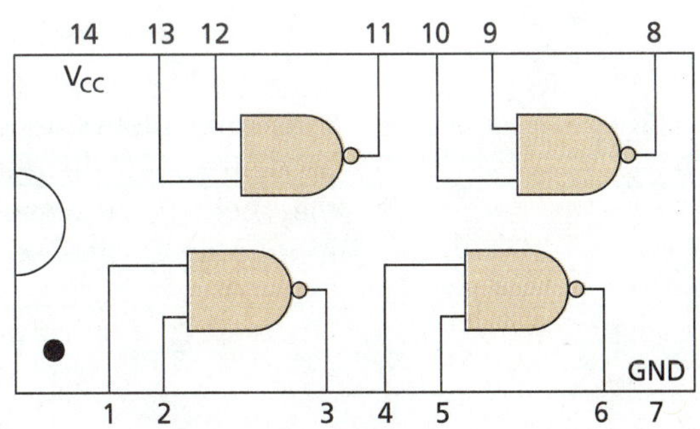

(d) 74LS00 QUAD Two-Input AND Gate IC

Figure 1.19 NAND Logic Gate

a new logic function. This causes the output of the AND gate to be complemented. By definition, the output of a NAND gate is a logic 1 if any input is 0. The output of the NAND is a logic 0 only when all inputs to the gate are logic 1. The truth table for the NAND is included in Figure 1.19. To form the logic expression for the NAND gate, place an inverting bar over the AND expression so it becomes a NOT AND:

$$Y = \overline{A \cdot B}$$

This is stated as "Y equals not A and B." The 74LSxx series IC for the NAND gate is the first device in the series, the 74LS00, whose pin out is included in Figure 1.19.

EXAMPLE 1.4

Develop a truth table, circuit, and expression for a logic circuit that prevents a person from entering a room if all the following characteristics match an unauthorized person:

1. HAIR color
2. EYE color
3. HEIGHT

Matches are indicated by a logic 1. Access to the room is logic 1, denied access is a logic 0.

SOLUTION Access is granted if any one or more of the features do not match. Access is denied when all three match. A truth table describing these conditions is

HAIR	EYE	HEIGHT	ACCESS
0	0	0	1
0	0	1	1
0	1	0	1
0	1	1	1
1	0	0	1
1	0	1	1
1	1	0	1
1	1	1	0

This truth table satisfies the definition of a NAND function in that for access to be denied (0) all inputs had to have matched the user (1). If anything mismatched (0), access is allowed (1). The gate to achieve this result is a three-input NAND gate (see Figure 1.20).

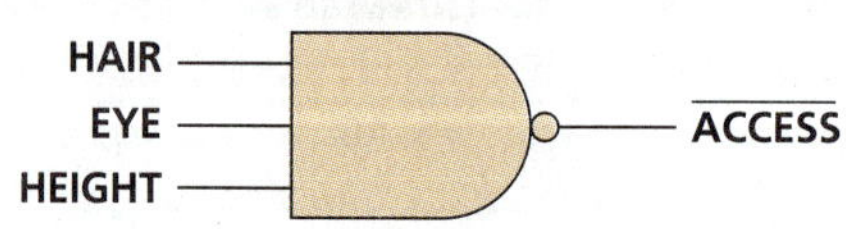

Figure 1.20 Three-Input NAND Gate

The logic expression may be written in one of two ways:

$$\text{ACCESS} = \overline{\text{(HAIR)(EYE)(HEIGHT)}} \qquad \overline{\text{ACCESS}} = \text{(HAIR)(EYE)(HEIGHT)}$$

The first says that access is granted (active high) if any characteristic does not match (active low). The second states that access is denied (active low) if all three characteristics are a match (active high).

NOR *not OR. Logic function where the output state is 0 if any input is a logic 1.*

The opposite or complement of the OR function is a NOT OR or **NOR** gate, shown in Figure 1.21, which includes the 74LSxx series pin diagram for the 74LS02 quad 2-input NOR gate. The NOR gate like the NAND gate is formed by combining an OR function with an inverter logic function. The output of an OR gate is connected to the input of an inverter, which causes the OR function to be complemented. By definition, the output of a NOR gate is a logic 1 when all inputs are logic 0. If any input is a logic 1, then the output is a logic 0. A truth table describing the NOR function is shown in Figure 1.21. The expression for the NOR gate is the same as the OR with a NOT bar over one side of the expression and is stated as "Y equals not A OR B":

$$Y = \overline{A + B}$$

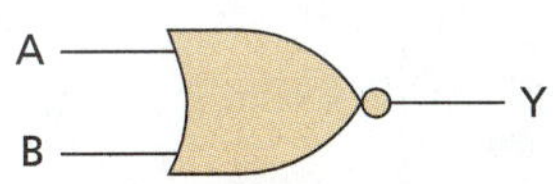

(a) Logic Symbol

A	B	Y
0	0	1
0	1	0
1	0	0
1	1	0

(b) Truth Table

$Y = \overline{A + B}$

(c) Logic Expression

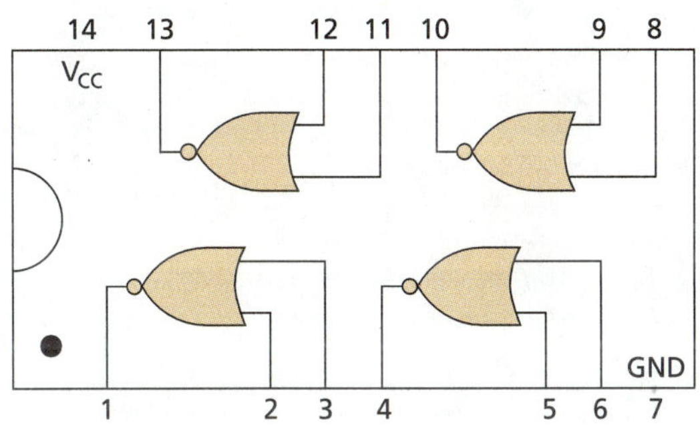

(d) 74LS02 QUAD Two-Input NOR Gate IC

Figure **1.21** NOR Logic Gate

EXAMPLE 1.5

Develop a truth table, circuit, and logic expression for a circuit that shuts off a water valve if any of the following conditions occur:

1. It is raining.
2. The ground is wet enough.
3. The water pressure gets too low.

These three stated conditions are logic 1's, and the water valve is a logic 0 when it is shut off and a logic 1 when it is open.

SOLUTION The valve is left open (1) as long as it isn't raining (0), the ground is still dry (0), and water pressure is sufficient (0). If any of these conditions are not met (1), then the valve is shut off (0), stopping the water flow. The NOR logic function suits these requirements, as shown by this truth table:

RAIN	WET	PRESSURE	VALVE
0	0	0	1
0	0	1	0
0	1	0	0
0	1	1	0
1	0	0	0
1	0	1	0
1	1	0	0
1	1	1	0

The logic expression for this three-input NOR gate is

$$\text{VALVE} = \overline{\text{RAIN} + \text{WET} + \text{PRESSURE}}$$

This says that the valve is open if it isn't raining, ground is not wet and pressure is OK or

$$\overline{\text{VALVE}} = \text{RAIN} + \text{WET} + \text{PRESSURE}$$

This states that the valve is not open if it is raining or the ground is wet or there is not enough water pressure. These expressions describe a three-input NOR gate:

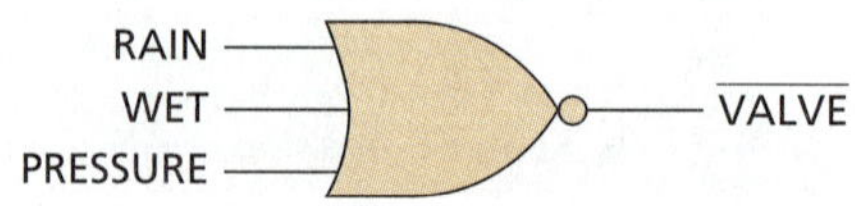

Figure 1.22 Three-Input NOR Gate

Contrast the truth tables in Examples 1.4 and 1.5, and observe that like the AND and OR gates, the NAND and NOR gates are not opposite to each other, but are, rather, differing functions entirely.

Section 1.10

Practice Problems

1. Section 1.9 Practice Problem 1 can be rewritten to state: Those annoying car buzzers, that warn us when a problem occurs, shut off if our seat buckle is buckled, the key is out of the ignition when the door is open, and the oil pressure is OK. If each condition is a logic 1 and the buzzer being off a logic 0, create the truth table, gate, and logic expression to turn the BUZZER off.
2. Section 1.7 Practice Problem 1 also can be rewritten as follows: The READY signal in a floppy disk circuit is made inactive (0) if any of the following conditions are met (logic 1):
   a. The disk DOOR is open.
   b. A disk is not inserted (IN) into the drive.
   c. The disk is inserted with the wrong SIDE up.

   READY is set to a 1 to indicate that the disk drive is ready for use, when all conditions are correct (all logic 0). Create the truth table, gate, and logic expression for the READY signal.

1.11 Combinational Logic

As previously stated, the four logic building blocks—buffer, inverter, AND, and OR gates—are combined to make other functional blocks that can be used to make yet additional logic circuits or blocks. In Figure 1.23 an AND gate is combined with an OR gate to form a new logic circuit, which can be represented by a truth table and a logic expression. Let's start by creating the truth table for the circuit. The inputs to the table are A, B, and C, and the final output is labeled Y. The output of the AND gate is an intermediate point and appears on the output side of the truth table as A·B. This intermediate output becomes one of the inputs to the OR gate, whose other input is C. Start by listing the inputs to the truth table in ascending numerical order and noting the outputs:

C	B	A	A·B	Y
0	0	0		
0	0	1		
0	1	0		
0	1	1		
1	0	0		
1	0	1		
1	1	0		
1	1	1		

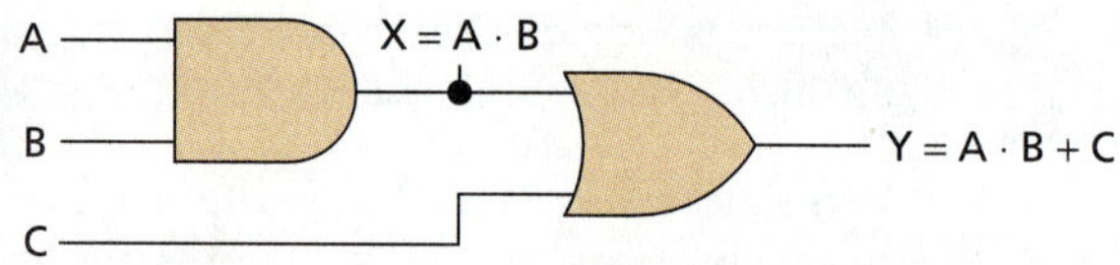

Figure 1.23 Combining Gates: Y = AB + C

Output A·B is derived from the AND function of A and B, which means that its state is low if either A or B is low and is high when A and B are both high:

C	B	A	A · B	Y
0	0	0	0	
0	0	1	0	
0	1	0	0	
0	1	1	1	
1	0	0	0	
1	0	1	0	
1	1	0	0	
1	1	1	1	

Finally, output Y is high whenever input A·B or input C is high, as defined by the OR function:

C	B	A	A · B	Y
0	0	0	0	0
0	0	1	0	0
0	1	0	0	0
0	1	1	1	1
1	0	0	0	1
1	0	1	0	1
1	1	0	0	1
1	1	1	1	1

Notice that the truth table of the final results at Y do not take the form of any of the previous truth tables discussed. This is because the output of this circuit is a combination of logic functions and is unique to this circuit. No other circuit or gate has been developed yet that matches this circuit's output. That does not mean that another combination of gates couldn't yield the same results. It simply means that this circuit is not an inverter, buffer, AND, OR, NAND, or NOR gate by itself.

Now for the logic expression that describes output Y. We have already determined that the expression for the intermediate output is A·B. This output becomes one of two inputs to the OR gate in the circuit, whose other input is C. This means that output Y is an OR function of those two inputs, or

$$Y = A \cdot B + C$$

EXAMPLE 1.6

Develop the truth table and logic expression for the circuit of Figure 1.24. Is there another logic expression that describes this circuit?

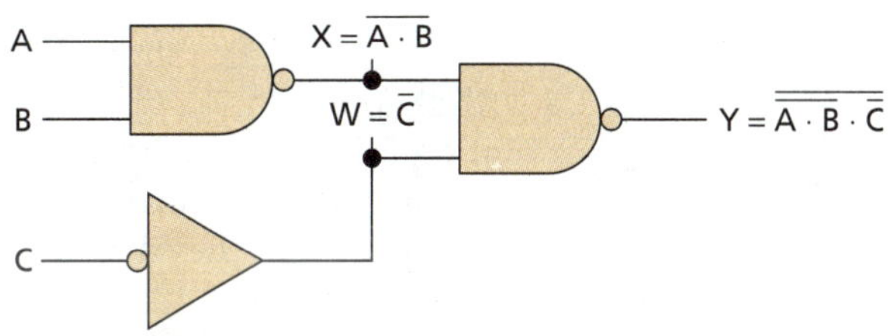

Figure 1.24 Combining Gates: $Y = \overline{\overline{AB} \cdot \overline{C}}$

Start by developing the logic expression. Intermediate output X is a NAND result of combining inputs A and B, or $X = \overline{A \cdot B}$. Intermediate output W of the inverter gate is input C inverted: $W = \overline{C}$. Output Y results from the NAND logic function of X and W, $Y = \overline{X \cdot W}$. Substituting for X and Y so the expression can be written in terms of inputs A, B, and C yields the final expression of

$$Y = \overline{\overline{A \cdot B} \cdot \overline{C}}$$

Next the truth table. The inputs are A, B, and C, and since there are three inputs, then the total number of input combinations is $2^3 = 8$. There are two intermediate outputs, $\overline{A \cdot B}$ and $\overline{C}$, and one final output, Y:

C	B	A	$\overline{A \cdot B}$	$\overline{C}$	$Y = \overline{\overline{A \cdot B} \cdot \overline{C}}$
0	0	0			
0	0	1			
0	1	0			
0	1	1			
1	0	0			
1	0	1			
1	1	0			
1	1	1			

The first step in completing the truth table is to fill in the intermediate outputs, NAND gate $\overline{A \cdot B}$ and inverter $\overline{C}$:

C	B	A	$\overline{A \cdot B}$	$\overline{C}$	$Y = \overline{\overline{A \cdot B} \cdot \overline{C}}$
0	0	0	1	1	
0	0	1	1	1	
0	1	0	1	1	
0	1	1	0	1	
1	0	0	1	0	
1	0	1	1	0	
1	1	0	1	0	
1	1	1	0	0	

Notice that the $\overline{A \cdot B}$ column shows a 0 whenever A and B are both 1, which describes a NAND function. Note that $\overline{C}$ is the complement of input C, and the states in that column are the inverse of C. The final output at Y is derived by the NAND operation of inputs $\overline{A \cdot B}$ and $\overline{C}$, which means that when either one is a 0, Y is a 1, and when they are both 1, Y is a 0:

C	B	A	$\overline{A \cdot B}$	$\overline{C}$	$Y = \overline{\overline{A \cdot B} \cdot \overline{C}}$
0	0	0	1	1	0
0	0	1	1	1	0
0	1	0	1	1	0
0	1	1	0	1	1
1	0	0	1	0	1
1	0	1	1	0	1
1	1	0	1	0	1
1	1	1	0	0	1

Look at the results for Y in this table and in the table that preceded this example. Notice that they are identical and therefore describe the same logic function. This means that either circuit will produce the same result for the same set of inputs. That makes both circuits equivalent, along with their logic expressions. So another expression for this circuit is

$$Y = A \cdot B + C$$

In Example 1.6, after the truth table was developed it was compared to the table of the circuit described in the text preceding Example 1.6. Both tables described the same logic function. This means that both circuits are logically the same, and the logic expressions that describe both circuits produce equivalent results. The expression in the discussion is

$$Y = A \cdot B + C$$

The expression in Example 1.5 is

$$Y = \overline{\overline{A} \cdot \overline{B} \cdot \overline{C}}$$

Based on the observation made by comparing the two tables, we can conclude that

$$A \cdot B + C = \overline{\overline{A \cdot B} \cdot \overline{C}}$$

In Figure 1.25, AND, OR, and inverter gates are combined to produce a unique result that will lead to a new logic function, called the exclusive OR or XOR. Let's first develop the logic expression for this logic circuit. The intermediate outputs of two inverter gates, supply complemented inputs $\overline{A}$ and $\overline{B}$ to a pair of AND gates. The other inputs to the AND gates are the opposite non-complemented inputs, B and A. This yields intermediate outputs of $X = A \cdot \overline{B}$ and $W = \overline{A} \cdot B$. These two outputs are the inputs to the final OR gate, producing an output of

$$Y = A \cdot \overline{B} + \overline{A} \cdot B$$

The truth table for this circuit is constructed with four combinations of the two inputs A and B, intermediate outputs $\overline{A}$, $\overline{B}$, $A \cdot \overline{B}$, and $\overline{A} \cdot B$, and output Y:

A	B	$\overline{A}$	$\overline{B}$	$A \cdot \overline{B}$	$\overline{A} \cdot B$	$Y = A \cdot \overline{B} + \overline{A} \cdot B$
0	0					
0	1					
1	0					
1	1					

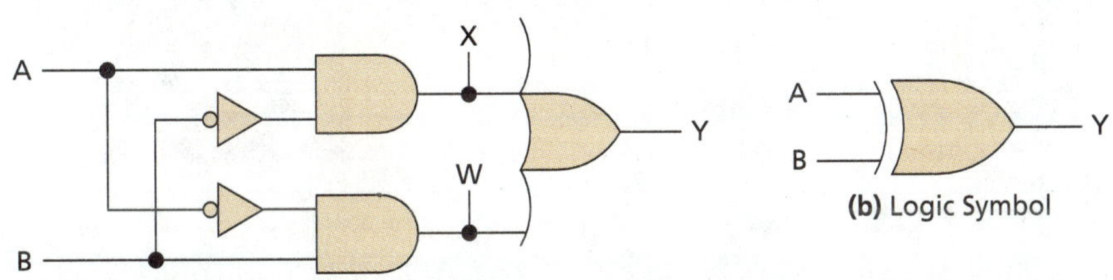

(a) Logic Function $Y = A \cdot \overline{B} + \overline{A} \cdot B$

(b) Logic Symbol

A	B	Y
0	0	0
0	1	1
1	0	1
1	1	0

(c) Truth Table

$Y = A \oplus B$

(d) Logic Expression

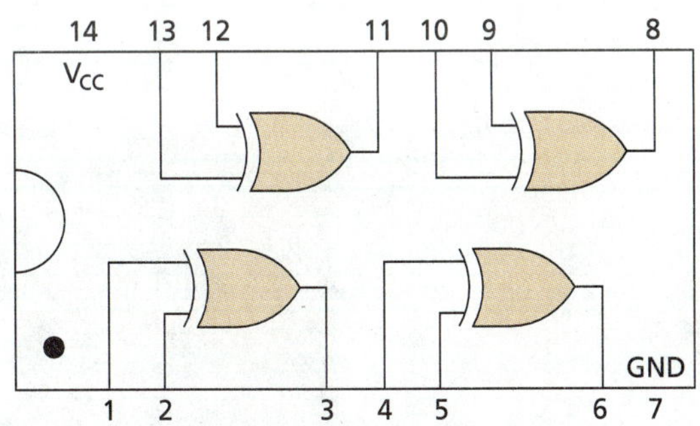

(e) 74LS86 QUAD Two-Input XOR Gate IC

Figure 1.25 Exclusive OR Logic Gate

$\overline{A}$ and $\overline{B}$ are the complements of inputs A and B:

A	B	$\bar{A}$	$\bar{B}$	$A \cdot \bar{B}$	$A \bar{\cdot} B$	$Y = A \cdot B \bar{+} A \cdot \bar{B}$
0	0	1	1			
0	1	1	0			
1	0	0	1			
1	1	0	0			

Formulate intermediate outputs $A \cdot \overline{B}$ and $\overline{A} \cdot B$ by doing the AND function on one noncomplemented input and the complemented input for the intermediate output:

A	B	$\bar{A}$	$\bar{B}$	$A \cdot \bar{B}$	$A \cdot \bar{B}$	$Y = A \cdot \bar{B} + \bar{A} \cdot B$
0	0	1	1	0	0	
0	1	1	0	0	1	
1	0	0	1	1	0	
1	1	0	0	0	0	

Last, the intermediate AND outputs supply the inputs to the OR gate whose output is a 1 if either of the AND outputs is a 1:

A	B	$\bar{A}$	$\bar{B}$	$A \cdot \bar{B}$	$A \cdot \bar{B}$	$Y = A \cdot \bar{B} + \bar{A} \cdot B$
0	0	1	1	0	0	0
0	1	1	0	0	1	1
1	0	0	1	1	0	1
1	1	0	0	0	0	0

***Exclusive OR** or **XOR** logic function where the output is a 1 if the total number of inputs that are logic 1 is odd. Uniquely, the two-input XOR function yields a logic 0 out when both inputs match, and a logic 1 when they do not match.*

By removing the intermediate output terms and presenting the abbreviated truth table, we describe a new logic function called an **exclusive OR** or **XOR.** The function is represented by a plus symbol in a circle, $Y = A \oplus B$, and has the logic symbol shown in Figure 1.25b. The final form of truth table for the exclusive OR gate is

A	B	$Y = A \oplus B$
0	0	0
0	1	1
1	0	1
1	1	0

The 74LSxx family includes the 74LS86 quad exclusive OR integrated circuit shown in Figure 1.25e.

Comparator *logic circuit that compares the states of binary signals.*

The first property of an exclusive OR is that its output is 0 when both inputs are the same (0 and 0 or 1 and 1) and a 1 when they are not (0 and 1 or 1 and 0). This property allows this circuit to be used as a digital **comparator** for comparing the state of one binary input to another binary input. If both inputs have the same state, then the output of this circuit is a low. Dissimilar inputs are indicated by a high state.

A second property of the exclusive OR defines the basic function for the XOR gate: When the number of inputs that are a logic 1 state are even, the output of an XOR gate is low. An odd number of logic 1 inputs forces the XOR output high. For the two-input XOR gate, when both inputs are in the same state the number of logic 1 inputs are even and the output is low. When the inputs are in opposite states, the number of logic 1 inputs are odd and the output is high. Integrated-circuit versions of the exclusive OR are limited to two inputs, but the exclusive OR function can be expanded by combining exclusive OR gates, as in Figure 1.26.

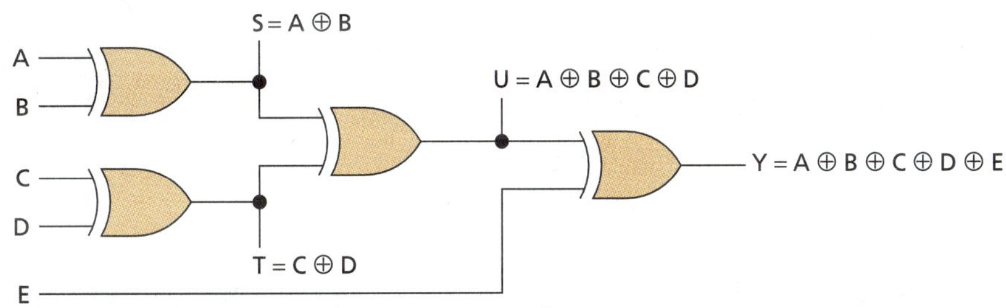

Figure **1.26** Multiple Exclusive OR Function

Intermediate outputs S and T are the XOR results of A and B, C and D: $A \oplus B$ and $C \oplus D$. Intermediate output U combines the two previous intermediate outputs through a third XOR gate to yield the exclusive OR expression $U = A \oplus B \oplus C \oplus D$. This output then becomes the input to the last XOR gate, along with input E, so the final expression appears as a five-input XOR function:

$$Y = A \oplus B \oplus C \oplus D \oplus E$$

By definition, the output of this circuit is low when the number of inputs that are a logic 1 are even and high when that number is odd. If inputs A, B, and C are high and D and E are low, then the number of inputs that are logically high is odd, and output Y is also high. Changing input E to a high now makes the number of logically high inputs even, and causes output Y to go low. Note that the order in which the exclusive OR function is applied to the inputs for this function does not matter. The result at output Y will be the same.

Parity generator *a circuit that creates a parity bit used for error checking digital data.*

One use of the exclusive OR, based on the comparator function, is to get an odd or even count of the number of 1's in a string of binary digits. A circuit that performs this operation is called a **parity generator**. When digital data are moved from one place to another, it is nice to know that the data arrived correctly and without errors. This is a major function of parity. A set number of bits of the binary data to be sent are selected as a group for the purpose of using parity. The number of logical 1's in that group are checked for an odd or even count, and a parity bit is generated and appended to the end of the group of bits currently being sent. The actual state of this parity bit depends on which type of parity system, odd or even, is going to be used. If the system used is an even parity system, the generator sets the parity bit high if the count is odd and low if the count is even. In an odd parity system, the opposite takes place: The parity bit is set high if the count is even and low if it is odd.

For instance, let's use an eight-bit group of bits with the binary value of 11001110 to illustrate the process. If the system in use is even parity, then the parity bit is set high and appended at the end of the group in the most significant bit position: **1**11001110. The group of bits, including the parity bit, are sent to wherever they need to go. At the receiving end, another parity generator counts the number of logic 1's in the eight-bit group and creates a second parity bit. The parity bit sent is then compared to the parity bit created at the receive side. If they do not match, then an error exists in one of the data bits. If the two parity bits match, then there is a good possibility that the data arrived correctly. However, if an even number of bits are incorrect, then the parity state will be identical for the incorrect and correct data. In the example data used, 11001110, the even parity bit state is high. Suppose the last two bits are incorrect, making the data 11001101. Once again, the parity bit would take a high state.

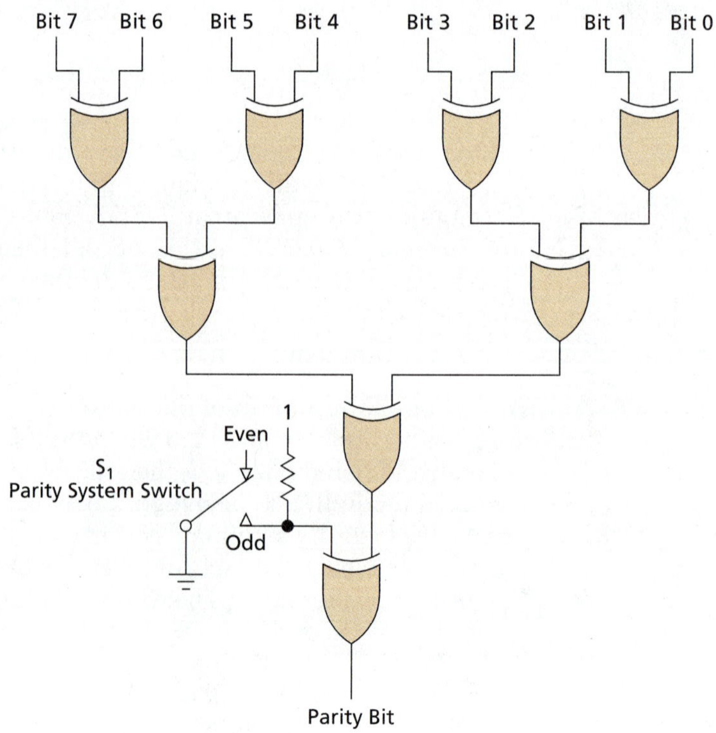

Figure 1.27 Parity Generator

Since the exclusive OR function produces two different results based on the total number of logic 1 levels at its input, it is used to generate the parity bits. The parity generator circuit for the eight-bit data word is shown in Figure 1.27. Each pair of data bits are compared, using a two-input XOR gate. If the inputs match, the number of logic 1's are even (0,0 and 1,1) causing the XOR output to go low. An odd count of logic 1's (0,1 and 1,0) forces the XOR output high. The outputs of the XORs are then compared, using additional XORs, until a single output is finally generated. This final output is the state of the parity bit in an even parity system. Causing the parity output to be the result of an odd parity system requires the even parity result to be complemented or inverted. This inversion takes place when a signal is sent to the parity generator requesting the use of odd parity. This is accomplished using another XOR because of an additional property of the exclusive OR function, controlled inversion.

A third property of a two-input exclusive OR gate uses input A as a control input and input B as binary data input. Control inputs are used to cause the output of a logic circuit to have a desired effect on another input. For this circuit, the control has the effect of causing the data to be passed to the output in either noncomplemented or complemented form. Notice on the exclusive OR truth table, when the control input, say input A, is a 0, the output has the same state as the binary data on input B. When the control input is a 1, the output has the opposite state of the input data. Therefore, this circuit can be used as a controlled inverter.

The even parity result is converted to an odd parity system result by setting the control input of the last XOR high, causing the parity results to be inverted.

Section 1.11

Practice Problems

1. Develop the truth table for the combinational circuit in Figure 1.28. What is the logic expression for this circuit? What is the equivalent gate for this circuit? *Hint:* Compare the truth table to the truth tables in Section 1.10.

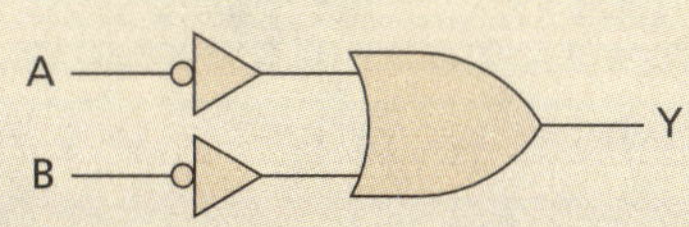

Figure 1.28 Practice Problem 1.11(1)

2. Design a logic circuit that compares the state of input A with the state of B. If they are the same, the output is high; if not alike, the output is low. Write out the truth table and logic expressions on which you base your design.

1.12 Sum-of-Products (SOP) and Product-of-Sums (POS) Logic Expressions

Sum of products (SOP) *form of logic expression that logically combines a number of AND terms using an OR function.*

Product of sums (POS) *form of logic expression that logically combines a number of OR terms using an AND function.*

Logic expressions can be written in one of three forms, the **sum of products (SOP),** the **product of sums (POS),** and a combination of SOP and POS forms. SOP and POS terms derive their name from associating Boolean algebra logic OR functions with arithmetic addition and logic AND functions with multiplication. Boolean algebra, its uses, and applications are described in Chapter 2.

The SOP format is a logic expression that includes a number of AND terms that are logically connected through an OR function. For example, the logic expression for an exclusive OR, $A\overline{B} + \overline{A}B$, is in the SOP format. The first term is an AND function of the variables A AND $\overline{B}$, and the second is an AND combination of $\overline{A}$ AND B. These two AND or product terms logically combined through an OR function to form the sum portion of the SOP expression.

Another example expression for the SOP format is

$$AB\overline{C} + \overline{A}CD + A\overline{B}C\overline{D} + \overline{B}D$$

Ignoring the NOT bars, this expression looks like the mathematical sum of several products. Each AND function is similar in appearance to a product term in mathematics. The product terms then appear to be summed together by the use of the OR plus symbols.

The POS format is an opposite form of logic expression that appears as a number of sums that are being multiplied together. In logic terms, it is a number of OR terms that are combined through the use of an AND function. As an example,

$$(A + B + \overline{C})(\overline{A} + C + D)(A + \overline{B} + \overline{C} + D)(\overline{B} + D)$$

is in a POS format. Each term within parentheses appears as if it is the sum of algebraic variables. A parenthesis is another format used in mathematics to represent multiplication or product functions. Chapter 2 shows how one form of an expression can be changed into the other form using DeMorgan's theorems.

SOP and POS forms are often mixed in a single logic expression as required by the context of the logic expression. An example of a mixed format is

$$[(A + B)(A \cdot B + A \cdot \overline{C} + \overline{C} \cdot D) + A \cdot B + C \cdot D](\overline{A} \cdot B + B \cdot \overline{C})$$

The overall expression is in a POS form of two terms. To see the formats being used, let's simplify the expression by substituting X for the expression $A + B$, and Y for the expression $A \cdot B + A \cdot \overline{C} + \overline{C} \cdot D$. The revised expression with the substitutions becomes

$$(X \cdot Y + A \cdot B + C \cdot D)(\overline{A} \cdot B + B \cdot \overline{C})$$

which is in the POS form. Both of the POS terms are, individually, in the SOP form, as are the expressions substituted by X and Y. The term $X \cdot Y$, once the original expressions are reinstated, is in the POS format:

$$(A + B)(A \cdot B + A \cdot \overline{C} + \overline{C} \cdot D)$$

Many combinational logic expressions result in a mixed SOP and POS format. In Chapter 2 we will see how to change the mixed format into a pure SOP or pure POS expression.

Minterms *terms in a sum-of-products logic expression.*

Maxterms *terms of a product-of-sums logic expression.*

Two additional terms are associated with the SOP and POS forms of logical expressions. They are **minterms** for the terms of a SOP expression and **maxterms** for the terms of a POS expression. They seem to suggest that an expression written in SOP form contains minimum-size terms and those written in POS form are maximum-size expressions. I present these two terms as a matter of historical reference and will not use them in the text. The reason I feel they are inaccurate is shown by these two logical expressions which represent the same circuit:

$$Y = (A + B)(C + D)$$

$$Y = A \cdot B + A \cdot C + B \cdot C + B \cdot D$$

In the POS "maxterms" expression, there are two OR functions and one AND function. In the SOP "minterms" expression, there are three OR functions and four AND functions. Which form is minimum, and which is maximum? Nevertheless, when you are reading information on digital circuits and expressions are referred to as containing minterms, you know they are discussing a SOP expression and for maxterms, a POS expression.

Section 1.12

Practice Problems

Which of these expressions is an SOP, POS, or mixed form?

1. ABC + DEF + AC + CE + BDE
2. (AB + C + E)(AB + CDE)(ACE + BDE)
3. (ABC + BDE)[(AB + ACE + DE) + (ABC + AD)(A + B + C)]

1.13 Logic Expressions from Truth Tables

Combinational logic uses the four basic building blocks, the AND, OR, buffer, and inverter, along with the first level of developed logic circuits, the NAND, NOR, and XOR, to form logic circuits based on needed criteria. For instance, suppose a logic circuit is required to satisfy this control situation: A pattern-cutting machine designated M can be operated if (1) the operator is qualified (Q) to operate it or (2) a trainee (T), scheduled to operate the machine, is accompanied by a foreman (F). A truth table can be developed from the statement to satisfy the requirements. Inputs are qualified operator (Q), trainee (T), and foreman (F), and the output is "operate the machine" (M):

Q	T	F	M
0	0	0	
0	0	1	
0	1	0	
0	1	1	
1	0	0	
1	0	1	
1	1	0	
1	1	1	

One condition under which the machine can be operated (M = 1) is by a qualified operator (Q = 1), so the lower half of the truth table where the Q input is high yields an output high state for M. The only other time that the machine may be operated is by a trainee (T = 1) when a foreman is present (F = 1) to observe how the trainee does. On the truth table, this condition occurs when both T and F are high. The remaining input combinations indicate when the machine may not be operated and are shown as a logic 0 state for M:

Q	T	F	M
0	0	0	0
0	0	1	0
0	1	0	0
0	1	1	1
1	0	0	1
1	0	1	1
1	1	0	1
1	1	1	1

The problem is to come up with a logic expression and diagram for the circuit that produces the result indicated on the truth table. One way that always works is to interpret the truth table directly. In this case, we can see that as long as Q is high, the output is high. When Q is low, we have to look at the remaining inputs to determine what combinations of T and F make M go high. On the upper half of the truth table, the only time M is high is when T and F are both high. Basically, if T and F are high, M is high. An AND gate is required for this function and is used as an intermediate output to drive one input of an OR gate. The other input of the OR gate in the circuit is connected to Q. The circuit for this problem is shown in Figure 1.29.

The logic expression shows a sum-of-products expression that describes the truth table and logic circuit:

$$M = T \cdot F + Q$$

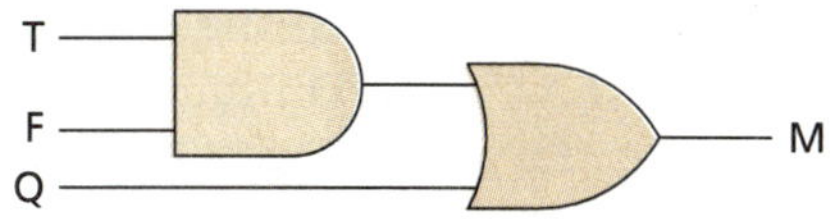

Figure 1.29 M = Q + TF

As long as the design criteria remain comparatively uncomplicated, you can derive the circuit using this process. However, for more complex logic circuits and their truth tables, a different approach is made to develop the logic expression from the truth table of the circuit. To read an expression from a truth table, first consider the states of the inputs. They are 0 or 1, or, in effect, they are in their active high state or their complement. For instance, the first line of the pattern machine truth table shows all the inputs low, can be represented as Q, T, and F = 0, 0, 0 or $\overline{Q}$, $\overline{T}$, and $\overline{F}$ = 1, 1, 1. The second line can be shown as Q, T, and F = 0, 0, 1 or $\overline{Q}$, $\overline{T}$, and F = 1, 1, 1 and so forth. Instead of writing the input side of the table in terms of 1's and 0's, it can be rewritten showing the states of the inputs that produce an active high condition. When this is done, the state of each output is derived from the AND combination of the input states. Return to the first line of the truth table that we just developed for the pattern cutter. This line can be rewritten as

INPUTS			M
$\overline{Q}$	$\overline{T}$	$\overline{F}$	0

which simply says that when all three inputs are low, the output is low. The entire table, replacing 1's and 0's with input variable is

INPUTS			M
$\overline{Q}$	$\overline{T}$	$\overline{F}$	0
$\overline{Q}$	$\overline{T}$	F	0
$\overline{Q}$	T	$\overline{F}$	0
$\overline{Q}$	T	F	1
Q	$\overline{T}$	$\overline{F}$	1
Q	$\overline{T}$	F	1
Q	T	$\overline{F}$	1
Q	T	F	1

The benefit of creating this type of table is that a SOP logic expression can be formed by placing an OR function between each group of AND inputs that produce a 1 state at the output. In this example, the last five lines of the truth table show a high output at M. The product terms that produce these 1 states are $\overline{Q} \cdot T \cdot F$, $Q \cdot \overline{T} \cdot \overline{F}$, $Q \cdot \overline{T} \cdot F$, $Q \cdot T \cdot \overline{F}$, and $Q \cdot T \cdot F$. These are summed together for the sum-of-products expression for the circuit represented by this truth table:

$$M = \overline{Q} \cdot T \cdot F + Q \cdot \overline{T} \cdot \overline{F} + Q \cdot \overline{T} \cdot F + Q \cdot T \cdot \overline{F} + Q \cdot T \cdot F$$

This expression for the pattern cutter can now be simplified by applying the rules, laws, and axioms of Boolean algebra that will be discussed in Chapter 2.

EXAMPLE 1.7

What is the logic expression for the circuit that generates this truth table?

A	B	C	Y
0	0	0	0
0	0	1	1
0	1	0	1
0	1	1	1
1	0	0	0
1	0	1	0
1	1	0	1
1	1	1	0

SOLUTION Rewrite the truth table to show product terms:

INPUTS			Y
$\overline{A}$	$\overline{B}$	$\overline{C}$	0
$\overline{A}$	$\overline{B}$	C	1
$\overline{A}$	B	$\overline{C}$	1
$\overline{A}$	B	C	1
A	$\overline{B}$	$\overline{C}$	0
A	$\overline{B}$	C	0
A	B	$\overline{C}$	1
A	B	C	0

and then sum all the product terms that yield a logic 1 level at output Y to form the SOP expression of the circuit that produced the truth table result:

$$Y = \overline{A} \cdot \overline{B} \cdot C + \overline{A} \cdot B \cdot \overline{C} + \overline{A} \cdot B \cdot C + A \cdot B \cdot \overline{C}$$

Section 1.13

Practice Problems

Develop the logic expression of the logic circuit that generates the truth table in Figure 1.30.

A	B	C	Y
0	0	0	0
0	0	1	0
0	1	0	1
0	1	1	1
1	0	0	1
1	0	1	1
1	1	0	0
1	1	1	0

Figure 1.30 Practice Problem 1.13

1.14 Troubleshooting Logic Circuits

Static combinational logic *logic circuits that produce outputs based on logic state inputs that do not change state with respect to time.*

The types of logic circuits that have been currently discussed are called **static combinational logic** circuits. A steady state, unvarying voltage is applied to inputs that represent a logic 1 or a logic 0 level. The output result is also a steady state logic level. The term *combinational* means that a combination of logic gates or devices is used in the circuit described by a logic expression that shows the logic relationship of the output to the inputs. Static logic circuits are comparatively easy to troubleshoot. A selected set of inputs are applied, and the output is checked. If the output state is correct for the first set of inputs, change the input conditions and recheck the output. Repeat this process for all possible combinations of inputs until you are satisfied that they all work correctly or until an incorrect result appears. Once an incorrect output is detected, intermediate states are checked to find out where the problems lies.

To illustrate the process of troubleshooting a logic circuit, let's go through the steps to check out the circuit in Figure 1.31. The first step in troubleshooting a logic circuit is for the troubleshooter to know what the conditions are supposed to be throughout the circuit. Known conditions are compared to the circuit as it is currently performing to determine if there is something wrong with the circuit. The known conditions for this logic circuit are presented on its truth table, with four inputs having a total of 16 different input combinations and output results. For purposes of showing the troubleshooting process, we only need to use the first two lines of the truth table. The first two lines of the truth table of the logic circuit in Figure 1.31 are

A	B	C	D	AB	$\bar{C}$	$AB + \bar{C}$	$\bar{C}D$	$Y = \overline{(AB + \bar{C})(\bar{C}D)}$
0	0	0	0	0	1	1	0	1
0	0	0	1	0	1	1	1	0

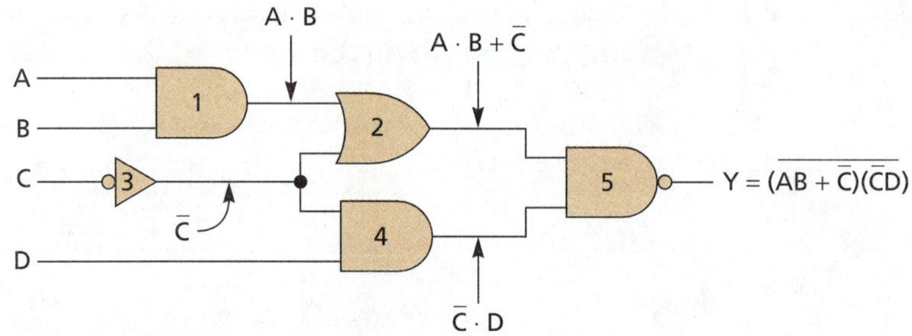

A	B	C	D	AB	$\overline{C}$	AB + $\overline{C}$	$\overline{C}$D	Y
0	0	0	0	0	1	1	0	1
0	0	0	1	0	1	1	1	0
0	0	1	0	0	0	0	0	1
0	0	1	1	0	0	0	0	1
0	1	0	0	0	1	1	0	1
0	1	0	1	0	1	1	1	0
0	1	1	0	0	0	0	0	1
0	1	1	1	0	0	0	0	1
1	0	0	0	0	1	1	0	1
1	0	0	1	0	1	1	1	0
1	0	1	0	0	0	0	0	1
1	0	1	1	0	0	0	0	1
1	1	0	0	1	1	1	0	1
1	1	0	1	1	1	1	1	0
1	1	1	0	1	0	1	0	1
1	1	1	1	1	0	1	0	1

Figure 1.31 Troubleshooting Example

The output of gate 1 is AB, which at the first line is 0 (a 0 into an AND gate forces the output of that AND gate to 0). $\overline{C}$ is the complement of C and is a 1 on the first line. When AB is ORed with $\overline{C}$, the result on the first line is 1, since any 1 into an OR gate produces a 1 out. The next point in the circuit is the result of ANDed $\overline{C}$ with input D. Since D is a 0 on line 1, the output of AND gate 4 is a 0. Output Y is derived from the result of NANDing AB + $\overline{C}$ with $\overline{C}$D, which is a high, because $\overline{C}$D is low and any low into a NAND gate forces the output high. The states of each part of the circuit in the second line are determined in a similar fashion.

On testing the output of Y with the inputs as shown on the second line of the table, the output appears to be a high instead of the expected low level. Something is wrong. You can proceed to check out the circuit from one of two directions. You may begin at the last gate, the NAND, in this example. We already know that the output is incorrect, so we must check its inputs. In this scenario, the output of AND gate 4 is 0 and OR gate 2 is a 1. Checking the truth table, OR gate 2, which is AB + $\overline{C}$, should be a 1. This means that at this point AND gate 1 and inverter 3 are probably all right. The output of AND gate 4 checks out to be incorrect, so the problem lies with this gate. Does this mean AND gate 4 is necessarily bad? No, there could be a problem with the inputs to the AND gate, and they are the next points to check. On checking these inputs, though, we find them to be correct, so AND gate 4 turns out to be the culprit.

An alternate plan of attack is to start at the inputs and work your way toward output Y. In this example, inputs A, B, C, and D are all correct. Input C drives an inverter gate, which is a good point to check next, because the inverter output is used as inputs to two additional gates. Since we know what the problem is going to be from the preceding paragraph, we expect the inverter gate to be good. On to the next level—check the output of the AND gate supplied by the inverter and input D. Again, as with the previous paragraph, we find this AND gate to be bad.

By comparing both procedures, it may look as if doing the checking from the inputs toward the outputs is a quicker method. Don't be easily fooled—in this case starting at the beginning would find the trouble sooner. Just suppose the trouble is with NAND gate 5 at the output—then starting the search at the output and working toward the inputs is quicker. There is no hard and set rule as to which direction you should use at any given time. It depends a lot on the circuit and on which place in the circuit is causing problems.

Logic probe *a portable testing device that detects logic states.*

The logic states of the inputs and outputs can be checked using several different types of test equipment. Some types of equipment provide the troubleshooter with more information than others. Other types may be easier and faster to use, but do not present the entire picture. One of the more popular types of troubleshooting devices used for logic circuits is the **logic probe**, shown in Figure 1.32. The red (V_{CC}) lead attached to the probe is connected to the positive voltage source for the logic chips being used (usually equal to a logic 1 level) and the black lead (gnd) to the negative supply (which is usually the ground and equal to a logic 0). The tip of the probe is then placed on the circuit point to be tested. Lights on the probe indicate if the state at that point is a logic 1 or a logic 0, or is switching or pulsing between these two logic levels.

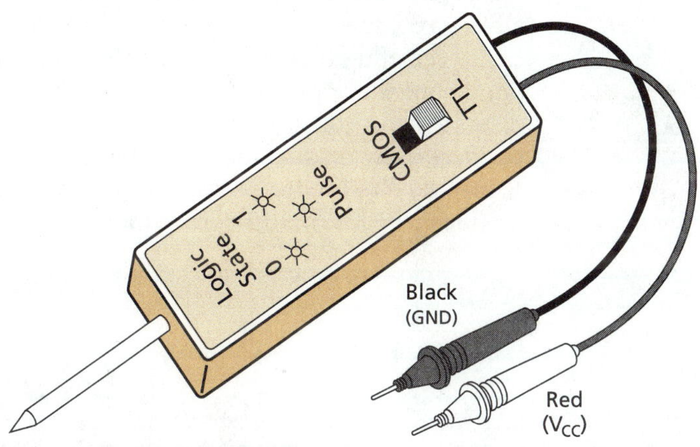

Figure 1.32 Logic Probe

Logic probes are easy to use and can help trace a signal through a circuit, like the one in the previous example. There is one problem with logic probes: They do not present the entire picture of what is going on and can sometimes give an incorrect indication of a logic state. For instance, if a circuit uses 5 V for a logic 1 and ground (0 V) for a logic 0, a probe will indicate when an input or output is in the wrong state. However, if the input or output point under test is at a voltage level between the 5 V and 0 V—say 2.0 V, which, for some logic devices is neither a logic 1 or a logic 0—the indication on the probe could be a logic 1, which may be the anticipated state of that point.

This problem becomes more pronounced depending on the technology of the logic device being tested. One technology, which uses high-input impedance FETs, called MOS, requires very small input currents to switch logic states. If you are testing a MOS logic circuit that has an open lead to an input, the terminating impedance of the logic probe will cause the MOS device to sense a logic input state at the open lead. For example, suppose you were looking for a logic 1 input level to a MOS device, and you placed the probe onto an open lead, which now terminates that input and the MOS device thinks it sees a logic 1 level. The probe indicates the logic 1, and you happily go to the next point in the circuit, looking for your trouble elsewhere, when actually the probe failed to indicate the open lead. The point here is that although you are not yet familiar with the different types of logic families of devices, you need to be aware of the limitations of logic probes that you may be using for experiments in a laboratory course you are taking concurrently with a lecture course.

Dynamic logic *logic circuits that continually change state with respect to time and usually synchronized to another logic signal.*

Pulse *dynamically changing logic signal.*

Logic probes are useful but limited in their application. Depending on the quality of the logic probe, their degree of limitation may increase once you start to probe around **dynamic logic** circuits. Dynamic circuits are those that are constantly changing state. This change is usually directed by a signal that switches from a logic 1 to a logic 0 at a specified rate. Dynamic logic circuits are discussed later. Logic probes indicate an input or output that is changing dynamically, called a **pulse** on the probe. The problem with this indication is that it doesn't tell you at what rate the logic level is changing, or what the change looks like. There could be problems with how quickly the change takes place, or what actual voltage levels the signal is switching between. Are they really correct logic states for a high and a low?

Oscilloscope *test device that shows a visual picture of a signal.*

A piece of test equipment that gives the troubleshooter more detailed information about the activity within a logic circuit is the **oscilloscope.** This test device paints a picture of what the point under test looks like electrically. Voltage levels can be checked by reading the vertical displacement from a central line on the oscilloscope's screen. Inputs or outputs that are dynamic show the changes as a dot that produces the line on the screen moves horizontally from left to right. Figure 1.33

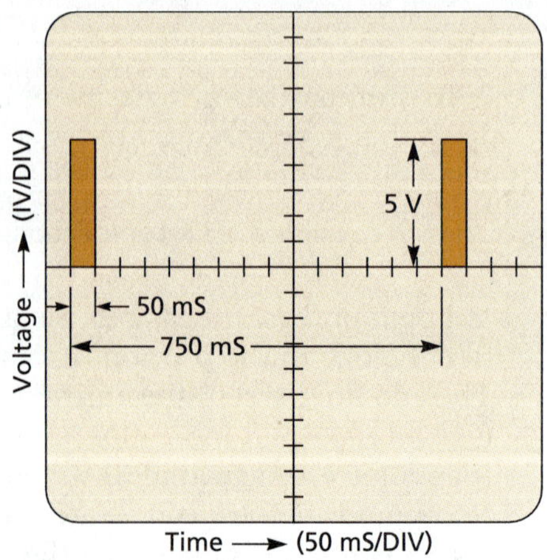

Figure 1.33 Oscilloscope Display

shows a representation of an oscilloscope picture of an AND gate output that is changing dynamically. The vertical deflection, scaled at 1 V per division, shows a change in state from a 0 V level to a 5 V level. The horizontal sweep, scaled in 50 millisecond (mS) divisions, shows the signal going high for 50 ms and returning low. This 50 mS high period repeats every 750 mS in the sample depicted.

The horizontal sweep rate can be changed so that the center line goes across the screen in 50 mS instead of 1 second, by changing the time per division scale from 50 ms per division to 2.5 mS per division. When the oscilloscope is set up this way, the 50 mS pulse can be examined in more detail, as shown in Figure 1.34. The voltage levels are still the same: 5 V for logic 1 and 0 V for logic 0, but you can now see that the output takes some time to change from one state to the other, and there is even some undershoot or ringing on the negative edge. These factors could be important in determining why a logic circuit acts in a particular fashion.

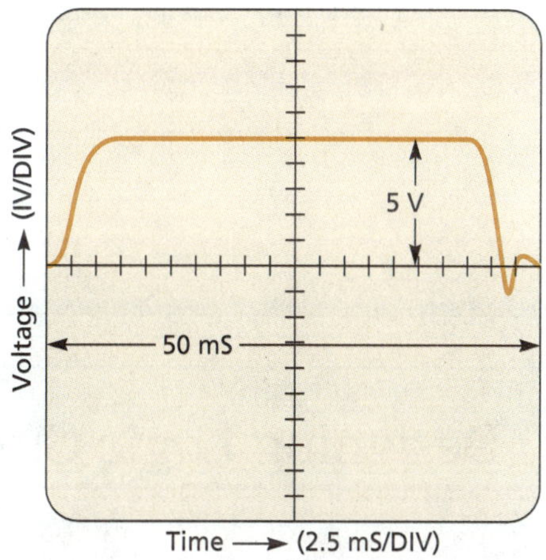

Figure 1.34 Expanded Horizontal Sweep

To summarize, a logic probe is a handy device for quick circuit tracing through a logic circuit, but when detailed information is needed to help analyze a circuit, a more refined type of test equipment, such as an oscilloscope, is required.

Section 1.14

Practice Problems

1. In the circuit of Figure 1.35, setting all the inputs to a 1 forces the output to a 1. Determine if this is correct, and if it is not, determine what is wrong given the logic levels shown at each point in the circuit.

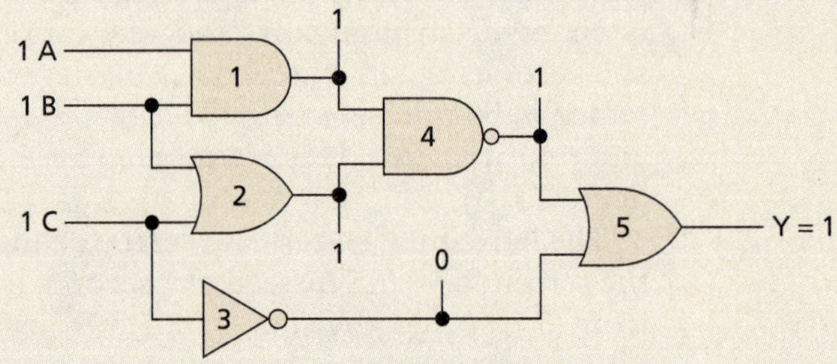

Figure 1.35 Practice Problem 1.14(1)

2. Repeat the instructions of Problem 1 for the circuit in Figure 1.36.

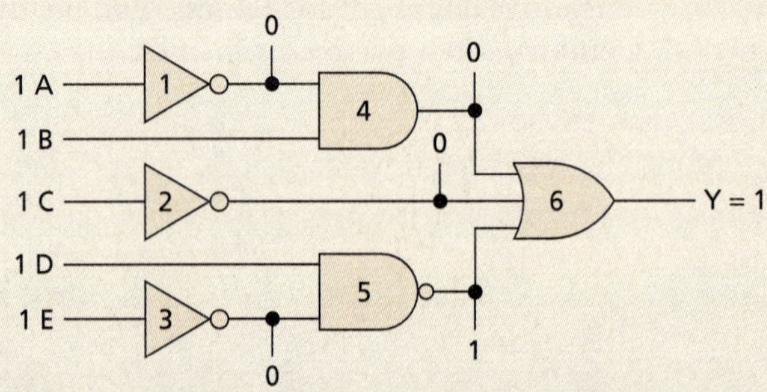

Figure 1.36 Practice Problem 1.14(2)

1.15 Designing Logic Circuits

We have already done some basic logic design in some of the preceding sections of this chapter. Let's now formalize the process. Designing logical circuits usually begins with the statement of a logic problem that requires resolution. As I discuss the process, I will use the following example of a problem requiring a logic circuit solution:

Statement of the Problem

We have a new security system that is capable of sounding an alarm and calling either the police or fire department as needed. The system monitors the windows and doors of the home against breakins. It also senses smoke, heat, and carbon monoxide (CO) to detect the possibility of fire. The system will only operate if it is armed.

Criteria

Once the problem is stated, the next step in the process is to clearly understand the exact criteria needed for the circuit's operation. In our security example, the criteria are as follows:

1. The alarm will sound if the system is armed and
   a. Someone is breaking in a window or
   b. Jimmying open a door or
   c. Smoke or heat or CO is detected
2. The police are called if the system is armed and there is a breakin.
3. The fire department is called if the system is armed and smoke, heat, or CO (carbon monoxide) is detected.

Once the problem is stated and criteria are established, it is now time to select logic levels to represent the states of the variables within the criteria. At this time, you also should establish which are input variables and which are outputs. Continuing with our example, the following states are chosen:

Selecting Logic States

1. Inputs:

A System:	1 = armed	0 = off
B Window:	1 = broken	0 = OK
C Door:	1 = open	0 = closed
D Smoke, heat, CO:	1 = present	0 = absent

2. Outputs:

E Alarm:	1 = 0	0 = off
F Police:	1 = call	0 = don't call
G Fire department:	1 = call	0 = don't call

Now that logic states have been assigned to the variables to be used in the circuit, it is time to create a truth table showing the relationship between the inputs and the outputs. Our table has these variables:

INPUTS						OUTPUTS		
ARMED	WINDOW	DOOR	FIRE	SMOKE	CO	ALARM	POLICE	FIREMEN

The criteria we have specified state that the system does not operate unless it is armed. Another way to make this statement is, the system does not operate, regardless of the other input states, as long as the system is not armed (ARMED = 0), the alarm is off (ALARM = 0), and neither the police or firemen are called (both 0). To shorten the length of the truth table, we shall use a new symbol, X, which is interpreted, in a truth table, to mean, "don't care." It indicates that the state of a variable, with respect to the output, does not matter. It can be a logic 1 or a logic 0, and the outcome will be the same. We will use the X symbol on the first line of the truth table to show that all the inputs are "don't cares" when the system is unarmed.

ARMED	WINDOW	DOOR	FIRE	SMOKE	CO	ALARM	POLICE	FIREMEN
0	X	X	X	X	X	0	0	0

This first line states that as long as the system is not armed, the alarm will not sound and no calls are made regardless if there is a breakin or fire. The remaining lines of the truth table reflect the output conditions once the system is armed. On the second line, we show the results of no alarm or calls if the system is armed and there is no breakin or fire. Next we must decide what actually causes the alarm to sound and the calls to be made. If a window is broken in or a door forced open, the alarm sounds and the police are called. This occurs whether or not there is a fire. On our table, the next two lines reflect these conditions.

ARMED	WINDOW	DOOR	FIRE	SMOKE	CO	ALARM	POLICE	FIREMEN
0	X	X	X	X	X	0	0	0
1	1	X	X	X	X	1	1	*
1	X	1	X	X	X	1	1	*

Once again, the "don't care" symbol is used, this time for fire, smoke, and CO inputs. An asterisk (*) is placed in the table under the FIREMAN column for these two inputs, because the state of that signal will depend on other entries in the table. Lastly, if there is smoke, heat, or CO, sound the alarm and call the fire department regardless if a breakin is occurring or not occurring at the same time:

ARMED	WINDOW	DOOR	FIRE	SMOKE	CO	ALARM	POLICE	FIREMEN
0	X	X	X	X	X	0	0	0
1	0	0	0	0	0	0	0	0
1	1	X	X	X	X	1	1	*
1	X	1	X	X	X	1	1	*
1	X	X	1	X	X	1	*	1
1	X	X	X	1	X	1	*	1
1	X	X	X	X	1	1	*	1

*Stage determined by other lines on the table.

Once we have completed the truth table, it is time to determine the logic expressions for each of the outputs. "Don't care" places for variables can be ignored in the product terms for the SOP expressions. To cause the alarm to go on, the expression read from the table is

$$\text{ALARM} = \text{ARMED} \cdot \text{WINDOW} + \text{ARMED} \cdot \text{DOOR} + \text{ARMED} \cdot \text{SMOKE} + \text{ARMED} \cdot \text{HEAT} + \text{ARMED} \cdot \text{CO}$$

To call the police, the logic expression is

$$\text{POLICE} = \text{ARMED} \cdot \text{WINDOW} + \text{ARMED} \cdot \text{DOOR}$$

To call the fire department, the expression is

$$\text{FIREMEN} = \text{ARMED} \cdot \text{SMOKE} + \text{ARMED} \cdot \text{HEAT} + \text{ARMED} \cdot \text{CO}$$

The design can be constructed and tested to prove it meets the criteria and solves the problem statement with which we began. The actual logic circuit derived from the logic expression is shown in Figure 1.37.

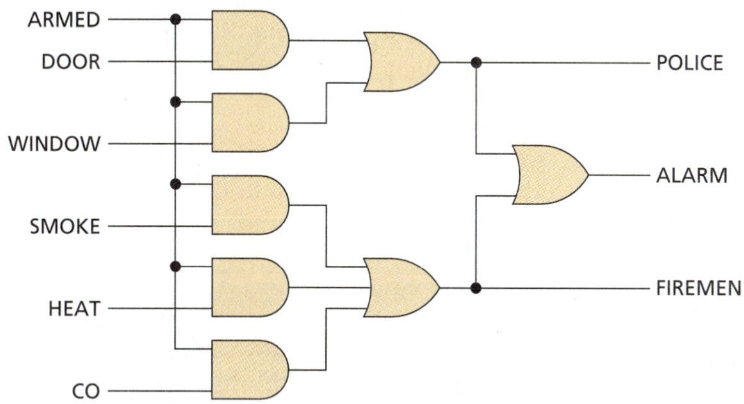

Figure 1.37 Design Circuit

The two circuits that call the police and fire department also form the two halves of the circuit that sets off the alarm. These two halves are brought together by an OR function of POLICE and FIREMEN.

Section 1.15

Practice Problems

1. Design a logic circuit that allows a computer to "BOOT" or come into a usable state if a SYSTEM disk is in DRIVE A or if DRIVE C has the boot-up program. This boot-up occurs on POWER or a RESET signal being applied to the computer.
2. Design a soft drink dispenser that yields a can of DRINK if two QUARTERS or a DOLLAR bill is inserted into the money slot and there is a can of drink (SELECTION) in the machine to dispense. Change is RETURNed if a DOLLAR is inserted, there is CHANGE left in the machine, and a can of drink (SELECTION) is available to be dispensed.

Summary

Digital electronic circuits are used extensively in all forms of commercial, industrial, and home electronics equipment. The more familiar uses are microwave ovens, television sets, VCRs, personal computers, and automobile engine control. Industry finds use for these circuits in automation applications and computerized work stations.

The basis of digital logic used to develop digital circuits is the binary number system, in which crucial quantities are represented by one of two states. These states are direct opposites such as On and Off, True and False, 1 and 0, or 5 V and 0 V. Truth tables are diagrams used to show the relationship between input states and output states of a logic circuit.

The four basic building blocks of digital electronic circuitry are the BUFFER, INVERTER, AND, and OR logic gates. Each has its own unique truth table and logic expression. These building blocks are combined to form other logical functions, the first of which is the exclusive OR. Two AND gates, two inverters, and one OR gate are combined to satisfy the expression $A\overline{B} + \overline{A}B$. This function is unique in that it has many direct applications. As a binary bit comparator, the output signals when inputs match or are different. The XOR can also be used as a controlled inverter where one input controls whether or not the other input is inverted.

Logic expressions defining a digital circuit can be written in either a sum-of-products form which shows ANDed variable groups ORed together, or a product-of-sums format, which shows ORed variable groups ANDed together. The sum-of-products form for a logic expression can be derived from a truth table that describes the logic circuit. The table is converted from 1's and 0's to the active state form. From this revised form, product terms that result in a logic 1 output state are read and ORed together to form the sum-of-products logic expression.

Logic circuit troubleshooting requires that the troubleshooter knows or can determine what the circuit is supposed to do. Only then can he or she determine if there is a problem with the actual circuit, and where that problem is. Test equipment such as logic probes and oscilloscopes by the troubleshooter to locate the problem. Each has its benefits and drawbacks. Logic probes are quick and easy but do not give full information about the point under test. Probes occasionally give the user a false or misleading indication. Oscilloscopes give a lot more information but are more complex to use and not as portable or handy.

Designing logic circuits starts with a problem statement and a list of logic criteria to solve the problem. Variable names and logic states are selected so that a truth table defining the relationship between inputs and outputs can be established. From the truth table, a logic expression and circuit is made to complete the design process. The final circuit must be tested and, if necessary, altered to make sure it meets the criteria and solves the original problem.

Glossary

active the logic state used to turn on a circuit or function.

amplitude measure of the quantity of an electrical characteristic, such as voltage or current, in a circuit.

analog branch of electronics dealing with circuits that operate with and/or produce electrical quantities, such as voltage, current, or power, that regularly vary in value.

AND logic function where the result is logic 1 if all the input states are logic 1's and a logic 0 for all other input combinations.

assert apply a signal or logic state that turns on a circuit or function.

base or **radix** defines the number system in use and has a value equal to the number of different values in that system, including the lowest value, zero. The decimal number system is a base 10 number system.

binary a number system based on two distinct values, 1 and 0.

buffer logic function where the output result is the same state as the input. This logic circuit is used when there is a requirement to change logic levels between circuits or to increase the drive capability of a gate's output.

central processing unit (CPU) the part of the computer that directs the actions of the remaining parts and is responsible for executing mathematical and logical operations.

combinational logic the use of different logic gates in a logic circuit to form a different logic function.

comparator logic circuit that compares the states of binary signals.

complement the opposite or inverted binary state.

computer electronic system that systematically executes a series of instructions to perform a given task.

digital system, circuit, or device that operates using binary bits and logic.

digital logic system that relates the output logic state of a digital circuit to the logic states of its inputs.

dynamic logic logic circuits that continually change state with respect to time and usually synchronized to another logic signal.

exclusive OR or **XOR** logic function where the output is a 1 if the total number of inputs that are logic 1 is odd. Uniquely, the two-input XOR function yields a logic 0 out when both inputs match, and a logic 1 when they do not match.

frequency the rate at which a cycle of an analog signal's amplitude varies within a specified time, usually 1 second.

gate or **logic gate** a single logic circuit whose output is determined by the logic states of its inputs.

inactive logic state that turns off a circuit or function.

inverter or **NOT** logic function where the output result is the opposite logic state or complement of the input.

integrated circuit a digital device that contains a number of electronic logic circuits.

logic 1 and **logic 0** represent the two opposing logic conditions of digital circuits and numbers.

logic probe a portable testing device that detects logic states.

maxterms terms of a product-of-sums logic expression.

microprocessor central processing unit on a single IC chip.

minterms terms in a sum-of-products logic expression.

NAND not AND. Logic function where the output state is a logic zero if all inputs are logic 1 and a logic 1 state for all other input conditions.

negate in terms of a logic signal, negate means to turn off a logic circuit or function.

negative logic binary logic system in which a logic 1 state has a lower voltage value than a logic 0 state.

NOR not OR. Logic function where the output state is 0 if any input is a logic 1.

NOT bar bar over a logic variable or term indicating a logic inverted or "NOT" state.

operator symbol for a logic function such as '+' for the OR function.

OR logic function where the output state is a logic one if any input is a logic one.

oscilloscope test device that shows a visual picture of a signal.

parity generator a circuit that creates a parity bit used for error checking digital data.

positive logic binary logic system in which a logic 1 state has a higher voltage value than a logic 0 state.

product of sums (POS) form of logic expression that logically combines a number of OR terms using an AND function.

pulse dynamically changing logic signal.

signal flow circuit schematic showing desired output and input logic states for correct circuit operation.

solid state devices composed of semiconductor material such as transistors.

state the current condition of an entity. As an example, the state of a light switch can be ON or OFF, UP or DOWN.

static combinational logic logic circuits that produce outputs based on logic state inputs that do not change state with respect to time.

sum of products (SOP) form of logic expression that logically combines a number of AND terms using an OR function.

truth table a diagram of columns and rows that illustrates the relationship between an entity's input and output states.

weight value of a digit based on its position in a number.

Solutions to Practice Problems

Section 1.1

Avoid static charges that can damage MOS devices.

Section 1.3

1.

ANSWER	SOUND
RIGHT	MUSIC
WRONG	BUZZER

2.

KEY	BELT	SIGN
OUT	UNFASTENED	OFF
OUT	FASTENED	OFF
IN	UNFASTENED	LIT
IN	FASTENED	OFF

3.

ANSWER	SOUND
0 = right	0 = music
1 = wrong	1 = buzzer

KEY	BELT	SIGN
0	0	0
0	1	0
1	0	1
1	1	0

Key, 0 = out, 1 = in
Belt, 0 = unfastened, 1 = fastened
Sign, 0 = off, 1 = on

4.

DOOR	KEY	HORN
0	0	0
0	1	0
1	0	1
1	1	0

Door, 0 = closed, 1 = open
Key, 0 = in, 1 = out

Section 1.6

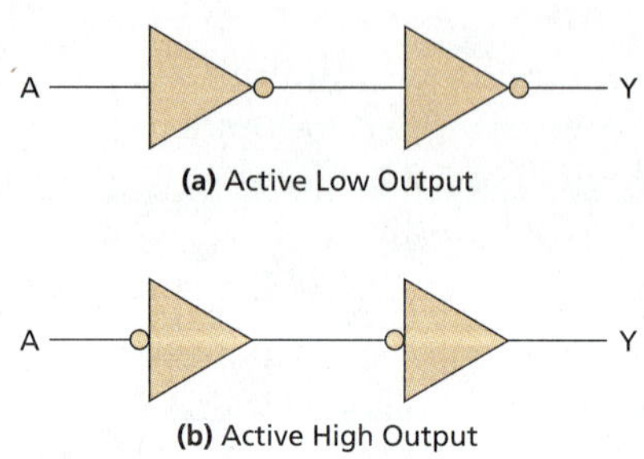

Figure 1.38 Practice Problem 1.6 (solution)

Section 1.7

1. $10011 = 16 + 2 + 1 = 19$
$11001 = 16 + 8 + 1 = 25$

2. $2^4 = 16 \quad 2^6 = 64$

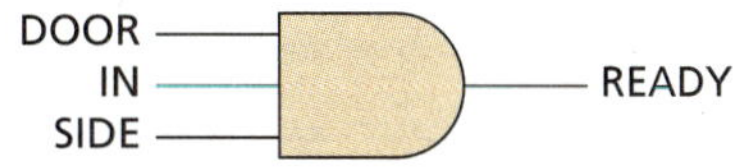

Figure 1.39 Practice Problem 1.7(1) (solution)

Section 1.8

DOOR	IN	SIDE	READY
0	0	0	0
0	0	1	0
0	1	0	0
0	1	1	0
1	0	0	0
1	0	1	0
1	1	0	0
1	1	1	1

DOOR 0 = open
1 = closed
IN 0 = disk not in
1 = disk in
SIDE 0 = wrong
1 = right

Section 1.9

1.

BUCKLE	KEY	OIL	BUZZER
0	0	0	0
0	0	1	1
0	1	0	1
0	1	1	1
1	0	0	1
1	0	1	1
1	1	0	1
1	1	1	1

2. $Y = A + B + C$

BUCKLE 0 = open
1 = closed
KEY 0 = out
1 = in
OIL 0 = o.k.
1 = low
BUZZER 0 = off
1 = on

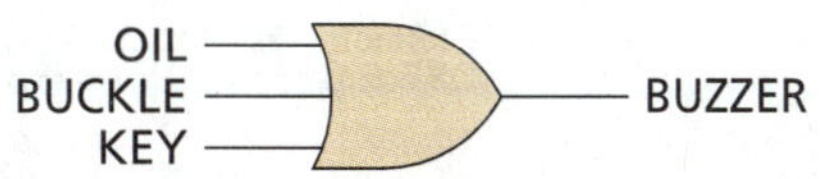

Section 1.10

1.

BUCKLE	KEY	OIL	BUZZER
0	0	0	1
0	0	1	1
0	1	0	1
0	1	1	1
1	0	0	1
1	0	1	1
1	1	0	1
1	1	1	0

$\text{BUZZER} = \overline{\text{BUCKLE} \cdot \text{KEY} \cdot \text{OIL}}$

Symbol is a three-input NAND gate.

2.

DOOR	IN	SIDE	READY
0	0	0	1
0	0	1	0
0	1	0	0
0	1	1	0
1	0	0	0

$\text{READY} = \overline{\text{DOOR} + \text{IN} + \text{SIDE}}$

1	0	1	0
1	1	0	0
1	1	1	0

Symbol is a three-input NOR gate.

Section 1.11

1. $Y = \overline{A} + \overline{B}$ is equivalent to $Y = \overline{AB}$

A	B	$\overline{A}$	$\overline{B}$	$Y = \overline{A} + \overline{B}$
0	0	1	1	1
0	1	1	0	1
1	0	0	1	1
1	1	0	0	0

2.

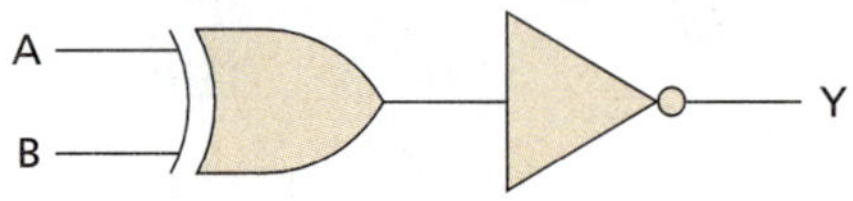

Figure 1.40 Practice Problem 1.11(2) (solution)

$Y = \overline{A \oplus B}$

A	B	Y
0	0	1
0	1	0
1	0	0
1	1	1

Section 1.12

1. SOP **2.** POS **3.** mixed

Section 1.13

$Y = \overline{A} \cdot B \cdot \overline{C} + \overline{A} \cdot B \cdot C + A \cdot \overline{B} \cdot \overline{C} + A \cdot \overline{B} \cdot C$

Section 1.14

1. One possible trouble is that Gate 4, the NAND gate, is bad. Output should be 0.

2. There is nothing wrong with this circuit as shown.

Section 1.15

1. BOOT = [(DRIVEA)(SYSTEM) + DRIVEC](POWER + RESET)

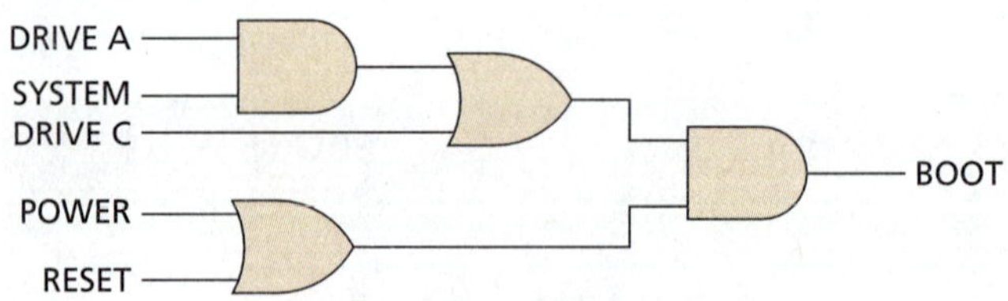

Figure 1.41 Practice Problem 1.13(1) (solution)

2. DRINK = (QUARTERS + DOLLAR)(SELECTION)
RETURN = (DOLLAR)(CHANGE)(SELECTION)

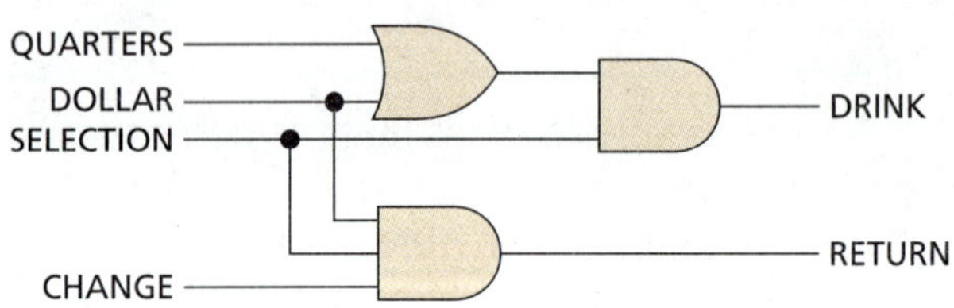

Figure 1.42 Practice Problem 1.13(2) (solution)

Problems

1.1 What advantages were realized when the transistor replaced the vacuum tube?

1.2 List two disadvantages of vacuum tubes.

1.3 What are the benefits of field effect transistors as compared to regular transistors?

1.4 List an advantage and a disadvantage of field effect transistors compared to bipolar transistors.

1.5 What benefit has been gained by the use of integrated circuits (ICs)?

1.6 How have integrated circuits made it possible to decrease computers in physical size?

1.7 Give three examples of digital applications that are not cited in the text.

1.8 Show at least one example of an analog signal that has been "digitized." What advantage is gained by digitizing the analog signal?

1.9 Compare an analog signal to a digital signal.

1.10 Give a benefit and a disadvantage of using digital signals compared to analog signals.

1.11 What does the term *state* mean in relationship to digital circuits and signals?

1.12 What are the two states of a digital signal?

1.13 Write a truth table for a part of a soft drink machine that dispenses a can of soft drink if two quarters are inserted or if a dollar bill is inserted. Make a second truth table that gives change if the dollar is inserted but not if the two quarters are inserted. Let one of the inputs to the table be QUARTERS and the other DOLLAR. The outputs are DRINK and CHANGE.

1.14 Rewrite the truth table of Problem 1.13 to use 1's to represent QUARTERS inserted, DOLLAR inserted, DRINK dispensed, and CHANGE returned, and 0's to represent no quarters or dollar, no drink, and no change.

1.15 Compare the buffer and inverter logic functions.

1.16 How do buffer and inverter logic gates differ?

1.17 Draw the logic symbols and truth tables for the buffer and inverter gates.

1.18 Draw the alternate symbols for the buffer and inverter gates using signal flow analysis. Describe what these symbols are actually telling the person reading them.

1.19 Draw the symbol, write the logic expression, and develop the truth table for a three-input AND gate.

1.20 Draw the symbol, write the logic expression, and develop the truth table for a four-input AND gate.

1.21 Define the AND logic function in terms of active high logic states.

1.22 Define the AND function in terms of active low logic states.

1.23 A combination lock is made from four wheels labeled WHEEL0 through WHEEL3. Inside the lock, each of the 10 numbers on each wheel has a pin that when pulled out generates a logic 1 (WHEEL = 1) and when in generates a logic 0 (WHEEL = 0). Develop the logic expression, truth table, and logic diagram that opens the lock (UNLOCK = 1) when the correct numbers are set on the wheels (when all WHEELs = 1).

1.24 An automobile approaches a railroad crossing equipped with flashing lights and a moving gate. Write a truth table that says it is OK to cross the tracks (OK = 1) when there is no TRAIN (1), no flashing LIGHTS (1), and the GATE is up (1). A train on the tracks, flashing lights, and a down gate are all represented by logic 0 levels. Draw the logic diagram for this truth table.

1.25 Draw the logic symbol, and create the logic expression and truth table for a three-input logic OR gate.

1.26 Draw the logic symbol, create the logic expression, and truth table for a four-input logic OR gate.

1.27 Describe the OR function in terms of active high logic levels.

1.28 Describe the OR function in terms of active low logic levels.

1.29 A defense plant has a high-security vault that can only be entered when four authorized people insert their ID cards. If any ID CARD is incorrect (logic 1), then an ALARM sounds and the DOOR does not open (logic 1). Design a logic circuit gate that allows the door to be opened (logic 0) when the correct cards are inserted (all four, logic 0) and sounds an ALARM (logic 1) if any card is incorrect (logic 1).

1.30 A store's security system sets off an ALARM and dials 911 if the front DOOR is opened after hours, a WINDOW is broken in, or SMOKE from a fire is detected anywhere in the store. Design a logic circuit that sounds the ALARM (logic 1) if any of these events occur.

1.31 Draw the logic symbols, and write down the truth tables and logic expressions for three-input NAND and three-input NOR logic gates.

1.32 Draw the logic symbols, and write down the truth tables and logic expressions for four-input NAND and four-input NOR logic gates.

1.33 A mileage counter is to be RESET to 0 when a count of 15 miles is reached. The count is a binary number (B_3, B_2, B_1, and B_0 = 1111) and the RESET signal is active low. Write down the logic expression and truth table, and draw the logic gate for this RESET signal.

1.34 In a computer, many IC chips are selected for use by detecting combinations of signals. A memory device in a computer is selected by setting signal SEL low. To make this signal low, four signals have to be set high. They are RAMHI, RAMLO, MEM, and PICK. Write down the truth table and logic expression, and draw the gate for the active low SEL signal.

1.35 Develop the truth table and logic expression for the circuit in Figure 1.43. Compare the truth table with those for the AND, OR, NAND, and NOR gates. What conclusion can be drawn from this comparison?

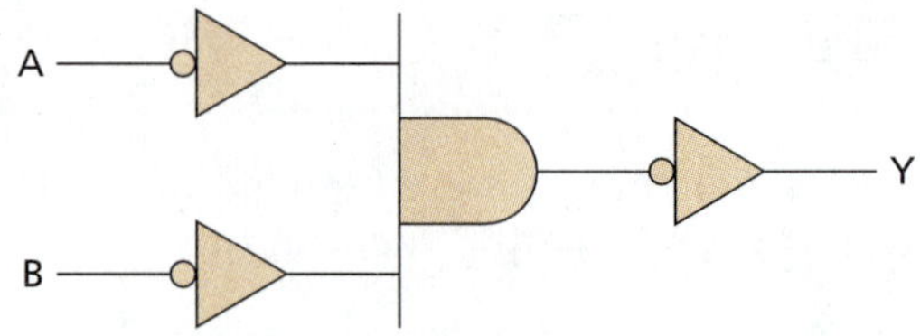

Figure 1.43 Problem 1.35

1.36 Develop the truth table and logic expression for the circuit in Figure 1.44. Compare the truth table with those for the AND, OR, NAND, and NOR gates. What conclusion can be drawn from this comparison?

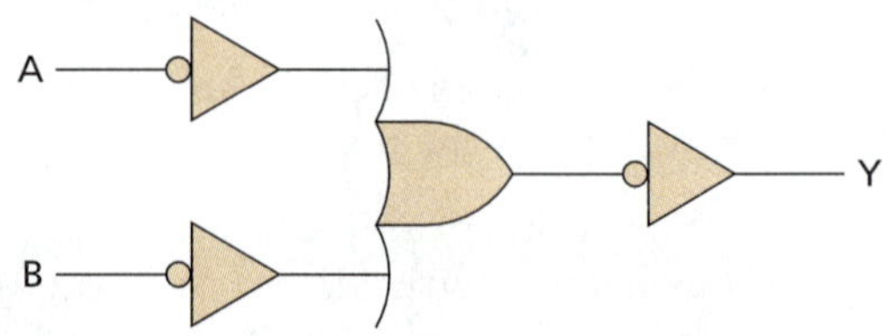

Figure 1.44 Problem 1.36

1.37 Develop the logic expression and truth table for the circuit in Figure 1.45.

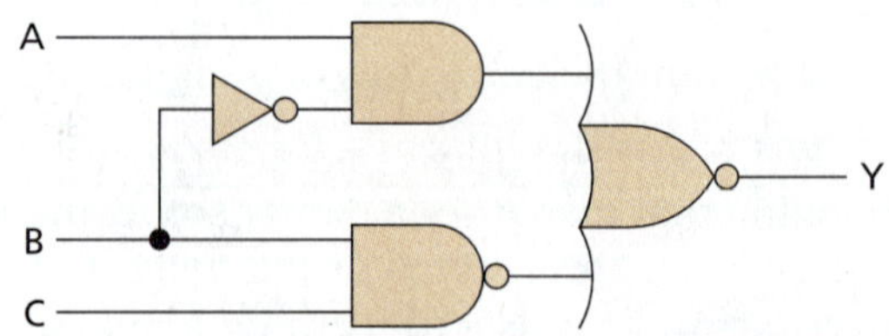

Figure 1.45 Problem 1.37

1.38 Develop the logic expression and truth table for the circuit in Figure 1.46.

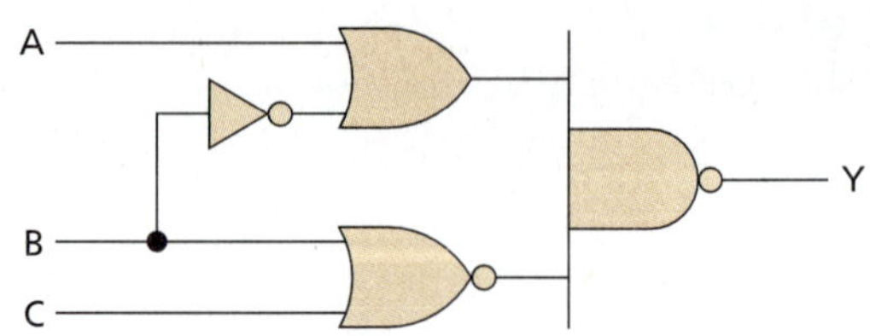

Figure 1.46 Problem 1.38

1.39 Develop the sum-of-products expression for the circuit in Figure 1.47.

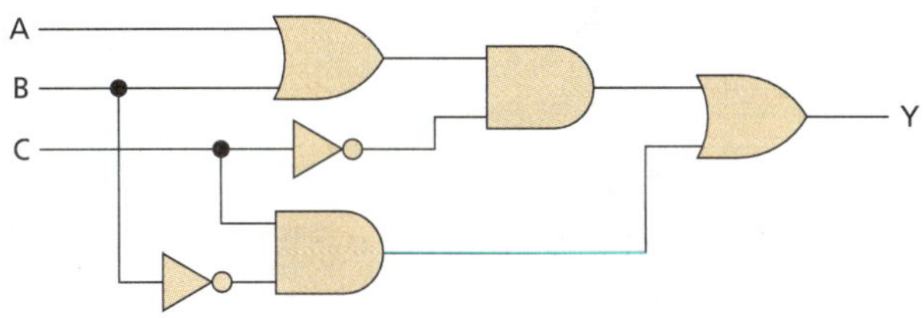

Figure **1.47** Problem 1.39

1.40 Develop the sum-of-products expression for the circuit in Figure 1.48.

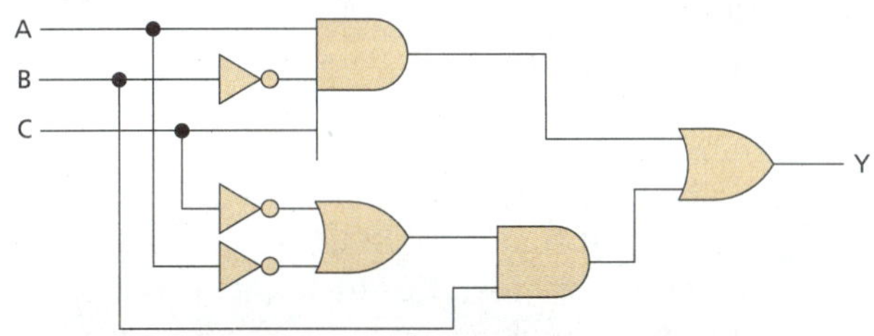

Figure **1.48** Problem 1.40

1.41 Draw a logic symbol, develop the truth table, and write the logic expression for a three-input exclusive OR function. Draw the actual combinational exclusive OR circuit using the 74LS86 two-input exclusive OR chip.

1.42 Draw the logic symbol, develop the truth table, and write the logic expression for a four-input exclusive OR function. Draw the actual combinational exclusive OR circuit using the 74LS86 two-input exclusive OR chip.

1.43 For the circuit in Figure 1.49,

a. Create its truth table.
b. Write its logic expression.
c. Determine what is wrong with the circuit based on the states of the gates shown in the circuit.

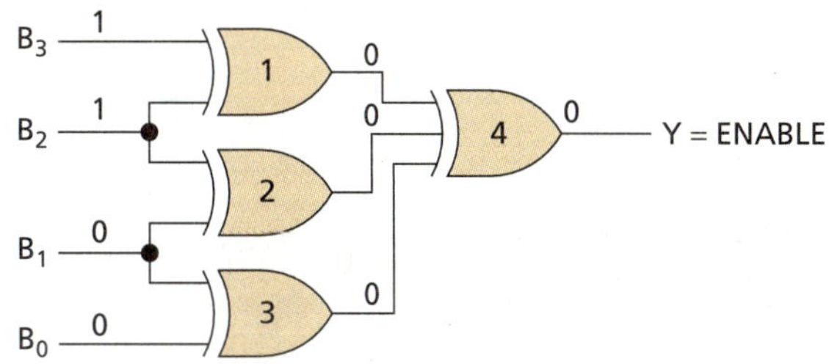

Figure **1.49** Problem 1.43

1.44 For the circuit in Figure 1.50,

a. Create its truth table.
b. Write its logic expression.
c. Determine what is wrong with the circuit based on the states of the gates shown in the circuit.

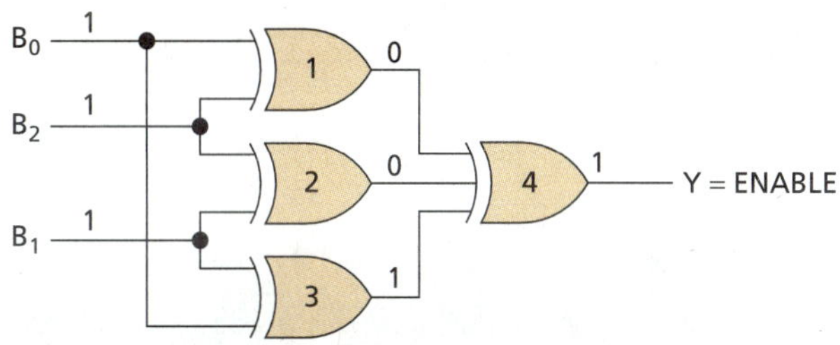

Figure **1.50** Problem 1.44

1.45 Explain the benefits and limitations of using a logic probe to troubleshoot digital circuits.

1.46 What benefits are gained by using an oscilloscope instead of a logic probe to troubleshoot logic circuits?

1.47 A logic circuit uses 5 V to represent a logic 1 state and 0 V to represent a logic 0 state. What might be wrong with the output of a logic circuit that shows the oscilloscope pattern of Figure 1.51?

1.48 A logic circuit uses 5 V to represent a logic 1 state and 0 V to represent a logic 0 state. What might be wrong with the input to a logic circuit that shows the oscilloscope pattern of Figure 1.52?

1.49 For the logic circuit in Figure 1.53, determine which gate is bad, based on the logic states shown.

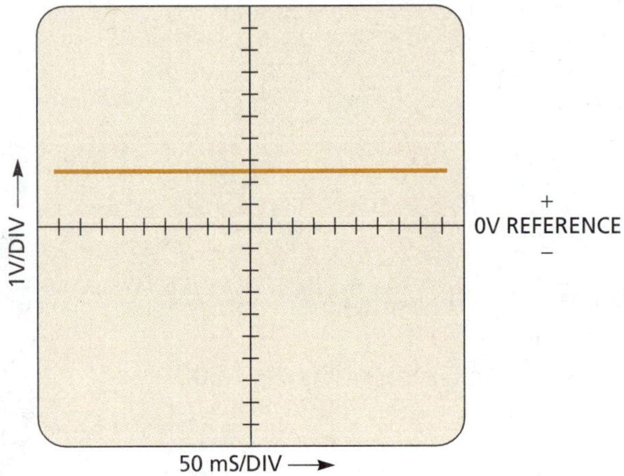

Figure 1.51 Problem 1.47

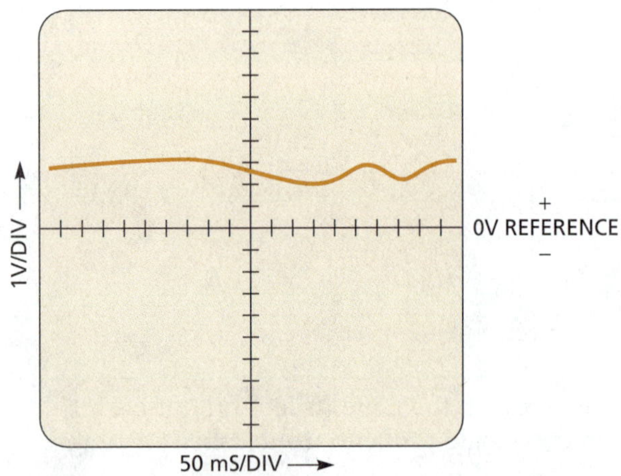

Figure 1.52 Problem 1.48

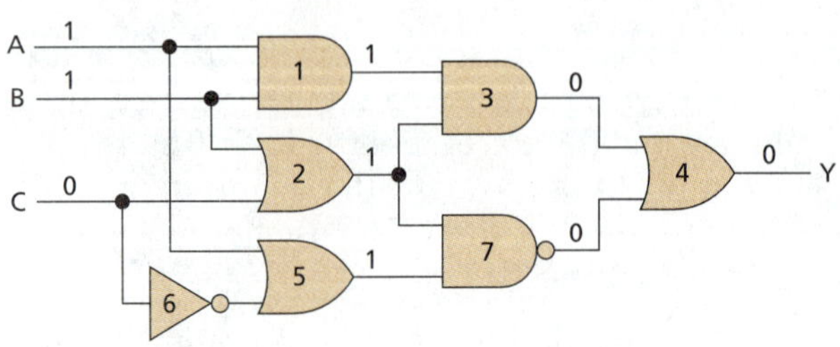

Figure 1.53 Problem 1.49

1.50 For the logic circuit in Figure 1.54, determine which gate is bad, based on the logic states shown.

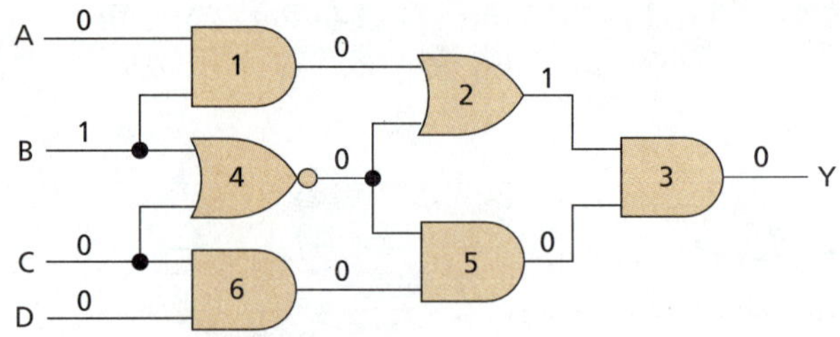

Figure 1.54 Problem 1.50

1.51 Design the logic circuit that turns on an active low ENABLE signal when four binary bits, A, B, C, and D, have the same state (all 0's or all 1's). Draw the schematic, develop the truth table, and write the logic expression for this circuit.

1.52 Design the logic circuit that operates a slot machine that pays out a JACKPOT whenever one of three identical symbols, LEMON, BAR, or BELL, appear. Each occurrence of a match is indicated by a logic 1 for the variable of that symbol while the other two variables are low. For instance, three lemons cause logic variable LEMON to go high and variables BAR and BELL to be low. Output JACKPOT is set to a 1 whenever a single variable is high when the other two variables are low. Draw the schematic, develop the truth table, and write the logic expression for this circuit.

1.53 Bonus Design Problem: Design a juke box that plays a particular song if the following conditions are met:

a. The SONG number matches the selection number. SONG = 1.

b. 50 CENTS (two quarters) or a DOLLAR is inserted into the money slot. CENTS = 1 or DOLLAR = 1.

c. CHANGE is returned if a DOLLAR was used, but not for 50 CENTS. CHANGE = 1.

d. If coins and bills are inserted at the same time, no SONG is selected, an ERROR message is displayed, and the money is returned. ERROR = 1.

two

Combinational Logic Solutions

KEY TERMS

associative law
Boolean algebra
DeMorgan's theorem
distributive law
Karnaugh map

OBJECTIVES

After studying this chapter, the student should be able to

- State the associative and distributive laws, which are used to expand or contract logic expressions in a manner similar to that of mathematical algebra.
- State and explain DeMorgan's theorem, which uses the NOT logic operator to alter logic expressions.
- Describe and illustrate Karnaugh mapping, which is a graphical method to aid in reducing logic expressions.

OVERVIEW

This chapter is about DeMorgan's theorem, Boolean algebra, and Karnaugh map-tools used for altering logic expressions to aid in designing logic circuits with a minimum number of gates. These tools may also be used to create circuits that use the same type of logic gate to perform different logic functions to reduce overall circuit cost. A third use of these tools is to aid in developing a combination of logic gates for a logic circuit or function that reduces operating time or complexity of the original logic circuit.

2.1 Introduction

In Chapter 1 logic expressions for devices and some basic combinational circuits were developed using truth tables. Truth tables can also be used for determining similarities among different logic expressions. For instance, let's examine the truth tables for a NOR gate whose expression is $Y = \overline{A + B}$ and for a combinational logic circuit using an AND gate whose inputs are inverted. The expression for this circuit is $Y = \overline{A} \cdot \overline{B}$. The gates and their truth tables are shown in Figure 2.1.

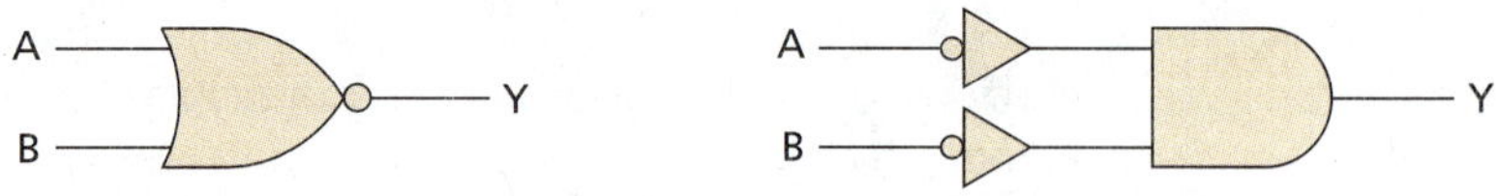

A	B	Y
0	0	1
0	1	0
1	0	0
1	1	0

(a) NOR Gate
$Y = \overline{A + B}$

A	B	$\overline{A}$	$\overline{B}$	Y
0	0	1	1	1
0	1	1	0	0
1	0	0	1	0
1	1	0	0	0

(b) AND/Inverter
$Y = \overline{A} \cdot \overline{B}$

Figure 2.1 NOR and AND/Inverter Comparison

As can be seen on the truth tables, the outputs of the NOR gate and the AND circuit are identical. From this, we can draw the conclusion that the two circuits produce the same logical results and are therefore logically identical. In other words, their expressions describe the same logic function and are equivalent, or $\overline{A + B} = \overline{A} \cdot \overline{B}$. Either circuit can be used to achieve the same result. Truth tables are used, in like manner, to establish theorems of Boolean algebra and the rules behind DeMorgan's theorem.

2.2 Boolean Algebra

Boolean algebra *a set of theorems to aid in reducing logic expressions.*

George Boole (1815–1864), a mathematician and clergyman in Cambridge, England, is credited with associating logical statements with symbols. Although his interest lay more with human thought and logic, the principles he developed have been put to good use in a mathematics of logic called **Boolean algebra.**

Boolean algebra is a statement of logical identities and theorems used to facilitate digital logic circuit design. Complex problems are often presented, for digital logic solution, that require what appears to be a complicated and involved logic circuit. However, when Boolean algebra identities and theorems are applied, the circuit can be reduced to a much simpler and more understandable form. In many cases, this reduction also reduces the cost of building the circuits themselves, by reducing gate count or allowing the use of a single gate type to perform the required logic task. Using the same gate type in large quantities reduces the cost per gate. We will start by establishing a set of basic identities and theorems for Boolean algebra and proving their validity by using truth tables for the gates represented by each identity or theorem.

Figure 2.2 contains the logic diagrams and truth tables for some of the basic theorems and identities of Boolean algebra that are going to be examined in this chapter. Some of these relationships are derived from the definition of the logic gates themselves. For instance, the first relationship, shown as Figure 2.2a, is an AND gate with both its inputs connected together. That makes input A and B both the same logic level or, in this instance, both equivalent to input A. From the truth table, we can see that there are only two conditions for both these inputs: 0 and 1. When both inputs to an AND gate are 1, then the output is a 1, and when they are both 0, the output is 0. As a result, we can surmise that the logic expression A AND A yields an output state equal to the state of input A or $A \cdot A = A$. A similar statement can be made for the OR function, whose inputs are the same. That is, as shown in Figure 2.2b, $A + A = A$.

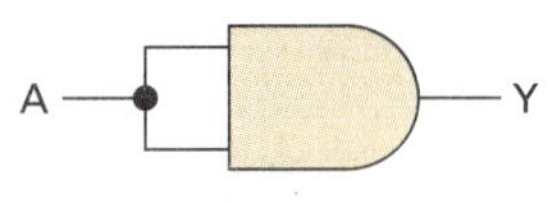

A	A	Y
0	0	0
1	1	1

(a) A · A = A

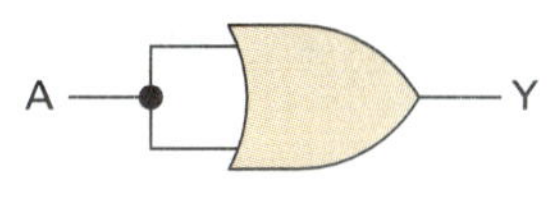

A	A	Y
0	0	0
1	1	1

(b) A + A = A

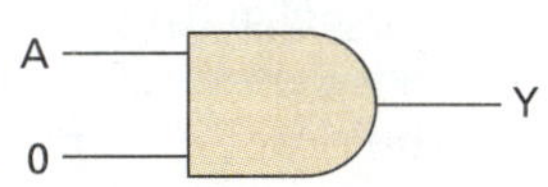

A	0	Y
0	0	0
1	0	0

(c) A · 0 = 0

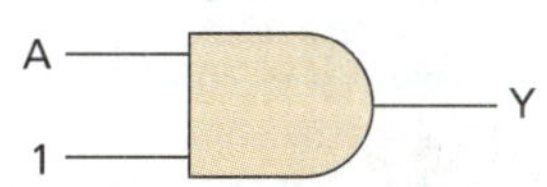

A	1	Y
0	1	0
1	1	1

(d) A · 1 = A

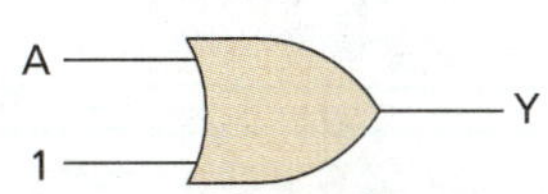

A	1	Y
0	1	1
1	1	1

(e) A + 1 = 1

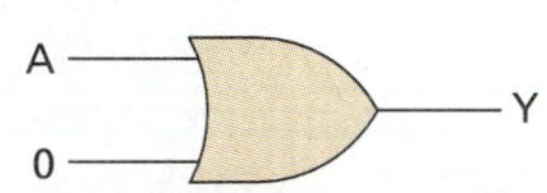

A	0	Y
0	0	0
1	0	1

(f) A + 0 = A

Figure 2.2 Boolean Algebra Identities

The next four relationships (Figure 2.2c–2.2f) illustrate what occurs when one of the inputs to an AND or OR gate is held at the constant logic state. In Figure 2.2c, input A is ANDed with a logic 0. By definition, a 0 input to an AND gate forces the output to a 0, and it doesn't matter what state the other input is. Thus A AND 0 is always 0 or $A \cdot 0 = 0$.

What happens when one input of an AND gate is connected to a 1? The AND gate swings open, passing the input to the output as described in Chapter 1. Using the truth table for an AND gate, we can see that the output in this situation is a copy of the other input. Thus, A AND 1 is A, or $A \cdot 1 = A$. These last two theorems help to define what it means to use a "forcing" input to a gate. In the case of an AND gate, a logic 0 input forces the output to a zero state, whereas placing a logic 1 on an input to the AND gate leaves its output dependent on the other input state.

Figure 2.2e and 2.2f repeat the process for an OR gate. In Figure 2.2e, one input is held to a 1 state, causing the output of the OR gate to be forced to a 1 regardless of the state of the other input. The expression for this is $A + 1 = 1$. Placing a 0 on one input, as in Figure 2.2f, causes the OR gate to act as an open gate and let the input pass to the output. The expression describing this function is $A + 0 = A$.

Before we can do much with these theorems to reduce Boolean expressions, we must define and become familiar with a few identities and formats. The first of these is called the associative law and has a set of rules that are similar to the distributive law in arithmetic.

2.3 Mathematical Laws for Logical Expressions

Associative law *the order of performing a logic operation by a single gate with many inputs does not matter.*

Boolean algebra, like other mathematical functions, contains a set of laws that deal with the order of performing logic functions. One law, called **associative law**, says that the order of performing the AND function using a number of inputs does not matter. This is equivalent, in arithmetic, to multiplying numbers in any order. For instance, performing the AND function using inputs A, B, and C can be done by ANDing A and B first, then ANDing that result to C as in the expression $(A \cdot B) \cdot C$. Or we could AND A with C first, and then AND B to that result, as in $(A \cdot C) \cdot B$. Or maybe we would prefer to AND B to C, and then AND A to that result, as in $(B \cdot C) \cdot A$. In all three instances, the logical results are the same, all three inputs are ANDed together to produce the final logic result.

EXAMPLE 2.1

Show that the order of ANDing three inputs does not matter.

SOLUTION To prove this associative law, let's compare the truth tables for $A \cdot (B \cdot C)$ and $(A \cdot B) \cdot C$. In the first case, we AND B and C and then AND A to that result. In the second, we AND A and B and then AND C to that result.

A	B	C	B · C	A · B	A · (B · C)	(A · B) · C	A · B · C
0	0	0	0	0	0	0	0
0	0	1	0	0	0	0	0
0	1	0	0	0	0	0	0
0	1	1	1	0	0	0	0
1	0	0	0	0	0	0	0
1	0	1	0	0	0	0	0
1	1	0	0	1	0	0	0
1	1	1	1	1	1	1	1

Intermediate output B · C is high twice when both inputs are high. The same can be said for A · B. As can be seen on the truth table, when input A is ANDed with B · C, the output of the AND gate is highly only when all three inputs are high, which describes the three-input AND function A · B · C. Looking at the results for (A · B) · C, we can also see that the output is high only when all three inputs are high. Thus both A · (B · C) and (A · B) · C not only produce the same result, but they are both equivalent to A · B · C.

The OR function, like the AND function, may be performed in any order. The output of the logic function A + B + C is the same regardless of the order in which the inputs are ORed together. This is much like adding a group of numbers together. The sum can be done in any order. Just as 1 + 5 + 6 can be done by adding 6 to the sum of 1 and 5 or by adding 5 to the sum of 1 and 6, OR inputs can be logically accomplished in any order. The same result is achieved by ORing A to (B + C) as with ORing C to (A + B) or, for that matter, ORing A and C first and then ORing B to the result.

***Distributive law** specifies the manner in which logic expressions can be combined similar to multiplying polynomials in mathematics. Also defines methods for factoring common variables from logic expressions.*

The **distributive law** is similar to multiplying two polynomials together. The similarity is incomplete, because we do not produce A^2 or A^3 as we would by multiplying A by A or A by A by A. Instead, we get A · A or A · A · A. The distributive law allows us to distribute the variable inputs of an OR gate into a set of inputs of a second OR gate when their outputs are combined through an AND function. This process is better understood by doing an example.

Let's take two OR gates, one whose output logic expression is A + B + C and another whose output logic expression is B + D + A, and combine them into a single expression using an AND function. Thus the complete logic expression for this example is Y = (A + B + C)(B + D + A), and its circuit is shown in Figure 2.3.

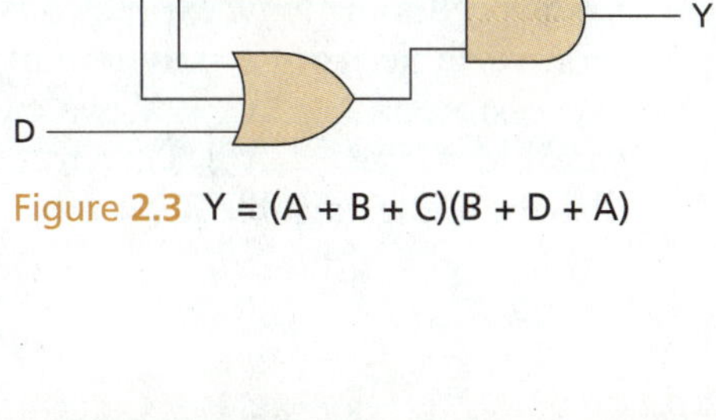

Figure 2.3 Y = (A + B + C)(B + D + A)

The distributive law says we can derive an equivalent expression by distributing the individual elements of one OR function into the elements of the second OR function. This is done by taking the first element of one OR gate and ANDing it with each element of the second gate:

AB + AD + AA

Do that with the second and third elements of the first OR expression:

BB + BD + BA and CB + CD + CA

Lastly, write it out as a full expression:

Y = (A + B + C)(B + D + A) = AB + AD + AA + BB + BD + BA + CB + CD + CA

This says that both expressions yield the same logical result and that the circuit in Figure 2.4, derived from this new expression, is equivalent to the one in Figure 2.3. Shortly, we will begin to apply some of the theorems from Section 2.2 to reduce this expression and create a smaller circuit that will produce the same logical results as both of these expressions.

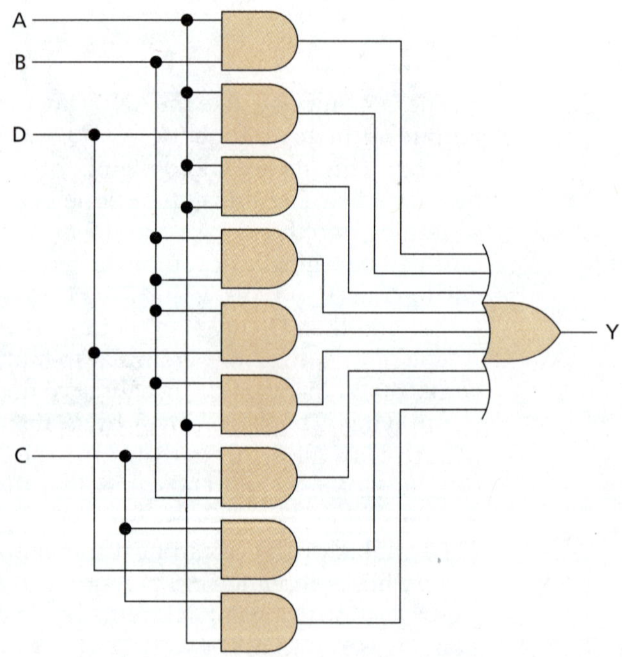

Figure 2.4 Y = AB + AD + AA + BB + BD + BA + CB + CD + CA

Let's look at one more part of the distributive law, and them we'll be ready to start reducing logic expressions. This law is the same idea as factoring a common variable from a polynomial in mathematics. If we have a mathematical expression $5X^3 + 3X^2 + X$, we can factor out the common X term from each variable: $X(5X^2 + 3X + 1)$. Why did the last term become 1? Basically, if we redistribute the variable X into the polynomial, we should have the same expression we started with. Factoring *X* from the last term must leave a 1 behind, so that the *X* can be redistributed to that term by multiplying *X* by 1.

Logically, the method is the same when applied to logic expressions: extract the common input variable from each AND term. For instance, let's factor $A \cdot B \cdot C + A \cdot D \cdot E + A \cdot B \cdot E + A$. The common variable that can be factored is A:

$$Y = A \cdot (B \cdot C + D \cdot E + B \cdot E + 1)$$

As with the math factoring, a single term factored out leaves a 1 behind so that if it is redistributed, the original term is replaced. In this example, A AND 1 is A.

How does this all help in making logic circuits smaller or more efficient? Let's go back to the example used previously and continue to work with it using the Boolean theorems from Section 2.2 and the distributive laws just described.

The original expression, shown in product-of-sums (POS) form, is

$$Y = (A + B + C)(B + D + A)$$

After distributing one OR function into the other, we get the sum-of-products (SOP) form of the same expression:

$$Y = A \cdot B + A \cdot D + A \cdot A + B \cdot B + B \cdot D + B \cdot A + C \cdot B + C \cdot D + C \cdot A$$

One of the theorems stated that if both input variables applied to an AND gate term are the same variable, the output of that term is equivalent to that input variable, or $A \cdot A = A$. We can substitute the single variable for any term in the SOP expression that has the same variable for each AND input. In our example, $A \cdot A = A$ and $B \cdot B = B$.

$$Y = A \cdot B + A \cdot D + A + B + B \cdot D + B \cdot A + C \cdot B + C \cdot D + C \cdot A$$

Now let's reorder the terms, because we can do the OR function in any sequence we like. Let's reorder the expression to group terms that contain an A input variable together and those with a B input variable together:

$$Y = A \cdot B + A \cdot D + A + B \cdot A + C \cdot A + B + B \cdot D + C \cdot B + C \cdot D$$

Next we factor the common variable A from the group with at least one A input and B from the group of terms containing a B input:

$$Y = A \cdot (B + D + 1 + B + C) + B \cdot (1 + D + C) + C \cdot D$$

Another of the theorems we looked at in Section 2.2 says that if an OR gate has an input that is a 1, its output is forced to a 1. After we factored out the common inputs in our groups, one of the OR inputs in each group is a 1, so the equivalent expression for both groups is a 1, or

$$Y = B + D + 1 + B + C = 1 \qquad \text{and} \qquad 1 + D + C = 1$$

Substituting 1 for both expressions, our expression now becomes

$$Y = A \cdot (1) + B \cdot (1) + C \cdot D$$

Redistributing A and B, the final reduced expression becomes

$$Y = A + B + C \cdot D$$

By using the distributive law and applying Boolean theorems, we have reduced the original expression (A + B + C)(B + D + A) whose circuit was shown in Figure 2.3, to a new expression, A + B + C · D, whose logic circuit is shown in Figure 2.5. The circuits in both figures yield the same logical result. How do we prove this? By now you know—do the truth table as shown in Figure 2.6.

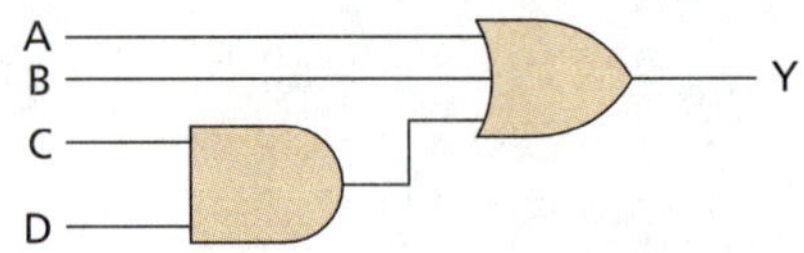

Figure 2.5 Y = A + B + CD

A	B	C	D	A + B + C	A + B + D	AB	AD	AA	BB	BD	BA	CB	CD	CA	(A + B + C) · (A + B + D)	AB + AD + AA + BB + BD + BA + CB + CD + CA	A + B + CD
0	0	0	0	0	0	0	0	0	0	0	0	0	0	0	0	0	0
0	0	0	1	0	1	0	0	0	0	0	0	0	0	0	0	0	0
0	0	1	0	1	0	0	0	0	0	0	0	0	0	0	0	0	0
0	0	1	1	1	1	0	0	0	0	0	0	0	1	0	1	1	1
0	1	0	0	1	1	0	0	0	1	0	0	0	0	0	1	1	1
0	1	0	1	1	1	0	0	0	1	1	0	0	0	0	1	1	1
0	1	1	0	1	1	0	0	0	1	0	0	1	0	0	1	1	1
0	1	1	1	1	1	0	0	0	1	1	0	1	1	0	1	1	1
1	0	0	0	1	1	0	0	1	0	0	0	0	0	0	1	1	1
1	0	0	1	1	1	0	1	1	0	0	0	0	0	0	1	1	1
1	0	1	0	1	1	0	0	1	0	0	0	0	0	1	1	1	1
1	0	1	1	1	1	0	1	1	0	0	0	0	1	1	1	1	1
1	1	0	0	1	1	1	0	1	1	0	1	0	0	0	1	1	1
1	1	0	1	1	1	1	1	1	1	1	1	0	0	0	1	1	1
1	1	1	0	1	1	1	0	1	1	0	1	1	0	1	1	1	1
1	1	1	1	1	1	1	1	1	1	1	1	1	1	1	1	1	1

Figure 2.6 Truth Table Proof

EXAMPLE 2.2

What is the equivalent Boolean expression for

$$Y = (A + B)(A + C)?$$

SOLUTION First distribute the variables as described earlier:

$$(A + B)(A + C) = A \cdot A + A \cdot C + A \cdot B + B \cdot C$$

Notice that we get an A · A term. This reduces to A by using the theorem, A · A = A:

$$A + AC + A \cdot B + B \cdot C$$

Next factor A from the first three terms:

$$(A)(1 + C + B) + B \cdot C$$

Using the theorem A + 1 = 1, the expression 1 + C + B is replaced by a 1:

$$(A)(1) + B \cdot C$$

Again, A · 1 = A, so, finally;

$$(A + B)(A + C) = A + B \cdot C$$

Section 2.3 Practice Problems

1. Reduce the expression (A + C + D)(B + C + D), and prove that your reduction is equivalent to the original expression.
2. Reduce the expression A · B · C + A(A + C + B) + B · C, and prove that your reduction is equivalent to the original expression.

2.4 More Boolean Theorems

Figure 2.7 illustrates what happens when you connect an input and its complement to an AND and an OR gate. For the AND case (Figure 2.5a), one input, A or its complement ($\overline{A}$) is a 0, because when one input is a 1, the other, being the complement of the first, must be a 0. Since one input or the other is always 0, the output of the AND gate, by definition, is always forced to a 0. The logic expression for this relationship is $A \cdot \overline{A} = 0$. The OR gate produces similar results, except the output is always a 1, since one input or the other is always a 1. The logic expression for this theorem is $A + \overline{A} = 1$.

A	$\overline{A}$	$Y = A \cdot \overline{A}$
0	1	0
1	0	0

(a) $A \cdot \overline{A} = 0$

A	$\overline{A}$	$Y = A + \overline{A}$
0	1	1
1	0	1

(b) $A + \overline{A} = 1$

Figure 2.7 Two More Boolean Theorems

To illustrate the use of these two theorems, let's reduce the logic expression $(\overline{A} + B)(A + B)$. Start by distributing the variables in one POS term into the other to create the equivalent SOP expression:

$$\overline{A} \cdot A + \overline{A} \cdot B + B \cdot A + B \cdot B$$

One of the new theorems applies to the first term, $\overline{A} \cdot A = 0$, which means that the term can be replaced by a 0:

$$0 + \overline{A} \cdot B + B \cdot A + B \cdot B$$

Another theorem, $A + 0 = A$, allows us to drop the 0 input from the SOP expression while another theorem allows us to reduce $B \cdot B$ to B:

$$\overline{A} \cdot B + B \cdot A + B$$

The variable B can be factored from each term in the expression

$$B \cdot (\overline{A} + A + 1)$$

Although $\overline{A} + A = 1$ works here, we can avoid doing this extra step by noticing that there is a 1 to one of the inputs of the OR expression. Whenever an OR input is a 1, the output is forced to a 1, so the entire expression $\overline{A} + A + 1 = 1$ becomes

$$B \cdot (1)$$

Finally, the theorem $A \cdot 1 = A$ applies to make the final result of the reduction B so that

$$(\overline{A} + B)(A + B) = B$$

EXAMPLE 2.3

Reduce the expression $\overline{A} \cdot B + A \cdot B + A \cdot \overline{C} + A \cdot C$.

SOLUTION Begin by factoring out the B variable in the first two terms:

$$B \cdot (\overline{A} + A) + A \cdot \overline{C} + A \cdot C$$

$\overline{A} + A$ reduces to a 1 according to one of the theorems, and then the 1 ANDed with B further reduces to B:

$$B \cdot (1) + A \cdot \overline{C} + A \cdot C = B + A \cdot \overline{C} + A \cdot C$$

Do the same for the AC terms:

$$B + A \cdot (\overline{C} + C) = B + A \cdot (1) = B + A$$

Table 2.1 summarizes the basic theorems and laws discussed to this point. Since they are all supported by truth tables based on the definitions of AND and OR gates, then each is a valid rule to assist in reducing logic expressions.

Table 2.1
Basic Boolean Rules

$A \cdot A = A$
$A \cdot 1 = A$
$A \cdot 0 = 0$
$A + A = A$
$A + 1 = 1$
$A + 0 = A$

Section 2.4

Practice Problems

Reduce each of the following logic expressions:

1. $A \cdot B \cdot C + A \cdot \overline{B} \cdot C + \overline{A} \cdot B \cdot C + \overline{A} \cdot \overline{B} \cdot C$
2. $(B + C)(A + \overline{C})(\overline{A} + B)$
3. $A \cdot C + B \cdot \overline{C} + \overline{A} \cdot C + B \cdot C$

2.5 The NOT Operator and DeMorgan's Theorem

DeMorgan's theorem *establishes the equality of logic expressions based on using complements (NOTs).*

Inverting variables and expressions is one of the handiest tools for reducing logic circuits. Using inversion at times is subtle and requires some practice and insight. However, before we get to that level, let's start by introducing Augustus DeMorgan. A contemporary of George Boole, he is responsible for much of the work and theory behind Boolean reduction. The concept DeMorgan is most remembered for bears his name, **DeMorgan's theorem,** which deals with the relationship between a logic expression and its complement. DeMorgan surmised that there was more than one way to arrive at the same logical result and that there were precise relationships between different logic expressions used to achieve a common goal. For instance, the expression for a NOR function, $\overline{A + B}$, can also be written as an AND function, $\overline{A} \cdot \overline{B}$. We have seen that is true through the use of a truth table proof. The principles of DeMorgan's theorem can be used to help reduce complex logic expressions.

A logic expression may be altered, using DeMorgan's theorem, by following this process:

1. Complement all the variables of the expression.
2. Change AND operators to OR operators.
3. Change OR operators to AND operators.

Before we explore the use of DeMorgan's theorem, we need to understand an additional application of the inverting NOT bar function. When looking at logic expressions, always keep in mind that the NOT bar indicates an inverting process. Examine the expression, $Y = \overline{\overline{A}}$. There is a "double NOT" over the variable A. In this example, this means that Y has the same logic as variable A inverted twice. The first inversion yields the complement of A, and the second inverts the complement back to the original state. In other words, if A is a logic 1, the A inverted twice is also a logic 1. Therefore, this relationship is true:

$$Y = \overline{\overline{A}} = A$$

Figure 2.8 is used to show how DeMorgan's theorem works, using both logic gates and truth tables. As an example, we will use the basic NAND gate, $\overline{A \cdot B}$. To use DeMorgan's theorem to obtain a similar logic circuit to the NAND gate, start by inverting each input variable and changing the logic operator between them. For this example, change the AND operator to an OR operator and invert the input variables, A and B: $\overline{A \cdot B} = \overline{A} + \overline{B}$. The truth table in Figure 2.8 bears this equality out. The steps are illustrated in Figure 2.8, showing the change of logic gate and the application of inverting circles to each of the gate's inputs. Also note the absence of the inverting circle on the converted gate's output.

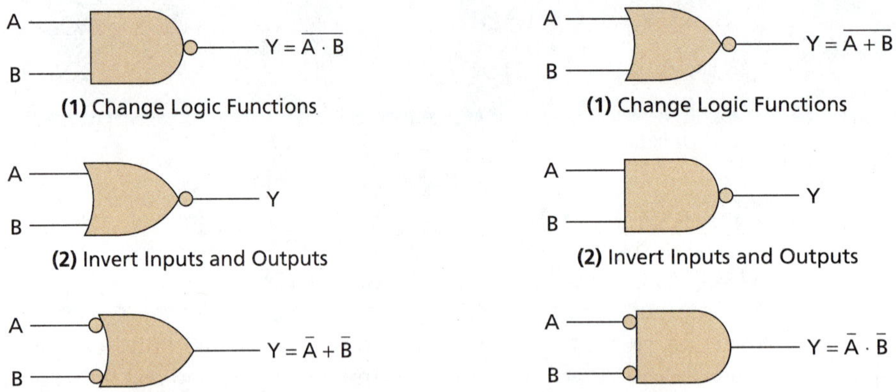

A	B	$A \cdot B$	$\overline{A \cdot B}$	$\overline{A}$	$\overline{B}$	$\overline{A} + \overline{B}$
0	0	0	1	1	1	1
0	1	0	1	1	0	1
1	0	0	1	0	1	1
1	1	1	0	0	0	0

Same

A	B	$A + B$	$\overline{A + B}$	$\overline{A}$	$\overline{B}$	$\overline{A} \cdot \overline{B}$
0	0	0	1	1	1	1
0	1	1	0	1	0	0
1	0	1	0	0	1	0
1	1	1	0	0	0	0

Same

Figure 2.8 DeMorgan's Theorem Proof

Applying the three steps of DeMorgan's theorem to the NOR gate results in the equality $\overline{A + B} = \overline{A} \cdot \overline{B}$. Notice that this is the equality we illustrated in the introduction to this chapter. Once again the truth table proves the equality is valid. A word of caution: When using DeMorgan's rules, be very careful of the grouping of variables under each NOT (inverting) bar in a logic expression. This will be emphasized as we explore the use of DeMorgan's theorem. Table 2.2 summarizes DeMorgan's theorem as applied to NAND and NOR gates.

Table 2.2
DeMorgan's Theorem

$\overline{A} \cdot A = 0$
$\overline{A} + A = 1$
$\overline{A \cdot B} = \overline{A} + \overline{B}$
$\overline{A + B} = \overline{A} \cdot \overline{B}$
$\overline{A} = A$

Now it is time to move up in applying DeMorgan's theorem to a more complex expression. Watch the groupings under each NOT bar when using the process behind the theorem. We will convert this logic expression using DeMorgan's theorem and then reduce it as far as possible:

$$\overline{(\overline{A} \cdot B + A \cdot C)\overline{(A + C)} + B}$$

It is usually best to start with the inner gates to get a feel for grouping. As you get more proficient at using this tool, you will begin to handle the process differently. NOR gate $\overline{A + C}$ can be converted to its AND equivalent by inverting the inputs $\overline{A} \cdot \overline{C}$ and the OR operator between them:

$$\overline{(\overline{A} \cdot B + A \cdot C)(\overline{A} \cdot \overline{C}) + B}$$

Recognize that the remaining logic expression is for a combinational circuit whose final output gate is a two-input NOR gate. To simplify matters, let's substitute a new variable, R for $(\overline{A} \cdot B + A \cdot C)(\overline{A} \cdot \overline{C})$. Now the NOR gate is more recognizable and has the logic expression $\overline{R + B}$, which changes to $\overline{R} \cdot \overline{B}$ after applying DeMorgan's theorem. Replacing the original expression for R, we have

$$\overline{(\overline{A} \cdot B + A \cdot C)(\overline{A} \cdot \overline{C})} \cdot \overline{B}$$

Next recognize that the expression under the long NOT bar is an AND gate with three inputs: (1) the expression $(\overline{A} \cdot B + A \cdot C)$, (2) $\overline{A}$, and (3) $\overline{C}$. Again, a substitution will make this clear: Let $S = (\overline{A} \cdot B + A \cdot C)$, and we substitute S into the expression:

$$\overline{S \cdot \overline{A} \cdot \overline{C}} \cdot \overline{B}$$

Notice that the last variable, $\overline{B}$, is not included under the NOT bar and is not affected by the next change. This is an example of carefully noticing proper grouping under a NOT (inverting) bar. Complement each of the variables of the NAND term, and change the operator (in this case, AND to OR):

$$(\overline{S} + A + C)\overline{B}$$

This is one place to emphasize the caution note that preceded this example. Do not lose sight of the fact that $\overline{B}$ is ANDed with the entire expression and not just the $\overline{C}$ term. Substitute the original expression for S back into the new expression:

$$(\overline{\overline{A} \cdot B + A \cdot C} + A + C) \cdot \overline{B}$$

The next NOT bar is over $\overline{A} \cdot B + A \cdot C$ only, so the remaining parts of the expression are not affected by the next action. This time, let's try to complete it all at once, keeping in mind the caution note about operator placement:

$$[(A + \overline{B})(\overline{A} + \overline{C}) + A + C] \cdot \overline{B}$$

Here is how it went: $\overline{A}$ was complemented to A and B to $\overline{B}$. The operator between them is an AND and was changed to an OR to yield the $A + \overline{B}$ portion of the new expression. This is repeated for AC, to make it become $\overline{A} + \overline{C}$. Finally, the OR operator between the two original AND terms was changed to an AND operator.

With the applications of DeMorgan's theorem completed, it remains to distribute the terms and apply Boolean theorems to reduce the expression. The first step is to distribute the POS terms into a SOP form:

$$(A \cdot \overline{A} + A \cdot \overline{C} + \overline{A} \cdot \overline{B} + \overline{B} \cdot \overline{C} + A + C) \cdot \overline{B}$$

$$A \cdot \overline{A} = 0$$

$$(0 + A \cdot \overline{C} + \overline{A} \cdot \overline{B} + \overline{B} \cdot \overline{C} + A + C) \cdot \overline{B}$$

0 + everything else = everything else. Input A is relocated to better illustrate the following steps:

$$(A \cdot \overline{C} + A + \overline{A} \cdot \overline{B} + \overline{B} \cdot \overline{C} + C) \cdot \overline{B}$$

Factor A from the first two terms:

$$(A(\overline{C} + 1) + \overline{A} \cdot \overline{B} + \overline{B} \cdot \overline{C} + C) \cdot \overline{B}$$

$\overline{C} + 1 = 1$:

$$(A(1) + \overline{A} \cdot \overline{B} + \overline{B} \cdot \overline{C} + C) \cdot \overline{B}$$

$A \cdot 1 = A$:

$$(A + \overline{A} \cdot \overline{B} + \overline{B} \cdot \overline{C} + C) \cdot \overline{B}$$

Distribute $\overline{B}$:

$$A \cdot \overline{B} + \overline{A} \cdot \overline{B} \cdot \overline{B} + \overline{B} \cdot \overline{B} \cdot \overline{C} + \overline{B} \cdot C$$

$\overline{B} \cdot \overline{B} = \overline{B}$:

$$A \cdot \overline{B} + \overline{A} \cdot \overline{B} + \overline{B} \cdot \overline{C} + \overline{B} \cdot C$$

Factor $\overline{B}$:

$$\overline{B}(A + \overline{A} + \overline{C} + C)$$

$A + \overline{A} = 1$ and $\overline{C} + C = 1$:

$$\overline{B}(1 + 1)$$

$1 + 1 = 1$:

$$\overline{B}(1)$$

$1 \cdot \overline{B} = \overline{B}$:

$$\overline{B}$$

This example illustrates one important procedural point: If your expression contains a single variable (A, $\overline{B}$, and so on) within a sum-of-products (SOP) expression, that variable becomes a powerful tool for reducing the number of terms in that SOP. All the product terms in that SOP that contain the single variable as one of their elements will be eliminated by factoring the common variable and leaving an expression that satisfies the rule of $1 + A = 1$.

EXAMPLE 2.4

Show how a single variable in a SOP expression can aid in reducing the expression

$$A \cdot B \cdot D + A \cdot B \cdot C \cdot D + A \cdot C \cdot D + A \cdot B \cdot C \cdot D \cdot E \cdot F \cdot G + C$$

SOLUTION Start by factoring C, the single variable, from as many product terms as possible:

$$A \cdot B \cdot D + C \cdot (A \cdot B \cdot D + A \cdot D + A \cdot B \cdot D \cdot E \cdot F \cdot G + 1)$$

When factored out, the single variable leaves a 1 in the remaining SOP expression. Since a 1 on any input into an OR gate forces its output to a 1, the SOP expression remaining in the parenthesis becomes a 1:

$$A \cdot B \cdot D + C \cdot (1)$$

C AND 1 = C, so the final expression is

$$A \cdot B \cdot D + C$$

What we could have done by recognizing the power of the single variable is look at the expression and cross out all the product terms that contained the single variable:

$$A \cdot B \cdot D + \cancel{A \cdot B \cdot C \cdot D} + \cancel{A \cdot C \cdot D} + \cancel{A \cdot B \cdot C \cdot D \cdot E \cdot F} \cdot G + C$$

The terms that remained after eliminating the terms that contained a C variable is the reduced expression

$$A \cdot B \cdot D + C$$

Section 2.5

Practice Problems

Reduce the following Boolean expressions:

1. $\overline{(A \cdot B \cdot \overline{C} + B \cdot C) \cdot C + \overline{B}}$
2. $\overline{\overline{A} \cdot B \cdot \overline{C} + B \cdot C \cdot \overline{D}} + \overline{A \cdot C \cdot \overline{D} + A \cdot \overline{B}}$
3. $(A \cdot B \cdot C + \overline{C})(\overline{\overline{C} \cdot D + D})$
4. $(\overline{AB + C})(\overline{BC}) + \overline{C}$

2.6 Applying DeMorgan's Theorem

One effect of DeMorgan's theorem is that AND and OR gates can be logically formed from their opposite members. In other words, the logic OR function can be derived from a NAND gate and the logic AND function from the NOR gate. Because of this effect, logic circuits that have different gates in them can be developed from one type of gate, the NAND or NOR. To illustrate this point, we will develop an exclusive OR using only NAND gates.

The logic expression for an exclusive OR is $Y = \overline{A} \cdot B + A \cdot \overline{B}$, whose circuit is shown in Figure 2.9. As with mathematical expressions, whatever action is

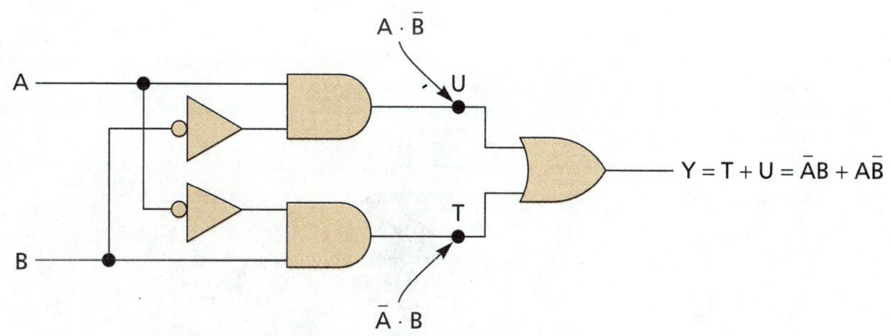

Figure 2.9 XOR Circuit

taken on one side of the expression must be taken on the other side to keep the equality of the expression. To begin the job of converting the exclusive OR logic expression from one that uses two AND gates, or OR gate, and two inverters, to one that uses only NAND gates, it is first helpful to recognize that the four basic building blocks of logical circuits, buffer, inverter, and AND and OR functions can be derived from NAND gates.

By applying DeMorgan's rule to a NAND gate, it can be redrawn as an OR gate with inverted inputs, as explained earlier and shown again here in Figure 2.10. Connecting both inputs to a two-input NAND gate together produces the inverting function, expressed by $Y = \overline{A \cdot A} = \overline{A}$. Placing the NAND inverter on the output of another NAND gate creates an AND gate, because $Y = \overline{\overline{A \cdot B}} = A \cdot B$. Figure 2.10 also shows how to use the NAND inverters to make the buffer function and OR function without inverted inputs. Since each of the basic logic building blocks can be formed from NAND gates, then NAND gates can be exclusively used to make any logic circuit.

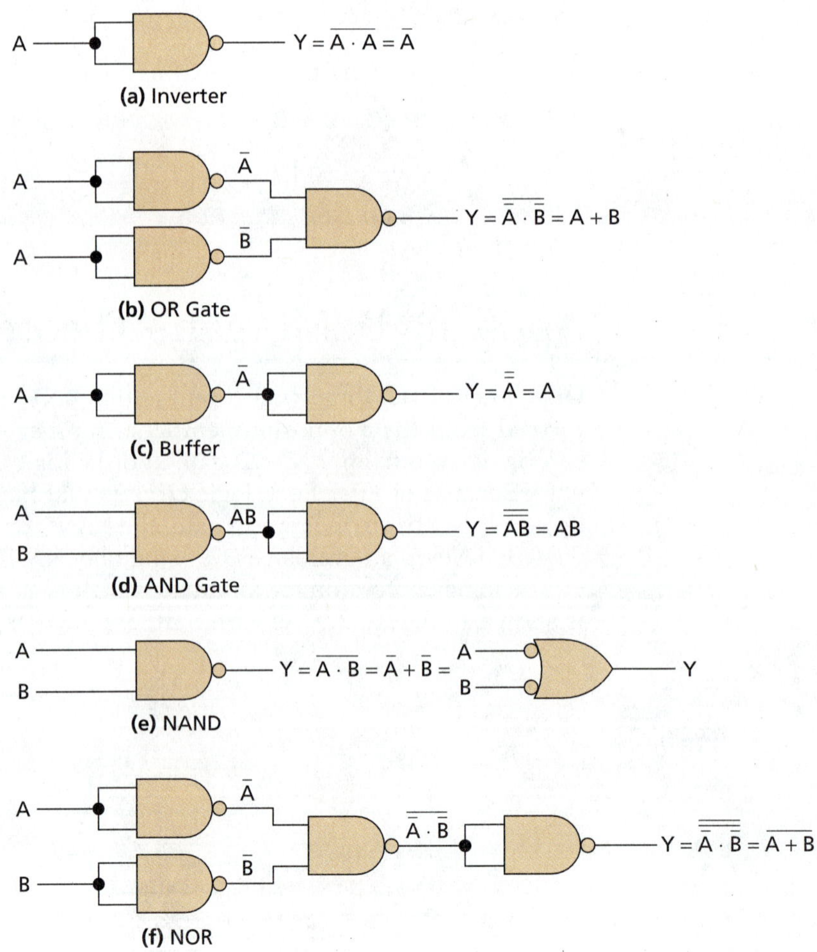

Figure 2.10 Logic Building Blocks

Going back to the exclusive OR, we start by inverting both sides of the expression:

$$\overline{Y} = \overline{\overline{A} \cdot B + A \cdot \overline{B}}$$

Realize that since we inverted both sides of the expression, the equality of the expression remains intact. The next step in the process of converting to NAND gates is to use DeMorgan's theorem on the right side of the expression:

$$\overline{Y} = (\overline{\overline{A} \cdot B}) \cdot (\overline{A \cdot \overline{B}})$$

Each input to the output OR gate is inverted, and the OR function is changed to an AND function. Notice that we have produced two NAND gates at this point, $\overline{\overline{A} \cdot B}$ and $\overline{A \cdot \overline{B}}$, but the final gate is an AND and not a NAND. Our last step is to make that AND gate a NAND by inverting both sides of the expression one more time. This puts a double inversion over Y. Recall that a double NOT or inversion is equivalent to the same variable without any inverting bars, or

$$\overline{\overline{Y}} = Y = \overline{(\overline{\overline{A} \cdot B}) \cdot (\overline{A \cdot \overline{B}})}$$

Figure 2.11 is the logic diagram of the exclusive OR function built using NAND gates. What remains to do is prove that this circuit performs the exclusive OR operation, by reducing the final expression we just derived to make sure it yields the original exclusive OR expression. Start by using DeMorgan's theorem again:

$$Y = (\overline{\overline{\overline{A} \cdot B}}) + (\overline{\overline{A \cdot \overline{B}}})$$

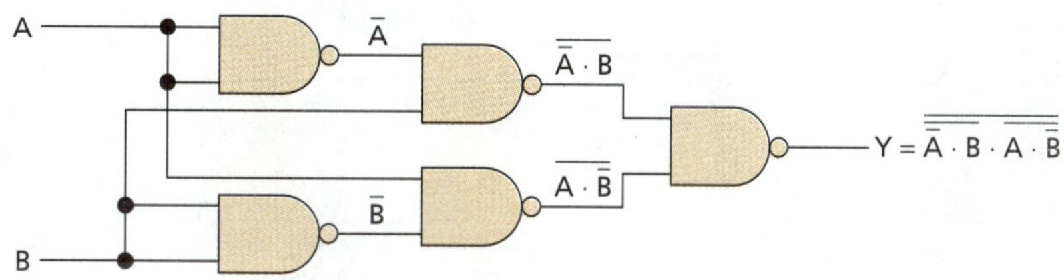

Figure 2.11 XOR Using NAND Gates

The double inversions cancel each other ($\overline{\overline{A}} = A$) to give us the original exclusive OR logic expression:

$$Y = \overline{A} \cdot B + A \cdot \overline{B}$$

In another example, we will make a logic circuit, using two-input NAND gates, for the logic expression $Y = A \cdot B + C + D$. Start with the circuit in Figure 2.12,

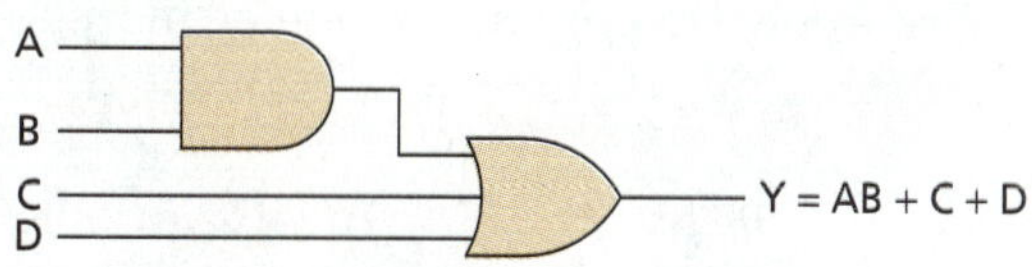

Figure 2.12 Y = AB + C + D

which is the original circuit created from the logic expression. The conversion is begun the same way used for the exclusive OR conversion, by inverting both sides of the expression:

$$\overline{Y} = \overline{A \cdot B + C + D}$$

Now apply DeMorgan's rules to produce our first NAND gate:

$$\overline{Y} = \overline{A \cdot B} \cdot \overline{C} \cdot \overline{D}$$

A second inversion of both sides of the expression does yield a NAND function, as it did with the exclusive OR, but this time the NAND has three inputs and we wish to restrict the circuit to two-input NAND gates:

$$\overline{\overline{Y}} = Y = \overline{\overline{A \cdot B} \cdot \overline{C} \cdot \overline{D}}$$

What we need now is a device to separate the three inputs without changing the logic function. That device is one we have used quite frequently already, the double inversion, this time over inputs $\overline{C}$ and $\overline{D}$:

$$Y = \overline{\overline{A \cdot B} \cdot \overline{\overline{\overline{C} \cdot \overline{D}}}}$$

Figure 2.13 shows the logic circuit for the new expression. Notice the exclusive use of two-input NAND gates and the development of the final expression. Also take note of the number of times NAND gates are used to perform the inverting function. To prove that this circuit and the one in Figure 2.12 are logically identical, let's reduce the final expression and make sure we again yield the original expression, Y = A · B + C + D.

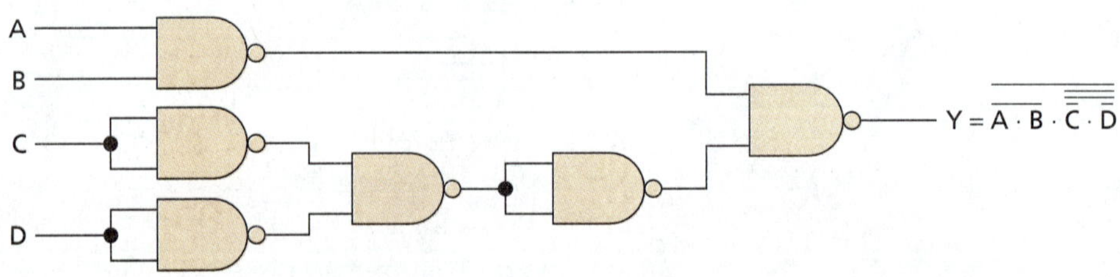

Figure 2.13 NAND Circuit

We begin the reduction by using DeMorgan's theorem:

$$Y = \overline{\overline{A \cdot B}} + \overline{\overline{\overline{\overline{C} \cdot \overline{D}}}}$$

Remove pairs of double inversion in the next step:

$$Y = A \cdot B + \overline{\overline{C} \cdot \overline{D}}$$

Apply DeMorgan's rule once more to finish the reduction:

$$Y = AB + C + D$$

This process of converting a logic circuit into one using only NAND gates allows these circuits to be produced using only one type of logic device, in this case the NAND. Large quantities of NAND gates can be purchased at a savings that make the effort worthwhile. A second use of the ability to develop circuits using one type of gate occurs in complex logic circuits that have leftover gates. Leftover gates result from using many logic devices in a circuit that have ICs with unused gates in them. For instance, a circuit may use 20 quad two-input NAND gate ICs in its design. Six or seven individual gates may not be used in various parts of the overall circuit. A designer can make use of them to perform a different logic function within the same circuit, possibly saving the need to add another IC to do the same function.

The same process for developing circuits using NAND gates can be used to create logic circuits using NOR gates, except that the core function is an OR instead of an AND function. As an example, let's create the exclusive OR function using two-input NOR gates only. The process starts the same way by inverting both sides of the expression and applying DeMorgan's theorem:

$$\overline{Y} = \overline{\overline{A} \cdot B + A \cdot \overline{B}} = (\overline{\overline{A} \cdot B}) \cdot (\overline{A \cdot \overline{B}}) = (A + \overline{B}) \cdot (\overline{A} + B)$$

We now have OR functions and, to make them NOR, invert both sides again:

$$\overline{\overline{Y}} = Y = \overline{(A + \overline{B}) \cdot (\overline{A} + B)} = (\overline{A + \overline{B}}) + (\overline{\overline{A} + B})$$

We are almost there, but the output gate is again an OR type and not a NOR type. We need a double inversion to finish the design expression:

$$Y = \overline{\overline{(\overline{A + \overline{B}}) + (\overline{\overline{A} + B})}}$$

The exclusive OR logic circuit using two-input NOR gates is shown in Figure 2.14. To prove that this circuit is logically the same as the ones in Figures 2.9 and 2.11, reduce this final expression and see if it yields the exclusive OR logic expression $Y = \overline{A} \cdot B + A \cdot \overline{B}$.

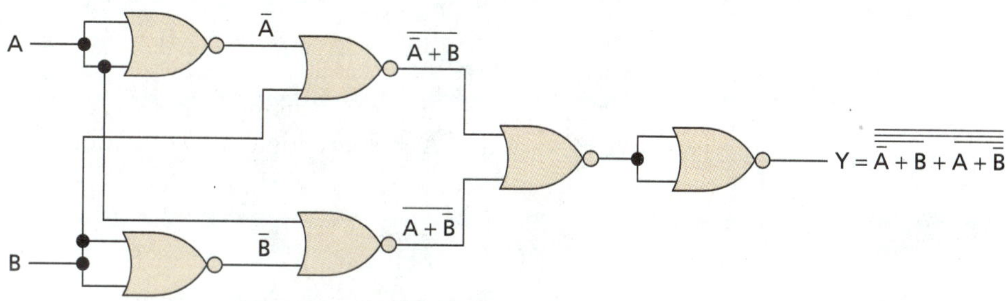

Figure 2.14 Exclusive OR Function Using NOR Gates

Section 2.6

Practice Problems

1. Implement the logic expression $Y = A \cdot B + C$, using two-input NAND gates only.
2. Implement the logic expression $Y = A \cdot B + C$, using two-input NOR gates only.

2.7 Two Additional Boolean Theorems

There are two additional logic relationships that appear, at first, not to be useful in reducing logic expressions. However, with a little Boolean algebra manipulation, they can be shown to be reducible and to form two new Boolean theorems. The original expressions and their reduced form can be further proven to be equal by using a truth table. Once their equality is established, they can be used as two additional theorems in our Boolean tool box. The first theorem states that if a SOP expression contains a single variable, such as A or $\overline{C}$, and another term in that SOP expression contains the complement of that variable ($\overline{A}$ or C), then the complemented variable can be canceled from that term. This theorem can better be understood by example. For instance, the SOP expression $\overline{A} + A \cdot B$ contains single variable $\overline{A}$ and its complement as a variable in the other term—the A in $A \cdot B$. This meets the requirements for this reduction, and the expression reduces to $\overline{A} + B$. The method to prove this reduction works is to double NOT the expression, which means that it looks like this:

$$\overline{\overline{\overline{A} + A \cdot B}}$$

Notice that the two NOT bars cancel each other, so there is no logical change to the expression. The reason for the double NOT is to allow us to use DeMorgan's theorem to alter the expression:

$$\overline{A \cdot \overline{(A \cdot B)}} = \overline{A \cdot (\overline{A} + \overline{\overline{B}})} = \overline{A \cdot \overline{A} + A \cdot \overline{\overline{B}}} = \overline{0 + A \cdot \overline{\overline{B}}} = \overline{A \cdot \overline{\overline{B}}} = \overline{A} + B$$

By using the double NOT and applying DeMorgan's theorem, we were able to reduce the expression $\overline{A} + A \cdot B$ to $\overline{A} + B$, thus eliminating the A variable from the term AB.

It should be also noted that the inverted variable could have been part of the AND function. In other words, $A + \overline{A} \cdot B = A + B$ applies equally as well as $\overline{A} + A \cdot B = \overline{A} + B$.

The theorem can be proven using a truth table to show that the two expression, $\overline{A} + A \cdot B$ and $\overline{A} + B$ are logically equivalent:

A	B	$\overline{A}$	$A \cdot B$	$\overline{A} + B$	$\overline{A} + A \cdot B$
0	0	1	0	1	1
0	1	1	0	1	1
1	0	0	0	0	0
1	1	0	1	1	1

The truth table illustrates that the output state for both expressions are identical, which proves that they represent the same logic function and are therefore equal, $\overline{A} + A \cdot B = \overline{A} + B$.

As an example of an application using this new theorem, we will reduce the Boolean expression $(A \cdot B + B \cdot C)(\overline{C} + B \cdot C)$. Start with the right-hand term $\overline{C} + B \cdot C$. This term satisfies the new theorem, since $\overline{C}$ is a single variable and its complement, C, is included in another term, $B \cdot C$. This reduces $\overline{C} + B \cdot C$ to $\overline{C} + B$. Next distribute the variables in one term into the other term:

$$(A \cdot B + B \cdot C)(\overline{C} + B) = A \cdot B \cdot \overline{C} + A \cdot B \cdot B + B \cdot C \cdot \overline{C} + B \cdot B \cdot C$$

Next apply Boolean theorems to change $B \cdot B$ to B and $\overline{C} \cdot C$ to 0:

$$A \cdot B \cdot \overline{C} + A \cdot B + 0 + B \cdot C$$

Apply $A + 0 = A$:

$$A \cdot B \cdot \overline{C} + A \cdot B + B \cdot C$$

Factor out B from each term:

$$B \cdot (\overline{A} \cdot C + A + C)$$

Reorder the variables in the parenthesis:

$$B \cdot (C + A \cdot \overline{C} + A)$$

There are two ways to continue and both yield the same result. Since we are dealing with a new theorem, we will use that method first. $C + A\overline{C}$ reduces to $C + A$:

$$B \cdot (C + A + A) = B \cdot (C + A) = B \cdot C + A \cdot B$$

We could have also factored an A from two of the terms instead:

$$B \cdot (C + A(\overline{C} + 1)) = B \cdot (C + A \cdot 1) = B \cdot (C + A) = B \cdot C + A \cdot B$$

EXAMPLE 2.4

Reduce the following logic expression:

$$Y = A \cdot B \cdot \overline{C} + \overline{A} \cdot C \cdot D + \overline{A} \cdot B \cdot C + A \cdot C$$

SOLUTION Begin by factoring the C variable from as many terms as possible:

$$Y = A \cdot B \cdot \overline{C} + C(\overline{A} \cdot D + \overline{A} \cdot B + A)$$

Now we can use the theorem $\overline{A} + A \cdot B = \overline{A} + B$ to reduce the term $\overline{A} \cdot B + A$ in the parenthesis to $B + A$:

$$Y = A \cdot B \cdot \overline{C} + C(\overline{A} \cdot D + B + A)$$

Use the associative law to rearrange the terms in the parenthesis, and recognize a second application of this new theorem:

$$Y = A \cdot B \cdot \overline{C} + C(B + A + \overline{A} \cdot D)$$

$$Y = A \cdot B \cdot \overline{C} + C(B + A + D)$$

Redistribute the C variable:

$$Y = A \cdot B \cdot \overline{C} + B \cdot C + A \cdot C + C \cdot D$$

Either factor A from two of the terms or B from two of the terms:

$$Y = A(B \cdot \overline{C} + C) + B \cdot C + C \cdot D \qquad \text{or} \qquad B(A \cdot \overline{C} + C) + A \cdot C + C \cdot D$$

Apply the theorem $\overline{A} + AB = \overline{A} + B$ once again:

$$Y = A(B + C) + B \cdot C + C \cdot D \qquad \text{or} \qquad B(A + C) + A \cdot C + C \cdot D$$

and redistribute the A or B variable:

$$Y = A \cdot B + A \cdot C + B \cdot C + C \cdot D \qquad \text{or} \qquad A \cdot B + B \cdot C + A \cdot C + C \cdot D$$

Notice that the results are identical, although in slightly different order.

The second theorem is a bit more difficult to state and is more involved. In a SOP expression with two variables in each product term, if two terms contain a variable and its complement and if the other variables are the two variables in the third term, the third term can be eliminated. Again, an example shows how this works—consider the SOP expression $\overline{A}B + AC + BC$. The first two terms contain A and its complement $\overline{A}$, which satisfies one condition of the theorem. The first term's other variable, B, and the second term's other variable, C, are used in the third term, BC, which satisfies the other criteria of the theorem. The conditions are met, so the third term can be dropped from the expression. Thus

$$\overline{A} \cdot B + A \cdot C + B \cdot C = \overline{A} \cdot B + A \cdot C$$

Proof lies in the truth table for both expressions:

A	B	C	$\overline{A}$	$\overline{A} \cdot B$	$A \cdot C$	$B \cdot C$	$\overline{A} \cdot B + A \cdot C + B \cdot C$	$\overline{A} \cdot B + A \cdot C$
0	0	0	1	0	0	0	0	0
0	0	1	1	0	0	0	0	0
0	1	0	1	1	0	0	1	1
0	1	1	1	1	0	1	1	1
1	0	0	0	0	0	0	0	0
1	0	1	0	0	1	0	1	1
1	1	0	0	0	0	0	0	0
1	1	1	0	0	1	1	1	1

Notice in the truth table that both expressions produce identical output states for all combinations of input states, proving that these two expressions are logically identical.

EXAMPLE 2.7

Reduce the logic expression $(\overline{A} + B)(A + C)$.

SOLUTION First distribute one term into the other:

$$\overline{A} \cdot A + \overline{A} \cdot C + A \cdot B + B \cdot C$$

Apply the rule $\overline{A} \cdot A = 0$:

$$0 + \overline{A} \cdot C + A \cdot B + B \cdot C$$

Then A + 0 = A:

$$\overline{A} \cdot C + A \cdot B + B \cdot C$$

Lastly, for the second new theorem $\overline{A}$ and A are variables for two terms whose other variables, B and C, form a third variable, BC, so that third variable can be omitted:

$$(\overline{A} + B)(A + C) = \overline{A} \cdot C + A \cdot B$$

Table 2.3 summarizes the Boolean algebra rules for reducing circuits presented in this chapter.

Table 2.3 Boolean Algebra Theorems

$$A \cdot A = A$$
$$A \cdot 1 = A$$
$$A \cdot 0 = 0$$
$$A + A = A$$
$$A + 1 = 1$$
$$A + 0 = A$$
$$\overline{A} \cdot A = 0$$
$$\overline{A} + A = 1$$
$$\overline{A \cdot B} = \overline{A} + \overline{B}$$
$$\overline{A} + \overline{B} = \overline{A} \cdot \overline{B}$$
$$\overline{\overline{A}} = A$$
$$A + \overline{A} \cdot B = A + B$$
$$A \cdot B + \overline{A} \cdot C + B \cdot C = A \cdot B + \overline{A} \cdot C$$
$$A \cdot B + A \cdot C = A(B + C)$$
$$(A + B)(C + D) = A \cdot C + A \cdot D + B \cdot C + B \cdot D$$

Section 2.7

Practice Problems

Reduce the following Boolean expressions:

1. $(A \cdot B \cdot C + \overline{A})(A \cdot \overline{C} + A \cdot B \cdot C)$
2. $(A \cdot B + B \cdot C)(\overline{A} \cdot C + B \cdot C)$
3. $\overline{A \cdot \overline{B} + B \cdot C + A \cdot C}$
4. $A \cdot \overline{B} \cdot C + B \cdot \overline{C} \cdot D + A \cdot \overline{C} \cdot D + \overline{A} \cdot \overline{C} \cdot D$

2.8 Karnaugh Mapping

Another aid in reducing logic expressions uses a graphical approach to represent logic circuits. This method is called Karnaugh (pronounced as in Car-No) mapping and involves converting a truth table into a diagram called a

Karnaugh map *graphical tool used in reducing logic expressions.*

Karnaugh map. Figure 2.15 is a two-variable truth table showing each combination of input states for the variables, along with equivalent decimal values for the binary numbers represented by the input variables, and with product terms for each line of the table. The two-variable inputs are arranged in ascending binary numerical order. Each combination of input states of the two variables is represented by a number of different methods according to application needs. The first method is the by-now-familiar listing of their 0 and 1 states.

A	B	Numerical Equivalent	Product Term
0	0	0	$\overline{A} \cdot \overline{B}$
0	1	1	$\overline{A} \cdot B$
1	0	2	$A \cdot \overline{B}$
1	1	3	$A \cdot B$

Figure **2.15** Truth Table Equivalences

A second way to represent each input combination is to use its numerical equivalent. For instance, line 1 of the truth table, where variables A and B are both low, is input 0. Line 2 is input 1, since A is 0, and B—the least significant bit of a two-bit equivalent binary number consisting of bits A and B—is a 1. Line 3 is input 2, and the last line is input 3.

A third method of defining inputs on a truth table is to list them according to the product term formed from their current active states as shown on the table. For the truth table in Figure 2.15, the first line where both input variables are low can be represented by the product term $\overline{A} \cdot \overline{B}$. The second line shows input A as a logic 0 or active low and input B as a logic 1 or active high, so the input for this line is $\overline{A} \cdot B$. The next line is $A \cdot \overline{B}$, and the last line of the table is $A \cdot B$. Sum-of-product (SOP) logic expressions for the circuit represented by the truth table is formed by summing all the product terms that cause the circuit's output to be in a logic 1 state.

In Figure 2.16, a two-input truth table shows inputs $\overline{A} \cdot \overline{B}$ and $A \cdot B$ both produce a logic 1 output result, while the other two inputs yield a logic 0 output state. The logic expression for output Y is found by creating a SOP expression using the product terms that show a logic 1 output state on the truth table. For this truth table, the expression is $Y = \overline{A} \cdot \overline{B} + A \cdot B$. A SOP logic expression can also be derived from those inputs that produce a 0, but this

Inputs		Product Term	Output
A	B		Y
0	0	$\overline{A}\overline{B}$	1
0	1	$\overline{A}B$	0
1	0	$A\overline{B}$	0
1	1	AB	1

Figure **2.16** Truth Table for $Y = \overline{A}\,\overline{B} + AB$

expression will be for the complement of the output, since each input product combination, in this case, yields the active low or logic zero output. In Figure 2.16, the active low SOP logic expression is $Y = \overline{\overline{A} \cdot B + A \cdot \overline{B}}$. To get the active high logic expression for Y, apply DeMorgan's theorem to reduce the expression.

$$Y = (\overline{\overline{A} \cdot B})(\overline{A \cdot \overline{B}}) = (A + \overline{B})(\overline{A} + B) = A \cdot B + \overline{A} \cdot \overline{B}$$

EXAMPLE 2.5

Derive the sum-of-products expression for the circuit represented by the two-input truth table shown in Figure 2.17. Reduce the circuit using Boolean algebra.

A	B	Y
0	0	1
0	1	1
1	0	0
1	1	1

Figure 2.17 Example 2.5 Truth Table

SOLUTION The inputs for the truth table are expressed in terms of their product terms, starting with $\overline{A} \cdot \overline{B}$ for inputs 00 and ending with $A \cdot B$ for inputs 11. A sum-of-products expression is formed by summing the product terms that produce a logic 1 output:

$$Y = \overline{A} \cdot \overline{B} + \overline{A} \cdot B + AB$$

This logic expression is reduced to the minimum SOP form by using Boolean algebra:

$$Y = \overline{A} \cdot \overline{B} + B(\overline{A} + A) = \overline{A} \cdot \overline{B} + B(1) = \overline{A} \cdot \overline{B} + B = \overline{A} + B$$

	$\overline{B}$	B
$\overline{A}$	$\overline{A}\overline{B}$ **0**	$\overline{A}B$ **1**
A	$A\overline{B}$ **2**	AB **3**

Figure 2.18 Two-Input Karnaugh Map with Product Terms and Equivalent Numbers

A two-variable Karnaugh map for the truth table in Figure 2.15 is shown in Figure 2.18. It is a graph with its sides labeled with input variables A and B and their complements $\overline{A}$ and $\overline{B}$. This labeling is used to form a matrix so that each box represents a different product term associated with a two-variable truth table. In the diagram, the upper left box represents the product term $\overline{A} \cdot \overline{B}$, found by combining $\overline{A}$ and $\overline{B}$ variable labels at the top and left side of the map. The box to its right contains product term $\overline{A} \cdot B$. The two bottom boxes work in the same manner to represent $A \cdot \overline{B}$ and $A \cdot B$.

The diagram has as many boxes as a truth table has input combinations, both of which can be determined by raising 2 to a power equal to the number of input variables (n) used in the table or the map (the number of boxes $= 2^n$). Each box must represent a unique product term, and adjacent boxes must contain product terms that do not differ by more than one variable. For instance, the box containing $\overline{A}B$ and AB are adjacent because only the A variable changed from $\overline{A}$ to A, while the B variable remained B.

The actual format that a Karnaugh map (or K-map, for short) takes is optional as long as the aforementioned criteria have been met. Two common formats are shown in Figure 2.19, the major difference being the actual labeling of the variables used by the map. The upper map is a copy of the one presented in Figure 2.18, and labels the edges of the map with variable names and states (A, B, $\overline{A}$, and $\overline{B}$). The other map places the variables names along a diagonal at the left corner. The edges are labeled with 1 and 0 to indicate the active state of the variable. For instance, along the top edge the box on the left has a label of $\overline{B}$, derived from the B label on the diagonal and the 0 state shown on the box edge. A standard method for labeling K-maps has not been formulated; it is left to the discretion of users to adopt a form that is comfortable for them or that they first learned when studying logic circuits and Boolean algebra.

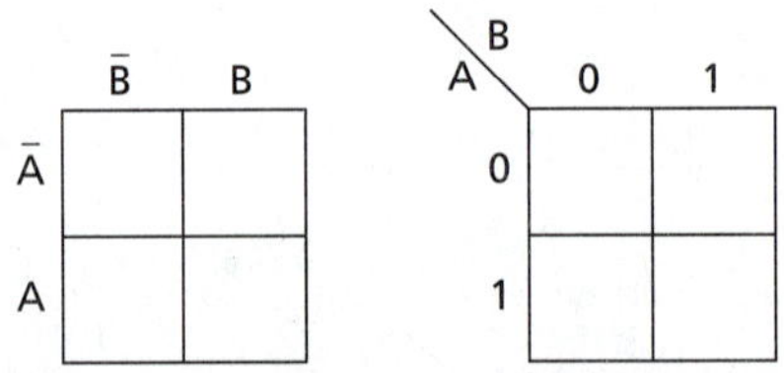

Figure 2.19 Alternate Two-Input K-map Formats

For each product term that produces a 1 output as indicated on a truth table, a corresponding box on a K-map is filled with a 1. Let's start with a simple example by making a K-map for an AND gate. From a truth table for the AND gate, only one product term, A · B, yields a 1 output. On the K-map for the AND gate in Figure 2.20, only one box gets a 1. This box is in the lower right corner of the map and represents the product term A · B. The remaining three boxes can be left blank or filled with zeros.

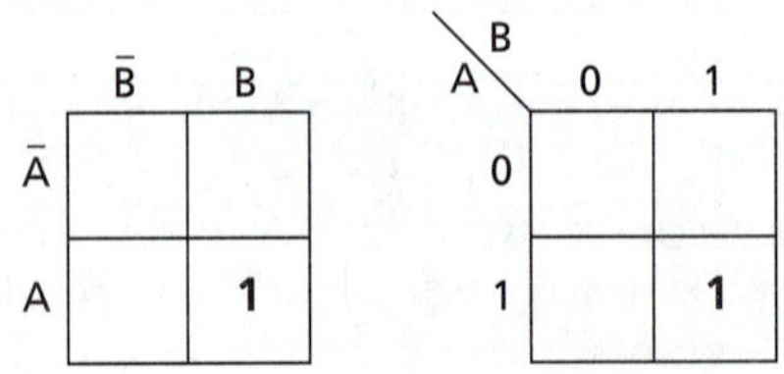

Figure 2.20 AND Gate Karnaugh Map

Figure 2.21 is a K-map for an OR gate that was filled by reading its truth table. Three product terms produce a 1 in the OR gate truth table. They are inputs $\overline{A}$ · B, A· $\overline{B}$ and A · B, and each has a corresponding box, filled with a 1, in the map. The last box, representing $\overline{A}$ · $\overline{B}$, is left blank or filled with a zero.

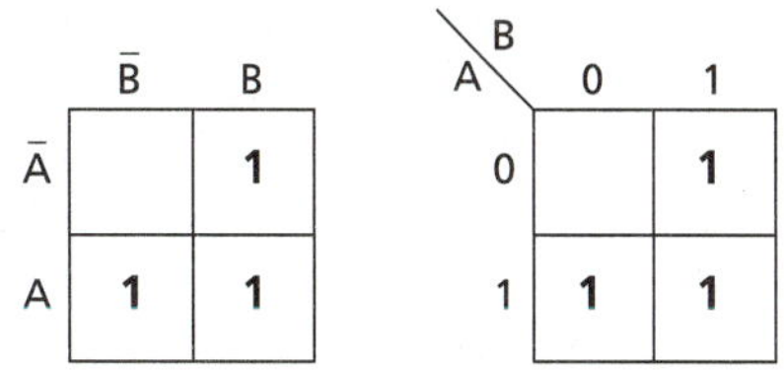

Figure 2.21 OR Gate Karnaugh Map

A K-map is read by first grouping adjacent boxes that represent product terms that yield logic 1 output results into rectangular groups or arrays, the sizes of which are based on the base 2 progression of 1, 2, 4, 8, or 16 boxes. For purposes of finding the minimum SOP logic expression for the circuit associated with the truth table or K-map in use, the larger the grouping the better. The reason for this will be shown as we work with K-maps and the expressions they generate. The largest possible group for a two-variable map contains four boxes, because this is the total number of boxes in the map. The other group sizes possible in the two-variable map are of one or two boxes. In the OR gate example, two groups contain two adjacent boxes that have product terms that yield a logic 1 output. Groups of boxes on a K-map may share as many individual boxes as needed with other groups in order to form the largest possible groupings. The problem is to avoid unnecessary redundancy, as will be shown on a bigger K-map later on.

In Figure 2.22, the OR gate K-map is shown with the two groups circled. Each group contains two adjacent boxes that contain a 1. The box common to both groups holds the product term $A \cdot B$. The next step in reading the map is to examine the boxes that are grouped together and to look for variables that do not change state from one box to the next in the group. Only these common variables are used in the product terms of the SOP expression read from the map. In the OR example of Figure 2.22, the two adjacent boxes in the vertical group contain the variable B common to both boxes. Variable A changes from $\overline{A}$ to A in this group, so this group represents the product term with the single variable B. Similarly the horizontal pair represent a product term with the single variable A, which is common to both boxes in that group. The product terms are summed together to complete the SOP expression for the logic circuit that produced the truth table and K-map—in this case, the OR gate, which has the reduced expression $Y = A + B$.

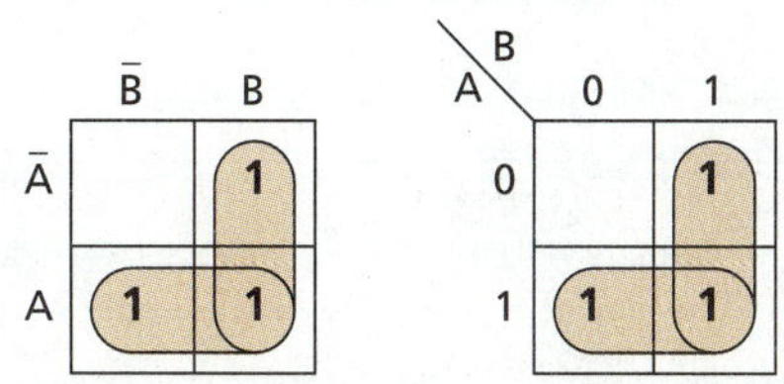

Figure 2.22 Grouping OR Gate Product Terms

The reason why the group of two boxes are represented by a single letter is best explained by forming the SOP expression for one of the groups itself, and applying Boolean algebra to reduce it to its minimum SOP form. The vertical group is actually the SOP expression taken from the two individual boxes in the group, $Y = A \cdot B + \overline{A} \cdot B$. Start by factoring the B variable from this SOP expression: $Y = B(\overline{A} + A)$. Next, realize that $\overline{A} + A = 1$ and $B \cdot 1 = B$, which is the same result achieved when reading the common variable from the group of two boxes on the K-map.

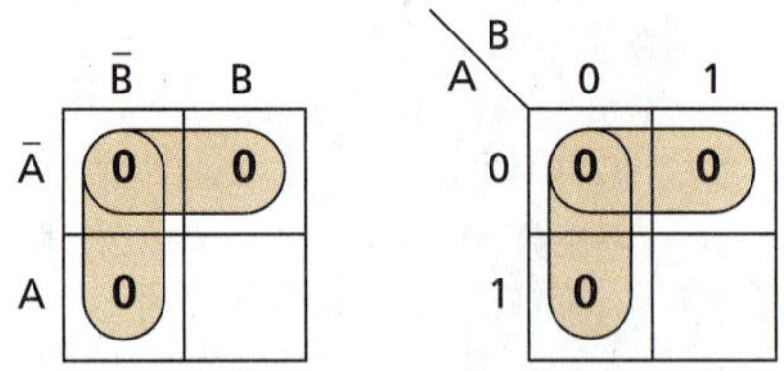

Figure 2.23 Reading Zero Entries from AND Gate Karnaugh Map

Zeros as well as logic 1's can be entered and read from the K-map. In the AND example of Figure 2.20, the product terms that produce a zero output on the truth table can be entered onto a K-map and arranged into groups of two adjacent boxes, as shown in Figure 2.23, and read in the same manner as reading groups of 1's. However, when reading groups of 1's from a K-map, you are deriving the sum-of-products expression that yields an active high or logic 1 output. When reading groups of zeros from the map, you are also creating a sum-of-products expression, but this time, for an active low or zero output. To write the SOP logic expression for the circuit, when reading groups of zeros, a NOT bar is placed over the entire expression after it is read from the map to show the active low state of the output. For the AND gate, the active low SOP expression as read from the K-map is

$$Y = \overline{\overline{A} \cdot \overline{B} + \overline{A} \cdot B + A \cdot \overline{B}}$$

The expression must now be further reduced using Boolean algebra before the final minimized SOP expression is found:

$$Y = \overline{\overline{A}(1 + B) + A \cdot \overline{B}} = \overline{\overline{A} + A \cdot \overline{B}} = \overline{\overline{A} + \overline{B}} = A \cdot B$$

The OR gate map has only one zero entered onto it, representing the product expression $\overline{A} \cdot \overline{B}$. The K-map in Figure 2.21 could have been read as

$$Y = \overline{\overline{A} \cdot \overline{B}}$$

DeMorgan's theorem is applied to derive the OR expression $Y = A + B$. In this case, doing the DeMorgan application does not require a significant effort. In most cases, K-maps are read using the boxes filled with 1's to avoid the need to use DeMorgan's theorem to get the minimized SOP expression for a logic circuit.

EXAMPLE 2.10 Derive the sum-of-products expression for this two-input truth table using Karnaugh maps, one diagram mapping 1's and a second one mapping 0's.

A	B	Y
0	0	1
0	1	1
1	0	0
1	1	1

SOLUTION First construct the Karnaugh map filling in the boxes with 1's for $\overline{A} \cdot \overline{B}$, $\overline{A} \cdot B$, and $A \cdot B$. Since we also want to read the active low expression as well as the active high one, also enter the 0 in the box for the product term $A \cdot \overline{B}$.

	$\overline{B}$	B
$\overline{A}$	**1**	**1**
A	**0**	**1**

Next group the adjacent boxes containing 1's into the smallest number of groups with the most adjacent 1's in them. In this example, there are two groups that contain two adjacent boxes each:

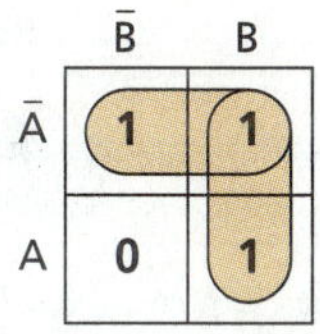

Next read the product terms from groups of boxes containing 1's. In this case, these terms are $\overline{A}$ from the two upper adjacent boxes and B from the two right vertical adjacent boxes. The minimum SOP expression for the circuit that generated the truth table is $Y = \overline{A} + B$.

A active low SOP expression comes from the single box containing a zero in it, $A \cdot \overline{B}$. The expression for the circuit is $Y = \overline{A \cdot \overline{B}}$, which reduces to $Y = \overline{A} + B$, the same expression derived from reading the minimum active high SOP expression.

Karnaugh maps can be used to reduce a SOP expression directly as well as creating an expression from a truth table. Boxes on a map are filled with 1's for each product term of the SOP expression. The boxes are grouped as before, and a minimum SOP expression is read from the map. As an example, earlier in the chapter the Boolean theorem, $\overline{A} + A \cdot B = \overline{A} + B$ was proven using Boolean algebra and a truth table. This theorem can also be verified using the K-map in Figure 2.24. Single variable terms in a SOP expression, like the $\overline{A}$ term in this expression, occupy all the boxes that contain that term. For a two-variable map, a single variable product term causes two boxes to be filled with 1's. For this theorem, the two boxes, representing product terms $\overline{A} \cdot \overline{B}$ and $\overline{A} \cdot B$ are filled with 1's because they both contain the variable $\overline{A}$. The box representing the other product term of the expression, $A \cdot B$, is filled with a 1. The filled boxes are then circled as groups of two boxes each and read off the map using the common variables for each group. The reduced expression is $Y = \overline{A} + B$, again showing that $\overline{A} + A \cdot B$ reduces to $\overline{A} + B$.

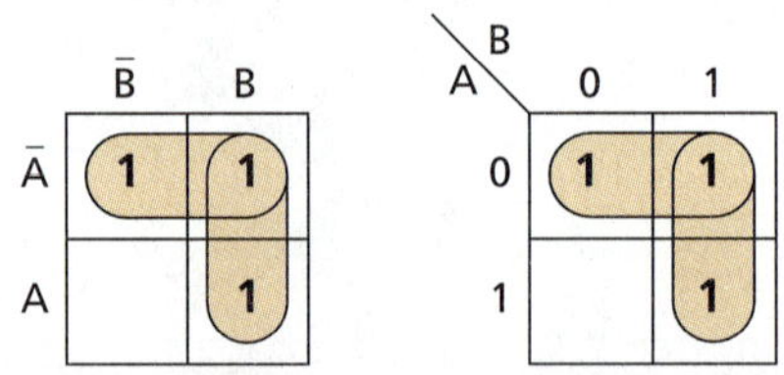

Figure 2.24 $\overline{A}$ + AB Karnaugh Map

Section 2.8

Practice Problems

1. Derive the minimum sum-of-products expression of output Y in this two-input truth table using a Karnaugh map.

A	B	Y
0	0	1
0	1	0
1	0	1
1	1	1

2. Map the following Boolean expressions on a Karnaugh map, and read off the reduced sum-of-products expressions:

(a) $Y = \overline{A}B + A\overline{B} + AB$ **(b)** $Y = \overline{A} \cdot \overline{B} + \overline{A} \cdot B + A \cdot B$

2.9 Three-Variable Karnaugh Maps

The design of the K-maps beyond two-inputs must meet exacting criteria. Each box must contain a unique combination of variable product terms. In addition, boxes that are on the outside of any diagram need to be logically adjacent to boxes on the opposite outside edges. That means they must meet the same criteria as other adjacent boxes in that only one variable in each of the product terms of two adjacent boxes can be different. Before looking at the form of a three-variable K-map, let's start with a three-variable truth table to formulate the product terms that will be used on the map.

The truth table in Figure 2.25 is a three-variable table showing equivalent numerical value and product term for each input combination. It is an extension of the two-variable table examined in the last section. We will reference this table to show the relationship between it and the three-variable K-map in Figure 2.26.

A	B	C	Numerical Equivalent	Product Term
0	0	0	0	$\overline{A} \cdot \overline{B} \cdot \overline{C}$
0	0	1	1	$\overline{A} \cdot \overline{B} \cdot C$
0	1	0	2	$\overline{A} \cdot B \cdot \overline{C}$
0	1	1	3	$\overline{A} \cdot B \cdot C$
1	0	0	4	$A \cdot \overline{B} \cdot \overline{C}$
1	0	1	5	$A \cdot \overline{B} \cdot C$
1	1	0	6	$A \cdot B \cdot \overline{C}$
1	1	1	7	$A \cdot B \cdot C$

Figure 2.25 Three-Variable Truth Table

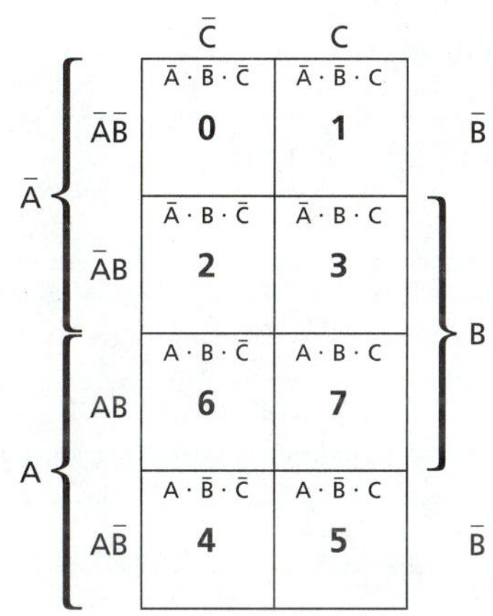

Figure 2.26 Three-Variable K-map

The upper edge of the K-map is labeled with variable $\overline{C}$ and C while the left edge uses product combinations of variables A and B. Product terms represented in each box are formed by combining the A and B product term on the left with variable C at the top. For example, the box in the upper left corner, numbered 0, represents the product term $\overline{A} \cdot \overline{B} \cdot \overline{C}$, pointed to by the combination of product term $\overline{A} \cdot \overline{B}$ on the left and $\overline{C}$ above the box. The box's number corresponds to the binary value of the active states of the term's variables, in this case 000. The box to its right represents the product term $\overline{A} \cdot \overline{B} \cdot C$, formed by combining $\overline{A} \cdot \overline{B}$ with C. These two boxes are adjacent to one another and meet the criteria because only one variable, C, changes state from one box to the next. The same can be said for the boxes in the upper and lower left corners. One box contains the product term $\overline{A} \cdot \overline{B} \cdot \overline{C}$ we have just mentioned. The other represents the term $A \cdot \overline{B} \cdot \overline{C}$. Notice that only the variable A has a different state from one box to the other. Essentially, this means that this K-map is a two-dimensional representation of a three-dimensional spherical form. The top edge is adjacent to or touching the bottom edge and the two sides are also touching. This means that the upper right box and lower right box both contain at least two common variables in their product terms, in this case $\overline{B}$ and C, and only one variable, A, changes state between the two boxes.

As with two-variable K-maps, 3-variable maps may be drawn in various ways, as shown in Figure 2.27. They may lie horizontally or vertically. They may be labeled with variable terms on their edges or with terms on a diagonal and logic states, 0 or 1, along their edges. I find this latter method adds a degree of awkwardness to mapping. Consider the map in the lower right of Figure 2.27. The diagonal shows that variable C lies on the left edge and the product of A and B along the top. The state of variable C is not too bad, it's either 0 ($\overline{C}$) or 1 (C). However, the four combinations of variables A and B, shown as combinations of 1 and 0, force the user to pause to interpret the product terms by mentally replacing the binary combination with its product equivalent and then combining that with the C variable. This map form is included here because it is used fairly frequently, and the student may find it used in other texts or applications.

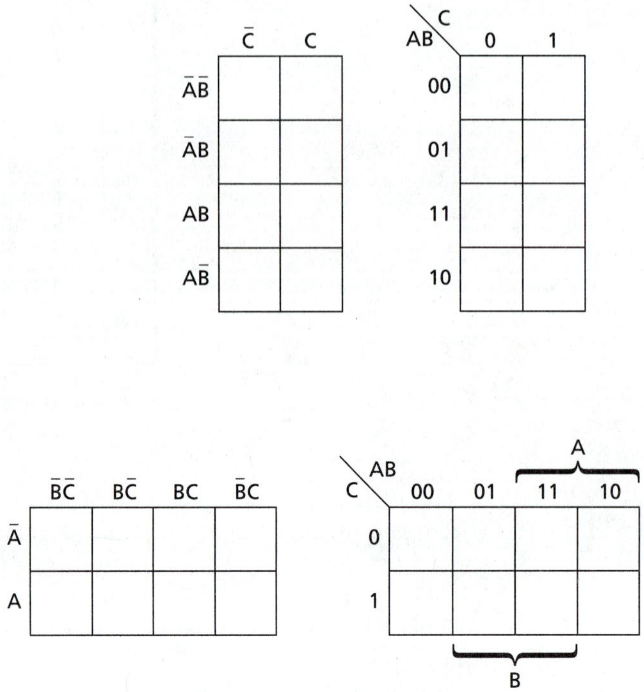

Figure 2.27 Variations of Three-Input K-map Formats

Three-variable maps are filled with 1's in the same manner as two-input maps by entering 1's into boxes that represent the product terms that produce 1's from a truth table or Boolean expression. A product term with three variables causes a single box to be filled with a 1, because each box represents a unique three-input product term. For instance, the term $\overline{A} \cdot \overline{B} \cdot C$ fills the box in the upper right corner with a 1. Product terms with two variables cause two boxes containing both of those variables to be filled with a 1. Product term $\overline{A} \cdot \overline{B}$ cause the upper two boxes of the map to contain a 1, since both boxes contain this term. Lastly, a product term with a single variable results in four boxes, each containing that variable is filled with a 1. If a logic expression contains the term A, the four boxes at the bottom of the map, each containing an A variable, are filled with 1's.

EXAMPLE 2.6

Map the following logic expression:

$$Y = \overline{A} \cdot B + A \cdot B \cdot C + \overline{C}$$

SOLUTION

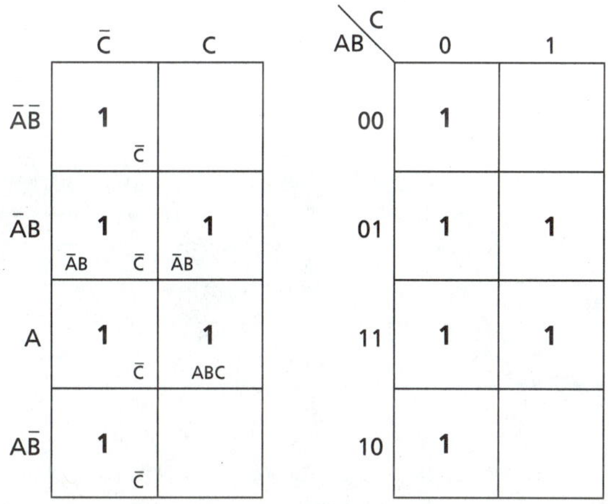

Figure 2.28 Mapping $\overline{A}B + ABC + \overline{C}$

The two-variable term $\overline{A} \cdot B$ causes 1's to be entered into two boxes in the second row from the top. Term $A \cdot B \cdot C$ fills one box in the third row, right, and the single-variable term $\overline{C}$ causes all the boxes on the left to receive a 1. Notice that both the terms $\overline{A} \cdot B$ and $\overline{C}$ cause a 1 to be placed in the left box second from the top.

Three-variable maps are read using a procedure similar to reading two-variable maps. Adjacent boxes are grouped into the largest possible groups of 1, 2, 4, and 8 boxes. The process of grouping stops when all boxes containing a 1 are accounted for. Groups can share boxes with 1's in them, but unnecessary redundancy should be avoided, which we will shortly illustrate by example. Then the variables that are common to the groupings are read off the map as the new product terms. If a three-variable map has a grouping of eight boxes, it means that all the boxes are full and the logic expression for the circuit is a steady state logic 1. A completely empty map means the circuit is producing a steady state 0. Groups of four boxes contain a single variable common to all four boxes. Two box groups have two common variables and single boxes represent product terms with three variables.

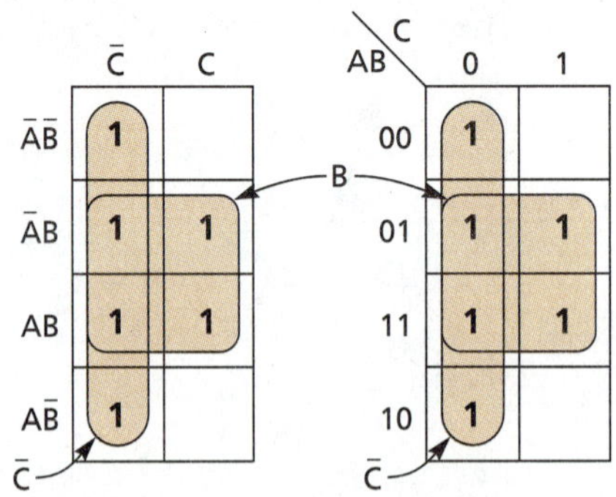

Figure 2.29 Reading B + $\overline{C}$ from Map

Figure 2.29 is the K-map produced in Example 2.6 for the logic expression $Y = \overline{A} \cdot B + A \cdot B \cdot C + \overline{C}$, with the groupings circled. Boxes are grouped into the largest possible size to minimize the number of variables of the product terms read from the map. Two groups of four adjacent boxes are found on this map, one along the left side of the map and the other containing the four boxes in the center of the map. Notice there are two boxes that are shared by both groups. Groups of two boxes each could have been formed instead of grouping by four boxes, but that would not yield the minimum SOP expression. Remember that boxes in groups of four represent product terms of one variable, whereas groups of two boxes represent product terms of two variables. In the current example, the group of four boxes along the left side all contain the common variable $\overline{C}$, whereas the group in the center all contain the variable B. The minimum SOP expression read from the map is $Y = B + \overline{C}$, which means that $\overline{A} \cdot B + A \cdot B \cdot C + \overline{C} = B + \overline{C}$. This can be proven using a truth table or Boolean algebra. Let's use the latter in the proof. Start by factoring the variable B from the first two terms:

$$Y = B(\overline{A} + A \cdot C) + \overline{C}$$

Reduce the middle term using the $\overline{A} + AB$ theorem, and redistribute the B variable:

$$B(\overline{A} + C) + \overline{C} = \overline{A} \cdot B + B \cdot C + \overline{C}$$

Repeat the theorem's use once again for $BC + \overline{C}$:

$$Y = \overline{A} \cdot B + B + \overline{C}$$

Factor variable B again, and complete the reduction:

$$Y = B(\overline{A} + 1) + \overline{C} = B(1) + \overline{C} = B + \overline{C}$$

Three-variable maps can be used to derive logic expressions from three-variable truth tables as well as SOP expressions. For example, the truth table in Figure 2.30 is mapped into the three-variable map shown in Figure 2.31. Each line of the truth table that causes the output to be high will fill in one box, because each line represents a product term of three variables. To read this map, first circle the groupings on the map as shown in Figure 3.32. The first of these is the pair of adjacent boxes in the upper right quadrant of the map representing product term $\overline{A} \cdot C$. Four adjacent boxes with a 1 in them form a group, using the top and bottom pairs of boxes whose edges actually touch. The common variable in the group of four is $\overline{B}$, so the minimum SOP expression for the circuit that is represented by the truth table of Figure 2.30 is

$$Y = \overline{A} \cdot C + \overline{B}$$

A	B	C	Y	Product Term
0	0	0	1	$\overline{A} \cdot \overline{B} \cdot \overline{C}$
0	0	1	1	$\overline{A} \cdot \overline{B} \cdot C$
0	1	0	0	
0	1	1	1	$\overline{A} \cdot B \cdot C$
1	0	0	1	$A \cdot \overline{B} \cdot \overline{C}$
1	0	1	1	$A \cdot \overline{B} \cdot C$
1	1	0	0	
1	1	1	0	

Figure 2.30 Three-Variable Truth Table Example

	$\overline{C}$	C
$\overline{A}\overline{B}$	**1** $\overline{A} \cdot \overline{B} \cdot \overline{C}$	**1** $\overline{A} \cdot \overline{B} \cdot C$
$\overline{A}B$		**1** $\overline{A} \cdot B \cdot C$
AB		
$A\overline{B}$	**1** $A \cdot \overline{B} \cdot \overline{C}$	**1** $A \cdot \overline{B} \cdot C$

AB \ C	0	1
00	**1** 000	**1** 001
01		**1** 011
11		
10	**1** 100	**1** 101

Figure 2.31 Mapping Truth Table

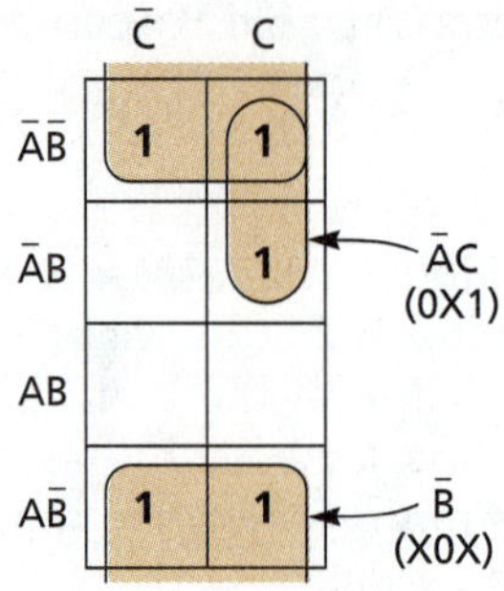

Figure 2.32 Reading $\bar{B} + \bar{A}C$ from Map

EXAMPLE 2.7

Show that the Boolean theorem $AB + \bar{A}C + BC = AB + \bar{A}C$, which was previously proven by using a truth table, can also be proven using a Karnaugh map.

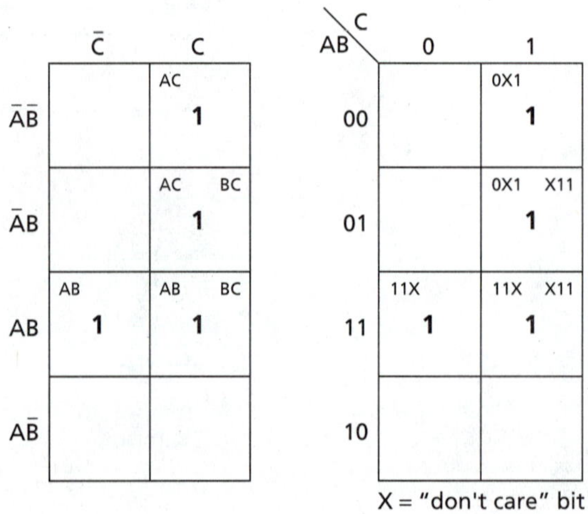

Figure 2.33 Mapping Boolean Algebra. Theorem $AB + \bar{A}C + BC$

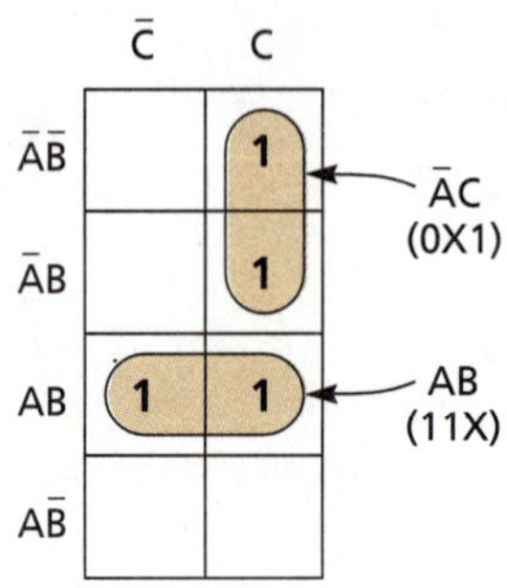

Figure 2.34 Reading Theorem Proof from Map Showing That $AB + \bar{A}C + BC = AB + \bar{A}C$

SOLUTION Enter 1's for each of the expression's sum-of-products terms in the appropriate boxes as indicated in Figure 2.33. Each term causes two boxes to be filled. Notice, as you do this, that one of the product terms actually fills two boxes already filled by the other two terms.

After the map is filled, circle the largest groupings required to cover all the boxes with 1's as shown in Figure 2.34. Avoid redundancy, which is the creation of an unnecessary group. Besides the two groups shown on this map, there is a third group of two boxes in the center right two boxes. This is a redundant group because both boxes in that group already make up other groups. The reduced expression is read by finding the common variables in each of two groups:

$$Y = \overline{A} \cdot C + A \cdot B$$

For some logic problems, a situation occurs where a product term has no influence on the output logic function. In other words, the output result does not depend on this product combination being a logic 1 or logic 0. For this case, as noted earlier, the product term is called a "don't care" term and is indicated as an "X" on a truth table and Karnaugh map. When reading a map with a "don't care" state in it, you can treat that box as if it contained either a 1 or a 0. If it helps to use this box to complete a group, do so, but if this box is not part of a group, ignore it.

In the K-map of Figure 2.35, the product terms $\overline{A} \cdot B \cdot C$ and $A \cdot \overline{B} \cdot \overline{C}$ have don't care states, indicated by an X in their boxes. The box containing the term $\overline{A} \cdot B \cdot C$ aids in forming the two groups, as shown in the figure, so we treat the box as if it had a 1 in it. The other don't care box, containing the term $A \cdot \overline{B} \cdot \overline{C}$, does not aid in forming any groups, so we treat it as if it contained a zero. The logic expression read from this map is $B + \overline{A} \cdot C$.

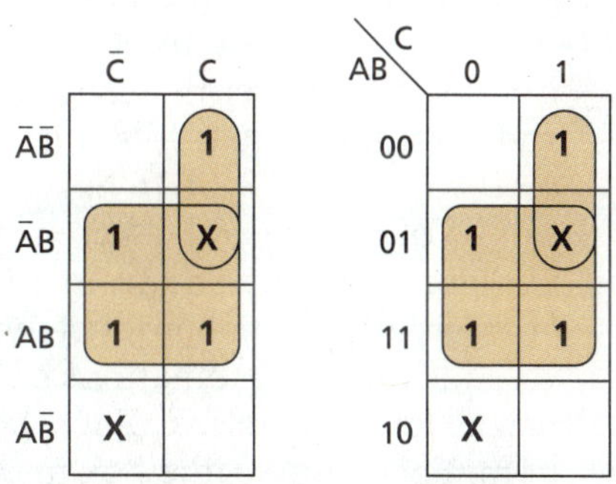

Figure 2.35 "Don't Care" Example

Section 2.9

Practice Problems

1. Reduce the following Boolean expressions using Karnaugh maps:

 (a) $Y = \overline{A} \cdot \overline{B} \cdot \overline{C} + B \cdot C + A \cdot \overline{C} + \overline{A} \cdot B \cdot \overline{C}$

 (b) $Y = A + \overline{A} \cdot B + B \cdot \overline{C} + \overline{A} \cdot C$

2. Derive the minimum sum-of-products expression for the circuit that generated the truth table in Figure 2.36.

A	B	C	Y
0	0	0	0
0	0	1	0
0	1	0	1
0	1	1	1
1	0	0	0
1	0	1	1
1	1	0	1
1	1	1	1

Figure 2.36 Practice Problem 2.9 (2)

2.10 Four-Variable Karnaugh Maps

Figure 2.37 shows a four-variable truth table with equivalent numerical values and product terms. A corresponding four-variable K-map is shown in Figure 2.37. This time, the labeling of the top and left side edges include two variables each, with A and B placed along the left edge and variables C and D at the top. The product term in each box is formed by combining the partial product terms from the labeling at the top with that from the left side. For example, the product term in the leftmost box that is second from the top is $\overline{A} \cdot B \cdot \overline{C} \cdot \overline{D}$, formed from the partial product terms $\overline{A} \cdot B$ at the left and $\overline{C} \cdot \overline{D}$ from the top. When entering 1 into the boxes of this map, product terms of four variables cause one box to be filled, three-variable product terms fill two boxes with 1's, two-variable product terms account for four boxes, single-variable terms fill 8 boxes, and a steady state 1 fills all 16 boxes. Adjacent boxes that contain 1's form groups of 1, 2, 4, 8, or 16 boxes. Four-variable K-maps, like their predecessors, can be drawn several different ways. Two commonly used methods, one used in Figure 2.37, are shown in Figure 2.38. The other method employs the diagonal with the partial product terms in the same manner as used with the three-variable map. Combinations of 1 and 0 signify the states of the variables of the partial product terms. Each box represents a unique four-variable product term. For example, the box in the upper left corner of the upper map is designated by the product term $\overline{A} \cdot \overline{B} \cdot \overline{C} \cdot \overline{D}$, whereas on the lower map it is referred to as the box containing 0000 ($A \cdot B = 00$ and $C \cdot D = 00$, or $\overline{A} \cdot \overline{B}$ and $\overline{C} \cdot \overline{D}$).

To illustrate how to use the four-variable map, let's use the map in Figure 2.37 to enter the logic expression:

$$Y = \overline{A} \cdot \overline{B} \cdot \overline{C} \cdot \overline{D} + \overline{A} \cdot B \cdot \overline{C} + A \cdot \overline{B} \cdot \overline{C} + B \cdot \overline{C} \cdot D + B \cdot C \cdot D + \overline{A} \cdot B \cdot C \cdot \overline{D}$$

onto the map and then read the map to find the minimum sum-of-products expression.

A	B	C	D	Numerical Equivalent	Product Term
0	0	0	0	0	$\bar{A}\bar{B}\bar{C}\bar{D}$
0	0	0	1	1	$\bar{A}\bar{B}\bar{C}D$
0	0	1	0	2	$\bar{A}\bar{B}C\bar{D}$
0	0	1	1	3	$\bar{A}\bar{B}CD$
0	1	0	0	4	$\bar{A}B\bar{C}\bar{D}$
0	1	0	1	5	$\bar{A}B\bar{C}D$
0	1	1	0	6	$\bar{A}BC\bar{D}$
0	1	1	1	7	$\bar{A}BCD$
1	0	0	0	8	$A\bar{B}\bar{C}\bar{D}$
1	0	0	1	9	$A\bar{B}\bar{C}D$
1	0	1	0	10 = A	$A\bar{B}C\bar{D}$
1	0	1	1	11 = B	$A\bar{B}CD$
1	1	0	0	12 = C	$AB\bar{C}\bar{D}$
1	1	0	1	13 = D	$AB\bar{C}D$
1	1	1	0	14 = E	$ABC\bar{D}$
1	1	1	1	15 = F	$ABCD$

(a) Four-variable Truth Table

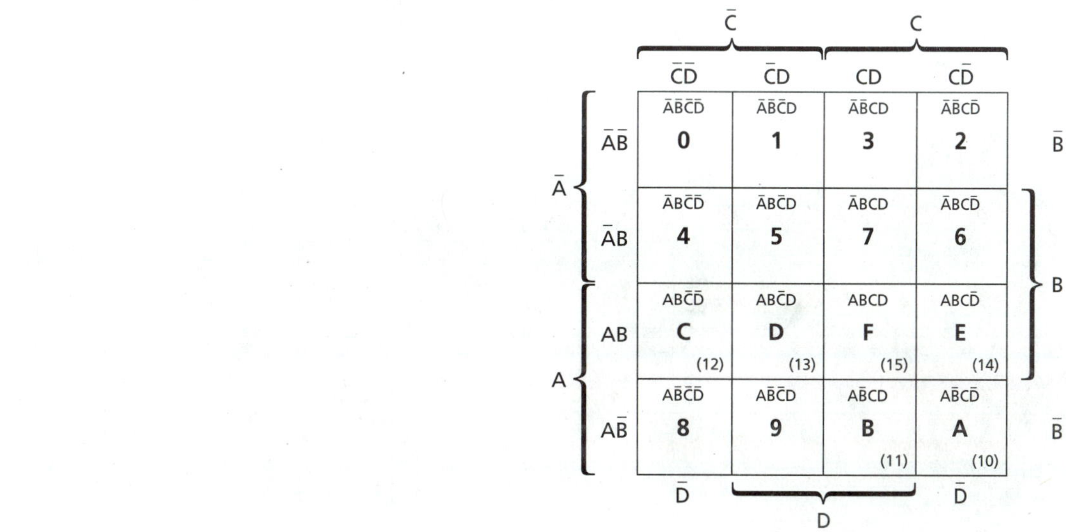

(b) Four-variable Map

Figure 2.37 Four-Variable K-map

The completed K-map for the expression is shown in Figure 2.39. Four input terms occupy one box each. First, $\bar{A} \cdot \bar{B} \cdot \bar{C} \cdot \bar{D}$ fills the box in the upper left corner, while $\bar{A} \cdot B \cdot C \cdot \bar{D}$ fills the second box from the top at the right side of the map. Three input terms occupy two boxes each. $\bar{A} \cdot B \cdot \bar{C}$ fills two leftmost boxes in the second row while $A \cdot \bar{B} \cdot \bar{C}$ fills the two lower left-hand boxes. Two more boxes are occupied by $B \cdot C \cdot D$ in the second and third rows in the third column from the left. Lastly, $B \cdot \bar{C} \cdot D$ accounts for two boxes, one in the third row, second column, and one above it that was already filled using the product

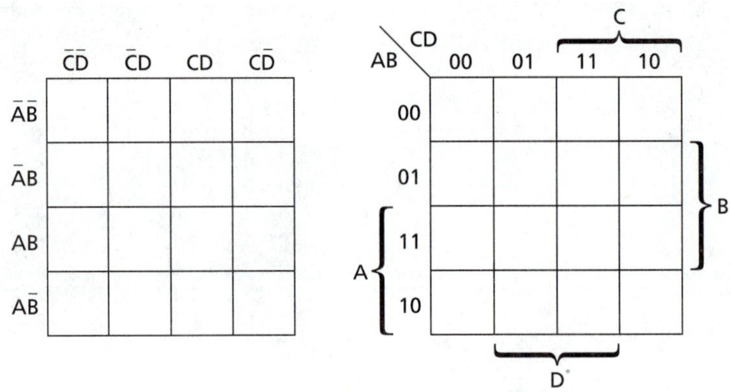

Figure 2.38 Variation on Four-Variable K-map Formats

term $\overline{A} \cdot B \cdot \overline{C}$. Identifying by location is cumbersome and awkward, so in the future each box will be referenced by its equivalent numerical value, as shown on the K-map in Figure 2.37 and the truth table of Figure 2.36. These numbers are included in the filled-in map of Figure 2.39 along the left edge of each filled box. This makes each box location easier to identify. For instance, we can now refer to "box 0," instead of "the box in the upper left corner," as the one representing the product term $\overline{A} \cdot \overline{B} \cdot \overline{C} \cdot \overline{D}$.

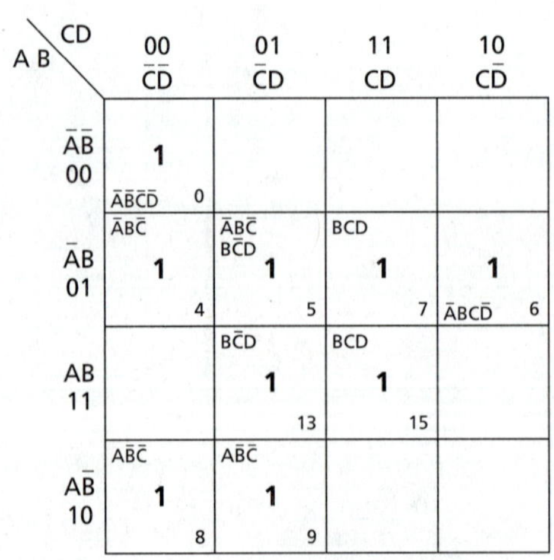

Figure 2.39 Mapping the Logic Expression $\overline{A}\overline{B}\overline{C}\overline{D} + \overline{A}B\overline{C} + A\overline{B}\overline{C} + B\overline{C}D + BCD + \overline{A}BC\overline{D}$

Figure 2.40 shows the four-variable map with groupings correctly circled. Two groups of four adjacent boxes exist, one consisting of boxes 4, 5, 6, and 7 representing the product term $\overline{A} \cdot B$, and the other, in the middle of the map, consisting of boxes 5, 7, 13, and 15, representing the product term $B \cdot D$. This leaves a number of double-box groups, each representing a three-variable prod-

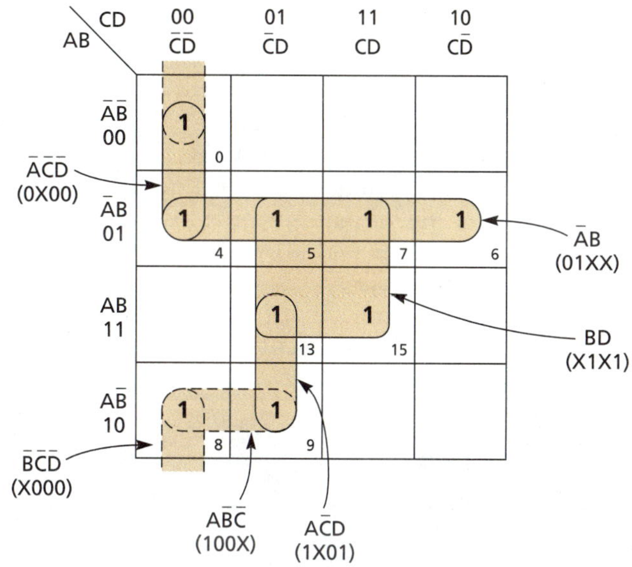

Three Alternate Expressions:

(1) $\overline{A}B + BD + \overline{B}\overline{C}\overline{D} + A\overline{B}\overline{C}$

(2) $\overline{A}B + BD + \overline{B}\overline{C}\overline{D} + A\overline{C}D$

(3) $\overline{A}B + BD + \overline{A}\overline{C}\overline{D} + A\overline{B}\overline{C}$

Figure 2.40 Reading Four-Variable Map

uct term. There are actually four groups of double boxes, providing us with an opportunity to observe what is meant by redundancy. To complete the minimized SOP expression, we have to select enough of the remaining two-box groups to account for all the remaining filled-in boxes on the map without creating extra product terms for our SOP expression. This often leads to choices between several SOP expressions, each of which is an optional valid expression for the circuit represented by the K-map. Let's see what this all means by first identifying the remaining groups.

A pair of adjacent boxes is formed, from box 0 and box 4, that represents the product term $\overline{A} \cdot \overline{C} \cdot \overline{D}$. A second pair includes boxes 0 and 8, because the upper and lower edges are touching and contain adjacent boxes. The product term for this group is $\overline{B} \cdot \overline{C} \cdot \overline{D}$. The next pair uses boxes 8 and 9 and represents the term $A \cdot \overline{B} \cdot \overline{C}$. The fourth and last grouping includes boxes 9 and 13 and represents the group $A \cdot \overline{C} \cdot D$. When reading the minimum SOP expression from the map, we start by reading the product terms that have no redundancy. In this example, they are the terms $\overline{A} \cdot B$ and $B \cdot D$. The remaining terms are selected to avoid unnecessary repetition of terms. We can select the terms $\overline{A} \cdot \overline{C} \cdot \overline{D}$ and $A \cdot \overline{B} \cdot \overline{C}$, which use boxes 0, 4, 8, 9 and account for all the remaining filled-in boxes. If we choose these two terms, then we do not need the terms $\overline{B} \cdot \overline{C} \cdot \overline{D}$ or $A \cdot \overline{C} \cdot D$, because each box that these terms cover has already been accounted for by using the other two product terms. The minimized SOP expression is $Y = \overline{A} \cdot B + B \cdot D + \overline{A} \cdot \overline{C} \cdot \overline{D} + A \cdot \overline{B} \cdot \overline{C}$, but this is only one of three possible solutions to this circuit.

A second solution opts to use the terms $\overline{B} \cdot \overline{C} \cdot \overline{D}$ and $A \cdot \overline{C} \cdot D$ instead of the previously used terms. These two also account for each of the remaining filled-in boxes, but produces a different SOP expression that still satisfies the original logic circuit that generated the original expression. This new expression is $Y = \overline{A} \cdot B + B \cdot D + \overline{B} \cdot \overline{C} \cdot \overline{D} + A \cdot \overline{C} \cdot D$. Still, there is one more optional solution that uses the terms $\overline{B} \cdot \overline{C} \cdot \overline{D}$ and $A \cdot \overline{B} \cdot C$ to account for the remaining filled-in boxes, yielding the minimized SOP expression $Y = \overline{A} \cdot B + B \cdot D + \overline{B} \cdot \overline{C} \cdot \overline{D} + A \cdot \overline{B} \cdot \overline{C}$.

Four-variable K-maps, like their three-variable counterparts, can be filled directly from a truth table. For example, let's map the following table and derive the minimum SOP expression from the map:

A	B	C	D	Y	BOX NUMBER
0	0	0	0	1	0
0	0	0	1	1	1
0	0	1	0	1	2
0	0	1	1	0	
0	1	0	0	0	
0	1	0	1	0	
0	1	1	0	1	6
0	1	1	1	1	7
1	0	0	0	1	8
1	0	0	1	1	9
1	0	1	0	0	
1	0	1	1	0	
1	1	0	0	1	12
1	1	0	1	1	13
1	1	1	0	0	
1	1	1	1	0	

The individual product terms that cause the output of the logic circuit responsible for the truth table to be a logic 1, represented by their K-map box number, are read from the truth table, and each fills one box with a 1, as shown on the map in Figure 2.41. Groups with the largest number of adjacent boxes are circled, and the map is read, yielding two possible minimum SOP solutions. Two groups of four adjacent boxes exist, each sharing boxes 8 and 9 and representing the product terms $\overline{B} \cdot \overline{C}$ and $A \cdot \overline{C}$. A group of two adjacent boxes, representing the product term $\overline{A} \cdot B \cdot C$, is formed from boxes 6 and 7.

The first solution, shown in Figure 2.41, makes use of one possible group of two adjacent boxes to account for the remaining filled box 2 by pairing that box with box 6 to form the product term $\overline{A} \cdot C \cdot \overline{D}$. Read from the map, this solution yields the minimum SOP expression $Y = A \cdot \overline{C} + \overline{B} \cdot \overline{C} + \overline{A} \cdot B \cdot C + \overline{A} \cdot C \cdot \overline{D}$. A second alternate solution, shown in Figure 2.42, combines box 2 with box 0 to form the product term $\overline{A} \cdot \overline{B} \cdot \overline{D}$, yielding the SOP expression $Y = A \cdot \overline{C} + \overline{B} \cdot \overline{C} + \overline{A} \cdot B \cdot C + \overline{A} \cdot \overline{B} \cdot \overline{D}$. Both SOP expressions represent the circuit that produces the result listed on the truth table and the K-map.

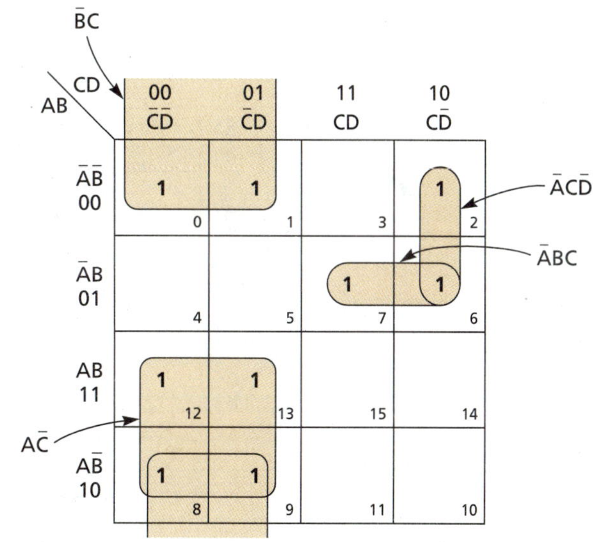

Figure 2.41 First Solution

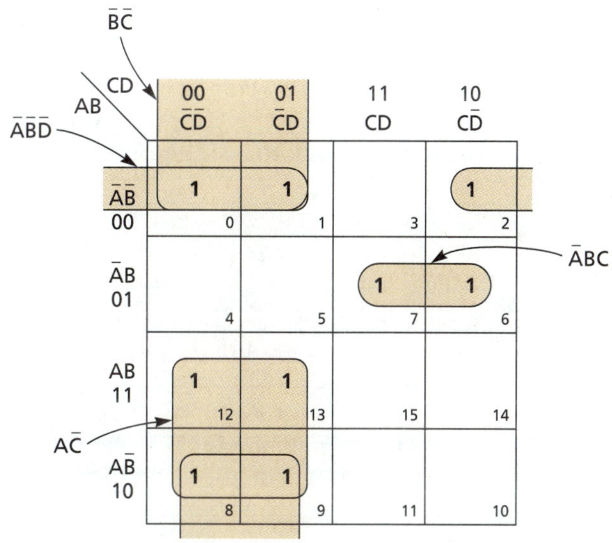

Figure 2.42 Second Solution

EXAMPLE 2.8

Reduce the following logic expression using a four-input Karnaugh map:

$$Y = \bar{A} \cdot B \cdot C + C \cdot \bar{D} + A + \bar{A} \cdot \bar{B} \cdot C \cdot D + \bar{A} \cdot B \cdot \bar{C} \cdot D$$

SOLUTION The map is filled as shown in Figure 2.43. The first term, $\bar{A} \cdot B \cdot C$, fills two boxes, 7 and 6. The second term, $C \cdot \bar{D}$, causes the four boxes on the right side of the map to be filled with 1's. The single term, A, fills all eight boxes on the bottom of the map. The last two terms, $\bar{A} \cdot \bar{B} \cdot C \cdot D$ and $\bar{A} \cdot B \cdot \bar{C} \cdot D$, fill

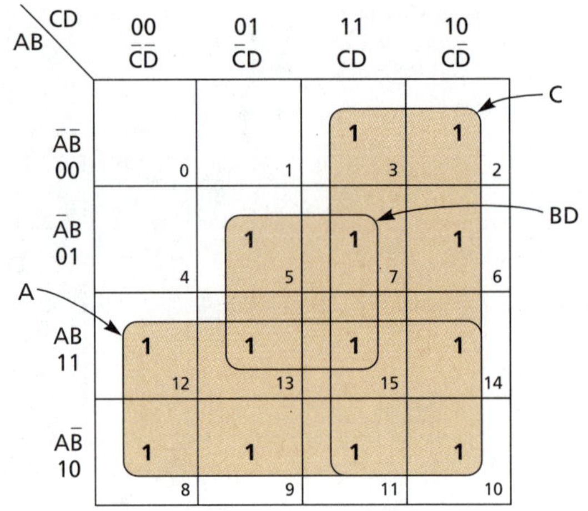

Figure 2.43 Example 2.8 Karnaugh Map

boxes 3 and 5 respectively. The map is then grouped as shown in Figure 2.43. Two large groups of eight are formed using the lower eight boxes for one, and the rightmost eight boxes for another. This leaves box 5 unaccounted for. It is included in a group of four boxes in the center of the map. The SOP expression read from the map comes from the product terms represented by these groupings: A for the lower eight boxes, C for the eight boxes on the right, and $B \cdot D$ for the four boxes in the center of the map: $Y = A + C + BD$.

Let's take one more look at why the product term for a group of boxes is a combination of the variables of the boxes within the group. In Example 2.8, there is a group of four boxes in the center of the map. The boxes represent the following four product terms: $\overline{A} \cdot B \cdot \overline{C} \cdot D$, $\overline{A} \cdot B \cdot C \cdot D$, $A \cdot B \cdot \overline{C} \cdot D$, and $A \cdot B \cdot C \cdot D$. This group could be read off the map as $\overline{A} \cdot B \cdot \overline{C} \cdot D + \overline{A} \cdot B \cdot C \cdot D + A \cdot B \cdot \overline{C} \cdot D + A \cdot B \cdot C \cdot D$. When we reduce this expression using Boolean algebra, we derive the minimum SOP expression for this group of four boxes:

$$\overline{A} \cdot B \cdot D(\overline{C} + C) + A \cdot B \cdot D(\overline{C} + C) = B \cdot D(\overline{A} + A) = B \cdot D$$

The expression $B \cdot D$ can also be read from the map by observing that variables B and D are the only two variables with the same active logic level in all four boxes.

The K-map in Figure 2.44 is an example of a logic circuit that has a number of don't care states indicated by X's in the boxes on the map. Two of the don't care terms, $\overline{A} \cdot \overline{B} \cdot C \cdot \overline{D}$ and $A \cdot \overline{B} \cdot \overline{C} \cdot D$, are used to form two groups and are treated as if they contain 1's in them, whereas the third term, $A \cdot B \cdot \overline{C} \cdot D$, is not part of any group and is treated as if it were a zero. The expression read from the map is $\overline{B} \cdot \overline{C} + \overline{A} \cdot C \cdot \overline{D}$.

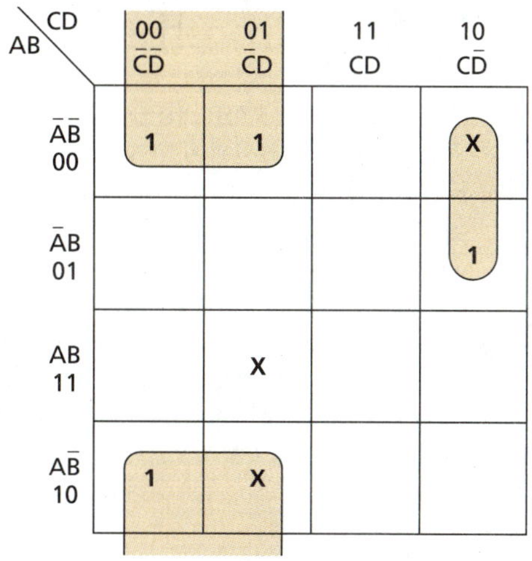

X = "don't care" state

Figure 2.44 Four-Variable Map with Don't Care States

Section 2.10

Practice Problems

1. Reduce the following Boolean expressions using Karnaugh maps:

   (a) $Y = \bar{A} \cdot \bar{B} \cdot \bar{C} \cdot \bar{D} + A \cdot \bar{C} \cdot \bar{D} + \bar{B} \cdot C \cdot \bar{D} + A \cdot \bar{D} + \bar{A} \cdot B \cdot C \cdot \bar{D}$

   (b) $T = W \cdot \bar{X} \cdot Y \cdot \bar{Z} + X \cdot \bar{Y} + W \cdot \bar{Y} \cdot \bar{Z} + \bar{W} \cdot Z + \bar{X} \cdot \bar{Y} \cdot \bar{Z} + \bar{W} \cdot \bar{X} \cdot Y \cdot \bar{Z}$

2. Derive the sum-of-products expression for the circuit that generated the truth table in Figure 2.45.

A	B	C	D	Y
0	0	0	0	0
0	0	0	1	0
0	0	1	0	1
0	0	1	1	1
0	1	0	0	1
0	1	0	1	0
0	1	1	0	0
0	1	1	1	0
1	0	0	0	0
1	0	0	1	0
1	0	1	0	1
1	0	1	1	1
1	1	0	0	1
1	1	0	1	1
1	1	1	0	1
1	1	1	1	0

Figure 2.45 Practice Problem 2.10 (2)

2.11 K-Mapping the Exclusive OR

Exclusive OR (XOR) functions appear on a Karnaugh map (K-map) in a special way. Recognizing the form a map takes for an XOR can save time in reading the K-map. Figure 2.46 shows two versions of the two-variable K-map with the XOR function mapped into it. Recall that an XOR logic function using A and B as the variables can be written in one of two ways, either $Y = A \oplus B$ or $\overline{A} \cdot B + A \cdot \overline{B}$. The second form is the sum-of-products (SOP) expression that can be placed into a K-map as shown in Figure 2.46. On those maps, notice that a unique form has evolved from mapping the XOR. Every other box is filled with a 1. The alternate boxes contain a 0. As a result, there are no groups larger than one box, so the original SOP expression cannot be reduced. However, because of the unique mapping, the resulting expression can be read as $A \oplus B$, the XOR function.

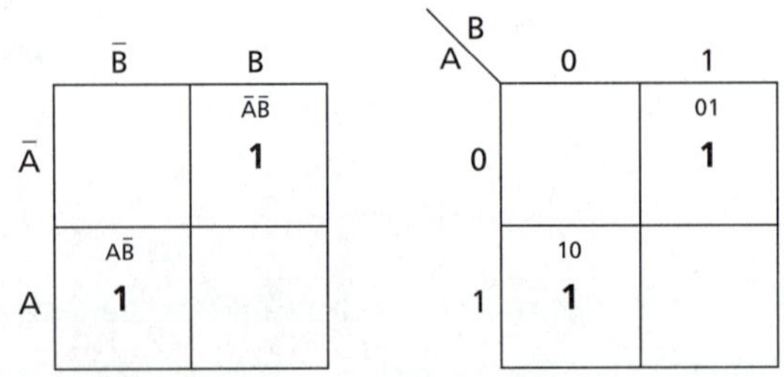

Figure 2.46 Mapping Exclusive OR $Y = \overline{A}B + A\overline{B}$

The so-called XNOR, which is actually the complement of the XOR, would find its K-map boxes filled with the opposite states of an XOR. A map so filled represents the expression $Y = A \cdot B + \overline{A} \cdot \overline{B}$, which is the complement of $\overline{A} \cdot B + A \cdot \overline{B}$.

A three-variable XOR function is mapped into the K-map of Figure 2.47. The SOP expression for this map is $\overline{A} \cdot B \cdot \overline{C} + A \cdot \overline{B} \cdot \overline{C} + A \cdot B \cdot C + \overline{A} \cdot \overline{B} \cdot C$, which is equivalent to $A \oplus B \oplus C$. Notice, again, how the boxes are filled alternately with 1's while the empty boxes hold 0's.

Figure 2.47 Mapping $Y = \overline{A}B\overline{C} + A\overline{B}\,\overline{C} + ABC + \overline{A}\,\overline{B}C$

One question you may have is how to recognize a two-variable XOR function on a three-variable map. Look at the map in Figure 2.47 again. This time, ignore the lower part of the map or consider that there aren't any boxes filled in that portion of the map. That would leave only the two upper boxes filled, as shown in the figure. The common variable to both those boxes is $\overline{A}$, so start by factoring out the $\overline{A}$ and make an XOR function out of the variables that remain: $Y = \overline{A}(B \oplus C)$. Let's prove that this is the expression for the upper portion of the map. Read the two boxes that contain 1's directly as

$$Y = \overline{A} \cdot \overline{B} \cdot C + \overline{A} \cdot B \cdot \overline{C}$$

Factor the common variable $\overline{A}$ from both terms:

$$Y = \overline{A}(\overline{B} \cdot C + B \cdot \overline{C})$$

The remaining expression left after factoring the common term is the logic XOR function of variables B and C:

$$Y = \overline{A}(B \oplus C)$$

This verifies the expression read from the map by using Boolean algebra.

In Figure 2.48, a four-variable map is shown for the logic expression $Y = A \oplus B \oplus C \oplus D$ whose SOP expression is

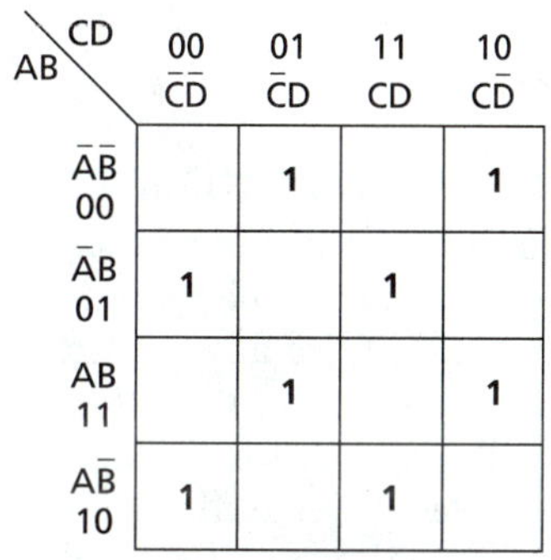

Figure 2.48 Mapping $Y = A \oplus B \oplus C \oplus D$

$$\begin{aligned} Y = {} & \overline{A}\cdot B\cdot \overline{C}\cdot \overline{D} + A\cdot \overline{B}\cdot \overline{C}\cdot \overline{D} + A\cdot B\cdot C\cdot \overline{D} + \overline{A}\cdot \overline{B}\cdot C\cdot \overline{D} \\ & + A\cdot B\cdot \overline{C}\cdot D + \overline{A}\cdot \overline{B}\cdot \overline{C}\cdot D + \overline{A}\cdot B\cdot C\cdot D + A\cdot \overline{B}\cdot C\cdot D \end{aligned}$$

Here is the Boolean algebra proof:

$$\begin{aligned} Y = {} & \overline{C}\cdot \overline{D}(\overline{A}\cdot B + A\cdot \overline{B}) + C\cdot \overline{D}(A\cdot B + \overline{A}\cdot \overline{B}) + \overline{C}\cdot D(A\cdot B + \overline{A}\cdot \overline{B}) \\ & + C\cdot D(\overline{A}\cdot B + A\cdot \overline{B}) \end{aligned}$$

$$Y = \overline{C}\cdot \overline{D}(A \oplus B) + C\cdot \overline{D}(\overline{A \oplus B}) + \overline{C}\cdot \overline{D}(\overline{A \oplus B}) + C\cdot D(A \oplus B)$$

Substitute X for $A \oplus B$:

$$Y = \overline{C}\cdot \overline{D}\cdot X + C\cdot \overline{D}\cdot \overline{X} + \overline{C}\cdot D\cdot \overline{X} + C\cdot D\cdot X$$

$$Y = \overline{D}(\overline{C}\cdot X + C\cdot \overline{X}) + D(\overline{C}\cdot \overline{X} + C\cdot X)$$

$$Y = \overline{D}(C \oplus X) + D(\overline{C \oplus X})$$

Make another substitution, R for $C \oplus X$:

$$Y = \overline{D}\cdot R + D\cdot \overline{R} = D \oplus R$$

To finish the proof, substitute the original expressions for R and X:

$$Y = D \oplus R = D \oplus C \oplus X = D \oplus C \oplus A \oplus B$$

What this all leads to is that if, when reading a K-map, you discover alternate boxes holding 1's after you have made your groupings, you may be able to read off the XOR terms into your final SOP expression.

2.12 Karnaugh Maps with More Than Four Variables

Karnaugh maps of up to four variables are practical and fairly easy to master with practice. Maps of logic expressions containing more than four variables become cumbersome and difficult to manage. Let's look at two additional maps, one with five variables and one with six, to see the problems that arise when using them. Figure 2.49 is a diagram of a five-variable Karnaugh map format. It is actually two four-input maps drawn one above the other. The top map includes the fifth variable, $\overline{E}$, while the map below uses the complement, E, as the fifth input variable. We shall use separate variable labeling to reduce the clutter caused by trying to label using partial product terms. Each box's product term is still a combination of all the variables labeled on the appropriate edges plus the diagonal fifth variable. Groups are created not only from those boxes that are adjacent to each other in either map, but also from those that are adjacent between the two maps. This is best understood by example, as we shall see in a moment. Attempts have been made to recreate the five-input map on a totally two-dimensional form, as shown in Figure 2.50, but you will see the complication this adds to their use.

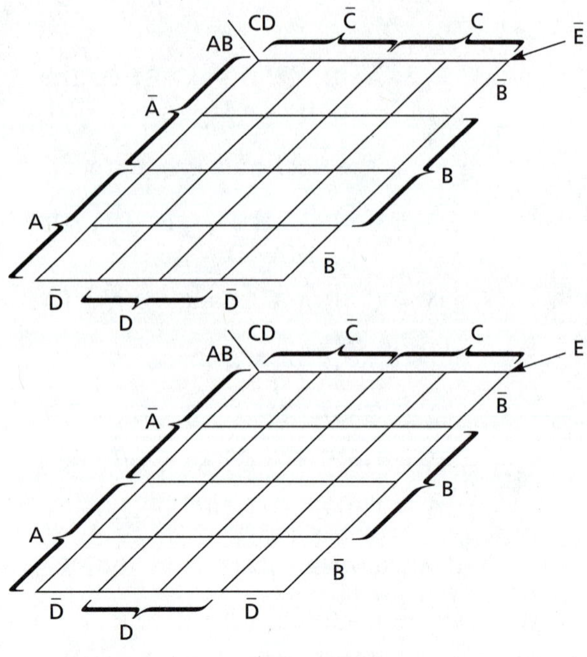

Figure 2.49 Five-Variable Karnaugh Map

Note, in the map of Figure 2.50, that the five-variable map is composed, again, of two four-variable maps that are joined together. Look carefully at this configuration, and you will notice that the four-variable map on the left is the mirror image of the one on the right. This is to facilitate the adjacent boxes for

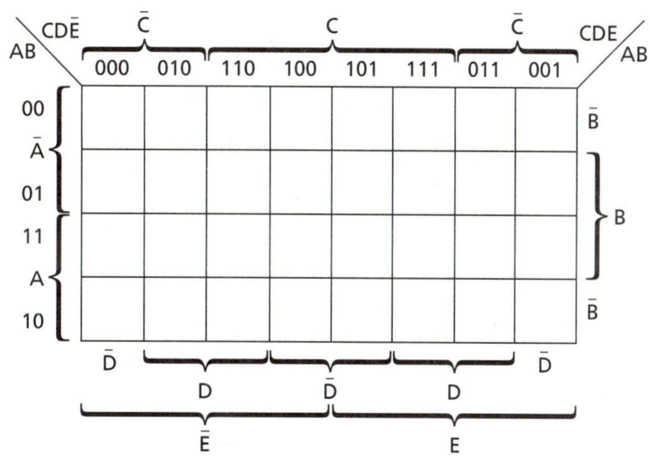

Figure 2.50 Two-Dimensional Form for Five-Variable Maps

variables $\overline{C}$ and D. The map on the left has $\overline{E}$ for a fifth variable, and the one on the right includes its complement, E.

Figure 2.51 shows a five-variable Karnaugh map that has been filled with 1's from a sum-of-products expression. The locations of the 1's in this example are chosen to illustrate how the map is read. Figure 2.52 shows the same map with the groups of adjacent boxes marked out. Some groups are familiar with those now used in four-variable maps. The pair of boxes in the lower right corner of the top map, combining $A \cdot \overline{B} \cdot C \cdot \overline{D} \cdot \overline{E}$ and $A \cdot \overline{B} \cdot C \cdot D \cdot \overline{E}$ into the term $A \cdot \overline{B} \cdot C \cdot \overline{E}$, is made the same way as it would on a separate four-input map. Of course, on that map the variable $\overline{E}$ would not be included in the term. There is also a group of four adjacent boxes on the lower map, as shown in Figure 2.52. This is read as the term $A \cdot \overline{D} \cdot E$. Now comes the group of boxes that are formed between the two maps. One group of four boxes is formed at the left top corners of both boxes. The variables common to these four boxes form the term $\overline{A} \cdot \overline{B} \cdot \overline{C}$. A second group of four is found almost in the center of both maps, and represents the term $B \cdot \overline{C} \cdot D$. The last group of four adjacent boxes marries the lower four corners of the maps for the term $A \cdot \overline{B} \cdot \overline{D}$. The three-dimensional aspect of this form of the five-variable Karnaugh maps is strange and a bit difficult to get used to, but watch what happens when you try to map the same expression on the two-dimensional form of the five-variable map.

The Karnaugh map in Figure 2.53 contains the same mapped expression as the one in Figures 2.51 and 2.52. Groups of boxes on each of the four-variable sections are, for the most part, still easy to find. There are three pairs that are fairly easy to read off the two four-variable sections of the five-variable map. They are the ones whose terms are $A \cdot \overline{B} \cdot C \cdot \overline{E}$, $B \cdot \overline{C} \cdot D \cdot \overline{E}$, and $B \cdot \overline{C} \cdot D \cdot E$. The first tricky part comes in realizing that two of these groups actually form a group of four boxes. Looking at the product terms that have just been listed, we realize that two of them can be reduced by Boolean algebra to $B \cdot \overline{C} \cdot D$. This means that two of the groups really form a four-box group, but how do we show this on the two-dimensional map?

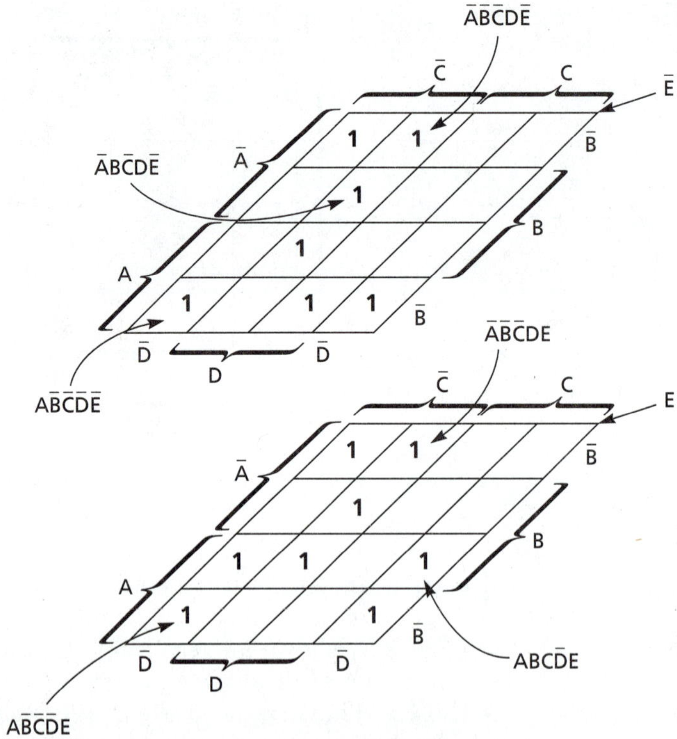

Figure 2.51 Filled-In Map

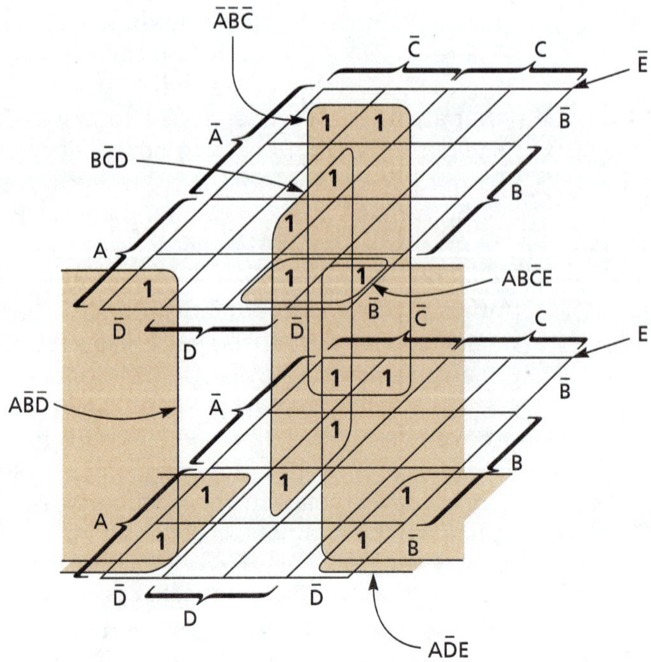

Figure 2.52 Making Groups

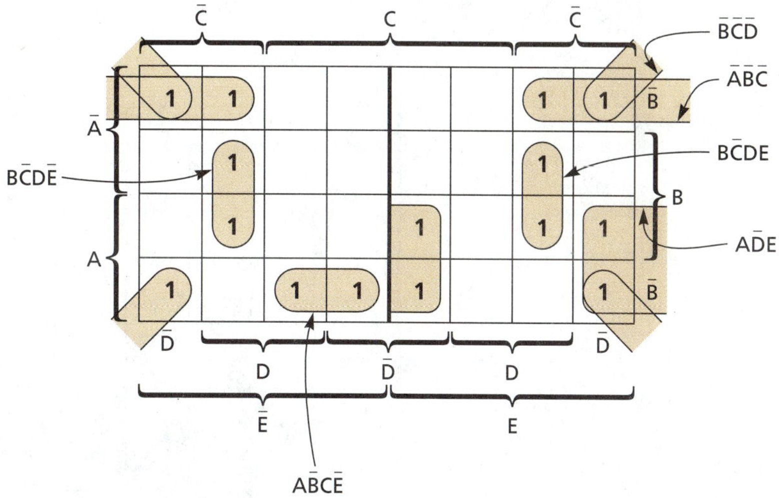

Figure 2.53 A Two-Dimensional Solution

Recall that the two four-variable maps that are attached to form the five-variable K-map are mirror images of each other. It's as if you could bend the map in the middle and fold one on top of the other. What this means is that all the boxes on half of the map are adjacent to their corresponding mirror image boxes on the other half of the map. As a result, the pair of boxes holding the product term $B \cdot \overline{C} \cdot D \cdot \overline{E}$ is adjacent to the pair representing the term $B \cdot \overline{C} \cdot D \cdot E$, and this group of four actually represents the product term $B \cdot \overline{C} \cdot D$. In the right half of the K-map, a group of four adjacent boxes represents the term $A \cdot \overline{D} \cdot E$. These four boxes conform to the method used with a four-variable map to form groups using boxes adjacent on the edges of the map. So not only are boxes adjacent on the mirror images, but they are also adjacent on the individual edges of each four-variable section of the five-variable map.

Another group of four boxes is formed at the four corner boxes of the complete map and is grouped using semiellipses shown at an angle from each box. Now look at the top of the map, and find the group of four formed by the upper pair in the left corner and the upper pair in the right corner of the complete map. Compare the three- and two-dimensional forms of the five-variable map in Figures 2.52 and 2.53. You may have to look between them several times until you connect the results of grouping boxes on each form. Notice that neither one is easy to work with.

As cumbersome as the five-variable Karnaugh map can be, it is nothing compared to the six-variable map shown in Figure 2.54. The six-variable map is made up of four adjoining four-variable maps that are mirrors of each other. It can also be thought of as joining two five-variable maps that are mirror images. An example of a filled map showing adjacent boxes set off in groups is depicted in Figure 2.55. No attempt is made to figure out if these groups are part of larger groups, although they *all* do except the group of four formed by the four corner boxes. Instead, the groups shown are read off and Boolean algebra is used to arrive at the final reduced expression.

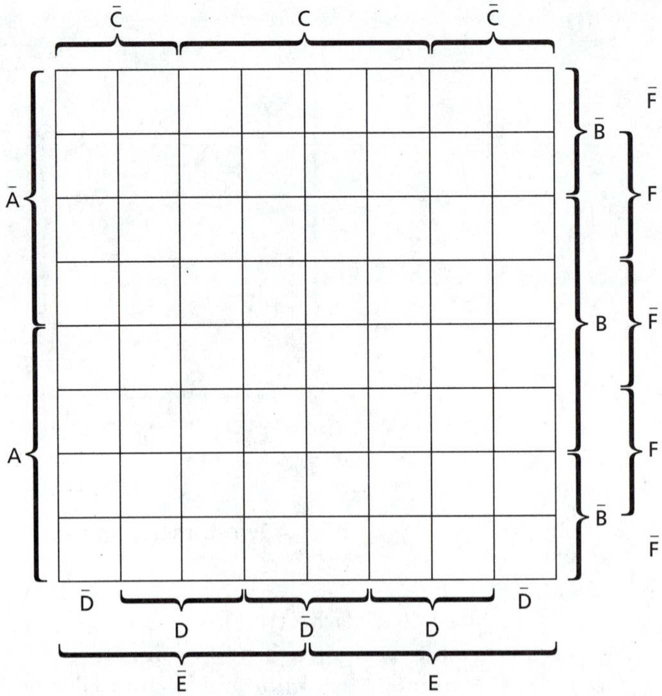

Figure 2.54 A Six-Variable Karnaugh Map

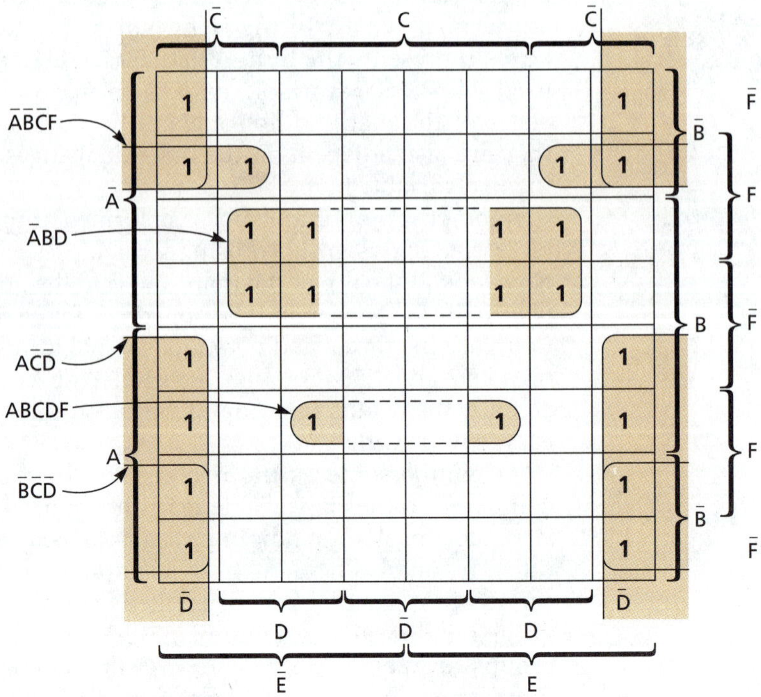

Figure 2.55 Groupings on a Six-Variable Map

Does this mean that we have gone beyond a practical limitation for the use of Karnaugh maps? Perhaps. In general, maps up to and including the four-variable size are most commonly used. The five-variable map finds minimal usage, and expressions with more variables are rarely, if ever, mapped onto a Karnaugh map.

Section 2.12

Practice Problem

What is the sum-of-products expression represented by the Karnaugh map of Figure 2.56?

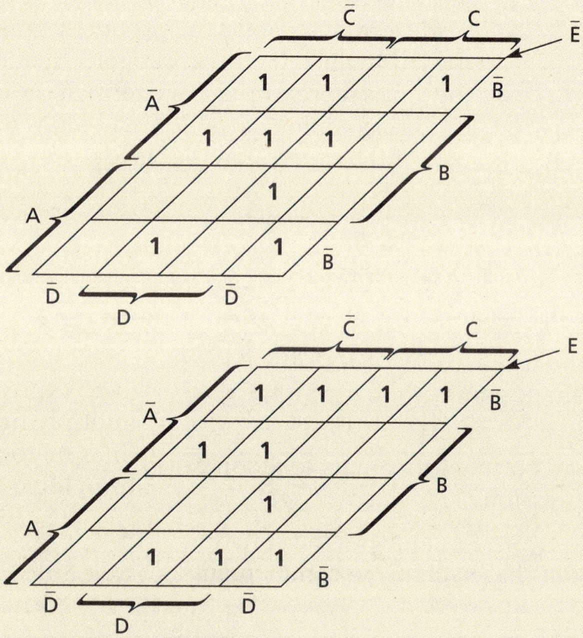

Figure 2.56 Practice Problem 2.11

Summary

Logic circuits can be documented in several different ways, including truth table, logic diagram, and logic expression. Complex combinational circuits can be refined and/or reduced using Boolean algebra, DeMorgan's theorem, and Karnaugh maps.

Boolean algebra methods apply specific theorems that allow logic expressions to be manipulated in much the same manner as mathematical algebra can be applied to algebraic expressions. In addition to the Boolean theorems summarized in Table 2.3, there is the associative law, which defines how variables and terms can be ordered. Another law, the distributive law, describes how one term can be distributed into another and the process of factoring common variables from a set of terms in a logic expression.

Complementing logic functions and variables is summarized in a set of processes defined by DeMorgan's theorem. A logic expression can be complemented by inverting the variables and changing AND logic functions to OR logic functions and vice versa. DeMorgan's theorem is also applied to the process of developing logic circuits using only NAND or NOR gates.

Karnaugh mapping, a graphical method of diagramming logic expressions, is used to aid in reducing these expressions to their minimal sum-of-products form. Product terms from a truth table can be entered onto a Karnaugh map by filling in 1's for boxes that represent terms that show a 1 on the truth table. The remaining boxes can be left empty or filled with 0's. Boxes are grouped into adjacent sets of 1, 2, 4, 8, or 16 boxes. The common variables in each group are read from the map as the product terms of the logic expression represented by the truth table. Logic expressions themselves can be mapped by entering 1's in the boxes representing each term of the expression. Adjacent boxes are grouped, and the common variables read off as a product in the minimum sum of products of the expression that is logically equivalent to the original expression.

Glossary

associative law the order of performing a logic operation by a single gate with many inputs does not matter.

Boolean algebra a set of theorems to aid in reducing logic expressions.

DeMorgan's theorem establishes the equality of logic expressions based on using complements (NOTs).

distributive law specifies the manner in which logic expressions can be combined similar to multiplying polynomials in mathematics. Also defines methods for factoring common variables from logic expressions.

Karnaugh map graphical tool used in reducing logic expressions.

Solutions to Practice Problems

Section 2.3

1. $(A + C + D)(B + C + D) = AB + AC + AD + BC + CC + CD + BD + CD + DD$

$CC = C,\ DD = D,\ CD + CD = CD$

$AB + AC + AD + BC + C + CD + BD + D = AB + C(A + B + 1 + D) + D(A + B + 1)$

$AB + C(1) + D(1) = AB + C + D$

A	B	C	D	A + C + D	B + C + D	AB	(A + C + D)(B + C + D)	AB + C + D
0	0	0	0	0	0	0	0	0
0	0	0	1	1	1	0	1	1
0	0	1	0	1	1	0	1	1

0	0	1	1	1	1	0	1	1
0	1	0	0	0	1	0	0	0
0	1	0	1	1	1	0	1	1
0	1	1	0	1	1	0	1	1
0	1	1	1	1	1	0	1	1
1	0	0	0	1	0	0	0	0
1	0	0	1	1	1	0	1	1
1	0	1	0	1	1	0	1	1
1	0	1	1	1	1	0	1	1
1	1	0	0	1	1	1	1	1
1	1	0	1	1	1	1	1	1
1	1	1	0	1	1	1	1	1
1	1	1	1	1	1	1	1	1

2. $ABC + A(A + C + B) + BC = ABC + AA + AC + AB + BC = ABC + A + AC + AB + BC = A(BC + 1 + C + B) + BC = A(1) + BC = A + BC$

A	B	C	ABC	A + B + C	A(A + B + C)	[BC	ABC + A(A + B + C) + BC]	A + BC
0	0	0	0	0	0	0	0	0
0	0	1	0	1	0	0	0	0
0	1	0	0	1	0	0	0	0
0	1	1	0	1	0	1	1	1
1	0	0	0	1	1	0	1	1
1	0	1	0	1	1	0	1	1
1	1	0	0	1	1	0	1	1
1	1	1	1	1	1	1	1	1

Section 2.4

1. $A \cdot B \cdot C + A \cdot \overline{B} \cdot C + \overline{A} \cdot B \cdot C + \overline{A} \cdot \overline{B} \cdot C = A \cdot C \cdot (B + \overline{B}) + \overline{A} \cdot C \cdot (B + \overline{B}) = A \cdot C \cdot (1) + \overline{A} \cdot C \cdot (1) = A \cdot C + \overline{A} \cdot C = C \cdot (A + \overline{A}) = C \cdot (1) = C$

2. $(B + C)(A + \overline{C})(\overline{A} + B) = (A \cdot B + B \cdot \overline{C} + A \cdot C + C \cdot \overline{C})(\overline{A} + B) = (A \cdot B + B \cdot \overline{C} + A \cdot C + 0)(\overline{A} + B) = A \cdot B \cdot \overline{A} + A \cdot B \cdot B + \overline{A} \cdot B \cdot \overline{C} + B \cdot B \cdot \overline{C} + A \cdot \overline{A} \cdot C + A \cdot C \cdot B = 0 + A \cdot B + \overline{A} \cdot B \cdot \overline{C} + B \cdot \overline{C} + 0 + A \cdot B \cdot C = A \cdot B \cdot (1 + \overline{C}) + B \cdot \overline{C} \cdot (\overline{A} + 1) = A \cdot B \cdot (1) + B \cdot \overline{C} \cdot (1) = A \cdot B + B \cdot \overline{C}$

3. $A \cdot C + B \cdot \overline{C} + \overline{A} \cdot C + B \cdot C = C \cdot (A + \overline{A}) + B \cdot (C + \overline{C}) = C \cdot (1) + B \cdot (1) = C + B$

Section 2.5

1. $\overline{(AB\overline{C} + BC)C + \overline{\overline{B}}} = \overline{BC + \overline{\overline{B}}} = (\overline{BC})B = (\overline{B} + \overline{C})B = B\overline{C}$

2. $\overline{\overline{A}B\overline{C} + BC\overline{D}} + \overline{ACD + A\overline{B}} = (A + \overline{B} + C)(\overline{B} + \overline{C} + D) + (\overline{A} + \overline{C} + \overline{D})(\overline{A} + B) = A\overline{B} + A\overline{C} + AD + \overline{B} + \overline{A} + B\overline{C} + B\overline{D} = \overline{B}(A + 1) + A\overline{C} + AD + \overline{A} + B\overline{C} + B\overline{D} = \overline{B} + A\overline{C} + AD + \overline{A} + B\overline{C} + B\overline{D}$

3. $(ABC + \overline{C})(\overline{CD + D}) = (AB + \overline{C})(\overline{D}) = AB\overline{D} + \overline{C} \cdot \overline{D}$ *Note:* This equation can be further reduced by using concepts in Section 2.7.

4. $(\overline{AB + C})(\overline{BC}) + \overline{C} = (\overline{A} + \overline{B})(\overline{C})(\overline{B} + \overline{C}) + \overline{C} = (\overline{A} \cdot \overline{C} + \overline{B} \cdot \overline{C})(\overline{B} + \overline{C}) + \overline{C} = \overline{A} \cdot \overline{B} \cdot \overline{C} + \overline{B} \cdot \overline{C} + \overline{A} \cdot \overline{C} + \overline{C} = \overline{C}(\overline{A} \cdot \overline{B} + \overline{B} + \overline{A} + 1) = \overline{C}(1) = \overline{C}$

Section 2.6

1.

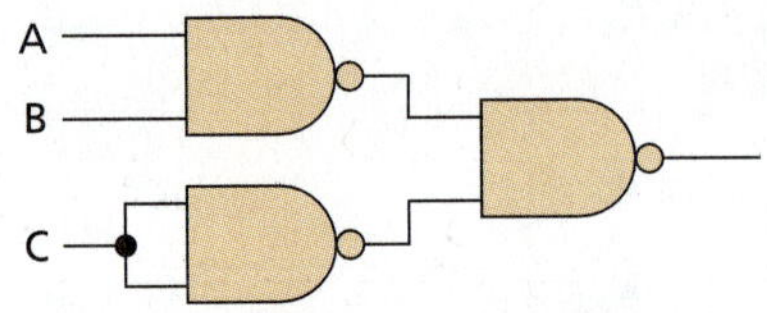

Figure 2.57 Practice Problem 2.6 (1) (Solution)

2.

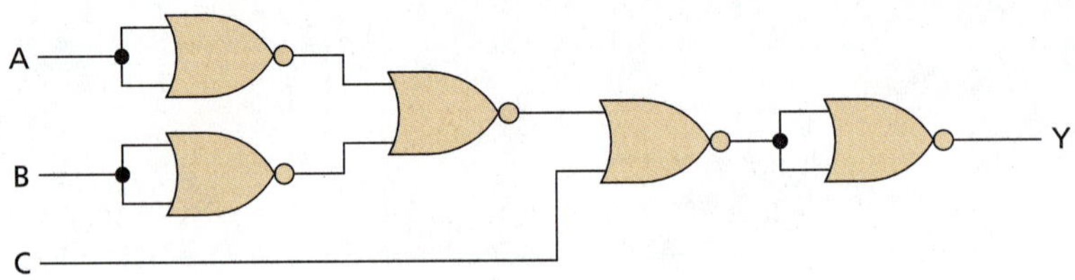

$$Y = \overline{\overline{\overline{\overline{A}+\overline{B}}+C}} = \overline{\overline{A}+\overline{B}}+C = AB + C$$

Figure 2.58 Practice Problem 2.6 (2) (Solution)

Section 2.7

1. $(ABC + \overline{A})(A\overline{C} + ABC) = (BC + \overline{A})(A)(\overline{C} + BC) = (BC + \overline{A})(A)(\overline{C} + B) = ABC(\overline{C} + B) = ABC\overline{C} + ABBC = 0 + ABC = ABC$

2. $(AB + BC)(\overline{A}C + BC) = ABC + \overline{A}BC + BC = BC(A + \overline{A} + 1) = BC(1) = B$

3. $\overline{AB + BC + AC} = \overline{AB + BC} = (\overline{A} + B)(\overline{B} + \overline{C}) = \overline{A}\cdot\overline{B} + \overline{A}\cdot\overline{C} + B\cdot\overline{C} + B\cdot\overline{B} = \overline{A}\cdot\overline{B} + \overline{A}\cdot\overline{C} + B\cdot\overline{C} + 0 = \overline{A}\cdot\overline{B} + B\cdot\overline{C}$

4. $A\cdot\overline{B}\cdot C + B\cdot\overline{C}\cdot D + A\cdot\overline{C}\cdot D + \overline{A}\cdot\overline{C}\cdot D = A\cdot\overline{B}\cdot C + \overline{C}\cdot D\cdot(B + A + \overline{A}) = A\cdot\overline{B}\cdot C + \overline{C}\cdot D\cdot(1) = A\cdot\overline{B}\cdot C + \overline{C}\cdot D$

Section 2.8

1.

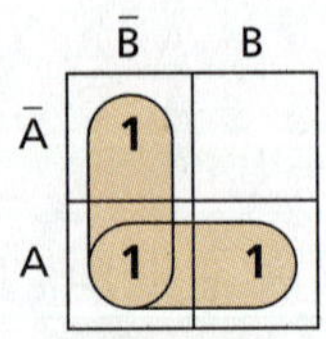

$Y = A + \overline{B}$

Figure 2.59 Practice Problem 2.8 (1) (Solution)

2. a.

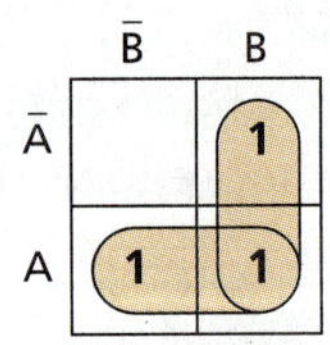

$Y = A + B$

Figure 2.60 Practice Problem 2.8 (2a) (Solution)

b.

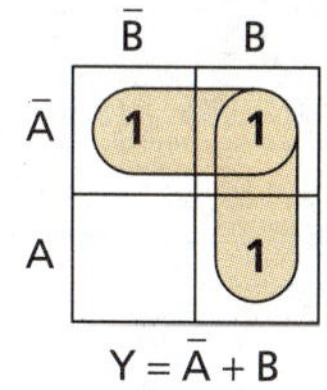

Figure 2.61 Practice Problem 2.8 (2b) (Solution)

Section 2.9

1. a.

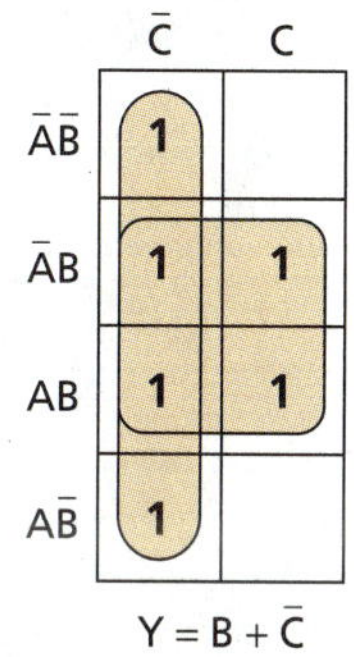

Figure 2.62 Practice Problem 2.9 (1a) (Solution)

b.

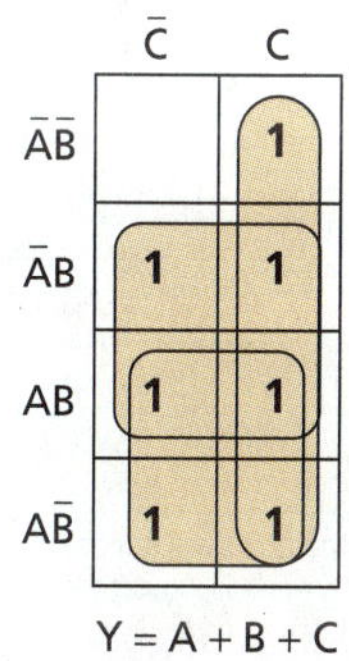

Figure 2.63 Practice Problem 2.9 (1b) (Solution)

2.

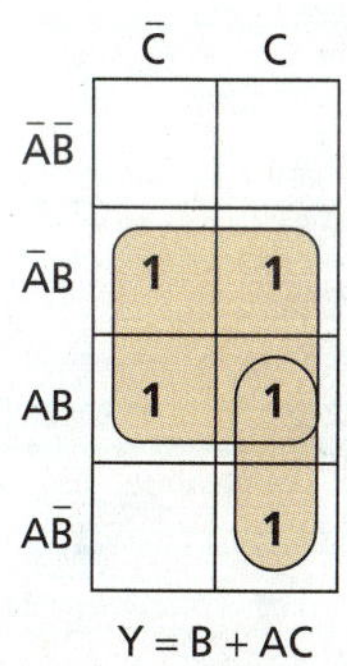

Figure 2.64 Practice Problem 2.9 (2) (Solution)

Section 2.10

1. a.

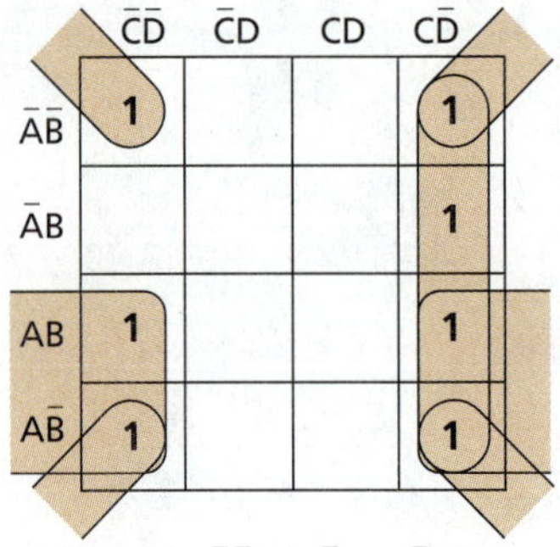

$$Y = \overline{B}\,\overline{D} + C\overline{D} + A\overline{D}$$

Figure 2.65 Practice Problem 2.10 (1a) (Solution)

b.

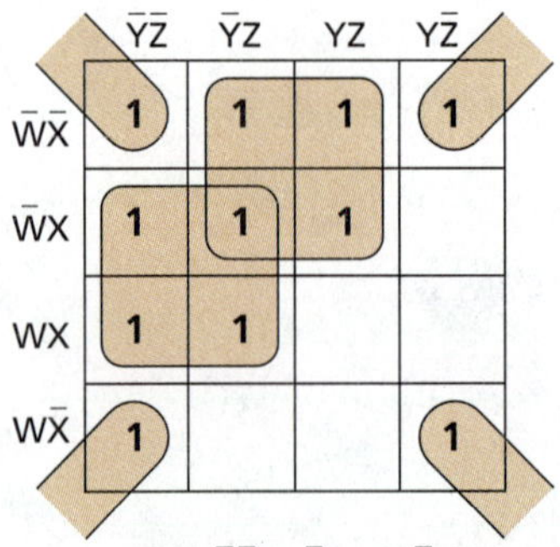

$$T = \overline{X}\,\overline{Z} + \overline{W}Z + X\overline{Y}$$

Figure 2.66 Practice Problem 2.10 (1b) (Solution)

2.

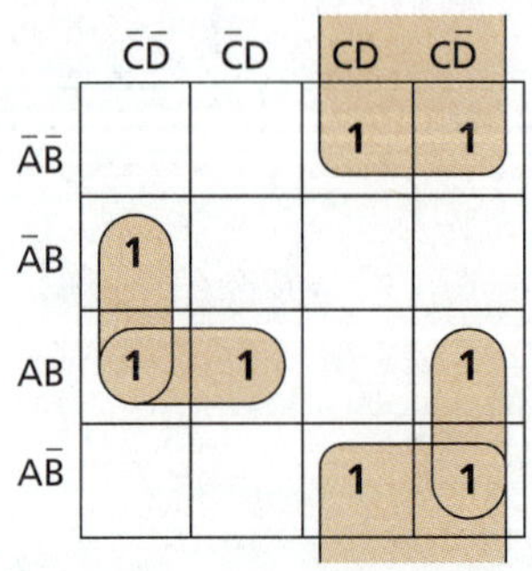

$$Y = \overline{B}C + B\overline{C}\,\overline{D} + AB\overline{C} + AC\overline{D}$$

Figure 2.67 Practice Problem 2.10 (2) (Solution)

Section 2.12

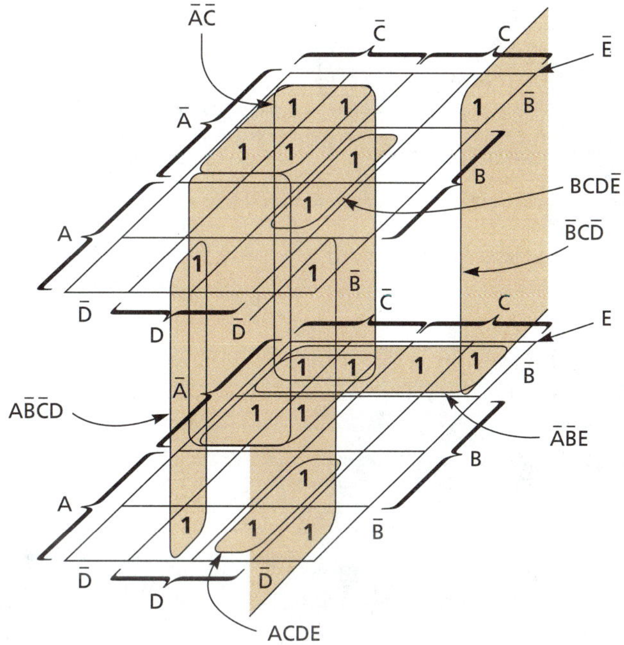

Figure 2.68 Practice Problem 2.11 (Solution) (not the *only* solution!)

$\bar{A} \cdot \bar{C} + \bar{A} \cdot \bar{B} \cdot E + \bar{B} \cdot C \cdot \bar{D} + B \cdot C \cdot D \cdot \bar{E} + A \cdot \bar{B} \cdot \bar{C} \cdot D + A \cdot C \cdot D \cdot E$
This is not the only solution—alternate groupings are possible.

Problems

2.1 Convert the following product-of-sums expressions to the sum-of-products form by using the distributive law.

a. $(A + B)(C + D)(\bar{A} + D)$
b. $(A\bar{B} + C)(ACD + B)$
c. $(\bar{A} + \bar{B} + \bar{C})(A + B + C)$
d. $(A + \bar{B} \cdot \bar{C} + ACD)(A + C)$
e. $(B + \bar{C})(\bar{A} + \bar{B} + \bar{C})(B)$
f. $(A\bar{C} + B\bar{C})(A + B)(A\bar{B}C + B)$

2.2 Convert the following product-of-sums expressions to the sum-of-products form, using the distributive law.

a. $(\bar{A} + B)(\bar{B} + D)(A + D)$
b. $(AB + \bar{C})(A\bar{B}D + C)$
c. $(\bar{A} + B + \bar{C})(A + \bar{B} + C)$
d. $(A + \bar{B} \cdot C + \bar{A}CD)(\bar{A} + C)$
e. $(\bar{B} + \bar{C})(\bar{A} + B + C)(B)$
f. $(BC + \bar{A}BC)(A + \bar{B})(\bar{A}BC + \bar{C})$

2.3 Factor the common variable(s) from the terms of each of these expressions.

a. $ABC + A\bar{B}C + ACD + A\bar{B}D$
b. $AB + A\bar{C}D + \bar{B} \cdot \bar{C} \cdot D + AD$
c. $A\bar{B}C\bar{D} + ABCD + A\bar{D} + BCD + AD$
d. $\bar{B}DE + ACE + BCE + ABCE + ACD\bar{E}$
e. RST + STU + BRS + BUST
f. THIS + IS + ITS + HIT

2.3 Factor the common variable(s) from the terms of each of these expressions.

a. $ABD + AB\bar{C} + AC\bar{D} + A\bar{B}D$
b. $AB\bar{D} + A\bar{C}D + \bar{B} \cdot \bar{C} \cdot \bar{D} + A\bar{D}$
c. $A \cdot \bar{B} \cdot \bar{C} \cdot \bar{D} + ABCD + A\bar{D} + BCD + BD$

d. $\overline{B}E + ACE + A\overline{B}ED + ABCE + AC\overline{D}E$
e. $REST + STU + BURST + BUST$
f. $TH\overline{E}M + TH\overline{A}T + H\overline{A}ST\overline{E} + HIT$

2.5 Simplify the logic expressions of Problem 2.1.

2.6 Simplify the logic expressions of Problem 2.2.

2.7 Simplify the logic expressions of Problem 2.3.

2.8 Simplify the logic expressions of Problem 2.4.

2.9 Reduce the following logic expressions to their minimum sum-of-products terms:

a. $Y = \overline{A} \cdot \overline{B} \cdot \overline{C} + \overline{A} \cdot \overline{B} \cdot C + \overline{A} \cdot B \cdot \overline{C} + A \cdot B \cdot \overline{C}$
b. $Y = (A + B)(A \cdot \overline{B} + C)(\overline{A} \cdot \overline{B} \cdot C + A \cdot C)$
c. $Y = (A \cdot \overline{B} + C)(A + \overline{C}) + A(\overline{B} \cdot C + A \cdot C)$
d. $T = \overline{W} \cdot \overline{X} \cdot Y \cdot Z + \overline{W} \cdot X + W \cdot \overline{X} \cdot Z + \overline{X} \cdot Z + W \cdot X \cdot Z$
e. $V = S(\overline{T} + U) + \overline{U} + U \cdot T$
f. $Y = A \cdot B \cdot (C + A \cdot \overline{D}) + (\overline{A} + C)(B + \overline{D}) + A \cdot B \cdot D$

2.10 Reduce the following logic expressions to their minimum sum-of-products terms:

a. $Y = \overline{A} \cdot B \cdot \overline{C} + \overline{A} \cdot \overline{B} \cdot \overline{C} + A \cdot B \cdot C + A \cdot \overline{B} \cdot C + A \cdot B \cdot \overline{C}$
b. $Y = (\overline{A} + \overline{C} + D)(\overline{A} + B \cdot C \cdot D) \cdot (B \cdot C + A \cdot D)$
c. $Y = A \cdot B \cdot C + \overline{A} \cdot \overline{D} \cdot (C + \overline{B}) + A \cdot (C \cdot D + \overline{B} \cdot C)$
d. $T = U \cdot \overline{V} \cdot X \cdot Y + \overline{U} \cdot \overline{V} \cdot (\overline{X} + \overline{Y}) + U \cdot (\overline{V} \cdot \overline{X} + \overline{V} \cdot \overline{Y})$
e. $V = \overline{S} \cdot \overline{T} \cdot (U + W) + \overline{W} + U \cdot \overline{W} \cdot T$
f. $Y = \overline{A} \cdot \overline{C} \cdot D \cdot (A + B + \overline{C}) + (A + \overline{B})(A + \overline{C})$

2.11 Derive the minimum sum-of-products expression for the circuits that generate each of these truth tables:

a.

A	B	C	Y
0	0	0	1
0	0	1	0
0	1	0	1
0	1	1	1
1	0	0	1
1	0	1	0
1	1	0	0
1	1	1	1

Figure 2.69 Problem 2.11a

b.

N	P	T	Z
0	0	0	0
0	0	1	0
0	1	0	1
0	1	1	1
1	0	0	1
1	0	1	1
1	1	0	1
1	1	1	0

Figure 2.70 Problem 2.11b

c.

A	B	C	D	Y
0	0	0	0	1
0	0	0	1	1
0	0	1	0	0
0	0	1	1	0
0	1	0	0	0
0	1	0	1	1
0	1	1	0	1
0	1	1	1	0
1	0	0	0	0
1	0	0	1	0
1	0	1	0	0
1	0	1	1	0
1	1	0	0	1
1	1	0	1	1
1	1	1	0	1
1	1	1	1	1

Figure 2.71 Problem 2.11c

d.

A	B	C	D	Y
0	0	0	0	0
0	0	0	1	0
0	0	1	0	0
0	0	1	1	1
0	1	0	0	1
0	1	0	1	1
0	1	1	0	1
0	1	1	1	1
1	0	0	0	0
1	0	0	1	1
1	0	1	0	1
1	0	1	1	0
1	1	0	0	0
1	1	0	1	0
1	1	1	0	1
1	1	1	1	1

Figure 2.72 Problem 2.11d

2.12 Derive the minimum sum-of-products expression for the circuits that generate each of these truth tables:

a.

A	B	C	Y
0	0	0	1
0	0	1	1
0	1	0	0
0	1	1	0
1	0	0	0
1	0	1	0
1	1	0	1
1	1	1	1

Figure 2.73 Problem 2.12a

b.

P	D	Q	N
0	0	0	0
0	0	1	1
0	1	0	0
0	1	1	1
1	0	0	0
1	0	1	1
1	1	0	0
1	1	1	1

Figure 2.74 Problem 2.12b

c.

A	B	C	D	Y
0	0	0	0	0
0	0	0	1	0
0	0	1	0	1
0	0	1	1	1
0	1	0	0	1
0	1	0	1	1
0	1	1	0	1
0	1	1	1	1
1	0	0	0	1
1	0	0	1	0
1	0	1	0	0
1	0	1	1	0
1	1	0	0	1
1	1	0	1	0
1	1	1	0	1
1	1	1	1	1

Figure 2.75 Problem 2.12c

d.

A	B	C	D	Y
0	0	0	0	1
0	0	0	1	1
0	0	1	0	0
0	0	1	1	0
0	1	0	0	1
0	1	0	1	1
0	1	1	0	0
0	1	1	1	0
1	0	0	0	1
1	0	0	1	1
1	0	1	0	0
1	0	1	1	0
1	1	0	0	1
1	1	0	1	1
1	1	1	0	0
1	1	1	1	0

Figure 2.76 Problem 2.12d

2.13 Reduce the following logic expressions to their minimum sum-of-products form:

a. $\overline{(\overline{A}\cdot B + C)(\overline{A\cdot C} + B\cdot C)}$

b. $W\cdot\overline{X\cdot Y} + \overline{\overline{X\cdot T} + (X + Y)(W + \overline{T})}$

c. $\overline{A\cdot\overline{B}\cdot\overline{C}\cdot D + \overline{A\cdot B}\cdot(A\cdot C + \overline{B\cdot D}) + \overline{A\cdot C\cdot D}}$

d. $\overline{\overline{R}\cdot\overline{S} + T\cdot U + R\cdot T}$

e. $\overline{\overline{(A\cdot B + C)(B\cdot C + A)} + (A\cdot C + B)\cdot\overline{(A\cdot B\cdot C)}}$

2.14 Reduce the following logic expressions to their minimum sum-of-products form:

a. $Y = \overline{(A + \overline{B})\overline{(B + A \cdot C)} + B \cdot \overline{(A \cdot \overline{C} + \overline{A} \cdot B \cdot C}}$

b. $V = \overline{\overline{W \cdot X} \cdot Z + \overline{X \cdot T} \cdot (W + Y \cdot \overline{T}) + W}$

c. $Y = A \cdot (\overline{B \cdot C}) + \overline{\overline{A} \cdot B \cdot \overline{C} + \overline{\overline{A} \cdot B \cdot C}} + (\overline{A \cdot B}) \cdot C$

d. $Z = \overline{\overline{R} \cdot S \cdot \overline{T} + U} + \overline{R \cdot T}$

e. $Y = \overline{\overline{A \cdot (B + \overline{C})} + B \cdot \overline{(C + \overline{A} \cdot \overline{C})}}$

2.15 Complete the Karnaugh maps for each of these logic expressions:

a. $A \cdot B + \overline{A} \cdot B + A \cdot \overline{B}$

b. $\overline{A} \cdot \overline{B} \cdot C + \overline{A} \cdot B \cdot \overline{C} + A \cdot \overline{B} \cdot \overline{C} + A \cdot B \cdot C$

c. $A \cdot \overline{B} \cdot \overline{C} \cdot D + \overline{A} \cdot \overline{B} \cdot \overline{C} \cdot D + A \cdot B \cdot \overline{C} \cdot \overline{D} + \overline{A} \cdot B \cdot C \cdot \overline{D}$

d. $A \cdot B + C + A \cdot \overline{C}$

e. $A \cdot \overline{B} \cdot \overline{C} + B \cdot C \cdot \overline{D} + A \cdot \overline{C} \cdot \overline{D} + A \cdot B \cdot C \cdot D + \overline{A}$

2.16 Complete the Karnaugh maps for each of these logic expressions:

a. $A \cdot \overline{B} + \overline{A} \cdot \overline{B} + A \cdot B$

b. $\overline{A} \cdot \overline{B} \cdot \overline{C} + \overline{A} \cdot B \cdot \overline{C} + A \cdot B \cdot \overline{C} + A \cdot \overline{B} \cdot \overline{C}$

c. $\overline{A} \cdot \overline{B} \cdot \overline{C} \cdot \overline{D} + \overline{A} \cdot \overline{B} \cdot C \cdot \overline{D} + A \cdot B \cdot \overline{C} \cdot \overline{D} + A \cdot \overline{B} \cdot C \cdot \overline{D} + A \cdot \overline{B} \cdot \overline{C} \cdot \overline{D}$

d. $A + \overline{A} \cdot B \cdot C + B \cdot \overline{C} + \overline{A} \cdot \overline{B} \cdot \overline{C}$

e. $A \cdot \overline{C} \cdot \overline{D} + C + B \cdot \overline{C} \cdot D + A \cdot \overline{B} \cdot \overline{C} \cdot D + A \cdot B$

2.17 What is the minimum sum-of-products (SOP) expression for each of the logic expressions in Problem 2.15?

2.18 What is the minimum sum-of-products (SOP) expression for each of the logic expressions in Problem 2.16?

2.19 Derive the minimum sum-of-products and product-of-sums expressions from each of the following Karnaugh maps.

a.

	$\overline{C}$	C
$\overline{A}\overline{B}$	1	
$\overline{A}B$	1	1
AB		
$A\overline{B}$	1	

Figure 2.77 Problem 2.19a

b.

	$\overline{X}$	X
$\overline{W}\overline{Y}$	1	1
$\overline{W}Y$		
WY		1
$W\overline{Y}$	1	1

Figure 2.78 Problem 2.19b

c.

	$\overline{C}\overline{D}$	$\overline{C}D$	CD	$C\overline{D}$
$\overline{A}\overline{B}$		1		
$\overline{A}B$	1	1	1	1
AB	1			1
$A\overline{B}$				

Figure 2.79 Problem 2.19c

d.

	$\overline{R}\overline{G}$	$\overline{R}G$	RG	$R\overline{G}$
$\overline{N}\overline{V}$	1	1	1	1
$\overline{N}V$		1	1	
NV				
$N\overline{V}$	1		1	1

Figure 2.80 Problem 2.19d

2.20 Derive the minimum sum-of-products and product-of-sums expressions from each of the following Karnaugh maps.

a.

	$\overline{B}$	B
$\overline{A}$	1	1
A		

Figure 2.81 Problem 2.20a

b.

	$\bar{C}$	C
$\bar{A}\bar{B}$	1	1
$\bar{A}B$	1	1
AB	1	1
$A\bar{B}$		

Figure 2.82 Problem 2.20b

c.

	$\bar{C}\bar{D}$	$\bar{C}D$	CD	$C\bar{D}$
$\bar{A}\bar{B}$	1			1
$\bar{A}B$	1	1	1	
AB	1			1
$A\bar{B}$	1			1

Figure 2.83 Problem 2.20c

d.

	$\bar{X}\bar{Z}$	$\bar{X}Z$	XZ	$X\bar{Z}$
$\bar{W}\bar{Y}$	1	1		
$\bar{W}Y$		1	1	
WY		1	1	1
$W\bar{Y}$	1			1

Figure 2.84 Problem 2.20d

2.21 Create the logic circuits for each of these logic expressions using only two-input NAND gates only:

a. $A \cdot \bar{B} \cdot \bar{C} + A \cdot C + \bar{A} \cdot \bar{B} \cdot C$

b. $(X \cdot Y)(\bar{W} + Y) + W \cdot X$

2.22 Create the logic circuits for each of these logic expressions, using only two-input NAND gates only:

a. $(X + T)(Y + X \cdot T)$

b. $A \cdot \bar{B} \cdot C + A \cdot B \cdot \bar{C}$

2.23 Repeat Problem 2.21, using only two-input NOR gates only.

2.24 Repeat Problem 2.22 using only two-input NOR gates only.

2.25 Reduce the following logic expressions to their minimum sum-of-products terms:

a. $(A \cdot \bar{B} + C)(A \cdot \bar{C} + \bar{B} \cdot \bar{D}) + \bar{A} \cdot B \cdot (\bar{C} \cdot D + B)$

b. $(\overline{A \cdot B \cdot C + D})(A \cdot \bar{C} + B \cdot D) + \overline{A \cdot B} \cdot (C + B \cdot \bar{D}) + \bar{A} \cdot \bar{B} \cdot \bar{C} \cdot \bar{D}$

c. $\overline{A \cdot C \cdot D + A \cdot B \cdot C \cdot D} + A \cdot B \cdot (\bar{C} + B \cdot \bar{D} + A \cdot C)$

2.26 Reduce the following logic expressions to their minimum sum-of-products terms:

a. $(A + B + C)(\bar{A} + B + \bar{C})(A + \bar{B} + \bar{C})$

b. $\bar{A} \cdot B \cdot \bar{C} \cdot D + \bar{A} \cdot B \cdot \bar{C} \cdot \bar{D} + \bar{A} \cdot \bar{B} \cdot \bar{C} \cdot D + A \cdot B \cdot \bar{C} \cdot \bar{D} + A \cdot \bar{B} \cdot C \cdot \bar{D} + A \cdot \bar{B} \cdot \bar{C} \cdot \bar{D} + A \cdot B \cdot C \cdot \bar{D}$

c. $\overline{A \cdot B \cdot C} \cdot D + \overline{(A \cdot C) \cdot (B + \bar{D}) + \bar{A} \cdot \bar{B} \cdot \bar{C}}$

2.27 For the logic circuit in Figure 2.85, do the following:

a. Derive its logic expression.

b. Reduce the expression to its minimum sum-of-products terms.

c. Draw the reduced circuit.

2.28 For the logic circuit in Figure 2.86, do the following:

a. Derive its logic expression.

b. Reduce the expression to its minimum sum-of-products terms.

d. Draw the reduced circuit.

2.29 From the truth table in Figure 2.87, create the Karnaugh map and read off the minimum SOP expression for the logic circuit represented by the truth table.

2.30 From the truth table in Figure 2.88, create the Karnaugh map and read off the minimum SOP expression for the logic circuit represented by the truth table.

2.31 Create the K-map that represents this expression:

$$Y = \bar{A} \cdot \bar{B} \cdot \bar{C} + A \cdot B \cdot \bar{C} + \bar{A} \cdot B \cdot C + A \cdot \bar{B} \cdot C$$

Read an equivalent expression from the map.

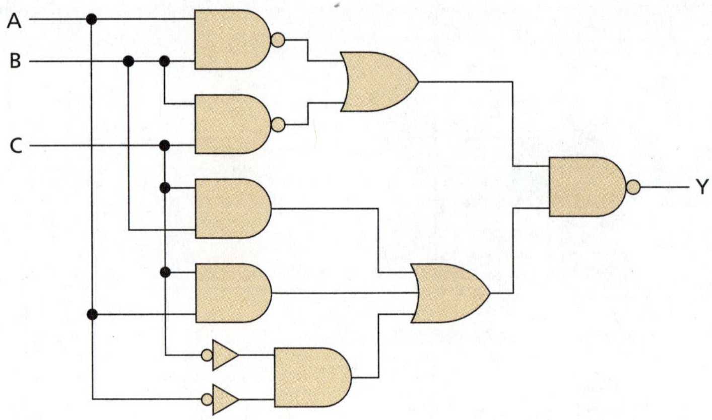

Figure 2.85 Problem 2.27

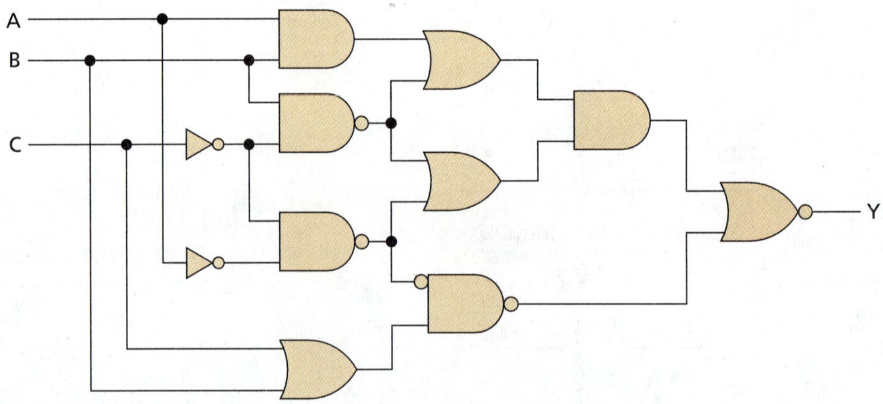

Figure 2.86 Problem 2.28

A	B	C	D	Y
0	0	0	0	1
0	0	0	1	1
0	0	1	0	0
0	0	1	1	0
0	1	0	0	X
0	1	0	1	1
0	1	1	0	1
0	1	1	1	0
1	0	0	0	X
1	0	0	1	X
1	0	1	0	1
1	0	1	1	1
1	1	0	0	1
1	1	0	1	0
1	1	1	0	0
1	1	1	1	0

X = "don't care" state

Figure 2.87 Problem 2.29

A	B	C	D	Y
0	0	0	0	X
0	0	0	1	X
0	0	1	0	0
0	0	1	1	0
0	1	0	0	0
0	1	0	1	1
0	1	1	0	1
0	1	1	1	1
1	0	0	0	1
1	0	0	1	0
1	0	1	0	1
1	0	1	1	X
1	1	0	0	1
1	1	0	1	0
1	1	1	0	0
1	1	1	1	1

X = "don't care" state

Figure **2.88** Problem 2.30

2.32 Create the K-map that represents this expression:

$$Y = \overline{A} \cdot \overline{B} \cdot \overline{C} \cdot \overline{D} + \overline{A} \cdot \overline{B} \cdot C \cdot D + \overline{A} \cdot B \cdot \overline{C} \cdot D + \overline{A} \cdot B \cdot C \cdot \overline{D} + A \cdot B \cdot \overline{C} \cdot \overline{D} + A \cdot B \cdot C \cdot D + A \cdot \overline{B} \cdot \overline{C} \cdot D + A \cdot \overline{B} \cdot C \cdot \overline{D}$$

Read an equivalent expression from the map.

three

The Binary Number System

KEY TERMS

address
ASCII and extended ASCII
base or radix
BCD (binary-coded decimal)
BCD (or decade BCD) adder
bit
bus
byte
carry
complement
complementing
data
data link characters
default
double word
EBCDIC (Extended Binary Coded Decimal Interchange Code)
flag or status flag
full adder
half adder
hex (hexadecimal)
invert
inverting
least significant digit
long word
LSB (least significant bit)
MSB (most significant bit)
nibble
nine's complement
octal
overflow
parallel adder
quad word
scale value
sign–magnitude system
signed number
ten's complement
unsigned number
weight
word

OBJECTIVES

After completing this chapter, the student should be able to

- Describe how number systems and arithmetic manipulation work using the familiar decimal or base 10 number system.
- Work with the binary or base 2 number system, including translation between binary and decimal numbers.
- Describe the hexadecimal or base 16 system, which is used as a shorthand method of representing binary numbers.
- Use concepts of binary addition to do arithmetic operation using binary numbers.
- Use one's complement to invert binary numbers, and perform binary subtraction by two's complement addition.
- Perform addition and subtraction of hexadecimal numbers.
- Develop and use digital circuits for addition and subtraction beginning with simple single-bit addition and continuing to multiple-bit arithmetic.
- Describe binary-coded decimal (BCD) character number code, and a method for adding BCD characters as decimal numbers.
- Demonstrate familiarity with the ASCII code character set as an application of a binary code.
- Troubleshoot and design adder and subtracter circuits.

OVERVIEW

A major task of most digital computers is to perform mathematical calculations. These are usually done on decimal numbers entered into the computer by a user. The computer converts these decimal numbers into binary numbers and performs the operation, returning an answer in binary form, which is then converted back to decimal for the user. In this chapter, the decimal and binary number base systems are defined and their relationship to each other explored. In addition, the student is shown how to perform the arithmetic operations of adding and subtracting in both of these number systems. It is essential to understand the process, so that we can demonstrate the design of digital circuits that perform these operations.

After the relationship between decimal and binary number systems is established, we explore two additional number systems that are used to represent binary number words: the hexadecimal, or base 16, and the octal, or base 8, number systems. The octal system is presented briefly for the sake of historic significance and its occasional use today. The hexadecimal system provides a way to manage binary numbers easily and efficiently. Converting between hexadecimal and binary numbers is quick and easy, facilitating the use of hexadecimal numbers to represent equivalent binary numbers.

An additional purpose of this chapter is to implement combinational logic circuits to perform binary addition and subtraction using discrete logic gates and medium-scale integrated devices. Truth tables and logic expressions based on binary addition and subtraction concepts are used to create the circuits to perform these arithmetic functions. From the basic adder circuits, a four-bit parallel adder is developed, which is used to implement two's complement addition for subtraction.

Binary codes such as binary-coded decimal, or BCD (used to represent decimal number characters) and ASCII (used to represent numbers, alphanumeric, and other characters) are presented, with applications for their use. The chapter ends with a section on troubleshooting arithmetic circuits and a design application for a four-bit BCD adder circuit.

3.1 Introduction

As noted in Chapter 1, digital electronic circuits and computer operations are based on a two-state operation—a high or 1 state and a low or 0 state. Also in Chapter 1, we presented a brief introduction to the binary number system to facilitate the use of truth tables. Now we are going to extend that discussion to complete detail as we explore the binary or base 2 number system. To see what is meant by a base 2 number system, let's first consider the familiar base 10 or decimal number system that we use daily.

3.2 Base 10 Fundamentals

Base** or **radix *a number system's base value.*

The decimal system uses 10 different symbols for numerical values: 0, 1, 2, 3, 4, 5, 6, 7, 8, and 9. The term describing the number system is called a **base** or **radix** number, which has a value equal to the number of symbols used in a particular number system. Hence the decimal number system is also known as a base 10 number system with a radix value of 10.

Least significant digit *rightmost digit of a number.*

Weight *the multiplier or scale value of a digit determined by its position in a number.*

Scale value *a weight value of a number.*

Let's start with the numeral 0, since it is the most misunderstood of the symbols. It has a value of nothing by itself, but consider its importance in a number such as 300. Each of the two zeros, alone, has a value of nothing, but placed along with the 3, they take on a significance—not to themselves, but to the 3. The number no longer has a value of three but instead, three hundred. What made the 3 take on so much more significance and value is its position in the number. Had the 3 appeared in the rightmost position, called the **least significant digit,** then the numeral 3 would have had a value of three. Move the 3 left one position and place a zero into the least significant digit position, and 3 takes a value of thirty or 30. Move the 3 once more to the left, tacking on another 0, and 3 now takes a value of three hundred or 300. The numeral 3 by itself, though, retains a value of three. What gave it a higher value was the numerical **weight** of the position it was in.

The least significant digit has a positional weight of 1. This weighted value is found by raising the base number (10) to an exponent value equal to the digit's position. In the case of the least significant digit, that position is position 0, so 10 raised to the 0 power is 1. The value of the number in that position is the product of the number multiplied by a weight of 1. The 1 is called a **scale value** and is used to determine the weight of the digit corresponding to its position within a number. A 3 in the least significant digit equals its value times a scale value of 1, for an actual value of 3×1, or 3. The next position to the left has a scale value or weight of 10. This is position 1, and raising 10 to a power of 1 yields a weighted value of 10. A 3 in that position becomes 3×10, or 30. Continuing to the next position with a weight of 10^2 or 100, 3 in that place is 3×100, or 300. Relating the position of a digit in a number to its weight yields these relationships:

POSITION	POWER	WEIGHT OR SCALE VALUE
Least significant digit =	10^0	1
Next =	10^1	10
Next =	10^2	100
Next =	10^3	1,000
	(and so on)	

Notice that as the position of the digit moves to the left, its weight increases by a factor of 10, which is the base value.

EXAMPLE 3.1

Express the number 1,538 in terms of each digit's value and weight.

SOLUTION Starting with the least significant digit, the number 1,538 can be written as

$$8 \times 1 = 8 \qquad 3 \times 10 = 30 \qquad 5 \times 100 = 500 \qquad 1 \times 1{,}000 = 1{,}000$$

How do we get the whole number from the product of the weights and digit values? Simply add the digit–weight products together. In Example 3.1,

$$8 + 30 + 500 + 1{,}000 = 1{,}538$$

The whole number is the decimal value of the original number. It is easily noted that this process has no usefulness, in this case, because the numbers we start with are the ones we end with. The usefulness lies in understanding the process that will be applied, later, when converting numbers from binary or other number systems to the more familiar decimal numbers.

EXAMPLE 3.2

Express the decimal number 84,762 in terms of weights and digit values. Show how the digit–weight products are combined to form the original decimal value.

SOLUTION Separate the number, and multiply each digit value by its weight:

$$8 \times 10{,}000 = 80{,}000$$

$$4 \times 1{,}000 = 4{,}000$$

$$7 \times 100 = 700$$

$$6 \times 10 = 60$$

$$2 \times 1 = 2$$

To find the decimal number, add these digit–weight products together:

$$80{,}000 + 4{,}000 + 700 + 60 + 2 = 84{,}762$$

Section 3.2

Practice Problems

1. Represent the following decimal numbers as products of their weights and digit values.
   a. 67,234 b. 123,456 c. 399 d. 4,001
2. Show how the digit–weight products of Problem 1 are used to find the decimal value of the original numbers.

3.3 Fractions—Values Less Than 1

Numbers with values less than 1, called *fractions* or *decimals,* are digits found to the right of the decimal point. In the number 45.23, the number 45 is the whole number integer component, while the number 23 is the fractional part of the complete number. This part of the number is actually the fraction 23/100, also known as the decimal value 0.23. Decimal numbers also have weights based on their position. The weight of the digit that is nearest to the decimal point has a weight of 1/10 or 0.1 expressed as a power of 10 as 10^{-1}. For the example value of 45.23, this gives the first digit to the right of the decimal point a value of 2×0.1, or 0.2. The next position to the right has a weight of 1/100 or 0.01 or 10^{-2} and so forth. The weights continue progressively smaller, with each succeeding weight having a value 1/10th of the preceding position. In the example, the number 3 has a value of 3×0.01 or 0.03. Summarizing fractional decimal weights as we did with whole number weights yields a progression like

POSITION	POWER	WEIGHT
Nearest to the decimal point	10^{-1}	= 0.1
Next right	10^{-2}	= 0.01
Next right	10^{-3}	= 0.001
Next right	10^{-4}	= 0.0001
(and so on)		

Notice that the weight or scale value of each succeeding position to the right reduces by a factor of 10, the base number.

EXAMPLE 3.3

Represent the number 654.92 as products of weights and digit values. Show how to combine these digit–weight products to form the original decimal number.

SOLUTION Separate the number into weight and digit products:

$$6 \times 100 = 600$$

$$5 \times 10 = 50$$

$$4 \times 1 = 4$$

$$9 \times 0.1 = 0.9$$

$$2 \times 0.01 = 0.02$$

To recover the original number, add the products together:

$$600 + 50 + 4 + 0.9 + 0.02 = 654.92$$

Application Note

Decimal Numbers

Actually, using a scale value or weight for calculating decimal values is not that unique in daily use. Consider what you do each time you go shopping—you gather your purchases and go to a checkout stand to pay for them. Essentially, when you gather the necessary bills and coins, you add them together to determine if they are enough to pay for your purchases. It seems obvious to say that you treat nickels and quarters and dollar bills differently, but consider why you do. You are actually using digits and weights to give you your sum. A quarter is actually 25×0.01 of a dollar. The dollar bill is 1×1. Our digits are limited by the coins and bills minted. In relation to the current discussion, our coins and bills can be equated using digits and weights:

1 penny = $1 \times \$0.01 = \$0.01 = 1¢$

1 nickel = $5 \times \$0.01 = \$0.05 = 5¢$

1 dime = $10 \times \$0.01 = \$0.10 = 10¢$

1 quarter = $25 \times \$0.01 = \$0.25 = 25¢$

1 half dollar = $50 \times \$0.01 = \$0.50 = 50¢$

1 dollar bill = $100 \times \$0.01 = 1 \times \$1 = \$1.00 = 100¢$

1 five-dollar bill = $5 \times \$1 = \$5.00 = 500¢$

(and so on)

We can also surmise that there is a scale factor of 100 between the \$ symbol and the ¢ symbol, or 100¢/\$. For example, \$.01 × 100¢/\$ = 1¢.

Section 3.3

Practice Problems

1. Represent each of these numbers as products of their weights and digit values.
   a. 123.456 b. 567,900.02 c. 39.04508
2. Show how the digit–weight products found in Problem 1 are used to find the decimal value of the original numbers.

3.4 The Binary or Base 2 Number System

Because of the exactness of mathematics, the mathematical rules that apply to one number base system, such as the decimal system, apply to all real number base systems. This is true for the binary or base 2 number system, as it is for the decimal or base 10 system. The binary system has only two numerical symbols, 1 and 0. These two numbers have the same value as they did in decimal (and any other system)—1 and 0. No other numerical symbols appear in binary numbers. So how do we get binary numbers that are equivalent to numbers greater than 1? The same way we got numbers greater than 9 in the decimal system, by using digit–weight products.

The weights for the binary system are found in the same manner used for decimal numbers—by raising the base value, 2, by a power equal to the value of the position of a digit in a binary number. Those weights, starting with the least significant digit, are

POSITION	WEIGHT
Least significant digit =	$2^0 = 1$
Next =	$2^1 = 2$
Next =	$2^2 = 4$
Next =	$2^3 = 8$
Next =	$2^4 = 16$
Next =	$2^5 = 32$
Next =	$2^6 = 64$
Next =	$2^7 = 128$
(and so on)	

Notice that each position change to the left increases the weight by a factor of 2, the base value. The sum of the products of the weights and the binary digit values yields the equivalent decimal value of a binary number. However, digit values in the binary system are only 1 and 0, so the products will be either 0 (weight $\times$ 0) or the weight value (weight $\times$ 1). As an example, we will convert the binary number, 1010101, to its decimal equivalent value.

First we need to determine the weight of each position in the binary number. A short-cut method to do this, instead of raising the base number of 2 to the power of each position, is to start with the rightmost digit, which has a weight of 1. As you move to the left, double the weight value for each position. In our example, the weights are

Binary number:	1	0	1	0	1	0	1
Weights:	64	32	16	8	4	2	1

The binary number is shown, as it appears, with the least significant digit on the right. Notice how the weights double as you move to the left.

The next part of the process is to get the digit–weight products. Again, since binary digits can only have a value of 0 or 1, these products will be 0 or the weight value of the digit position. Digit–weight products in our example are 64, 0, 16, 0, 4, 0, and 1. Finally, to get the base 10 equivalent value, add up the digit–weight products: $64 + 0 + 16 + 0 + 4 + 0 + 1 = 85$.

EXAMPLE 3.4

Find the base 10 equivalent value for the binary number 110011101.

SOLUTION Determine the weight values for all the binary digits, multiply them by the digit values of 1 or 0, and add them together:

$$256 + 128 + 0 + 0 + 16 + 8 + 4 + 0 + 1 = 413$$

Bit *a single binary digit.*

LSB (least significant bit) *the rightmost binary digit, which has the lowest weighted value.*

MSB (most significant bit) *the leftmost binary digit, which has the largest weighted value.*

Word *original definition: a fixed number of binary bits used at a given time. Second definition (based on de facto use by microprocessor manufacturers): 16-bit binary number. Also, 32-bit quantities are often designated as **double** or **long words.***

Nibble *a four-bit binary number.*

Binary digits are called **bits**. The least significant digit is called the **least significant bit** or **LSB**. The leftmost bit is the **most significant bit** or **MSB**. Binary values contained within a number of bits with a fixed length are called **words**. A word length of four bits is referred to as a **nibble** and one of eight bits is a **byte**. The usage of various binary word lengths has been standardized by the requirements of digital computers whose circuits are built supporting a microprocessor that uses binary numbers in specific word sizes. Early microprocessors were based on nibble sizes, but soon advanced to byte-size words. Currently, these devices manipulate binary word sizes of 16-, 32-, and 64-bit lengths. The de facto standard for a binary word length of 16 bits has become a word. This is unfortunate because of the confusion created from the use of this term *word*. It is important for the reader to know in exactly what context the term is being used. Is it for the general application for binary lengths of any size, or does it mean a binary length of 16 bits? After this paragraph, use of the term *word* will be clearly specified before its use, because both definitions are valid and have specific use and application. However, 32-bit binary numbers have been termed **long words** by some microprocessor manufacturers and **double words** by others, and 64-bit lengths have generally come to be known as **quad words**. No standard name, de facto or other, has yet arisen for 32-bit or larger-size binary numbers.

Section 3.4

Practice Problems

Convert the following base 2 numbers into decimal numbers.

1. 11011101 2. 1001001 3. 1111111 4. 10000011

3.5 Designating Different Number Bases

Convert the binary number 1011 to a base 10 number. The answer is $8 + 0 + 2 + 1 = 11$. It is not too difficult to place values on the two numbers because the problem was clearly stated—convert the binary number 1011 to a base 10 number. The answer, 11, in this case is understood to be eleven, because that was the result being asked for. However, if the number 11 appears without any reference, how would its value be determined as a decimal eleven or a bi-

Byte *a group of eight binary bits.*

Long word *a 32-bit binary number.*

Double word *a 32-bit binary number.*

Quad word *a 64-bit binary number.*

nary three? The number base system in use must be specified in some manner. There are several methods used, not the least of which is someone telling the reader which number system is in use. Another method, which doesn't require the person who wrote the number to be present, is the use of a subscript reference. The base value of the system is used as a subscript, appended after the least significant digit of the number. The example used at the beginning of this section is written, with subscripts, as

$$1011_2 = 11_{10}$$

Subscripts do clarify the number base being used, but you have to wait until you are finished reading the number digits before determining which number base is in use, since the subscript is appended at the end of the number. An alternate method is to use a *preceding* letter to indicate the number base used. This letter is usually the lowercase abbreviation for the base name, and it is surrounded by single quote marks. A 'b' is used for binary numbers and a 'd' for decimal. The preceding conversion example is written with leading letter designations as

'b'1011 = 'd'11

Default *quantity or form that is used when it is unspecified.*

Frequently the decimal base system is used as a **default** system, which does not require any notation. A number lacking a reference designation is assumed to be in the default base system. The example being used can be written, using base 10 or decimal as a default base system, as

'b'1011 = 11

Still another method for designating number base systems uses a trailing capital letter without quotation marks for a reference. A capital B designates binary numbers and a D, decimal numbers. Continuing with our example, the results could now be written as

1011B = 11D

As with the preceding reference letter, if the value does not have a trailing letter, the default base—usually decimal—is assumed. This text follows the convention of the more familiar form of the trailing capital letters. Decimal or base 10 numbers are the default number type. Default means that a designation is not required for these numbers, and any number without a designation is assumed to be decimal.

EXAMPLE 3.5

Give the decimal value for these numbers:

110011B and 1011

SOLUTION The first number is a binary number designated by the trailing B. Use the weight product method to get the decimal equivalent:

$$32 + 16 + 0 + 0 + 2 + 1 = 51$$

$$110011B = 51$$

The second number is a decimal number already, since it does not have a designating letter and does not require conversion. Its value is one thousand and eleven.

Application Note

Number Base Designations

The concept of using symbolic designations for numeric values is as old as time. Designations are used to differentiate between types of money, weights, and measures. Money amounts in the United States use the dollar sign ($) and cent sign (¢), while in England the chief symbol is the pound (£). In the United States, it's the foot (′) and inch (″), while in Europe it is centimeter (cm), and meter (M). In the United States, speed limits are in mph (miles per hour), and in Europe they are in kph (kilometers per hour). Even writing dates, numerically, requires understanding the symbolism used. What date is 2/7/93? In the United States, it is February 7, 1993, because the symbols are M/D/Y for month, day, and year. In Germany, those same symbols mean July 2, 1993, or 2 July 1993, because the symbols for the date are ordered D/M/Y. Whenever numbers are used, it is important to first know what the numbers represent.

Section 3.5

Practice Problems

1. Give the decimal values for the following numbers:
   a. 101101B b. 120021 c. 100011B d. 11101
2. Show how each of these binary numbers are written using the three methods described in this section:
   a. 100110 b. 1110011 c. 10101010

3.6 Fractional Binary Numbers

Like decimal numbers, binary numbers can have components that are fractional. Studying fractional binary numbers is important because complex numbers are used in computing systems. Keep in mind that all decimal values entered into a computer are converted to binary values so that the binary digital computer can perform requested arithmetic processes using those values. Finding their decimal value uses the same principles discussed for decimal fractions. Weights for binary decimal numbers are derived by raising the base 2 by negative exponents. The position closest to the decimal point has a weight of 2^{-1} or ½. Each succeeding position reduces the weight by a factor of 2:

POSITION	WEIGHT OR SCALE VALUE
Nearest to the decimal point	$2^{-1} = \frac{1}{2} = 0.5$
Next	$2^{-2} = \frac{1}{4} = 0.25$
Next	$2^{-3} = \frac{1}{8} = 0.125$
Next	$2^{-4} = \frac{1}{16} = 0.0625$
(and so on)	

EXAMPLE 3.6

Convert 101.011B and 11.0011B to decimal.

SOLUTION Same process—add up the products of weights and digit values:

$$101.011B = 4 + 0 + 1 + 0 + .25 + .125 = 5.375$$

and

$$11.0011B = 2 + 1 + 0 + 0 + .125 + .0625 = 3.1875$$

Section 3.6

Practice Problems

Convert the following binary numbers to decimal numbers:

1. 10111.0101B 2. 1.111B 3. 10001.1001B

3.7 Changing Decimal Numbers to Binary

The process to change a numerical quantity from a decimal value into another number system value is the same for all number base systems, including the decimal base system. Start by *dividing* the decimal number by the base value. The remainder from the division becomes the least significant digit of the new base number. Using the quotient from the division, repeat the process. The remainder is the next digit of the new base number. Again, take the quotient and divide it by the base value. Keep repeating the process until the division produces a quotient of zero. To see how this works, let's convert the decimal number 3,528 to its decimal equivalent. (We expect to get 3,528 as a result and are only performing the conversion to illustrate the process used.)

3,528 divided by 10 is 352 with a remainder of 8

352 divided by 10 is 35 with a remainder of 2

35 divided by 10 is 3 with a remainder of 5

3 divided by 10 is 0 with a remainder of 3

Reassemble the remainders with the first remainder, 8, as the least significant digit, the next remainder, 2, as the next digit and so forth: 3,528.

The process described to convert a decimal number into another base system value works with any base system. Use the base system value as the division, and organize the remainders in the same fashion, beginning with the least significant digit. Next we will convert a decimal number to its binary number equivalent.

Let's convert the decimal number 137 to binary. Use the same process as before, using the base number 2 for a divisor:

137 divided by 2 = 68 with a remainder of 1 LSB

68 divided by 2 = 34 with a remainder of 0

34 divided by 2 = 17 with a remainder of 0

17 divided by 2 = 8 with a remainder of 1

8 divided by 2 = 4 with a remainder of 0

4 divided by 2 = 2 with a remainder of 0

2 divided by 2 = 1 with a remainder of 0

1 divided by 2 = 0 with a remainder of 1 MSB

Now assemble the binary number, starting with the remainder of the first division as the least significant bit, continuing with the next remainder as the next binary digit and so forth:

137 = 10001001B

The fractional part of a number is converted from decimal to another number base by first *multiplying* the fractional part by the base value. Any whole value of the product becomes the digit next to the decimal point. The remaining fraction of the product is multiplied by the base value again. The whole part is the next digit to the right. The fraction is multiplied once more. The process is repeated until the product is 0 or a desired accuracy is achieved.

As an example, let's convert the decimal number 23.575 to a binary number. To do the conversion, start with the whole number and divide it by the base value, 2:

23 divided by 2 = 11 remainder 1 LSB

11 divided by 2 = 5 remainder 1

5 divided by 2 = 2 remainder 1

2 divided by 2 = 1 remainder 0

1 divided by 2 = 0 remainder 1 MSB

The whole number portion is 10111B. Next, to convert the fractional portion of the decimal number, multiply that fraction by the base value, 2. The whole numbers from the products become the binary fraction values and are deleted before the next multiplication is performed.

$0.575 \times 2 = 1.150$ digit = 1

$0.15 \times 2 = 0.3$ = 0

$0.3 \times 2 = 0.6$ = 0

$0.6 \times 2 = 1.2$ = 1

$0.2 \times 2 = 0.4$ = 0

$0.4 \times 2 = 0.8$ = 0

$0.8 \times 2 = 1.6$ = 1

$0.6 \times 2 = 1.2$ = 1 LSB

The whole numbers from the multiplications become the fraction binary values: 0.10010011B. Combine that with the whole binary number found from the division process, and the final answer is 23.575 = 10111.10010011B. Before we abandon this idea, let's prove that this conversion is correct by finding the decimal equivalent of the binary result. If the method works, we should get 23.575 back. Use the sum of the products of the weights and digits to convert the binary number back to decimal: 16 + 0 + 4 + 2 + 1 + 0.5 + 0 + 0 + .0625 + 0 + 0 + .0078125 + .00390625 = 23.57421875, which is reasonably close to 23.575.

Section 3.7

Practice Problems

1. Convert the following decimal numbers to binary numbers:
   a. 123.456 b. 97.05 c. 400.004 d. 511.0075
2. Prove that the conversions in Problem 1 are correct by using the digit–weight product method.

3.8 Other Useful Number Systems

Octal *the base 8 number system, each digit of which can be represented by a three-bit binary word.*

Hex (hexadecimal) *the base 16 number system, each digit of which can be represented by a four-bit word.*

Binary numbers are cumbersome to read and difficult to discuss. Consider a future discussion involving the sum of 1101010111010111010111 0110B and 11011110111101010101011111B. Go ahead and read that number—without mistaking a 1 for a 0 or vice versa! It is a mouthful, to say the least. One way to make the binary numbers more readable is to convert them to decimal and then deal with them. In this case, convert the two numbers to decimal, do the addition, and then convert the decimal sum back to binary. But converting binary values into decimal numbers is an involved process, particularly when considering large values. Try the two numbers in this discussion if you need to be convinced of that difficulty. Two other number systems can help you manage binary numbers more easily. The **octal** or base 8 system and the **hexadecimal** or base 16 number system are used to overcome the problem of dealing with large binary numbers. Conversions between these two systems and binary numbers are easy, facilitating their use. Being number systems, the same arithmetic rules used for decimal and binary systems also apply to the octal and hexadecimal (usually shortened to **hex** for convenience) number systems. The octal system has fallen into minimal use and is presented here only for historical context. The **hex** system is the current choice, for reasons we shall see shortly.

Data *binary words that represent information or numbers.*

The binary information or **data** in early computers were formed into binary words of three-bits in length. This was done so that the listing of binary codes of the programs run on these computers, as well as the data results produced by the programs could be printed out in a shorter format, using the octal number base system to represent the 1 and 0 program and data binary codes. For instance, which of the following do you find easier to read?

110011010101011110101101101010101010101111101111011011OB

or 1465275332525373666#

The octal number system uses a # sign at the end to signify its use rather than an *O*, because an *O* looks too much like a 0. The octal number system is a base 8 number system, which means that only eight numerical symbols, 0 through 7, are used. Each octal digit can be directly converted into a three-bit binary number, from 000 for 0 to 111 for 7. The relationship that defines the number of binary bits required to represent all the values of the digits, in another number system, is 2^n = the number of values, where n is the number of binary bits used. For example, the octal system requires a three-bit word because a word of three bits can represent eight different values: $2^3 = 8$. The eight binary values for the octal system are summarized in this table.

OCTAL NUMBER	BINARY NUMBER
0	000
1	001
2	010
3	011
4	100
5	101
6	110
7	111

EXAMPLE 3.7

How many numerical values can be represented by binary numbers with a length of 4, 6, and 8 bits?

SOLUTION Raise the base 2 to an exponent value equal to the number of bits in the length of the binary word to get the total number of values that can be represented by the binary word.

$2^4 = 16 \qquad 2^6 = 64 \qquad 2^8 = 256$

The highest decimal value that can be represented by a binary number is based on the binary bits in that number. This value can be found by using the equation

$$\text{Highest decimal value} = 2^n - 1$$

where n is the number of bits in the binary number.

EXAMPLE 3.8

How many different values are represented by a binary number of 12 bits? What is the largest decimal value that can be represented by this number?

SOLUTION To find the total number of values in a system of 12 bits, raise the base 2 to the 12th power:

$$2^{12} = 4{,}096$$

Subtracting 1 from that total yields the highest value usable in that system:

$$2^{12} - 1 = 4{,}095$$

The first requirement for developing or using a new number base system is to make it useful. With that in mind, converting numbers to and from binary numbers, using the new system, had to be done easily. In the case of the base 8 or octal system, this requirement is met. Converting from binary to octal begins by sectioning the binary number into groups of three bits starting from the least significant bit (LSB) position. As an example, let's use the number 1110111011011101B. Start with the LSB, and separate the bits into groups of three bits each: 1 110 111 011 101B. If necessary, use leading 0's to fill out the leading group to three bits: 001 110 111 011 011 101B. Next convert the binary number as if you were attempting to determine the decimal value of each group of three bits, using the same three weights for each group: 4, 2, and 1: 001B = 1 110B = 6 111B = 7 011B = 3 011B = 3 101B = 5. These numbers when grouped together form the octal base number equivalent of the binary number, so that 1110111011011101B = 167335#.

Admittedly, 167335# is easier to say than 1110111011011101, but is 167335# really a valid number? The proof is arrived at by converting both the binary and octal numbers to decimal to see if they are the same. The principle for making the conversions was explained earlier—add the products of the digits times the weights for both numbers. Recall that the weights for the binary number are derived by raising the base value to an exponent value equal to the position of each bit in the binary number. The same procedure is used for both binary and octal numbers using a base value of 2 for the binary numbers and 8 for the octal numbers.

First the binary-to-decimal conversion:

$$\begin{aligned} 1110111011011101\text{B} &= 32{,}768 + 16{,}384 + 8{,}192 + 0 + 2{,}048 + 1{,}024 \\ &\quad + 512 + 0 + 128 + 64 + 0 + 16 + 8 + 4 + 0 + 1 \\ &= 61{,}149 \end{aligned}$$

and then the octal-to-decimal conversion:

167335#: $1 \times 8^5 + 6 \times 8^4 + 7 \times 8^3 + 3 \times 8^2 + 3 \times 8^1 + 5 \times 8^0$

The individual digit numbers multiplied by their weight values are

$1 \times 32{,}768 = 32{,}768 \quad 6 \times 4{,}096 = 24{,}576 \quad 7 \times 512 = 3{,}584$

$3 \times 64 = 192 \quad 3 \times 8 = 24 \quad 5 \times 1 = 5$

and the final sum is

$32{,}768 + 24{,}576 + 3{,}584 + 192 + 24 + 5 = 61{,}149$

It is now safe to say that, since both the binary and octal numbers equal the same decimal value, they are equal to each other.

EXAMPLE 3.9

Convert the following binary numbers to octal and prove the binary and octal numbers are numerically equal:

100011101B 1110011B

SOLUTION Start by grouping the binary numbers into sets of three bits each, starting from the LSB:

100 011 101B 001 110 011B

Then convert each group of three bits to their decimal equivalent:

4 3 5 1 6 3

When regrouped, the individual decimal numbers form the octal number equal to the original binary number:

100011101B = 435# and 1110011B = 163#

Now prove they are equal by converting each number to its decimal value:

100011101B = $256 + 0 + 0 + 0 + 16 + 8 + 4 + 0 + 1 = 285$

and 435# = $5 \times 1 + 3 \times 8 + 4 \times 64 = 5 + 24 + 256 = 285$

1110011B = $64 + 32 + 16 + 0 + 0 = 2 + 1 = 115$

and 163# = $3 \times 1 + 6 \times 8 + 1 \times 64 = 3 + 48 + 64 = 115$

To convert from octal to binary, take each octal digit and convert it into its three bit binary value. For example, take the number 435# from the preceding example and convert it back to a binary number:

4 = 100, 3 = 011, and 5 = 101

435# = 100011101B

EXAMPLE 3.10

Convert the following octal numbers into binary numbers:

123# 555# 2007#

Prove that both numbers have the same decimal value.

SOLUTION Separate the octal digits, and convert each individually to their binary equivalent:

1	2	3	5	5	5	2	0	0	7
001	010	011	101	101	101	010	000	000	111

and 123# = 1010011B 555# = 101101101B

2007# = 10000000111B

Converting both octal and binary numbers of each example into decimal yields the same value:

$$123\# = 1 \times 64 + 2 \times 8 + 1 \times 3 = 83$$

$$1010011B = 64 + 16 + 2 + 1 = 83$$

$$555\# = 5 \times 64 + 5 \times 8 + 5 \times 1 = 365$$

$$101101101B = 256 + 64 + 32 + 8 + 4 + 1 = 365$$

and

$$2007\# = 2 \times 512 + 7 \times 1 = 1{,}031$$

$$10000000111B = 1{,}024 + 7 = 1{,}031$$

Section 3.8

Practice Problems

1. Convert the following binary numbers into octal numbers, and prove they have the same decimal value:
   a. 100011B b. 111001110B c. 100001110011B
2. Convert the following octal numbers to binary, and prove that they are equal to each other:
   a. 466# b. 15244# c. 3351#

3.9 The Hexadecimal or Base 16 Number System

The discussion of the octal system was brief because that number system has fallen into minimal use, having been replaced by the hexadecimal or HEX number system. The main reason that the hex system came into heavy use is the

Table 3.1 Hexadecimal Numbers

BINARY NUMBER					
b_3	b_2	b_1	b_0	Hexadecimal Number	Decimal Number
0	0	0	0	0	0
0	0	0	1	1	1
0	0	1	0	2	2
0	0	1	1	3	3
0	1	0	0	4	4
0	1	0	1	5	5
0	1	1	0	6	6
0	1	1	1	7	7
1	0	0	0	8	8
1	0	0	1	9	9
1	0	1	0	A	10
1	0	1	1	B	11
1	1	0	0	C	12
1	1	0	1	D	13
1	1	1	0	E	14
1	1	1	1	F	15

advent of computers built around a single integrated-chip central processing unit called a microprocessor. The CPU is responsible for gathering instructions from a program list and executing those instructions that include manipulating information or data and performing arithmetic and logic operations. For the purposes of our current discussion, the crucial aspect of the microprocessor is the size of the binary words used for data. Generally, the lengths of the binary data words are multiples of four: 4, 8, 16, and 32. Using the relationship of 2^n = number of different values, we can determine that a binary word of four bits, called a *nibble,* can represent 16 different numerical values: 0000 (0) to 1111 (15). The base 16 number system, called the *Hex* number system, is used to represent groups of four-bit binary numbers. The system uses 16 different numerical symbols, the first 10 being 0 to 9 and the remaining symbols, the first six letters of the alphabet, *A* through *F.* Table 3.1 shows the relationship between each group of four binary bits and their hexadecimal equivalent value.

Converting binary numbers to hex numbers follows the same principle used for converting to octal numbers, except that the binary numbers are now set into groups of four bits instead of three bits each for each hex digit. Hex values are derived by considering each group of four binary bits as having the same weight scale—the LSB is 1, next 2, next 4, and the MSB is 8. This relationship is sometimes referred to as the 1-2-4-8 system. For instance, the binary sequence 1011001101111001B is converted to its hex equivalent by first grouping the binary bits into groups of four beginning with the least significant bit:

1011 0011 0111 1001B

Then each group is converted into their hex digit equivalent using Table 3.1 or the 1-2-4-8 method. A trailing H is used to designate hex numbers:

1011B = 11 = BH 0011B = 3H 0111B = 7H

1001B = 9H

The hex digits are united to form the final hex value:

1011001101111001B = B379H

Converting hex numbers to decimal and back uses the same principles discussed for conversion between binary and decimal numbers. Add the digit–weight values, using 16 as the base instead of 2, 8, or 10 to convert hex numbers to decimal. A summary of hexadecimal weights are

$16^0 = 1$ $16^1 = 16$ $16^2 = 256$ $16^3 = 4{,}096$ and so on

The equivalent decimal value of a hex number is found by adding the products of the numbers times their weights. To convert decimal numbers to their hexadecimal equivalent values, divide by 16 in the same way you divided by 2 to convert decimal values to binary numbers. Use the remainders from the divisions to form the hex number.

Let's start by finding the decimal equivalent of 3C2H. First get the decimal equivalent for each hex digit from Table 3.1:

3 = 3 C = 12 2 = 2

Next multiply the decimal equivalents of each hex digit by its weight:

$3 \times 256 = 768$ $12 \times 16 = 192$ $2 \times 1 = 2$

Last, add up the products to get the final decimal equivalent:

3C2H = 768 + 192 + 2 = 962

Now that we have seen an example of converting hex to decimal, let's do one going the other way. We will convert the decimal number 397 to hex using the same principle as in converting from decimal to binary. Set up the problem, and divide by 16, using the remainders from the division as the new hex digits:

16)397 = 24, with a remainder of 13DH

16) 24 = 1, remainder 8

16) 1 = 0, remainder 1

The first remainder is the least significant digit, and each succeeding remainder fills the next digit position to the left of the preceding digit, making 397 = 18DH.

EXAMPLE 3.11

What is the hex equivalent of 100111011B? Prove they are of equal value.

SOLUTION Group the binary number into sets of four bits, starting at the LSB:

0001 0011 1011B

Note that leading zeros are used to make the leading group four bits long. Convert each group, using 8-4-2-1 weights:

0001B = 1H

0011B = 3H

1011B = 11 = BH

100111011B = 13BH

The proof is the same as before: add up the digit–weight products, and check to see if both numbers have the same decimal value.

100111011B = 256 + 32 + 16 + 8 + 2 + 1 = 315

13BH: 1 × 256 = 256   3 × 16 = 48   11 × 1 = 11

13BH = 256 + 48 + 11 = 315

and since 315 = 315, then 100111011B = 13BH = 315.

Application Note

Hexadecimal Numbers

Bus a group of wires that carries binary data.

Address binary code for a memory location that stores data.

The 8088 microprocessor built by INTEL Corporation was widely used in a very successful personal computer called the IBM PC. This microprocessor moved eight bits of binary information or data along a set of parallel wires called a **bus**. In essence, the microprocessor is said to have an eight bit data bus. These data were moved between the microprocessor and memory for storage. Each memory location is designated by an **address** unique to that location. As with the data bus, there is also an address bus. In the case of the 8088, this address bus is 20 bits wide. The range of data and addresses controlled by the 8088 microprocessor is represented by hexadecimal equivalents of the binary minimums and maximums dictated by the size of the binary numbers. An eight bit data value can range from 00000000B to 11111111B, which is easier to represent in hex as 00H to FFH. The range of addresses is 00000000000000000000B to 11111111111111111111B or better, 00000H to FFFFFH. As you can see, discussing the 8088 microprocessor in terms of its data and address sizes is easier in hex than in binary.

Section 3.9

Practice Problems

1. Convert the following binary numbers to hex numbers, and prove the binary and hex numbers are equal to each other by converting both numbers to decimal.
   a. 1001110101B b. 1110010001B c. 101011000011B
2. Convert the following hex numbers to binary, and prove, by converting both numbers to decimal, that the binary and hex numbers are equal to each other.
   a. 3C1H b. 15F2H c. 345H

3.10 Binary Addition

When decimal numbers are added by hand, the process is taken for granted strictly from long-time usage. Add the columns of numbers together, and if there is anything left over from the addition, carry it over to the next column and continue the process. For example, when adding 278 to 569 start by adding 8 to 9 and getting 17. The 7 is written down, and 1 is "carried" to the next column:

```
  1
278
569
---
  7
```

In the next column, 7 is added to 6 and to the 1 that was carried over from the previous column, getting 14. The 4 is written down, and the 1 is carried to the sum of the last two digits:

```
11
278
569
---
 47
```

Finally the most significant column of numbers is added, yielding a sum of 847:

```
11
278
569
---
847
```

Overflow a condition under which the results of a mathematical value produces a value too large to be represented by the number of digits available.

Carry overflow that occurs when the number of bits in the sum are more than can be held in the circuit that stores the result.

In the previous example, 8 was added to 9 for a sum of 17, as stated. However, what we really have is 8 added to 9, producing a sum of 7 in the same column, with an **overflow** value of 1 to the next column called a **carry.** Essentially, the value 17 is too large to fit into a single digit position, so we must carry over to the next digit. This carry is added to the next pair of digits, which in return produced a sum of 4 and an additional carry of 1 to the next column. Last, the second carry is added to the last column. This sum did not produce a carry, because its total fit into a single digit position. The carry value is 0, since none was produced, and the sum appears as

```
011  carries
 569
 278
 ---
 847 sum
```

Adding binary numbers is no different from adding decimal numbers, except that you are restricted to two values, 0 and 1. The total possible results from adding binary bits, restricted to values of 1 and 0 in the sum as well as the bits being added together, are

$0 + 0 = 0 \qquad 0 + 1 = 1 \qquad \text{and} \qquad 1 + 1 = 10$

Notice in the last case that the sum is binary 10 (pronounced "one, zero") which in actuality is a sum of 0 with a carryover of 1. When the carry is added to the sum of the next two bits, the total number of binary bits being added together is three and the possible number of results increases:

$$0 + 0 + 0 = 0 \quad 0 + 0 + 1 = 1 \quad 0 + 1 + 1 = 10 \quad \text{and} \quad 1 + 1 + 1 = 11$$

Using these results as a guide, let's see how the addition of two binary numbers works by adding 11001110B to 1001110B. We will start by setting up the addition problem like any addition problem:

$$\begin{array}{r} 11001110 \\ +\ \ 1001110 \\ \hline \end{array}$$

The process is begun as you would when adding two decimal numbers, by adding the least significant bits first, $0 + 0 = 0$:

$$\begin{array}{r} 11001110 \\ +\ \ 1001110 \\ \hline 0 \end{array}$$

Continue on to the next pair of digits, noting that you will get a sum of 0 with a carry of 1, which must be placed onto the next column, $1 + 1 = 10$:

$$\begin{array}{r} 1\ \ \\ 11001110 \\ +\ \ 1001110 \\ \hline 00 \end{array}$$

Now we have three bits to add together, whose sum is a 1 with a carryover to the next column, $1 + 1 + 1 = 11$:

$$\begin{array}{r} 11\ \ \ \\ 11001110 \\ +\ \ 1001110 \\ \hline 100 \end{array}$$

Keep repeating the process until the addition has been completed:

$$\begin{array}{r} 1\ \ \ 111\ \ \ \\ 11001110 \\ +\ \ 1001110 \\ \hline 100011100 \end{array}$$

How do we prove that the addition is correct? By now you know—convert the numbers and their sum to decimal, and check the results:

$$\begin{array}{rcr} 11001110 & = & 206 \\ +\ \ 1001110 & = & +\ \ 78 \\ \hline 100011100 & = & 284 \end{array}$$

EXAMPLE 3.12

Do the following additions of binary numbers:

11011001B + 1101111B and 10011101B + 1000011B

SOLUTION Set up the problems like any addition problem:

```
  11011001B       10011101B
+  1101111B     +  1000011B
```

Start with the least significant bits, and add them: 1 + 1 = sum of 0 and a carry of 1. Place the sum below, and carry any value over to the next column. Add the three binary bits in that column, 1 + 0 + 1, which, again, yields a sum of 0 and a carry of 1. Continue the process until the addition is completed:

```
 1111111       0011111   carryover
 11011001      10011101
+1101111      +1000011
101001000      11100000  sum
```

Section 3.10 Practice Problems

1. Add the following binary numbers:
   a. 100110B + 110111B b. 11110111B + 10011111B
   c. 101010B + 11110B d. 110000111B + 101110111B
2. Verify that each sum in Problem 1 is correct.

3.11 Binary Subtraction

What is different when considering subtraction is the concept of borrowing. In decimal numbers, when a borrow is required to complete a subtraction, 1 is deducted from one column and 10 is added to another. Since they are neighboring columns, the value of the original number is maintained. Take the number 345. Recall that it can be written, using weights, as $3 \times 100 + 4 \times 10 + 5 \times 1$. When a borrow is done for subtraction, for example, from the 4 to the 5, the 4 is reduced by 1 to 3 and 5 is increased to 15. Adding the new digit–weight products, $3 \times 100 + 3 \times 10 + 15 \times 1$, still adds up to 345. Of course, there are now two digits in a single position, but this is only temporary. Once the subtraction is performed, the results will be returned to a single digit. Otherwise a borrow would not have been necessary in the first place. Let's observe how the subtraction process applies, by subtracting 3,627 from 8,331.

We begin by trying to subtract 7 from 1, but we run into a problem immediately. This subtraction requires a borrow from the next digit whose current value is 3, which causes it to be reduced to 2. Borrowing the 1 from the 3 allows us to add 10 to the least significant digit value of 1, resulting in a new value of 11. Now 7 can be subtracted from 11 to yield 4:

```
      2
 8 3 3̸ 11
-3 6 2  7
        4
```

Next, 2 − 2 = 0. No trouble there, but the next subtraction, 6 from 3, requires a borrow from 8 to increase the 3 to 13 so that the subtraction can be completed:

```
 7   2
 8 13 3 11
-3  6 2  7
 4  7 0  4
```

The process for doing binary subtraction is the same, except the quantity that is borrowed is 2 instead of 10. Notice, however, that they are the same symbols: 10 (1 and 0) and represent the base or radix number of the system in use, 2 for binary and 10 for decimal. Before we get into the binary subtraction, though, an additional aspect must be understood—borrowing from zero.

What happens when we subtract 3,265 from 20,003? We start out as before and try to subtract 5 from 3; since 5 is greater than 3, we require a borrow, but the next digit is 0. You cannot borrow from nothing, so go to the next digit—0 again, and once more as well. Instead of borrowing from 0, we are going to borrow from 2,000. Borrowing 1 from 2,000 leaves 1,999. Adding the borrowed value to the 3 makes 13 and maintains the original decimal value, and thus the subtraction can be done:

```
 1 9 9 9
 2 0 0 0 13
-  3 2 6  5
 1 6 7 3  8
```

A 0 occurs in binary numbers more frequently than in decimal. After all, a 0 is one-half the symbols in the binary system, but only one-tenth the symbols in the decimal system. As with the addition of binary numbers, there are a limited number of possible differences when subtracting one binary bit from another:

```
  0    1    1
 -0   -0   -1
  0    1    0
```

```
and  0
    -1
```

which after borrowing yields a difference of

```
 10
- 1
  1
```

Of course, if you also include borrowing as a part of the subtraction, then the possible number of results changes. When borrow equals 0, the subtraction results are the same as before. Subtracting 0 (the borrow value) does not alter the differences. However, when a borrow is required, giving it a value of 1, the differences are affected (here the borrow is the third binary value in each example):

```
  0            requires a borrow:  10
 -0                                -0
 -1 borrow bit                     -1
                                    1

  0            requires a borrow:  10
 -1                                -1
 -1 borrow bit                     -1
                                    0

  1            no additional borrow required
 -0
 -1 borrow bit
  0

  1            requires a borrow:  11
 -1                                -1
 -1 borrow bit                     -1
                                    1
```

When borrowing from 0, the same problem arises when attempting to borrow from 0 in decimal—you cannot borrow from nothing. So once again you must borrow from a whole number. For example, when a 1 is borrowed from 10000B, the result is 1111B, which is one less than 10000B.

To see the process at work, let's subtract 101010B from 1110010B. Set up the subtraction problem, and do the first subtractions, which do not require a borrow:

```
  1110010
-  101010
      000
```

But then we run into our first and only borrow for this example. The borrow reduces 1 to 0 and adds 10 to the previous column. Now the remaining subtractions can be done:

```
     0
 1 1 1̸ 10 0 1 0
-  1 0  1 0 1 0
 1 0 0  1 0 0 0
```

A second example problem illustrates the difficulty when trying to borrow from 0. Subtract 1000111B from 11010000B. Immediately, we need to borrow, but the column we need to borrow from contains a 0, as do the next two columns. Instead of trying to borrow from 0, we will borrow from 1000B:

```
       0 1 1 1
 1 1 0 1̸ 0̸ 0̸ 0̸ 10
-1 0 0 0 1 1   1
 1 0 0 0 1 0 0  1
```

Borrowing from 1000B reduces it to 0111B and adds 10B to 0B, and the remaining subtractions can be performed without any further difficulty.

Another aspect should be discussed before moving on, namely a “double” borrow. What this actually means is that borrowing from one bit created the need to borrow when moving on to that bit’s position. By example, subtract 111B from 1110B. The first step requires a borrow:

```
     0
 1 1 1 10
-1 1   1
       1
```

Moving onto the next bit, borrowing for the first subtraction reduced the digit value in that column to 0 and caused the need to borrow again to do the second subtraction. Nothing different is required; do the borrow as before, by borrowing from the next column:

```
   0 10
 1 1  1 10
-1  1  1
    1  1
```

And once more:

```
0 10 10
1  1  1 10
-  1  1  1
   1  1  1
```

(The 1s beneath the borrowed values are printed struck through.)

Section 3.11

Practice Problems

1. Subtract the following pairs of numbers:
   a. 1100011B–1001110B b. 10011101B–1111110B
   c. 1111000B–1010101B d. 10000000B–100011B
2. Prove each of the subtractions is correct by converting each problem to decimal.

3.12 Half Adders

There are four possible results from adding two binary bits together, summarized in the truth table of Figure 3.1. This truth table uses A and B to designate the two binary bits that are added together to produce SUM and CARRY results. If the two bits both have a value of zero, then their sum is zero and no carry is produced. This condition is shown as the first line of the truth table in Figure 3.1. The remaining entries to the table show the results of adding bit A to bit B. If one of the bits is a one and the other a zero, then the SUM is 1

A	B	SUM	CARRY
0	0	0	0
0	1	1	0
1	0	1	0
1	1	0	1

Figure 3.1 Half Adder Truth Table

(0 + 1 = 1) and there is no CARRY. When both bits are high, the resulting SUM is 2 (1 + 1 = 10) which is shown as a SUM value of 0 and a CARRY of 1.

Half adder *an adder circuit that adds two binary bits producing a sum and carry out.*

From the truth table, we can derive the logic expressions for SUM and CARRY. Notice that the results for the SUM are a truth table for an exclusive OR and the results for CARRY are a truth table of an AND gate. This makes the logic expression for SUM $= A \oplus B$ and for CARRY $= A \cdot B$. Figure 3.2a is a logic diagram for the SUM and CARRY; Figure 3.2b shows a block diagram of the same circuit labeled as a **half adder.** This circuit performs the addition of binary bits A and B, but it does not consider what would happen if a carry from a preceding column were required in adding these two bits. That makes the addition process incomplete, because it cannot accommodate the carry inputs from preceding columns. Because of this, the circuit is called a *half adder,* since

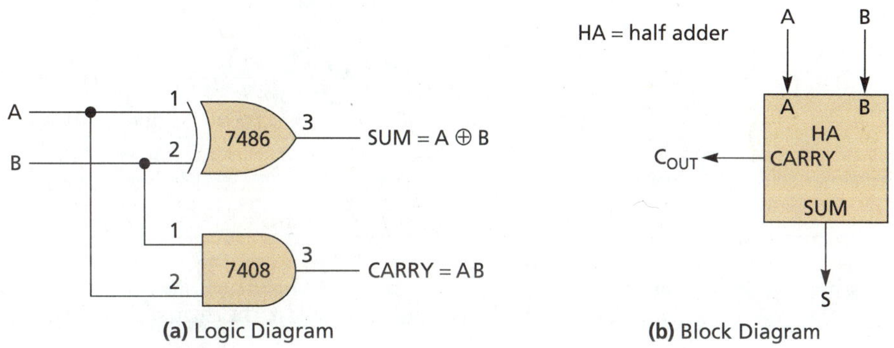

Figure 3.2 Half Adder

Section 3.12 Practice Problems

1. Derive the logic circuit for a half subtracter. Use the subtraction results in Section 3.11 to derive the truth table for the half subtracter.
2. Why is the circuit of Problem 1 only a "half" subtracter?

Full adder *an adder circuit that adds two binary bits and a carry-in bit and produces a sum and carry out.*

it only accomplishes half of the addition process. What is needed is a circuit that will do the complete addition of two binary bits and a carry bit. Such a circuit is called a **full adder.**

3.13 Full Adders

Figure 3.3 is the truth table for a full adder. On this table, inputs A and B are the same binary input bits as they were for the half adder. The third input, C, represents the carry bit generated from the binary addition of the preceding column. Notice that when C is low, indicating that there was no carry, the truth table is identical to that of the half adder. However, when C is high, the truth table results change. The entire table is developed by listing the results of adding three binary bits (A, B, and C) together and is used to derive the logic expressions for the SUM and CARRY outputs.

C_{IN}	B	A	SUM	C_{OUT}
0	0	0	0	0
0	0	1	1	0
0	1	0	1	0
0	1	1	0	1
1	0	0	1	0
1	0	1	0	1
1	1	0	0	1
1	1	1	1	1

Figure 3.3 Full Adder Truth Table

Start by "reading" the sum-of-products expression for SUM from the table:

$$\text{SUM} = A \cdot \overline{B} \cdot \overline{C} + \overline{A}B\overline{C} + \overline{A} \cdot \overline{B} \cdot C + ABC$$

Now apply Boolean algebra methods to reduce the expression to its fundamental form. Factor $\overline{C}$ and C from the SOP terms

$$\text{SUM} = \overline{C}(A\overline{B} + \overline{A}B) + C(AB + \overline{A} \cdot \overline{B})$$

Recognize that $A\overline{B} + \overline{A}B$ is the logic expression for the exclusive OR, $A \oplus B$. To recognize $AB + \overline{A} \cdot \overline{B}$ as that of the complemented exclusive OR (or exclusive NOR), $\overline{A \oplus B}$, complement the exclusive OR expression and reduce the result:

$$\begin{aligned}\text{Exclusive NOR} &= \overline{A \cdot \overline{B} + \overline{A} \cdot B} = \overline{A \cdot \overline{B}} \cdot \overline{\overline{A} \cdot B} = (\overline{A} + B) \cdot (A + \overline{B}) \\ &= \overline{A} \cdot A + \overline{A} \cdot \overline{B} + A \cdot B + \overline{B} \cdot B \\ &= 0 + \overline{A} \cdot \overline{B} + A \cdot B + 0 = \overline{A} \cdot \overline{B} + A \cdot B\end{aligned}$$

Substitute the exclusive OR forms into the original expression, and then let the letter *Z* represent $A \oplus B$ to help see how the expression further clarifies itself:

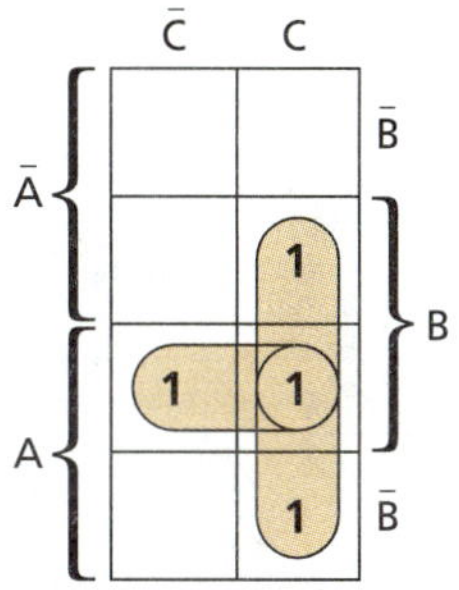

CARRY = AB + BC + AC

Figure 3.4 Carry Karnaugh Map

$$\text{SUM} = \overline{\text{C}}(\text{A} \oplus \text{B}) + \text{C}(\overline{\text{A} \oplus \text{B}})$$

$$\text{SUM} = \overline{\text{C}}\text{Z} + \text{C}\overline{\text{Z}}, \text{ which is the expression for } \text{C} \oplus \text{Z}$$

Substitute $\text{A} \oplus \text{B}$ back for Z, and end up with the final expression for SUM = $\text{A} \oplus \text{B} \oplus \text{C}$.

Now for CARRY. The Karnaugh map of Figure 3.4 is used to help simplify its expression:

Reading the reduced expression from the map yields the logic expression for the carry result:

$$\text{CARRY} = \text{AB} + \text{BC} + \text{AC}$$

Figure 3.5a is the logic circuit for the SUM and CARRY of the full adder, and Figure 3.5b is its equivalent block diagram. Although this circuit will work fine, another logic circuit accomplishes the same result with fewer gates. This circuit is shown in Figure 3.6.

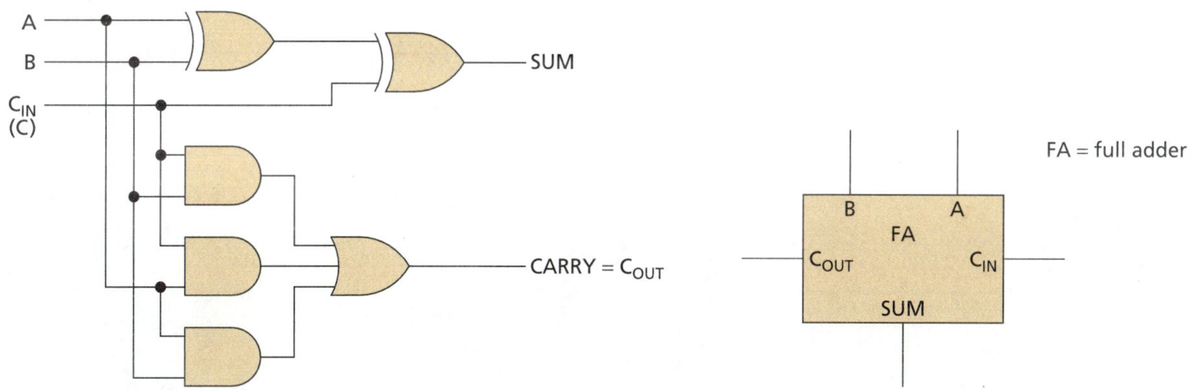

(a) Schematic

(b) Block Diagram

Figure 3.5 Full Adder

Notice that the circuit used for the SUM in Figure 3.6b is the same as it was in Figure 3.5a. What is different is the circuit for CARRY output. Before it is proven that this is a viable alternative for the CARRY output of the full adder, observe what this entire circuit actually is. Compare the circuit in Figure 3.6b to the half adder in Figure 3.2a. What is actually being done is that two half adders are combined to form a full adder. The only addition to the two half adders is the OR gate, which combines the two half-adder carry output bits into the full adder carry output. This is the output that needs to be proven.

The output of the first half adder carry is AB. The output of the second half adder carry is C ANDed with the sum output of the first half adder or C $(\text{A} \oplus \text{B})$, which is $\text{C}(\overline{\text{A}}\text{B} + \text{A}\overline{\text{B}})$. Distributing C yields the SOP expression $\overline{\text{A}}\text{BC} + \text{A}\overline{\text{B}}\text{C}$. The two half adder carry outputs are ORed together to form the logic expression for CARRY:

$$\text{CARRY} = \text{AB} + \overline{\text{A}}\text{BC} + \text{A}\overline{\text{B}}\text{C}$$

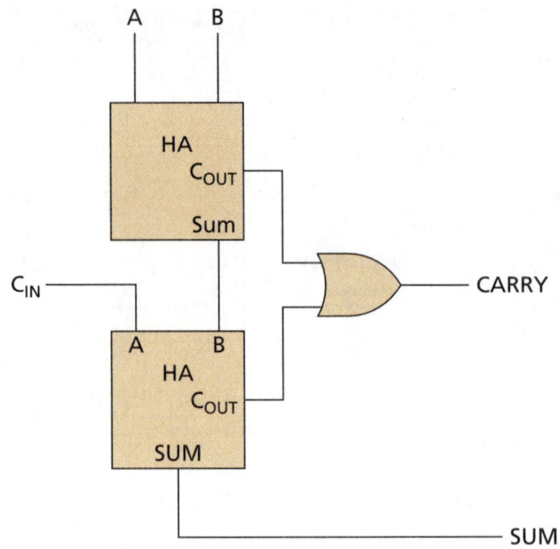

(a) Block Diagram

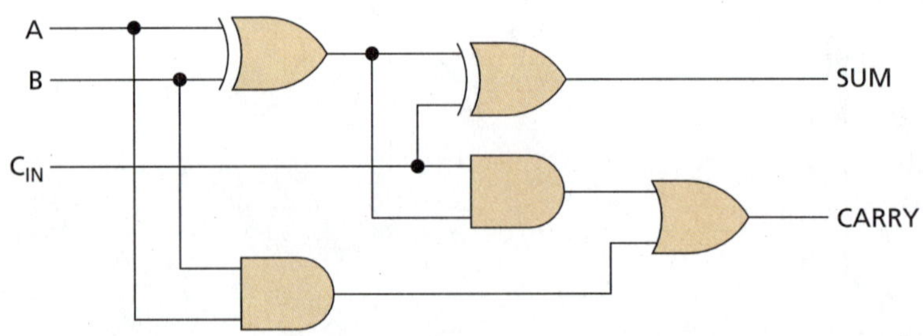

(b) Schematic

Figure 3.6 Full Adder Using Half Adders

Now for a little Boolean algebra magic. Factor variable B from the first two terms and then apply the Boolean axiom A + AC = A + C:

$$\text{CARRY} = B(A + \overline{A}C) + A\overline{B}C = B(A + C) + A\overline{B}C$$

Redistribute the B variable, and repeat the process once again:

$$\text{CARRY} = AB + BC + A\overline{B}C = BC + A(B + \overline{B}C)$$

$$\text{CARRY} = BC + A(B + C) = BC + AC + AB$$

which is the same expression developed for the CARRY in Figure 3.5, thus proving the two circuits produce the same result.

Section 3.13

Practice Problem

Derive the logic circuit for a full subtracter that subtracts binary bit B from binary bit A. The circuit becomes a full subtracter when you consider the effect of borrowing from the preceding binary bit column.

3.14 Four-Bit Parallel Adders

A four-bit parallel adder is constructed using four full adders connected in cascade, as shown in Figure 3.7. This circuit adds the four binary bits applied to the "A" inputs to the four binary bits supplied to the "B" inputs. A single carry in (C_{IN}) input is available to the first full adder, and a final carry out (C_{OUT}) is generated by the last full adder. Each full adder operates as described in Section 3.13. Inputs A, B, and C to each full adder are summed supplying a sum (S) and a carry (C_O) output, while each succeeding bit's adder receives the carry out from the preceding adder so that carry values are properly included in the addition process.

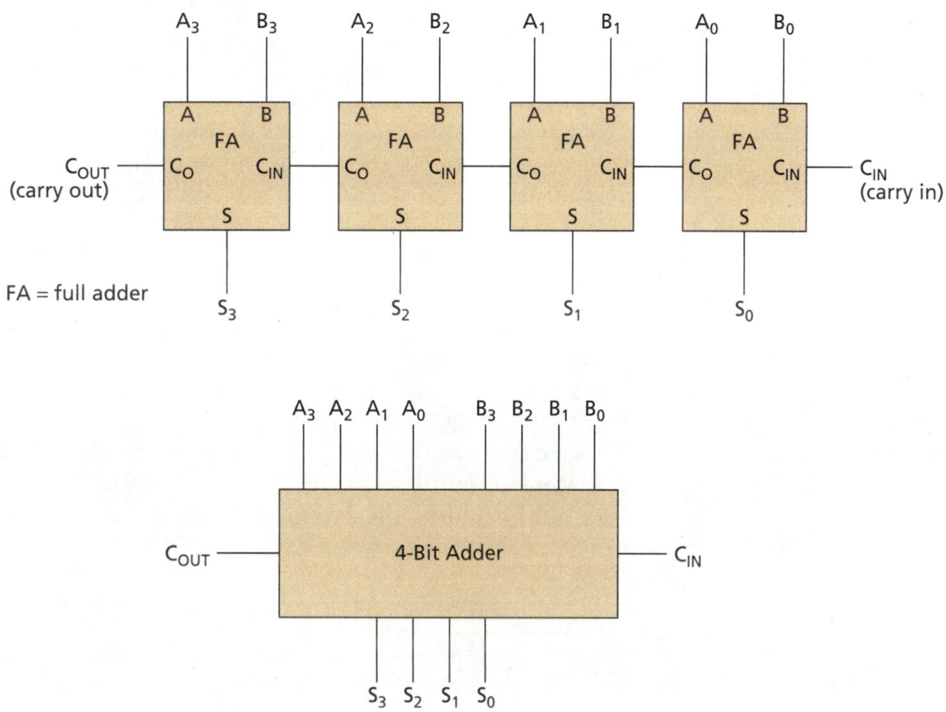

Figure 3.7 Four-Bit Parallel Adder

We will see how the four-bit adder successfully sums four-bit binary numbers by adding 1001B at input A to 1011B at input B. That is, A_3, A_2, A_1, A_0 = 1001B and B_3, B_2, B_1, B_0 = 1011B. Input C_{IN} is 0, since A_0 and B_0 are the least significant bits (LSBs) of both numbers and there would be no carry in to the LSB. The first adder that sums these LSBs produces a sum of 0 ($S_0 = 0$) and a carry of 1. The carry from this addition is added to the values in the next column, A_1 (0) and B_1 (1), using the next adder, which yields a S_1 sum of 0 and a carry of 1 to the next full adder. This carry is added to A_2 (0) and B_2 (0) in the third full adder, resulting in a S_2 sum of 1 and a carry of 0. This carry value is summed with the last two binary bits, A_3 and B_3, in the final adder to produce the last S_3 sum of 0 and a final carry out, C_{OUT}, of 1. This makes the sum word, S_3, S_2, S_1, S_0 = 0100B with a C_{OUT} = 1 or 1001B + 1011B = 10100B.

Parallel adder *a circuit that sums several binary bits simultaneously.*

This circuit is called a **parallel adder,** because the binary values to be summed are presented, to the A and B word inputs, at the same time and are added in parallel. Their sum is produced rapidly—only delayed by the time it takes for the logic states to propagate through the numerous AND and OR gates in the circuit and the output to react to these inputs and generate the SUM word and C_{OUT} results.

The four-bit parallel adder circuits are manufactured as medium-scale integrated (MSI) circuit devices such as the 74LS83 pictured in Figure 3.8. This diagram is a functional block diagram for the circuit Figure 3.7. You will note that data sheets do not always follow logical conventions in their diagrams. You must examine the drawings carefully and follow the truth tables supplied with the data sheets to follow the way the device is designed to work. For the 74LS83, inputs for binary number word A are grouped together, as are the values for input word B and the sum word S outputs. All three words are shown with their LSB on the left side of each four-bit group. The LSB are labeled with a subscript of 1. This paragraph continues the discussion with the designations as shown on the data sheet, but subsequent discussions revert to using a subscript of 0 for the LSB of all binary words and numbers. The carry in, C_{IN}, is shown on the left edge of the diagram and the carry out, C_{OUT}, on the right. This arrangement places the LSB of the input and output words at the left side of the diagrams, which is the opposite of where they would be as an actual binary number. The carry lines are added so that they reflect their relationship to the LSB and MSB bits of all three words. Interpreting the operation of this chip is straightforward: Number A + number B + C_{IN} = SUM and C_{OUT}.

For example, the sum and carry out results of a 74LS83 four-bit parallel adder, when number A is 1001B and number B is 1011B, with a carry in of 0, is found by adding these two numbers as you would on paper:

	0	C_{IN}
	1001	A
	+ 1011	B
C_{OUT}	1 0100	SUM

Since A_0 and B_0 are the least significant bits of the binary number words, why does the four-bit adder require a carry in (C_{IN}) and a full adder to sum these bits? Can C_{IN} be omitted and a half adder used to sum these bits? The answer to these two questions depends on the size of the binary numbers that are to be added together. If that size is always four bits and the four-bit adder is going to be used, then yes, C_{IN} can be discarded and the full adder replaced with a half adder. However, if it is desirable to add binary numbers greater than four bits, there needs to be a way to have the carry out from one 74LS83 supply a carry in to the next 74LS83. Figure 3.9 shows how two four-bit parallel adders are used to sum two eight-bit number words. The carry out from the lower four-bit adder is connected to the carry in of the upper adder so that the carry from the addition of A_3 and B_3 is added to the sum of A_4 and B_4.

FEATURES

- **High speed 4-bit binary addition**
- **Cascadeable in 4-bit increments**
- **LS83A has fast internal carry lookahead**
- **See "283" for corner power pin version**

FUNCTIONAL DESCRIPTION

The "83" adds two 4-bit binary words (A_n plus B_n) plus the incoming carry. The binary sum appears on the Sum outputs (Σ_1-Σ_4) and the outgoing carry (C_{OUT}) according to the equation:

$$C_{IN}+(A_1+B_1)+2(A_2+B_2)+4(A_3+B_3)+8(A_4+B_4)=\Sigma_1+2\Sigma_2+4\Sigma_3+8\Sigma_4+16C_{OUT}$$

Where (+) = plus

Due to the symmetry of the binary add function the "83" can be used with either all active HIGH operands (positive logic) or with all active LOW operands (negative logic)–see Table "A". With active HIGH inputs, C_{IN} cannot be left open; it must be held LOW when no "carry in" is intended. Interchanging inputs of equal weight does not affect the operation, thus C_{IN}, A_1, B_1, can arbitrarily be assigned to pins 10, 11, 13, etc.

LOGIC SYMBOL

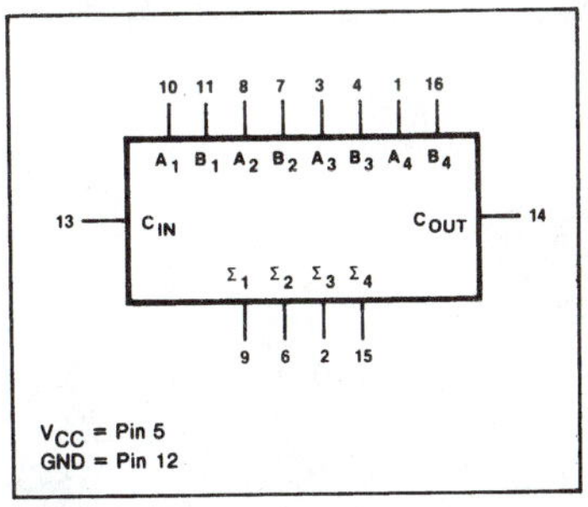

PIN CONFIGURATION

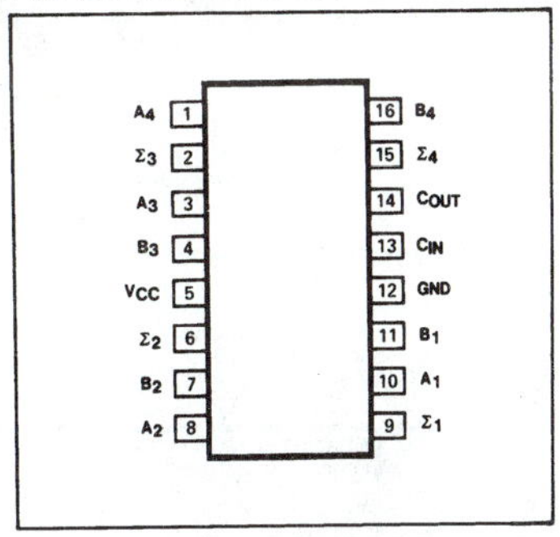

LOGIC DIAGRAM

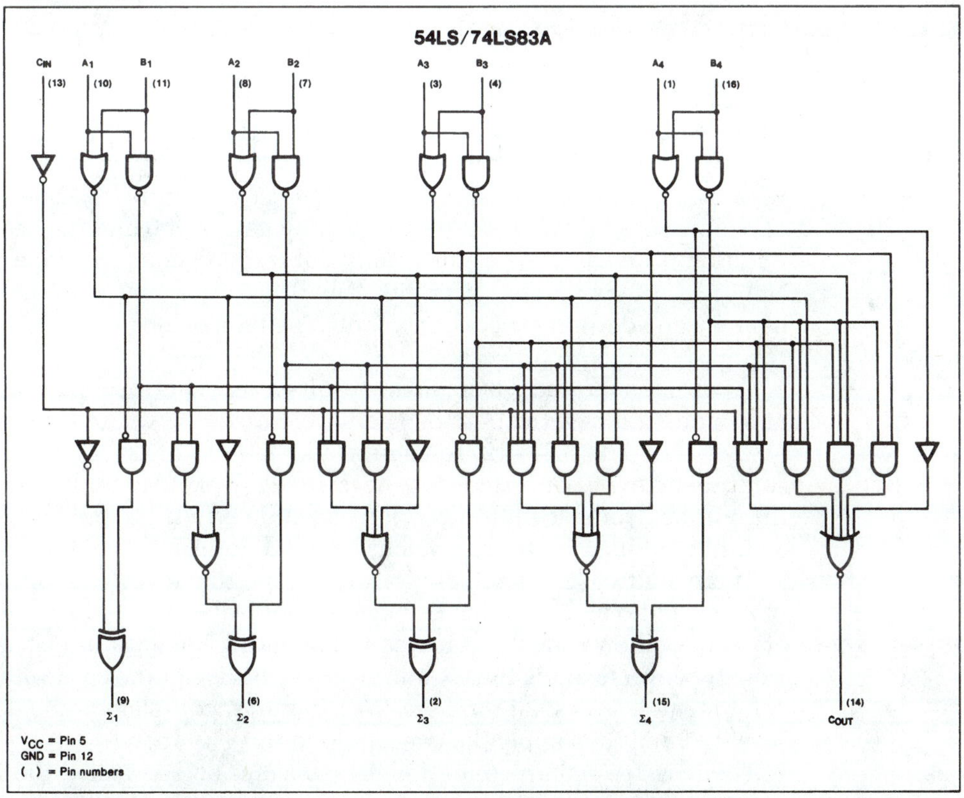

PINS	C_{IN}	A_1	A_2	A_3	A_4	B_1	B_2	B_3	B_4	Σ_1	Σ_2	Σ_3	Σ_4	C_{OUT}	
Logic Levels	L	L	H	L	H	H	L	L	H	H	H	L	L	H	
Active HIGH	0	0	1	0	1	1	0	0	1	1	1	0	0	1	(10+9=19)
Active LOW	1	1	0	1	0	0	1	1	0	0	0	1	1	0	(carry+5+6=12)

Figure 3.8 Four-Bit Adder (courtesy of Signetics Corp.)

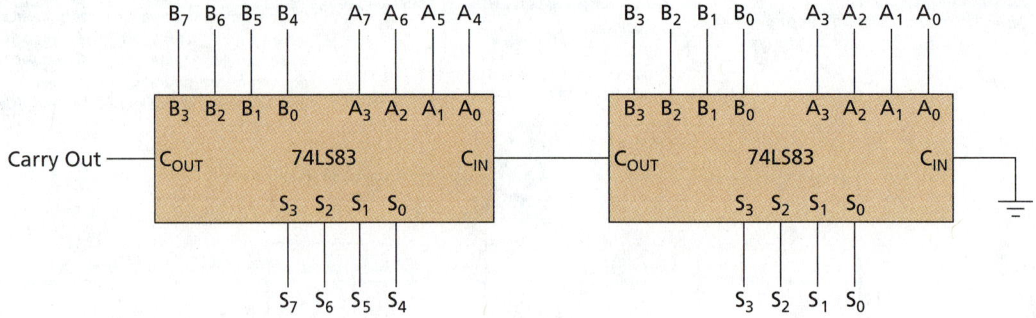

Figure **3.9** Eight-Bit Parallel Adder

Section 3.14

Practice Problem

Show how the cascaded 74LS83 adders of Figure 3.9 successfully add the two binary numbers 10011101B and 11100101B.

3.15 Subtraction by Complement Addition

Subtraction can be a nuisance—keeping track of borrows, especially with binary numbers, where there are so many instances of 0 and the need to borrow. Addition is an easier process. What if subtraction could be done by addition? Not only is the process easier, but if both addition and subtraction could be done by an addition process, then only one type of electronic circuit would be required to do both processes.

When 3 is subtracted from 5, the results are 2. This result can also be arrived at by adding a negative 3 to 5. The problem can be written one of two ways: $5 - 3 = 2$ or $5 + (-3) = 2$. To do subtraction by addition as shown requires the subtrahend to be converted to its negative equivalent so that it can be added to the number from which it was originally subtracted. The process of converting a number to its negative equivalent is called **complementing**.

Complementing *the process by which numbers are inverted.*

Complement *the inverted form or negative equivalent of a number.*

There are two forms of **complements**, one that gives you the "inverse" of a number and one that yields its negative equivalent. The word "inverse" is set off by quotation marks because complementing binary numbers does actually yield the inverse of the binary number, but this is not the case with other number systems, as we shall soon see. We'll begin by complementing decimal numbers first and then apply the concept to binary numbers.

Nine's complement *the first step in negating a decimal number.*

The "inverse" complement for decimal numbers is called the **nine's complement** and is found by subtracting each digit of the decimal number from 9. Note that this subtraction is straightforward and does not require any borrowing. To complement the number 473, subtract each digit from 9:

$$\begin{array}{r} 999 \\ -473 \\ \hline 526 \end{array}$$

Ten's complement *the negative equivalent of a decimal number.*

The nine's complement is the first step in converting a decimal value to its negative equivalent. The next step is to add 1 to the nine's complement. This produces the **ten's complement**, which is the negative value of the original number:

$$\begin{array}{r} 526 \\ \underline{+\quad 1} \\ 527 \end{array}$$

This means that 527 is the same as −473 when changing a subtraction problem to an addition problem. As an example, let's say we want to subtract 473 from 983. First we do the subtraction normally, to determine the difference for comparison purposes:

$$\begin{array}{r} 982 \\ \underline{-473} \\ 509 \end{array}$$

Now let's do the same problem using complement addition. This is accomplished by adding the negative equivalent or ten's complement number instead of subtracting the original value. In our case, add the ten's complement of 473, which is 527, to 982, effectively adding the negative value of 473:

$$\begin{array}{r} 982 \\ \underline{+527} \\ 1{,}509 \end{array}$$

The results are interpreted as 509 with the carryover of 1. Well, 1,509 is not 509, but the sum part of the results is correct—it is 509, which means we must interpret what the carry results are telling us. Since the sum results are positive and the correct value, then the carryover is *not* included as part of result, but instead indicates the nature of the results, in this case positive, and the actual difference value of 509. To further illustrate the significance of the carryover, let's subtract 473 from 200, which yields negative results:

$$\begin{array}{r} 200 \\ \underline{-473} \end{array}$$

One method to do this subtraction is to reverse the order of the subtraction and place a minus sign in front of the result:

$$\begin{array}{r} 473 \\ \underline{-200} \\ -273 \end{array}$$

What we are really saying here is that the subtrahend was larger than the minuend and to complete the subtraction requires borrowing from somewhere. Since there is no "somewhere" at the moment to borrow from, the need to do the borrowing is indicated by a negative difference. To show that this is true, do the subtraction normally:

$$\begin{array}{rrrr} 1 & 1 & 9 & \\ & 2 & 0 & 10 \\ \hline - & 4 & 7 & 3 \\ & 7 & 2 & 7 \end{array}$$

The problem is, where did the 10 come from that was borrowed to make the most significant digit 11? We had to borrow it from somewhere. Since there is no *actual* "somewhere," we would simply indicate that a borrow was required to complete the subtraction. The number 727 is certainly not the correct answer, since 473 is bigger than 200—meaning we expected a negative answer! We did get one, since, as we shall see shortly, 727 is the negative equivalent of 273. Take the ten's complement of 727:

$$\begin{array}{r} 999 \\ \underline{-727} \\ 272 \\ \underline{+\quad 1} \\ 273 \end{array}$$

and we see that 727 is actually −273. Now, we will do the same subtraction using ten's complement addition from the onset:

$$\begin{array}{rrl} 200 & 200 & \\ \underline{-473} & \underline{+527} & \text{ten's complement of 473} \\ & 727 & \text{ten's complement of 273} \end{array}$$

and we already know that 727 is equivalent to −273. The same difference result is produced by subtraction and ten's complement addition. Notice that in this second example there is no carryover from the addition operation. A lack of a carryover when doing subtraction by ten's complement addition signifies that the answer required a borrow and is in its negative form, requiring complementing to find its actual value.

EXAMPLE 3.13

Find the difference of 8,331 − 3,627 and 1,538 − 4,557 using ten's complement.

SOLUTION The first part of the task is to develop the ten's complement of 3,627 and 4,557 by subtracting each decimal digit from 9 and adding 1 to the result:

$$\begin{array}{rrl} 9{,}999 & 9{,}999 & \text{nines} \\ \underline{-3{,}627} & \underline{-4{,}557} & \text{numbers to be complemented} \\ 6{,}372 & 5{,}442 & \text{nine's complement} \\ \underline{+\qquad 1} & \underline{+\qquad 1} & \text{add 1} \\ 6{,}373 & 5{,}443 & \text{ten's complement (negative)} \end{array}$$

The next step is to add the complements to 8,331 and 1,538 and then interpret the results:

$$\begin{array}{rrl} 8{,}331 & 1{,}538 & \\ \underline{+6{,}373} & \underline{+5{,}443} & \\ 14{,}704 & 6{,}981 & \text{difference} \end{array}$$

For the first result, 14,704, the last carry is omitted, leaving the positive true difference of 4,704. In the second example, lack of a carryover indicates that the results are negative and must be complemented to yield the actual answer:

$$\begin{array}{rl} 9{,}999 & \text{nines} \\ -\underline{6{,}981} & \text{original number} \\ 3{,}018 & \text{nine's complement} \\ +\underline{1} & \text{add 1} \\ 3{,}019 & \text{ten's complement} \end{array}$$

Thus, 1,538 − 4,557 = −3,019.

The following list summarizes, in generic form, the steps to use the complement method for subtraction:

1. To get the negative complement form of a number, first subtract it from the highest value of the number base system in use to get the "inverse" of the number. Then add 1 to the result to get its negative equivalent.
2. Add the complement value to the other number to get the difference of the two original numbers.
3. If the result of the addition produces a carryover, the result is the correct positive difference.
4. If the result of the addition does *not* produce a carryover, the difference is negative and must be complemented to show the true value.

For the complement system to work properly, both numbers in the subtraction problem must contain the same number of digits. This is needed to maintain the negative quality of the complemented number. For instance, compare the results of trying to do subtraction by ten's complement for the problem 250 − 36, first without making the numbers equal in length and then lengthening the number 36 to three digits by adding a leading zero to it, which does not change the value of the number. In this case, 036 is equivalent to 36.

The complement of 36 as shown is 99 − 36 + 1 = 63 + 1 = 64. Adding that to the number 250 yields 315 without a carry. That means that the result is negative and requires complementing to get its actual value. It can be seen by the size of the two numbers that the result should have been positive, so this first result must be incorrect. Next try it again with the extended number 036. Its complement is 999 − 036 + 1 = 963 + 1 = 964. Now when it is added to 250, the results are 214 with a carryover. The results are positive and correct. Add a fifth step to the process of subtraction by complement addition:

5. Add leading 0's to the number to be complemented, the subtrahend, to assure that it has the same length as the minuend.

Section 3.15

Practice Problems

1. Do the following subtractions using ten's complement addition:
   a. 1,234 − 567 b. 520,034 − 187,099 c. 380,207 − 728,965
2. Show each result is correct by using normal subtraction.

3.16 Binary Subtraction Using Two's Complement Addition

Inverting the process of altering the logic state of a bit or binary quantity.

Invert the opposite value of a binary state.

Now that subtraction by ten's complement addition for decimal numbers has been shown, binary subtraction by two's complement addition can be demonstrated, using the same process. To perform the two's complement process, subtract each binary bit of the original binary number from 1, since 1 is the maximum value in the binary number system. This is called the one's complement and is the same thing as **inverting** (getting the opposite of) each binary digit. This is why this complement process is called an "inversion" process even though similar complements in other number systems (such as decimal) do not actually **invert** the original number. Add a 1 to the one's complement result to get the two's complement, which is the negative equivalent of the original number. To demonstrate the process, we will complement the number 10110011B.

Start by subtracting each digit of the number from 1:

11111111B	one's
−10110011B	original number
01001100B	one's complement

Note that the resulting binary number has bits that one's are the inverse of each of the original number's bit values (0's replaced 1's and 1's replaced 0's), which means that we could have taken a short-cut to this step. Instead of doing the subtraction, all we had to do was invert each bit of the original number to get the one's complement. To arrive at the two's complement or negative equivalent of the original number, add 1 to the one's complement:

01001100B	one's complement (inverse)
+ 1B	adding 1
−01001101B	two's complement (negative)

EXAMPLE 3.15

Find the one's and two's complement of 11100001000B.

SOLUTION To get the one's complement, invert each binary bit of the original number—1's to 0's and 0's to 1's:

One's complement of 11100001000B is 00011110111B

For the two's complement, simply add 1 to the one's complement:

00011110111B	one's complement
+ 1B	adding 1
−00011111000B	two's complement

When doing subtraction using two's complement addition, the process and interpretation of the results are similar to the ten's complement process for decimal numbers. To show this, we will first subtract 100110B from 11000100B and then subtract 11001111B from 10000101B using the two's complement addition method. After doing the subtraction, we will convert all numbers to decimal and prove the results are correct.

For any complement system to work, both numbers in the problem must have the same number of bits. Add leading 0's to the shorter number: 00100110B. Now it has the same number of bits, but its value is unchanged. To do the subtraction, first take the two's complement of the subtrahend:

Two's complement of 00100110B = 11011001B + 1 = 11011010B

Then add it to the second number:

11000100B	original minuend
+11011010B	two's complement subtrahend
110011110B	difference of numbers

Since there is a carryover from the addition, the difference is a positive one, so drop the carryover and you are left with the actual positive difference: 10011110B. To prove the results are correct, convert the two original numbers and difference to decimal, and verify the subtraction:

11000100B	=	196
− 100110B	=	− 38
10011110B	=	158

The solution to the second set of numbers begins by complementing the subtrahend as before. Invert the subtrahend for one's complement: 00110000B and add 1 for the two's complement negative equivalent of the subtrahend: 00110001B. Next perform the subtraction by adding the negative subtrahend to the original minuend:

10000101B	original minuend
+00110001B	two's complement subtrahend
10110110B	difference of numbers

Since the result lacks a carryover, a borrow was required, and the difference is negative and must be complemented to get the actual value:

10110110B original difference
−(01001001B + 1) = −01001010B actual negative value

The proof is to convert the original binary values and the final answer to decimal and see if the answer is correct:

10000101B	=	133
−11001111B	=	−207
− 1001010B	=	− 74

Section 3.16

Practice Problems

1. Do the following subtraction using two's complement addition. Give each answer in true value form:
   a. 110011101B − 10011101B b. 100001110B − 111000101B
   c. 1110111B − 100111B d. 101010010B − 110011001B
2. Prove each answer is correct by converting each problem to decimal numbers.

3.17 Two's Complement Adder/Subtracter Circuit

Circuits for half and full subtraction can be developed and created in the same manner that the half and full adders were done. However, when it comes to designing the arithmetic circuits for digital computers a decision is made to incorporate addition and subtraction using the same basic circuit. The benefit in doing this is to reduce the number of different circuits in the overall system. The down side of incorporating dual functions into a single circuit results in the need for some method of controlling which function is to be performed. A four-bit parallel adder is going to be used to accomplish subtraction, but it requires one of the four-bit number inputs to be complemented when subtraction is to be performed. However, do not forget that this same circuit is also required to do addition, at which time neither four-bit number is to be complemented.

The first step is to state the requirements for our circuit:

1. A control signal is required to determine when to add and when to subtract—call this control signal SUB, which is defined as causing the circuit to perform subtraction when the signal is asserted (SUB = 1) and addition when it is negated (SUB = 0).
2. To do the two's complement process, number B is first inverted and a 1 is added to it when SUB is asserted.
3. For addition, do *not* invert number B or add 1 when SUB is negated.
4. Add the two four-bit numbers—one number A and the other the complemented (SUB = 1) or noncomplemented (SUB = 0) number B.
5. Set CARRY high when adding and C_{OUT} is high, which is the carryover indication of addition.
6. Set BORROW high when subtracting and C_{OUT} is low, which indicates a borrow condition in two's complement addition for subtraction.
7. Combine the CARRY and BORROW indications into a single CARRY/BORROW indication.

If the circuit satisfies all these criteria, then it will be a circuit that adds and subtracts two four-bit binary numbers. The next step, in the design, is to resolve some of the requirements that have been listed. Start with the inverting problem. What is needed is a circuit that will invert when SUB = 1 (subtracting) and not invert when SUB = 0 (adding). As in any digital logic design, translate this requirement to a truth table as shown in Figure 3.10.

SUB	B_{input}	B_{word}
0	0	0
0	1	1
1	0	1
1	1	0

Figure 3.10 One's Complement Truth Table

This table looks very familiar—it is an exclusive OR. This turned out to be a comparatively easy solution to one of the requirements. Another aspect of the circuit requires the two numbers to be added together. That much will be accomplished by a 74LS83 four-bit adder. Number A is applied directly to the A inputs of the 74LS83. The outputs of the exclusive OR gates connected to the B number bits are connected to the B inputs of the 74LS83. This gives us the sum of numbers A and B when SUB is low and the addition of number A and the inverted value of number B when B is high. There is still the problem of adding 1 to complete the two's complement of number B. The SUB control line is connected to the carry in (C_{IN}) pin of the 74LS83. This adds a 1 to the inverted B number when SUB is high and adds 0 to the sum of A and B when SUB is low. To this point, we have the circuit of Figure 3.11.

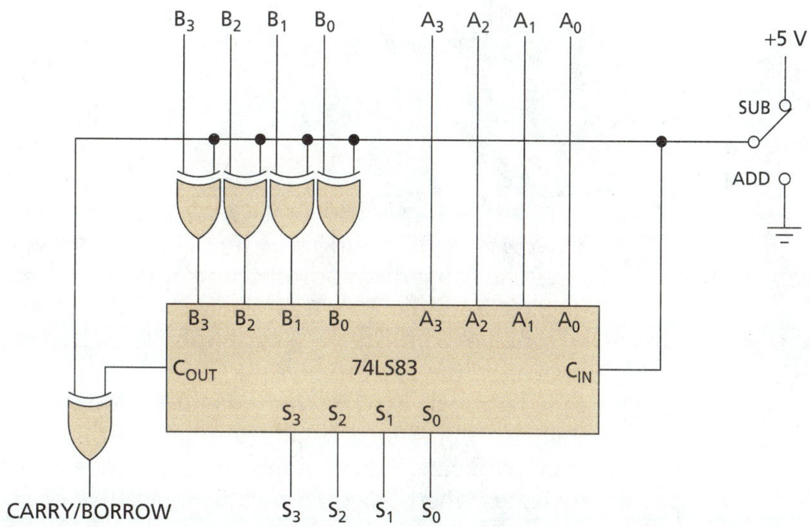

Figure 3.11 Two's Complement Adder/Subtracter

The last requirement for the CARRY/BORROW indication is also shown in Figure 3.11. Notice that an exclusive OR once more provides the solution to generating the CARRY/BORROW indicator. When SUB is a 0 and C_{OUT} is a 1, the CARRY/BORROW is a 1, indicating that a carryover from addition has occurred. Equally, when SUB is a 1 and C_{OUT} is a 0, a borrow condition exists and is indicated by the CARRY/BORROW being set to a 1. All other conditions cause the CARRY/BORROW to be a 0.

EXAMPLE 3.16

Show how the exclusive OR resolved the problem of creating a CARRY/BORROW indicator that shows both carry and borrow results of adding and subtracting.

SOLUTION By now the solution should be obvious—create a truth table representing the SUB and C_{OUT} signals as inputs and the CARRY/BORROW indication as an output. Recall that a carry out from addition indicates an unsigned overflow called carry and the absence of a carry out from two's complement subtraction indicates a borrow condition.

The truth table uses SUB and C_{OUT} as inputs and the CARRY/BORROW indicator as an output. Defining the circuit needs on the truth table does indicate that an exclusive OR fits the needs of the CARRY/BORROW signal:

SUB	C_{OUT}	CARRY/BORROW
0	0	0
0	1	1
1	0	1
1	1	0

Flag *or* ***status flag*** *an indication of the state of a condition.*

The carry/borrow indicator line is an example of one of many output signal lines used as indications of the results of an operation used frequently in digital circuits and computers. Since they show the status of a condition, they are referred to as **status flags** or simply **flags.** In our circuit, the CARRY/BORROW indicator line is shortened to CARRY, and as a status flag in most cases it is called a *carry flag.* The carry flag is understood to also indicate borrow conditions during subtraction. Examples of other flag indicators are a *zero flag,* which indicates when all the bits in the result are low (0), and a *sign flag,* which indicates whether a binary value is positive or negative.

What happens if we need to add and subtract numbers larger than four bits? A solution for an eight-bit adder/subtracter is shown in Figure 3.12. The circuit works because of the nature of doing two's complement. A 1 needs to be added to the inverted subtrahend regardless of the number of bits in that subtrahend. As a result, the 1 is added into the C_{IN} input of the circuit's least significant bit by using the SUB line. The C_{OUT} from the lower four bits is connected to the C_{IN} of the next four in much the same manner as done earlier with the eight-bit adder of Figure 3.9. Now the eight-bit subtrahend applied to the B inputs is two's complemented by the assertion of the SUB input.

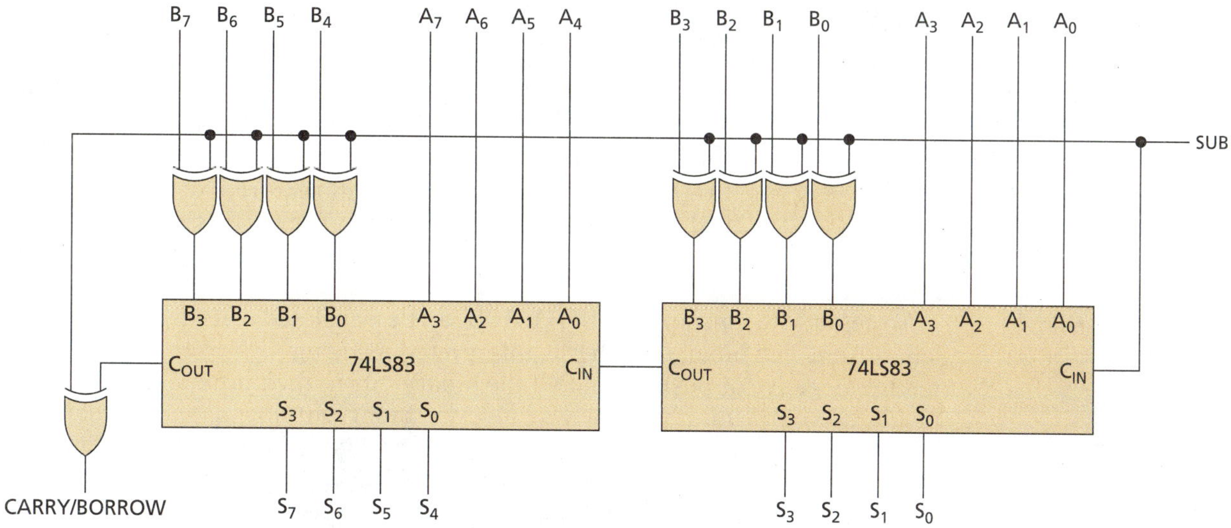

Figure 3.12 Eight-Bit Two's Complement Adder/Subtracter

EXAMPLE 3.17

Show how the eight-bit adder/subtracter circuit in Figure 3.12 subtracts the number 11011001B from 11101110B.

SOLUTION The number 11011001B is applied to the B word inputs and number 11101110B to the A word inputs. The SUB line is asserted high to put the circuit into the subtract mode. The B word is complemented through the eight exclusive ORs and sent to the B inputs of the cascaded four-bit adders. There it is added to the A word. The two's complement function is completed by adding 1 to the least significant bit through the carry input of the right four-bit adder. Any carry from the lower nibble of the eight-bit addition to the upper nibble is accomplished by the connection of C_{OUT} to C_{IN}, as shown in Figure 3.12. The eight-bit difference appears at the S outputs, while the carry flag indicates, in this case, that a borrow was not required and the difference is a positive true value result.

Section 3.17

Practice Problems

1. Apply the following numbers to the A and B inputs of the eight-bit adder/subtracter of Figure 3.12, and verify that they can be successfully added and subtracted:

   A = 11001101B B = 10001110B

2. If number B is larger than number A, a borrow is indicated from the adder/subtracter. What does this tell you about the value at the S outputs?

3.18 Signed Binary Numbers

In Section 3.16, we did binary subtraction using two's complement addition. At that time we mentioned that the two's complement process was a means to derive the negative equivalent of a binary number. To actually indicate that a number is negative, though, a minus sign is still required. Minus signs are cumbersome to deal with in a system that uses only two symbols. Therefore other methods were developed to indicate negative and positive binary numbers.

Unsigned number *magnitude-only number with no reference to negative or positive quantity.*

Signed number *a negative or positive number.*

Sign–magnitude system *the representation of a binary number whose MSB is the sign of the number and whose remaining bits are its actual magnitude.*

To begin with, the number 10011010B can be translated to its decimal equivalent of 154. The binary number is interpreted as being **unsigned.** That is, it is neither negative or positive and all the binary bits in the number represent the magnitude of the number. One way to give a binary number a **signed** quality is to assign one of the binary bits to represent the sign condition. The bit usually assigned this task is the most significant bit (MSB).

One system, called the **sign–magnitude system,** uses the MSB to signify the sign of the number. A positive number has a MSB of 0 and a negative one, a MSB of 1. The remaining bits of the number represent the magnitude of the number. For the preceding example, 10011010B, the decimal equivalent is −26, since the MSB = 1 and the remaining bits equate to 26.

EXAMPLE 3.18

Give the decimal equivalent of the following numbers in the sign–magnitude system:

0011001B and 1011100B

SOLUTION The first number is positive (MSB = 0), and the remaining bits represent the number's magnitude (B11001 = 25), so the decimal value of 0011001B = +25. For the second number, the sign is negative (MSB = 1) and the magnitude is 11100B = 28, so that 1011100B, in the sign–magnitude system, is a decimal −28.

The most common method of representing signed numbers is the two's complement system. In this system, the MSB is still used to represent the sign of the number and the remaining bits are the magnitude. The difference lies in the value of the magnitude. Magnitude bits for positive numbers (MSB = 0) are the actual value of the number as they are in the sign–magnitude system. Negative magnitude bits (MSB = 1), however, are in two's complement form. That is, to find the actual binary value, the entire number, including the sign bit, has to be complemented and a minus sign placed in front of the complemented value. Consider the number we used as an example for magnitude only,

10011010B, and interpret its value in the two's complement system. The number is negative, since the MSB is a 1, and we must complement the number to find its actual value.

```
  10011010B
  01100101B   one's complement
+        1B
 -01100110B   actual binary number
```

The decimal equivalent of the number is now −102 instead of the 154 value it had as an unsigned number of −26 as a sign–magnitude number. In each case, the binary number was the same, but took on different decimal equivalent values based on the type of system in use.

Consider any binary number with a leading 1 followed by a series of zeros such as 10000B. The number is negative because the sign bit is a 1. However, when you take the two's complement of this number to find its true binary magnitude, you will get the exact same number:

```
  01111B   invert number
+     B   add 1 to complete complement
  10000B   two's complement value
```

Keep in mind that the purpose of taking the two's complement is to derive the number's magnitude. You will need to precede the complemented value with a minus sign regardless of the appearance of the result. In this example, the value of the number is −10000B or −16.

EXAMPLE 3.19

What are the decimal values of the binary numbers in Example 3.18 using the two's complement system?

SOLUTION The value of the first number, since it is positive (MSB = 0) is the same in both systems, +25. The second number in the two's complement system is negative as before (MSB = 1), but its magnitude value requires complementing to get the decimal equivalent:

```
  1011100B   original number
  0100011B   one's complement
+       1B   add 1
  0100100B   two's complement
```

Finally, convert the complemented result to decimal. Do not forget to place a minus sign in front of the complemented result: −0100100B = −36. This is different from the −28 value of the sign–magnitude result in Example 3.17.

Section 3.18

Practice Problems

1. Give the decimal equivalent of the following numbers in the sign–magnitude system:
   a. 001101111B b. 11111111B c. 011111111B
   d. 100000000B e. 10011110B
2. What are the decimal equivalent values for the binary numbers in Practice Problem 1 if they are in the two's complement form? Be careful of part d.
3. What are the decimal equivalent values for the binary numbers in Practice Problem 1 if they are unsigned (magnitude only) binary numbers?
4. What are the minimum and maximum decimal equivalents of the signed and unsigned numbers that can be represented by byte-size binary numbers? *Hint:* Eight-bit binary numbers range from 0000000B to 11111111B.

3.19 Hexadecimal Addition

To understand and use the basic machine language of today's computers, it helps to be able to perform some of the basic arithmetic processes using hexadecimal numbers. Having a working knowledge of addition, subtraction, and complementing using hex numbers provides a good foundation for the study of microprocessors and microprocessor-based computers.

Hex numbers convert to binary easily. They also conform to the process for conversion to decimal numbers, and since the hex system is an actual number system, arithmetic processes of addition and subtraction can be performed in hex with dependable results. Adding hex numbers follows all the arithmetic rules used for decimal and binary numbers, except that carry and borrow values are 16 and not 10 for decimal or 2 for binary additions and subtractions. The other part that takes getting used to is manipulating letters as if they are numbers. You need to get used to handling A, B, C, D, E, and F as you do 0–9. The key, always, is the base value of 16.

As an example, we will add the hex numbers 3EB6H and 52A3H. The problem is set up like any other addition problem and the least significant digits are summed, 6 + 3 = 9. The sum is placed under the least significant digit:

```
  3EB6H
 +52A3H
 ------
     9H
```

The next column of digits is added together, but this time they are letter symbols. Keep in mind these letters stand for a numerical quantity, B is 11 and A is 10 in decimal. Adding B to A is the same as adding 11 to 10, yielding a sum of 21. However, a second difficulty arises in that 21 is larger than 16 and has no separate hex symbol. The number 21 must be converted to its hex equivalent, yielding a sum and carryover condition. To complete the addition of these two numbers, subtract 16 from 21, place the 5 under the second column, and carry the 16 you just subtracted as a carryover to the next column:

```
  1
 3EB6H
+52A3H
   59H
```

Continue the process, keeping in mind that E has a value of 14. The sum of the carry and the digits in the next column is 1 + E + 2 = 1 + 14 + 2 = 17, which is bigger than 16, so again subtract 16 from the sum and carry 1 to the next column:

```
 11
 3EB6H
+52A3H
  159H
```

Finally, in the most significant column, 1 + 3 + 5 = 9, which does not generate a carryover:

```
 11
 3EB6H
+52A3H
 9159H
```

EXAMPLE 3.20

Add A2CH to 59EH, and verify that the result is valid.

SOLUTION First do the addition:

CH + EH = 12 + 14 = 26 26 − 16 = 10 (AH) and carry 1 **1**

1 + 2 + 9 = 12 = CH **2**

AH + 5 = 10 + 5 = 15 = FH **3**

```
  1
 A2CH
+59EH
 FCAH
```

Now convert each hex number to decimal, and check if the addition is correct:

AH = 10 CH = 12 FH = 15 **1**

10 × 1 = 10 12 × 16 = 192 15 × 256 = 3,840 **2**

The sum: FCAH = 10 + 192 + 3,840 = 4,042 **3**

59EH = 5 × 256 + 9 × 16 + 14 = 1,438 **4**

A2CH = 10 × 256 + 2 × 16 + 12 = 2,604 **5**

and . . . 2,604 + 1,438 = 4,042 . . . which is the decimal equivalent of the sum.

Section 3.19

Practice Problems

Add the following pairs of hex numbers, and verify that the sums are correct.

1. 123H + ABCH
2. C01H + 800H
3. 44EH + 1FFH
4. 6DAH + A3CH

3.20 Hexadecimal Subtraction

Subtraction of hex numbers can be done two ways, just as subtraction of decimal or binary numbers can be done two ways—straight and complement addition. Straightforward hex subtraction is rarely ever used and will not be demonstrated here. Using hex complements to accomplish subtraction by addition is the method used. Keep in mind that the hex system is being used to actually represent binary numbers in an easier-to-use form. Doing hex arithmetic performs binary arithmetic using hex digits.

To negate a hex number so that subtraction can be performed by addition, subtract each digit from FH, the largest value in the base 16 system. This yields the fifteen's complement of the original number. Adding 1 to the fifteen's complement produces the sixteen's complement, which is the negative value of the original hex number. Since hex numbers are used as a method to represent binary numbers easily, the fifteen's complement performs the one's complement and the sixteen's complement performs the two's complement of the binary number equivalent to the original hex number.

To illustrate how this works, we will find the fifteen's and sixteen's complement of the hex number 3E5H and show that they represent the one's and two's complement of the binary equivalent number, 001111100101B. Complement both the hex number and its binary equivalent by subtracting both from their system's maximum value of FB and 1B and adding 1 to the result.

FFFH	111111111111	
−3E5H	−001111100101	
C1AH	110000011010	fifteen's and one's complement
+ 1H	+ 1	add 1
C1BH	110000011011	sixteen's and two's complement

Notice how each hex number and binary number are equivalent.

EXAMPLE 3.21

Find the fifteen's and sixteen's complements of 200D0H, and show that they are the same as the one's and two's complements.

SOLUTION First convert the hex number to its binary equivalent:

200D0H = 00100000000011010000H

and then complement both hex and binary numbers:

FFFFFH		
−200D0H	00100000000011010000	
DFF2FH	11011111111100101111	invert
+ 1H	+ 1	
DFF30H	11011111111100110000	sixteen's and two's complement

To do subtraction by sixteen's complement addition, after complementing the subtrahend, the differences of the two hex numbers can be found by adding the complement to the minuend. To find the difference of the hex numbers 5C20H − 3BBBH, complement the subtrahend:

FFFFH	F's
− 3BBBH	original subtrahend
C444H	fifteen's complement
+ 1H	add 1
C445H	sixteen's complement

Then add that result to the minuend:

11	carries
5C20H	
+C445H	
12065H	

The difficult part of this addition is CH + 4H = 12 + 4, which is 16 = 0H plus a carry. Then 1H + 5H + CH = 1 + 5 + 12 = 18 = 2H plus a carryover. Recall that when doing subtraction by complement addition, a final carryover is discarded and indicates that the result is positive, so the difference in this example is

5C20H − 3BBBH = 2065H

EXAMPLE 3.22

Find the difference of 200BH − 9CE5H using complement addition.

SOLUTION Start by getting the sixteen's complement of the subtrahend, 9CE5H:

FFFFH	F's
−9CE5H	original subtrahend
631AH	fifteen's complement
+ 1H	add 1
631BH	sixteen's complement

Now add the complemented subtrahend to the minuend to get the difference:

1
200BH
+631BH
8326H

In this example, there is no final carryover, so the result is negative and must be complemented to get the actual value:

$$\begin{array}{r} \text{FFFFH} \\ -\ \underline{8326\text{H}} \\ 7\text{CD9H} \\ +\ \underline{\quad 1\text{H}} \\ 7\text{CDAH} \end{array}$$

therefore 200BH − 9CE5H = −7CDAH

Section 3.20

Practice Problems

1. Find the sixteen's complement of each of these hex numbers. Verify that they represent the two's complement of the binary equivalent of the original hex values.
   a. 34CDAH b. ABCDOH c. 001234H d. 80000H
2. Do the following hex subtractions using sixteen's complement addition:
   a. 2389H − 1001H b. 60D00H − 4EB01H
   c. 5E0B2H − 99CCFH

3.21 Binary-Coded Decimal

BCD (binary-coded decimal) *a binary code representing decimal (0–9) characters.*

One thing that should be duly noted is that converting from a number base to decimal requires a certain amount of effort. The quest for a quicker way of representing decimal numbers in binary form led to the development of the **binary-coded decimal** or **BCD** code. First and foremost, let it be fully understood that the following set of binary combinations are a code and *not* a number base system. Since this is a code and not a number system, arithmetic functions cannot be done directly using the BCD code. The BCD code uses combinations of binary bits, similar to the binary combinations in the hexadecimal number system, to represent decimal characters. Groups of four binary bits are used to represent the decimal characters 0–9. The first 10 hex numbers have the same binary combinations as the BCD code. Table 3.2 lists the BCD code. Compare this table with the hex Table 3.1 earlier in the chapter.

Take a look at this binary number: 100101001000B. As binary numbers, the 1's and 0's tend to run together as you attempt to read them. Since a good deal of digital theory uses hex and BCD values to represent binary quantities, this text will adopt a slightly different way of writing binary numbers. The purpose is to make the binary numbers more readable but does not affect their value. From this point on, binary numbers will be written as groups of four bits with a space between each group. For example, the binary sequence under discussion now appears as 1001 0100 1000B, which is easier to read. Again, the spacing is for convenience and makes the binary sequence easier to deal with.

Table 3.2 BCD Decimal Codes

BINARY CODE				
b_3	b_2	b_1	b_0	Decimal Character
0	0	0	0	0
0	0	0	1	1
0	0	1	0	2
0	0	1	1	3
0	1	0	0	4
0	1	0	1	5
0	1	1	0	6
0	1	1	1	7
1	0	0	0	8
1	0	0	1	9

Note: The remaining codes are invalid.

The binary sequence 1001 0100 1000B can be converted to either BCD or hex: 948 or 948H. Same thing with a second number: 0110 0111 1001 0011B. This binary sequence translates to 6,793 or 6793H. The difference comes when they are added together. In binary, the results are the same: 1001 0100 1000B + 0110 0111 1001 0011B = 0111 0000 1101 1011B. These results can only be converted to hex as 70DBH since "D" and "B" are characters that do not lie within the BCD code. Therefore attempting to get a BCD sum from this binary sequence using the BCD code yields non-BCD results and therefore has no meaning. At this time, it is noted that microprocessor instruction sets include an instruction to allow two BCD "numbers" to be added to produce a BCD "sum." These instructions, however, adjust the results of adding hexadecimal numbers limited to the values of 0 to 9 so that their sum appears to resemble a BCD addition. The method that is used to do this is explored in Section 3.22.

The preceding example emphasizes the difference between the hexadecimal number base system, which can be used to do arithmetic, and the BCD code, which is not a number system, but a binary code. The BCD code is used so that decimal values can be translated to a binary sequence fairly easily. It takes advantage of the 8-4-2-1 weight system that the hex base uses when translating to binary. Because of this, the BCD code is often referred to as the 8-4-2-1 code. A binary sequence representing BCD characters is designated by appending the letters BCD at the end, much like appending a B for binary or H for hex; for example, 76 = 0111 0110BCD.

EXAMPLE 2.23

Convert the decimal number 1,234 to BCD code and to a binary number. Compare the two results.

SOLUTION Converting to BCD requires that each decimal digit be converted to its 8-4-2-1 code:

1 = 0001BCD   2 = 0010BCD   3 = 0011BCD   and   4 = 0100BCD

Together, 1,234 = 0001 0010 0011 0100BCD.

The trailing letters BCD are used to denote a binary sequence as a BCD code in the same manner that a B denotes the sequence of bits as a binary number.

Converting 1,234 to a binary number requires the decimal value to be divided by 2 and to assemble the remainders from successive divisions as done in Section 3.7. The conversion is 1,234 = 0100 1101 0010B.

As can be seen by comparing the two binary sequences, they are not the same. Since the binary value is an actual number, it can be used to do arithmetic operations directly, whereas BCD, being a code to represent decimal characters instead of numbers, cannot be used to do arithmetic operations directly.

What good is the BCD code if it can only be used to change decimal characters to binary but cannot be used to do arithmetic operations? BCD-coded decimal numbers serve many purposes in allowing easy ways for decimal numbers to be entered into a computer or calculator and later shown on a display screen. The key is that this is a code and although direct arithmetic operations cannot be performed using a binary code, they can be done by performing a little manipulation.

Section 3.21

Practice Problems

1. Convert the following decimal numbers into BCD code:
   a. 472 b. 9,876 c. 192,837
2. Convert the following BCD codes to decimal numbers:
   a. 1000 0110 0011 0111BCD b. 0010 0111 1001 0001BCD
   c. 0001 0100 1001 0111 0110 0011BCD

3.22 BCD Addition

BCD sums can be achieved by adding BCD codes as if they were hexadecimal numbers and then adjusting the addition to make the result appear as if a BCD addition took place. The purpose of creating a system for BCD addition is to allow for an easy conversion of decimal numbers to a binary code and provide a way for that code to be used in an arithmetic addition. The result of the addition is also a BCD code and can be easily converted into decimal characters. Consider the problem of entering decimal numbers, converting them into equivalent binary (or hex) numbers, having the digital computer do the addition, and converting the result back to a decimal number. Instead, using BCD codes makes the conversion of a decimal numerical character to a binary sequence, usable by a computer, faster and easier.

The basic requirements to do BCD additions are that the numbers you begin with are BCD codes and that all additions are performed as if the codes were hex numbers. Keep in mind that when hex addition is used, the same results can be achieved using binary addition, since hex values are used to represent binary numbers. Hex is preferred because of the smaller values and applications that arise when studying microprocessors and computers later on. Two rules apply to the hex addition to correctly adjust the results to appear as if the BCD number codes were added together:

1. When the hex sum of two BCD digits produce a value greater than 9H (1001B), add 6H (0110B) to that sum.
2. When the sum of two BCD digits produces a carry, add 6H (0110B) to that sum.
3. The adjustments themselves will not produce a value greater than 9H, and any carries resulting from making the adjustment are treated as part of the regular hex addition. No additional adjustment is required because of the addition of the 6H.

Let's add 0100 1001 0110 0111BCD to 0110 1000 0001 0101BCD, which in decimal is 4,967 + 6,815, to see how the adjust works. Set up the problem like any other addition problem, and add the numbers as hex numbers:

```
  1
  4,967
 +6,815
  B17C
```

There are two sums that are greater than 9H and one sum that produced a carry. A 6H must be added to each of these sums:

```
   1
   4967
+  6815
   B17C
+  66 6
  11782
```

When 6H is added to the original sum, subsequent carries are added as normal carries in hex addition and do not cause additional additions of 6H. However, if a sum from adding 6H produces a number greater than 9H, 6H will be added again. For instance, suppose the preceding example was modified to add 4967 to 6835. The initial hex addition yields

```
  1
  4967
 +6835
  B19C
```

Now let's do the adjustment by adding 6H as expected:

```
   1
   4967
+  6835
   B19C
+  66 6
  117A2
```

Notice the "A" in the adjusted result, which is *not* a BCD character, since it is beyond 9 (1001BCD). This will require a second adjustment to create the correct results:

```
   1
  4967
+ 6835
  B19C
+ 66 6
 117A2
+    6
 11802
```

Check that the addition is correct by adding the two decimal numbers represented by the BCD characters: 4,967 + 6,835 = 11,802. Look at these two examples carefully. All the additions are done using hex numbers, but the end result appears as if two BCD numbers were added together: 4,967 + 6,815 = 11,782, which in BCD code is

0100 1001 0110 0111BCD + 0110 1000 0001 0101BCD =
0001 0001 0111 1000 0010BCD

and 4,967 + 6,835 = 11,802, which in BCD code is

01000 1001 0110 0111BCD + 0110 1000 0011 0101BCD =
0001 0001 1000 0000 0010BCD

Section 3.22

Practice Problems

Add the following decimal equivalents of BCD number codes using hex addition and showing the adjustments that are required to complete the BCD addition.

1. 391 + 5,239 **2.** 57,189 + 59,228 **3.** 4,567 + 9,876

3.23 The American Standard Code for Information Interchange (ASCII)

Along with the advent and use of computers, the need to transfer digital information from one place to another became a common necessity. To accomplish this, alphabet, numeral, punctuation, and other various characters had to be coded into binary sequences. One of the first codes developed was called the *Baudot code,* after one of the pioneering engineers in the field. Combinations of five binary bits are used to represent letters, numbers, and other characters, but five bits proved quite limiting. Raising the binary base 2 to the number of bits per character word, which for the Baudot code is 5, shows us the number of characters that a code can represent. The Baudot code is limited to representing 32 different characters, one for each group of five bits. This allows hardly enough combinations to cover the alphabet (26 letters) and numbers (10 numbers), let alone any additional characters. To facilitate this shortcoming, the Baudot code uses the code for a shift key to almost double the number of characters the code can represent. The same set of binary combinations is used to represent letters and then numbers. The difference is that the shift key had to be pressed to switch from one group to another. This is similar to what was needed on a typewriter to shift from numbers to punctuation characters or from lower- to uppercase letters. By the way, the Baudot code only allowed for uppercase letters.

EBCDIC (Extended Binary-Coded Decimal Interchange Code) *IBM's character code.*

***ASCII** and **extended ASCII** abbreviation for American Standard Code for Information Interchange, which is a seven-bit binary code used to represent alpha, numeric, and other characters. An extended ASCII set, which includes an eighth binary bit, also includes many graphic and specialty characters.*

An improvement over the Baudot code came in two forms, one code developed by the International Business Machines (IBM) company and the other by the American National Standards Institute (ANSI). IBM developed the **Extended Binary Coded Decimal Interchange Code** or **EBCDIC** (pronounced *eb-ci-dic*) and ANSI is responsible for the **American Standard Code for Information Interchange** or **ASCII** (pronounced *as-key*). EBCDIC uses combinations of eight bits and ASCII, seven bits, to represent various characters. ASCII has emerged as the predominant code and is the one emphasized in this text. An eighth bit has been added to the ASCII code to create the current **extended ASCII code** to include foreign letters, mathematical symbols, and other special characters. Table 3.3 summarizes the extended ASCII code used in today's personal and other computers. In the table, each binary combination is shown in its hex equivalent, with the relevant character or symbol shown to the right, first in regular, and then in extended ASCII.

The table is used by looking up the characters to be converted into the binary ASCII code and writing them out as a string of hex digits. For instance, the ASCII code for characters in the word *Vacation!* are looked up from the table and assembled in order starting with *V* in *Vacation!* In hex, the translated code is

56 61 63 61 74 69 6F 6E 21H

These codes work for the regular seven-bit ASCII and for the eight-bit extended ASCII. However, when converting hex codes to binary, the one-bit-per-character difference is crucial. Since standard ASCII uses seven bits per character, the actual binary sequence for *Vacation!* is

1010110 1100001 1100011 1100001 1110100 1101001 1101111
1101110 0100001B

Using the eight-bit extended ASCII yields a binary sequence of

01010110 01100001 01100011 01100001 01110100 01101001
01101111 01101110 00100001B

Spaces between characters in the hex and binary sequences are placed for clarity and are not actually included in the sequences. Note that the extended ASCII of necessity is longer than the ASCII sequence.

Several different types of characters are represented by the binary sequences in the ASCII code. There are the more familiar alphabet, number, and punctuation symbol characters. Still somewhat familiar are the functional characters, such as shift, line feed, and carriage return, which respond to typewriter and computer function keys.

Table 3.3 ASCII Codes Including Extended ASCII Set

	ASCII			ASCII	
Hex Code	**Regular**	**Extended**	**Hex Code**	**Regular**	**Extended**
00	NUL	NUL	30	0	0
01	SOH*	☺	31	1	1
02	STX*	☻	32	2	2
03	ETX*	♥	33	3	3
04	EOT*	♦	34	4	4
05	ENQ*	♣	35	5	5
06	ACK*	♠	36	6	6
07	BELL	●	37	7	7
08	BKSP	◘	38	8	8
09	HTAB	○	39	9	9
0A	LF	◙	3A	:	:
0B	VTAB	♂	3B	;	;
0C	FF	♀	3C	<	<
0D	CR	♪	3D	=	=
0E	SFTO	♫	3E	>	>
0F	SFTI	✲	3F	?	?
10	DLE*	▶	40	@	@
11	DC1*	◀	41	A	A
12	DC2*	↕	42	B	B
13	DC3*	!!	43	C	C
14	DC4*	¶	44	D	D
15	NACK*	§	45	E	E
16	SYNC*	–	46	F	F
17	ETB*	↨	47	G	G
18	CAN*	↑	48	H	H
19	END*	↓	49	I	I
1A	SUB*	→	4A	J	J
1B	ESC*	←	4B	K	K
1C	FSP*		4C	L	L
1D	GSP*	↔	4D	M	M
1E	RSP*	▲	4E	N	N
1F	USP*	§	4F	O	O
20	SPACE	SPACE	50	P	P
21	!	!	51	Q	Q
22	"	"	52	R	R
23	#	#	53	S	S
24	$	$	54	T	T
25	%	%	55	U	U
26	&	&	56	V	V
27	'	'	57	W	W
28	(	(	58	X	X
29	)	)	59	Y	Y
2A	*	*	5A	Z	Z
2B	+	+	5B	[	[
2C	,	,	5C	\	\
2D	–	–	5D	]	]
2E	.	.	5E	^	^
2F	/	/	5F	—	—

Table 3.3 Continued

	ASCII			ASCII	
Hex Code	**Regular**	**Extended**	**Hex Code**	**Regular**	**Extended**
60	`	`			
61	a	a			
62	b	b			
63	c	c			
64	d	d			
65	e	e			
66	f	f			
66					
67	g	g			
68	h	h			
69	i	i			
6A	j	j			
6B	k	k			
6C	l	l			
6D	m	m			
6E	n	n			
6F	o	o			
70	p	p			
71	q	q			
72	r	r			
73	s	s			
74	t	t			
75	u	u			
76	v	v			
77	w	w			
78	x	x			
79	y	y			
7A	z	z			
7B	{	{			
7C	|	|			
7D	}	}			
7E	~	~			
7F	DEL	⌂			

The remaining ASCII codes apply to the extended ASCII code only.

80	Ç	81	ü	82	é	83	â	84	ä	85	á
86	å	87	ç	88	ê	89	ë	8A	è	8B	ï
8C	î	8D	ì	8E	Ä	8F	Å	90	É	91	æ
92	Æ	93	ô	94	ö	95	ò	96	û	97	ù
98	ÿ	99	ö	9A	Ü	9B	¢	9C	£	9D	¥
9E	Pt	9F	ƒ	AO	á	A1	í	A2	ó	A3	ú
A4	ñ	A5	Ñ	A6	a	A7	o	A8	¿	A9	
AA		AB	½	AC	¼	AD	¡	AE	«	AF	»
BO	░	B1	▒	B2	▦	B3	│	B4	┤	B5	╡
B6	╢	B7	╖	B8	╕	B9	╣	BA	║	BB	╗
BC	╝	BD	╜	BE	╛	BF	┐	CO	└	C1	┴
C2	┬	C3	├	C4	─	C5	┼	C6	╞	C7	╟
C8	╚	C9	╔	CA	╩	CB	╦	CC	╠	CD	═

Table 3.3 Continued

	ASCII			ASCII	
Hex Code	Regular	Extended	Hex Code	Regular	Extended
CE ╬	CF ╧	D0 ╨	D1 ╤	D2 ╥	D3 ╙
D4 ╘	D5 ╒	D6 ╓	D7 ╫	D8 ╪	D9 ┘
DA ┌	DB █	DC ▄	DD ▌	DE ▐	DF ▀
E0 α	E1 β	E2 Γ	E3 π	E4 Σ	E5 σ
E6 µ	E7 τ	E8 Φ	E9 Θ	EA Ω	EB δ
EC ∞	ED φ	EE ε	EF ∩	F0 ≡	F1 ±
F2 ≥	F3 ≤	F4 ⌠	F5 ⌡	F6 ÷	F7 ≈
F8 °	F9 •	FA ·	FB √	FC ^{n}	FD ^{2}
FE ■	FF blank				

Data link characters *part of a character code set that defines control functions for establishing a communication link between two stations.*

Other characters in the extended ASCII set, that have a code with the most significant bit equal to 0, make up the balance of the standard seven-bit ASCII code subset. These characters are used to facilitate the actual transfer of data and the interpretation of those data. They are called **data link characters,** and include such functions as Start of Text (STX), End of Transmission (EOT), and End of Text (ETX). The standard ASCII set codes in the extended ASCII set are replaced by printable characters and such symbols as ☺, ♫, and ♥.

The eighth bit of the extended ASCII set was added to double the total number of characters that the ASCII code could represent. Additional extended ASCII characters all have that eighth bit, the most significant bit, high. The characters within this portion of the ASCII set include special foreign letters that require special characters like the umlaut above some German vowels, such as ä. Mathematical symbols such as π and $\div$ are included in this set along with symbols to create graphical drawings like ╣ and ░. Part of the Greek letter set is also included in the extended ASCII code, which allows for ease of use in mathematical, electrical, scientific, and engineering documentation and messages.

Application Note

ASCII

Binary character codes such as EBCDIC and ASCII have found heavy use in computers and in the field of *digital* and *data communications.* This field deals with the numerous ways that people and machines communicate with each other over vast distances. Many aspects of our daily living are digitized—that is, converted into some form of binary sequence—and transmitted from one place to another. In some cases, these digitized signals are sent across cables from one computer to another using a *local area network* (*LAN*). In other cases, these digitized signals are used to modulate or alter a high-frequency signal and transmitted from one site via a radio or satellite to another site. When messages containing words and numbers are sent, codes such as ASCII are used to digitally represent these messages.

Section 3.23

Practice Problems

Convert the following "messages" into extended ASCII codes, in hex. Do not forget that a space is a separate ASCII character with a code of 20H.

1. a. π is = to 3.14. **b.** Put on a ☺!

3.24 Troubleshooting Adder Circuits

The method used in troubleshooting adder circuits starts the same way as troubleshooting any logic circuit—know what the circuit is expected to do. Only then can you tell if something is wrong with it. Take the two's complement adder/subtracter in Figure 3.13, which includes various test points throughout the circuit. Since there are four bits to both input words, all 16 combinations of number A added to a like number for word B would have to be checked to make sure the circuit works 100 percent. Fortunately, we do not normally have to check all the combinations. A bad circuit usually comes to us with a note, from someone, as to an indication that something is wrong with a given set of input conditions that produced an incorrect result.

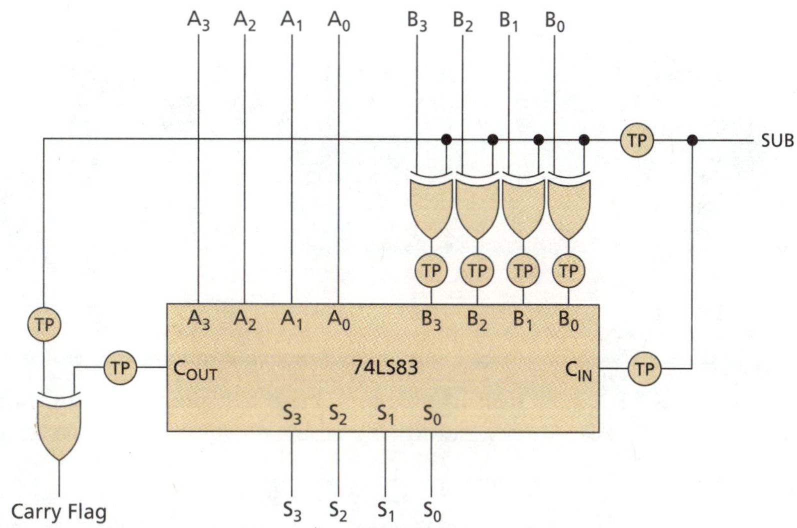

Figure 3.13 Two's Complement Adder/Subtracter with Test Points (TP)

Suppose we receive an adder/subtracter circuit that adds the two words 1010B and 0011B correctly, but gives us an error when subtracting 0011B from 1010B. We can already tell quite a bit about the circuit from this information. The four-bit adder portion of the circuit is functioning correctly, or the sum result would have an error in it. Basically, the XOR gates operate OK as long as SUB is low or, again, the sum result would not be correct. To find out what is wrong with the circuit, we apply the inputs that are yielding the incorrect result and try to uncover where in the circuit things go wrong. In this example, when the inputs are applied, the difference appears as 0110B instead of the expected 0111B. This should make you suspect already where the problem is. A quick check using two other sets of inputs helps to verify the suspicion. When 0001B is subtracted from 0111B, the result in our bad circuit is 0111B instead of 0110B. Why did we use the second set of numbers to verify the problem? The reason lies in the result of the first subtraction—the answer was off by a value of 1. The second set quickly verified that the same problem exists with another set of inputs. So what is the suspected problem? Looks like the 1 added into the least significant bit at C_{IN} to the four-bit adder is not being done. Out comes the

Section 3.24

Practice Problems

1. Given the states of the test points in the adder/subtracter circuit of Figure 3.14, determine what might be wrong with the circuit.

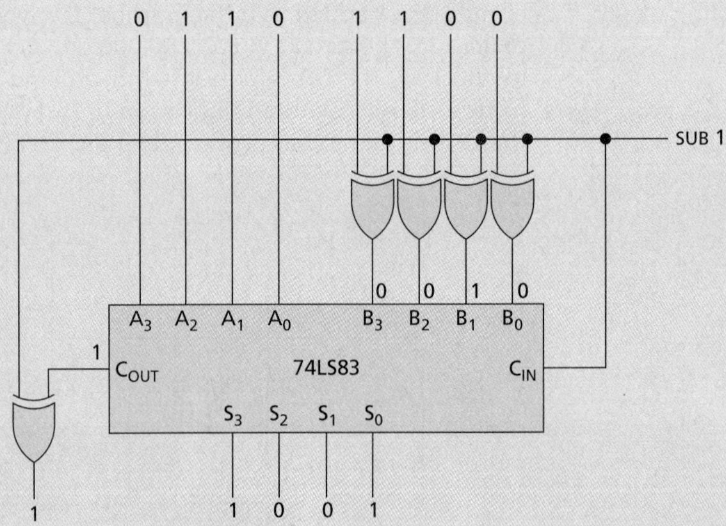

Figure 3.14 Practice Problem 3.24(1)

2. Given the states of the test points of the four-bit parallel adder of Figure 3.15, determine what, if anything, is wrong with the circuit.

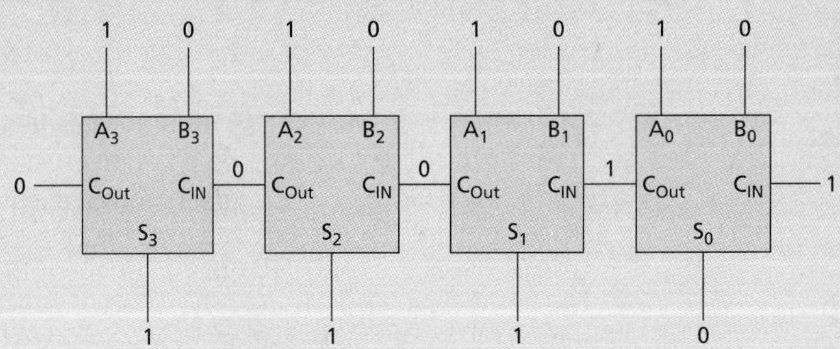

Figure 3.15 Practice Problem 3.24(2)

oscilloscope, and the C_{IN} test point is checked. Success! C_{IN} is always low, regardless of the logic state of the SUB control line. The actual cause of the problem may be a short from C_{IN} input to ground or something internally wrong with the four-bit adder that holds C_{IN} at OV regardless of the input level applied to C_{IN}.

3.25 Designing a BCD Adder

We do our arithmetic using the base 10 decimal system with digits 0–9. Computers work with binary values 0 and 1 that are sometimes represented by hexadecimal numbers (each hexadecimal digit is equivalent to four binary bits). Both the binary and hexadecimal systems are true number base systems. That is, you can do arithmetic with them directly and receive desired results. Presenting decimal values in binary form led to the creation of the binary-coded decimal code discussed in Section 3.22, which included a method to change the results of binary or hex addition to make it appear as if a BCD sum was produced. Recall that this required the addition of 6H or 0110B when the sum produced a value greater than 9H or 1001B or a carry out. Our design task is to create the BCD addition using the four-bit adder.

As with earlier design problems, we begin by stating the circuit requirements and then derive truth tables and logic expressions to meet those requirements:

1. Add 0110B to the sum if it is greater than 1001B.
2. Add 0110B if the sum produced a carry out.
3. Do not add 0110B if the sum is equal to or less than 1001B and there is no carry out.
4. Produce a final carry if either the original sum or the result of adding 0110B results in a carry out.

S_3	S_2	S_1	S_0	C_{OUT}	ADD6
0	0	0	0	0	0
	to				
1	0	0	1	0	0
1	0	1	0	0	1
1	0	1	1	0	1
1	1	0	0	0	1
1	1	0	1	0	1
1	1	1	0	0	1
1	1	1	1	0	1
X	X	X	X	1	1

X = "don't care" state

Figure 3.16 BCD Adder Adjust Truth Table

The truth table generated from these requirements is shown in Figure 3.16. The table shows the 16 possible results of adding two numbers with a four-bit adder. The table indicates that as long as the results are 1001B or less, no adjustment to the result is needed. Adjusting the result by adding that 0110B is shown with results 1010B to 1111B or if a carry out is generated by the original addition process. From the truth table, we develop a logic expression for ADD6, which is the signal that causes the adjustment to be made:

$$\text{ADD6} = S_3 \cdot S_2 + S_3 \cdot S_1 = C_{OUT} = S_3(S_2 + S_1) + C_{OUT}$$

As for the final carry out, it is set whenever a carry out is produced from the original addition. The final carry out is also generated from the addition of 0110B to numbers greater than 1001B. Since 0110B is added when a carry out is produced from the original addition and from numbers greater than 1001B, the same line that causes 0110B to be added to the original sum is also the final carry out. Generation of the final carry out is summarized as follows:

1. A 0110B is added when a carry results from adding two numbers. This is also a final carry out.
2. Each time 0110B is added to a sum greater than 1001B, it results in a final carry out: 6H + AH = 10H; 6H + BH = 11H; and so on.
3. Both conditions cause 0110B to be added to the sum and result with a final carry out.
4. Result—the signal that causes 0110B to be added also generates the final carry out. The same circuit produces both results.

Figure 3.17 is the circuit that generates the sum of two BCD numbers at A and B. Inputs A and B must be BCD values between 0000BCD and 1001BCD (0

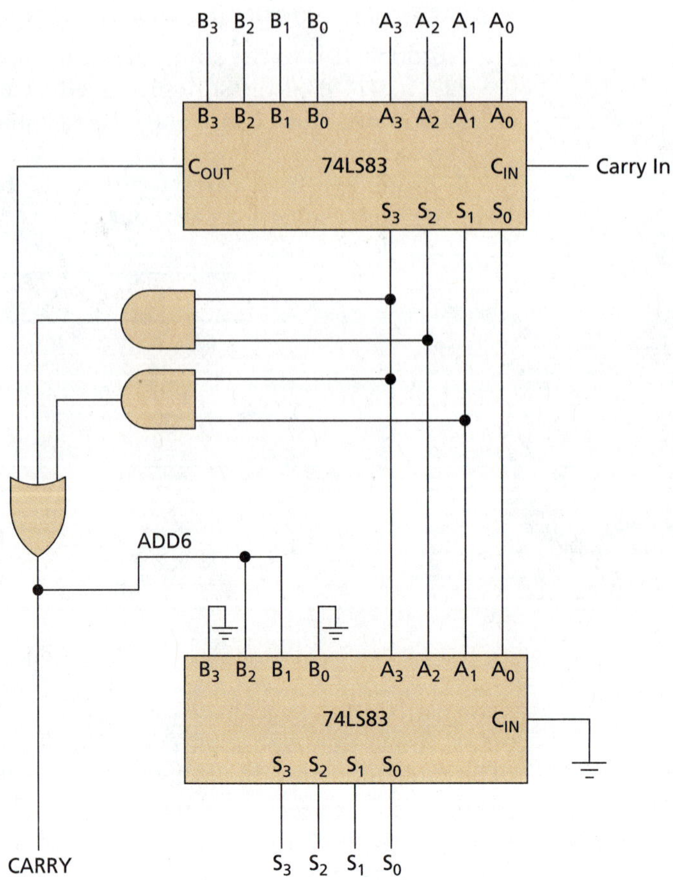

Figure 3.17 Single-Decade BCD Adder

and 9). Their original sum appears at the outputs of the top four-bit adder. Here the logic circuit we just designed, $S_3(S_2 + S_1) + C_{OUT}$, is used to sense when that sum is greater than 1001B (9) or a carry out is generated from the original addition. A high output from this circuit causes the value 0110B (6) to be added to the original sum in the lower four-bit adder by causing bits A_2 and A_1 to go high while A_3 and A_0 remain permanently low. The actual BCD sum appears at the output of the lower adder. CARRY, the final carry out, is taken from the output of the logic circuit that produces the 0110B adjustment.

BCD (or decade BCD) adder *an adder circuit that sums two binary-coded decimal (BCD) numbers.*

The BCD circuit just developed is often called a **decade BCD adder.** This circuit can be cascaded with others to allow addition of larger BCD numbers. Figure 3.18 illustrates a three-decade BCD adder that can be used to add values from 000 to 999. The process of connecting the adders together is similar to that used with the four-bit binary and two's complement adders. That is, the C_{OUT} of the previous decade adder is connected to the C_{IN} of the next decade adder.

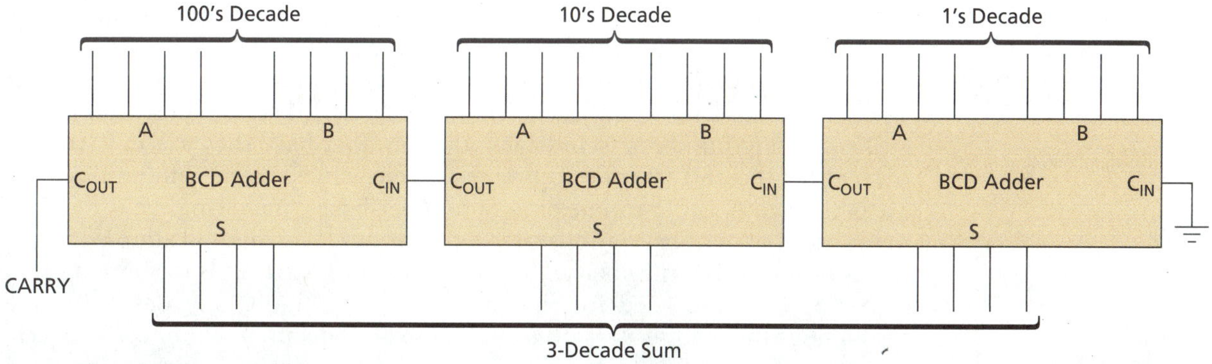

Figure 3.18 Cascading BCD Adders

Section 3.25

Practice Problems

1. Show that the numbers 1000B and 0101B produce the correct BCD result when applied to the A and B inputs of a BCD adder.
2. Describe how the BCD adder shows that using non-BCD numbers as inputs to words A and B does not produce a correct result.

Summary

Exploring any new number base system requires a thorough understanding of the decimal system, which most readers use daily. The principles behind the decimal system can be applied to any number systems, including the binary and hexadecimal systems. Converting values from one number system into the decimal base system requires the summing of the products of digit values and weights. This is applicable for both whole numbers and fractional components of these numbers. To convert from a decimal number into another base system, divide the whole part of the decimal number by the base value of the new system, then use the remainders of the division as the digits of the new base number. The process is repeated until the division produces a quotient of 0. To convert the fractional part of the decimal number, multiply it by the base value. Whole numbers of the products are used as the fractional digits of the new base number. Repeat the process until the product is 0 or a desired level of accuracy is achieved.

Binary and hex arithmetic is best understood once the principles behind decimal arithmetic are understood. The concepts of carrying over and borrowing apply to decimal, hexadecimal, and binary numbers. The main difference is the value of the carry and borrow, which is always 10, but has different numerical interpretations that depend on the radix for the number system in use. For decimal, 10 has a value of 10; for binary, it has a value of 2; and for hex, a value of 16.

Circuits to add and subtract binary numbers are designed around a single integrated circuit called the four-bit parallel adder which uses four full adder circuits as its code. Exclusive ORs are used to complement the subtrahend when using the four-bit adder in a two's complement adder/subtracter circuit.

Designating numbers to indicate what number base they are in is very important for the correct interpretation of their values. Many different methods are employed to differentiate among number bases, including the use of trailing subscripts or abbreviations such as H for hex. The method adopted for this text is to follow the number with a base indicator, such as B for binary, which identifies the base used for that number. The default—the base presumed throughout this text, unless otherwise noted—is decimal. Numbers appearing without a designation are assumed to be using the decimal number system.

The sign of a binary number is usually indicated by the most significant bit of that number. A 0 in that bit position indicates that the number is positive, and a 1 means it is negative. The most common system used to deal with signed numbers in binary is the two's complement system, in which a positive number is shown in true binary form and a negative number in its two's complement form.

Working with binary numbers of any significant size can become cumbersome. Add to this the possibility of making an error when dealing with a great number of binary bits, and you can see that we need a way to represent these binary numbers in a less cumbersome fashion. Initially, decimal representation of binary numbers would seem convenient, because most people are used to working with the base 10 system. However, converting between binary and decimal is not easy. The octal and hexadecimal number systems are easy to convert to binary numbers and are much less cumbersome to use than binary. However, the need to be able to operate with decimal numbers in a binary format gave rise to the binary-coded decimal code to represent decimal numbers. Doing arithmetic using this code requires special adjustments to the initial hexadecimal (or binary) results from adding BCD numbers.

To facilitate the transfer of messages between computers and distant sites, we employ binary character codes such as Baudot, EBCDIC, and ASCII to represent letters, numbers, punctuation, and a host of other characters.

Troubleshooting adder circuits requires, as with all electronic circuits, knowing what results are expected, to determine what is wrong with the circuit. For adder and subtracter circuits, the first expectations are sums and differences of the input numbers. The next level is the individual outputs of the circuit sections within the adder or subtracter. Many times, the area of difficulty becomes obvious from the result of testing the circuit with a couple of input combinations. At other times, each stage of the circuit must be checked until the trouble is found.

Designing adder/subtracter circuits follows the general procedure for designing any logical circuit. Begin by listing the circuit requirements and developing truth tables for the logic circuits that will be used to meet those requirements. Once the design is completed, it must be checked for accurate operation. The BCD adder was used in this chapter to illustrate the process. The BCD adder circuit generates an adjustment value of 0110B whenever the original sum of two BCD digits exceeds 1001B or yields a carry out. Once the adjustment is added to the original result, the output of the adder appears as if the two BCD characters had been added together.

Glossary

address binary code for a memory location that stores data.

ASCII and **extended ASCII** abbreviation for American Standard Code for Information Interchange, which is a seven-bit binary code used to represent alpha, numeric, and other characters. An extended ASCII set, which includes an eighth binary bit, also includes many graphic and specialty characters.

base or **radix** a number system's base value.

BCD (binary-coded decimal) a binary code representing decimal (0–9) characters.

BCD (or **decade BCD) adder** an adder circuit that sums two binary-coded decimal (BCD) numbers.

bit a single binary digit.

bus a group of wires that carries binary data.

byte a group of eight binary bits.

carry overflow that occurs when the number of bits in the sum are more than can be held in the circuit that stores the result.

complement the inverted form or negative equivalent of a number.

complementing the process by which numbers are inverted.

data binary words that represent information or numbers.

data link characters part of a character code set that defines control functions for establishing a communication link between two stations.

default quantity or form that is used when it is unspecified.

double word a 32-bit binary number.

EBCDIC (Extended Binary-Coded Decimal Interchange Code) IBM's character code.

flag or **status flag** an indication of the state of a condition.

full adder an adder circuit that adds two binary bits and a carry-in bit and produces a sum and carry out.

half adder an adder circuit that adds two binary bits producing a sum and carry out.

hex (hexadecimal) the base 16 number system, each digit of which can be represented by a four-bit binary word.

invert the opposite value of a binary state.

inverting the process of altering the logic state of a bit or binary quantity.

least significant digit rightmost digit of a number.

long word a 32-bit binary number.

LSB (least significant bit) the rightmost binary digit, which has the lowest weighted value.

MSB (most significant bit) the leftmost binary digit, which has the largest weighted value.

nibble a four-bit binary number.

nine's complement the first step in negating a decimal number.

octal the base 8 number system, each digit of which can be represented by a three-bit binary word.

overflow a condition under which the result of a mathematical value produces a value too large to be represented by the number of digits available.

parallel adder a circuit that sums several binary bits simultaneously.

quad word a 64-bit binary number.

scale value a weight value of a number.

sign–magnitude system the representation of a binary number whose MSB is the sign of the number and whose remaining bits are its actual magnitude.

signed number a negative or positive number.

ten's complement the negative equivalent of a decimal number.

unsigned number a magnitude-only number with no reference to negative or positive quantity.

weight the multiplier or scale value of a digit determined by its position in a number.

word original definition: a fixed number of binary bits used at a given time. Second definition (based on de facto use by microprocessor manufacturers): 16-bit binary number. Also, 32-bit quantities are often designated as **double** or **long words.**

Solutions to Practice Problems

Section 3.2

1. a. 67,234: $4 \times 1 = 4$
$3 \times 10 = 30$
$2 \times 100 = 200$
$7 \times 1{,}000 = 7{,}000$
$6 \times 10{,}000 = 60{,}000$

b. 123,456: $6 \times 1 = 6$
$5 \times 10 = 50$
$4 \times 100 = 400$
$3 \times 1{,}000 = 3{,}000$
$2 \times 10{,}000 = 20{,}000$
$1 \times 100{,}000 = 100{,}000$

c. 399: $9 \times 1 = 9$
$9 \times 10 = 90$
$3 \times 100 = 300$

d. 4,001: $1 \times 1 = 1$
$0 \times 10 = 0$
$0 \times 100 = 0$
$4 \times 1{,}000 = 4{,}000$

2. a. 60,000 + 7,000 + 200 + 30 + 4 = 67,234
b. 100,000 + 20,000 + 3,000 + 400 + 50 + 6 = 123,456
c. 300 + 90 + 9 = 399
d. 4,000 + 0 + 0 + 1 = 4,001

Section 3.3

1. a. 123.456: $6 \times 0.001 = 0.006$
$5 \times 0.01 = 0.05$
$4 \times 0.1 = 0.4$
$3 \times 1 = 3$
$2 \times 10 = 20$
$1 \times 100 = 100$

b. 567,900.02: $2 \times 0.01 = 0.02$
$0 \times 0.1 = 0$
$0 \times 1 = 0$
$0 \times 10 = 0$
$9 \times 100 = 900$
$7 \times 1{,}000 = 7{,}000$
$6 \times 10{,}000 = 60{,}000$
$5 \times 100{,}000 = 500{,}000$

c. 39.04508: $8 \times 0.00001 = 0.00008$
$0 \times 0.0001 = 0$
$5 \times 0.001 = 0.005$
$4 \times 0.01 = 0.04$
$0 \times 0.1 = 0$
$9 \times 1 = 9$
$3 \times 10 = 30$

2. a. 100 + 20 + 3 + 0.4 + 0.05 + 0.006 = 123.456
b. 500,000 + 60,000 + 7,000 + 900 + 0 + 0 + 0 + 0.02 = 567,900.02
c. 30 + 9 + 0 + 0.04 + 0.005 + 0 + 0.00008 = 39.04508

Section 3.4

1. 221 **2.** 73 **3.** 127 **4.** 131

Section 3.5

1. a. 45 **b.** 120,021 **c.** 35 **d.** 11,101

2. a. 100110_2 100110B 'b'100110
b. 1110011_2 1110011B 'b'1110011
c. 10101010_2 10101010B 'b'10101010

Section 3.6

1. 23.3125 **2.** 1.875 **3.** 17.5625

Section 3.7

1. a. 1111011.011101B **b.** 1100001.000011B
c. 110010000.00000001B **d.** 111111111.00000001111B

2. a. 64 + 32 + 16 + 8 + 2 + 1 + .25 + .125 + .0625 + .015625 = 123.453
b. 64 + 32 + 1 + .03125 + .015625 = 97.047
c. 256 + 128 + 16 + .0039 = 400.004
d. 256 + 128 + 64 + 32 + 16 + 8 + 4 + 2 + 1 + .00390625 + .001953125 + .000965625 + .00048828125 = 511.007313

Section 3.8

1. a. 43# = 4 × 8 + 3 × 1 = 35
100011B = 32 + 2 + 1 = 35
b. 716# = 7 × 64 + 1 × 8 + 6 = 462
111001110B = 256 + 128 + 64 + 8 + 4 + 2 = 462
c. 4163# = 4 × 512 + 1 × 64 + 8 × 8 + 3 = 2,163
100001110011B = 2048 + 64 + 32 + 16 + 2 + 1 = 2,163

2. a. 100110110B = 256 + 32 + 16 + 4 + 2 = 310
466# = 4 × 64 + 6 × 8 + 6 × 1 = 310
b. 001101010100100B = 4,096 + 2,048 + 512 + 128 + 32 + 4 = 6,820
15244# = 1 × 4,096 + 5 × 512 + 2 × 64 + 4 × 8 + 4 × 1 = 6,820
c. 011011101001B = 1,024 + 512 + 128 + 64 + 32 + 8 + 1 = 1,769
3351# = 3 × 512 + 3 × 64 + 5 × 8 + 1 × 1 = 1,769

Section 3.9

1. a. 1001110101B = 512 + 64 + 32 + 16 + 4 + 1 = 629
275H = 2 × 256 + 7 × 16 + 5 × 1 = 629
b. 1110010001B = 512 + 256 + 128 + 16 + 1 = 913
391H = 3 × 256 + 9 × 16 + 1 × 1 = 913
c. 101011000011B = 2,048 + 512 + 128 + 64 + 2 + 1 = 2,755
AC3H = 10 × 256 + 12 × 16 + 3 × 1 = 2,755

2. a. 3C1H = 3 × 256 + 12 × 16 + 1 × 1 = 961
1111000001B = 512 + 256 = 128 + 64 + 1 = 961
b. 15F2H = 1 × 4,096 + 5 × 256 + 15 × 16 + 2 × 1 = 5,618
1010111110010B = 4,096 + 1,024 + 256 + 128 + 64 + 32 + 16 + 2 = 5,618
c. 345H = 3 × 256 + 4 × 16 + 5 × 1 = 837
1101000101B = 512 + 256 + 64 + 4 + 1 = 837

Section 3.10

1. a. 1011101B **b.** 110010110B **c.** 1001000B **d.** 1011111110B

2. a. 32 + 4 + 2 = 38 32 + 16 + 4 + 2 + 1 = 55
38 + 55 = 93
64 + 16 + 8 + 4 + 1 = 93
b. 128 + 64 + 32 + 16 + 4 + 2 + 1 = 247
128 + 16 + 8 + 4 + 2 + 1 = 159 247 + 159 = 406
256 + 128 + 16 + 4 + 2 = 406
c. 32 + 8 + 2 = 42 16 + 8 + 4 + 2 = 30 42 + 30 = 72
64 + 8 = 72
d. 256 + 128 + 4 + 2 + 1 = 391
256 + 64 + 32 + 16 + 4 + 2 + 1 = 375 391 + 375 = 766
512 + 128 + 64 + 32 + 16 + 8 + 4 + 2 = 766

Section 3.11

1. a. 10101B **b.** 11111B **c.** 100011B **d.** 1011101B

2. a. 64 + 32 + 2 + 1 = 99 64 + 8 + 4 + 2 = 78
99 − 78 = 21 16 + 4 + 1 = 21
b. 128 + 16 + 8 + 4 + 1 = 157 64 + 32 + 16 + 8 + 4 + 2 = 126
157 − 126 = 31 16 + 8 + 4 + 2 + 1 = 31
c. 64 = 32 + 16 + 8 = 120 64 + 16 + 4 + 1 = 85
120 − 85 = 35 32 + 2 + 1 = 35

d. 128 = 128   32 + 2 + 1 = 35   128 − 35 = 93
64 + 16 + 8 + 4 + 1 = 93

Section 3.12

1.

A	B	DIFFERENCE (A − B)	BORROW
0	0	0	0
0	1	1	1
1	0	1	0
1	1	0	0

DIFFERENCE = $A \oplus B$ BORROW = $\overline{A}B$

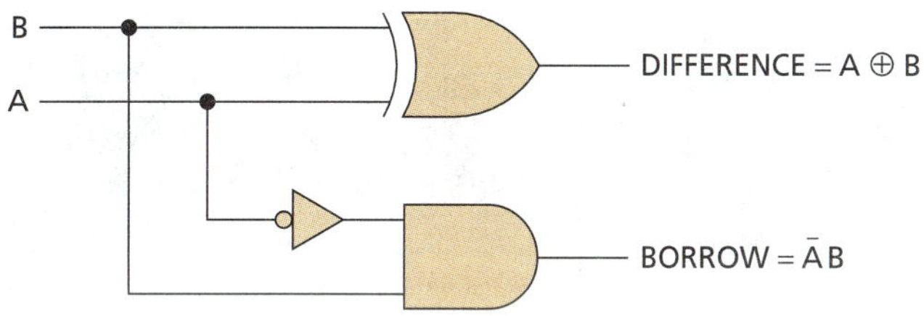

Figure 3.19 Practice Problem 3.12(1) (Solution)

2. A half subtracter does not consider the need to borrow because of the subtraction of the previous digit column.

Section 3.13

Borrow In (C)	A	B	A − B	Borrow Out
0	0	0	0	0
0	0	1	1	1
0	1	0	1	0
0	1	1	0	0
1	0	0	1	1
1	0	1	0	0
1	1	0	0	0
1	1	1	1	1

Difference = $A \oplus B \oplus C$ Borrow out = $\overline{A} \cdot B + B \cdot C + \overline{A} \cdot C$
Borrow in, requested borrow from preceding binary column
Borrow out, borrow required to do the subtraction of A − B − C

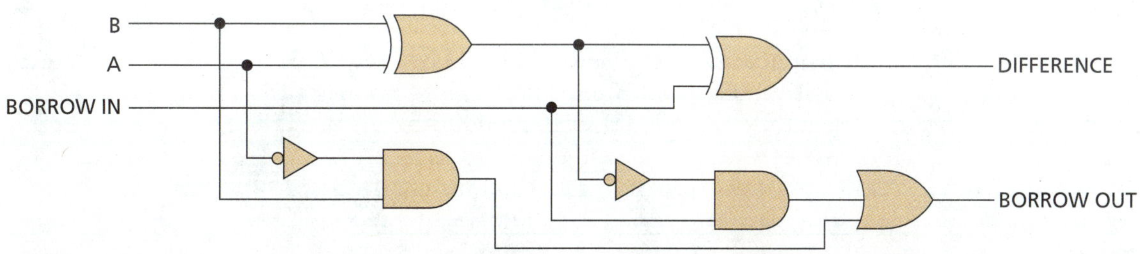

Figure 3.20 Practice Problem 3.13 (Solution)

Section 3.14

The lower four bits are added in the right 74LS83 four-bit adder; 1101B + 0101B = sum of 0010B and carry = 1. The carry is added to the sum of the upper four bits in the other 74LS83; 1001B + 1110B + 1 = sum of 1000 and a carry = 1. Combined, the eight-bit addition is 10011101B + 11100101B = 10000010B with a carry = 1.

Section 3.15

1. a.
$$\begin{array}{rcr} 1234 & = & 1234 \\ \underline{-567} & = & \underline{+9433} \\ & & 10667 \end{array}$$

b.
$$\begin{array}{rcr} 520034 & = & 520034 \\ \underline{-187099} & = & \underline{+812901} \\ & & 1332935 \end{array}$$

c.
$$\begin{array}{rcr} 380207 & = & 380207 \\ \underline{-728965} & = & \underline{+271035} \\ & & 651242 \end{array} \qquad \begin{array}{r} 999999 \\ \underline{-651242} \\ 348757 \\ \underline{+\qquad 1} \\ -348758 \end{array}$$

2. a. 1234 − 567 = 667 **b.** 520034 − 187099 = 332935
c. (−728965 + 380207) = −398758

Section 3.16

1. a. 100000000B **b.** −10110111B **c.** 1010000B **d.** −1000111B

2. a. 413 − 157 = 256 **b.** 270 − 453 = −183
c. 119 − 39 = 80 **d.** 338 − 409 = −71

Section 3.17

1. With input word A = 1001101B and word B = 10001110B and the SUB input 0 for addition, the output of the XORs is the same as the B word. The words at the adder inputs are summed to yield a result of 01011011B with C_{OUT} = 1. When SUB is brought high, the XOR outputs are the inverse of the input word B: 01110001B. This inverted number and word A are added together, along with a C_{IN} of 1, to yield a difference of 00111111B with C_{OUT} = 1. A high at C_{OUT} indicates that the difference is a positive real value.

2. When SUB is high for subtraction and C_{OUT} is low, the difference result is the two's complement negative value of the true difference.

Section 3.18

1. +111 **b.** −127 **c.** +127 **d.** −0 **e.** −30

2. a. +111 **b.** −1 **c.** +127 **d.** −256 **e.** −98

3. a. 111 **b.** 255 **c.** 127 **d.** 256 **e.** 158

4. Unsigned: 00000000B to 11111111B = 0 to 255
Positive: 00000000B to 01111111B = +0 to +127
Negative: 11111111B to 10000000B = −1 to −128

Section 3.19

1.
$$\begin{array}{rcr} 123H & = & 291 \\ \underline{+ABCH} & = & \underline{+2{,}748} \\ BDFH & = & 3{,}039 \end{array}$$

2.
$$\begin{array}{rcr} C01H & = & 3{,}073 \\ \underline{+800H} & = & \underline{+2{,}048} \\ 1401H & = & 5{,}121 \end{array}$$

3.
$$\begin{array}{rcr} 44EH & = & 1{,}102 \\ \underline{+1FFH} & = & \underline{+\ 511} \\ 64DH & = & 1{,}613 \end{array}$$

4.
$$\begin{array}{rcr} 6DAH & = & 1{,}754 \\ \underline{+A3CH} & = & \underline{+2{,}620} \\ 1116H & = & 4{,}374 \end{array}$$

Section 3.20

1. a. 34CDAH = 0011 0100 1100 1101 1010B
CB325H = 1100 1011 0011 0010 0101B fifteen's & one's
CB326H = 1100 1011 0011 0010 0110B sixteen's & two's

b. ABCDOH = 1010 1011 1100 1101 0000B
5432FH = 0101 0100 0011 0010 1111B fifteen's & one's
5433OH = 0101 0100 0011 0011 0000B sixteen's & two's

c. 001234H = 0000 0000 0001 0010 0011 0100B
FFEDCBH = 1111 1111 1110 1101 1100 1011B fifteen's & one's
FFEDCCH = 1111 1111 1110 1101 1100 1100B sixteen's & two's

d. 80000H = 1000 0000 0000 0000 0000B
7FFFFH = 0111 1111 1111 1111 1111B fifteen's & one's
80000H = 1000 0000 0000 0000 0000B sixteen's & two's

2. a.

2389H	2389H
−1001H	+EFFFH
	11388H

b.

60D00H	60D00H
−4EB01H	+B14FFH
	1121FFH

Leading 1 as final carry out indicates numbers are positive. Carry is discarded, 2389H − 1001H = 1388H and 60D00H − 4EB01H = 121FFH

c.

5E0B2H	5E0B2H
−99CCFH	+66331H
	C43E3H

No final carry out, so the answer must be complemented to ARRIVE at its actual negative value:

FFFFFH
−C43E3H
3BC1BH
+ 1
3BC1DH

Section 3.21

1. a. 0100 0111 0010BCD **b.** 1001 1000 0111 0110BCD
c. 0001 1001 0010 1000 0011 0111BCD

2. a. 8,637 **b.** 2,791 **c.** 149,763

Section 3.22

1.	**2.**	**3.**
391	57189	4567
+5239	+59228	+9876
55CA	B03B1	DDDD
+ 66	+66 66	+6666
5630	116417	14443

Section 3.23

1. E3 20 69 73 20 3D 20 74 6F 20 33 2E 31 34H
1110 0011 0010 0000 0110 1001 0111 0011 0010 0000 0011
1101 0010 0000 0111 0100 0110 1111 0010 0000 0011 0011
0010 1110 0011 0001 0011 0011 0100B

2. 50 75 74 20 6F 6E 20 61 20 01 21H
0101 0000 0111 0101 0111 0100 0010 0000 0110 1111 0110 1110 0010 0000 0110 0001 0010 0000 0000 0001 0010 0000 0100 0001B

Caution: Spaces are used for clarity and are not actually in the sequence. Notice the difference between hex and binary in length and readability.

Section 3.24

1. There are two problems with this circuit—two bad XOR gates, one fed by C_{OUT} and one driving B_0.
2. Carry out of the full adder with bits A_1 and B_1 is a 0 when it should be a 1. The remaining full adders are working correctly.

Section 3.25

1. Word A = 1000B and word B = 0101B. This produces a sum of 1101B. S_3 and S_2 are both high, which generates a 1 to the lower adder, causing it to add 0110B to the sum result, producing a final result of 0011B and a carry out = 1. This makes the results appear as if this addition were performed: 1000BCD + 0101BCD = 1 0011BCD or 8 + 5 = 13.
2. Applying non-BCD numbers to the inputs of a BCD adder will yield incorrect results. Look what happens when 1010B is added to 1110B. Both numbers are non-BCD types. The sum from the first adder is 0000B with a carry out = 1. The carry out causes 0110B to be added in the lower circuit. The final results are 0110B with a carry out of 1. This makes no sense, if you interpret the results as being BCD characters, because 1010B + 1110B does not equal 1 0110B (10 + 14 does not equal 22 or 16).

Problems

3.1 Convert the following decimal numbers into digit–weight products. Show how these products are used to develop the original decimal values.
a. 3,497 b. 255.396 c. 6,003.205
d. 34.00678 e. 98,207.5102

3.2 Convert the following decimal numbers into digit–weight products. Show how these products are used to develop the original decimal values.
a. 475 b. 1,234.56 c. 858,323.454
d. 999,999 e. 10,000.005

3.3 Convert the following numbers into equivalent decimal values.
a. 1234# b. 8355H c. 010111B
d. 65CDH e. 10101110B

3.4 Convert the following numbers into equivalent decimal values.
a. 10001101B b. 111100011B c. 444#
d. ABCDH e. 5A6BH

3.5 Which binary number in each pair has the greater unsigned value? Prove it by converting each pair into their decimal values and comparing them.
a. 10011B and 10001B
b. 11101101B and 11100101B
c. 100001100001B and 100011000001B

3.6 Which binary number in each pair has the higher signed value in a sign–magnitude system? Consider positive numbers to be higher than negative ones. Prove it by converting each pair into their decimal values and comparing them.
a. 10011B and 10001B
b. 11101101B and 11100101B
c. 100001100001B and 100011000001B

3.7 Determine the decimal value for each of these numbers:
a. 101.111011B b. 1234.567D
c. 11.011D d. 111011.11B
e. 1001.1001B

3.8 Determine the decimal value for each of these numbers:
a. 1111.111B b. 110011.110011B
c. 11.0101D d. 10000001.00001B
e. 11100011.00001111B

3.9 Convert the following decimal numbers to their binary values:
a. 1,298 b. 628.35 c. 59.875 d. 31,298
e. 1,024.75

3.10 Convert these decimal numbers to their base 2 equivalent values:
a. 123 b. 56.25 c. 100.052 d. 555.55
e. 6.252

3.11 Add the following pairs of binary numbers:
a. 1011101B + 100011B
b. 111011B + 111011B
c. 100111B + 1110001B
d. 1101.1101B + 1110.0011B
e. 100011.1B + 101.0011B

3.12 Do the following binary additions.
a. 11101B + 10011B
b. 100111B + 100001B
c. 110000.011B + 11101.001B
d. 110100B + 101101B
e. 1010.0101B + 1111.1111B

3.13 Verify that each addition in Problem 3.11 is correct by converting the binary values into decimal numbers and checking that the decimal sums are the same as the binary sums.

3.14 Verify that each sum in Problem 3.12 is correct by converting each binary value into a decimal number and checking the decimal addition.

3.15 Subtract the following pairs of binary numbers:
a. 1011101B − 100011B
b. 111011B − 111011B
c. 100111B − 1110001B
d. 1101.1101B − 1110.0011B
e. 100011.1B − 101.0011B

3.16 Do the following binary subtractions:
a. 11101B − 10011B
b. 100111B − 100001B
c. 110000.011B − 11101.001B
d. 110100B − 101101B
e. 1010.0101B − 1111.1111B

3.17 Verify that each subtraction in Problem 3.15 is correct by converting the binary values into decimal numbers and checking that the decimal differences are the same as the binary differences.

3.18 Verify that each difference in Problem 3.16 is correct by converting each binary value into a decimal number and checking the decimal subtraction.

3.19 What is the ten's complement of these decimal numbers?
a. 726 b. 1,032 c. 53.78 d. 59,123
e. 812,569

3.20 What is the ten's complement of these decimal numbers?
a. 7020 b. 21,000 c. 883.708 d. 3,193
e. 810,023

3.21 What is the two's complement of these binary numbers?
a. 1100100B b. 100011B c. 10011.011B
d. 100000B e. 101101.101B

3.22 What is the two's complement of these binary numbers:
a. 101001B b. 11100011B c. 111.1011B
d. 10011000B e. 10111011.111001B

3.23 Repeat the subtractions in Problem 3.15 using two's complement addition. Show how the process is done for each pair of numbers.

3.24 Repeat the subtractions in Problem 3.16 using two's complement addition. Show how the process is done for each pair of numbers.

3.25 Convert the following sign–magnitude binary numbers to their decimal equivalents.
a. 0011011B b. 1101101B
c. 100011011B d. 0111100011B
e. 1111111B

10AM TOMORROW

3.26 Convert the following two's complement signed binary numbers into their decimal equivalents.
a. 0110110B **b.** 10011011B
c. 11111111B **d.** 0111010B
e. 100110100B

3.27 What are the maximum and minimum unsigned, positive, and negative decimal values that can be represented by a 16-bit binary number?

3.28 A computing system requires over one million memory locations, each holding binary data. The binary data length is to be large enough to accommodate equivalent maximum decimal values of −15,000 and +15,000.
a. How many binary bits are needed to select among the one million memory locations?
b. How many binary bits are needed to represent the range of data values?
c. How many actual memory locations can be selected using the binary bits in the answer to Problem a?
d. What are the actual maximum and minimum positive and negative values that are represented by the binary bits in the answer to Problem b?
e. What are the minimum and maximum unsigned values represented by the number of bits in the answer to Problem b?

3.29 Convert the following binary numbers into hexadecimal numbers, and verify that they have the same value.
a. 1101111101110B
b. 1001111001101B
c. 11111110110001B
d. 110000111111101101B
e. 10101100111000B
f. 1110111101110011B

3.30 Convert the following hexadecimal numbers to binary numbers, and verify that their values are the same.
a. 3CE2H **b.** 7FB05H **c.** 500AH
d. 2076EH **e.** E77AH **f.** BEADH

3.31 Find the sum of each of the following hexadecimal pairs, and verify that these sums are correct.
a. 3C0EH + A37EH **b.** D20C7H + 1FBE3H
c. 5B0H + AAAH

3.32 Find the sum of each of the following hexadecimal pairs, and verify that these sums are correct.
a. BB0H + 275H **b.** 12EFH + 9ACH
c. 10A2H + 2C59H

3.33 Find the fifteen's and sixteen's complement of the following hexadecimal numbers.
a. 2E0C5H **b.** A32000H **c.** C115BBH

3.34 Find the fifteen's and sixteen's complement of the following hexadecimal numbers.
a. 600BE0H **b.** AB12H **c.** 002C1EH

3.35 Find the difference for each of these hexadecimal pairs using sixteen's complement addition.
a. AC9H − 25BH **b.** C005H − 2E71H
c. 350H − 25AH **d.** C50H − 86BH
e. 9000H − 7A28H **f.** 13F2H − C00H

3.36 Do the following subtractions using sixteen's complement addition.
a. 27C3H − 8B3EH **b.** 5E2H − AC8H
c. 1C2EEH − 9F00AH

3.37 Convert the following decimal numbers into BCD codes.
a. 9,376 **b.** 500,785 **c.** 70,663
d. 84,093 **e.** 7,921 **f.** 78,029

3.38 Convert the following BCD codes into decimal characters.
a. 1000 0110 0000 0111 0010BCD
b. 0011 0101 0111 0100 0101BCD
c. 0001 0000 0000 1001 0011 0101BCD
d. 0111 1000 0011 0111 0000 0010BCD

3.39 When is the value 6H added to the original sum when using hexadecimal addition to get BCD sums?

3.40 Using hexadecimal addition, and adjusting for BCD, find the BCD sums of the following pairs of decimal numbers.
a. 3,095 + 7,689
b. 59,318 + 25,991
c. 7,625,803 + 9,747,189
d. 660,983 + 847,705
e. 987,123 + 996,388
f. 370,089 + 397,092

3.41 Convert the characters in the following messages to the hexadecimal representation of their extended ASCII codes. Do not forget the space character (20H) where needed.
a. 37 + 52 = 89 **b.** ←THIS WAY→

3.42 Convert the characters in the following messages to the hexadecimal representation of their extended ASCII codes. Do not forget the space character (20H) where needed.
a. Date: 8/23/93 **b.** The End!

3.43 Compare the truth tables for the sum and carry of a half adder and a full adder. Why is a half adder only doing "half" of the addition process?

3.44 Explain why the logic expression for the carry out of a full adder is different from the expression for the carry out of the half adder.

3.45 Explain the meaning of a high state on the carry out of a two's complement adder/subtracter when SUB is 0 and when SUB is 1. Show how this information leads to the design of the carry flag circuit.

3.46 Explain how a two's complement adder/subtracter does the actual two's complement process.

3.47 In the four-bit adder of Figure 3.21, applying the binary value 1100B to A and 0110B produces the correct sum. However, when number A is 0111B and B is 1001B, the sum is 1110B instead of 0000B. What is wrong with the adder and why?

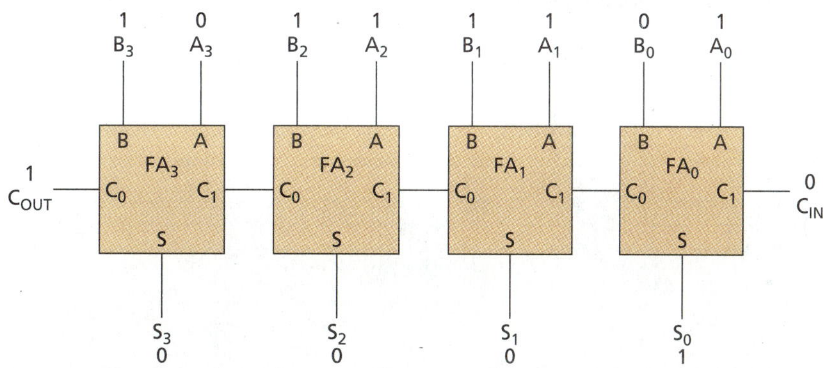

Logic States Shown for Problem 3.50

Figure 3.21 Problems 3.47 and 3.50

3.48 For the two's complement adder/subtracter of Figure 3.22 applying the inputs shown, are the results correct? If not, what is possibly wrong with the circuit?

3.49 What additional function is being performed by the circuit in Figure 3.23? When does the circuit perform the function? Logic states shown are for Problem 3.48 and are not necessarily all correct. Those states do not apply to this question.

3.50 Given the inputs and logic states shown in Figure 3.21 used in Problem 3.47, what is wrong with the circuit?

3.51 What is wrong with the BCD circuit of Figure 3.24, given the logic states shown?

3.52 The following pairs of BCD values were applied to a BCD adder like the one shown in Figure 3.24. What is possibly wrong with the circuit, given the results shown?

A =	1000	0110	0001	0011
B =	+0111	+0001	+1001	+0110
S =	1 0101	0111	1010	1001

3.53 Design the circuit required to generate an active high zero flag indicator for a four-bit parallel adder. This indicator is a 1 when all four sum outputs are 0 and is 0 when any one sum output is a 1.

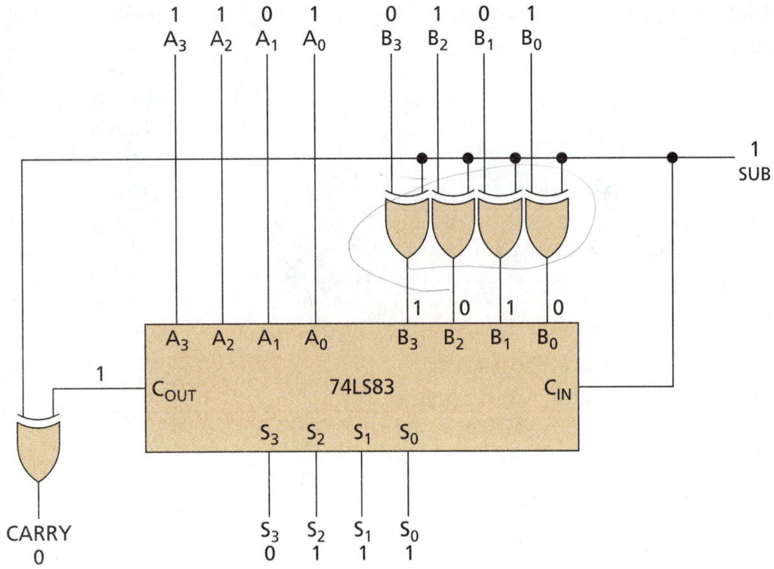

Figure 3.22 Problem 3.48

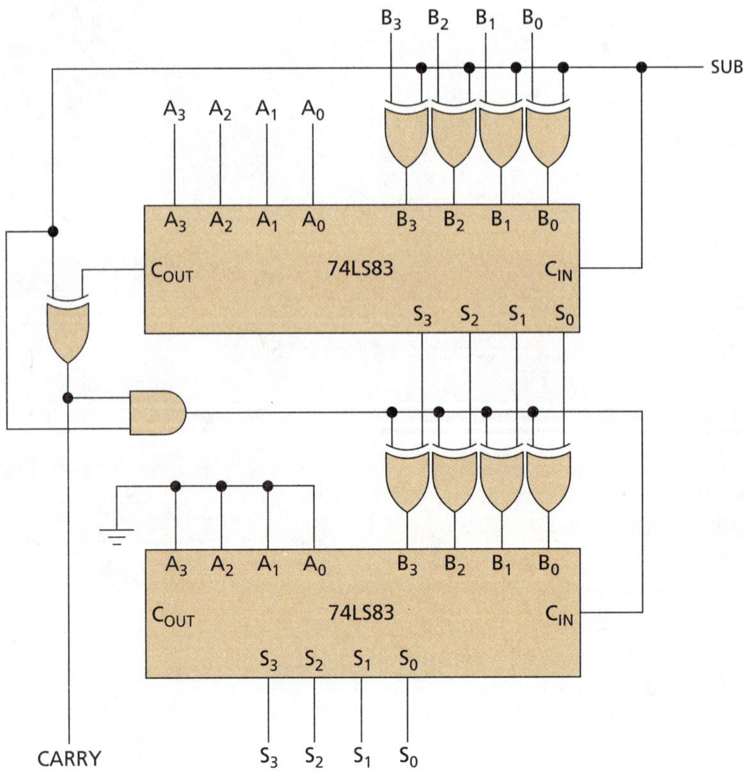

Figure 3.23 Problem 3.49

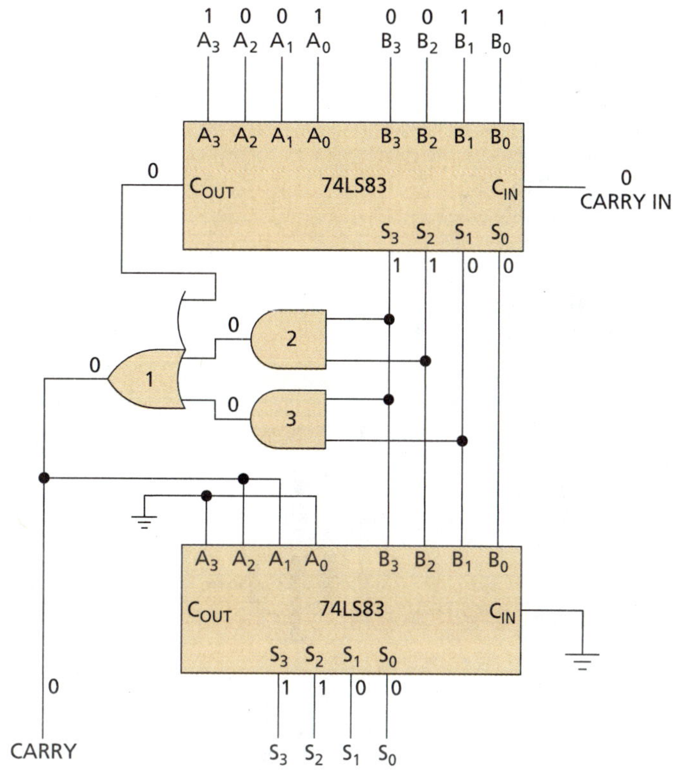

Figure 3.24 Problems 3.51 and 3.52

3.54 For the eight-bit adder of Figure 3.25, add the circuitry required to generate a sign flag. This indicator is in a high state for negative numbers and a low state for positive numbers.

3.55 *Bonus Design Problem:* Design a ten's complement BCD adder/subtracter.

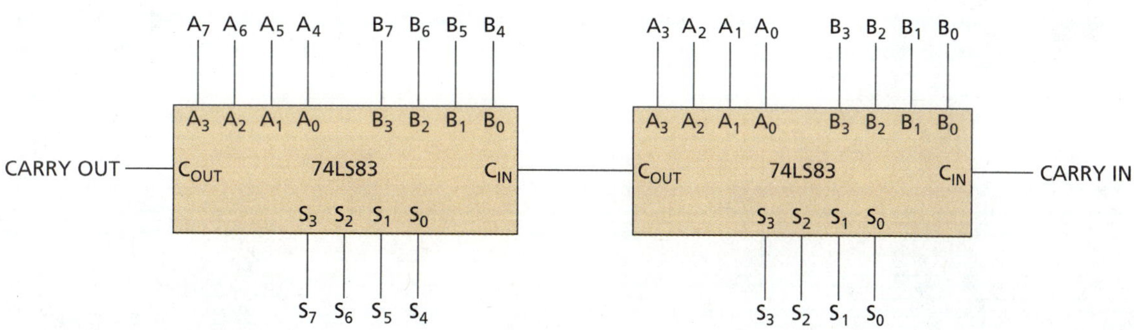

Figure 3.25 Problem 3.54

four

Flip-Flops and Counters

KEY TERMS

astable or free-running multivibrator (MVB)
asynchronous control signals
asynchronous counters
binary ripple counter
branch
CLOCK
control signal
counter
current or present state
data
data or D flip-flop
decoder
duty cycle
edge-triggered circuit
ENABLE
flip-flop
freeze
glitch
J–K flip-flop
Johnson counter
label
latch
level-triggered circuit
memory
MOD or modulo number
next state
no change
program counter
propagation delay
race condition
R–C time constant
register
RESET
ring counter
ripple count or RC
set
setup time
state diagram
state table
synchronous
terminal count or TC
timing diagram
toggle or T flip-flop
undefined or invalid condition

OBJECTIVES

After completing this chapter, the student will be able to

- Describe how the basic flip-flop latch is constructed from NAND or NOR gates and holds or stores a logic state.
- Describe the operation of the more common types of flip-flops, the R–S, the clocked R–S, the data, and toggle flip-flops, using truth tables and timing diagrams.
- Show how the universal J–K flip-flop can be used as any of the other types of flip-flop circuits.
- Describe a method to create a clock timing signal using a 555 timer circuit.
- Describe state tables and diagrams used to describe synchronous circuits, and apply them to several types of flip-flops.
- Use state tables, state diagrams, timing diagrams, and so on to illustrate the operation of counting circuits built from different types of flip-flop devices.
- Explain several methods used to troubleshoot counters, including signal tracing.
- Describe counter design using state tables and Karnaugh maps.

OVERVIEW

In this chapter logic circuits that are used for latching digital signals, counting binary numbers, and storing binary information are developed, explained, and illustrated with applications. These circuits respond to a set of inputs and retain the resulting output until a new set

Flip-flop *a logic circuit that changes from one state to another state, latching or holding the current state between changes.*

Register *a temporary data storage circuit.*

Counter *logic circuit that sequences through a set of binary values.*

of inputs are applied. The outputs remain latched even though the signals that generated those outputs have been removed. Commonly referred to as **flip-flops,** these circuits are used as the basis for counting and memory-type logic operations.

This chapter continues with an exploration of the concepts, design, and troubleshooting techniques involving circuits constructed from flip-flops. The main two types of such circuits are **registers** and **counters.** Registers are *any* digital circuit that can temporarily hold or store binary information. When you type in a number on your calculator and it is displayed on the readout, registers are at work. They hold the binary equivalent of the number you just entered into the calculator. They hold the information that drives the display so that you can see the number entered. These storage elements are definitely temporary, because when you enter the next number into the calculator, all these registers change to hold the new number. Register circuits are discussed in detail in Chapter 5.

Control signal *an input to a logic circuit that enables or disables some activity within that circuit.*

Counters are logic circuits that change in a sequential manner when a clock signal is applied. For instance, a circuit capable of sequencing through a binary-coded decimal (BCD) sequence (0000BCD to 1001BCD) is a decade counter. The starting point and the number of cycles these circuits sequence through can be controlled using binary signals called **control signals.**

After exploring the basic theories of individual counter circuits, integrated chip (IC) examples of some of the counters are given. Troubleshooting techniques and counter design using state tables and diagrams complete the chapter.

4.1 Introduction

Propagation delay *the time between applying an input signal to a device and getting a change of state condition at the output.*

Latch *a logic circuit that changes to a state and then holds that state until a new set of signals is applied.*

Memory *digital devices or circuits used to retain or store binary information (data).*

Until now our study of logic circuits has involved static elements. That is, the outputs of previous circuits were determined solely by the inputs that were being applied immediately to the circuit. Shortly after the inputs were applied, and after device propagation delays, outputs were generated. **Propagation delay** is the time required for a circuit or device to sense a change of input state and cause its output to change. Removing inputs generally resulted in outputs changing or becoming indeterminable. For instance, imagine a full adder with input A = 1, input B = 1, and CARRY IN = 0. The sum output is 0 with a carry of 1. If inputs A, B, and CARRY IN are then removed—that is, the inputs are left open—what happens to the outputs? Depending on the type of logic devices used, the SUM and CARRY OUT outputs take on either a high or low state. The inputs were not latched, so the circuit changed in reaction to what it thought the inputs were. If a new set of inputs are applied, once again the SUM and CARRY OUT take on their appropriate states. The change is nearly instantaneous and does not depend on the influence of another signal. In contrast, flip-flop circuit outputs are generated by the given inputs, but retain their states after the inputs are removed. The outputs can also be changed by a different set of inputs, but the element gained by using flip-flops is the **latch** or **memory** ability of the circuit to hold the current information after the inputs have been removed.

4.2 The Basic Flip-Flop Latch

Figure 4.1 illustrates the two forms of the basic latch or flip-flop circuit. The term *latch* refers to the circuit's ability to latch or remain in a state once inputs are removed. The word *flip-flop* refers to these circuits' ability to change from one latched state to another determined by inputs applied to them, thus flip-flopping from one state to the other. They are called the cross-coupled *NAND latch* (gates 1 and 2 in Figure 4.1a) and cross-coupled *NOR latch* (gates 3 and 4 in Figure 4.1b). For the NAND latch, the inputs are active low and labeled $\overline{S}$ for $\overline{SET}$ and $\overline{R}$ for $\overline{RESET}$, and the outputs are Q and $\overline{Q}$ (not-Q). The difference with the NOR latch is that the S and R inputs are active high instead of active low and are labeled S and R. The cross-coupled latch is characterized by the outputs of the gates being coupled across to the input of one of the opposite gates.

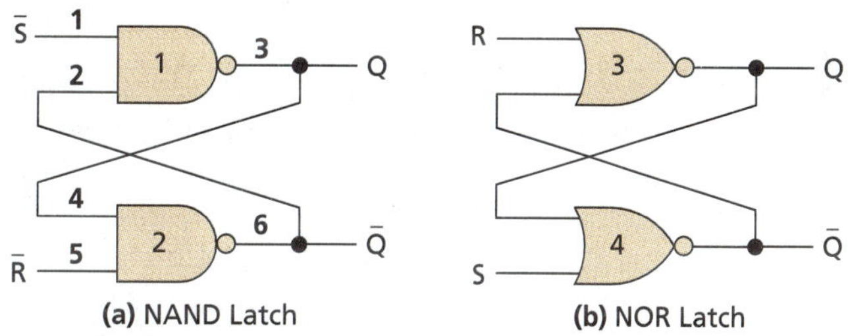

Figure 4.1 Cross-Coupled Latches

Race condition *an undeterminable state of a logic circuit whose outcome is based more on the propagation delay of the individual circuit parts than the logic of the circuit.*

With nothing connected to the $\overline{R}$ and $\overline{S}$ inputs of the NAND latch, power is applied to the circuit and a **race condition** begins. That is, both gates, believing that their inputs are 1, try to go low. One of the two gates will win the race and reach a logic 0 state before the other one. This logic 0 is cross-coupled to the input of the opposite gate and, as you can expect, a 0 into the NAND gate, forces its output to a 1 state. The 1 is coupled back to the input of the first gate, and the circuit is latched into a steady state condition with one output low and the other high.

EXAMPLE 4.1

Show the initial state of the NAND flip-flop if gate 1 of Figure 4.1 wins the race.

SOLUTION Pin 3 of gate 1 goes low to a zero state before pin 6 does. The zero state from pin 3 is coupled to pin 4 of gate 2, causing its output to go to a one state. The one state at pin 6 is connected back to pin 2 of gate 1 and, along with the one at pin 1, keeps NAND output pin 3 at a zero state. The zero coupled back to pin 4 keeps pin 6 at a one state. Thus the flip-flop is latched with output Q = 0 and $\overline{Q}$ = 1.

The basic characteristic of cross-coupled latches is that they remain in the latched state until new logic signals are applied to the inputs. A second characteristic of the latch and all flip-flops is that the Q and $\overline{Q}$ outputs are always opposite to one another. Whenever this situation is not true, then the circuit is not acting as a flip-flop and is in an undefined state.

Set *the logic 1 condition of the Q output of a flip-flop.*

RESET *as applied to counters, the signal that returns a counter to its initial condition. As applied to flip-flops, it is the logic 0 condition of the Q output of a flip-flop.*

Applying a 0 to the $\overline{S}$ input of the NAND flip-flop and a 1 to the $\overline{R}$ input causes the flip-flop to go into a **set** condition characterized by a 1 at Q and a 0 at $\overline{Q}$. The Q output goes high and couples the 1 logic state to input at pin 4. The 1's at pins 4 and 5 force the $\overline{Q}$ to go low. The 0 is then coupled back to pin 2, holding the Q output high. Reversing the input states causes the latch to flip to the **reset** state characterized by a 0 at the Q output and 1 at $\overline{Q}$. The 0 on the $\overline{R}$ input forces the $\overline{Q}$ output to go high. This high is coupled back to pin 2, which, along with the high at pin 1, forces Q to a 0. This 0 is coupled back to pin 4, holding $\overline{Q}$ at a 1.

The terms *set* and *reset* are convenient to use to describe the output states of the flip-flops, but these terms are not limited to that definition. The term *set* does refer to the condition of the flip-flop when Q = 1, but it is also defined as the input conditions that place the flip-flop in that state. Thus, the cross-coupled latch can be said to be set when the SET input is asserted. The purpose is not to confuse the user, but to present a consistent indication for the operation of the flip-flop. One could ask, "What conditions does it take to set a cross-coupled latch?" The answer would be "Asserting the SET input and negating the RESET input." A similar statement can be made for the term *reset.*

No change *condition in which a flip-flop or circuit retains its current state.*

Placing a logic 1, the inactive or nonforcing state, on both $\overline{R}$ and $\overline{S}$ inputs of the NAND latch does not change the output states of the flip-flop. This **"no change"** condition is used in logic circuits as the "memory" input condition. The flip-flop remembers the last state it was set in when both inputs are inactive. Let's look back at the set condition of the flip-flop that was achieved by placing a 0 on $\overline{S}$ and a 1 on $\overline{R}$. While still in the set condition, change the input on $\overline{S}$ to a 1. What does the flip-flop do? Nothing! The 0 from $\overline{Q}$, fed back to the other input of NAND gate number 1, forces that gate to keep a 1 at its output. This 1, coupled with the 1 on the $\overline{R}$ input, keeps its NAND gate output low at Q holding the latch condition. The only way to change the state of the flip-flop is to place a 0 on $\overline{R}$ while $\overline{S}$ is held high.

EXAMPLE 4.2

Illustrate the no-change condition of the cross-coupled NAND.

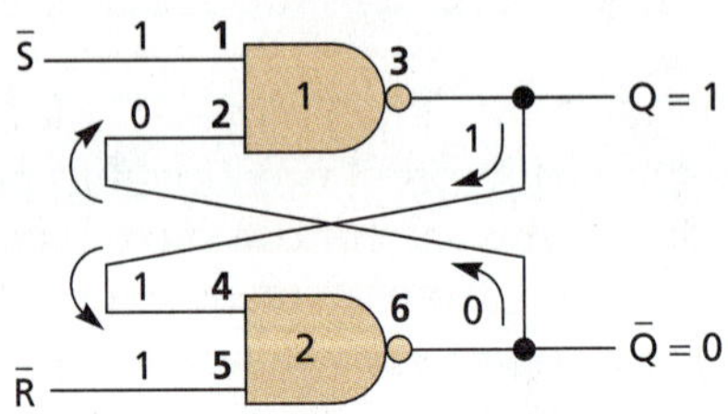

Figure 4.2 No Change Inputs

SOLUTION Suppose the inputs to the flip-flop are $\overline{S} = 0$ and $\overline{R} = 1$. This causes Q to be a 1 and $\overline{Q}$ to be 0. If we now make $\overline{S} = 1$ and hold $\overline{R}$ to a 1, nothing changes. The 0 from $\overline{Q}$ still causes pin 2 to be low and keeps the Q output high. The high from Q, coupled with the high input from $\overline{R}$, holds $\overline{Q}$ low. The result is that 1's on both inputs do not cause the state of the flip-flop to change, and the circuit is latched as shown in Figure 4.2.

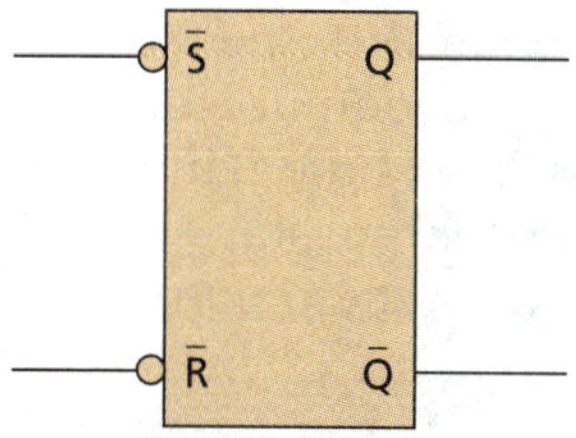

$\overline{S}$	$\overline{R}$	Q	$\overline{Q}$	State
0	0	1	1	Undefined
0	1	1	0	Set
1	0	0	1	Reset
1	1	—	—	No Change

Figure 4.3 NAND Flip-Flop Truth Table

The input and output states of the NAND latch are summarized in the truth table based on the discussion to this point and shown in Figure 4.3. This table also gives us an opportunity to define some terms involved with flip-flops. The condition of a flip-flop when Q is a 1 and $\overline{Q}$ is a 0 is called the **set** state of the flip-flop. The **reset** state of a flip-flop occurs when the outputs are in the opposite state, $Q = 0$ and $\overline{Q} = 1$. The S and R inputs have a bar across them because their active state is a low. That is a 0 on $\overline{S}$ forces the set state, and a 0 on $\overline{R}$ forces the reset state, so that the active state is often referred to as the *forcing state.* As shown in the previous paragraphs, making both inputs inactive (1 inputs for the NAND latch) places the flip-flop in a "no change" condition. Last, the state of the flip-flop when both inputs are 0 is **undefined** or **invalid.** The reason for this is that a 0 on $\overline{R}$ and a 0 on $\overline{S}$ forces both Q and $\overline{Q}$ to be high, which violates one of the conditions of a flip-flop that defined its operation as a latch. That condition is that Q and $\overline{Q}$ are always opposite in state. Also illustrated in Figure 4.3 is the logic block symbol for the cross-coupled NAND latch. A more common name for this circuit is the *reset–set* or *R–S flip-flop.*

Undefined** or **invalid condition *the condition of a flip-flop when Q and $\overline{Q}$ outputs are the same state, which violates the basic definition of flip-flops.*

R–S flip-flops can also be made from NOR gates, as shown in Figure 4.1b. Note that one main difference is that with the NOR gate the active state of the control signal is high and, as a result, there are no bars over the R and S inputs. The truth table and logic block for the NOR latch are shown in Figure 4.4. The

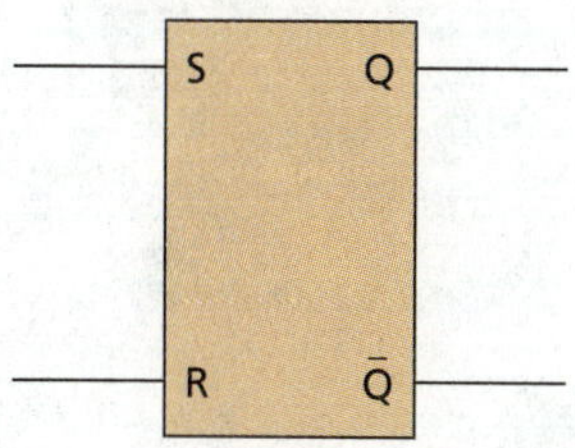

S	R	Q	$\overline{Q}$	State
0	0	X	X	No Change
0	1	0	1	Reset
1	0	1	0	Set
1	1	0	0	Undefined

X = "don't care" state

Figure 4.4 NOR Latch Truth Table

operation of the circuit is similar to that of the NAND except that the active state of the logic inputs is a 1 rather than a 0.

EXAMPLE 4.3 Show how the NOR is latched into the set state.

SOLUTION To set the NOR latch, a 1 is applied to the S input and a 0 to the R input. The 1 input to a NOR gate forces its output low. The 0 coupled to the opposite gate, in conjunction with the 0 on the R input, causes the output of that NOR gate to go high. That output is coupled back to the first NOR gate and the flip-flop is latched. Bringing the S input low while maintaining a 0 on the R input has no effect on the output states of the flip-flop, because it will be held latched by the 1 coupled from the Q output to the set NOR gate.

Application Note

NAND Latch

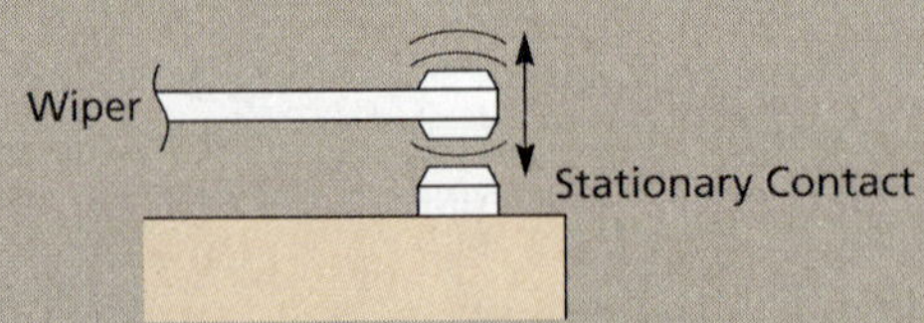

Figure 4.5 Switch Bounce

Mechanical switches have an inherent problem with contact bounce. That is, when the movable wiper strikes a stationary contact, as shown in Figure 4.5, that wiper physically bounces off the stationary contact, breaking the connection. The wiper returns to make contact with the stationary pole and then bounces again. In between making contact on each bounce, the wiper is in an open or no-contact state. This occurs until the bounces dampen out and the wiper finally rests on the stationary contact.

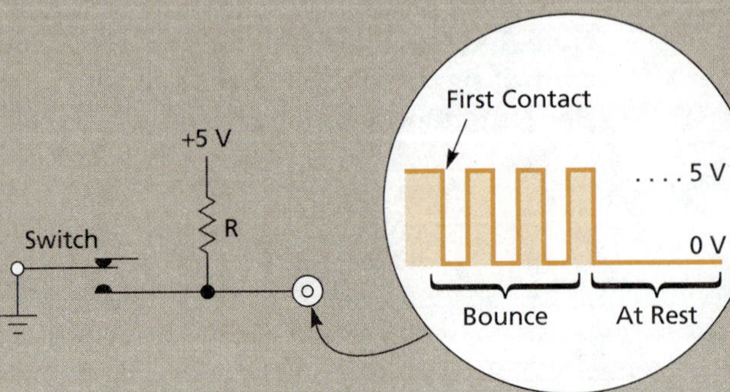

Figure 4.6 Electrical Switch Bounce

Electrically, the bounce can be seen on an oscilloscope connected to the wiper as in Figure 4.6. When the wiper first makes contact, the voltage goes low. When the first break occurs due to the bounce, the voltage returns high through the pull-up resistor. The wiper does not bounce far enough to remake connection with the opposite stationary contact. Instead, when a bounce occurs the moving wiper breaks connection and acts as an open circuit. The high level seen on the oscilloscope is a result of the resistor between the wiper and ground.

When contact is made again, by the wiper returning to the stationary contact, the voltage drops. Another bounce occurs, and again the voltage rises. This bounce can be removed by connecting the switch to a latch, as shown in Figure 4.7. This time, when the switch first makes contact it sets the latch. An open state due to the bounce does not change the latch. The latch must wait until the switch is returned to the opposite contact, causing the latch to be reset.

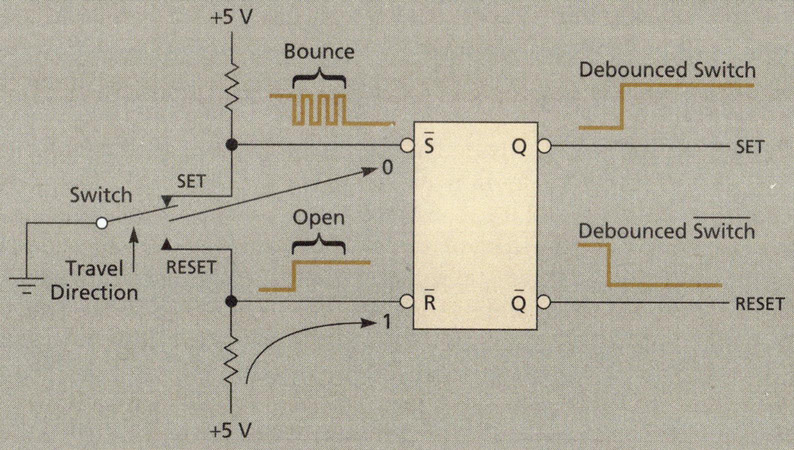

Figure 4.7 Debounced Switch

Section 4.2

Practice Problem

Explain why a cross-coupled latch cannot be made using AND gates.

4.3 Synchronous Flip-Flops

Asynchronous control signals *input and output control signals that are independent of any synchronizing CLOCK signal.*

The R–S flip-flop finds many uses in applications requiring a logic state to be held for a given period of time. It is called an **asynchronous** flip-flop because there are no additional signals used to synchronize the setting or resetting of the flip-flop. Other applications require the latching process to be synchronized with a given occurrence or periodic signal. In essence, these circuits require a set-up time for inputs R and S to be applied before they are latched into the flip-flop. This timing signal is most frequently a recurring pulse or square wave signal called a **clock** and is used for synchronizing changes in the state of a flip-flop. A flip-flop that requires a clocking signal to synchronize changes in its output state is called a **synchronous** flip-flop or latch.

CLOCK *a square wave or repetitive pulse signal that regularly alternates between a logic 1 and logic 0; used for synchronizing circuit elements and control signals.*

Synchronous *with respect to counters, synchronous circuits are those in which all elements are clocked or triggered at the same time.*

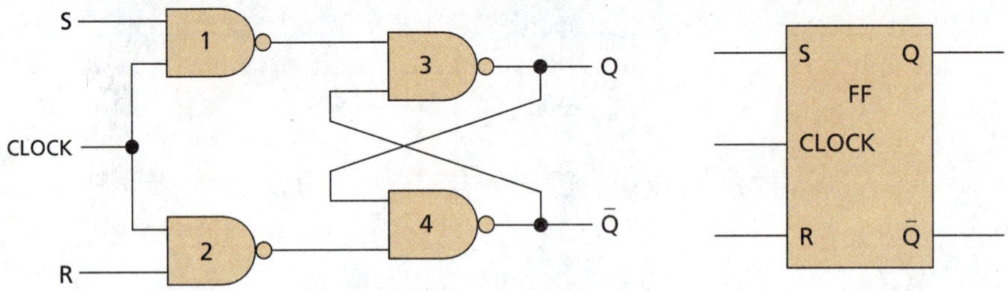

Figure 4.8 Clocked R–S Flip-Flop (FF)

Illustrated in Figure 4.8 is the basic form of a synchronous latch, the *clocked R–S flip-flop*. NAND gates 1 and 2 of the clocked R–S flip-flop pass the states of S and R to the NAND latch whenever the clock input (CLOCK) is high. This gating effect of the CLOCK allows for some control of when the flip-flop will change state. Essentially, the change of the R–S latch is synchronized to the high state of the CLOCK. The designation CLOCK is derived from the most common use of the clocked R–S flip-flop: to effect changes of the latch portion in coincidence with a periodic square wave. The regularity of the square wave signal suggests a similarity to the regularity of a clock. The clock records the steady flow of time just as the square wave continues in a regular progression with time, as shown in Figure 4.9. The high and low state alter at a regular rate or frequency. The amount of time between repetitions of one square wave clock cycle is the time period of the clock signal. These two quantities are reciprocally related to one another by the formula $f = 1/t_p$, where f is the square wave frequency in Hertz (Hz) and t_p is the time period in seconds.

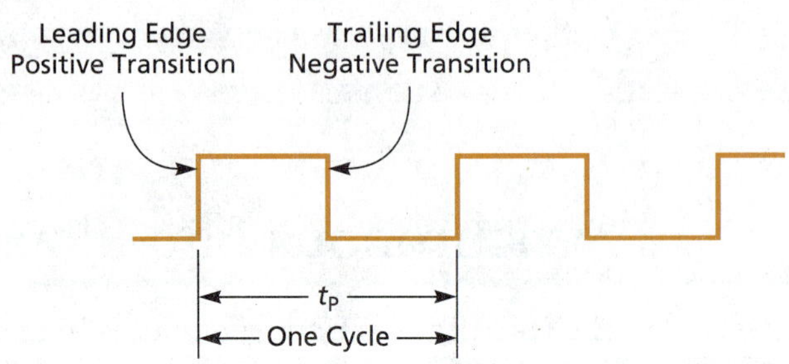

Figure 4.9 Square Wave CLOCK Signal

EXAMPLE 4.4 At what rate does a square wave clock signal with a time period of 325 μS change?

SOLUTION The frequency (f) of the square wave is the reciprocal of the time period, so

$$f = 1/325\ \mu\text{S} = 3.077\ \text{KHz}$$

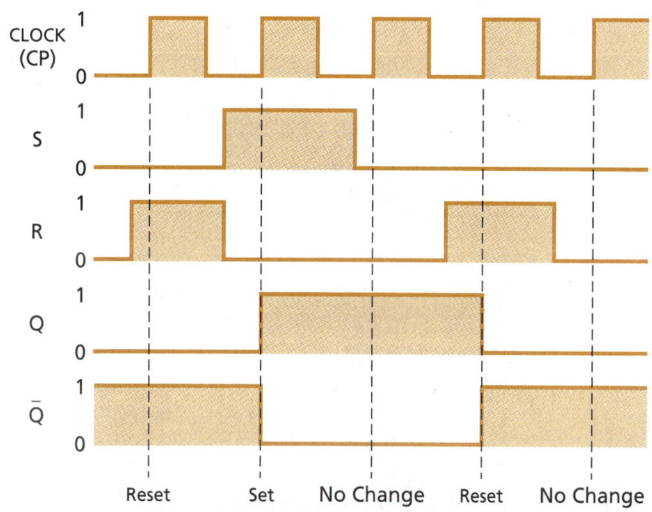

Also, no change occurs on each negative level of CLOCK.

(a) Timing Diagram

CLOCK	S	R	Q $\overline{Q}$	
0	X	X	No Change	
1	0	0	No Change	
1	0	1	0 1	Reset
1	1	0	1 0	Set
1	1	1	Undefined	

(b) Truth Table

X = "don't care" state

Figure **4.10** Clocked R–S Flip-Flop Analysis

Analysis of the action of the clocked R–S flip-flop is undertaken using a timing diagram to illustrate the changes of the flip-flop in synchronization with the CLOCK. The operation of the clocked R–S is then summarized into the truth table shown with the timing diagram in Figure 4.10. **Timing diagrams** are developed by showing the synchronizing CLOCK at the top of the diagram followed by asynchronous inputs and the resulting synchronized outputs, one on each line under the CLOCK signal. Dotted lines are frequently drawn from the vertical edges where changes of state occur in CLOCK, down through the other signal lines. These dotted lines make it easier to detect changes in inputs and outputs with relationship to changes in the clock state. In the case of the clocked R–S flip-flop, the CLOCK signal heads the timing diagram followed by inputs R and S and outputs Q and $\overline{Q}$ as shown in Figure 4.10. Vertical dotted lines from the clock edges are used to show the start and stop times of each half clock cycle.

Often it is necessary to arbitrarily select a starting state for an output at the beginning of a timing diagram, because it is generally unknown. For the clocked R–S flip-flop, the starting state will place the flip-flop in the reset condition ($Q = 0$ and $\overline{Q} = 1$). Initially a low level is applied to S and a high level to the R inputs, as indicated by the first change of state on the R line. As long as the CLOCK is in a low state, the outputs of NAND gates 1 and 2 are held high, since one input of each gate is held low by the CLOCK signal. These ones at the inputs of the NAND latch (gates 3 and 4) are defined in the R–S latch truth table as the no-change state of the latch. Whichever state the flip-flop was latched in, when power is first applied, that is the state it remains in until CLOCK goes high. Given $S = 0$ and $R = 1$, the clocked R–S flip-flop will be reset when CLOCK goes high. In our case, since we arbitrarily selected the flip-flop to be initially in the reset condition, there will be no change from applying reset inputs at the first CLOCK cycle. The state of the clocked R–S flip-flop is no longer arbitrary, though. It is now definitely reset. This happens because a high on the CLOCK input passes the inverted states of R and S to the NAND latch. A high on CLOCK and R inputs makes that NAND output low and the low on S keeps the other NAND output high. Applying a low to the reset input and high to the set input of a NAND latch puts it in the reset condition. On the timing diagram in Figure 4.10, the input and output states, for this first condition, are shown below the high state of the first CLOCK cycle.

Reversing the states of the R and S inputs will cause the flip-flop to change state when the CLOCK again goes high. At that time, the input NAND gates 1 and 2 pass the set inputs to the NAND latch, gates 3 and 4, causing the flip-flop to go to the set condition. It is important to note that although the operation of the flip-flop appears to show instantaneous changes of state, that this is not the actual case. It takes time for a NAND gate to react to the inputs and change its output. This time between a gate detecting an input change and achieving a change at the output state is known as the gate's propagation time and is a chief factor in setting the flip-flop's maximum operating rate. For instance, if it takes 1 μS propagation time for a R–S flip-flop to change state, then the highest CLOCK frequency that can be used is 1/1 μS = 1 MHz. Propagation time or delay also explains what happens when we first apply power to a flip-flop and get an arbitrary state. Although a gate may have a 1 μS propagation delay specification, the actual time for each gate will be slightly different. For instance, one NAND gate may have a delay time of 1.05 μS and another 1.13 μS. The one with the shorter delay will switch states faster and win the "race" on power up, setting its output low and placing the latch in a specific state.

Applying 0's to both R and S causes the outputs of NAND gates 1 and 2 to be high. This is the no-change condition for the latch and is independent of the CLOCK state. The set and no-change condition are shown on the timing diagram under cycles 2 and 3 respectively in Figure 4.10a.

The remaining combination of inputs to the R–S flip-flop is to apply 1's to both R and S. When the CLOCK goes high, this causes NAND gates 1 and 2 to both go to a low state. The undefined state of the latch is 0's on both inputs to the NAND latch. These inputs should not be used, because the results of applying these inputs defies the definition of the flip-flop latch, which requires the Q and $\overline{Q}$ outputs to always be opposite in state. This condition is not shown on the timing diagram, because it is an invalid state for the flip-flop. A truth table, summarizing the activity of a clocked R–S flip-flop, is included in Figure 4.10b.

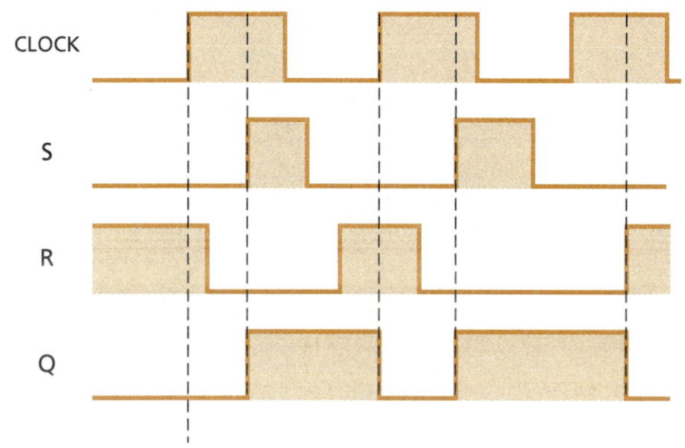

Figure 4.11 Level-Trigger R–S Flip-Flop Timing Example

Level-triggered circuit *logic circuits that change state when a repetitive pulse is at a specific logic level.*

One very important aspect of this latch is that the synchronization occurs whenever the CLOCK signal is high. Because of this, this type of clocking is often called **level-triggered** clocking. Although this affords a measure of synchronization, it is relatively sloppy. The flip-flop can change any time CLOCK is high. For instance, if the inputs were to change while CLOCK is still high, then ideally the output changes immediately, as shown on the timing diagram in Figure 4.11. The timing diagram, at this point, is an ideal depiction, because it does not take propagation delay into consideration. To gain a more precise control over timing, the amount of time CLOCK is high must be restricted.

Edge-triggered circuit *logic circuits that change state at the transition time of a repetitive pulse.*

One method to restrict the time that CLOCK is high is to convert the CLOCK signal to an **edge-triggered CLOCK signal,** as shown in Figure 4.12. An illustration of one way to accomplish this is included in the resistor/capacitor circuit in Figure 4.12. As the leading edge of the square wave is applied, the capacitor charges rapidly, resulting in a large voltage across the resistor. As the capacitor charges, the current through the resistor decreases, along with the voltage across it. The result is a "spike" of voltage at the leading edge of the CLOCK and occurs with every clock cycle. The width of this "spike" is determined by the product of the resistor (R) and capacitor (C) values, called the **R–C time constant.** On the trailing edge, a negative "spike" is formed, but this voltage is below a 0 logic level and can be eliminated by placing a diode across the resistor as shown.

R–C time constant *the time it takes a capacitor to charge to a specific voltage level.*

The operation of an edge-triggered R–S flip-flop is the same as that of a level-triggered flip-flop except that changes in the output state of the flip-flop occur only at the leading edge of the CLOCK signal. Figure 4.13 shows the

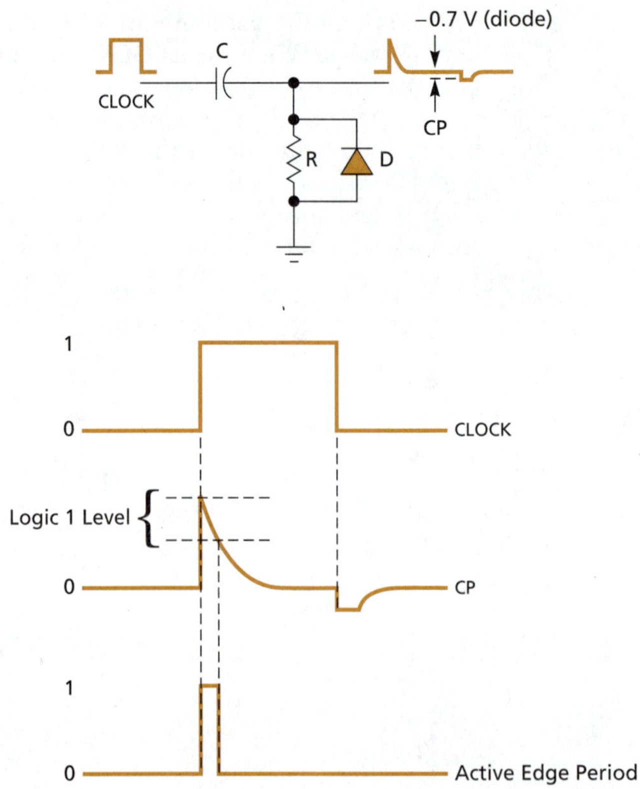

Figure 4.12 Development of Edge-Triggered Clock Pulse

edge-triggered flip-flop with various inputs applied. Note that the dotted lines used for illustrating synchronization are at the leading edge or positive-going transition of the square wave only, since that is the only time the flip-flop will change state. Timing diagrams often include arrows and circles to enhance the description of the synchronization activity in a circuit. A filled-in circle at the base of an arrow indicates the source signal that signifies the time when changes are to occur. In Figure 4.13, this is the CLOCK signal. From the source, arrows are drawn to the new states of the outputs that are affected at the specified time by the base of the arrow. Open circles along the body of an arrow indicate the state or condition of other signals that are required to make the change occur. For instance, look at the second clock cycle in Figure 4.13. The filled-in circle notes that a change occurs on the leading edge of that clock cycle. The changes that occur are shown by the arrowheads at the end of arrows from the source. In this case, the changes are Q = 1 and $\overline{Q}$ = 0. The inputs needed for this to happen, at this time, are indicated by open circles along the arrow's body. For this instance, S = 1 and R = 0 to allow the change to take place.

Figure 4.14 shows a comparison of the logic block diagrams for the level-triggered and edge-triggered clocked R–S flip-flops. Edge triggering is

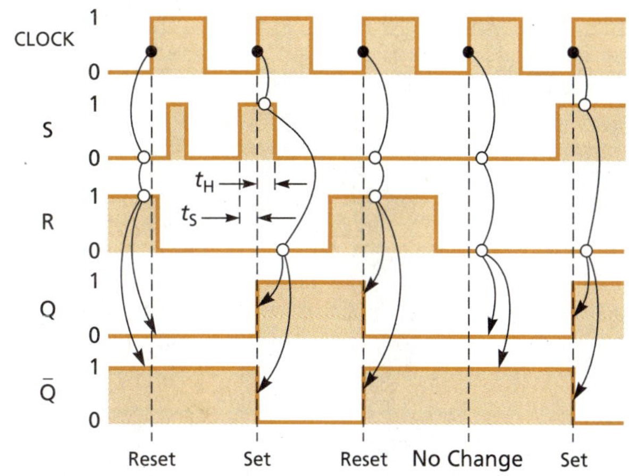

Figure 4.13 Edge-Triggered R–S Flip-Flop Timing Diagram

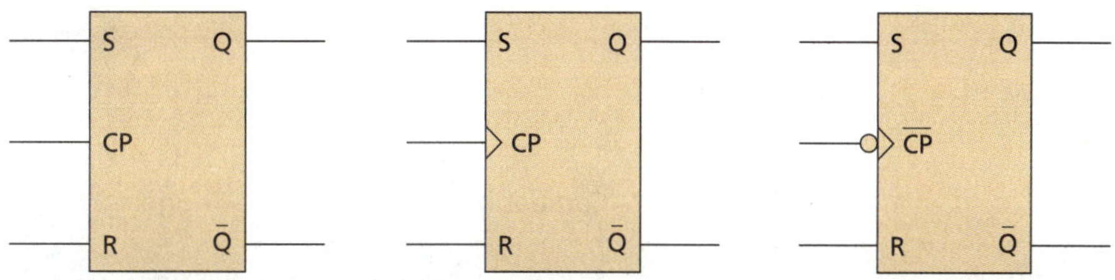

(a) Clocked R–S Flip-Flop **(b)** Positive-Edge-Triggered R–S Flip-Flop **(c)** Negative-Edge-Triggered R–S Flip-Flop

Figure 4.14 Logic Block Diagrams of R–S Flip-Flops

indicated by a triangle at the clock input, while the level-triggered flip-flop has no triangle. CP, standing for clock pulse, is a frequently used designation for clock inputs. Numerous edge-triggered flip-flops are triggered on the negative or trailing edge of the CLOCK signal instead of the leading or positive edge. Negative edge-triggered flip-flops are indicated by an inverting bubble in front of the triangle, as shown on the third block diagram in Figure 4.14c.

Timing diagrams for positive-level and edge-triggered R–S flip-flops, using the same inputs on R and S, are shown in Figure 4.15. Changes for the positive-edge-triggered flip-flop are governed by the leading or positive edge of the CLOCK signal. These changes only occur during the transition from a low level to a high level as indicated at the edge of the square wave. For the positive-level-triggered flip-flop, changes in the flip-flop occur whenever the CLOCK signal is in a high state. The most significant difference in both outputs is that changes for the edge-triggered flip-flop occur regularly, in synchronization, with the start of each clock cycle, while the level-triggered flip-flop outputs occur as long as the CLOCK signal is high.

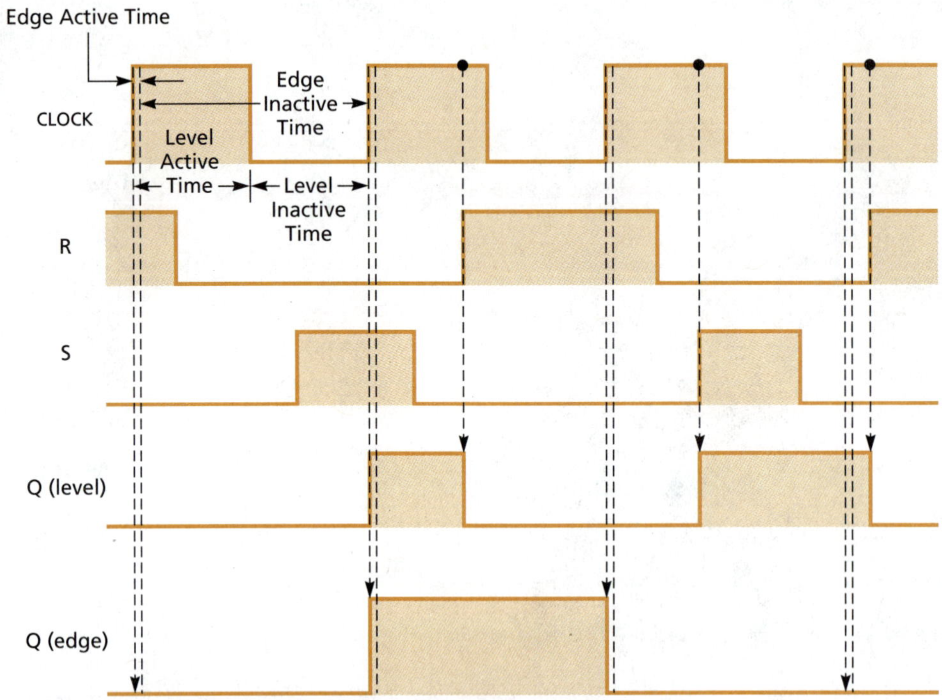

Figure 4.15 Comparing Level- and Edge-Triggered R–S Flip-Flops

Section 4.3

Practice Problems

1. Complete the timing diagram for the negative-level-triggered flip-flop in Figure 4.16.

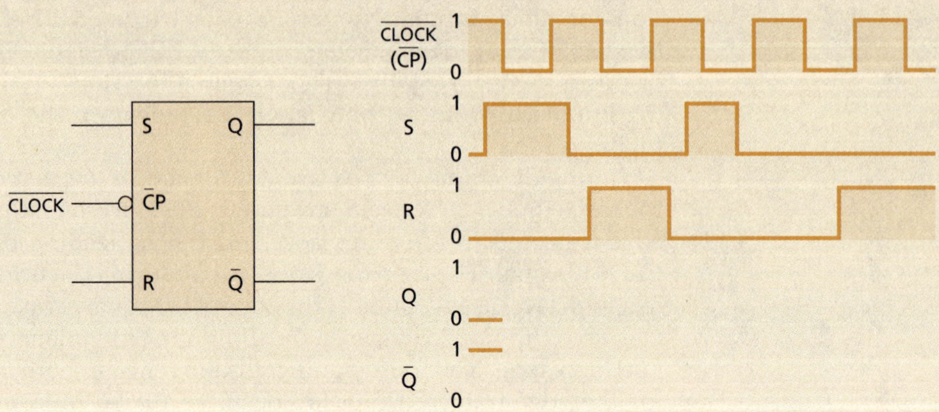

Figure 4.16 Practice Problem 4.3(1)

2. Complete the timing diagram for the edge-triggered flip-flop shown in Figure 4.17.

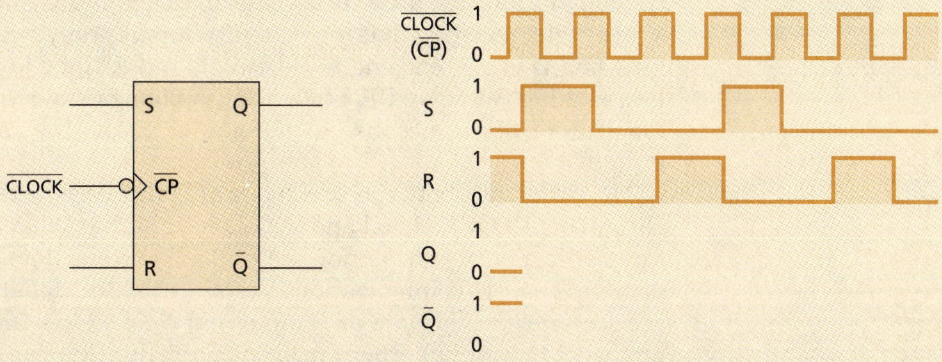

Figure 4.17 Practice Problem 4.3(2)

4.4 Data and Toggle Flip-Flops

Data* or *D flip-flop *a flip-flop whose output changes, when clocked, to the state of the input at D.*

Two specialized forms of flip-flop latches are used for storing binary information, called *data,* and for binary counting. The first of these two types, the **data** or **D flip-flop** comes in two varieties, asynchronous without an additional CLOCK input and synchronous which uses a synchronizing CLOCK signal. Both are shown with their truth tables in Figure 4.18. Data in digital electronics

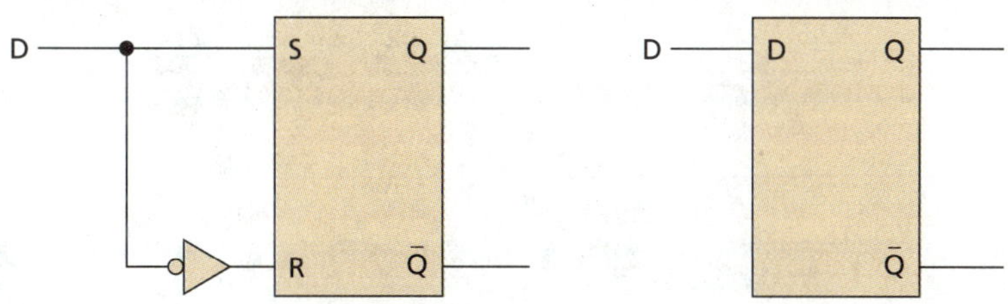

(a) Logic Diagram

(b) Block Symbol

D	Q	Q̄
0	0	1
1	1	0

(c) Truth Table

(I) D Latch Flip-Flop (asynchronous)

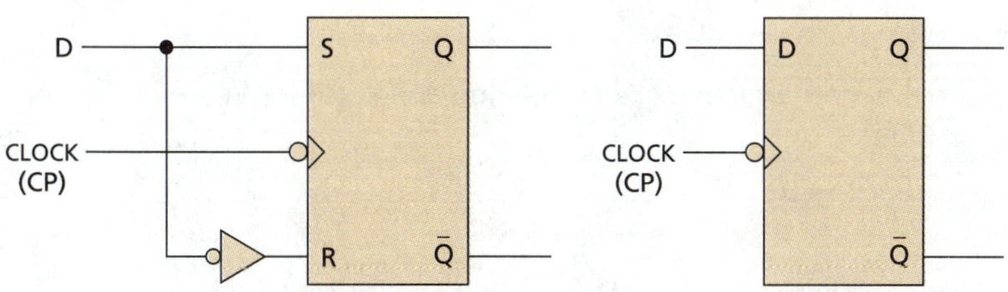

(a) Logic Diagram

(b) Block Symbol

CLOCK	D	Q	Q̄
Steady State	X	No Change	
↑	0	0	1
↑	1	1	0

(c) Truth Table

x = "don't care" state

(II) D Latch Flip-Flop (synchronous)

Figure 4.18 Data (D) Flip-Flops

are binary information stored as bits or groups of bits. Data flip-flops are the storage elements that hold digital data. They are formed from R–S and clocked R–S flip-flops by connecting the data or D input to the S input and to the R input through an inverter. This forces inputs S and R to always be opposite in state, meaning that the data flip-flop has only two operating conditions: set when D is a 1, causing S to be high and R to be low, and reset when D is low, making S low and R high. The no-change state when S and R are both low cannot occur, because of the action of the inverter between the S and R inputs.

The difference between two types of D flip-flops is that one requires a synchronizing CLOCK signal and the other does not. When triggered by a CLOCK signal, the binary information, a 1 or 0, is saved or stored into the synchronous data flip-flop. This information is retained by the data flip-flop until new data are presented at the data or D input and clocked in with the correct transition on the CLOCK input. The synchronized D flip-flop has a no-change condition when the CLOCK input is held at a steady state.

EXAMPLE 4.5

Complete the timing diagram for the D flip-flops of Figure 4.18 using the inputs in Figure 4.19.

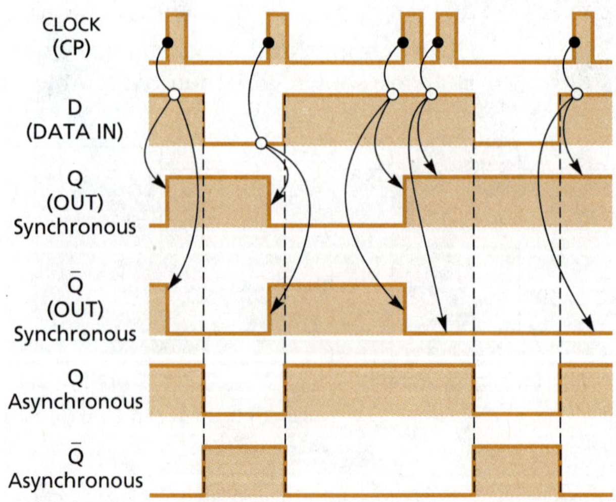

Figure 4.19 Example 4.5, D Flip-Flop and Latch Input Signals

SOLUTION Since the synchronized D flip-flop of Figure 4.18 is positive-edge-triggered, the Q output will match the D input whenever the CLOCK input changes from a low to a high. At all other times, the Q output remains unchanged. For the asynchronous D flip-flop, the Q output will match the D input and change whenever the D input changes. The complete timing diagram is shown in Figure 4.20.

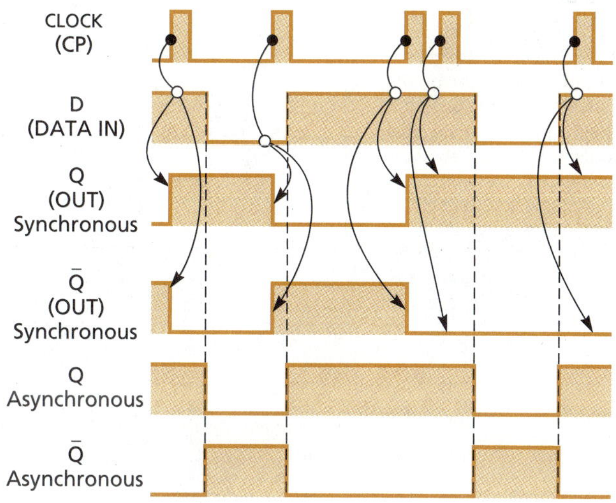

Figure 4.20 Example 4.5 Timing Diagram (solution)

Digital data are stored in D flip-flops with one flip-flop required for each binary bit. Storing four binary bits requires four D flip-flops, as shown in Figure 4.21. Data are presented on all four D inputs and then "clocked" into the flip-flops by a negative transition on the CLOCK line. Circuits constructed in this manner with the purpose of temporarily storing binary data are called

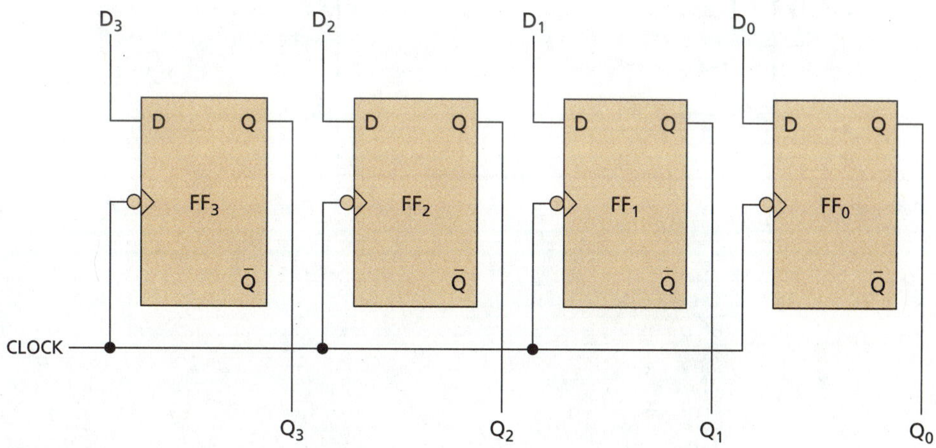

Figure 4.21 Four-Bit Data Register

DESCRIPTION

The "74" is a Dual Positive Edge-Triggered D-Type Flip-Flop featuring individual data, clock, set and reset inputs; also complementary Q and $\overline{Q}$ outputs.

Set ($\overline{S}_D$) and Reset ($\overline{R}_D$) are asynchronous active LOW inputs and operate independently of the clock input. Information on the data (D) input is transferred to the Q output on the LOW-to-HIGH transition of the clock pulse. The D inputs must be stable one setup time prior to the LOW-to-HIGH clock transition for predictable operation. Although the clock input is level sensitive, the positive transition of the clock pulse between the 0.8V and 2.0V levels should be equal to or less than the clock to output delay time for reliable operation.

PIN CONFIGURATION

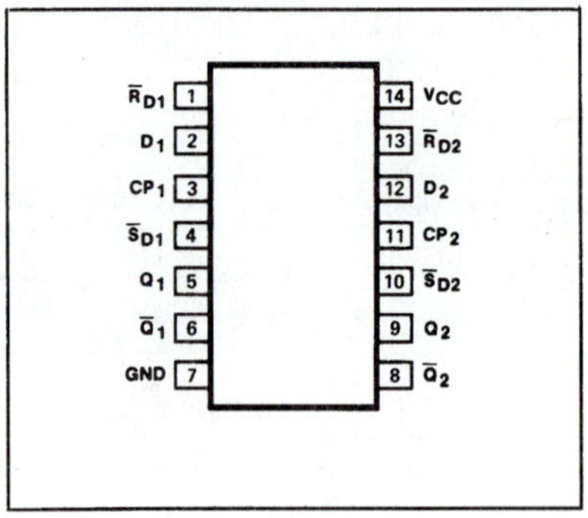

LOGIC SYMBOL

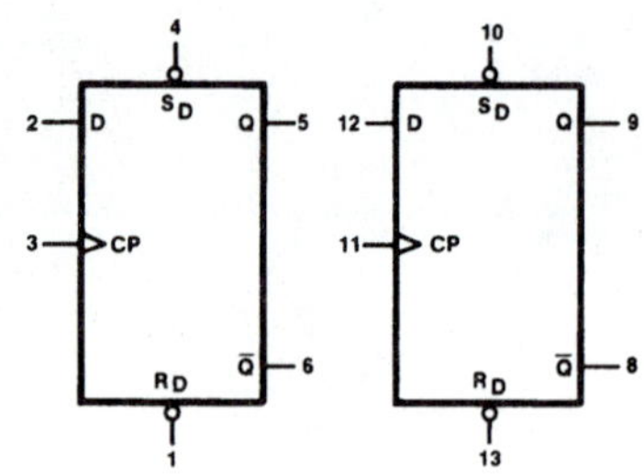

Pin numbers for pin configuration A

LOGIC DIAGRAM

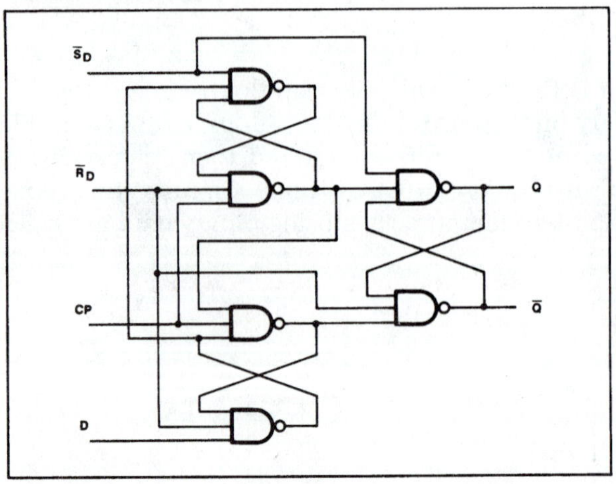

MODE SELECT—TRUTH TABLE

OPERATING MODE	INPUTS				OUTPUTS	
	$\overline{S}_D$	$\overline{R}_D$	CP	D	Q	$\overline{Q}$
Asynchronous Set	L	H	X	X	H	L
Asynchronous Reset (Clear)	H	L	X	X	L	H
Undetermined (c)	L	L	X	X	H	H
Load "1" (Set)	H	H	↑	h	H	L
Load "0" (Reset)	H	H	↑	l	L	H

H = HIGH voltage level steady state.
h = HIGH voltage level one setup time prior to the LOW to HIGH clock transition.
L = LOW voltage level steady state.
l = LOW voltage level one setup time prior to the LOW to HIGH clock transition.
X = Don't care.

Figure 4.22 74LS74 D Flip-Flop

registers. In actuality, any circuit, regardless of what else it is designed to do, that temporarily stores digital data, also functions as a register. For instance, a circuit that adds digital data and holds the sum for later use is an adder and also a register. In the case of the D flip-flop circuit in Figure 4.21, its sole purpose and design is that of a register.

Figure 4.22 shows a 74LS series TTL version of a dual-D flip-flop, the 74LS74, with its pin out and truth table. The determination of whether a flip-flop is synchronous or asynchronous is based on whether outputs changed in synchronization to the CLOCK input or not. The 74LS74 chip is a synchronous-D flip-flop, which also has asynchronous control inputs that are independent of the CLOCK (CP) input. The D input, which is synchronized to the CP input, and Q and $\overline{Q}$ outputs, this chip also includes active low asynchronous set ($\overline{S}$) and reset ($\overline{R}$) inputs. These two inputs take precedence over the D function of the flip-flop and operate like the NAND latch described first in this chapter and shown on the truth table in Figure 4.22. The D function is a synchronized function and will operate as long as both set and reset inputs are held inactive.

EXAMPLE 4.6

Complete the timing diagram in Figure 4.23 for the 74LS74 D flip-flop of Figure 4.22.

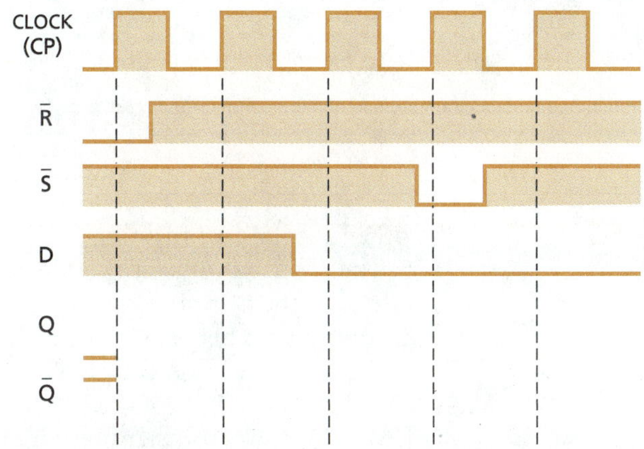

Figure 4.23 Example 4.6

SOLUTION At the start of the timing diagram in Figure 4.23, the reset input, $\overline{R}$, is asserted low, causing the flip-flop to become reset, Q = 0. By the time of the second clock cycle, the reset input is negated and the D flip-flop Q output follows the D input synchronized to the positive edge of the CP signal. When the set input, $\overline{S}$, is brought low, the flip-flop is forced into a set condition with

Q = 1 and resumes following the D input after the set input is negated. The completed timing diagram is shown in Figure 4.24.

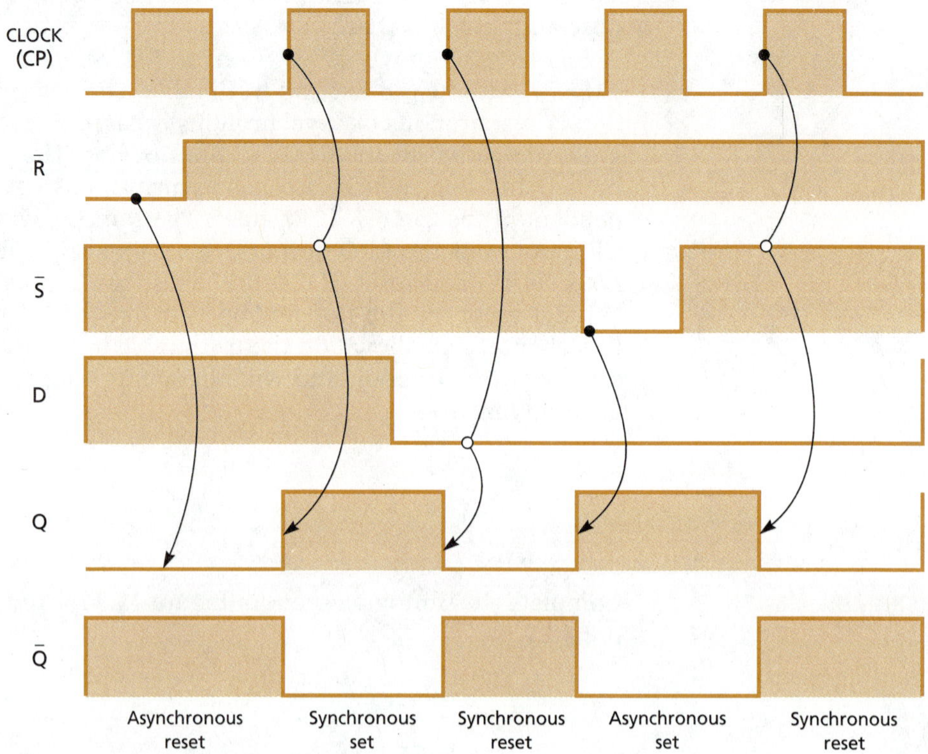

Figure 4.24 Example 4.6 (solution)

***Toggle** or **T flip-flop** logic circuit that alters states between two stable conditions whenever it is clocked.*

The other specialized flip-flop is the **toggle** or **T flip-flop**, illustrated in Figure 4.25. The Q output of a clocked R–S flip-flop is connected to the R input and the $\overline{Q}$ output to the S input. Power is applied to the flip-flop, and after the race is over it is either set or reset. For purposes of discussion, let's say it is reset—Q is 0 and $\overline{Q}$ is a 1. The cross-coupled connections from the outputs to the inputs place the inputs into a set condition (S = 1 and R = 0). When a CLOCK signal is applied, the flip-flop goes set, after which the cross-coupled connections present reset input states to the S and R inputs of the flip-flop (S = 0 and R = 1). Since the toggle flip-flop is edge-triggered, the output will only change on the beginning of the clock pulse. The next clock cycle causes the flip-flop to reset, and the process keeps repeating, with the outputs toggling or changing state with each clock cycle. A timing diagram for the action of the T flip-flop is included in Figure 4.25. Compare the frequency of the CLOCK signal with that at the Q output—the Q output rate is one-half that at the clock input. Because of this frequency division, a toggle flip-flop is often referred to as a "divide-by-two" flip-flop.

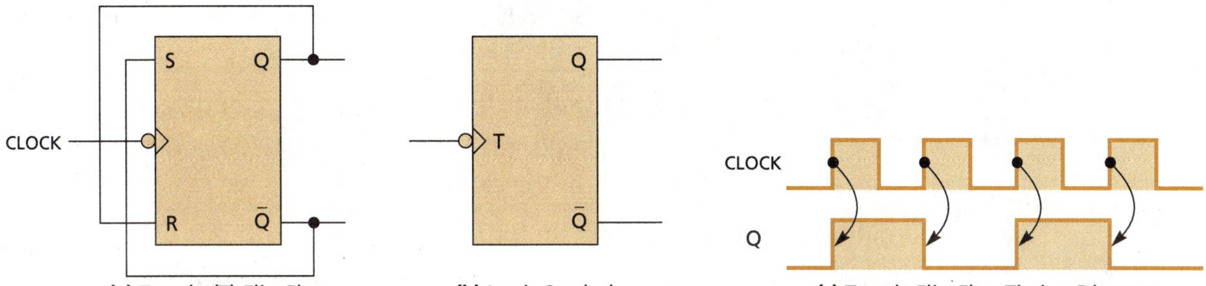

(a) Toggle (T) Flip-Flop **(b)** Logic Symbol **(c)** Toggle Flip-Flop Timing Diagram

Figure **4.25** Toggle Flip-Flop

Duty cycle *the ratio of the time a repetitive pulse is in the high state to the total time of a single cycle.*

Figure 4.26 shows what happens at the Q output of a toggle flip-flop when a repetitive pulse is applied to the CP input. Essentially, a repetitive pulse is a rectangular shape alternating signal that has a duty cycle other than 50 percent. A **duty cycle** (DC) is a measure of the time that a repetitive pulse is at a high level during a complete cycle of the wave form, compared with the time period of that complete cycle. The duty cycle, computed as a percentage, is found by dividing the time the repetitive wave is high by the total time period of a single cycle and then multiplying the result by 100 percent.

$$\text{Duty cycle} = \frac{\text{time high}}{\text{total time}} \times 100\%$$

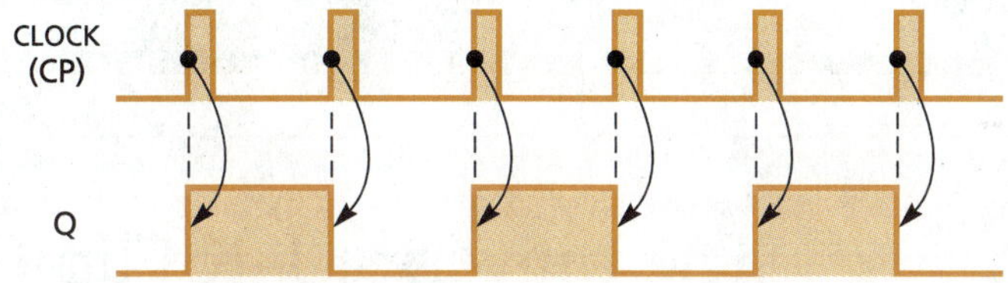

Figure **4.26** Fifty Percent Duty Cycle of Toggle Flip-Flop Output

As long as the time period of the positive time period and negative period of a repetitive wave are the same, the waveform will have a duty cycle that is 50 percent. Any other combination of time periods yield duty cycles other than 50 percent.

Examining the output of the toggle flip-flop, you will note that the frequency of that output is still one-half of the input at CP, but the Q output has a duty cycle of 50 percent. This happens because the Q output, in this example, changes on the leading edge of the clock pulse applied to CP without regard to the duty cycle of that clock, resulting in the state of the output to be high the same amount of time that it is low.

Section 4.4

Practice Problems

1. Draw a circuit using D flip-flops that can store a byte of data.
2. Design a circuit, using T flip-flops that will produce a square wave that is one-eighth the frequency of the clock input.
3. Complete the timing diagram for Figure 4.27.

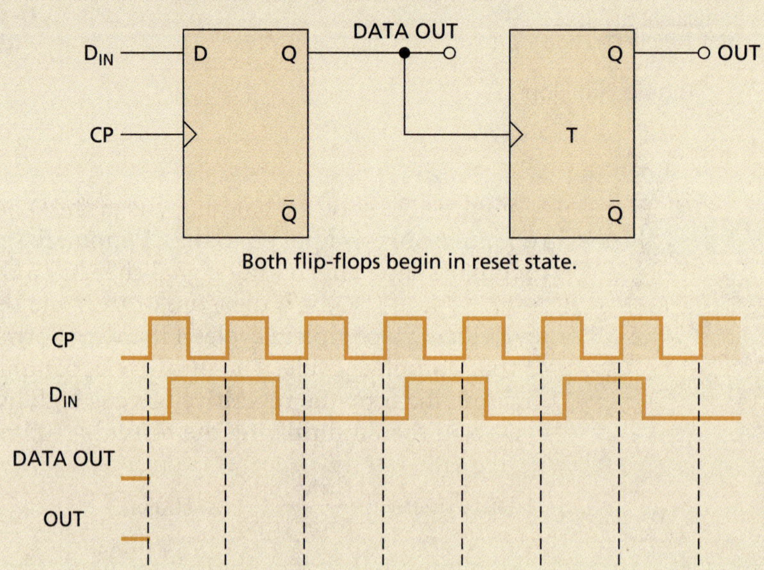

Figure 4.27 Practice Problem 4.4(3)

4.5 The Universal J–K Flip-Flop

***J–K flip-flop** a universal flip-flop that can mimick all the other types of flip-flops with minimal additional circuitry.*

The most versatile flip-flop is the **J–K flip-flop**, which can perform the functions of all the previous flip-flops with minimal wiring or additional circuitry. Each section of the dual 74LS series J–K, the 74LS76, in Figure 4.31, includes asynchronous reset and set inputs, synchronous J and K inputs, and separate $\overline{CP}$ inputs. To understand the effect of these inputs on the Q output, let's develop the truth table for the 74LS76 J–K flip-flop that can be used as a general reference for most J–K flip-flops. We begin the analysis by examining the use of the asynchronous inputs first. The first line of the truth table shows what happens when the reset input, $\overline{R}$, is asserted low. The asynchronous set input, $\overline{S}$, must be held high, although many J–K flip-flops are designed so that applying an active reset overrides *all* other inputs, including the set input. Since this is not universal, we will make sure that $\overline{S}$ and $\overline{R}$ are not asserted low simultaneously. When using the asynchronous inputs, the states of the remaining inputs are each a "don't care." "Don't care" does not mean that we do not care what is applied to those inputs, but it does mean that they may be a logic 1 or

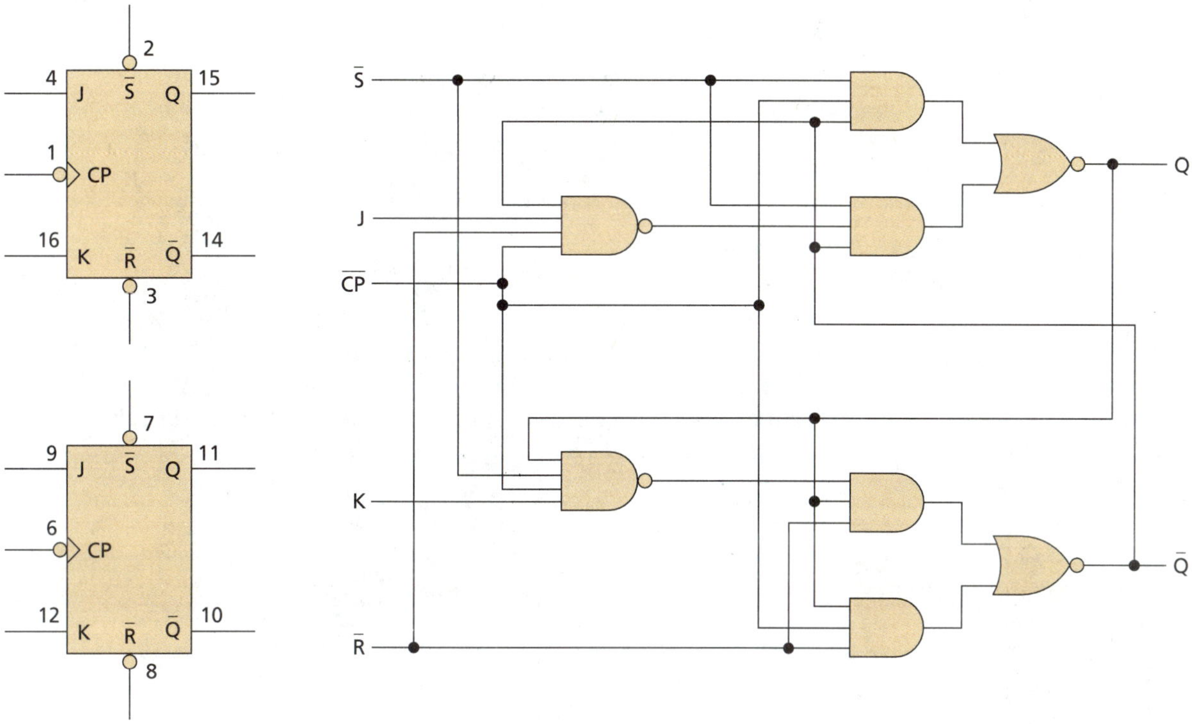

Figure 4.28 74LS76 Dual J–K Flip-Flop

a logic 0—it literally does not matter which state they are in. An active RESET signal causes the J–K flip-flop to be placed into the reset state, and an active set input puts it into the set condition. The first two lines on a truth table for the 74LS76 J–K flip-flop are

Function	$\overline{R}$	$\overline{S}$	J	K	$\overline{CP}$	Q	$\overline{Q}$
Reset	0	1	X	X	X	0	1
Set	1	0	X	X	X	1	0

Note: X = "don't care" state.

These first two lines show the J–K flip-flop operating like an asynchronous R–S latch. The remaining lines of the table require both the set and reset to be in an inactive (high) state to remove their influence on the outcome at Q and $\overline{Q}$. Applying a logic 0 to both J and K inputs with the set and reset inputs high keeps the flip-flop in a no-change state regardless of the signal applied to the clock input. Another way to keep the J–K flip-flop from changing state is to keep the set and reset inputs high and a steady state on the $\overline{CP}$ input. Add these two conditions as the next two lines on the truth table:

Function	$\overline{R}$	$\overline{S}$	J	K	$\overline{CP}$	Q	$\overline{Q}$
Reset	0	1	X	X	X	0	1
Set	1	0	X	X	X	1	0
No change	1	1	0	0	X	—	—
No change	1	1	X	X	ss	—	—

Note: ss = steady state.
X = "don't care" state.

Setting J low and K high and applying a CLOCK signal to $\overline{CP}$ will cause the flip-flop to go into the reset state. Reversing the input states on the J and K inputs places the flip-flop into the set condition on the next clock pulse. Thus the J–K flip-flop can also be used as a synchronous clocked R–S flip-flop by using the J, K, and CP inputs.

Function	$\overline{R}$	$\overline{S}$	J	K	$\overline{CP}$	Q	$\overline{Q}$
Reset	0	1	X	X	X	0	1
Set	1	0	X	X	X	1	0
No change	1	1	0	0	X	—	—
No change	1	1	X	X	ss	—	—
Reset	1	1	0	1	↓	0	1
Set	1	1	1	0	↓	1	0

Note: ss = steady state.
↓ = negative clock transition.
X = "don't care" state.

When the inputs to the R–S flip-flop were placed at a high state, the outputs of Q and $\overline{Q}$ were the same state and this condition was referred to as an undefined state for a flip-flop. For the J–K flip-flop, when inputs J and K are set high, the flip-flop acts as a toggle flip-flop, changing state on each clock cycle. This completes the truth table with the toggle flip-flop function as its last line:

Function	$\overline{R}$	$\overline{S}$	J	K	$\overline{CP}$	Q	$\overline{Q}$
Reset	0	1	X	X	X	0	1
Set	1	0	X	X	X	1	0
No change	1	1	0	0	X	—	—
No change	1	1	X	X	ss	—	—
Reset	1	1	0	1	↓	0	1
Set	1	1	1	0	↓	1	0
Toggle	1	1	1	1	↓	I	I

Note: ss = steady state.
↓ = negative clock transition.
X = "don't care" state.
I = inverse or complement of previous state.

The truth table for the J–K flip-flop can be used to analyze the output result of a J–K flip-flop given a set of input signals like those shown in Figure 4.29. It is important to keep in mind that asynchronous inputs, $\overline{R}$ and $\overline{S}$, cause changes

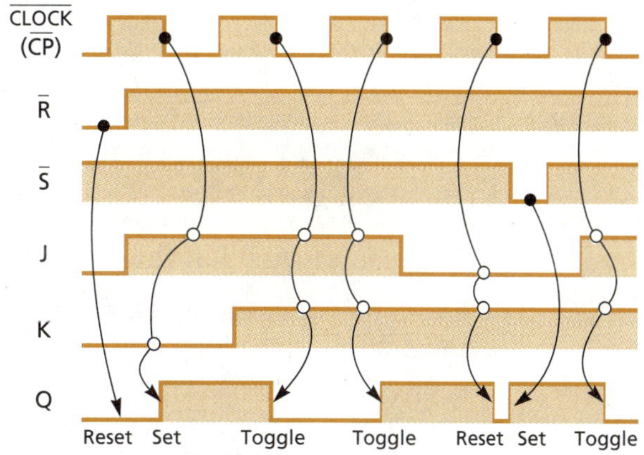

Figure 4.29 J–K Timing Example

in the Q output immediately while synchronous inputs J and K must wait for the application of the correct clock edge before the change occurs. In Figure 4.29, an asynchronous reset signal is applied to $\overline{R}$, causing Q to go to 0. Once $\overline{R}$ is negated, the state of the flip-flop depends on the remaining input signals. For instance, on the first negative clock edge after the reset is removed, a 1 is applied to J and a 0 to K while $\overline{S}$ and $\overline{R}$ inputs are both high. Between the negating of the reset and the application of the negative clock edge, the clock is in a steady state condition, keeping the flip-flop from changing state. On the first clock cycle, the flip-flop goes to the set condition, since J is 1 and K is 0. All three of these results can be verified from the truth table: An active R causes the flip-flop to reset, a steady state (ss) on the CP input is a no-change condition, and applying a 1 to J and 0 to K causes the flip-flop to set on the negative clock edge.

The remaining timing diagram is developed by checking the input conditions to the flip-flop with the J–K flip-flop truth table, and selecting the correct output state for each set of input conditions.

What about emulating the D flip-flop using a J–K flip-flop, because all the other flip-flop types are represented by one function or another of the J–K flip-flop? Allowing the J–K flip-flop to perform as a data or D flip-flop requires the addition of an inverter gate between the J and K inputs, as shown in Figure 4.30.

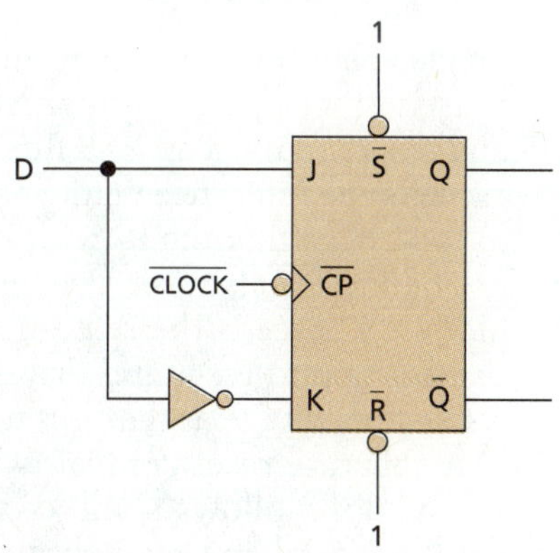

Figure 4.30 J–K Flip-Flop as a D Flip-Flop

This results in opposite J and K states and applying either set or reset inputs based on the state of the single synchronous input, D.

Section 4.5

Practice Problems

1. Describe why the J–K flip-flop is known as the universal flip-flop.
2. Given the input timing diagram in Figure 4.31 for the J–K flip-flop of Figure 4.28, complete the Q output time line.

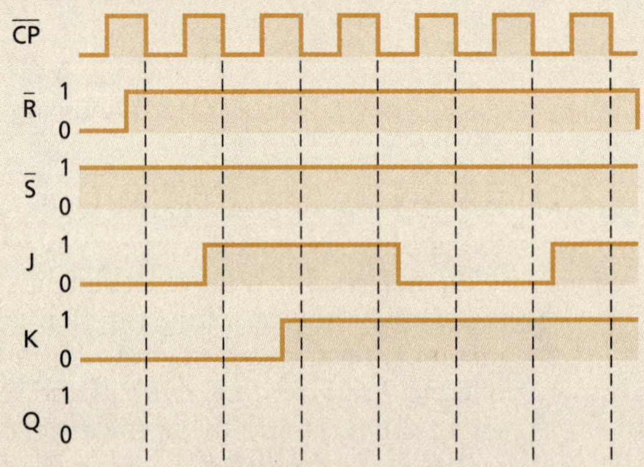

Figure 4.31 Practice Problem 4.5(2)

4.6 Generating a Clock Signal with a 555 Timer

Square waves used as clock signals for CP inputs can be created in many ways. One integrated circuit device used for this purpose is the 555 timer, illustrated in Figure 4.32. The circuit's name is derived from the use of an internal voltage divider between V_{CC} and ground using three 5 kΩ resistors. This divider chain is used to set a pair of reference voltages for two comparators that drive the set and reset inputs of an R–S flip-flop. A voltage applied to trigger input pin 2 that exceeds the reference voltage to the lower comparator causes that comparator to switch, setting the R–S flip-flop, making output pin 3 high.

The flip-flop is reset, driving output pin 3, when an input voltage to threshold pin 6 exceeds the reference voltage to the upper comparator. This causes the flip-flop to reset and drives the transistor connected to discharge pin 7 to cause the voltage at that pin to be near ground (about 0.1 V).

A separate reset control is available to the 555 timer through reset pin 4. Lastly, pin 5 allows some control over the reference voltages for both comparators. Applying a voltage to this pin replaces the reference voltages supplied

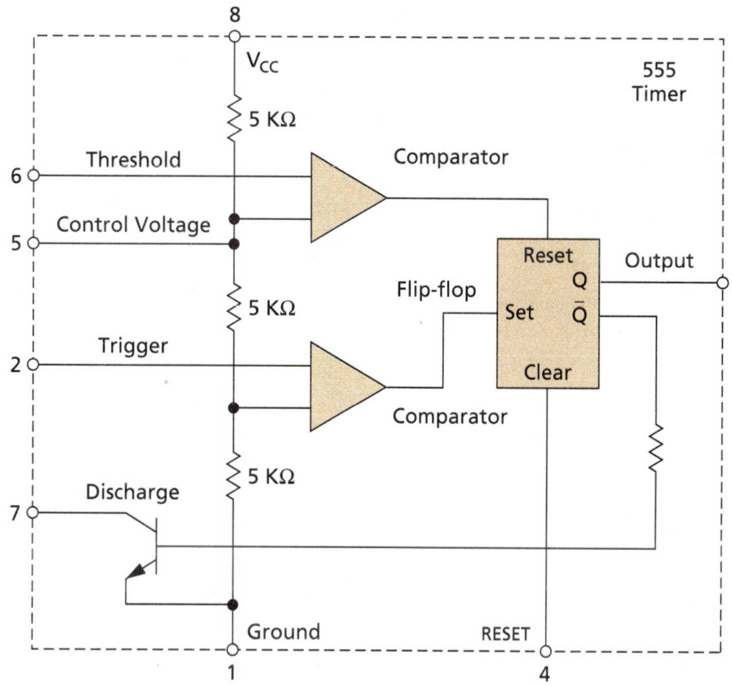

Figure 4.32 555 Timer Functional Diagram

by the 5 K voltage divider chain. When not used for that purpose, a noise-decoupling capacitor of 0.01 μf is connected to pin 5.

***Astable** or **free-running multivibrator (MVB)** a circuit that generates a CLOCK signal.*

To use the 555 to generate a clock signal, it must be connected to some external timing components, as shown in Figure 4.33. This configuration causes the 555 timer to act as an **astable** or **free-running multivibrator (MVB).** The term *multivibrator* stems from the days when a square wave voltage was generated by vibrating a movable contact between two stationary contacts, one at one voltage level and one at another voltage level. The signal on the movable contact would be a repetitive alternating voltage between the two voltages on the stationary contacts.

In the 555 circuit of Figure 4.33, when power is first applied a large surge of current takes place as the timing capacitor (C_T) begins to charge through timing resistors R_A and R_B. This causes the voltage across C_T to be initially low and begin to increase as the capacitor charges. Eventually the charge on C_T reaches the trigger level and the 555 flip-flop is set, forcing output pin 3 high. The voltage across the capacitor continues to build up until it reaches the threshold level. At this point, the flip-flop is reset, causing output pin 3 to go low and driving pin 7 low, creating a discharge path for C_T through R_B to the ground. The capacitor starts to discharge until the voltage drops low enough to cause the flip-flop to set, removing the low at pin 7 and setting output pin 3 high again. Timing capacitor, C_T, again charges up to the threshold level, causing the process to repeat as long as power remains applied to the circuit. Timing diagrams for the voltage across C_T and the associated change in voltage at pin 3 are included in Figure 4.33.

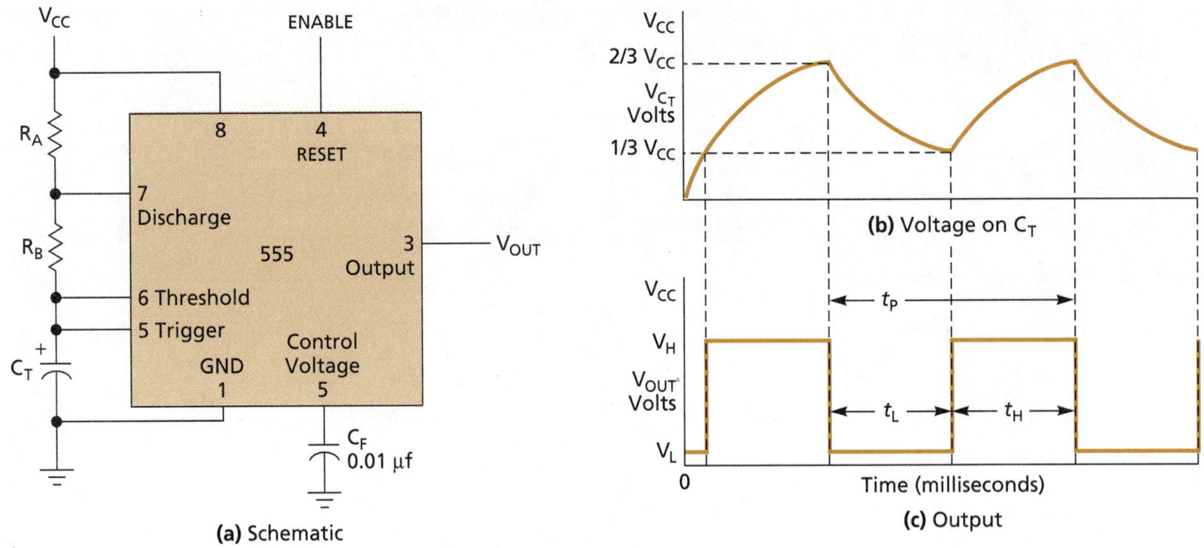

Figure **4.33** Astable or Free-Running Multivibrator

A set of formulas are used to aid in selecting values for R_A and R_B with a given capacitor value for C_T and square wave frequency. The time that pin 3 is high is determined from the formula $t_H = 0.693(R_A + R_B)(C_T)$ and for the time that pin 3 is low: $t_L = 0.693(R_B)C_T$. The total time or period of a single clock cycle (t_P) is equal to the sum of t_H and t_L. Finally, the frequency of the square wave is the reciprocal of the period, or

$$f = 1/t_P = 1/(t_H + t_L) = 1/[0.693(R_A + 2\,R_B)(C_T)]$$

The duty cycle (DC) of the square wave is found by dividing the high time, t_H, by the period, t_p, times 100 percent, or

$$DC = (100\%)(t_H)/t_P = (100\%)(t_H)/(t_H + t_L)$$

We shall see by doing a little algebra that the duty cycle for the square wave from this circuit will always be greater than 50 percent:

$$DC = \frac{(100\%)(0.693)(R_A + R_B)(C_T)}{(0.693)(R_A + 2\,R_B)(C_T)}$$

$$DC = \frac{(100\%)(R_A + R_B)}{R_A + 2\,R_B}$$

As a general rule, the square wave output can approximate a 50 percent duty cycle by selecting R_B to be 10 times the value of R_A. When 10 R_A is substituted for R_B in the previous formula, the calculation for duty cycle becomes

$$DC = (100\%)(11\,R_A)/(21\,R_A) = 52.4\%$$

EXAMPLE 4.7

What are the values for R_A and R_B for a 555 timer circuit to generate a 50 KHz square wave with an approximate duty cycle of 50 percent?

SOLUTION Select a starting value for C_T of 0.01 µf. Substitute 10 R_A for R_B, and calculate a value for R_A:

$$\text{Frequency} = 1/[(0.693)(R_A + 2\ R_B)(C_T)] = 1/[(0.693)(21\ R_A)(C_T)]$$

$$50\ \text{KHz} = 1/[(0.693)(21\ R_A)(0.01\ \mu)]$$

$$R_A = 1/[(0.693)(21)(0.01\ \mu)(50\ \text{KHz}) = 137.4\ \Omega$$

and for $R_B = 10\ R_A = (10)(137.4) = 1.374\ \text{K}\Omega$.

Suppose you wish to generate a CLOCK signal with less than 50 percent duty cycle. You may, for instance, want a repetitive pulse of 1 µS at rate of 150,000 pulse per second (PPS). This means that the total time period between pulses is 1/150,000 = 6.67 µS. The resulting duty cycle is the pulse width divided by the period, or

$$\text{Duty cycle} = 1\ \mu\text{S}/6.67\ \mu\text{S} \times 100\% = 15\%$$

To achieve such an output using the 555 timer, the external circuit has to be modified as shown in Figure 4.34. The diodes in the circuit cause the charging circuit to be dependent on R_A and C_T, bypassing R_B. The discharge path, as before, is from C_T through R_B to ground. This changes the time the waveform is high from $0.693(R_A + R_B)C_T$ to $0.693R_A \cdot C_T$ and the duty cycle expression to

$$DC = \frac{R_A}{R_A + R_B}$$

This configuration of the astable multivibrator permits duty cycles from 5 to 95 percent.

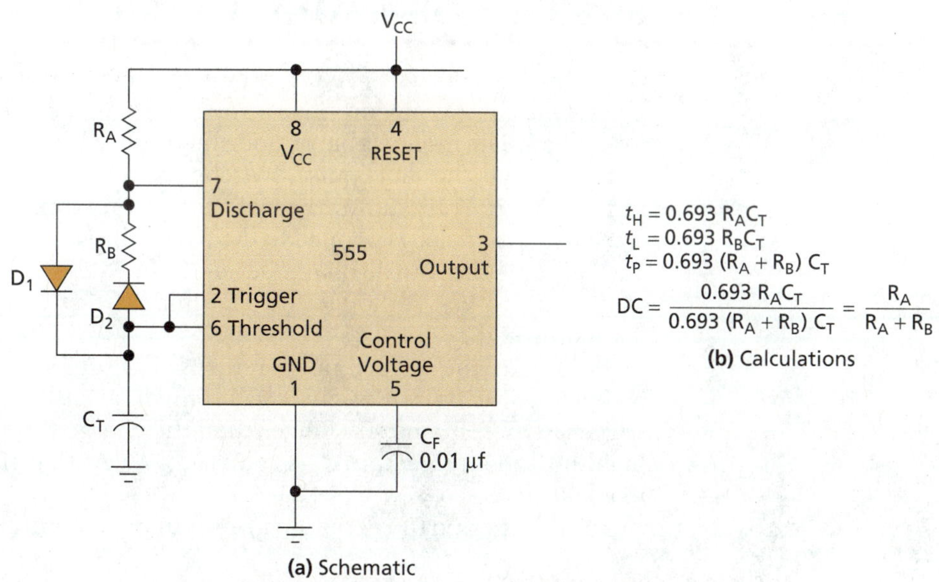

Figure 4.34 Astable Operation for Duty Cycles from 5 to 95 Percent

For stable operation of the 555 in the free-running mode, manufacturers recommend adherence to the following specifications, using a V_{CC} of 5 V:

Operating frequency: 2 to 500 KHz

Minimum value for R_B: 3 KΩ

Maximum value for R_B: 6.6 MΩ

Value for C_T between 0.01 μf and 10 μf

This does not mean that you cannot use values beyond these recommendations. However, for stable operation of the 555 timer it is best to work within these parameters.

Section 4.6

Practice Problem

Design a 555 timer circuit to produce a CLOCK signal with these characteristics:

1. Approximately 50 percent duty cycle
2. Frequency of 15 KHz
3. Timing capacitor, C_T, of 0.0047 μf

4.7 Using the 555 Timer as a One-Shot Multivibrator

On occasion, you will need a circuit that can "stretch" the time period of a pulse. One circuit may produce a short pulse of 3 μS to turn on another circuit for 2.5 mS. The monostable or one-shot multivibrator is ideal for the task. This circuit accepts the short pulse, switches, and holds its output for the longer duration. Figure 4.35 shows the 555 timer configured as a one-shot circuit. A short negative pulse is applied to the trigger input, which sets output pin 3 high and removes the low discharge level at pin 7, allowing capacitor C_T to charge through R_A to V_{CC}. Although the charging process continues, the output at pin 3 remains high.

As long as the trigger pulse returns high before the charge time of C_T is completed, the output at pin 3 will return low at that time. If the level on the trigger input remains low longer than the charge time of C_T, pin 3 will remain high until the trigger, pin 2, is returned low. At that time output pin 3 will go to a low state.

The time period (t_P) of the single output pulse computed using the expression

$$t_P = 1.1\ R_A \cdot C_T$$

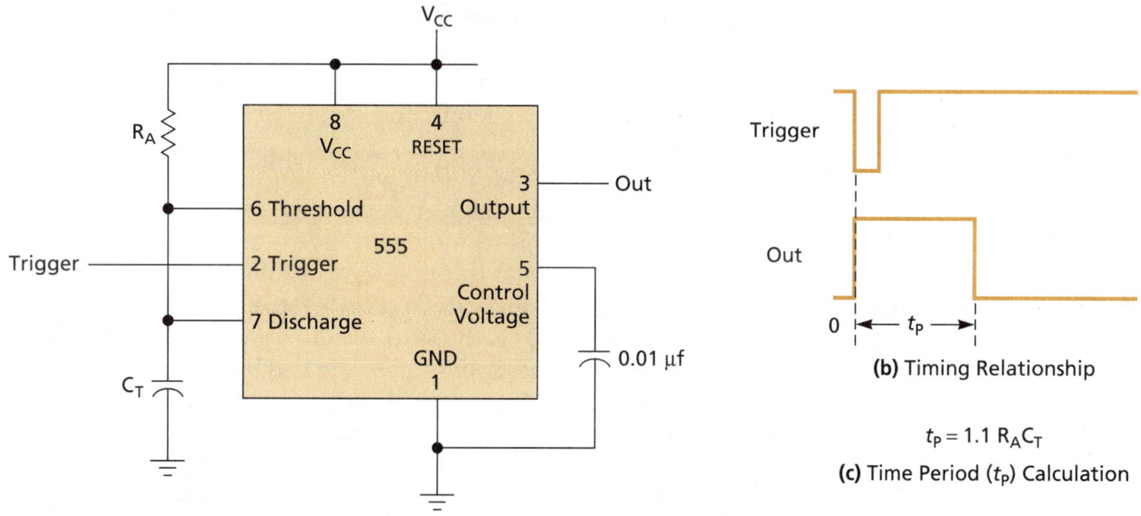

(a) Schematic

$t_P = 1.1\ R_A C_T$

(c) Time Period (t_P) Calculation

Figure 4.35 555 as a Monostable (one-shot) Multivibrator

EXAMPLE 4.8

Find the value of R_A required to produce a single pulse of 13 mS duration using a 555 timer with a timing capacitance of 0.47 μf.

SOLUTION Substitute the known values into the formula, and solve for R_A:

$13\ \text{mS} = 1.1(0.47\ \mu\text{f}) \cdot R_A \qquad R_A = 13\ \text{m}/(1.1 \cdot 0.47\ \mu) = 25\ \text{K}\Omega$

Section 4.7

Practice Problem

A one-shot pulse of 2.45 mS is required. Find the value of R_A to be used in a 555 timer circuit with a timing capacitance of 0.033 μf.

4.8 State Diagram Graphical Representation of Synchronous Devices

State diagram *a pictorial representation of a synchronous circuit.*

Truth tables and timing diagrams are two ways to represent a synchronous device, one static (truth table) and one dynamic (timing diagram). **State diagrams** are a third way to show flip-flop operation by combining the static and dynamic characteristics of truth tables and timing diagrams. The concept of

State table *similar to a truth table, it includes both present and next states.*

Current or ***present state*** *an indication on a state diagram of the present condition of a circuit before a clock pulse is applied.*

Next state *on a state diagram, it is the condition of a circuit after a CLOCK signal has been applied.*

state diagrams is an extension of **state tables,** which are truth tables that express two output states, the **current** or **present** state and the **next** state. Present state signifies the condition of a flip-flop or circuit *before* the application of a CLOCK signal. The next state details the condition of the output *after* the clock pulse has arrived.

In Figure 4.36 there is an R–S flip-flop with a timing diagram of the set input condition. Below the timing diagram is the state table entry for the set input condition. A small n is used to indicate the present state time and $n + 1$ indicates the next state time—that is, the state that follows the present state after one clock transition. Therefore, R_n and S_n are the input states before the CLOCK signal is applied, and Q_n is the state of the flip-flop at the present time. Q_{n+1} is the resulting state of the output after the CLOCK is applied and given the n time input states. Relate the state table entry to the timing diagram to get a feel for exactly what the table is telling you that a standard truth table did not.

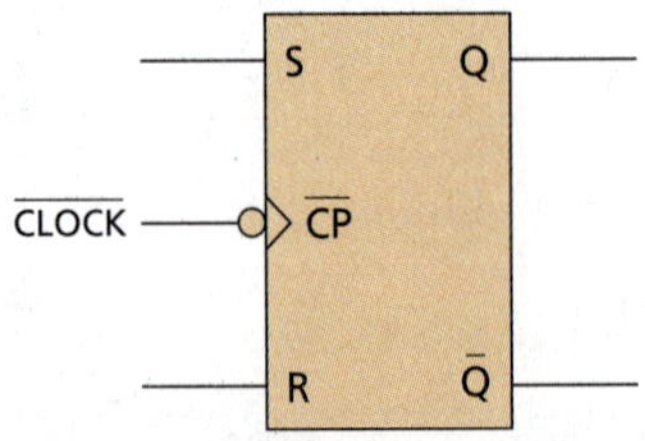

(a) Clocked R–S Flip-Flop

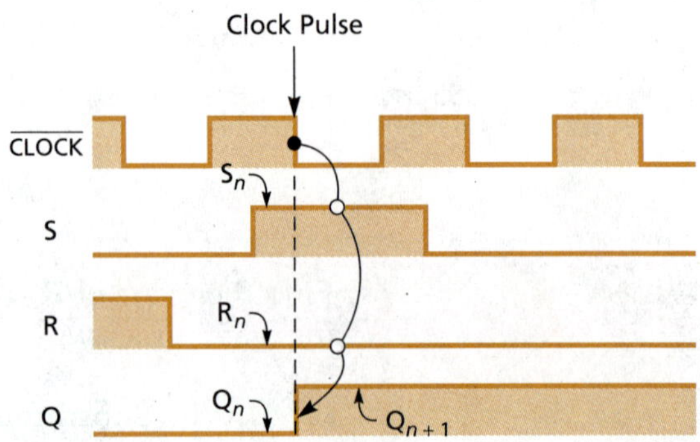

(b) Timing Example

S_n	R_n	Q_n	Q_{n+1}
1	0	0	1

(c) State Table Entry

Figure **4.36** State Table Entry Example

To get a feel for the difference between a timing diagram, truth table and state table, let's compare the three for the reset inputs to an R–S flip-flop all shown in Figure 4.37. The timing diagram shows when the flip-flop changes

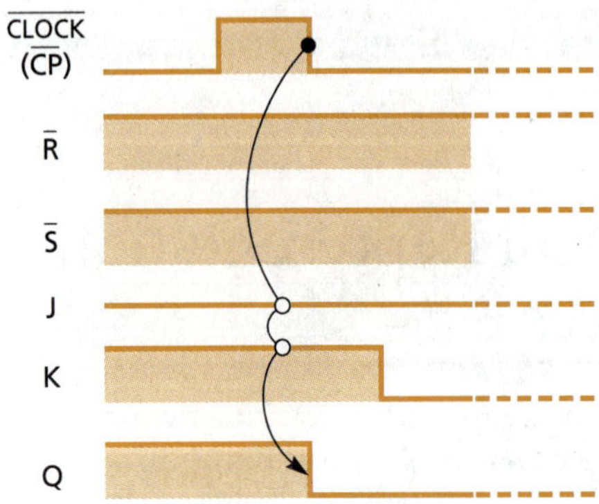

(a) Timing Diagram

$\overline{R}$	$\overline{S}$	$\overline{CP}$	J	K	Q	Mode
1	1	↑	0	1	0	Synchronous Reset

(b) Truth Table Entry

$\overline{R}$	$\overline{S}$	J	K	Q_n	Q_{n+1}	Mode
1	1	0	1	1	0	Synchronous Reset

(c) State Table Entry

Figure **4.37** Synchronous Reset Function of J–K Flip-Flop

state. You can see the R, S, and CP inputs and exactly when Q changes state. On the timing diagram, the change of output state is expressed as a change of level from a 1 to a 0. A truth table shows the input conditions that cause the output condition to occur. For reset, it shows R = 1, S = 0, and a transition on the CP input to cause Q to go to 0. In the state table, input conditions are shown as in the truth table. Next the present state of the output, *before* the application of the clock to CP, is shown as Q_n. Lastly, the next output state at Q, *after* the clock has been applied, is indicated under Q_{n+1} on the state table.

Branch *a line on a state diagram that shows the transition from the present to the next state.*

Label *defines the control signals on a state diagram that cause a transition to occur.*

State diagrams are a graphical extension of state tables. Before we can use a state diagram, we must understand the symbols used in the diagram. An output state for a flip-flop or any combinational synchronous logic circuit is indicated by a circle containing the value of the state of the device or circuit. In Figure 4.38a, two circles show how a flip-flop's reset (0) and set (1) states are described in a state diagram. Changes in state of a device or circuit are illustrated by an arrow, called a **branch,** originating from the present state circle and terminating at the next state circle. In Figure 4.38b there are two illustrations of this process. A no-change branch originates from the present state and terminates back to itself as the next state. That says that there was no change from the present to the next state. A second branch is made from the reset present state to the set next state. This branch says there is a change from the present state to the next state when a CLOCK signal is applied. **Labels** are used to mark the input conditions that cause the branch to occur. Branches also include information about additional asynchronous inputs and outputs that affect the change from present to next state or are affected by the change from present to next state of the device or circuit. These inputs and outputs are other circuits connected to the device or circuit input or output. This may sound a little confusing at this point, but what this all means will become clear once we go through several examples. In Figure 4.38c, the label on the no-change branch indicates that a low input is required for no change to occur. The label on the branch, affecting the change from the present to the next state, indicates that a high input is required for the change to occur. An asynchronous output indicated on the right side of the label's slash line (/) indicates that it follows the state of the Q output.

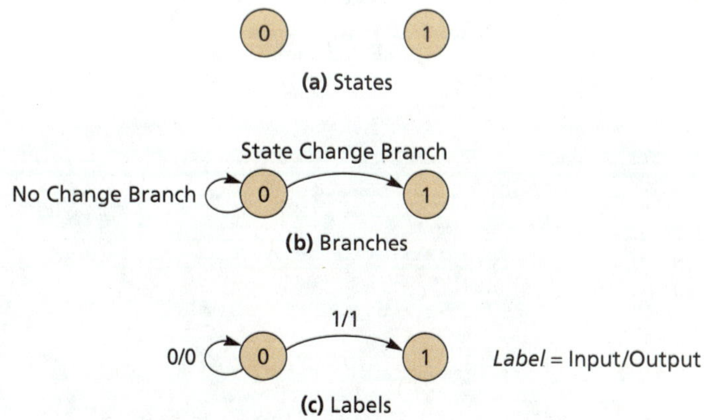

Figure 4.38 State Diagram Symbols

EXAMPLE 4.9 Which flip-flop is described by the state diagram of Figure 4.39?

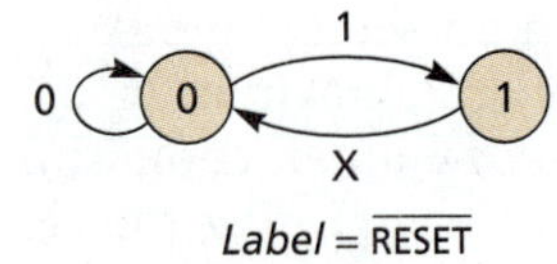

Figure 4.39 Example 4.8 State Diagram

SOLUTION The upper branch in the state diagram shows the flip-flop changing from a 0 to a 1 as long as $\overline{\text{RESET}}$ is high or inactive. The lower branch states that the state of the $\overline{\text{RESET}}$ does not matter; the flip-flop will return to the reset state from the set state if $\overline{\text{RESET}}$ is active or not. Essentially, it returns to the 0 state if the present state is 1 and $\overline{\text{RESET}}$ is high, and it also returns to the 0 state if $\overline{\text{RESET}}$ is low. Also shown on the diagram is what happens when $\overline{\text{RESET}}$ is asserted and the flip-flop is already reset—it remains reset. This diagram describes a toggle flip-flop.

In Figure 4.40, state diagrams are shown describing different types of synchronous devices and circuits. The first one, in Figure 4.40a, is for a clocked

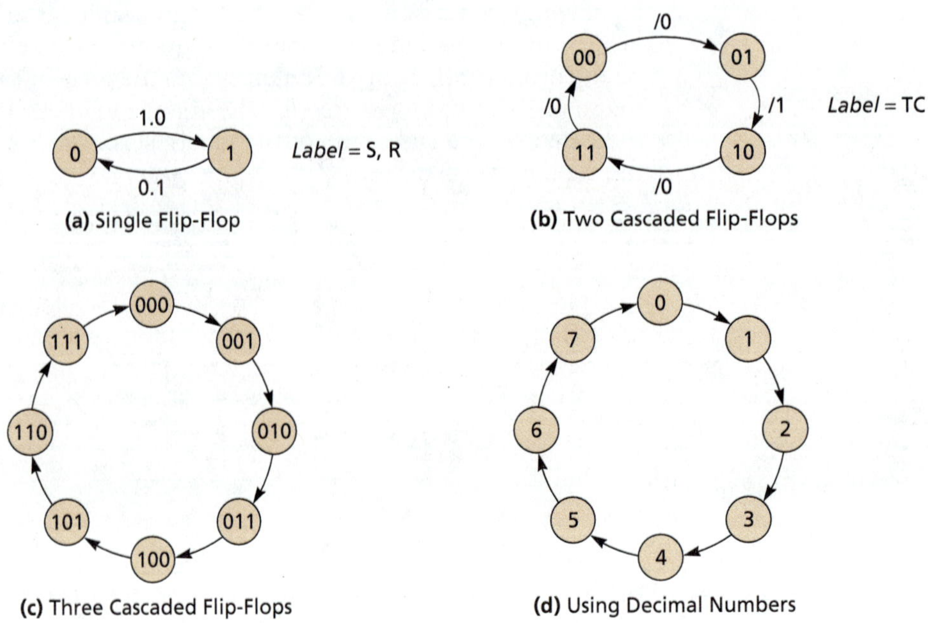

Figure 4.40 State Diagram Examples

R–S latch. The label indicates that there are two inputs, S and R, and that there is no other output except the state output. Label 0,1 on the branch originating at the set state and going to the reset state, says that it is required for input S to be a 0 and input R a 1 in order for the transition from the present set state to the next reset state to happen. The label on the branch in the other direction indicates that input S must be a 1 and input R a 0 to effect the change to a 1 or set state.

A free-running two-bit counting circuit (counter) has two conditions in each state, indicated in each of the four state circles of the counter diagram shown in Figure 4.40b. The states are 00, 01, 10, and 11. Since there are no inputs indicated in the branch labels, these are unconditional branches, making this a free-running counter. There is an additional output, called *terminal count,* or TC, shown in the state diagram as a label on the branches. This output indicates, when it is high, that the counter state is 01.

In Figure 4.40c, three flip-flops are cascaded to produce a free-running three-bit counter whose states go from 000B to 111B and back to 000B. It should be noted that these states do not have to change in a specific order, but whatever order is shown on the diagram is the order that particular changes in the circuit states occur (more on this when we discuss counters later in this chapter). Lastly, states can be shown as other conditions besides binary ones and zeros. In Figure 4.40d, decimal numbers are used to indicate the different states. Hexadecimal or other number systems can be employed as well as names and other descriptive labels. To show some additional aspects of state diagrams, we will make a state diagram for an edge-triggered J–K flip-flop having active low asynchronous reset and set inputs.

Figure 4.41a shows the block diagram, and Figure 4.40b shows the state table, for the J–K flip-flop. The state diagram in Figure 4.40c shows all the possible input conditions and resulting output states for the J–K flip-flop. An asynchronous input, such as $\overline{\text{RESET}}$, in Figure 4.41, is shown as an unterminated branch pointing to the state achieved by asserting that input signal. The principle associated with this is that the state indicated by the asynchronous branch results from asserting the input regardless of the present state of the circuit or synchronizing clock signal. This indication is frequently used with synchronous circuits rather than individual devices, because these circuits usually possess many states, as opposed to only two states of a device such as a D flip-flop.

In Figure 4.41, a reset state, Q = 0, is achieved by setting $\overline{\text{RESET}}$ to 0 and $\overline{\text{SET}}$ to 1 as shown by the unterminated branch to the zero state of the flip-flop. The label on the branch shows the states of the asynchronous signals and the synchronous J and K inputs. Note that the synchronous inputs are "don't cares" when applying either of the asynchronous inputs. The branches indicating changes from one present state to a next state are labeled the same as the unterminated branch. Note that there are two branches showing both asynchronous inputs, $\overline{\text{SET}}$ and $\overline{\text{RESET}}$ inactive. In the top branch, the J input is a 1 and the K input is a "don't care" when the flip-flop changes from the 0 state to the 1 state. This says that the change will occur on the next clock cycle if J = 1 and K = 0 (synchronous set inputs) or J = 1 and K = 1 (toggle inputs). A similar analysis can be made for the lower branch showing a change from the set to reset state of the flip-flop.

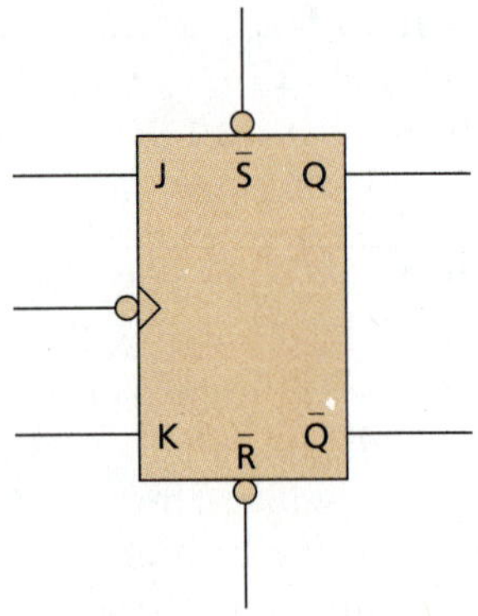

(a) Block Diagram

$\overline{R}$	$\overline{S}$	J	K	Q_n	Q_{n+1}	Mode
1	0	X	X	X	1	Asynchronous Set
0	1	X	X	X	0	Asynchronous Reset
1	1	0	1	1	0	Synchronous Reset
1	1	1	0	0	1	Synchronous Set
1	1	0	0	Q_n	Q_n	No Change
1	1	1	1	0	1	Toggle
1	1	1	1	1	0	Toggle

(b) State Table

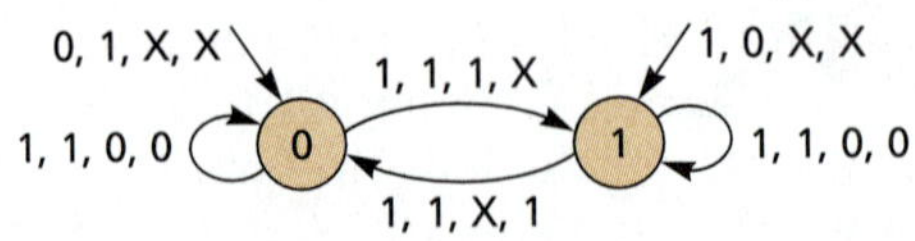

State = Q

Branch = $\overline{\text{RESET}}$, $\overline{\text{SET}}$, J, K

(c) State Diagram

X = "don't care" state

Figure **4.41** J–K Flip-Flop Example

Section 4.8

Practice Problems

1. Create a state table and state diagram for a data (D) flip-flop.
2. Make a state diagram for a three-bit counter that counts in this sequence: 000B 011B 101B 001B 110B 000B. It counts when an enable is set high and stops counting when enable is low. The counter can also be set to 000B by setting a reset input low. A terminal count (TC) output signal goes high when the counter state goes to 101B. *Hint:* This diagram requires labels with two inputs and one output and each state has three binary bits. Use an unterminated arrow for the asynchronous reset input.

4.9 Binary Counters

Set-up time the minimum time an input must be supplied before a CLOCK signal is applied.

Flip-flops are used as the building blocks for counter circuits. In most cases, the versatile J–K flip-flop is used as the basis for these circuits. At other times, when less functionality is required, the simpler data or D flip-flop is employed. It is essential, when analyzing these circuits, that you keep in mind the timing needs of the flip-flops. The primary care to be taken, particularly for an edge-triggered flip-flop; is that inputs must be applied before a flip-flop is triggered or clocked. This time is referred to as **set-up time** or t_S and is the basis for the following two rules:

1. The outputs of any flip-flop are determined by the states of the inputs at the time the clock trigger is applied.
2. Outputs will not change again until the next clock trigger is applied.

Keep these in mind as each circuit is presented and discussed.

Synchronous devices are defined as those which require a clock input to operate. In actuality, some flip-flops have both synchronous and asynchronous modes of operation. The J–K flip-flop, for instance, operated synchronously using the J and K inputs along with the clock pulse (CP) and operated asynchronously using the set (S) and reset (R) inputs, which do not require triggering from CP. The same applies to counter circuits—some are synchronous and some are asynchronous. The trouble is that both types of counters require clock signals to allow counting to happen. Synchronous counters are those circuits in which all the flip-flop elements are clocked at the same time. In asynchronous counters, flip-flops are not clocked simultaneously, but are driven by the resulting output state changes of other flip-flops within the circuit. In effect, the changes in the states of the flip-flops of an asynchronous counter appear to ripple from one stage to the next. Such an asynchronous circuit, the **binary ripple-up counter**, is shown in Figure 4.42 and will be used to illustrate this ripple effect.

Binary ripple counter an asynchronous counter whose output sequence either counts up or down.

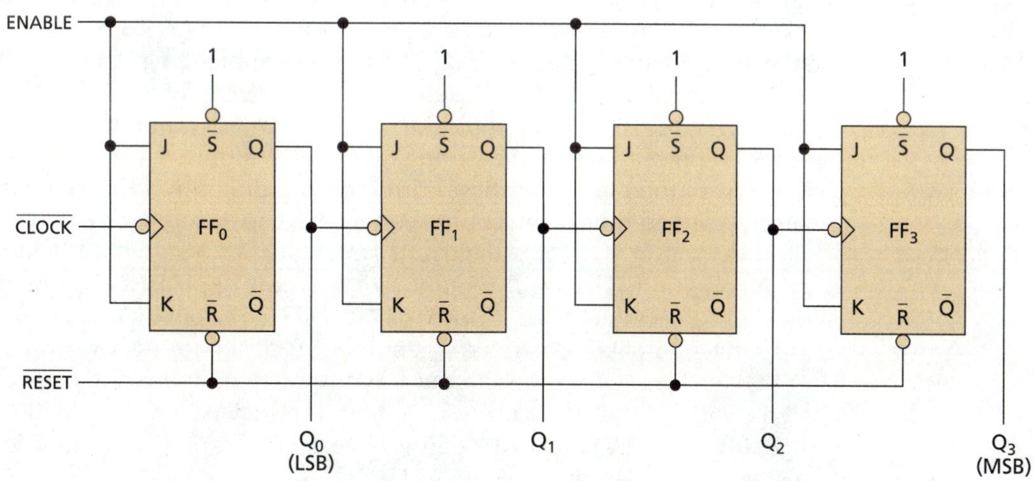

Figure 4.42 Four-Bit Binary Ripple Up Counter

ENABLE a signal that controls when a counter can count.

Freeze to hold the current count output, usually by negating the ENABLE signal.

Notice, right off the bat, that the circuit is an asynchronous counter according to the definition in the preceding paragraph. The CLOCK signal is brought only to the first flip-flop (FF_0). The Q output of that flip-flop drives the clock input of the second flip-flop (FF_1), and so forth. The J–K inputs of these flip-flops are all connected together and brought out as an **ENABLE** control signal. This signal is used to allow the counter to count or to stop and **freeze** the count. The ENABLE signal is connected to all the J and K inputs so that the J–K flip-flops can only operate in the no-change mode (J and K both 0) or the toggle mode (J and K both 1). You can see that if this line is brought low, none of the flip-flops will change state, causing the output to be "frozen" at its current state.

A second control signal, $\overline{\text{RESET}}$, when asserted, presets the counter to a value of 0000 ($Q_3 = 0, Q_2 = 0, Q_1 = 0, Q_0 = 0$). The $\overline{\text{RESET}}$ signal is connected to the reset, $\overline{\text{R}}$, input of each J–K flip-flop. When that line is asserted low, each flip-flop is placed in the reset state (Q = 0) regardless of the states on inputs J and K. Initially, this is what will be done to the circuit—the $\overline{\text{RESET}}$ input is brought low momentarily to initialize the counter outputs all to 0.

For the remaining discussion, the outputs will be referred to as a binary word of four bits (a nibble) in the order of Q_3, Q_2, Q_1, and Q_0, with Q_0 being the least significant bit (LSB). For instance, if the Q outputs of the counter are $Q_3 = 1$, $Q_2 = 1$, $Q_1 = 0$, and $Q_0 = 1$, then the output is 1101B.

When we bring both $\overline{\text{RESET}}$ and ENABLE high, the flip-flops are taken out of the reset mode and placed into the toggle mode, awaiting application of the CLOCK signal to FF_0. Applying the CLOCK signal to FF_0 causes its output to toggle each time the clock has a negative transition, since the 74LS76 J–K flip-flop is a negative-edge-triggered flip-flop. The Q output of FF_0 is applied to the clock input of FF_1, which will toggle whenever Q_0 goes low. The first clock cycle to FF_0 causes Q_0 to toggle to a set state. This low to high transition does not affect FF_1, which remains in a reset state. The second clock transition to FF_0 causes it to toggle to a reset condition, which is now impressed onto FF_1 CP input as a negative transition, which causes FF_1 to toggle to a set state. This does not affect FF_2, since its CP input sees a low to high transition. FF_3 does not change either, since its CP input sees a low steady state condition. FF_2 will only toggle when FF_1 changes from a set to a reset state, which occurs on the fourth clock pulse. FF_3 must wait even longer for FF_2 to change from a high to a low state before it can toggle. The outputs seem to ripple down from one flip-flop to the next because of the way each is driven, in turn, from the preceding flip-flop, which gives rise to the ripple part of the counter's name. The action of this counter is summarized in the timing diagram of Figure 4.43.

The reset mode is shown first on the timing diagram, followed by one no-change cycle and finally the count sequence. Keep in mind, that if ENABLE is brought low at any time, the current states of the counter are held in a no-change condition. They will remain that way until ENABLE is brought high, allowing the counter to continue or if $\overline{\text{RESET}}$ is brought low returning the counter back to 0000. Consider Q_0 as the LSB of the four-bit word composed of Q_3, Q_2, Q_1, and Q_0, and write out the output word following each clock cycle. You will see that the output counts in a binary sequence, 0000B, 0001B, 0010B, 0011B, . . . 1111B. This count is going up from a minimum low value of 0000B to the highest value of 1111B.

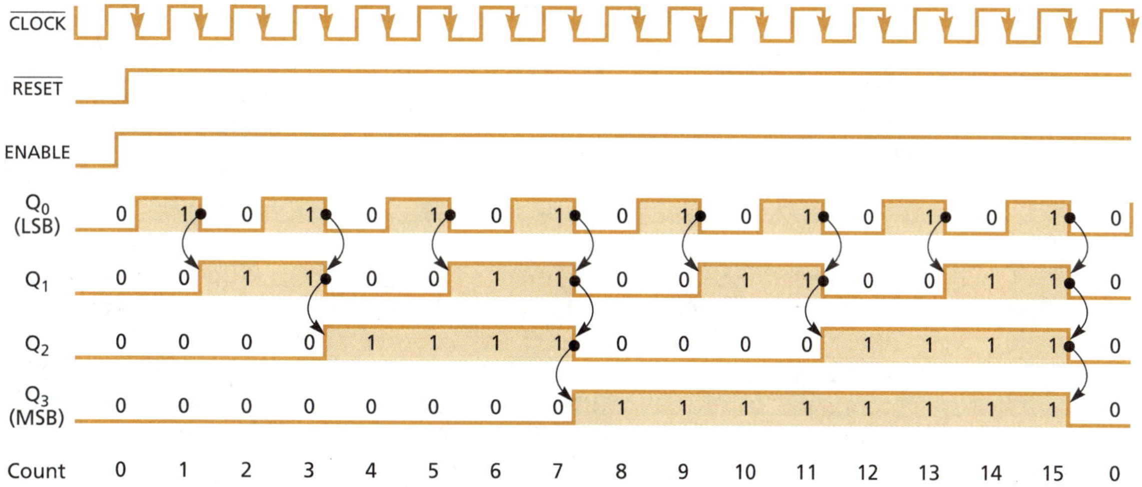

Figure 4.43 Up Counter Count Sequence

EXAMPLE 4.10

What is the four-bit word from the counter in Figure 4.40 if ENABLE is brought low 5 clock cycles after $\overline{\text{RESET}}$ is negated and ENABLE had been made active?

SOLUTION After reset, the counter starts from 0000B and continues to count until it is disabled by negating the ENABLE signal. The first clock cycle brings the counter output to 0001B; the second, 0010B; third, 0011B; fourth, 0100B, and the fifth to 0101B. At this point, it is disabled and the output will remain at 0101B until the count is resumed or $\overline{\text{RESET}}$ asserted.

__Mod__ or __modulo number__ the number of different conditions or states that a set of outputs can have.

From the timing diagram, notice that the output of FF_0 changes at a rate that is half the CLOCK rate. FF_1's output is half the frequency of the Q output of FF_0 and, as well, one-quarter of the CLOCK signal frequency. FF_2 toggles based on Q_1's output and further divides the rate in half again, as does FF_3. As well as counting, this circuit can be used to divide the clock frequency by 2, 4, 8, or 16. The highest rate of division is also known as the counter's **mod** or **modulo number**. This number actually reflects the number of different output state combinations the circuit can have. For the asynchronous up counter, the mod number is found by raising 2 to a power equal to the number of flip-flops in the counter. For the four-bit counter in Figure 4.41, the mod number is $2^4 = 16$.

EXAMPLE 4.11

What is the modulo number for a three-bit counter?

SOLUTION The number of different states for an asynchronous ripple-up counter (mod) is found by raising the base 2 by an exponent value equal to the number of flip-flops in the circuit (n): mod = 2^n. For a three-bit counter, the modulo number is $2^3 = 8$.

Decoder *a logic circuit that translates a binary number into a single logic signal.*

By relating the modulo number with the number of flip-flops in the counter, counters with mod numbers of 2, 4, 8, 16, 32, and so on can be easily made. What about a mod-10 counter? This counter will count from 0000B to 1001B, reset, and repeat the sequence. What needs to be done is to reset a binary up counter when it tries to increment from 9 (1001B) to 10 (1010B). This is accomplished by generating a RESET signal when the counter output reaches 1010B, as shown in Figure 4.44. The type of circuit used to generate that RESET signal is called a **decoder.** In general, decoder circuits convert a combination of binary input signals into a single output signal. In this case, the decoder is interpreting or decoding the binary input 1010B and converting it into a single RESET signal to cause the counter to return to 0000B when it tries to go to the 1010B state. Decoders are, basically, AND gates whose output only goes to a high logic state when all its inputs are high.

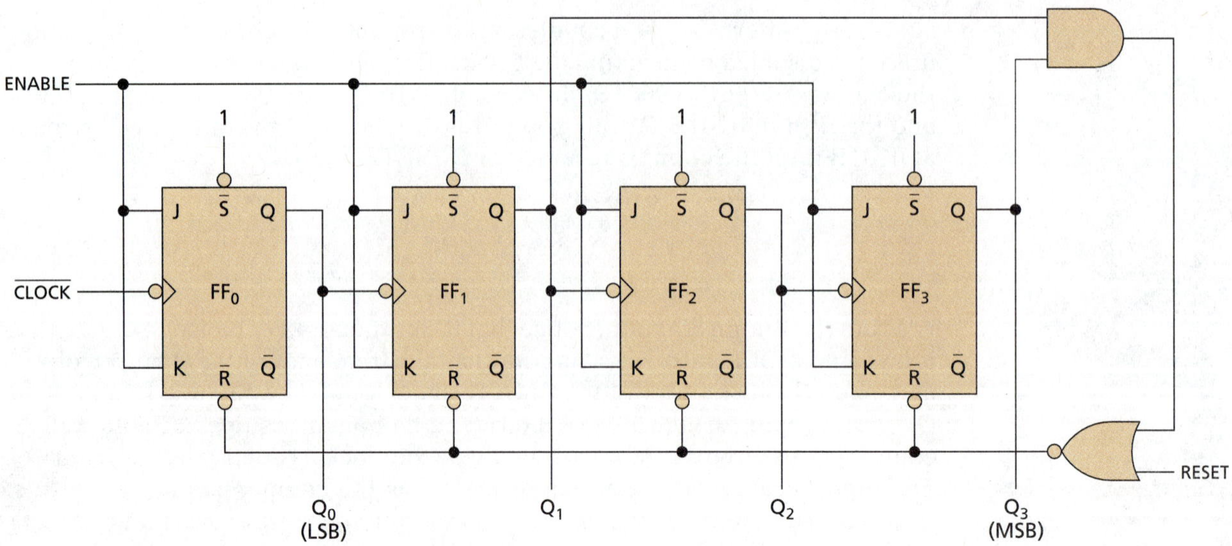

Figure 4.44 Mod-10 Binary Up Counter

The counter counts normally from 0000B to 1001B, but when it tries to go to the next count, the AND gate output is forced high and makes the output of the NOR gate go low, causing the counter to reset back to 0000B. One caution to keep in mind is that there is a certain amount of propagation time in each flip-flop for the flip-flop to change state from one level to another. First the flip-flops have to switch to the 1010B state, which is followed by the AND and NOR gate propagation times, before the reset line is asserted. This results in the presence of a short time between the counter starting to go to the 1010B count and being reset to 0000B. For a short period of time, the counter is in the 1010B state. This condition is called a **glitch,** which is any undesired output condition that exists for a very brief period of time. Unfortunately, in some applications this time period could be crucial and cause other circuits dependent on the counter to malfunction. Care must be taken in using asynchronous counters that use count decoding to adjust for a different mod number.

Glitch *a transient condition created by the propagation time of a counter changing from a maximum count to a zero condition.*

Another use for up counters is frequency division—dividing the frequency of the CLOCK signal applied to CP. A single toggle flip-flop divides the input frequency in half. That is, the output changes at a rate one-half of that of the input. It has already been shown that the process continues with each additional stage adding a division factor of 2. The counter discussed in the previous paragraph is a mod-10 counter having 10 different output states whose sequence repeats every 10 clock cycles. The output of the AND gate goes low at a rate that is one-tenth the frequency of the CLOCK signal at CP and can be used as a divide-by-10 output signal.

EXAMPLE 4.12

Design a mod-5 or divide-by-5 counter.

SOLUTION A mod-5 up counter counts from 000B to 100B, requiring three flip-flops. Therefore we start with a three-bit up counter. Output 101B supplies the inputs to a AND whose output drives a NOR gate that supplies the reset inputs so that when the counter attempts to count from 100B to 101B, it is forced to reset back to 000B and repeat the count sequence. *Note:* The mod number or divide-by number is also the binary value used to generate the reset signal.

Figure 4.45 shows the binary ripple down counter. Notice that the only difference between this counter and the up counter in Figure 4.42 is that the clock input of a flip-flop is driven by the $\overline{Q}$ output instead of the Q output and the $\overline{\text{RESET}}$ signal is connected to the $\overline{S}$ set inputs instead of the $\overline{R}$ reset inputs of the flip-flops. Asserting $\overline{\text{RESET}}$ low causes the flip-flops to be set and the output of the counter to be 1111B. Now when the clock is first applied to FF_0, it toggles as before, bringing its Q output to 0 and the output word to 1110B. At this time, $\overline{Q}_0$ goes high, which is applied to FF_1's CLOCK input. This low to high transition does not affect the state of FF_1, which requires a negative transition as before. The next CLOCK signal into FF_0 causes it to toggle high, forcing its $\overline{Q}$ to a low and triggering FF_1. FF_1 toggles low, bringing the output word to 1101B. The process continues as it did with the up counter as long as the clock remains on and ENABLE and $\overline{\text{RESET}}$ are both high. Each flip-flop's output again divides the previous output rate by one-half, as it did with the up counter.

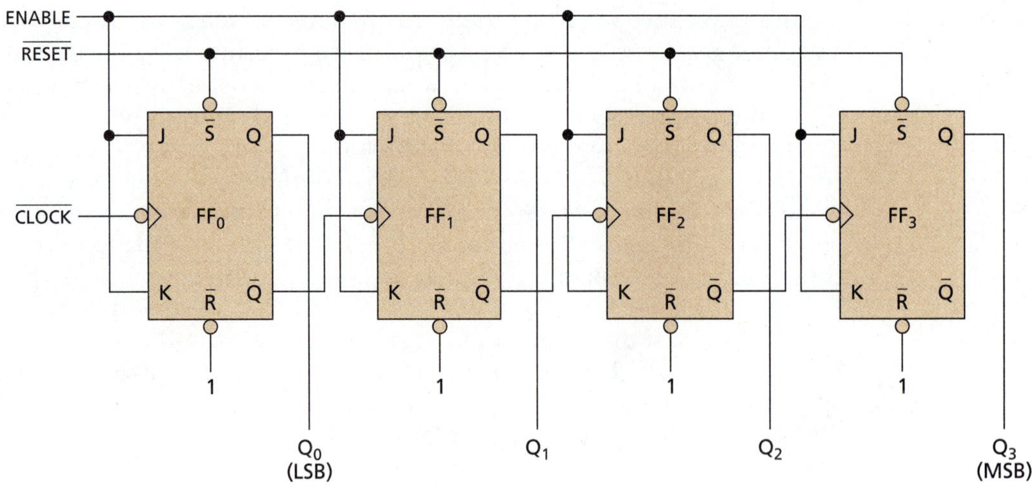

Figure 4.45 Four-Bit Binary Ripple Down Counter

The count sequence and timing diagram for the down counter is summarized in Figure 4.46.

Mod numbers for down counters are the same as they are for up counters, a three-bit down counter has a mod number of 8, a four-bit has a mod number of 16, and so on. However, setting up down counters for other mod numbers takes a little careful thought. For instance, let's design a mod-10 down counter. The counter will start at 1001B or 9 and count down to 0000B or 0.

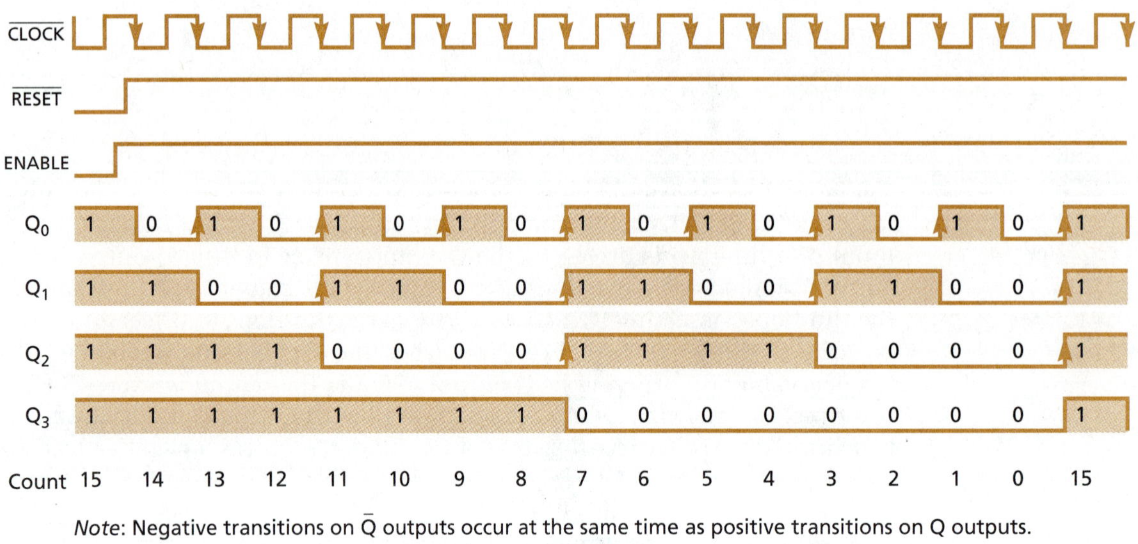

Figure 4.46 Down Counter Timing and Count

As with the up counter, the output that causes the counter to return to the beginning of the count sequence is detected using an AND gate. For this example, the output detected is 0000B and counter output that begins the count sequence is 1001B.

Figure 4.47 shows the circuit for the mod-10 down counter. Counter output 0000B causes all four flip-flop $\overline{Q}$ outputs to go high. The AND gate connected to these outputs is driven high, causing the NOR gate to supply an active low input to FF_0 and FF_3 set inputs and FF_2 and FF_1 reset inputs. This signal resets the counter to 1001B and allows the sequence to be repeated. The NOR gate is used to allow the application of a RESET signal to set the initial conditions for the counter. As with the up counter, the down counter can also be used for frequency division. The output of the mod-10 down counter AND gate goes low at a rate that is one-tenth of the $\overline{\text{CLOCK}}$ signal at CP.

Binary counters are manufactured in single IC packages like the 74LS93 pictured in Figure 4.48. A single D flip-flop configured to operate as a divide-by-2 toggle flip-flop is driven by a $\overline{\text{CLOCK}}$ signal applied to $\overline{CP}_0$. The output of that flip-flop, Q_0, changes at a rate that is one-half the frequency rate of the clock input. This flip-flop can be used independently, or it can be connected to input $\overline{CP}_1$ to extend the mod-8 up counter, comprised of the remaining three flip-flops, into a mod-16 counter. The main difference between this counter and the previously described binary ripple up counter is that there is no ENABLE control signal included with this counter. The only way to "freeze" the count is to place a steady state on the $\overline{CP}$ inputs. A pair of master reset inputs (MR_0 and MR_1) are included to place all the flip-flops into reset states when they are both asserted high.

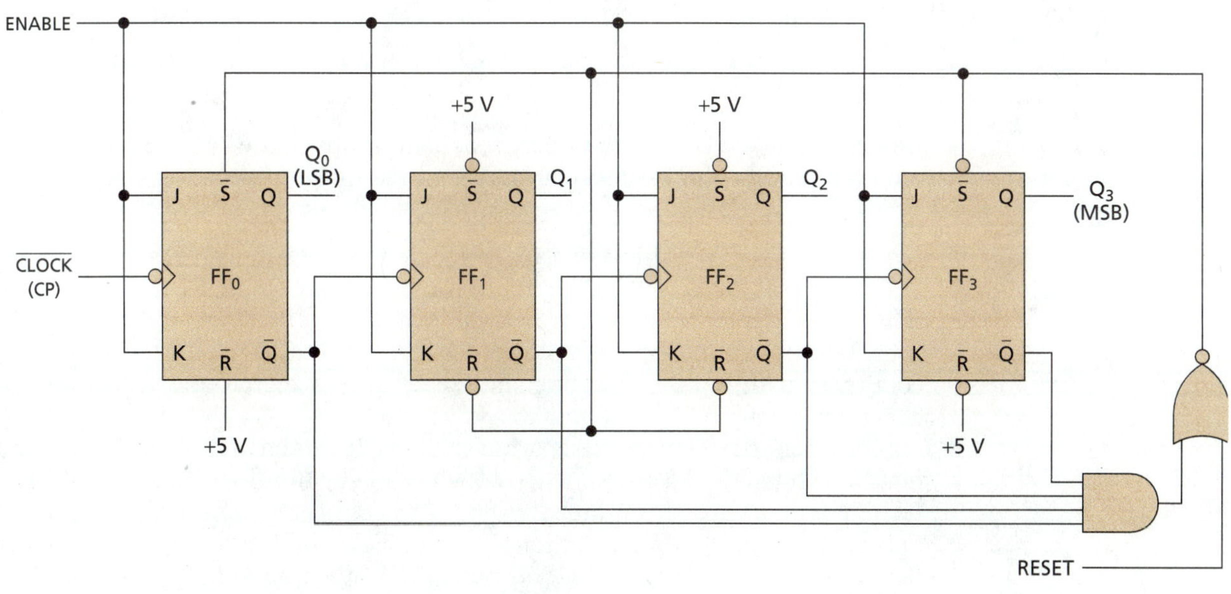

Figure 4.47 Mod-10 Binary Ripple Down Counter

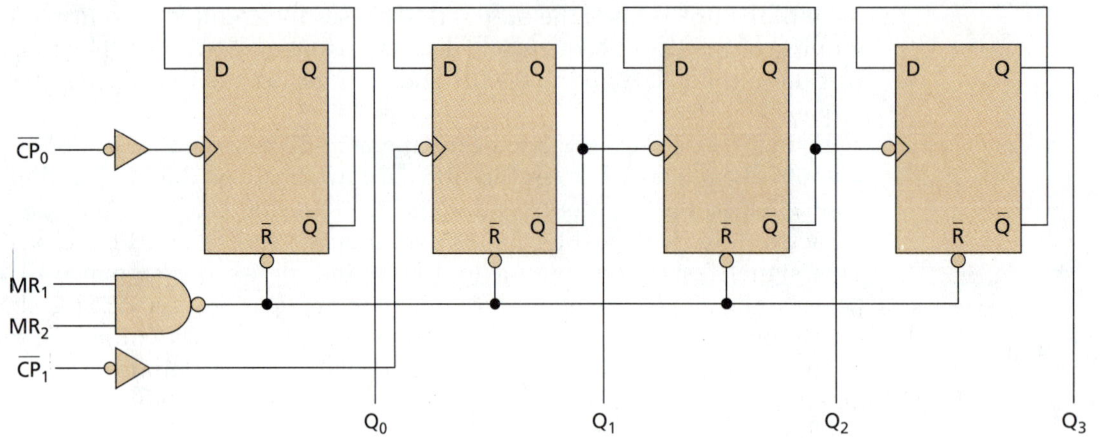

(a) Schematic

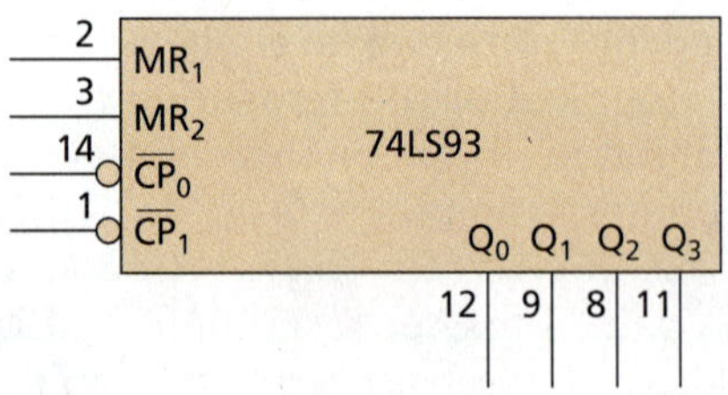

(b) Symbol

MR_1	MR_2	$\overline{CP}_0$	$\overline{CP}_1$	Q_0	$Q_3 Q_2 Q_1$
1	1	X	X	0	0 0 0
0	X	↓	↓	Toggle	Count Up
X	0	↓	↓	Toggle	Count Up
0	0	S	↓	No Change	Count Up
0	0	↓	S	Toggle	No Change

(c) Truth Table

X = "don't care" state
S = steady state

Figure 4.48 74LS93 Ripple Up Counter

EXAMPLE 4.13

Describe how the D flip-flops operate as T flip-flops in the 74LS93. What is the mod number for the 74LS93 counter?

SOLUTION $\overline{Q}$ of the D flip-flops are connected to the D inputs. Start with a flip-flop reset, making $\overline{Q} = 1$. When there is a transition on the $\overline{CP}$ input, the flip-flop sets, forcing $\overline{Q} = 0$, which is brought back to the D input. The next $\overline{CP}$ transition causes the flip-flop to reset, changing $\overline{Q}$ and D in to a 1. The process continues, with flip-flop toggling states each time a high to low transition occurs on $\overline{CP}$.

Since the 74LS93 has two separate sections, it can have three different mod numbers. It is a mod-2 circuit when considering the flip-flop driven by $\overline{CP}_0$. The second section consists of three flip-flops in a count-up configuration. The outputs range from 000B to 111B for a total of eight different states. This makes that section a mod-8 counter. If output Q_0 is connected to the $\overline{CP}_1$ input, then the entire circuit is a mod-16 counter.

The 74LS93 binary up counter can be used as a mod-2, mod-8, or mod-16 counter directly; other mod number counters can also be achieved by using the master reset (MR) inputs. To reset the 74LS93, both master resets, MR1 and MR2, have to be brought high. To make a counter with a different mod number from 2, 8, or 16, the MR inputs are asserted at a certain count to cause the counter to reset and repeat its count sequence.

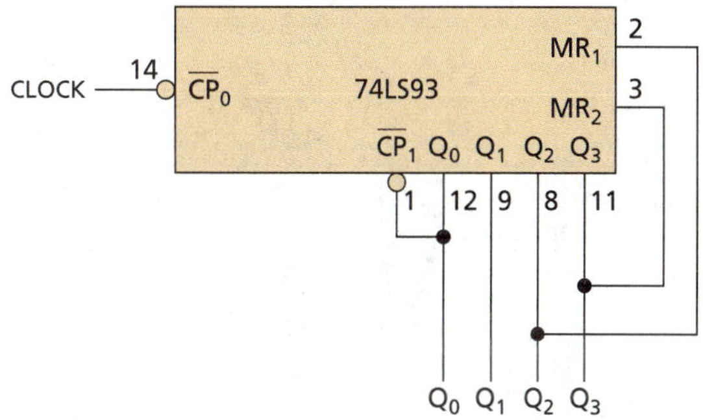

Figure 4.49 Mod-12 Binary up Counter Using 74LS93

The configuration of the 74LS93 counter in Figure 4.49 is a mod-12 counter. Output Q_0 is connected to input $\overline{CP}_1$ connecting the mod-2 section to the mod-8 section of the counter. Output Q_2 is connected to MR_1, and output Q_3 is connected to MR_2. The counter counts from 0000B to 1011B (0H to 11H) normally. When it tries to increase its count to 12H (1100B), both master resets are asserted and the counter is returned to 0H and resumes counting. Notice how direct the connections are made to change the mod number from 16 to 12. The following table lists all the possible mod number counters that can be created using the 74LS93 with jumpers only to the master resets.

Connections Made When Q = 1				
Q_3	Q_2	Q_1	Q_0	MOD NUMBER
0	0	0	1	1
0	0	1	0	2
0	0	1	1	3
0	1	0	0	4
0	1	0	1	5
0	1	1	0	6
1	0	0	0	8
1	0	0	1	9
1	0	1	0	10
1	1	0	0	12

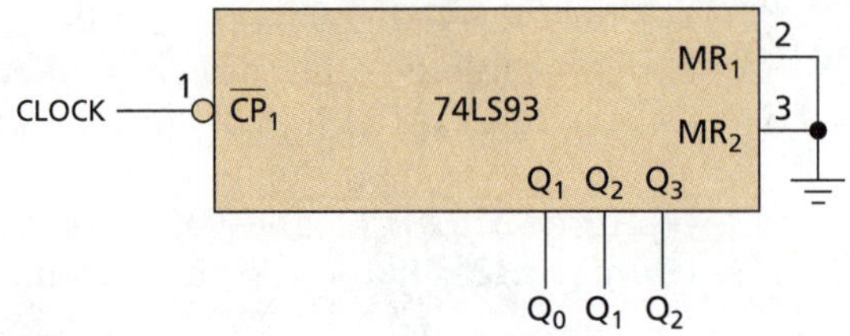

(a) Three-Stage Counter

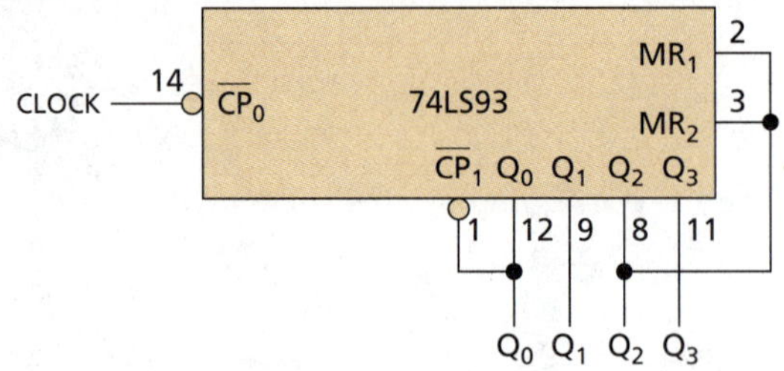

(b) Four-Stage Counter

Figure **4.50** Mod-8 Binary Up Counter Using 74LS93

Figure 4.50 shows two ways to make a mod-8 counter using the 74LS93. Figure 4.50a uses the three-stage counter driven by CP_1 to produce a mod-8 counter. Figure 4.50b shows an alternate method, which would not be used, because the other counter is easier to implement. However, the counter in Figure 4.50b does illustrate the use of connecting the master resets to the counter outputs to change the counter's mod number. In this case, output Q_3 is connected to both master reset inputs. The counter counts from 0H to 7H as expected, and when it tries to count to 8H, Q_3 goes high and asserts the master reset inputs, returning the count to 0H.

Figure 4.51 shows examples of the requirement to make mod number counters that are not on the previous table. Figure 4.51a is a mod-11 counter that employs an AND gate to assert MR_1 when Q_0 and Q_1 are both high. MR_2 is turned on when Q_3 is set high. The counter is reset when both master reset in-

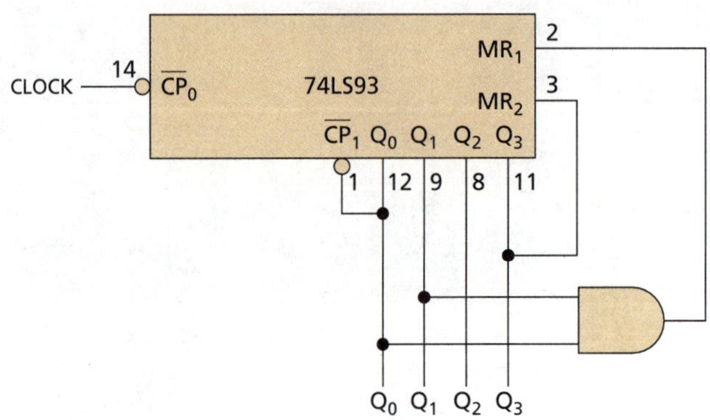

(a) Mod-11 Counter

Figure **4.51** Mod-11 and Mod-15 Binary Up Counters Using 74LS93

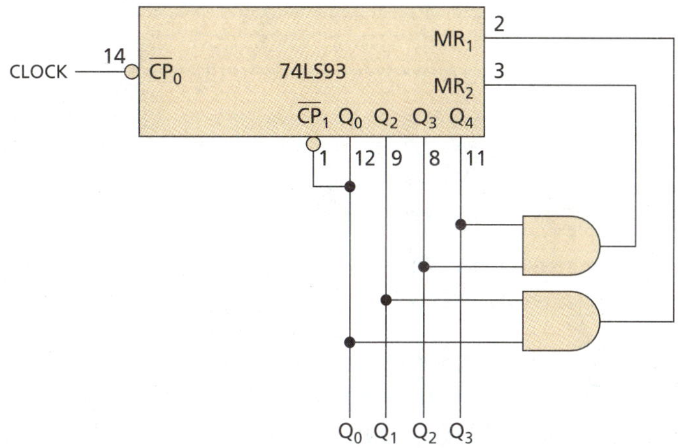

(b) Mod-15 Counter

Figure **4.51** Mod-11 and Mod-15 Binary Up Counters Using 74LS93

Application Note

Computer Application for Binary Up Counter

Program counter a binary up counter used to keep track of program instruction usage in a computer or microprocessor.

Programs executed by digital computers are stored in the computer's memory. Each instruction and data word is referenced by a binary address specifying the location of that instruction or data word. To execute a program, the computer must first fetch the instruction and data from their memory locations. These are fetched sequentially, one program item following another. Keeping track of where the computer is to correctly fetch these instructions and data is the job of the **program counter**, a binary up counter. During the fetch cycle, the contents of the program counter are used to "point" to the memory location that holds the program part that is to be fetched. After the current instruction or data word is fetched, the program counter counts up by one (increments) so that it points to the next instruction or data word. After the current instruction is executed (performed), a new fetch cycle is begun, allowing the computer to fetch the next part of the program now pointed to by the program counter.

Section 4.9

Practice Problems

1. As long as a CLOCK signal is applied to the CP input of the binary up counter in Figure 4.42, the Q outputs will show an up count sequence. What sequence appears at the $\overline{Q}$ outputs?
2. The Q outputs of the counter in Figure 4.45 show a countdown sequence. What sequence is detected at the $\overline{Q}$ outputs?
3. What is the mod number of a six-bit asynchronous ripple binary down counter?

4. Design and draw a mod-5 asynchronous binary ripple down counter that counts from 100B to 000B.

5. One way of making a mod-16 up counter using the 74LS93 is to connect Q_0 to input $\overline{CP_1}$. What alternate connection can be made to expand the mod-8 counter into a mod-16 counter using one 74LS93 and one jumper wire?

puts are high, which in this case occurs when Q_0, Q_1, and Q_3 are all high. This takes place when the counter attempts to count to 11H.

The counter in Figure 4.51b is a mod-15 counter configuration using two AND gates to cause the counter to reset on a count of 15H. The counter counts from 0H to 14H, is reset and repeats the count as long as a CLOCK signal is applied.

4.10 Synchronous Binary Counters

The counters that have been discussed up to this point are asynchronous circuits because the flip-flops that make up the counters are not clocked simultaneously. Instead, the clock input of one flip-flop is driven by the Q or $\overline{Q}$ output of the previous flip-flop. The advantage of doing this is the simplicity of the circuit involved. The main drawback is the ripple effect. It takes time for the change of states to ripple through the counter. Consider the time it takes for a counter to change from 0111B to 1000B. The first flip-flop is reset in response to the transition on CP. That takes one propagation delay time between CP input and the change at Q. The second flip-flop now changes in response to the change in the Q output from the first flip-flop, and a second propagation time is needed before its output changes state. This recurs for the next two flip-flops, making a total of four propagation delay times from the application of the clock to the first flip-flop CP input to the change of the last flip-flop's output from a reset to a set state. If the propagation time of the flip-flop is 15 nS, then the total time for the change is 15 ns × 4 = 60 nS. The minimum time period for the clock input to CP must take this total time into consideration, making the period for a clock cycle 60 nS. The maximum frequency the circuit can operate at is equal to the reciprocal of the period or 1/60 nS = 16.7 MHz.

EXAMPLE 4.14

What is the maximum operating frequency for 8-bit and 16-bit asynchronous up counters using flip-flops with 20 nS propagation delay times?

SOLUTION Basically, the total propagation delay is the number of flip-flops times the propagation delay of one flip-flop:

$$8 \times 20 \text{ nS} = 160 \text{ nS} \qquad \text{and} \qquad 16 \times 20 \text{ nS} = 320 \text{ nS}$$

The operating frequency is the reciprocal of these times:

1/160 nS = 6.25 MHz and 1/320 nS = 3.125 MHz

Synchronous counters are counters in which all the flip-flops are clocked simultaneously, reducing the overall operating time considerably. The tradeoff is that additional logic is required to make them function correctly.

A four-bit mod-16 synchronous up counter is pictured in Figure 4.52. A 0 applied to the ENABLE line forces the J–K inputs of the first flip-flop low and sets AND gate A's output low. The output from the AND gate sets the second flip-flop J–K inputs low and forces AND gate B's output low. This repeats for the last two stages of the counter. Since all J–K inputs are low, all the flip-flops are in the no-change mode and the counter is disabled and does not count. Applying a low to the $\overline{\text{RESET}}$ input forces the counter outputs to 0000B. Bringing both ENABLE and $\overline{\text{RESET}}$ high removes the reset inputs to the flip-flops and allows the counter to start counting. The minimum operating time is restricted by the propagation delay time of the AND gates needed to propagate the condition of the Q outputs to the J–K inputs. Much of the propagation time for the flip-flops occurs during the propagation time for these AND gates. The worst condition is the propagation of a 0 or a 1 from the first flip-flop (FF_0) through all three AND gates to the J–K inputs of the last flip-flop (FF_3). If the delay of one AND gate is 15 nS, then the total operating time is 45 nS and the maximum operating frequency is 1/45 nS or 22.22 MHz, which is higher than the asynchronous four-bit counter frequency of 16.7 MHz.

One very important point to keep solidly in mind when analyzing synchronous counters—the state of a flip-flop *after* the clock is applied depends solely on the state of the inputs *before* that clock signal is applied. With ENABLE and

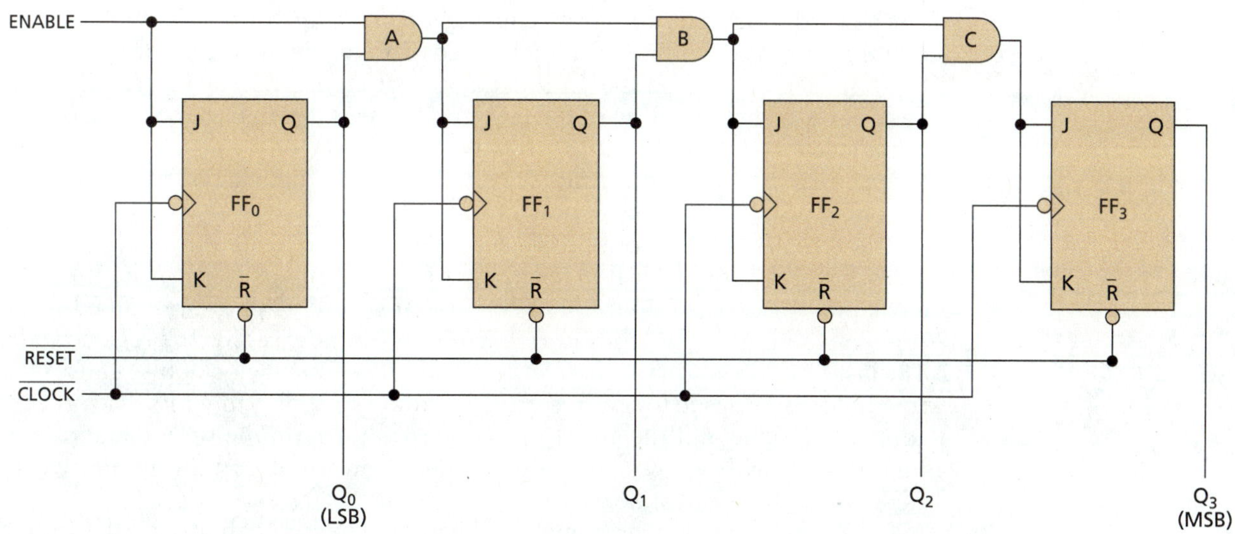

Figure 4.52 Synchronous Binary Up Counter

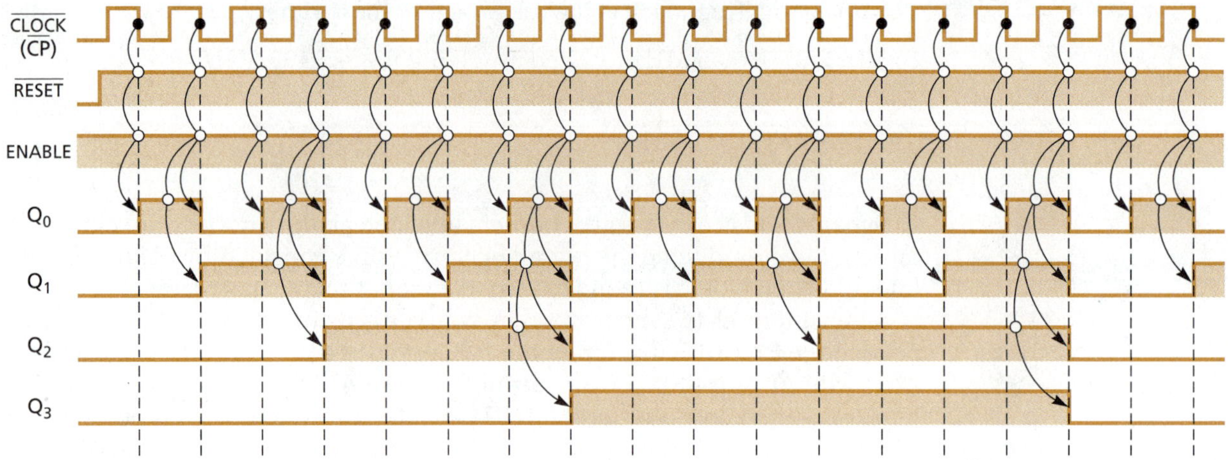

Figure 4.53 Synchronous Up Counter Timing Diagram

$\overline{\text{RESET}}$ high, FF_0 will toggle with each clock pulse, as shown on the timing diagram in Figure 4.53. What happens to the remaining flip-flops depends on the state of its J–K inputs before CP changes from a high to a low. Initially, as a result of a reset, the outputs of FF_1, FF_2, and FF_3 are all low. These zeros are propagated through the AND gates to the J–K inputs of the flip-flops.

Setting ENABLE high places the J–K inputs of FF_0 high, putting that flip-flop into the toggle mode while the remaining flip-flops experience no-change inputs from the zero outputs of the AND gates. When the CLOCK signal is applied to all the CP inputs simultaneously, FF_0 toggles and the others are held in a no-change mode and stay reset. The set condition of FF_0 places a 1 at its Q output and, through AND gate A, to the J–K inputs of FF_1. Since FF_1 is still reset at this time, the zero on its Q output is propagated through AND gates B and C to the J–K inputs of the last two stages.

The next clock pulse arrives, causing FF_0 and FF_1 to toggle while the remaining two flips-flop remain unchanged. The toggle operation of the second flip-flop is illustrated on the timing diagram of Figure 4.53 by an arrow, whose base starts at the transition of the clock and passes through the set state of FF_0, as indicated by the circle in the arrow body. The head of the arrow points to the change resulting from the conditions indicated by the base of the arrow, and the circle points along the arrow's body.

The zero on the Q output of FF_0 is once again propagated through the AND gates, so that the only flip-flop to change on the next clock is FF_0. After the third clock pulse, FF_0 and FF_1 are both set while FF_2 and FF_3 have remained reset. Inputs to all but the last flip-flop's J–K inputs are now set for the toggle mode. The last flip-flop is still in a no-change mode. The fourth clock pulse causes the first three flip-flops to toggle, resetting FF_0 and FF_1 and setting FF_2. The process continues tracing out the count-up sequence shown on the timing diagram of Figure 4.53, with arrows indicating source and input conditions needed to cause each change of state.

Counting down, synchronously, is accomplished by connecting the inputs of the AND gates to the $\overline{Q}$ outputs instead of the Q outputs. The reset control

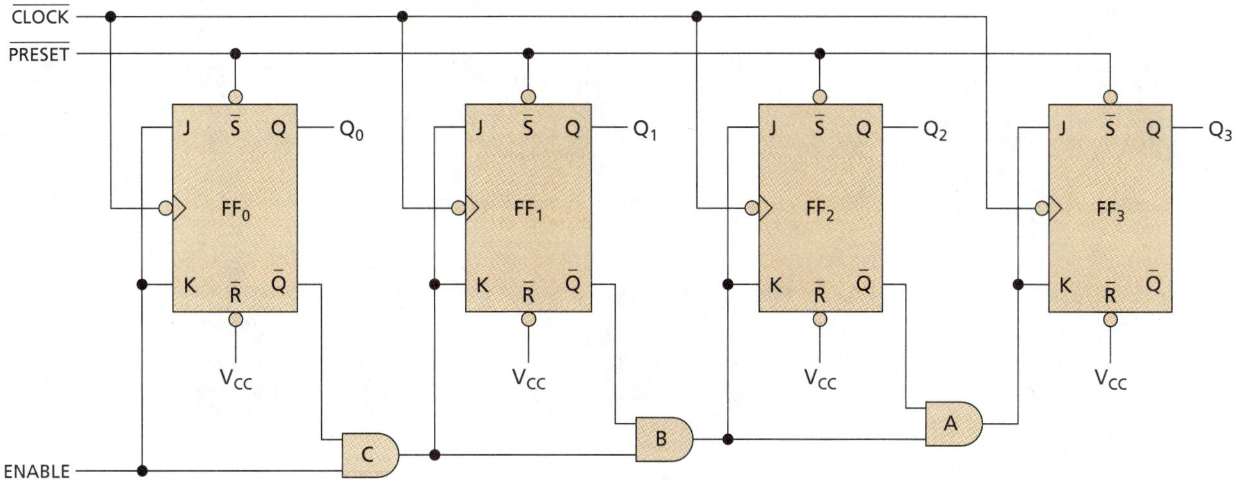

Figure 4.54 Synchronous Binary Down Counter

line is connected to the set inputs of the J–K flip-flops instead of the reset inputs as shown in Figure 4.54. When the counter is first reset, all Q outputs go to 1 and the $\overline{Q}$ outputs go to a 0. The lows from the $\overline{Q}$'s are passed to the J–K inputs of each flip-flop, except the first one, FF_0, via the AND gates. On the first clock pulse, the only flip-flop that toggles is the first one, changing the output of the counter from 1111B to 1110B. The $\overline{Q}$ output of FF_0 is now high, so both of the first two flip-flops toggle with the next clock pulse and the count goes to 1101B. The process continues with the counter counting down to 0000B, back to 1111B and repeating the sequence as long as ENABLE remains high.

EXAMPLE 4.15

What is the binary equivalent value of the outputs of a four-bit down counter if the ENABLE is brought low after six clock cycles?

SOLUTION Bringing the ENABLE line low "freezes" the current count value in the counter. The results of the counter after six clock cycles is 1001B:

RESET: 1111B
Clock 1: 1110B
Clock 2: 1101B
Clock 3: 1100B
Clock 4: 1011B
Clock 5: 1010B
Clock 6: 1001B

Application Note

Down Counter Applications

Counting down sequences appear in many places. Many TV game shows use time limits for contestants to answer questions or perform some mildly amusing task. Sometimes the countdown to no time left is shown on the screen and sometimes, as with *Jeopardy,* a musical tune is played until the time is reached. Microwave oven timers use down counters to turn off the oven when it "times out." VCR counters are actually up/down counters, counting up when the tape is running forward and down when it is rewinding. There are many applications for down counters, so—10, 9, 8, 7, 6, 5, 4, 3, 2, 1, 0—lift off and continue . . .

Section 4.10

Practice Problem

Make a timing diagram for a three-bit synchronous down counter that triggers on a negative transition at the CLOCK input. Place arrows on the diagram to indicate the set-up states required for each flip-flop state change.

4.11 State Diagram Representation of Synchronous Counters

Earlier in this chapter you were introduced to the concept of state diagrams to represent various types of flip-flops. Now let's extend those concepts to synchronous counters. These diagrams will assume a CLOCK signal is required to make the counter count at all. The discussion will begin by developing a progression from timing diagram to actual circuit for a two-bit synchronous up counter.

The timing diagram for the counter is shown in Figure 4.55a. It will be a mod-4 counter, with a sequence count of 00B, 01B, 10B, and 11B. A signal called TERMINAL COUNT ($\overline{TC}$) is arbitrarily selected to be asserted low each time the counter output is 10B. From the timing diagram, a state table, Figure 4.55b, was constructed. The inputs are $\overline{RESET}$, ENABLE, and the current states of the flip-flop outputs (Q_{1n} and Q_{0n}). The outputs are the next state of the flip-flop outputs (Q_{1n+1} and Q_{0n+1}) and $\overline{TC}$.

The first line of the state table shows the conditions resulting from applying an active $\overline{RESET}$ signal. The remaining inputs are "don't cares," as is to be expected. The flip-flops are placed in the reset state (Q outputs low) and TC is forced high. The next line details the no-change or no-count condition of the counter, when $\overline{RESET}$ and ENABLE are both inactive. All the outputs have the same state as their current states. The last four lines of the state table show the count sequence, indicating the current and next states after each CLOCK signal is applied. $\overline{TC}$ indicates an active low condition when the count is $Q_1 = 1$ and $Q_0 = 0$.

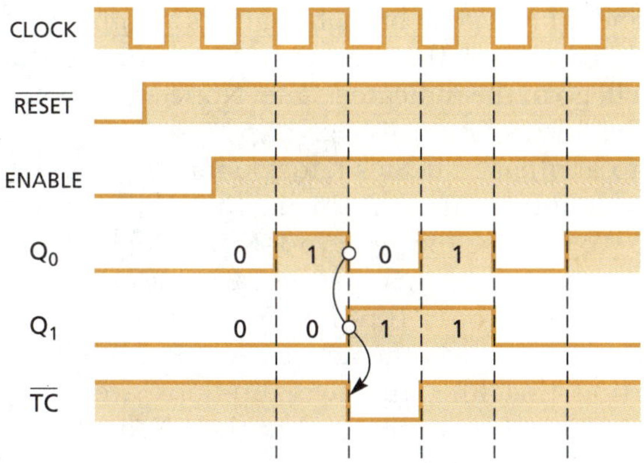

(a) Timing Diagram

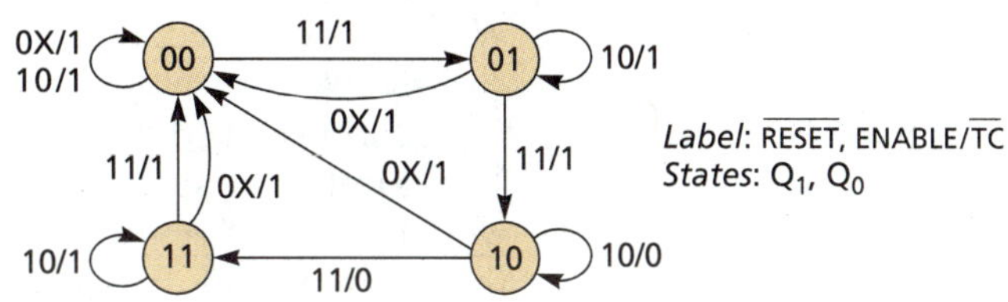

(c) State Diagram

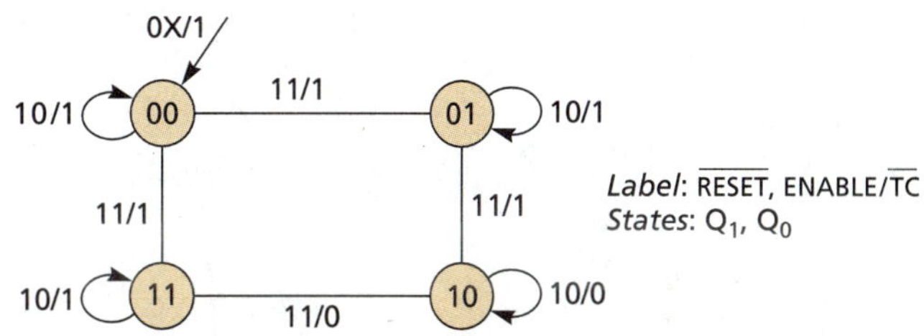

(d) Asynchronous Branch

$\overline{RESET}$	ENABLE	Q_{1n}	Q_{0n}	Q_{1n+1}	Q_{0n+1}	$\overline{TC}$
0	X	X	X	0	0	1
1	0	X	X	Q_{1n}	Q_{0n}	$\overline{TC}_n$
1	1	0	0	0	1	1
1	1	0	1	1	0	0
1	1	1	0	1	1	1
1	1	1	1	0	0	1

(b) State Table

X = "don't care" state

Figure 4.55 Two-Stage Up Counter Descriptors

The state table differs from a truth table by showing both current and next state conditions, but not showing the assumed CLOCK signal input. It tends to be more informative than a regular truth table. The last part of Figure 4.55 contains the state diagram corresponding to both the state table and the timing diagram.

Four circles with connecting arrow branches show the normal count sequence states of the counter. The states within the circles represent Q_1 and Q_0, in that order. A label used on each branch is defined in the diagram as representing inputs $\overline{RESET}$ and ENABLE, to the left of the slash and output $\overline{TC}$ to the right of the slash. For instance, look at the branch connecting the two top state circles. The label for that branch, 1,1/1, indicates that the state of the counter will change from 0,0 to 0,1 if $\overline{RESET}$ and ENABLE are both 1 and that $\overline{TC}$ for the next state (0,1) is inactive (1).

A number of branches point to the 00 state from each of the other states. They all have the label 0X/1, which means that the 00 state will result whenever $\overline{RESET}$ is set low, regardless of the state of the ENABLE signal or of the outputs of both flip-flops. On this diagram, these branches do not cause much of a problem, but if the counter had many more stages then drawing each of these branches might make a very full and difficult diagram to read. Another approach to adding a signal such as RESET into a state diagram, is to use an asynchronous branch. This is an open-ended branch looking like an arrow with a label on it, as shown in Figure 4.55d. This type of branch signifies an input that is independent of the CLOCK signal and of the present state of the circuit. When it is asserted, the counter will go to the indicated state. Notice how much clearer this diagram looks than the one in Figure 4.55c.

Each branch from one circle to another indicates a toggle for this counter. The number of states in each circle inform us that this is a two-bit counter, which will require two flip-flops. We will use the 74LS76 J–K flip-flop to build this counter. The first input, shown in Figure 4.56, is the easy one—it is the asynchronous $\overline{\text{RESET}}$ signal that causes both flip-flops to reset regardless of the other inputs. This signal is connected to the active low reset inputs of the 74LS76 flip-flops. By asserting this line, the two flip-flops are reset. When $\overline{\text{RESET}}$ is set high, the counter is free to respond to the ENABLE and CLOCK inputs.

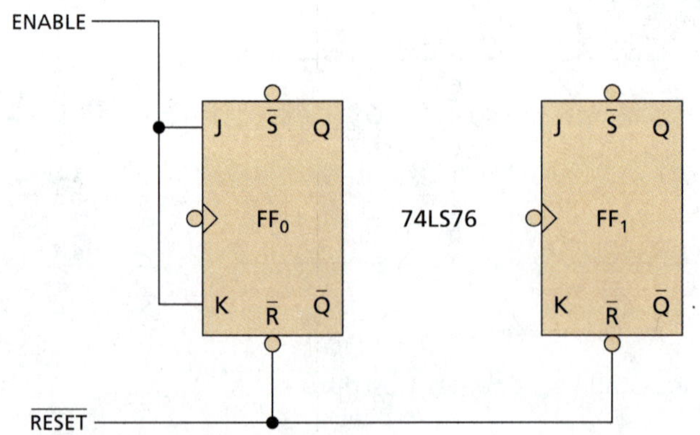

Figure 4.56 Asynchronous Inputs

Looking at the state diagram, notice that the state of Q_0 toggles as long as ENABLE and $\overline{\text{RESET}}$ are high. There are no other restrictions applied to the counter. As a result, we will connect the ENABLE signal to the J–K inputs of FF_0 as shown in Figure 4.56. Next examine the state of Q_1. It toggles twice, once when the current state is 01B and once when the current state is 11B. Since both J and K inputs must be high for the flip-flop to toggle and low for a no-change condition, a logic expression can be written to express the logic required to drive Q_1's J–K inputs. A state of 01B is $\overline{Q_1}$ and $Q_0 = 1$ or $\overline{Q_1} \cdot Q_0$. State 11B is Q_1 and Q_0, both high, or $Q_1 \cdot Q_0$. The expression for Q_1 J–K inputs is the sum of these product terms:

$$J_1 = K_1 = (\text{ENABLE}) \cdot (Q_1 \cdot Q_0 + \overline{Q_1} \cdot Q_0) = (\text{ENABLE}) \cdot (Q_0)$$

The last consideration in this counter design is for output $\overline{\text{TC}}$, which goes low when the counter output is 10B. A NAND gate is used with inputs Q_1 and $\overline{Q_0}$ and output $\overline{\text{TC}}$. The logic expression for this gate is

$$\text{TC} = \overline{Q_1 \cdot \overline{Q_0}}$$

The completed circuit is illustrated in Figure 4.57. The two-bit synchronous counter gave us a chance to see how the counter can be represented by a timing diagram, state table, and state diagram, leading to the actual circuit in Figure 4.57.

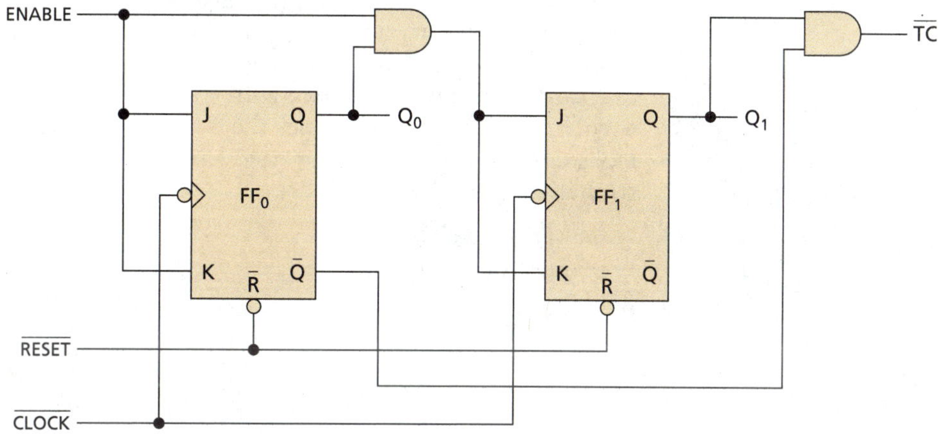

Figure 4.57 Two-Bit Synchronous Up Counter

Section 4.11 Practice Problems

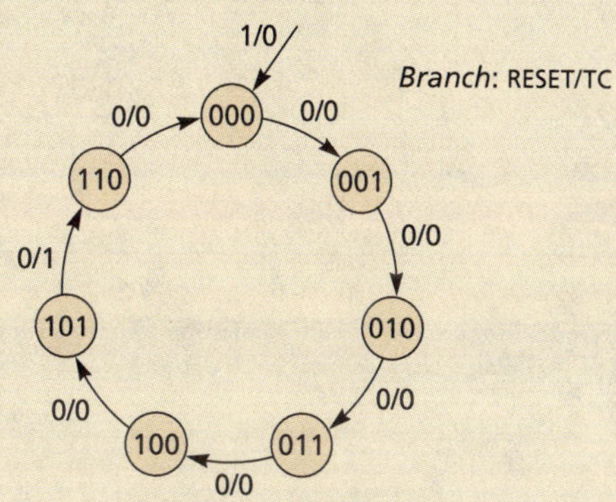

Figure 4.58 Practice Problem 4.10(1)

1. What is the mod number for the counter shown in the state diagram of Figure 4.58?
2. What input of Figure 4.58 is asynchronous?
3. What state is the counter of Figure 4.58 in, to assert the TC signal?

4.12 Ring and Johnson Counters

Binary counters have basic flaw. In order for them to control the occurrence of different events, their binary outputs have to be translated, using AND or NAND gates, into single signal outputs. For instance, if an event is to occur after six clock cycles, output 110B, from the counter, must be converted into a single signal to turn on the event. One way to eliminate this difficulty is to develop a counter that produces a single active output for each clock cycle. A state diagram for a mod-6 counter with that capability is shown in Figure 4.59. An asynchronous reset line causes the counter to begin at a state of 000001B. Each clock causes the counter output to shift left one bit position, rather than count, so that succeeding outputs are

000001B Reset state
000010B First count
000100B
001000B
010000B
100000B

and then returns to 000001B and repeats the sequence. The counter that corresponds to this state diagram is the **ring counter** of Figure 4.60.

Ring counter *a synchronous counter all of whose outputs except one are low.*

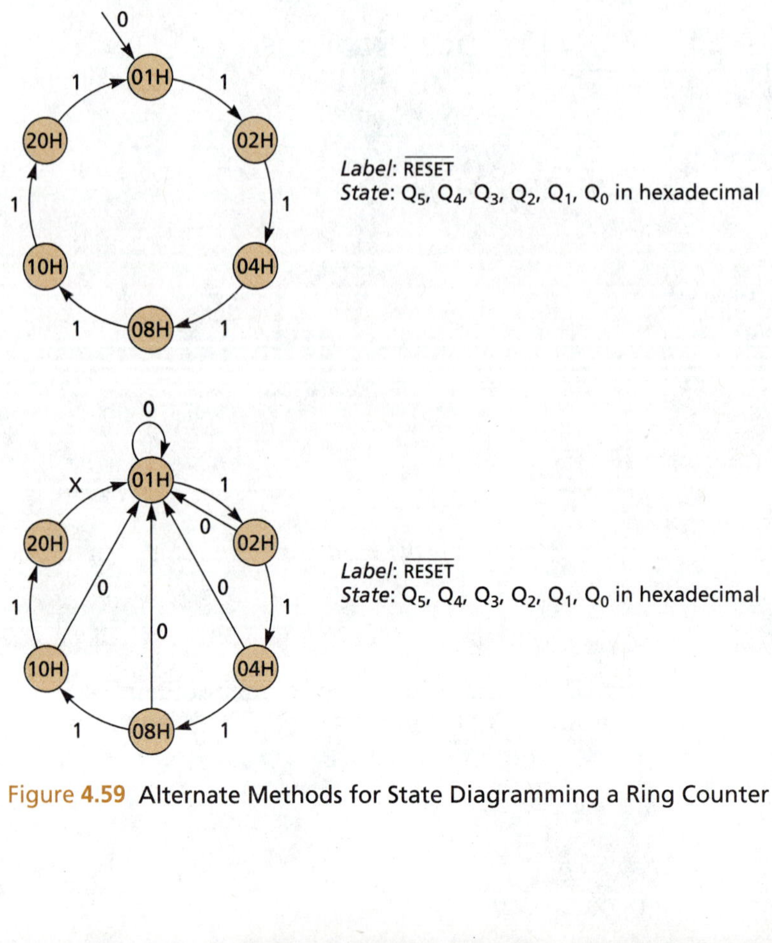

Figure 4.59 Alternate Methods for State Diagramming a Ring Counter

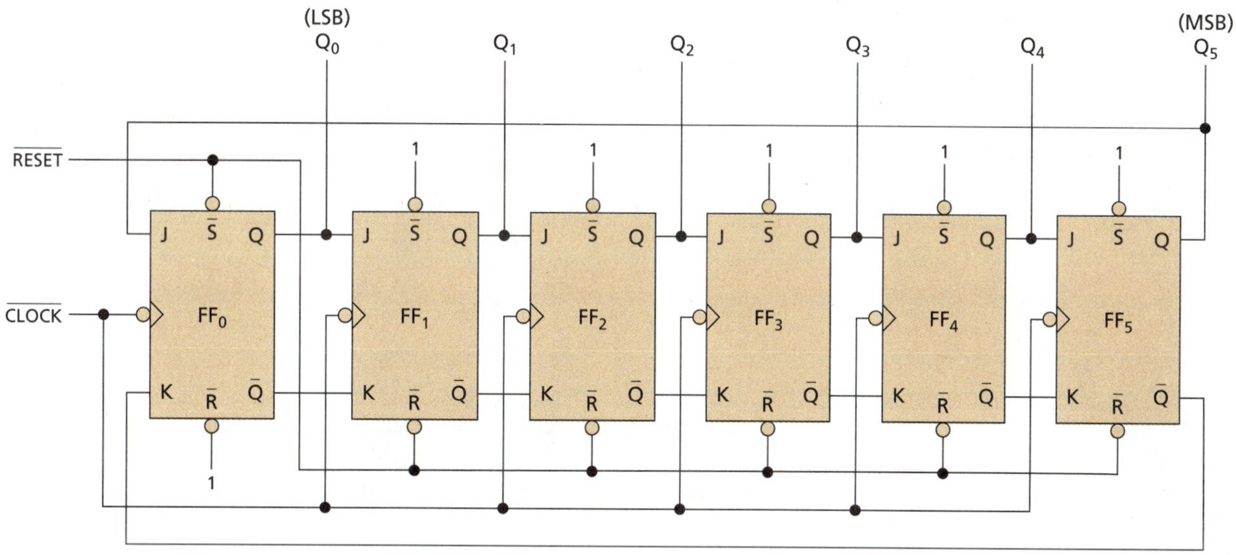

Figure **4.60** Six-Bit Ring Counter

The counter is initialized by asserting the $\overline{\text{RESET}}$ line, which causes FF_0 to be set and the remaining flip-flops to be reset. This produces an output of 000001B with FF_0 holding the least significant bit. The inputs to each flip-flop are based on the output of the preceding flip-flop. This makes the inputs to FF_1 such that it will cause FF_1 to set when the first cycle of the CLOCK signal is applied to the counter. The remaining flip-flops have inputs that keep them or place them into the reset state at the time of that same clock cycle. This includes FF_0, which receives its inputs from the last flip-flop, FF_5. In effect, the first clock cycle will cause the set output to shift left from FF_0 to FF_1. This may seem odd when looking at the way the diagram is drawn. Looking at the ring counter in Figure 4.60, it is drawn with the convention of input signals on the left and outputs toward the right side of the diagram. However, when discussing the direction binary bits move when shifting, a different convention is used. The movement of bits from the least significant bit position (FF_0 in this circuit) toward the most significant bit (FF_5 in the ring counter) is considered as shifting left. The reason for this lies in treating the outputs of counters, and, later, registers, as binary numbers. When these numbers are read, the least significant bit is always in the rightmost position, much the same as decimal numbers, with the units digit on the right. Using the convention dictated by the form numbers take, movement of data from the least significant position, which is the rightmost position, toward any other bit position, is shifting left. Thus the ring counter, starting at FF_0, as the least significant bit position, shifts left despite the appearance of the schematic.

When the shift of FF_0 state to FF_1 occurs, input state to FF_2 is changed from reset (J low and K high) to set (J high and K low). This allows the set state to be shifted to that flip-flop on the next clock cycle. A timing diagram illustrating the process is shown in Figure 4.61. This diagram agrees with the state

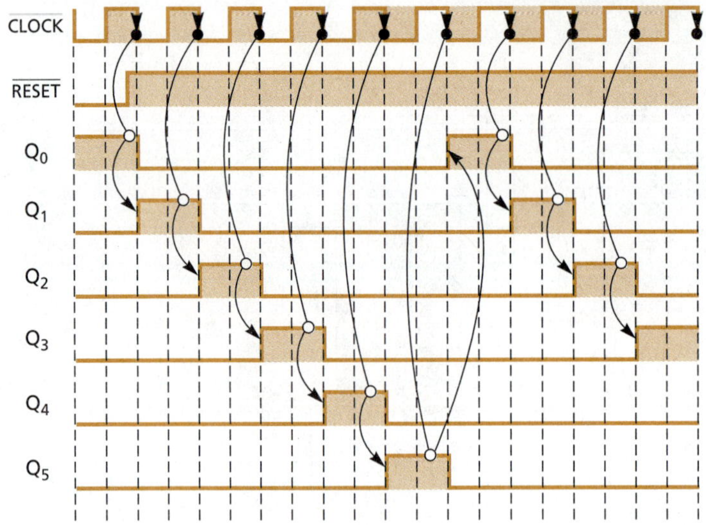

Figure **4.61** Ring Counter Timing Diagram

diagram of Figure 4.59, both indicating that a single flip-flop will be set while the remaining flip-flops are in a reset state.

EXAMPLE 4.16

Six lamps are controlled by a ring counter so that they turn on in sequence, one following the other. They are of different colors and are connected to the Q outputs as follows:

Brown to FF_0
Red to FF_1
Orange to FF_2
Yellow to FF_3
Green to FF_4
Blue to FF_5

Which color light is illuminated after 22 clock cycles following the assertion and release of a $\overline{\text{RESET}}$ signal?

SOLUTION The $\overline{\text{RESET}}$ signal causes the brown light to come on. Since this is a mod-6 counter, the brown light will be lit every six clock cycles following the reset state. This accounts for 18 clock cycles, leaving four remaining cycles. The lamp lit in order by these four clock cycles are red, orange, yellow, and green. The one that is lit after 22 clock cycles is green.

The advantage of a ring counter is that only one flip-flop is set at any given time. This removes the need for logic circuits to convert multiple counter outputs to a single output for use as synchronized control signals. The down side, however, is the large number of flip-flops required for high mod numbers. It takes a single flip-flop for each increase in mod number. For example, a mod-12 ring counter requires 12 flip-flops. Compare the number of flip-flops in the mod-16 binary up counter in Figure 4.42 with a mod-16 ring counter. A mod-16 binary up counter uses four flip-flops, whereas the same mod number ring counter requires 16 flip-flops. Any attempt to reduce the number of flip-flops to achieve a given mod number requires some form of additional logic to convert the outputs into single control signals. One way to improve on the ratio of flip-flop count to mod number, while minimizing the use of logic circuits for generating single control signals for each count, is the use of another synchronous counter called a **Johnson counter.**

Johnson counter *a synchronous counter whose mod number is equal to twice the number of flip-flops in the counter.*

At first glance, the Johnson counter in Figure 4.62 looks very much like the ring counter with its tell-tale "ring" connection from the output of the last flip-flop to the input of the first. However, on close examination of the Johnson counter, you will note that the last output connections are swapped back to the first set of inputs. That is, FF_5–Q is connected to FF_0–K instead of its J input and FF_5–$\overline{Q}$ is connected to FF_0–J. The other major difference between the two counters is that the $\overline{RESET}$ input of the Johnson counter is connected to reset inputs on *all* the flip-flops while the $\overline{RESET}$ input of the ring counter goes to the set input of FF_0 and the reset inputs of the remaining flip-flops.

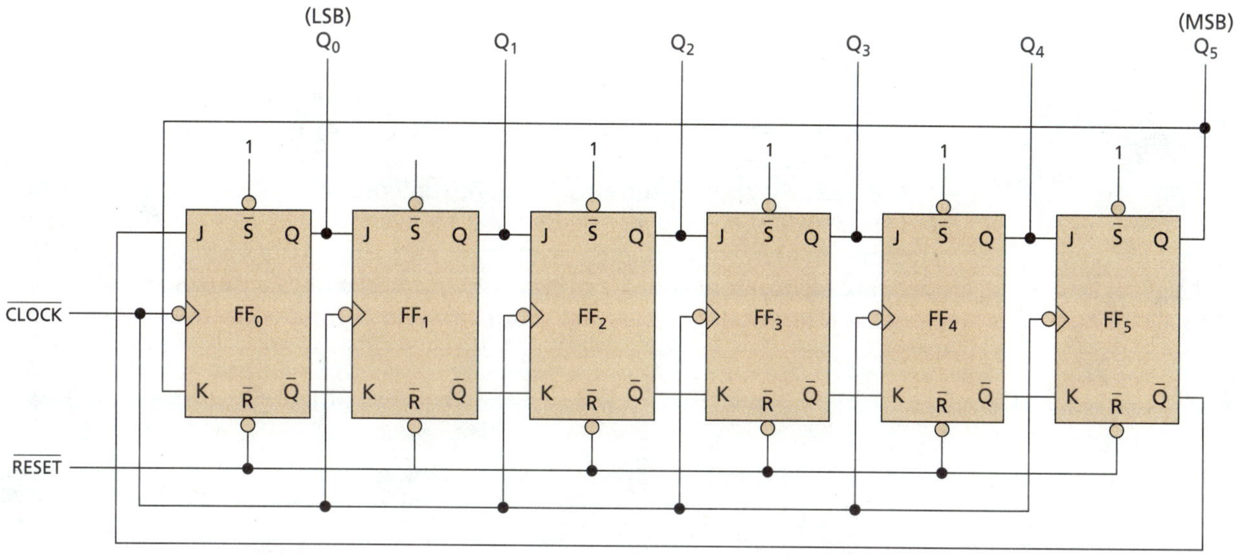

Figure 4.62 Six-Bit Johnson Counter

Refer to the timing diagram of Figure 4.63 during the discussion of the operation of the Johnson counter. As expected, asserting the $\overline{RESET}$ input causes all the flip-flops to be placed into the reset state. This places reset inputs to the J–K lines of all of the flip-flops except FF_0, which "sees" set inputs from the cross-coupled outputs of FF_5. The first clock cycle causes FF_0 to set and keeps the remaining flip-flops in the reset state. Now there are set inputs to both FF_0 and FF_1. The second clock cycle causes FF_1 to set while retaining the current

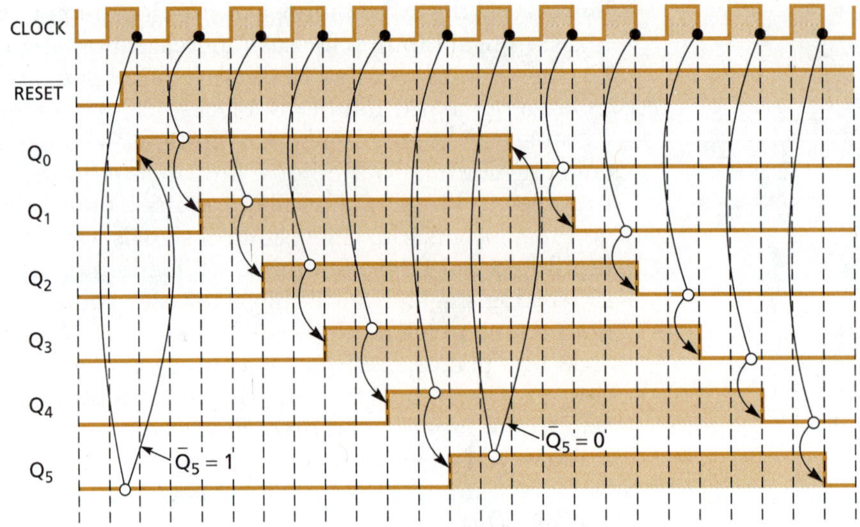

Figure **4.63** Johnson Counter Timing Diagram

states of the remaining flip-flops, including the set condition of FF_0. This increases the number of set inputs, for the next clock cycle, to three (FF_0, FF_1 and FF_2). The process continues until all the flip-flops have been set. When FF_5 goes set, its cross-coupled outputs send reset inputs to FF_0, causing FF_0 to reset with the next clock cycle. The process is similar, except that each flip-flop will be reset instead of set until they are all reset. This brings us back to the first state and the sequence repeats.

The state diagram describing the Johnson counter is shown in Figure 4.64. The states are presented as hexadecimal equivalents using FF_0 as the least significant bit. Match the changing states with the timing diagram, and verify that there are 12 different states for this counter. In effect, the Johnson counter configuration has doubled the mod number when compared to the ring counter.

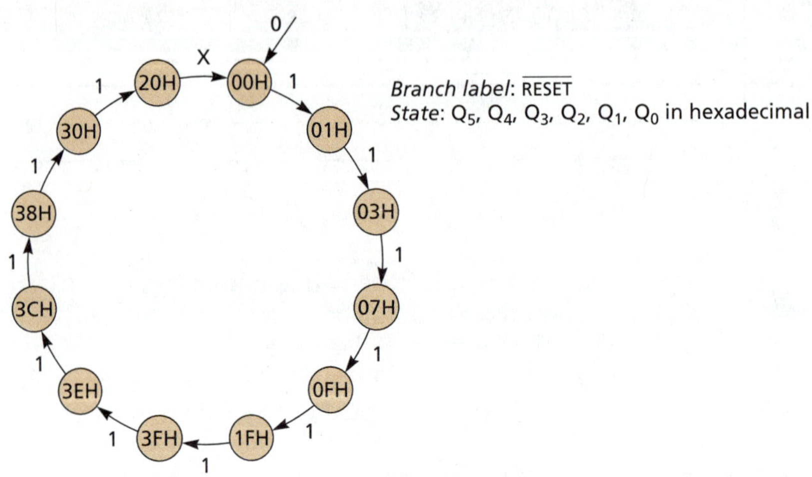

Figure **4.64** Johnson Counter State Diagram

Both used six flip-flops, but the ring counter was only a mod-6 counter, while the Johnson counter is a mod-12 counter. Of course, the Johnson counter outputs now require logic circuits to convert the output combinations into single control signal formats. For instance, if a signal needs to be generated after five clock cycles have been applied to this counter, output 01 1111B has to change to a single signal. This is accomplished using a six-input AND gate whose inputs are connected to the $\overline{Q}$ output of FF_5 and the Q outputs of the remaining flip-flops. Now when the counter reaches the 01 1111 state, the output of the AND gate goes high. At all other times, it is low.

Another difference between the two synchronous counters is that the "on" or high state of a Q output in the ring counter is of much shorter duration than that of a Johnson counter. To understand the effect of this, let's consider an event counter application for these counters. The objective of an event counter is to cause events to occur in sequence by turning them on and off in order. A ring counter is ideal for this job, because only one output is high at any given time. Since the flip-flops of the Johnson counter are set in order, why is it not readily usable as an event counter? Although it is true that Johnson counter outputs turn on and off in sequence, they are also on for longer periods of time. For the ring counter, each output is on for a single clock cycle only. In the mod-12 Johnson counter, each output is set and remains set for six clock cycles and then off for six clock cycles. This means that an event would be turned on, but would also remain on while five succeeding events are turned on before it finally is turned off. This is a handy device when a circuit requires events to be turned on in sequence and *remain* on until all the events are turned on.

One difference in using a ring counter and a Johnson counter is illustrated by using these counters to illuminate a sign spelling the word *ENTRANCE*. A ring counter would be used if you wanted to turn on each letter by itself in sequence. First the E is lit, and all other letters are out. Then the E is turned off and the first N is on, followed by the N being turned off, and the T on and so forth. When using the Johnson counter to turn on the sign, the E would come on first, as it did with the ring counter, but it would remain on while the N came on and the T came on, until the whole sign was lit. Then the E would go out while the remainder of the sign stayed lit, followed by the N and so forth. The sequence over the cycle of the Johnson counter appears as

```
E
EN
ENT
ENTR
ENTRA
ENTRAN
ENTRANC
ENTRANCE
 NTRANCE
  TRANCE
   RANCE
    ANCE
     NCE
      CE
       E
```

Section 4.12

Practice Problems

1. Draw a mod-8 ring counter and its state diagram, using D flip-flops.
2. Draw a mod-8 Johnson counter and its state diagram, using D flip-flops.
3. Compare the two counters in Problems 1 and 2.

4.13 Troubleshooting Counter Circuits

The largest problem faced when analyzing a counter circuit is that its outputs are dynamic in nature. That is, they must be constantly changing, or else they would not count. Most counters count in a logical sequence making expected outputs fairly easy to determine. If a counter is not working correctly, the first thing to determine, after you have verified that source power is connected to the circuit, is whether the clocking input is applied and correct. Never assume that a good clock at one CP input means that the CLOCK signal is present at all CP inputs. Check each one in turn to make sure that each flip-flop or circuit that is supposed to receive a CLOCK signal, does, indeed have one.

If the CLOCK signal is present, then it is, basically, a matter of starting at one end of the counter and checking outputs, one stage after another until an incorrect output is detected. The best instrument to use for this exploration is a dual-trace oscilloscope. Two outputs can be observed simultaneously on the oscilloscope to see the relationship between them. For instance, if a mod-16 binary up counter is being checked, one vertical trace can be used to monitor the output of the first flip-flop. Its output should be half the frequency of the CLOCK signal that is applied to the input of the first flip-flop. Monitoring the clock with the second trace of the oscilloscope lets you compare the input and output signals of the first flip-flop, as shown in Figure 4.65.

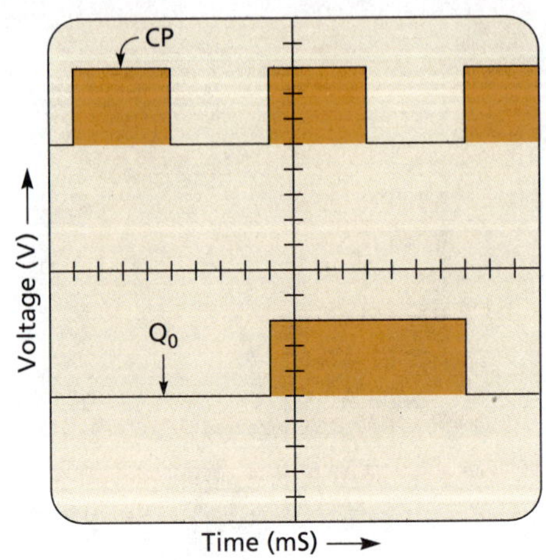

Figure 4.65 Dual-Trace Example

The lower trace monitoring the output of the first flip-flop shows a square wave that has a time period that is twice as long as that of the CLOCK signal on the upper trace, if it is working correctly. You can also check the clock and output voltage levels to determine if they are correct. Keeping the upper trace on the clock, the other can be moved to the next flip-flop output, which should be one-quarter the frequency of the CLOCK signal. The time period of this output is four times as long as the time period of the CLOCK signal. You could also leave one trace on the output of the first flip-flop and move the other the second flip-flop. In this case, the time periods would be in a ratio of 2:1 instead of 4:1. Repeat this process until all the stages are checked or until you find the bad output.

When checking a ring or Johnson counter, the sequence of the outputs is as important as their appearance at the flip-flop's Q outputs. To check these counters, one output is used to synchronize the horizontal sweep so that the sequence can be compared. Figure 4.66 is an illustration of a dual-trace oscilloscope used to monitor two stages of a ring counter. The horizontal sweep is synched to the output of the first stage of the ring counter. That means that the sweep begins when the output of that stage first goes high. On the top trace, it shows that the output of the second stage goes high one clock cycle after the sweep begins, which is correct. The bottom trace shows the output of the third stage going high much later than it should. There is a problem with this ring counter in the third stage.

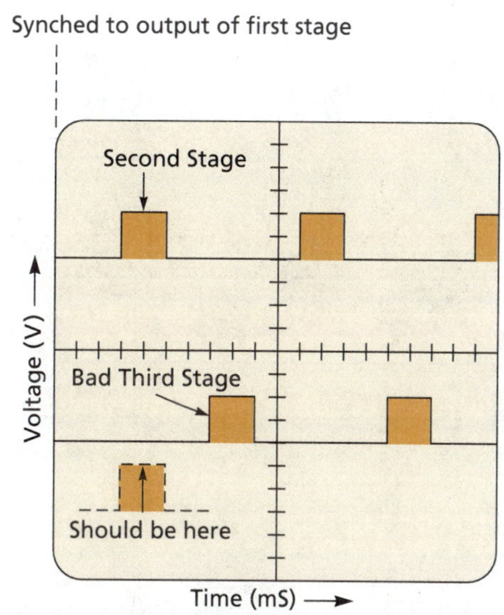

Figure 4.66 Ring Counter Display

Multitrace oscilloscopes are available to allow several stages of a counter to be monitored at one time. Another helpful addition to an oscilloscope is a storage function. These allow a single sweep of all the traces to be saved on the screen of the oscilloscope until another sweep is manually triggered. This allows the user to carefully diagnose the results by allowing him or her to look at the results for an extended period of time.

Section 4.13

Practice Problems

1. From the four-trace oscilloscope pattern shown in Figure 4.67, determine which stage of the mod-16 up counter is bad.

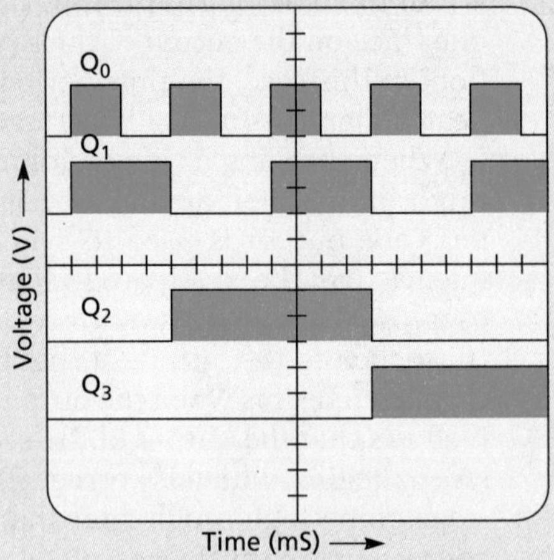

Figure 4.67 Practice Problem 4.13(1) Mod-16 Up Counter

2. Determine what is wrong with the Johnson counter that generated the four-trace pattern of Figure 4.68.

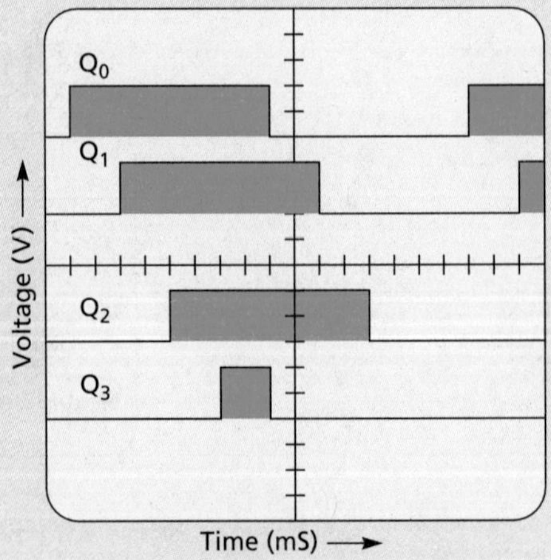

Figure 4.68 Practice Problem 4.13(2): Johnson Counter

4.14 Designing an Asynchronous UP/DOWN Counter

The Q outputs of the binary up counter of Figure 4.45 in Section 4.9 describe an up count sequence. Simultaneously, because the $\overline{Q}$ outputs are the complement of the Q outputs, they show a down count sequence. The same type of relationship can be drawn for the down counter. Suppose that you have a circuit in which it is crucial that a binary count sequence is to count from 0H (0000B) up to 9H (1001B) and then back down to 0H (0000B). One way to solve this problem is to switch connections from the Q outputs of a counter to its $\overline{Q}$ outputs by using an AND/OR logic combination, shown in Figure 4.69. When UP/$\overline{\text{DOWN}}$ is high, the up count sequence from the Q outputs is switched to the AND/OR circuit outputs. When UP/$\overline{\text{DOWN}}$ is made low, the down counting sequence appears at the AND/OR outputs. A more direct and simpler method is to design a circuit that can switch the direction of the counter itself.

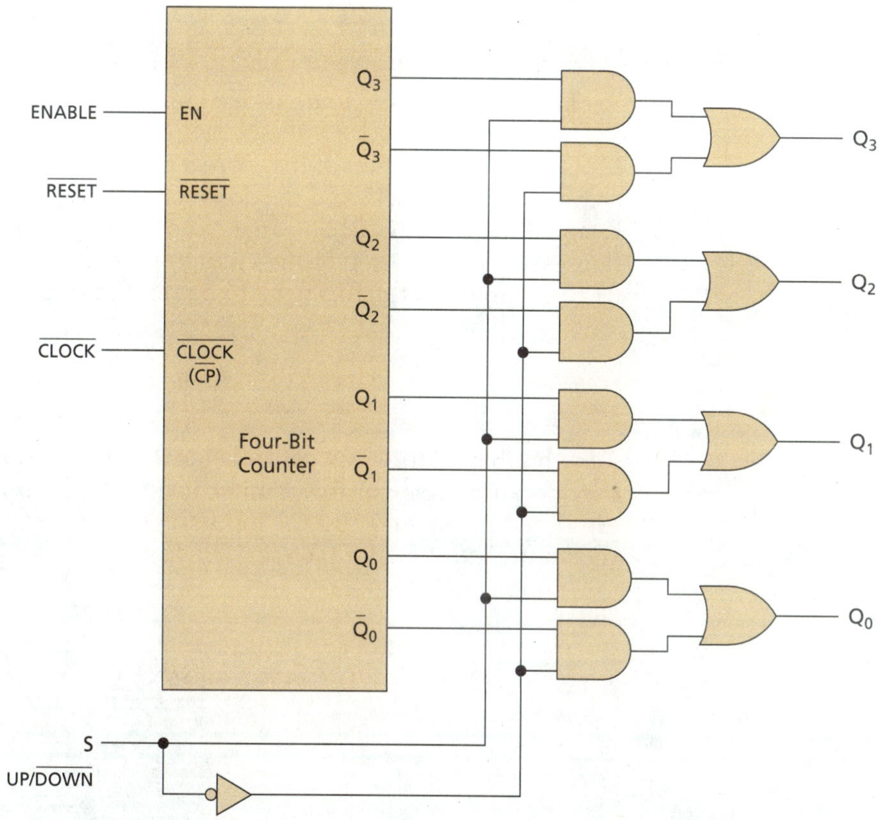

Figure 4.69 Binary Up/Down Counter Using AND/OR Select Circuit

To begin the design problem, decide what you wish the circuit to do, set out a list of objectives, and then decide how to resolve those objectives. For the UP/DOWN counter a typical list might be

1. Count up when control signal UP/$\overline{\text{DOWN}}$ = 1, and the count is enabled by ENABLE = 1 and $\overline{\text{RESET}}$ = 1.
2. Count down when UP/$\overline{\text{DOWN}}$ = 0, ENABLE = 1, and $\overline{\text{RESET}}$ = 1.
3. Freeze count when ENABLE = 0 and $\overline{\text{RESET}}$ = 1.
4. Output of counter = 0000B when $\overline{\text{RESET}}$ = 0.

Once the requirements for the circuit are listed, it is time to decide how to achieve some of those requirements. This is the brainstorming that goes on before an actual design is developed. For the UP/DOWN counter, the requirements are met by satisfying the needs for counting up and down with binary counters:

1. Use J–K flip-flops with active high CP and active low $\overline{\text{RESET}}$ inputs.
2. Connect all J and K inputs to ENABLE to control when the counter counts.
3. Connect all $\overline{\text{R}}$ inputs to $\overline{\text{RESET}}$.
4. To count down, connect the Q outputs of the flip-flops to positive edge CP inputs of the each succeeding flip-flop.
5. To count up, connect the $\overline{\text{Q}}$ outputs to the positive edge CP inputs.

EXAMPLE 4.17

Explain how the "how to do it" list suits the needs of the UP/DOWN counter.

SOLUTION Connection of the correct outputs to the CP inputs describes the up counter of Figure 4.42 and the down counter of Figure 4.45. Tying all the J and K inputs to the ENABLE signals allows us to control whether the J–K flip-flops are toggling or in a no-change mode. Connecting all the $\overline{\text{R}}$ inputs to $\overline{\text{RESET}}$ allows all the flip-flops to be reset at once.

A truth table relating the control signals $\overline{\text{RESET}}$, UP/$\overline{\text{DOWN}}$, and ENABLE to the operating mode of the counter including the interconnection between the flip-flop outputs and the CP inputs is established from the requirements as shown in Figure 4.70.

$\overline{\text{RESET}}$	ENABLE	UP/$\overline{\text{DOWN}}$	CLOCK	Mode
0	X	X	X	Reset (Q = 0)
1	X	X	Steady State	No Change
1	0	X	$\downarrow$	No Change
1	1	1	$\downarrow$	Count Up
1	1	0	$\downarrow$	Count Down

X = "don't care" state

Figure 4.70 UP/DOWN Counter Truth Table

The first line of the table describes the reset state of the counter when $\overline{\text{RESET}}$ is brought low. The second line shows the no-change condition of the counter when ENABLE is negated. Both of these modes exist regardless of the UP/DOWN sequence selection. The second two lines in the table describe the conditions required to allow the counter to count up or down.

The details of the connections between the flip-flop outputs and the CP clock inputs are shown in Figure 4.71. When UP/$\overline{\text{DOWN}}$ is low, the counter will count down if the Q outputs are channeled to the CP inputs. When UP/$\overline{\text{DOWN}}$ is set high, $\overline{\text{Q}}$ outputs are required to be connected to the CP inputs. As can be seen from the table in Figure 4.70, we can produce the necessary logic connections by using an exclusive OR gate whose inputs are the flip-flop Q outputs and the UP/$\overline{\text{DOWN}}$ signal and whose output drives the CP input of the next flip-flop.

UP/$\overline{\text{DOWN}}$	Q	CP
0	0	0 (Q)
0	1	1 (Q)
1	0	1 ($\overline{\text{Q}}$)
1	1	0 ($\overline{\text{Q}}$)

Figure 4.71 Connections to CP

Figure 4.72 takes all the parts just described and combines them into an operational binary ripple up/down counter. When UP/$\overline{\text{DOWN}}$ is high, the Q outputs are inverted and look like $\overline{\text{Q}}$ outputs. These are sent to the CP inputs, and the circuit counts up. When UP/$\overline{\text{DOWN}}$ is brought low, the state of the Q outputs are sent logically unchanged to the CP inputs, and the circuit counts down.

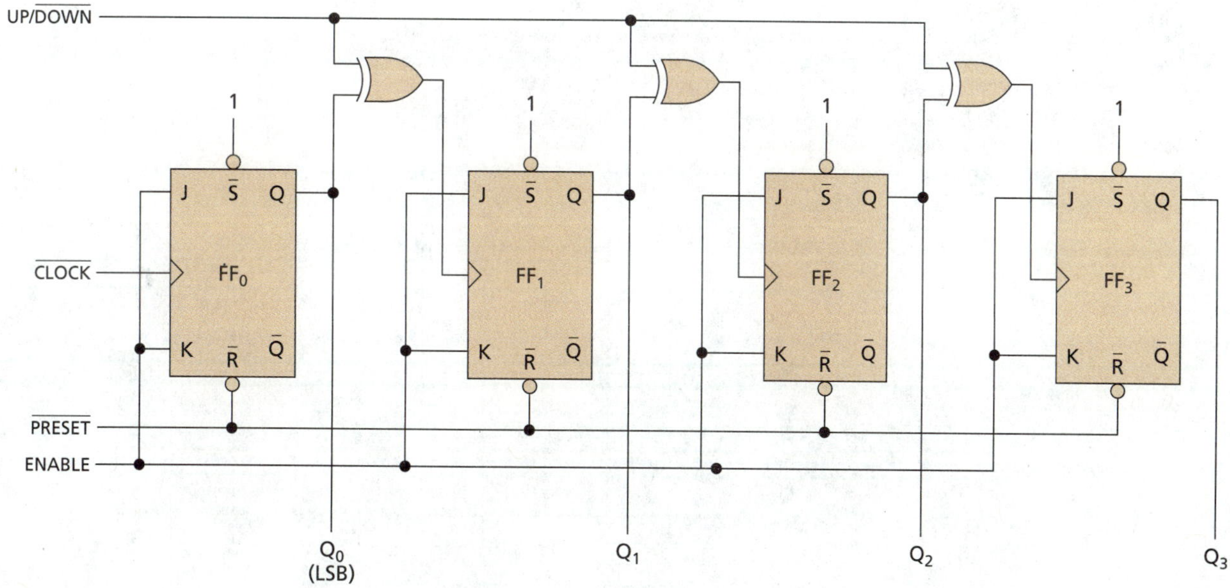

Figure 4.72 Binary UP/DOWN Counter

EXAMPLE 4.18

Complete the timing diagram in Figure 4.73 for the UP/$\overline{\text{DOWN}}$ counter of Figure 4.72.

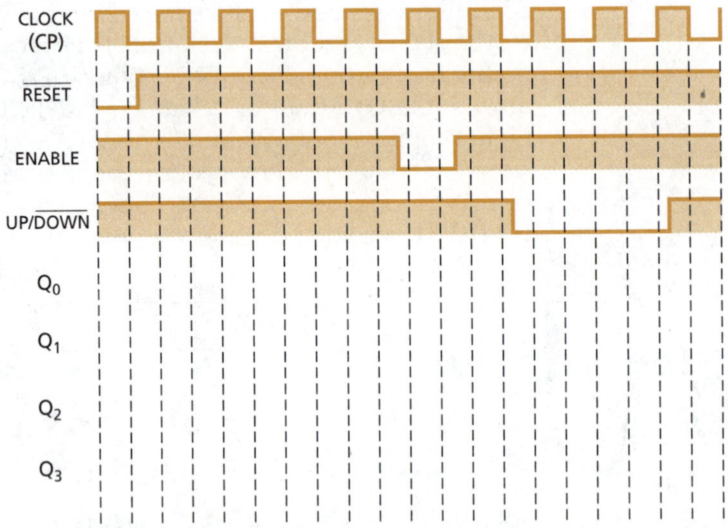

Figure **4.73** Example 4.16: UP/DOWN Timing Problem

SOLUTION Apply the input signals, and trace the output results as shown in Figure 4.74. Asserting the $\overline{\text{RESET}}$ signal causes the counter to start at 0000B. Setting $\overline{\text{RESET}}$ inactive, asserting the ENABLE signal, and making UP/$\overline{\text{DOWN}}$ selects the counter to count up. Setting the UP/$\overline{\text{DOWN}}$ signal low causes the counter to reverse direction and count down.

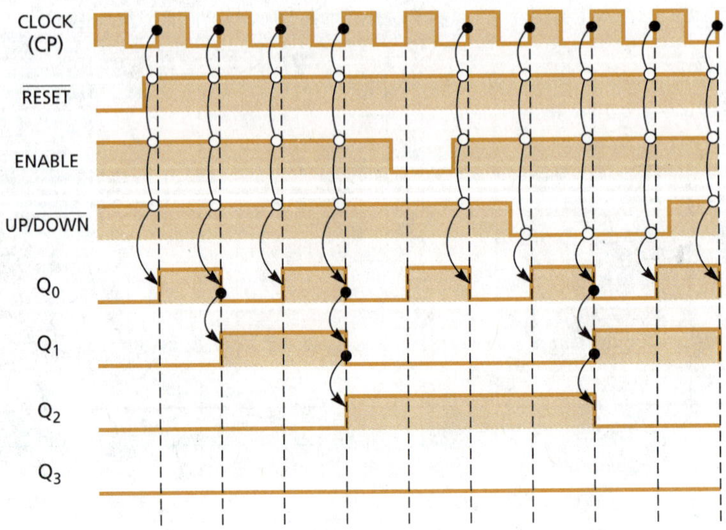

Figure **4.74** Example 4.16 (solution)

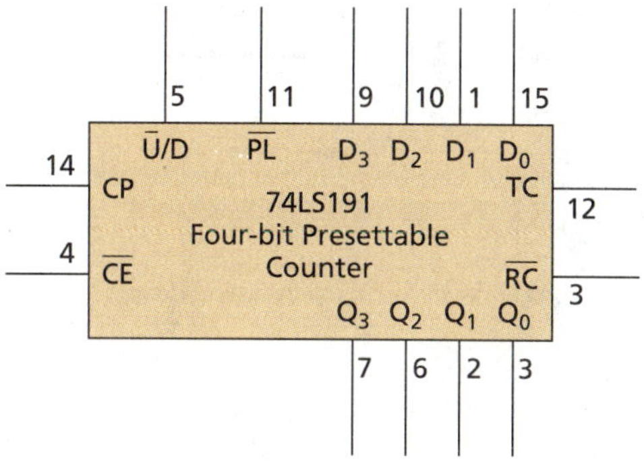

(a) Symbol

$\overline{PL}$	$\overline{U}$/D	$\overline{CE}$	CP	D_n	Q_n
0	X	X	X	D_n	D_n
1	0	0	↑	X	Count Up
1	1	0	↑	X	Count Down
1	X	1	X	X	No Change
1	X	X	Steady State	X	No Change

(b) Truth Table

X = "don't care" state

Figure **4.75** 74LS191 Presettable UP/DOWN Counter

The counter in Figure 4.75 is a 74LS191 presettable four-bit binary up/down counter with a truth table describing its control signals and modes of operation. To load a starting value into the counter, place that value in binary onto the four parallel input pins, D_0 to D_3 and assert the parallel-load control pin, $\overline{PL}$. The binary states on the input pins are latched into the four counter flip-flops. To make the counter count up or down, input $\overline{PL}$ must be negated by setting its line high. The count direction is selected by using the $\overline{U}$/D control pin. A low on this pin allows the counter to count up, and a high sets it to count down. Enabling the counter to count in either direction is controlled by the count enable ($\overline{CE}$) pin, which must be asserted low for the counter to count. The last input is the clock pulse (CP), which is a positive edge CLOCK signal. The outputs include four Q states from the counter flip-flops and two end-of-count indications called **TERMINAL COUNT (TC)** and **RIPPLE COUNT (RC)**. TC is normally low and goes high when the counter counts down to 0000B or up to 1111B. TC returns low if the counter continues to count or if the state of $\overline{U}$/D is reversed. TC will also return to a low state if a new value is loaded into the counter using the PL signal. The second output is the $\overline{\text{RIPPLE COUNT}}$ ($\overline{RC}$) and is used to cascade counters. $\overline{RC}$ is connected to the CP input of another 74LS191 counter. Other control signals are connected in parallel so that the two cascaded counters act in unison.

***Terminal count** or **TC** an indication of when a counter has reached its maximum (up) or minimum (down) count value.*

***Ripple count** or **RC** a short pulse generated by a binary counter when it reaches its maximum (up) or minimum (down) count value.*

EXAMPLE 4.19

Show how to connect four 74LS191 counters to make a 16-bit up counter that begins at an equivalent hex value of 084CH. What can be done to make this counter count continuously from 084CH to FFFFH, returning to 084CH and resume counting?

SOLUTION The $\overline{\text{RIPPLE COUNT}}$ ($\overline{RC}$) outputs from each counter are connected to the CP input of the following counter to cascade the four 74LS191 counters. Control line $\overline{CE}$ from each 74LS191 is connected to the ENABLE control signal so that all the counters are enabled or disabled simultaneously.

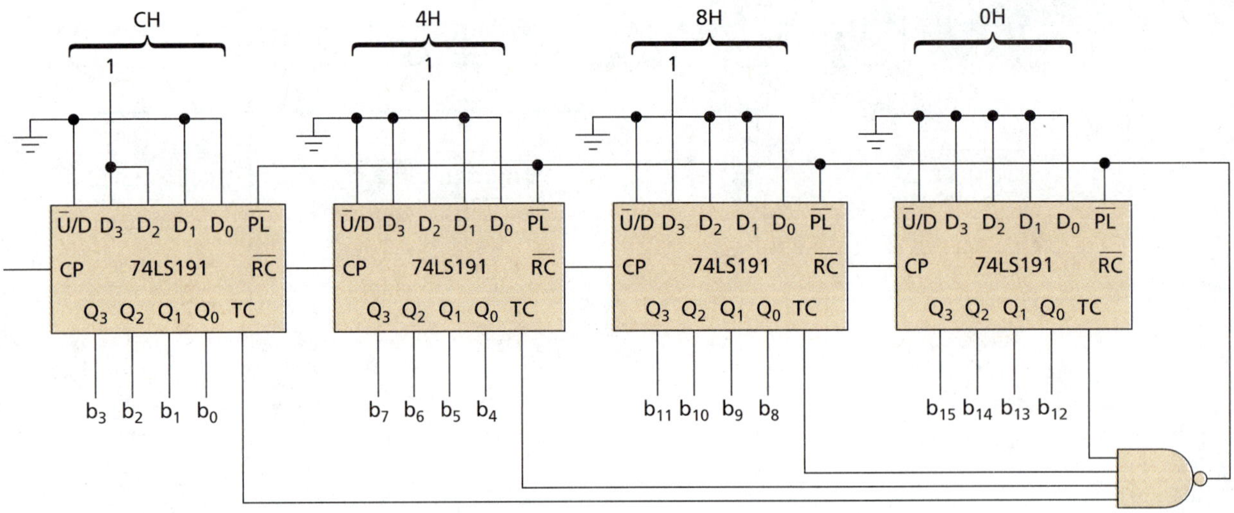

Figure 4.76 Sixteen-Bit Up Counter Using 74LS191 ICs

The $\overline{PL}$ inputs are wired together and brought to a NAND gate whose inputs are the TC outputs of all four counters. On the final count from FFFEH to FFFFH, counter 1, the low four bits of the 16-bit counter, reaches 1111B and

Section 4.14

Practice Problems

1. Complete the timing diagram of Figure 4.77 for the UP/DOWN counter of Figure 4.72.

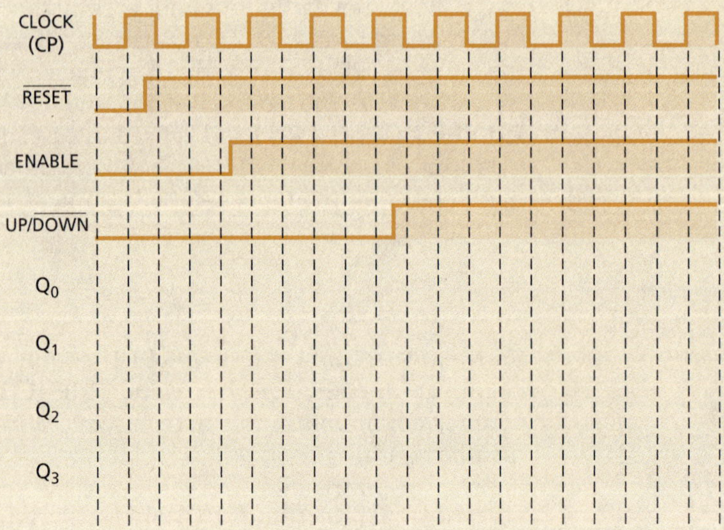

Figure 4.77 Practice Problem 4.14(1)

2. Complete the timing diagram of Figure 4.78 for the 74LS191 UP/DOWN counter.

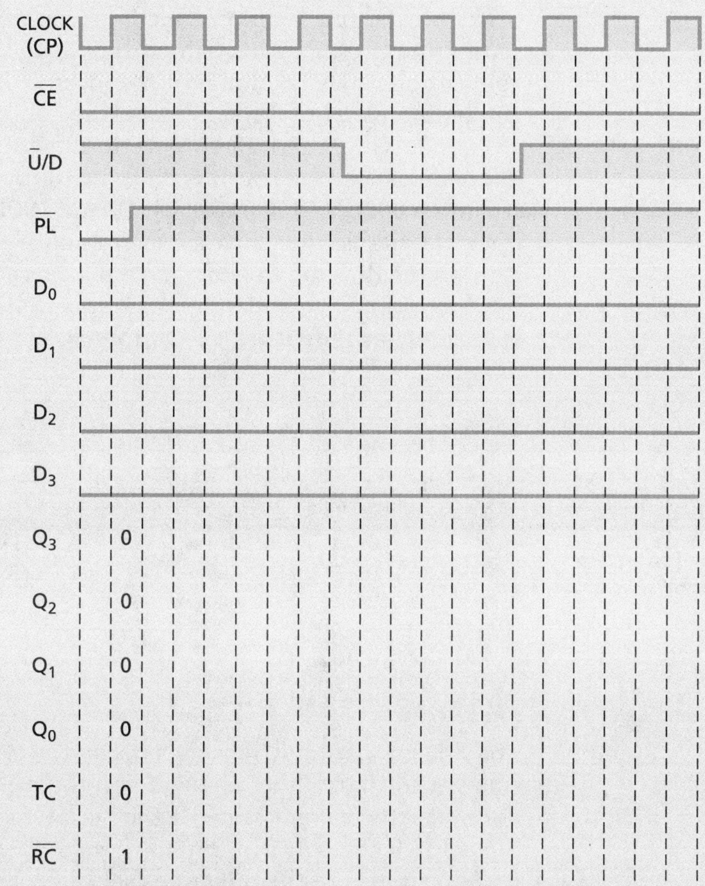

Figure 4.78 **Practice Problem 4.14(2)**

sets its TC high. The other three counter TC outputs are already high, having previously reached their maximum count. With all four TC outputs high, the NAND gate goes low, forcing the $\overline{PL}$ inputs to be asserted and reloading 084CH into the counter. Once the counter is no longer FFFFH, the TC outputs are negated and the NAND gate returns high, negating the $\overline{PL}$ inputs and allowing the counter to resume counting.

4.15 Designing Synchronous Counters

When designing an asynchronous up counter, the maximum value of the counter was used as the mod number. This output was converted by an AND or NAND gate and returned as the reset input so that when the count was reached, the counter was immediately reset so that it could restart the count. It was a fairly simple and straightforward method of design. The main drawback was the propagation time between reaching the decoded count and resetting the counter. This "glitch" count is highly undesirable in many digital applications. Synchronous counter design, in contrast, does not experience the glitch difficulty, but it involves a more complex design process.

Let's develop a mod-5 counter to illustrate the steps in the design process. The sequence of the count is 000B to 100B, yielding five different states of the counter—000B, 001B, 010B, 011B, and 100B. One approach to the design is to use a state table that shows current and next state conditions. The point here is that for a flip-flop to change from a current state to a *different* next state, the J–K inputs must be high, and to remain in the same state they must be low. The state table for the mod-5 counter is

Current State			Next State			J–K Inputs		
Q_{2n}	Q_{1n}	Q_{0n}	Q_{2n+1}	Q_{1n+1}	Q_{0n+1}	FF_2	FF_1	FF_0
0	0	0	0	0	1	0	0	1
0	0	1	0	1	0	0	1	1
0	1	0	0	1	1	0	0	1
0	1	1	1	0	0	1	1	1
1	0	0	0	0	0	1	0	0

The J–K inputs show the states needed to change the counter from the current state on the left to the next state on the right. For instance, in the first line of the table the current state of the counter is 0,0,0 and the next state is 0,0,1. Since FF_0 is the only flip-flop to change state from the present to the next state, its J–K inputs are the only ones set high. The other two flip-flops remain in a no-change condition by keeping their J–K inputs low. Looking at all the J–K input conditions for the first flip-flop (FF_0) on the table, we see that they are high except when the current state of the counter is 100. From the table, we can write an expression for the high state of J_0 and K_0 inputs or for the low state of these inputs. Since they are low only once, the low state would probably be a preferred expression. It is

$$\overline{J_0} = \overline{K_0} = (Q_2)(\overline{Q_1})(\overline{Q_0})$$

which says that the J–K inputs of the first flip-flop are low when the counter is 100 and at no other time. However, looking at the state table, we notice that J_0 and K_0 are high whenever Q_{2n} is low, so another expression for J_0 and K_0 that works equally as well is

$$J_0 = K_0 = \overline{Q_2}$$

Examining the state table closely yields the logic expression for J_1 and K_1 as well. Flip-flop FF_1 toggles whenever the first flip-flop's current state is a 1. From this, the logic expression for J_1 and K_1 can be written as $J_1 = K_1 = Q_0$.

The last flip-flop toggles twice, first when the counter is 011B, and then when the counter is 100B. The unique property for the 011B count is that that is the only time when the first two flip-flops of the counter are both set. The other count has the unique property of Q_2 being in a set state. If either of these conditions are met as a current state, then FF_2 toggles. Therefore the logic expression for J_2 and K_2 can be written as an OR function describing the present state conditions that would result in FF_2 toggling:

$$J_2 = K_2 = Q_2 + Q_1 \cdot Q_0$$

Once the logic expressions for all the inputs have been derived, the counter circuit can be drawn as shown in Figure 4.79.

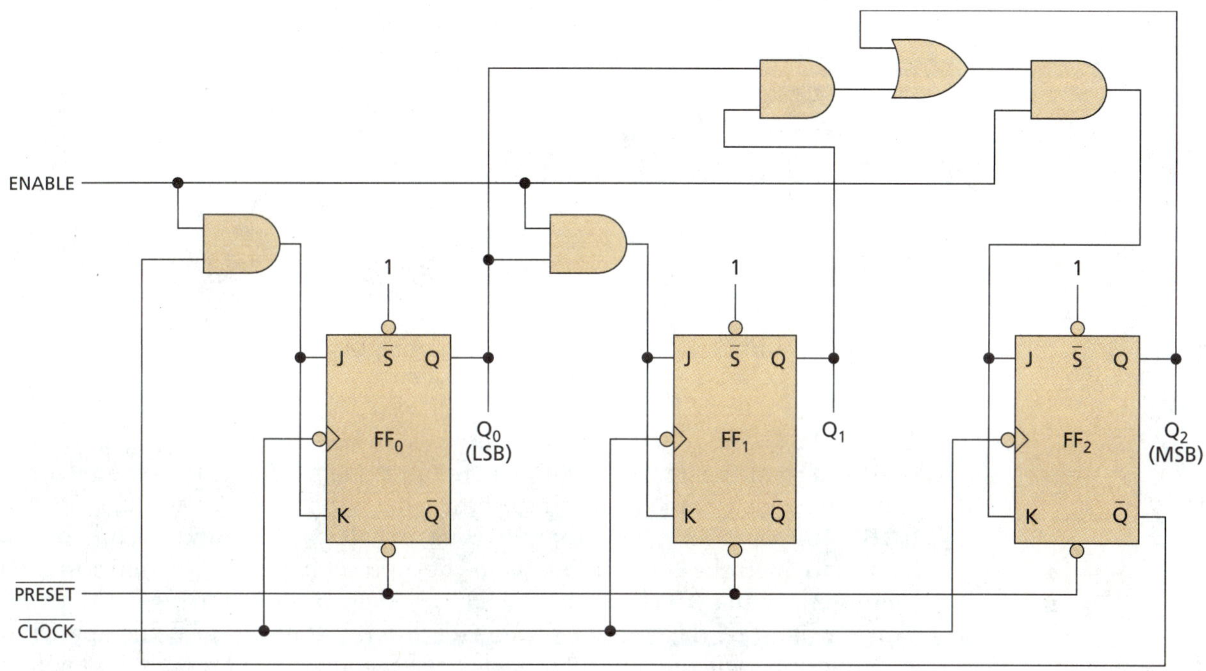

Figure 4.79 Mod-5 Synchronous Counter

There are three flip-flops in this counter. The inputs to the first flip-flop (J_0 and K_0) are connected to $\overline{Q_2}$. As long as FF_2 remains reset, FF_0 will toggle. J_1 and K_1, the inputs to the second flip-flop, are driven from Q_0 so that whenever FF_0 is set FF_1 will toggle. The third set of inputs are connected to the output of an OR gate. One input to the OR gate comes from Q_2 and the second input from an AND gate whose inputs are driven by the Q outputs of FF_0 and FF_1, satisfying the last logic expression. To make the counter complete, $\overline{\text{RESET}}$ and ENABLE inputs are added as shown in Figure 4.79. The $\overline{\text{RESET}}$ input is connected to all the reset inputs on the flip-flops. The ENABLE drives AND gates similar to those in the mod-16 synchronous counter of Figure 4.79.

The mod-5 counter was developed from its state table, but can the same type of circuit be designed using a state diagram instead? To investigate this, there will be an added wrinkle to the design: The mod-5 counter will be self-starting. That simply means that instead of using a $\overline{\text{RESET}}$ signal to start the counter at 000B and assure that the counter will count properly, this counter enters the count sequence regardless of where it starts on being powered up. Looking at the state diagram in Figure 4.80, note that the regular count sequence is as you would expect it—counting up from one current state to the next. The counter is always enabled, and there is no $\overline{\text{RESET}}$ input. Each branch shows a single label for output $\overline{\text{TC}}$, which this time will be asserted when the counter enters its highest count at 100B.

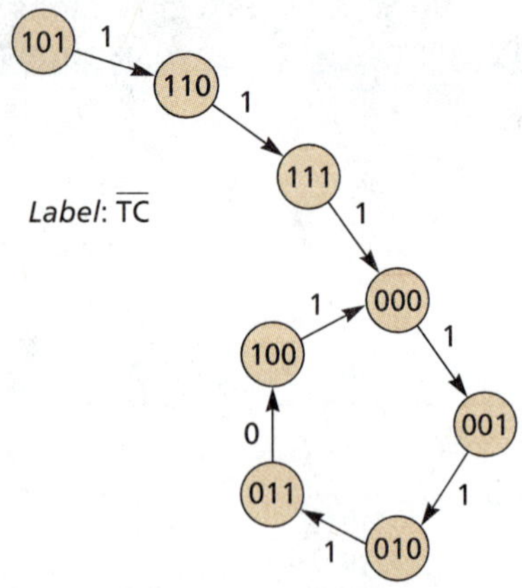

Figure 4.80 Self-Starting Mod-5 Counter State Diagram

What has been added to this diagram is an open-ended branch starting at 101B and counting up until 000B, where the sequence enters the regular count. This open-ended branch is used to show what happens if the counter starts at 101B, 110B, or 111B. It says that the counter will start to count up, and once it returns to 000B the counter begins the desired sequence, never returning to a count greater than 100B. This is what is meant by self-starting—it does not matter what state the counter starts with as a result of power being applied to it. The counter will quickly enter its designed sequence and remain in that loop as long as power stays on and the counter is allowed to count.

The circles indicate the current state of Q_2, Q_1, and Q_0 in that order as the counter progresses through its sequence. Because there are no asynchronous inputs such as $\overline{\text{RESET}}$ or ENABLE, the next state of the counter only depends on the current state that precedes it. Looking at the states for Q_0, notice that it toggles every time except when the current state is 100B. This differs from the previous mod-5, because you did not have to consider states 101B to 111B. The logic expression for J_0 and K_0 based on the state diagram is $\overline{J_0} = \overline{K_0} = Q_2 \cdot \overline{Q_1} \cdot \overline{Q_0}$, which says that FF_0 should not toggle when the current state of the counter is 100B, but should toggle every other time.

Next examine the progression of Q_1 with respect to the current states of the counter: Q_1 toggles whenever the current state of Q_0 is a 1. You can see this by looking for the change of state of Q_1 and then backing down to the state that preceded it. For instance, the state of 010B indicates a change of state of Q_1 from the previous state of 001B. This previous state, as a current state to next state 010B, shows FF_0 set. For each change Q_1, the current (preceding) state always shows Q_0 set, so the logic expression for J_1 and K_1 is simply Q_0.

Q_2 goes through three changes of state from 011B to 100B, from 100B to 000B and from 111B to 000B. From each current state, a product term can be derived for the sum-of-product terms for J_2 and K_2:

$$J_2 = K_2 = \overline{Q_2} \cdot Q_1 \cdot Q_0 + Q_2 \cdot \overline{Q_1} \cdot \overline{Q_0} + Q_2 \cdot Q_1 \cdot Q_0$$

Factoring $Q_1 \cdot Q_0$ from the first and last terms reduces the expression to

$$J_2 = K_2 = Q_1 \cdot Q_0 + Q_2 \cdot \overline{Q_1} \cdot \overline{Q_0}$$

All that remains is to draw the logic circuit of Figure 4.81, using these derived logic expressions:

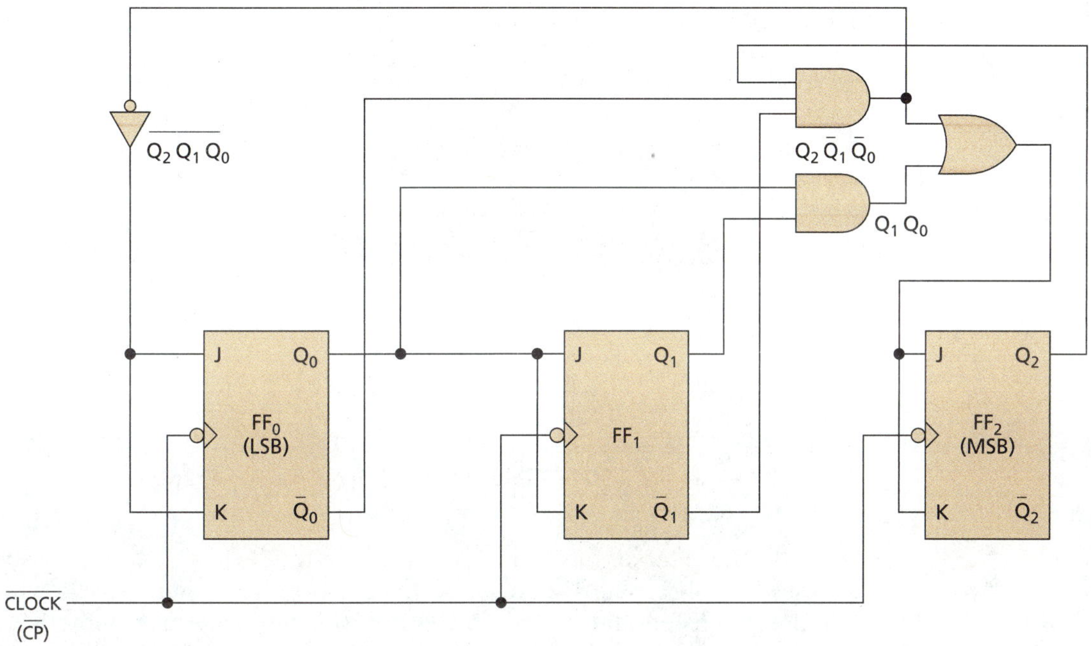

Figure 4.81 Self-Starting Mod-5 Counter

You can see that using a state table is clearer than using a state diagram. In doing the design process, you might consider generating a state table from a state diagram and then proceeding with the design of the counter.

Section 4.15

Practice Problems

1. Create a state table for a mod-6 synchronous down counter. $\overline{TC}$ goes active low when zero count is reached. The counter is preset to 101B using an active low $\overline{PRESET}$ signal. The counter counts continuously as long as a clock is applied and the ENABLE control line is high.
2. Translate the counter in Problem 1 into an actual circuit.

4.16 Synchronous Counter Design Using Karnaugh Maps

Another tool used in designing synchronous counters is the Karnaugh map (K-map) introduced in Chapter 2. To see how it is implemented, let's return to designing a mod-5 counter whose state table is reproduced in Figure 4.82. The table is laid out a little differently, to assist in illustrating the process of translating the diagram onto a K-map. The current and next states of the counter are shown as expected, but added to the table is a listing of expected input conditions to the flip-flops used in the counter. Two sets of conditions are shown, one for D-type flip-flops and one for J–K flip-flops used in the toggle configuration. We will start by using the J–K set so that the resulting design is that of the mod-5 counter in Section 4.15.

Q_n			Q_{n+1}			High D Inputs			Toggle J–K's		
Q_2	Q_1	Q_0	Q_2	Q_1	Q_0	D_2	D_1	D_0	J_2	J_1	J_0
0	0	0	0	0	1	–	–	1	–	–	1
0	0	1	0	1	0	–	1	–	–	1	1
0	1	0	0	1	1	–	1	1	–	–	1
0	1	1	1	0	0	1	–	–	1	1	1
1	0	0	0	0	0	–	–	–	1	–	–

Figure 4.82 State Table for Mod-5 Synchronous Counter

On the state table, the states of current outputs Q_2, Q_1, and Q_0 associated with setting J–K inputs high to cause the flip-flops to toggle are read whenever a circled 1 is listed under the J_n columns. For example, input J_2 and K_2 are set high (toggle flip-flop mode) when outputs Q_2, Q_1, and Q_0 are 011 or 100 respectively. These are entered onto the K-map of Figure 4.83 for all three sets of

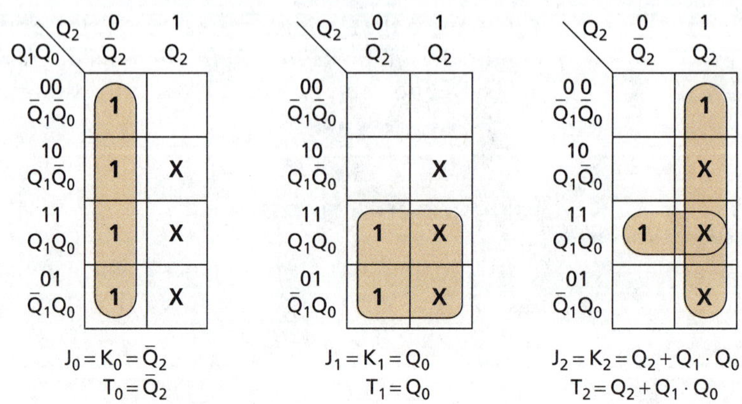

Figure 4.83 K-Map Counter Design for Mod-5 Counter Using Toggle Configuration Flip-Flops

J and K inputs. One map holds the required Q states for J_0 and K_0, another for J_1 and K_1 and a third for J_2 and K_2. Five boxes are accounted for by the states through which the mod-5 counter cycles. The remaining three boxes are "don't cares" because the counter will not produce these states.

The maps are then read to generate the minimum sum-of-products (SOP) expressions for each set of J–K inputs:

$$J_0 = K_0 = \overline{Q_2}$$

$$J_1 = K_1 = Q_0$$

$$J_2 = K_2 = Q_2 + Q_1 \cdot Q_0$$

Notice how the "don't cares" are used to aid in obtaining the minimum SOP expression for each set of J–K inputs. The actual counter is shown in Figure 4.84.

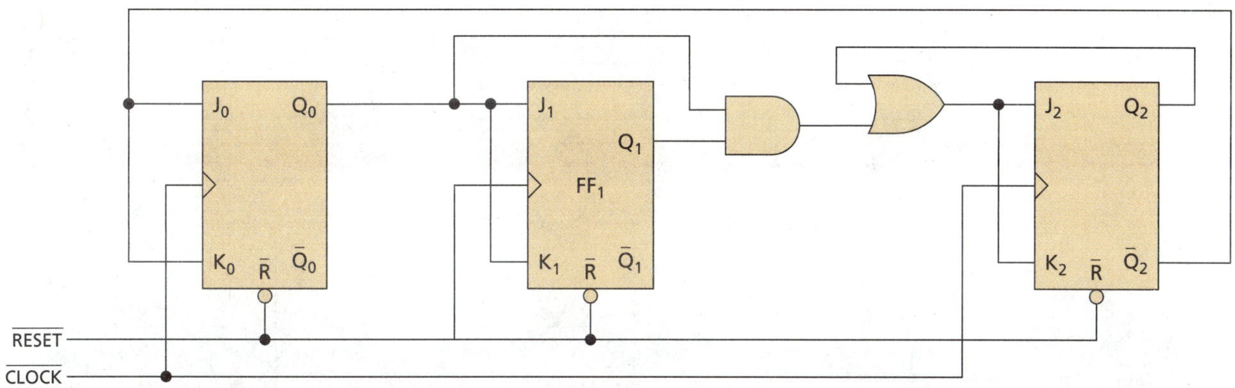

Figure 4.84 Mod-5 Synchronous Counter Using J–K Flip-Flops in Toggle Configuration

A synchronous counter can also be designed using D flip-flops. In Figure 4.85, K-maps are used to determine the minimum SOP expression that sets each individual D input high for the mod-5 counter using D flip-flops. From the

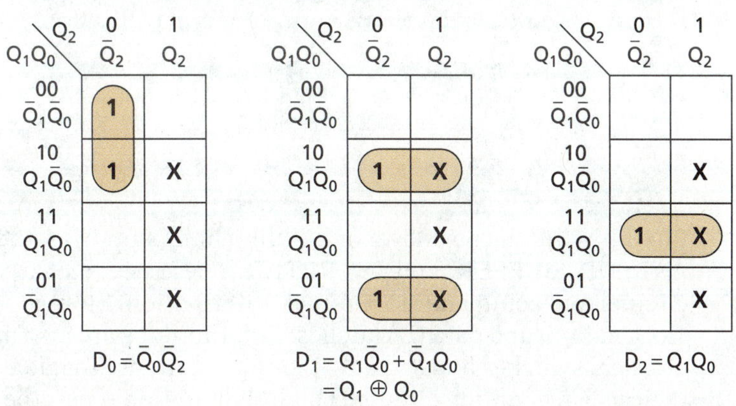

Figure 4.85 K-Map Counter Design for Mod-5 Counter Using D Flip-Flops

state table, the current state Q outputs are again translated onto the set of K-maps. In one respect, designing with D flip-flops is a little easier, because a high input to the D inputs is reflected in the next state condition of the Q outputs. Look on the state table of Figure 4.82. For the next state of Q_2, Q_1, and Q_0 to be 0, 0, and 1 respectively, the D inputs of the current state must have those same states, D_2, D_1, and D_0 = 0, 0, and 1.

After the K-maps are filled, they are read for the minimum SOP expressions and the counter is built. Notice on the left side of the map for D_1. The boxes that hold 1's are alternating, which indicates the possibility of an XOR function, as described in Chapter 2. When the map is read, it confirms this notion:

$$D_1 = Q_1 \cdot \overline{Q_0} + \overline{Q_1} \cdot Q_0 = Q_0 \oplus Q_1$$

The actual counter using D-type flip-flops is shown in Figure 4.86.

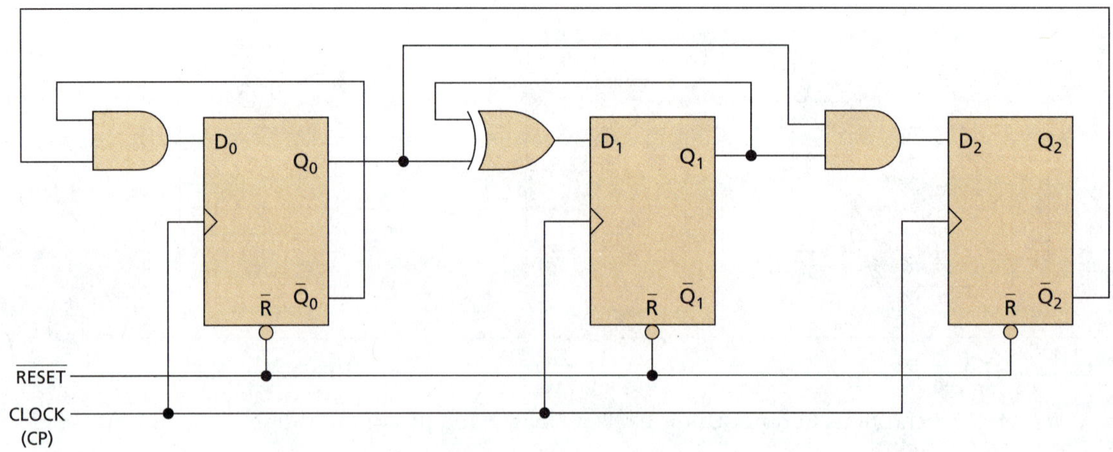

Figure 4.86 Mod-5 Synchronous Counter Using D Flip-Flops

Section 4.16

Practice Problem

Design a mod-6 synchronous counter using D flip-flops and K-maps.

Summary

Flip-flops are the next level of building blocks up from the basic four blocks—AND, OR, BUFFER, and INVERTER. These new blocks will be used to build circuits that count and store binary information. Storage flip-flops are the R–S, clocked R–S, and data (D) latches. R–S flip-flops are also used to debounce mechanical switches and several other applications. Toggle flip-flops find most of their use in counting circuits. Of all the flip-flop types, the J–K is the most universal, capable of operating as any of the other types with minimal wiring or additional logic circuits.

State diagrams are a graphical representation of the combined static and dynamic operations of synchronous devices and circuits. Circles indicate the output states of the device or circuit, while branches indicate transitions from one state to another. These branches may include labels that identify asynchronous input influences on these transitions and additional output conditions not specified in the state circles.

Sequential counting circuits have numerous applications. They are used for controlling the sequence of events and synchronizing activity to a timing or clock signal. Counters come in either asynchronous or synchronous operation depending on whether their individual flip-flops are clocked simultaneously or not. Counters can be made to count up or down in binary or simple to rotate a high state through the sequence of its flip-flops. Today many counter configurations are manufactured in medium- and large-scale integrated chips.

Troubleshooting counter circuits is not much different from troubleshooting other logic circuits. The main difference is that counter circuits are dynamic in nature, while the logic circuits we have looked up to now are static in nature. This means that counter circuits are constantly changing. Still, initial exploration of these circuits involves knowing what to expect and then hunting for where the results are incorrect. With counters, though, you may have outputs at all points of the circuit, but they may be in the wrong sequence, or have incorrect durations as well as logic levels.

Designing counters can be accomplished by using truth tables, state tables, or state diagrams to illustrate the count sequence desired. Logic expressions, representing the input states required to cause the counter to change from present or current states to next states, are derived from the state tables or diagrams. Karnaugh maps are also used to determine the logic expression required to drive the D inputs or J–K inputs of the counter being designed.

Glossary

astable or **free-running multivibrator (MVB)** a circuit that generates a CLOCK signal.

asynchronous control signals input and output control signals that are independent of any synchronizing CLOCK signal.

asynchronous counters counters whose flip-flop components are not clocked simultaneously.

binary ripple counter an asynchronous counter whose output sequence either counts up or down.

branch a line on a state diagram that shows the transition from the present to the next state.

clock a square wave or repetitive pulse signal that regularly alternates between a logic 1 and logic 0; used for synchronizing circuit elements and control signals.

control signal an input to a logic circuit that enables or disables some activity within that circuit.

counter logic circuit that sequences through a set of binary values.

current or **present state** an indication on a state diagram of the present condition of a circuit before a clock pulse is applied.

data binary information stored in single bits or in a series of bits called a data word.

data or **D flip-flop** a flip-flop whose output changes, when clocked, to the state of the input at D.

decoder a logic circuit that translates a binary number into a single logic signal.

duty cycle the ratio of the time a repetitive pulse is in the high state to the total time of a single cycle.

edge-triggered circuit logic circuits that change state at the transition time of a repetitive pulse.

ENABLE a signal that controls when a counter can count.

flip-flop a logic circuit that changes from one state to another state, latching or holding the current state between changes.

freeze to hold the current count output, usually by negating the ENABLE signal.

glitch a transient condition created by the propagation time of a counter changing from a maximum count to a zero condition.

J–K flip-flop a universal flip-flop that can mimick all the other types of flip-flops with minimal additional circuitry.

Johnson counter a synchronous counter whose mod number is equal to twice the number of flip-flops in the counter.

label defines the control signals on a state diagram that cause a transition to occur.

latch a logic circuit that changes to a state and then holds that state until a new set of signals are applied.

level-triggered circuit logic circuits that changes state when a repetitive pulse is at a specific logic level.

memory digital devices or circuits used to retain or store binary information (data).

mod or **modulo number** the number of different conditions or states that a set of outputs can have.

next state on a state diagram, it is the condition of a circuit after a CLOCK signal has been applied.

no change condition in which a flip-flop or circuit retains its current state.

program counter a binary up counter used to keep track of program instruction usage in a computer or microprocessor.

propagation delay the time between applying an input signal to a device and getting a change of state condition at the output.

race condition an undeterminable state of a logic circuit whose outcome is based more on the propagation delay of the individual circuit parts than the logic of the circuit.

R–C time constant the time it takes a capacitor to charge to a specific voltage level.

register a temporary data storage circuit.

RESET as applied to counters, the signal that returns a counter to its initial condition. As applied to flip-flops, it is the logic 0 condition of the Q output of a flip-flop.

ring counter a synchronous counter all of whose outputs except one are low.

ripple count or **RC** a short pulse generated by a binary counter when it reaches its maximum (up) or minimum (down) count value.

set the logic 1 condition of the Q output of a flip-flop.

set-up time the minimum time an input must be supplied before a CLOCK signal is applied.

state diagram a pictorial representation of a synchronous circuit.

state table similar to a truth table, it includes both present and next states.

synchronous with respect to counters, synchronous circuits are those in which all elements are clocked or triggered at the same time.

terminal count or **TC** an indication of when a counter has reached its maximum (up) or minimum (down) count value.

toggle or **T flip-flop** logic circuit that alters states between two stable conditions whenever it is clocked.

undefined or **invalid condition** the condition of a flip-flop when Q and $\overline{Q}$ outputs are the same state, which violates the basic definition of flip-flops.

Solutions to Practice Problems

Section 4.2

A 0 input to an AND gate forces its output low. When this low is cross-coupled to a second AND gate in a latch configuration, its output also goes low. Having both outputs occur in the same state defies the basic characteristic of a flip-flop in that Q and $\overline{Q}$ are always opposite in state.

Section 4.3

1.

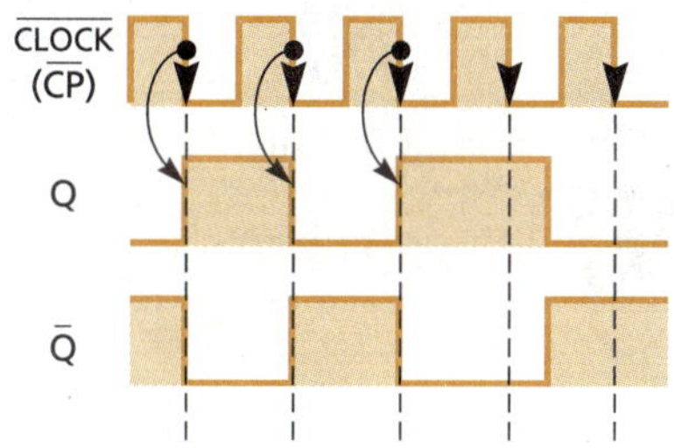

Figure 4.87 Practice Problem 4.3(1) (solution)

2.

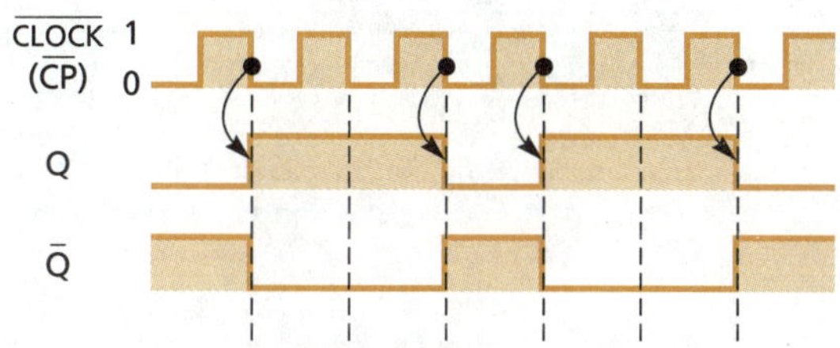

Figure 4.88 Practice Problem 4.3(2) (solution)

Section 4.4

1.

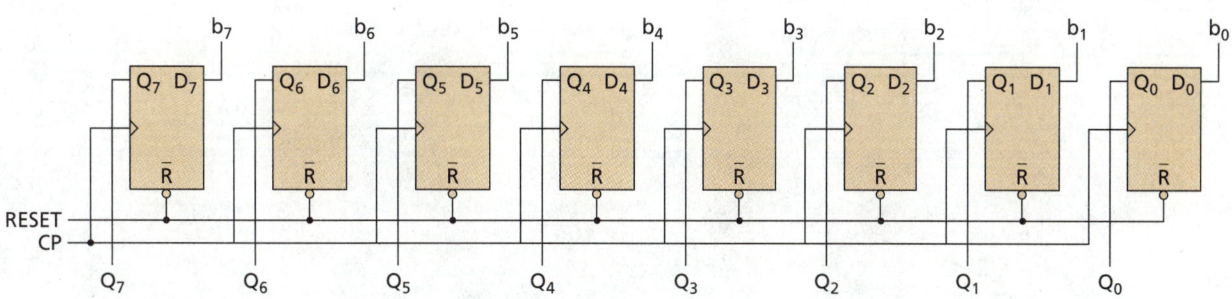

Figure 4.89 Practice Problem 4.4(1) (solution)

2.

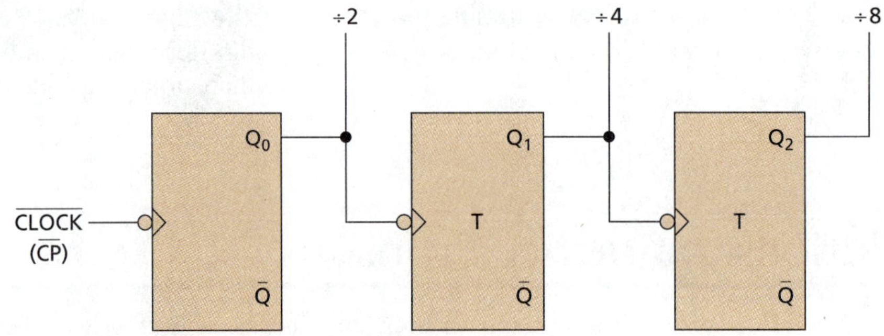

Figure 4.90 Practice Problem 4.4(2) (solution)

3.

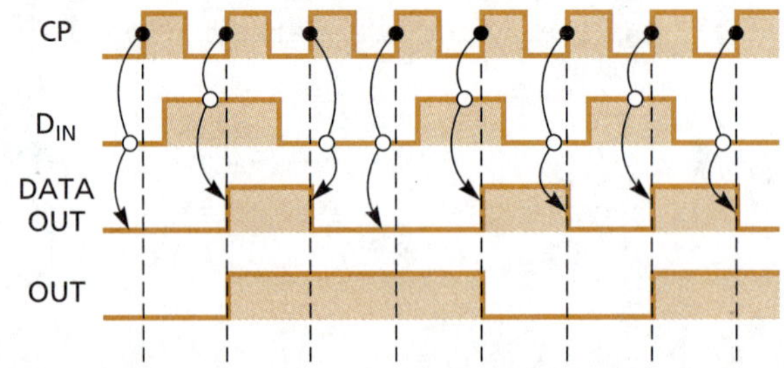

Figure 4.91 Practice Problem 4.4(3) (solution)

Section 4.5

1. The J–K flip-flop is usable as R–S, clocked R–S, toggle, or D flip-flop with minimal wiring or logic circuitry.

2.

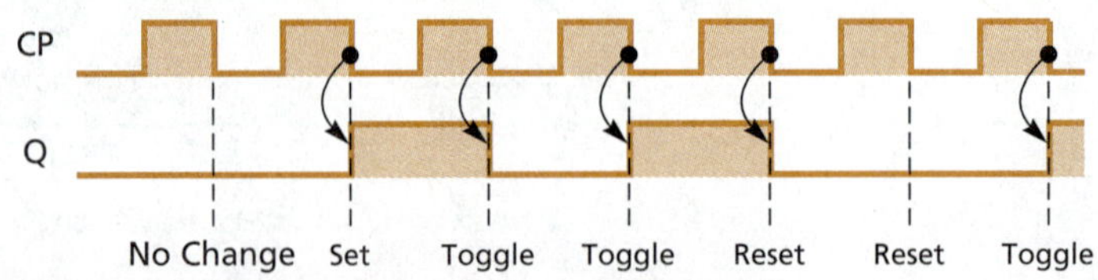

Figure 4.92 Practice Problem 4.5(2) (solution)

Section 4.6

$R_A = 975\ \Omega \qquad R_B = 9.75\ K\Omega$

Section 4.7

$R_A = 67.5\ K\Omega$

Section 4.8

1.

D	Q_n	Q_{n+1}
0	0	0
0	1	0
1	0	1
1	1	1

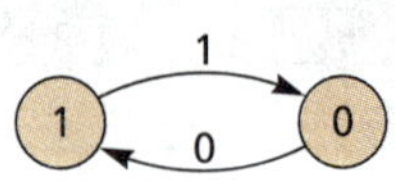

2.

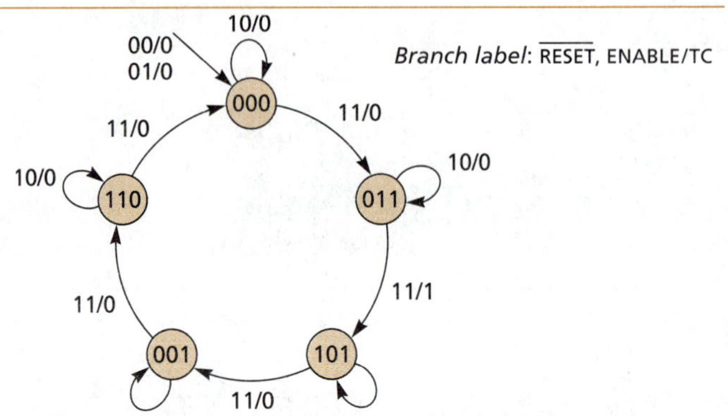

Figure 4.93 Practice Problem 4.8(2) (solution)

Section 4.9

1. Down count **2.** Up count **3.** 64

4. Max count − mod number = decoded reset: 111B − 101B = 010B

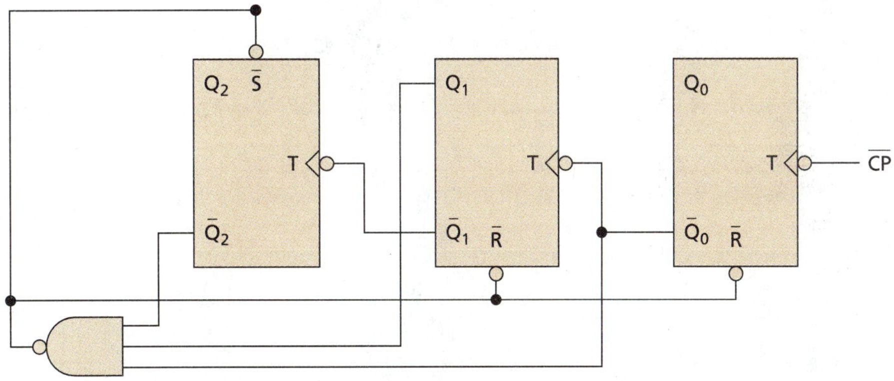

Figure 4.94 Practice Problem 4.9(4) (solution)

5.

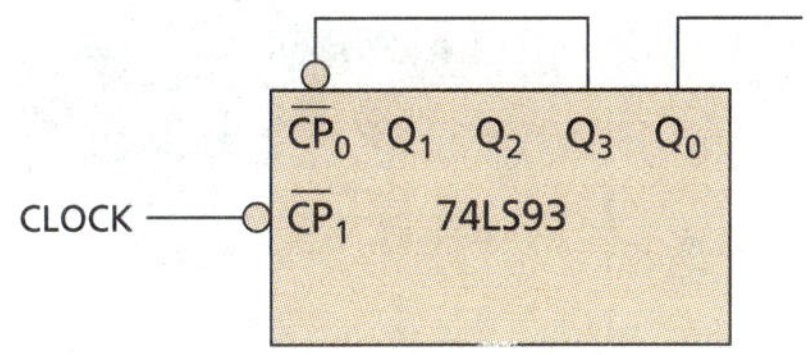

Figure 4.95 Practice Problem 4.9(5) (solution)

Section 4.10

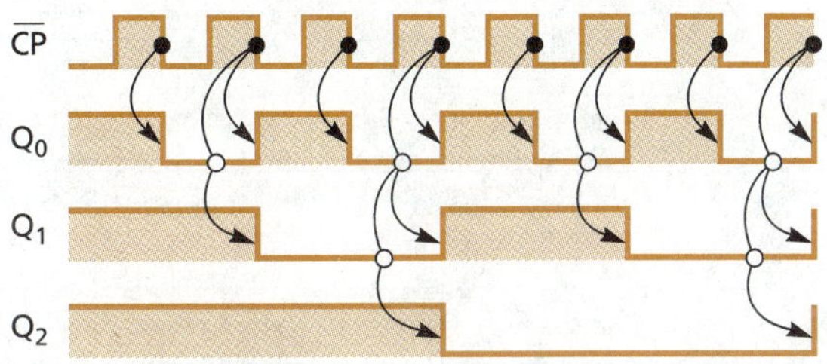

Figure 4.96 Practice Problem 4.10 (solution)

Section 4.11

1. 7 **2.** RESET **3.** 110

Section 4.12

1.

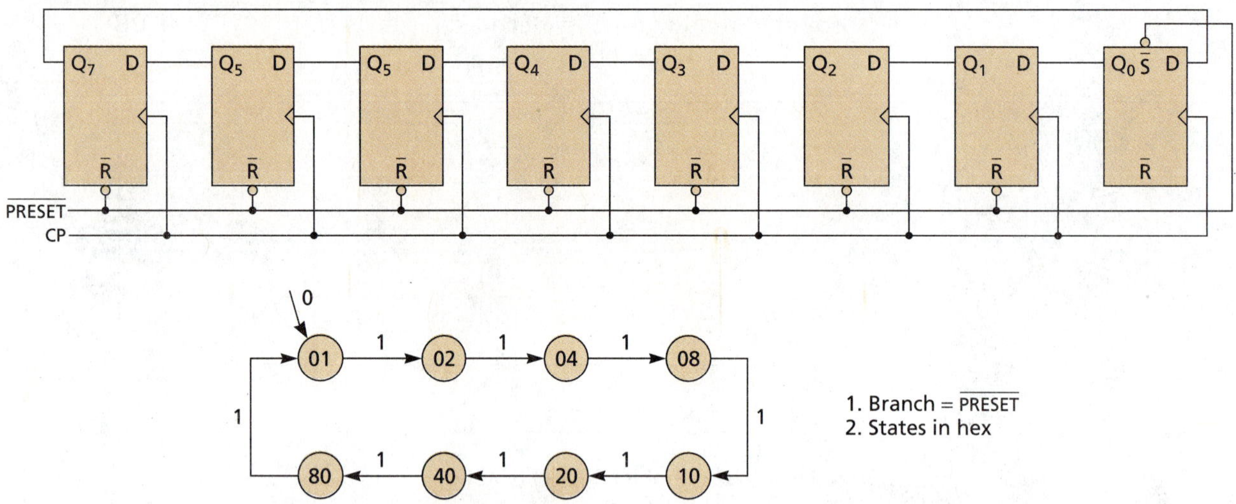

Figure 4.97 Practice Problem 4.12(1) (solution)

2.

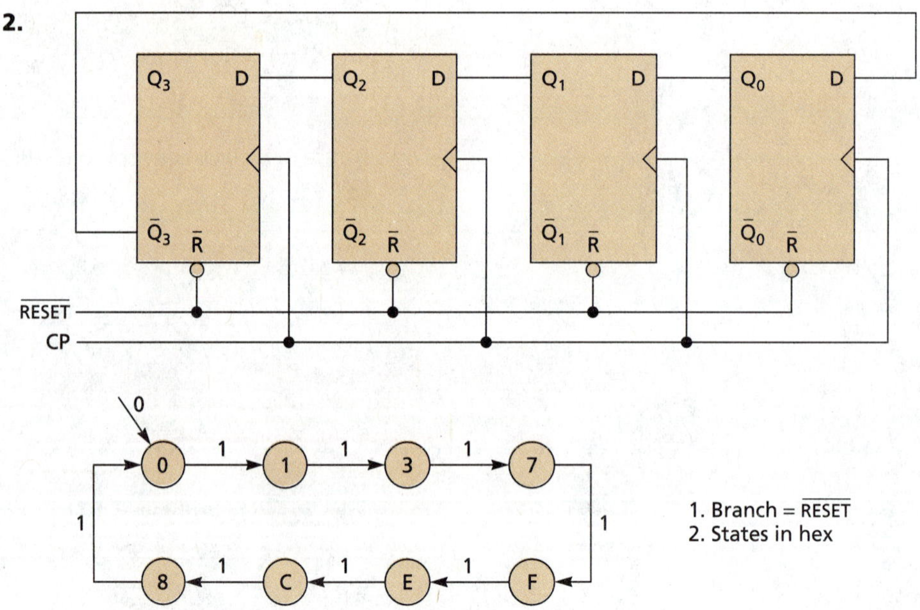

Figure 4.98 Practice Problem 4.12(2) (solution)

3. The mod-8 ring counter requires eight flip-flops and no decoding, whereas the mod-8 Johnson counter only requires four flip-flops, but needs some form of decoding circuitry.

Section 4.13

1. The first stage is acting as a count down rather than a count up stage.
2. Output Q_3 acts like a ring-counter-type output rather than a Johnson counter output.

Section 4.14

1.

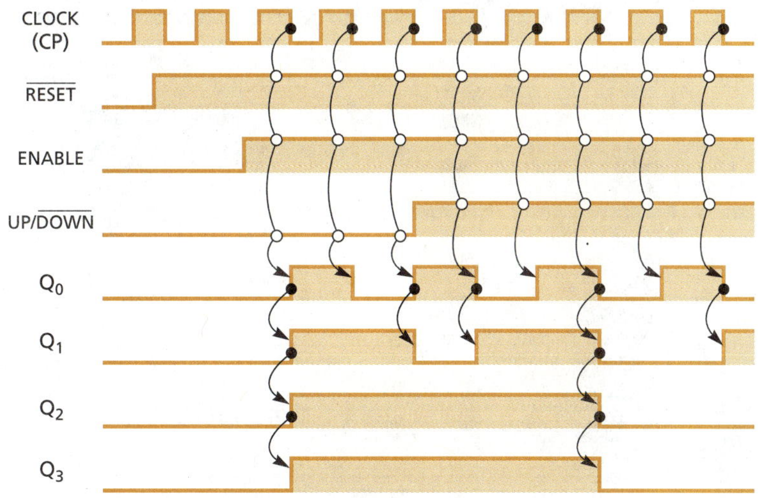

Figure 4.99 Practice Problem 4.14(1) (solution)

2.

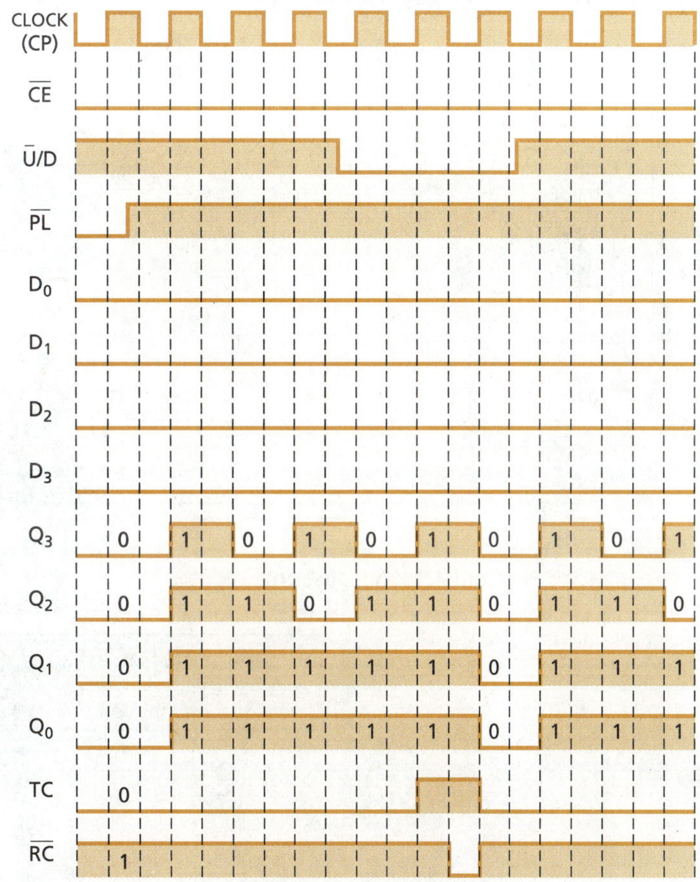

Figure 4.100 Practice Problem 4.16(2) (solution)

Section 4.15

1.

$\overline{PRESET}$	ENABLE	Q_n			Q_{n+1}			$\overline{TC}$
		Q_2	Q_1	Q_0	Q_2	Q_1	Q_0	
0	X	X	X	X	1	0	1	1
1	0	X	X	X		Q_n		$\overline{TC}$
1	1	1	0	1	1	0	0	1
1	1	1	0	0	0	1	1	1
1	1	0	1	1	0	1	0	1
1	1	0	1	0	0	0	1	1
1	1	0	0	1	0	0	0	0
1	1	0	0	0	1	0	1	1

X = "don't care" state

Figure 4.101 Practice Problem 4.15(1) (solution)

2.

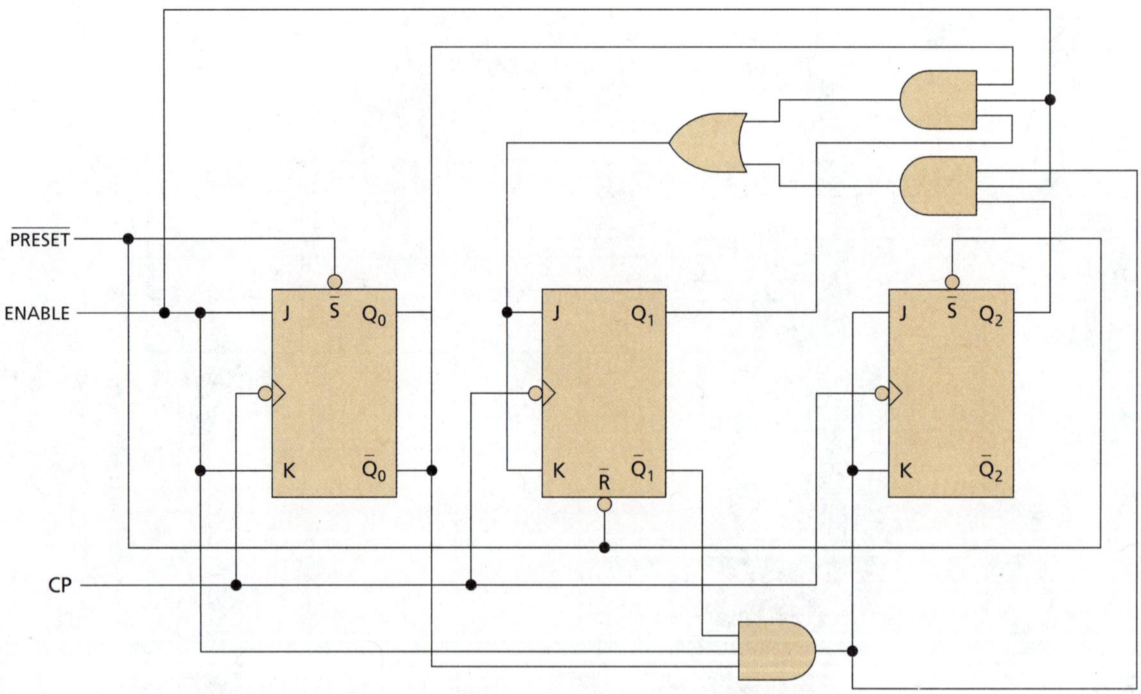

Figure 4.102 Practice Problem 4.15(2) (solution)

Section 4.16

Q_n			Q_{n+1}			High D Inputs		
Q_2	Q_1	Q_0	Q_2	Q_1	Q_0	D_2	D_1	D_0
0	0	0	0	0	1	–	–	(1)
0	0	1	0	1	0	–	(1)	–
0	1	0	0	1	1	–	(1)	(1)
0	1	1	1	0	0	(1)	–	–
1	0	0	1	0	1	(1)	–	(1)
1	0	1	0	0	0	–	–	–

Figure 4.103 Practice Problem 4.16 (solution)
State Table

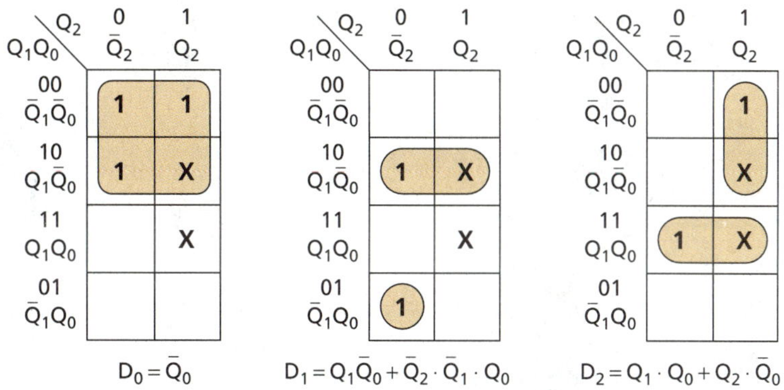

Figure 4.104 Practice Problem 4.16 (solution) Karnaugh Maps

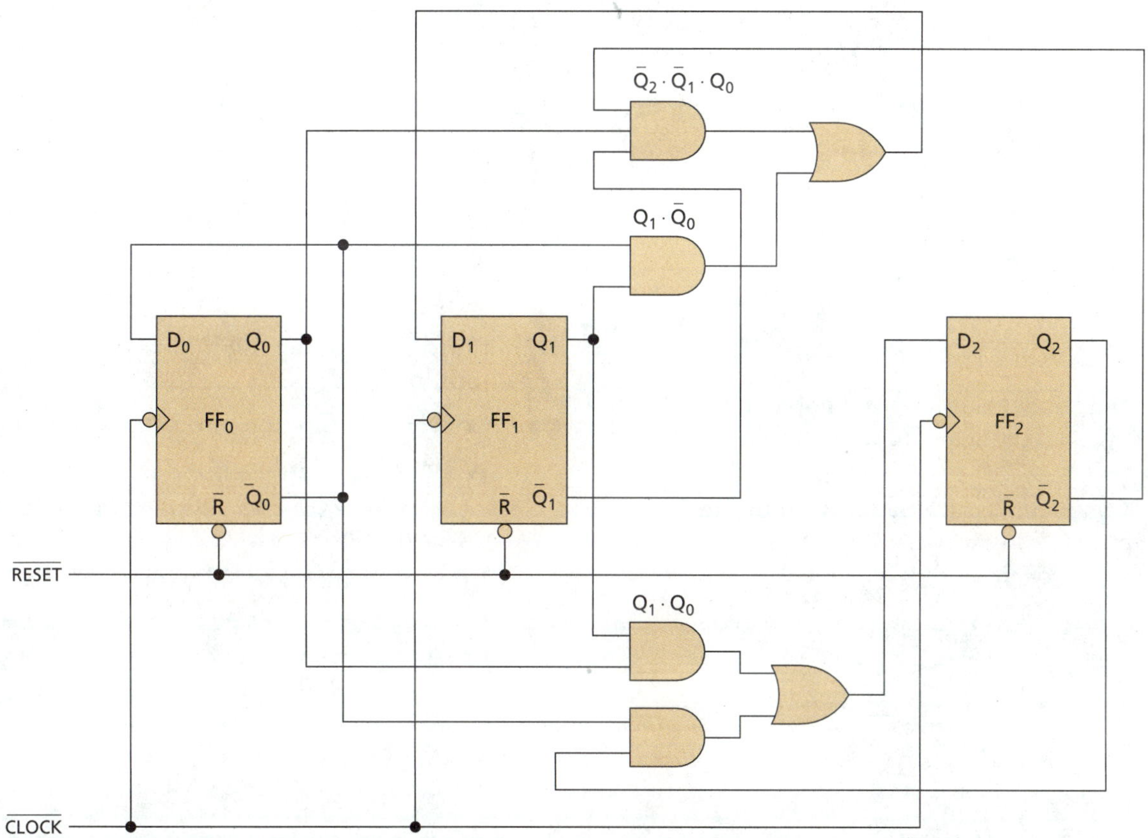

Figure 4.105 Practice Problem 4.16 (solution): Mod-6 Counter

Problems

4.1 What is the active state of the inputs to a cross-coupled NAND latch?

4.2 What is the active state of the inputs to a cross-coupled NOR latch?

4.3 Complete a truth table that describes the operation of the cross-coupled latch in Figure 4.106.

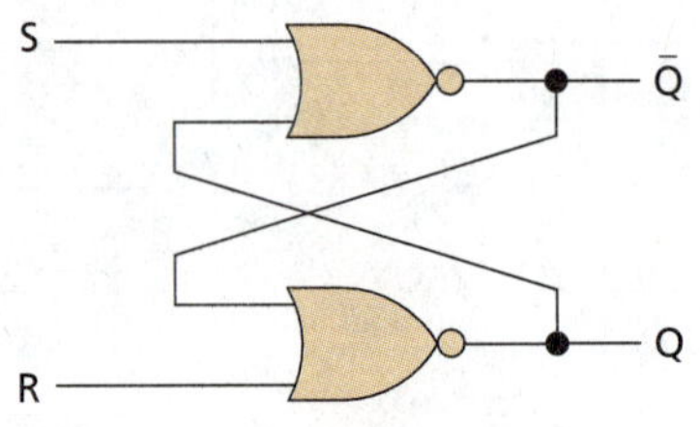

Figure 4.106 Problem 4.3

4.4 Complete the timing diagram in Figure 4.107 for the latch in Figure 4.106.

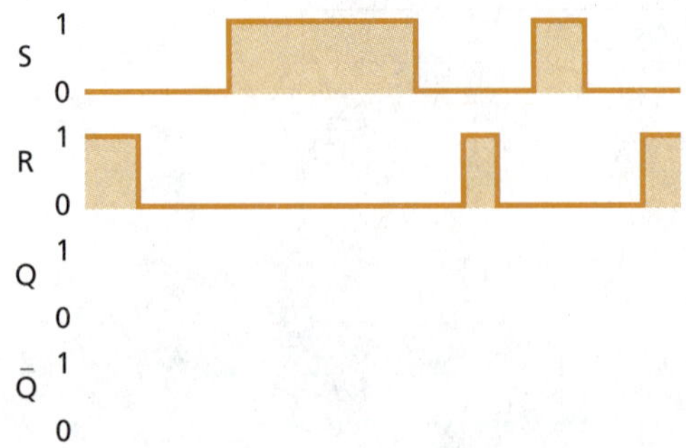

Figure 4.107 Problem 4.4

4.5 Describe the difference in operation between a level-triggered and an edge-triggered R–S flip-flop.

4.6 Complete the timing diagram for the R–S flip-flop in Figure 4.108.

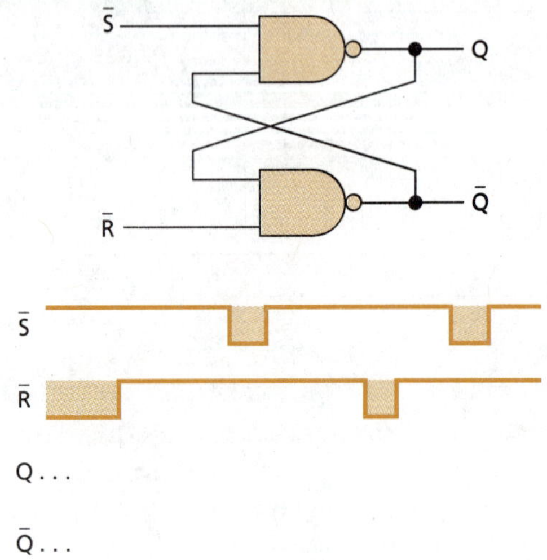

Figure 4.108 Problems 4.6, 4.7, and 4.8

4.7 Applying 0's to the $\overline{R}$ and $\overline{S}$ inputs of the R–S flip-flop in Figure 4.108 is considered as an undefined condition for that flip-flop. Why?

4.8 Describe how the R–S flip-flop of Figure 4.108 remains latched or in a no-change condition when 1's are applied to both R and S inputs.

4.9 What makes the Q output of a D flip-flop copy the D input when the CLOCK signal is applied?

4.10 Complete the timing diagram in Figure 4.109 for the 74LS74 D flip-flop.

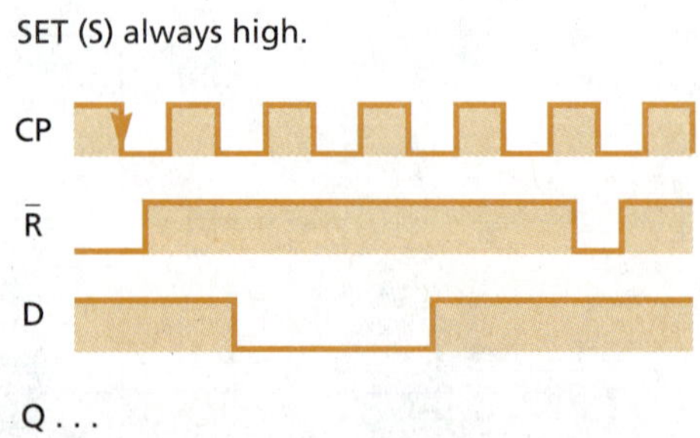

Figure 4.109 Problem 4.10

4.11 Convert a clocked R–S flip-flop into a toggle flip-flop.

4.12 Convert a D flip-flop into a toggle flip-flop.

4.13 What is the relationship of the frequency rates between the clock input and Q output of a toggle flip-flop?

4.14 Draw a circuit whose final output has a frequency that is 1/16th of the CLOCK input.

4.15 Create the truth table for the J–K flip-flop in Figure 4.110. Label all the other flip-flop

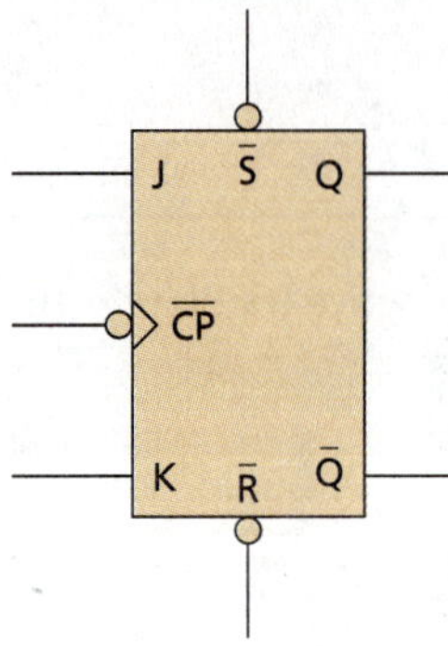

Figure 4.110 Problem 4.15

functions that the J–K can emulate without additional circuitry.

4.16 Complete the timing diagram of Figure 4.111 for the 74LS76 J–K flip-flop.

Figure **4.111** Problem 4.16

4.17 Determine the values for R_A and R_B used in a 555 timer circuit to generate a square wave whose frequency is 36 KHz with a duty cycle of approximately 50 percent. Use a timing capacitor value of 0.0033 μf.

4.18 Determine the values for R_A and R_B used in a 555 timer circuit to generate a square wave whose frequency is 64 KHz with a duty cycle of 66 percent. Use a timing capacitor value of 0.01 μf.

4.19 Compute the value of R_A for a one-shot pulse of 1.2 mS duration using a 555 timer with a timing capacitance of 0.1 μf.

4.20 Compute the value of R_A for a one-shot pulse of 456 μS duration using a 555 timer with a timing capacitance of 0.025 μf.

4.21 What type of information is used in labels on branches in a state diagram?

4.22 What information about a synchronous device or circuit do state circles describe?

4.23 What is the purpose of branches in a state diagram?

4.24 What is meant by present (or current) state in a state table? Compare this to the meaning of the next state.

4.25 What additional information is conveyed by a state table that is not included in a truth table?

4.26 Create a state table and state diagram for a toggle flip-flop that has an asynchronous active low reset input.

4.27 Draw a state diagram for an eight-state counter (000B to 111B) that has the following control signals:

a. Active low RESET

b. Output TERMINAL COUNT that goes high when the counter reaches 111B and is low otherwise.

4.28 Draw a state diagram for a mod-5 counter (000B to 100B) that has the following control signals and conditions:

a. Active high RESET (sets counter to 000B)

b. Output TERMINAL COUNT that goes low when the counter reaches 000B and is low otherwise.

4.29 What is the basic definition of a counter circuit?

4.30 Describe the two important considerations when analyzing counters.

4.31 What is the difference between asynchronous and synchronous binary counters?

4.32 Complete the timing diagram for the counter in Figure 4.112. What is the count sequence of this counter?

4.33 In general terms, what is the relationship between the Q output of one stage and the CLOCK input of the next stage, which determines whether a binary ripple counter counts up or counts down?

4.34 What two modes do the J–K flip-flops in Figure 4.112 operate in as controlled by the ENABLE signal?

4.35 What is the mod number of a ripple counter with eight flip-flops?

4.36 What is the relationship of the frequency between the output of the last stage of a six-bit ripple counter and the CLOCK input?

4.37 Draw an asynchronous mod-11 (0000B–1010B) binary ripple up counter.

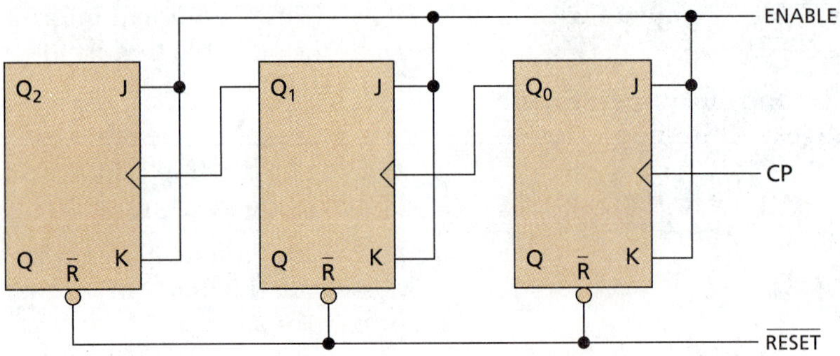

Figure **4.112** Problem 4.37

4.38 Draw an asynchronous mod-13 (1100B–0000B) binary ripple down counter.

4.39 Complete the timing diagram of Figure 4.113 for the 74LS93 counter. Q_0 is connected to $\overline{CP}_1$.

4.40 Complete the timing diagram for the UP/DOWN counter in Figure 4.114.

4.41 Complete the timing diagram in Figure 4.115 for the 74LS191 presettable four-bit binary UP/DOWN counter.

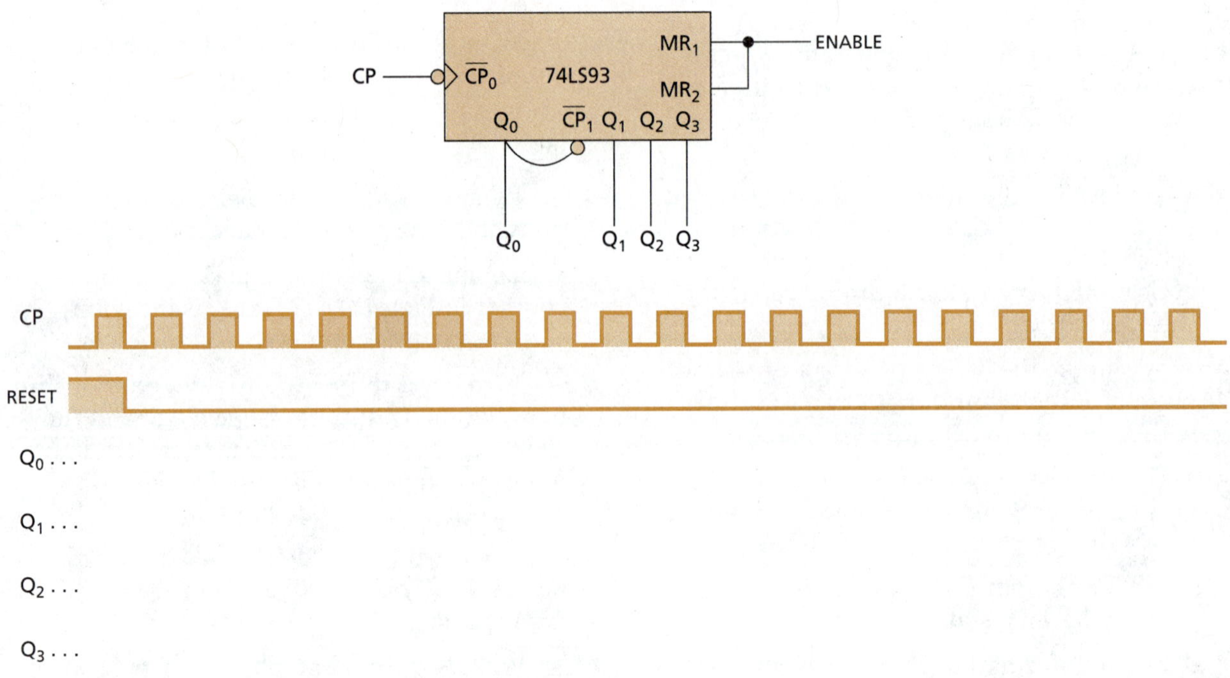

Figure **4.113** Problem 4.38

4.42 Complete the timing diagram in Figure 4.116 for the 74LS191 presettable four-bit binary UP/DOWN counter.

4.43 Complete the timing diagram for the synchronous up counter in Figure 4.117.

4.44 Create the state table and state diagram for the counter in Figure 4.117.

4.45 Compare an eight-stage ring counter to an eight-stage Johnson counter using D flip-flops.

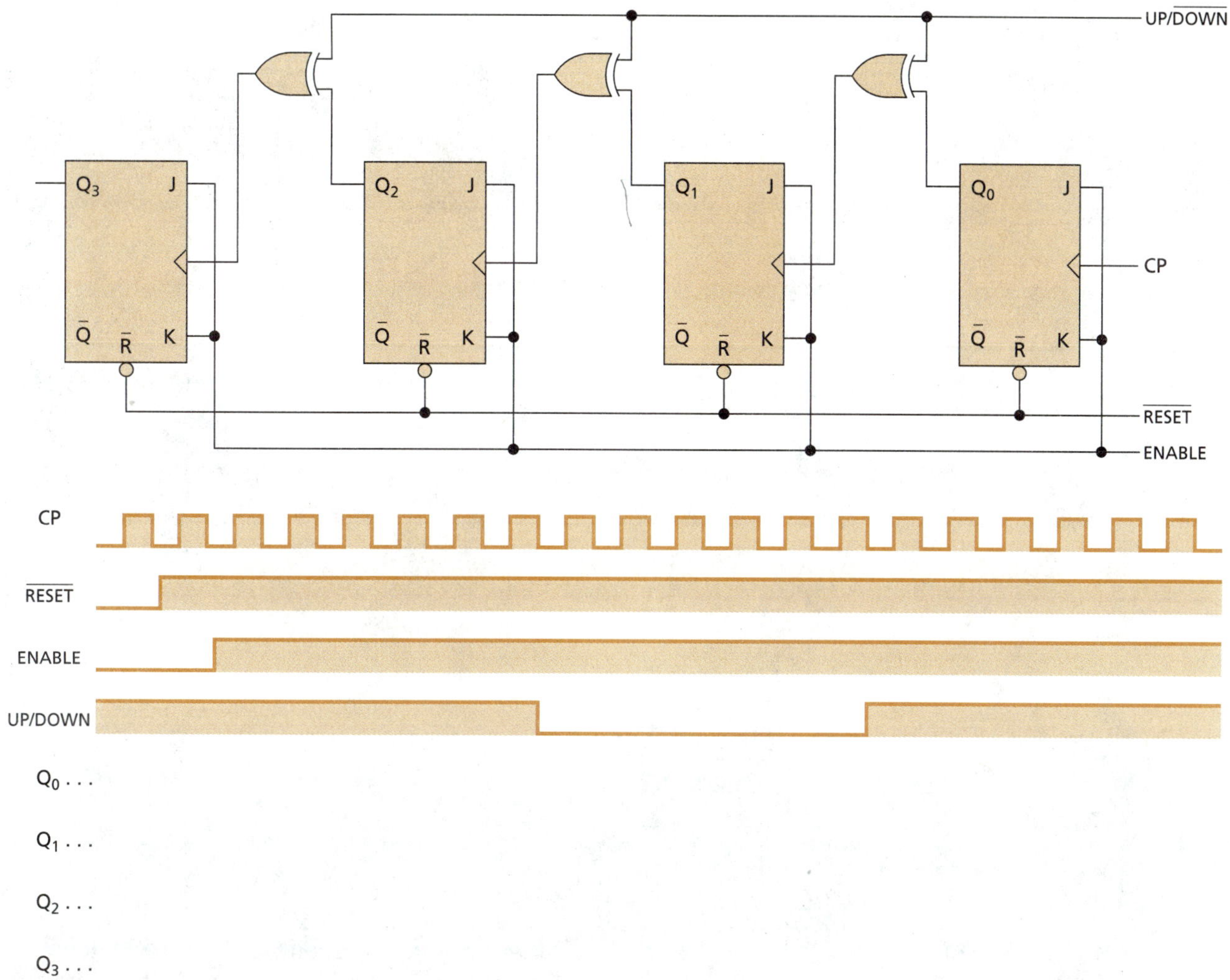

Figure 4.114 Problem 4.40

Give a solid advantage of each compared to the other.

4.46 Create state diagrams for the two counters in Problem 4.45. Both counters have an active low $\overline{\text{RESET}}$ to initiate the count sequence.

4.47 Determine which stage of the counter in Figure 4.118 is bad, given the timing sequence in Figure 4.119.

4.48 Determine which stage of the counter in Figure 4.120 is bad, given the timing sequence in Figure 4.121.

4.49 Design a mod-13 (0000B–1100B) synchronous up counter. Include a reset input to start the counter. You may use truth tables, state tables, or state diagrams in your design. Show all your work.

4.50 Design a mod-11 synchronous down counter that starts at 1010B. Include a PRESET signal to start the counter. You may use truth tables, state tables, or state diagrams in your design. Show all your work.

4.51 Develop the state diagram for a mod-9 synchronous UP/DOWN counter. The count sequence is 0000B to 1000B up and 1000B to 0000B down. The control signals included in the circuit are

Figure 4.115 Problem 4.41

a. Active low RESET
b. Active high count enable
c. Active high up/active low down control
d. A terminal count output that goes high when counting up and reaching 1000B or counting down and reaching 0000B. This output is low for all other count values.

4.52 Develop the state diagram for a mod-11 synchronous UP/DOWN counter (0000B–1010B up or 1010B–0000B down). The control signals included in the circuit are
a. Active low RESET
b. Active high count enable
c. Active high down/active low up control
d. A terminal count output goes low when counting up and reaching 1010B or counting down and reaching 0000B. This output is high for all other count values.

4.53 Design a self-starting mod-10 ring counter. This counter does not have a PRESET control line.

4.54 Design a mod-10 Johnson counter. Include the logic needed to decode the 10 output states into single output control signals.

4.55 Design a mod-10 synchronous up counter using D flip-flops and Karnaugh maps.

4.56 Design a mod-13 synchronous up counter using D flip-flops and Karnaugh maps.

Figure 4.116 Problem 4.42

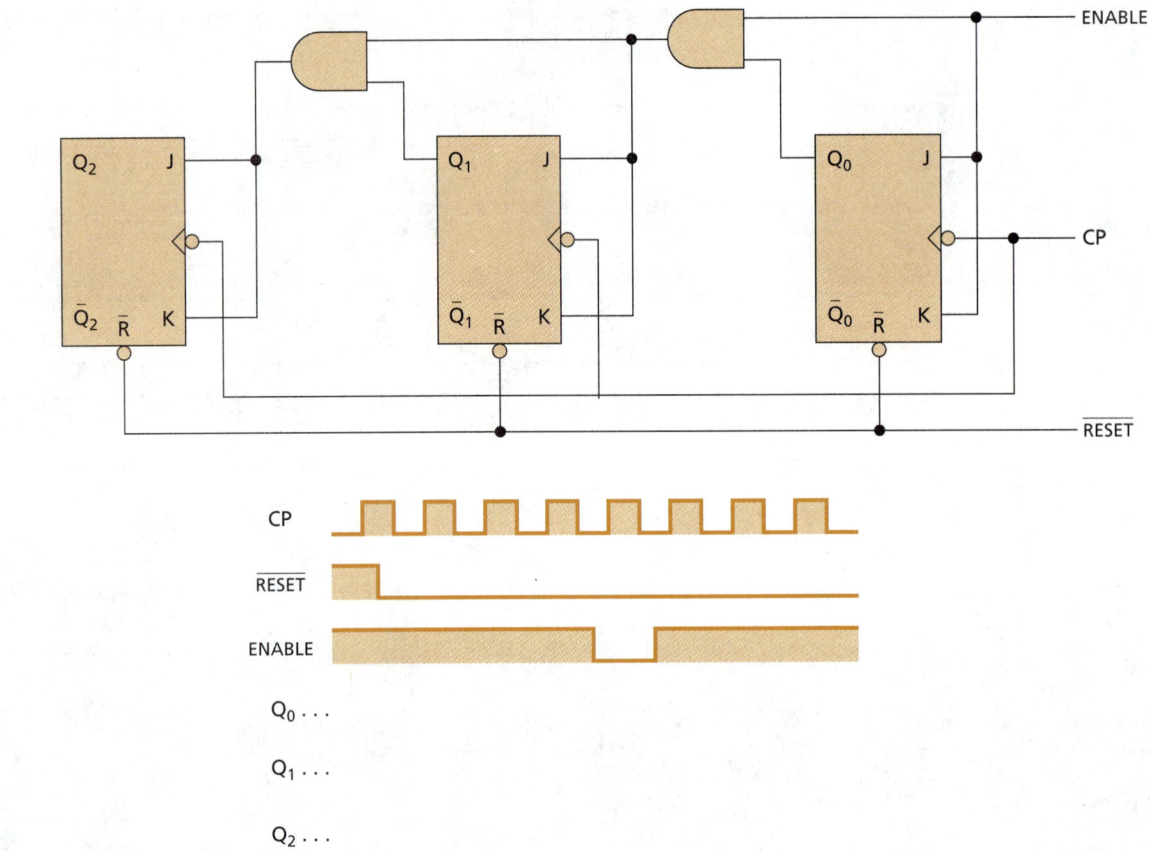

Figure **4.117** Problem 4.44

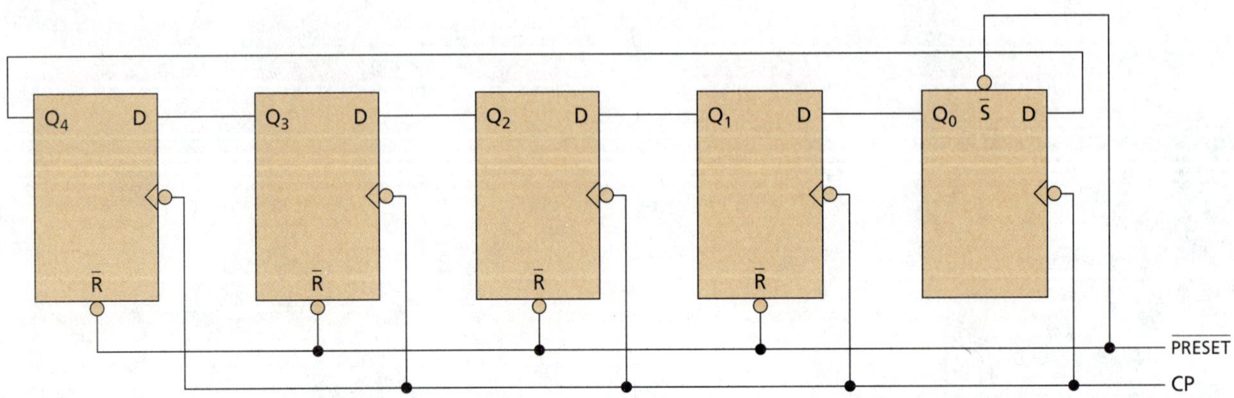

Figure **4.118** Problem 4.47 Counter

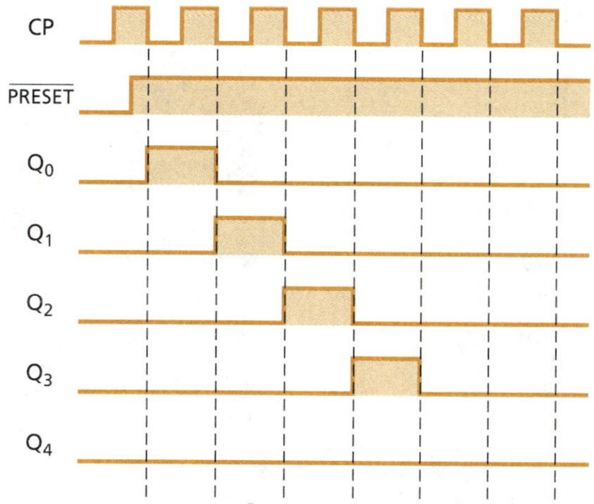

Figure 4.119 Problem 4.47 Timing Diagram

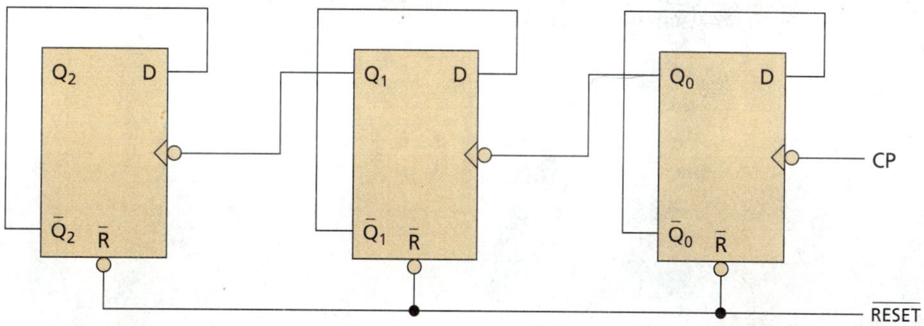

Figure 4.120 Problem 4.48 Counter

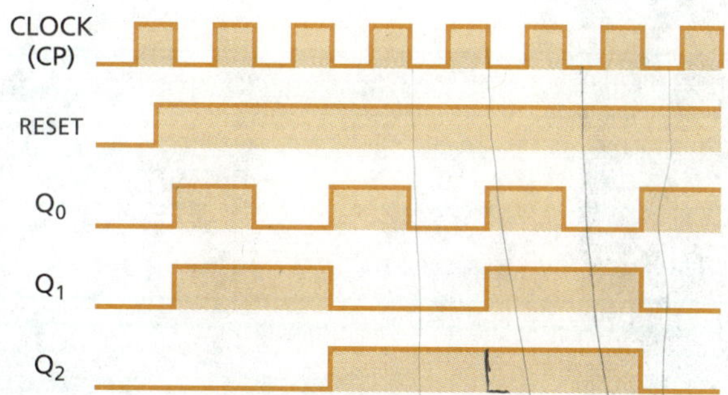

Figure 4.121 Problem 4.48 Timing Diagram

five

Memory Elements

KEY TERMS

access time
address
address bus
address decoding
bus
capacity
CAS
cell
data bus
decoder
demultiplexer (demux)
DRAM
encoder
EEPROM
EPROM
full address decoding
LSI (large-scale integration)
memories
memory map
MSI (medium-scale integration)
multiplexer (mux)
nondestructive read
parallel
partial address decoding
PROM
RAM
RAS
read cycle
refreshing
register
ROM
SAM
serial
shift left
shift right
SSI (small-scale integration)
strobe
TTL (transistor–transistor logic)
volatile
write cycle

OBJECTIVES

After reading this chapter, the students should be able to

- Describe registers that store data serially, one data bit following another and others that store data in parallel (store and hold an entire data word at one time).
- Describe decoder circuits that sense a binary input and deliver a single signal for each binary combination sensed and encoders that do the reverse process.
- Explain multiplexers, which are circuits that select among many inputs and direct that selection to a single output, and demultiplexers, which direct a single input to one of many outputs.
- Discuss the concepts and circuits for static and dynamic memory devices used for longer storage times.
- Discuss random access memory (RAM) and read-only memory (ROM), that are the two basic types of memory devices, and their internal organization.
- Explain the purpose behind memory address decoding, which is a method to select which device in a system is being used at any given time.
- Describe methods for troubleshooting register circuits. Describe the design processes for memory systems.

OVERVIEW

This chapter begins by covering the last major area of digital circuitry that uses flip-flops, the register. Registers are defined as those circuits that hold or store data temporarily. This definition applies to all circuits that perform this function, regardless of any other function they may fill. As an example, the counter circuits discussed in Chapter 4 are also registers. By disabling the count enable signal to most counters, the current count remains "frozen" or held until the counter is allowed to resume counting. Because of this ability to temporarily hold a value, counters are also considered registers.

For larger-capacity storage or for longer storage periods, various types of memory devices are used. The basic types of memory devices we will discuss are the semiconductor types of memories used in most computers today. We will also examine the circuits that are used to select where in the memory data are located and the methods to access those locations. These circuits include decoders, encoders, multiplexers, and demultiplexers.

5.1 Introduction

The need to store and retrieve data is essential to the operation of digital circuits and computers. Storage requirements may be temporary, as when you enter a value into a calculator, or long term, as when a program is held in a computer's memory so that it can be run. Some forms of memory devices and circuits are responsible for holding a relatively small amount of data, while others store vast quantities of information. In addition, some memories must be permanent and always available for access, while others are memories that can be readily changed. The first type of memory element we are going to explore is used for temporary storage of a small quantity of binary data and is called a **register.**

Register *a logic circuit that temporarily stores digital data.*

5.2 Moving Data in and out of a Register

Serial *the movement of binary data one bit at a time.*

Parallel *the movement of all the bits of a binary data word at one time.*

Binary data are placed into and taken from a register in one of two basic manners, **serial** or **parallel.** Serial data make up a stream of binary bits that move into or out of a circuit, one bit at a time. Imagine that you are sitting at a railroad crossing watching a train pass by. Each railroad car passes by you, one at a time, until the whole train has passed. If the train is stopped at a station to let passengers on and off, it can be said to be held at that position. Once the conductor yells "All aboard!" the train again begins to move, leaving the station one car at a time. Serial data move into a circuit one bit at a time, much like the train entered the railway station. Once all the bits are entered into the register, the data can be held or stored there for as long as they are needed. After other circuits are done with the data, the register can be emptied, like the train, one bit at a time.

When every bit of a data word moves from one circuit to another at the same time, this movement of data is said to be "in parallel." Consider a roll of pennies: There are 50 pennies in a single roll. When you hand a roll of pennies to another person, you are moving all 50 pennies at one time—that is a parallel movement. Binary data words made of individual bits can be moved, like the pennies, all at once from one logic circuit to another, in parallel.

A register that accepts parallel data in and supplies parallel data out is shown in Figure 5.1. This eight-bit parallel-in/parallel-out register is made up of eight flip-flops. Data are presented to the data or D inputs of all eight flip-flops at one time. When a clock signal is applied to the clockpulse (CP) input, the eight bits are stored into the register. Since each element of the register is a D flip-flop, the Q outputs will copy the D inputs when CP is triggered. A short signal pulse on the CP line is sometimes called a **strobe.** Data can be said to be *strobed* into the register. Outputs can be taken from the Q outputs at any time after the data have been loaded into the register. Loading data into the register is called a **write** operation, because data are being written into the register. Taking the data from the Q outputs and using them elsewhere is called a **read** operation, because the data are being read from the register.

Strobe *a signal of short duration that initiates an event or process.*

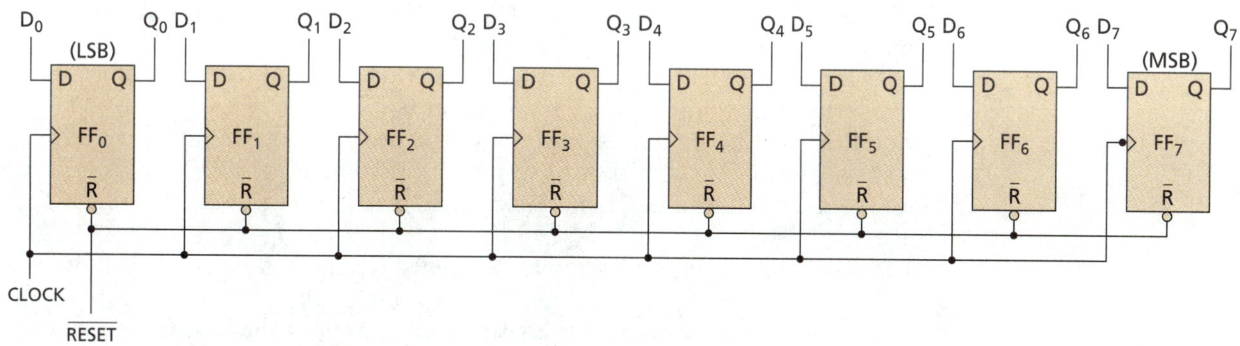

Figure 5.1 Parallel-In/Parallel-Out Register

EXAMPLE 5.1

Initially, the binary data word 10010110B was written into the register. Considering D_0 as the least significant bit of the data word, what are the data presently read from the register? At a later time, a new word, 00111100B, is placed onto the D inputs and strobed in. What is now read at the Q outputs of the register?

SOLUTION This example illustrates the temporary nature of the data stored in a register. The original data read were those which had been initially stored into it, 10010110B. When the new data were strobed in, they replaced the initial data word, totally destroying the old data. The second time the register was read, it showed the new data word, 00111100B.

Keep in mind a number of things about the parallel register:

1. The register is updated with whatever binary values are placed on the D lines and strobed into the register.
2. The data originally in the register are lost when the new data are written in. This is due to the temporary nature of a register.
3. Data in the register can be read and used at any time as long as new data have not replaced them.

A disadvantage of parallel registers is that they require individual data input and output connections and lines for each data bit to allow all bits of a data word to be loaded into or read from the register. The plus side is that the loading and reading of data are rapid, taking only the application of a single clock signal along with the propagation of the flip-flops to load a complete data word into the register. Transfers of data in parallel are most efficient in circuits where the transfer takes place between circuits in close proximity to each other. However, when you must send data over a long distance, it is preferable to reduce the number of data lines needed to facilitate the transfer. The most common method used is to send the data serially, one bit at a time. For example, the data word 10110110B can be sent from circuit A to circuit B in parallel. The entire word is sent at once, requiring a single clock pulse on CP and a separate data line for each bit. In a serial transfer, either the least significant bit (LSB, which is a 0 in our sample data) or the most significant bit (MSB, which is a 1 in our data) is sent first, depending on the direction of the serial shift that is used. This first bit is followed by the next bit (1 following the LSB in our data sample) and the next (1) continually until all the bits are sent, completing the transfer. Only a single data path is required to move the data, but it takes eight clock pulses on CP to send the data, one bit at a time.

The movement of data through a shift register is best illustrated by showing a shift register circuit and constructing a timing diagram for data moving through it. The sample data 1101B shifted in and out of the four-bit register in Figure 5.2 are used to show the shifting process. The data are presented on the D_{IN} line one bit at a time and enter and move through the register on succeeding clock pulses. Since there are four bits in the register and data stream, it takes four clock cycles to shift the data in, and four cycles to completely shift the data out through Q_3. Following a RESET signal, on the first clock cycle, the most significant bit of the data is latched into FF_0. This causes Q_0 to go high. On the next clock pulse, the high state of Q_0, presented to D_1, causes Q_1 to set, and the second bit of data on D_{IN} keeps FF_0 set. On the third clock cycle, the state of FF_1 is shifted to FF_2, that of FF_0 to FF_1 and the third bit of the data word is latched into FF_0. The process is repeated once more to shift the least significant bit into FF_0 and the other three bits into each following flip-flop. Reading the contents of the register at the Q outputs of the four D flip-flops after these four clock cycles, shows the data word 1101B is in the register. Figure 5.3 shows the progression of data word 1101B through the register, as just described, in a set of timing diagrams, showing the shift process for each clock cycle.

Holding D_{IN} low, and applying four additional clock cycles, shifts the data out through Q_3 and clears the register. Figure 5.4 shows the progression of shifting the data out and shifting in zeros.

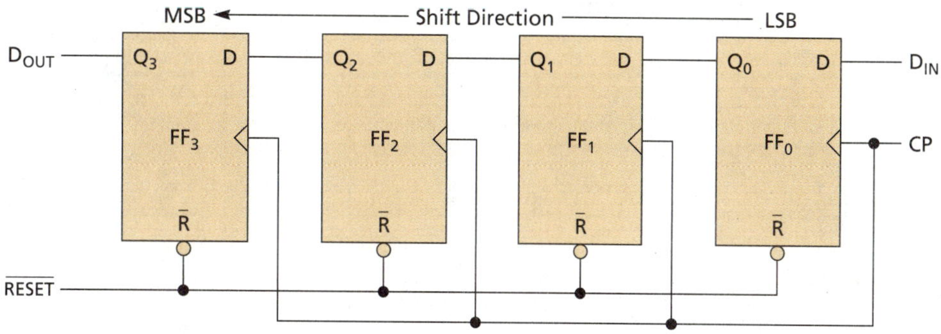

Figure 5.2 Shift Register

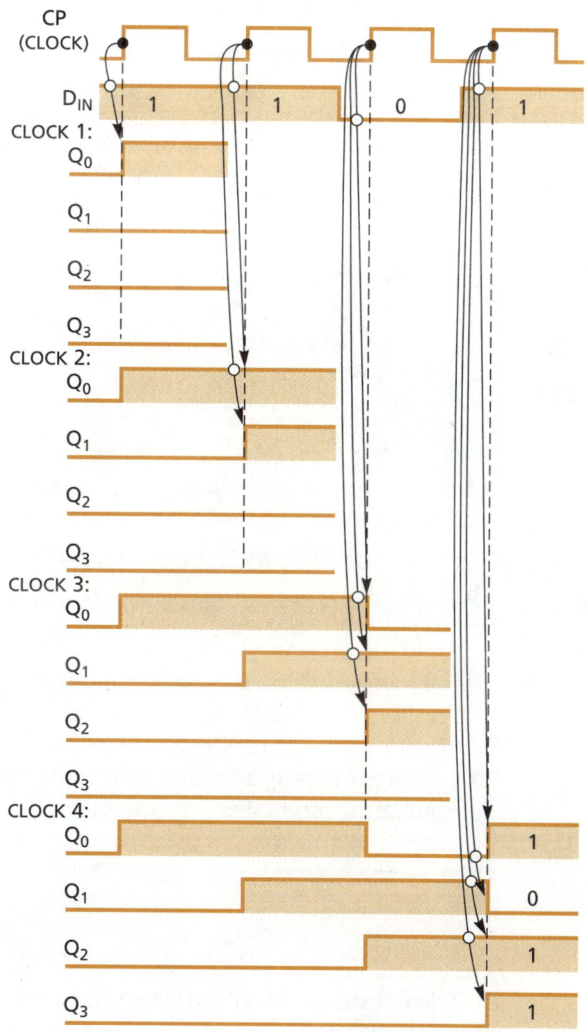

Figure 5.3 Shifting 1101B into a Four-Bit Register

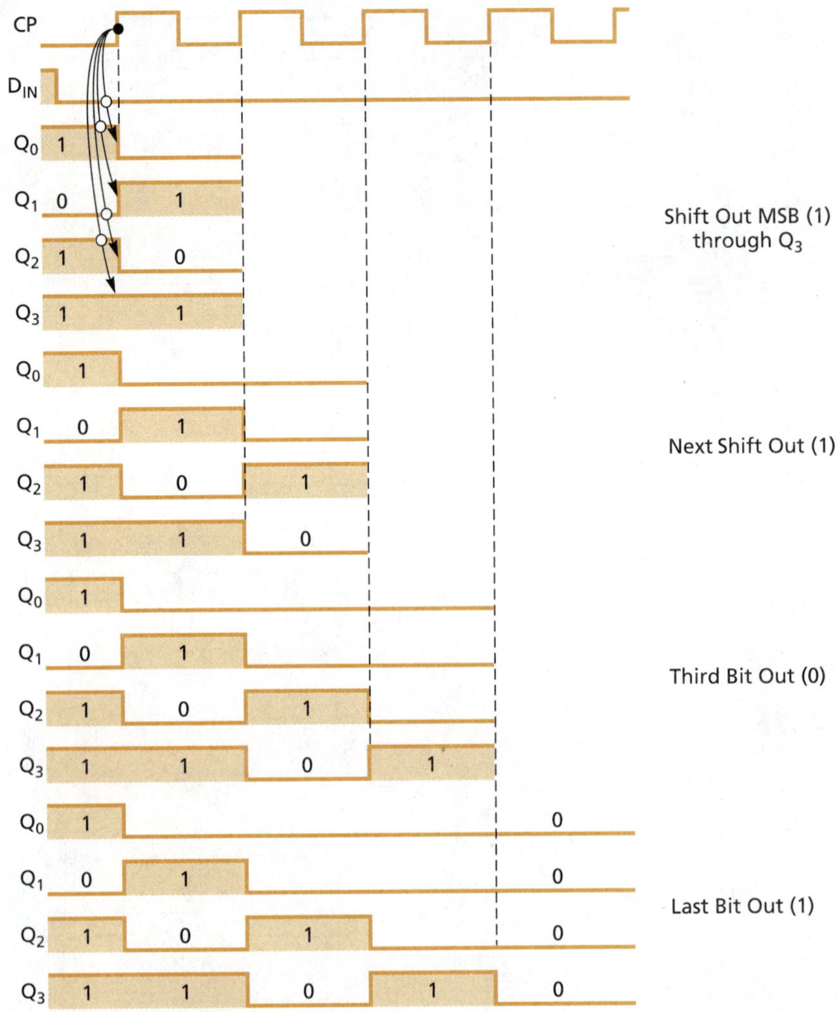

Figure 5.4 **Shifting Data Out**

The size of the data word is not significant when moving data serially, but the amount of data that can be held by the register at any instance depends on the number of flip-flops in the register. For example, an eight-bit data word can be shifted through the four-bit register of Figure 5.2, but only half of the word (four bits) can be held at any given time.

EXAMPLE 5.2

The data word 10011101B is shifted through the four-bit register of Figure 5.2, starting with the most significant bit first. What are the contents of that register after six clock cycles?

SOLUTION Six bits, 100111B, are shifted into the shift register after six clock cycles. Two of those bits are also shifted out. The remaining four bits, 0111B, are held in the register until the shift process is continued.

Shift left *the serial movement of data from the least significant bit position in a register toward the most significant bit.*

Shift right *the serial movement of data from the most significant bit position in a register toward the least significant bit.*

Shift registers are defined by the direction of data movement through them. A **shift-left** register shifts serial data starting at the least significant bit (LSB) flip-flop and shifting toward the most significant bit (MSB) flip-flop. The shift direction is not based on the physical arrangement of the flip-flops in the register, but rather on the relationship of the position of the bits of binary data with respect to a numerical or character code equivalent of the data. In other words, the LSB of the data is determined by how the data are to be used. For instance, the LSB of the binary word 10011010B is on the right if this binary word is interpreted as a number. The LSB could be on the left if the interpretation of the word is a code instead of a number.

To **shift right**, the process is reversed. Data are entered in through the MSB position and shifted toward the LSB position. The shift register circuits that implement this are shown in Figure 5.5. Notice that they are very similar to each other; only the connections between the Q outputs and D inputs differ. In the shift-right register of Figure 5.5b, data are entered into FF_3, whose output,

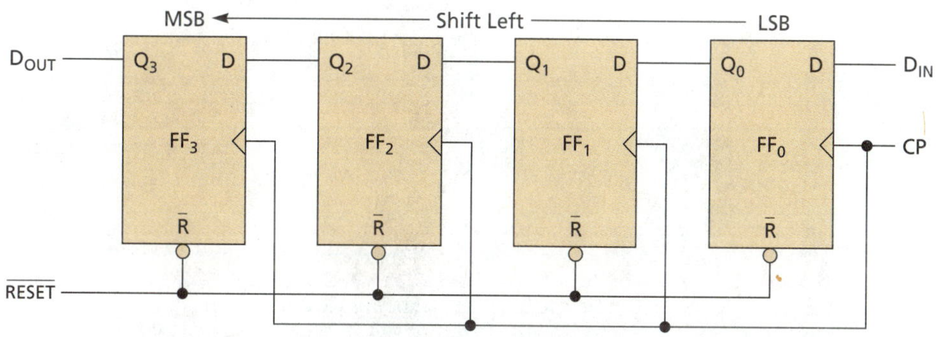

(a) Shift left

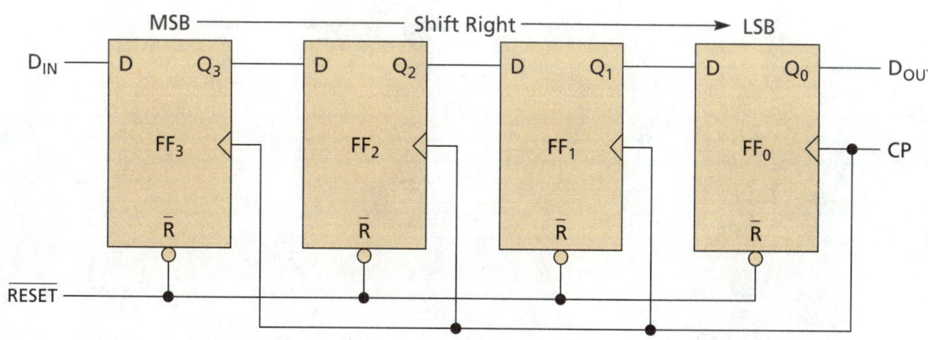

(b) Shift right

Figure **5.5** Shift Registers

Q_3, is connected to D_2 so that the state of FF_3 is shifted right to FF_2 on succeeding clock cycles. The same follows for the remaining flip-flops in the circuit. The shift-left register in Figure 5.5a shows data entering through the LSB, FF_0, and shift left into FF_1, FF_2, and out FF_3 on succeeding clock pulses.

The registers discussed thus far are loaded by pulsing the CP (clock pulse) line. That is, the register is updated each time CP is triggered. As with counters, using the CP input to control this activity invites possible difficulties with unexpected, accidental clocking of the CP line due to noise or other unwanted signal. It is preferred to apply a constant CLOCK signal to CP and to use other control inputs to manage the shift activity. The logic used to control when a reg-

Section 5.2

Practice Problems

1. Complete the timing diagram in Figure 5.6 for the shift-left register shown in Figure 5.5.

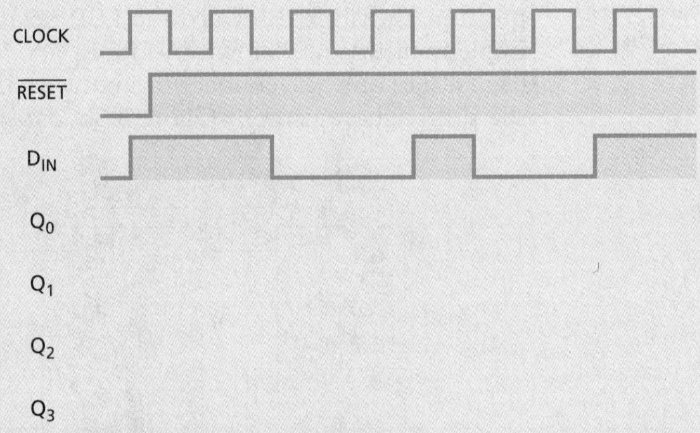

Figure 5.6 Practice Problem 5.2(1)

2. Complete the timing diagram in Figure 5.7 for the shift-right register shown in Figure 5.5.

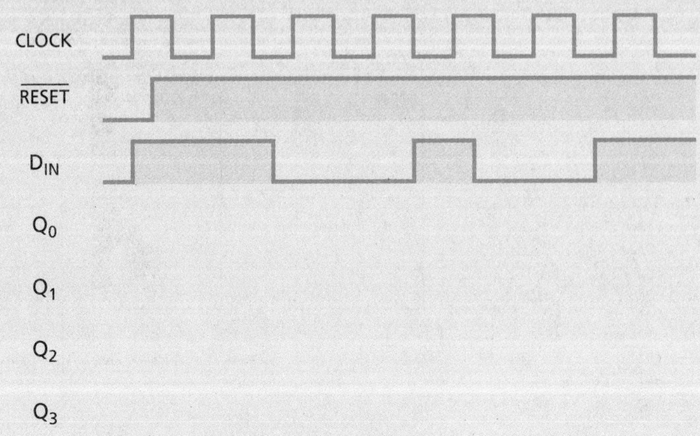

Figure 5.7 Practice Problem 5.2(2)

ister loads or shifts is formed by a combination of logic circuits called *decoders* and *multiplexers*.

5.3 Decoders and Encoders

Decoder *a logic circuit that translates binary codes into individual outputs.*

Encoder *a logic circuit that translates numerous individual inputs into a binary code.*

Two combinational logic circuits are used to translate between binary numbers and single signals. They are **decoders,** which generate a single signal for each binary combination placed at its inputs: and **encoders,** which reverse the process and yield a binary number or code for each input signal. These circuits are helpful in interpreting binary numbers and in causing an action to occur based on a specific binary value.

Figure 5.8 illustrates the basic circuitry for a two-bit decoder. Notice that this decoder has two inputs and four outputs. Only one of the outputs is active for any given input combination. Thus decoders are often designated by their decoding relationship of input versus output. In the two-bit example, this decoder is known as a 2-to-4 decoder because of the two-bit input and four possible outputs. Since only one of those four outputs is active at any given time, it is also known as a 1-of-4 decoder. The basic circuit for the decoder is the AND gate. The two binary bits and their complements supply the inputs to the four AND gates. Two-bit inputs supply four different possible binary input values, 00B, 01B, 10B, and 11B. Each AND gate receives a unique pair of inputs. For instance, applying 00B to inputs A and B causes two 1's from the inverters to be applied to gate 1. Each of the three other gates receives at least one 0 input. As a result, the only gate with a high output is gate 1—the three other gate outputs are forced low. We have decoded the binary input, 00B, into a single high output at gate 1.

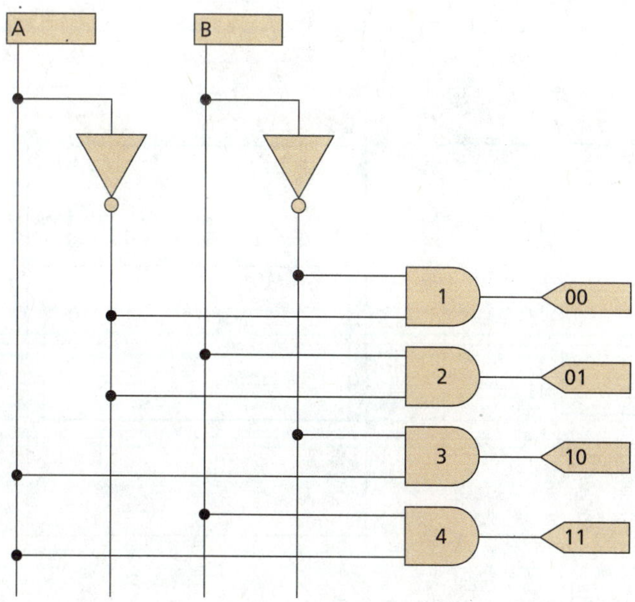

Figure 5.8 Two-bit Decoder

EXAMPLE 5.3

Show how the remaining three input combinations to the two-bit decoder also result in unique outputs.

SOLUTION As each binary value combination is applied to the inputs of the two-bit decoder, only one AND gate will have 1's on both its inputs. The remaining three gates in each case will have at least one input 0, causing their outputs to be 0, while the one gate with both inputs 1 has a 1 output. Input 01B asserts gate 2 output high; 10B, gate 3; and 11B, gate 4.

Figure 5.9 shows a 74LS138 1-of-8 decoder. It has three binary inputs, labeled A_0, A_1, and A_2 and eight individual active low outputs, $\overline{O}_0$ to $\overline{O}_7$. Since NAND gates are used to do the actual decoding, active outputs are low while inactive outputs are high. The device also has three enabling inputs: one active high, E_2, and two active low, $\overline{E}_0$ and $\overline{E}_1$. All three enable inputs must be asserted in order for *any* output to be set low. Disabling any single enable input forces all eight outputs to remain inactive or high. The decoding action of the 74LS138 decoder works in the same manner as it did with the two-bit decoder, only on a larger scale. One of eight binary combinations from 000B to 111B is placed at the binary inputs, A_2, A_1, and A_0. These inputs and their complements supply the inputs to the NAND gates. Each NAND gate receives a unique input combination for each binary input, with only one NAND gate receiving all high inputs for any one binary combination. Thus one, and only one, NAND gate output goes low for any given input. The remaining NAND gates remain in a

FEATURES

- **Demultiplexing capability**
- **Multiple input enable for easy expansion**
- **Ideal for memory chip select decoding**
- **Direct replacement for Intel 3205**

PIN CONFIGURATION

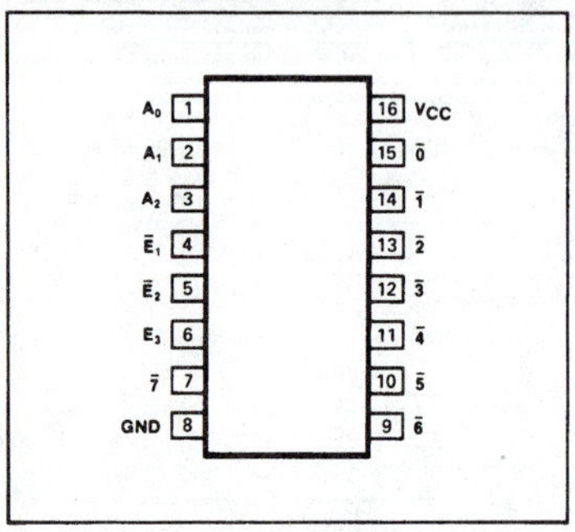

LOGIC SYMBOL

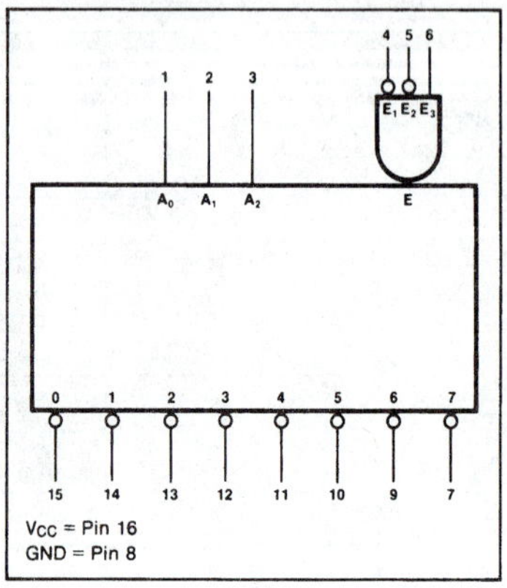

FUNCTIONAL DESCRIPTION

The "138" decoder accepts three binary weighted inputs (A_0,A_1,A_2) and when enabled provides eight mutually exclusive active LOW outputs ($\overline{0}$-$\overline{7}$). The device features three Enable inputs: two active LOW ($\overline{E}_1$,$\overline{E}_2$) and one active HIGH (E_3). Every output will be HIGH unless $\overline{E}_1$ and $\overline{E}_2$ are LOW and E_3 is HIGH. This multiple Enable function allows easy parallel expansion of the device to a 1-of-32 (5 lines to 32 lines) decoder with just four "138's" and one inverter.

The device can be used as an eight output demultiplexer by using one of the active LOW Enable inputs as the data input and the remaining Enable inputs as strobes. Enable inputs not used must be permanently tied to their appropriate active HIGH or active LOW state.

Figure 5.9 74LS138 Decoder/Demultiplexer

TRUTH TABLE

INPUTS						OUTPUTS							
$\overline{E}_1$	$\overline{E}_2$	E_3	A_0	A_1	A_2	$\overline{0}$	$\overline{1}$	$\overline{2}$	$\overline{3}$	$\overline{4}$	$\overline{5}$	$\overline{6}$	$\overline{7}$
H	X	X	X	X	X	H	H	H	H	H	H	H	H
X	H	X	X	X	X	H	H	H	H	H	H	H	H
X	X	L	X	X	X	H	H	H	H	H	H	H	H
L	L	H	L	L	L	L	H	H	H	H	H	H	H
L	L	H	H	L	L	H	L	H	H	H	H	H	H
L	L	H	L	H	L	H	H	L	H	H	H	H	H
L	L	H	H	H	L	H	H	H	L	H	H	H	H
L	L	H	L	L	H	H	H	H	H	L	H	H	H
L	L	H	H	L	H	H	H	H	H	H	L	H	H
L	L	H	L	H	H	H	H	H	H	H	H	L	H
L	L	H	H	H	H	H	H	H	H	H	H	H	L

NOTES

H = HIGH voltage level
L = LOW voltage level
X = Don't care

LOGIC DIAGRAM

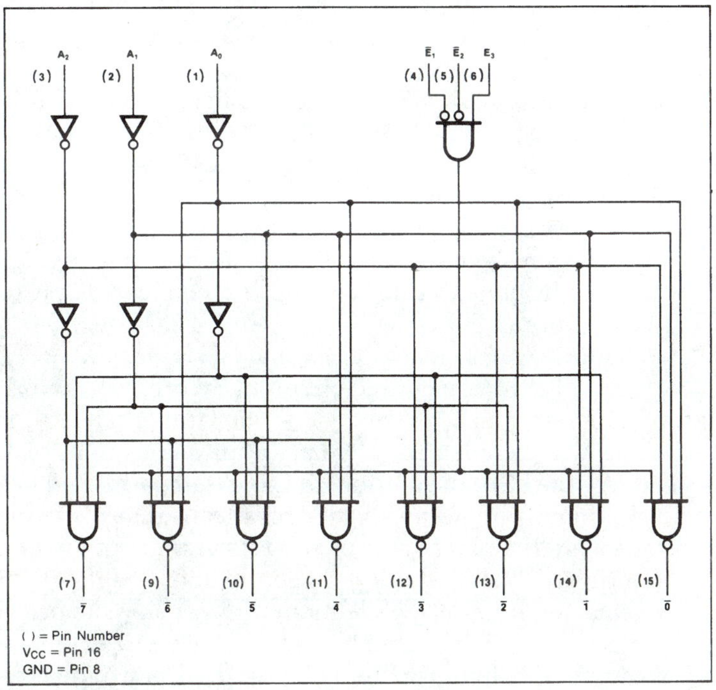

Figure 5.9 Continued

high state. Figure 5.9 includes a pin diagram for the 74LS138 and a truth table that summarizes the decoding process.

Encoders reverse the process of decoders. These circuits interpret single-signal inputs and generate binary outputs for each asserted input. A two-bit encoder, which has four individual input signals and two binary outputs, is shown in Figure 5.10. Like the two-bit decoder, the encoder is designated by the relationship between its inputs and outputs; in this case, a 4-to-2 encoder. The heart of an encoder is the OR gate, which asserts its output to be high whenever one of its inputs is set high. There is one OR gate for each binary

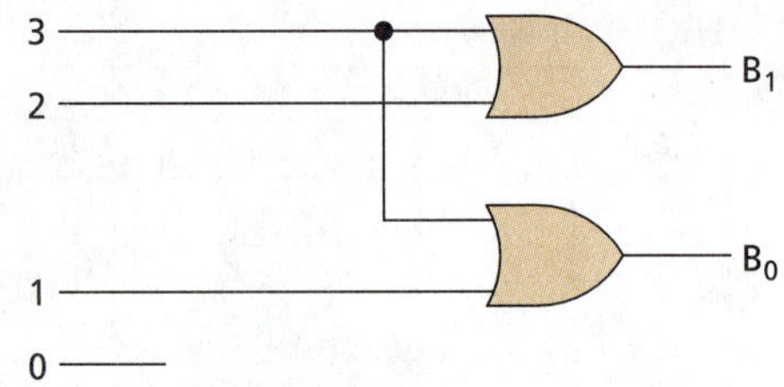

Figure 5.10 Two-bit Encoder

output. The inputs are connected in such a way that when an input is applied, the corresponding binary outputs go high. In Figure 5.10, the inputs are designated 0, 1, 2, and 3, which is the decimal equivalent that the binary outputs generated for each input. Outputs are labeled B_0 and B_1, with B_0 as the LSB of the two-bit binary output. When input line 0 is asserted, the binary output is expected to be 00, so this input line is not connected to either OR gate. Only one input is to be set high at any given time. For instance, when line 0 is asserted high, the other three lines must be held low. Keep in mind that the purpose of this circuit is to convert a single signal input into a binary value. For this encoder, setting line 0 high leaves both OR gate outputs low, because line 0 is not connected to either gate. Thus input 0 produces the binary number 00B as expected.

Setting line 1 high causes B_0 to go to a 1 and B_1 to stay at 0, because line 1 is only connected to the B_0 OR gate. Considering B_0 as the LSB, this relates input 1 to the binary value 01B. Input 2, connected only to the B_1 OR gate causes the output to be a binary 10B when it is asserted. Finally, input 3, connected to both OR gates, generates a binary 11B output.

The 74LS148 integrated circuit of Figure 5.11 is a priority encoder. It is first and foremost an 8-to-3 encoder, translating active low inputs $\overline{I_0}$ to $\overline{I_7}$ into their respective complemented binary equivalent values of 111B to 000B at $\overline{A_2}$, $\overline{A_1}$, and $\overline{A_0}$, the LSB. In addition, the inputs are arranged in ascending priority with $\overline{I_7}$ having the highest priority. This means that if two inputs are applied simultaneously, the binary output will reflect the input with the higher binary value or priority. Two lines are provided to allow multiple 74LS148s to be cascaded (connected in series) while maintaining complete priority. They are the active low enable-in ($\overline{EI}$) and enable out ($\overline{EO}$) pins. To let the binary output correspond to any of the inputs, $\overline{EI}$ must be asserted low. Whenever one of the inputs are applied and $\overline{EI}$ is asserted, $\overline{EO}$ is negated or set high. If $\overline{EI}$ is set low and none of the inputs are asserted, then $\overline{EO}$ is forced to an active low state. Figure 5.12 shows how three 74LS148's are connected in cascade fashion.

If $\overline{EI}$ of IC_2 is asserted and, assuming that no input is selected, then its $\overline{EO}$ is low, causing IC_1 $\overline{EI}$ to be asserted. In turn, its $\overline{EO}$ is forced low, making IC_0 $\overline{EI}$ active and, in turn, its $\overline{EO}$ low. The $\overline{GS}$ outputs from the three encoders signify when an output from that encoder is valid. The $\overline{GS}$ line is high whenever none of the inputs are selected and goes low whenever one of the inputs and the $\overline{EI}$ line are active. What is the result of enabling the circuit in Figure 5.12 and applying a low to each of the $\overline{I_3}$ inputs of the three encoders?

With the $\overline{EI}$ signal on, applying a low to input $\overline{I_3}$ of IC_2 causes the three binary outputs from this encoder to be 100B (the complement of 011B, or 3) and

FUNCTIONAL DESCRIPTION

The "148" 8-input priority encoder accepts data from eight active LOW inputs and provides a binary representation on the three active LOW outputs. A priority is assigned to each input so that when two or more inputs are simultaneously active, the input with the highest priority is represented on the output, with input line $\overline{I_7}$ having the highest priority.

A HIGH on the Input Enable ($\overline{EI}$) will force all outputs to the inactive (HIGH) state and allow new data to settle without producing erroneous information at the outputs.

A Group Signal output ($\overline{GS}$) and an Enable Output ($\overline{EO}$) are provided with the three data outputs. The $\overline{GS}$ is active level LOW when any input is LOW; this indicates when any input is active. The $\overline{EO}$ is active level LOW when all inputs are HIGH. Using the output enable along with the input enable allows priority coding of N input signals. Both $\overline{EO}$ and $\overline{GS}$ are active HIGH when the input enable is HIGH.

TRUTH TABLE

INPUTS									OUTPUTS				
$\overline{EI}$	$\overline{I_0}$	$\overline{I_1}$	$\overline{I_2}$	$\overline{I_3}$	$\overline{I_4}$	$\overline{I_5}$	$\overline{I_6}$	$\overline{I_7}$	$\overline{GS}$	$\overline{A_0}$	$\overline{A_1}$	$\overline{A_2}$	$\overline{EO}$
H	X	X	X	X	X	X	X	X	H	H	H	H	H
L	H	H	H	H	H	H	H	H	H	H	H	H	L
L	X	X	X	X	X	X	X	L	L	L	L	L	H
L	X	X	X	X	X	X	L	H	L	H	L	L	H
L	X	X	X	X	X	L	H	H	L	L	H	L	H
L	X	X	X	X	L	H	H	H	L	H	H	L	H
L	X	X	X	L	H	H	H	H	L	L	L	H	H
L	X	X	L	H	H	H	H	H	L	H	L	H	H
L	X	L	H	H	H	H	H	H	L	L	H	H	H
L	L	H	H	H	H	H	H	H	L	H	H	H	H

H = HIGH voltage level
L = LOW voltage level
X = Don't care

LOGIC DIAGRAM

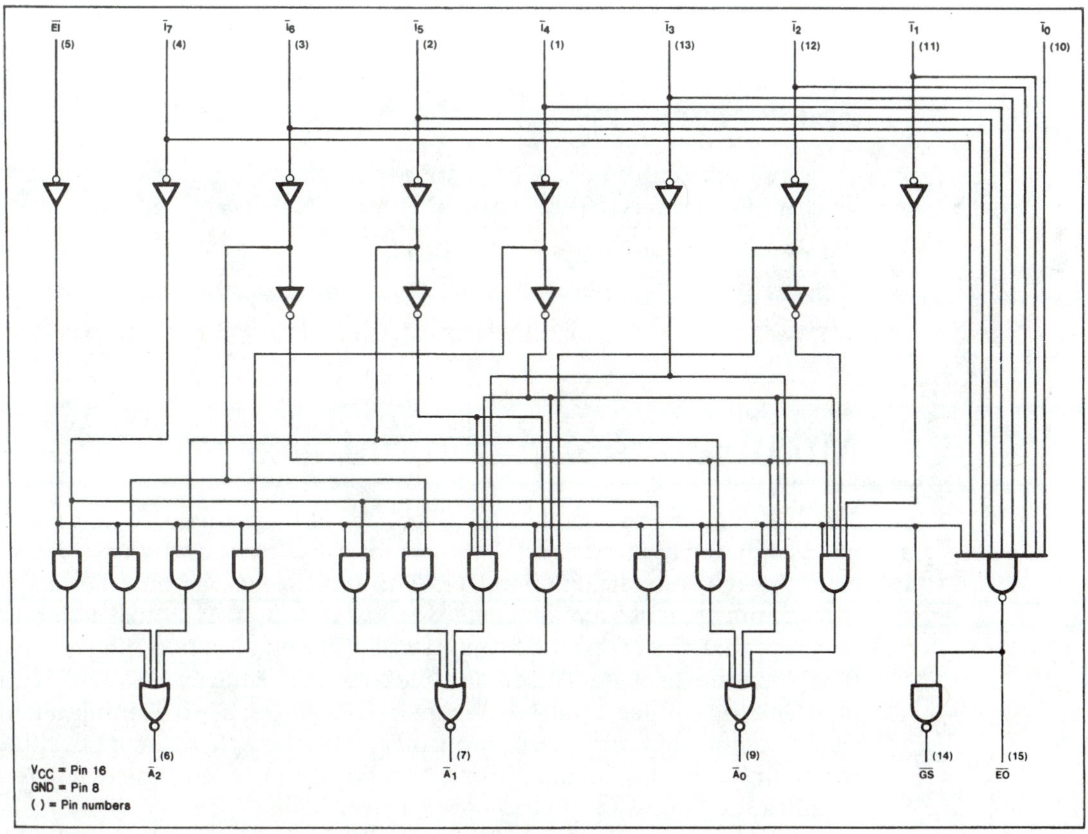

Figure 5.11 74LS148 Priority Encoder

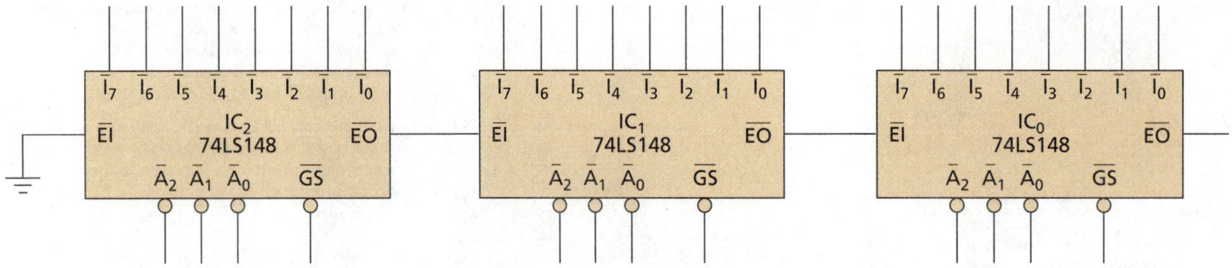

Figure 5.12 Cascaded Priority Encoders

forces its $\overline{EO}$ line high. Nothing else changes, because this high is applied to the $\overline{EI}$ input of the next encoder and propagated through to the $\overline{EI}$ of the third encoder, which disables both encoders. While $\overline{I_3}$ of IC_2 remains low, applying a signal to $\overline{I_3}$ of IC_1, which has a lower priority than any input of IC_2, has no effect on IC_1's outputs. When $\overline{I_3}$ of the first encoder is set inactive, it causes $\overline{EO}$ of IC_2 to go low, enabling IC_1. This allows its outputs to respond to the input on its $\overline{I_3}$, resulting in an output of 100B from IC_1. IC_2 at this point now has an output of 111B, reflecting no asserted inputs, and IC_0 shows 111B also, because its $\overline{EI}$ is held high by IC_1's EO. To get 100B from IC_0, all the inputs to IC_1 and IC_2 must go inactive.

Section 5.3

Practice Problems

1. Draw the circuit to decode the following binary numbers into individual active low signals: 0011B, 01010B, 1000B, 1010B, and 1111B.
2. Draw the encoder that translates eight inputs, which are asserted individually, into three active high binary outputs.

5.4 Multiplexers and Demultiplexers

Multiplexer (mux) *a logic circuit that selects an output from many inputs.*

Demultiplexer (demux) *a logic circuit that channels an input into one of many outputs.*

Just as decoders and encoders have very specific jobs to do, so do another type of combinational circuit called a **multiplexer (mux)** and its companion circuit, the **demultiplexer (demux)**. The multiplexer or mux selects among many inputs and delivers its selection to its output. It is similar to a busy traffic artery that accepts traffic from several different directions and channels the flow into a single route. The demultiplexer reverses the process by directing an input into one of many outputs. Using traffic control as an example again, consider a freeway interchange. Coming off one freeway, there are many other freeway entry ramps to choose from depending on where you want to go. A demultiplexer does this type of control electrically.

The 74LS153 of Figure 5.13 is a dual four-line to one-line multiplexer. It has separate enables ($\overline{E_a}$ and $\overline{E_b}$) for each of the two muxes. There are four

FEATURES

- **Non-inverting outputs**
- **Separate Enable for each section**
- **Common Select inputs**
- **See "253" for 3-State version**

FUNCTIONAL DESCRIPTION

The "153" is a Dual 4-input Multiplexer that can select two bits of data from up to four sources under control of the common Select inputs (S_0, S_1). The two 4-input multiplexer circuits have individual active LOW Enables ($\overline{E}_a$, $\overline{E}_b$) which can be used to strobe the outputs independently. Outputs (Y_a, Y_b) are forced LOW when the corresponding Enables ($\overline{E}_a$, $\overline{E}_b$) are HIGH.

The device is the logical implementation of a 2-pole, 4-position switch, where the position of the switch is determined by the logic levels supplied to the two Select inputs. The logic equations for the outputs are shown below.

$$Y_a = \overline{E}_a \bullet (I_{0a} \bullet \overline{S}_1 \bullet \overline{S}_0 + I_{1a} \bullet \overline{S}_1 \bullet S_0 + I_{2a} \bullet S_1 \bullet \overline{S}_0 + I_{3a} \bullet S_1 \bullet S_0)$$

$$Y_b = \overline{E}_b \bullet (I_{0b} \bullet \overline{S}_1 \bullet \overline{S}_0 + I_{1b} \bullet \overline{S}_1 \bullet S_0 + I_{2b} \bullet S_1 \bullet \overline{S}_0 + I_{3b} \bullet S_1 \bullet S_0)$$

The "153" can be used to move data to a common output bus from a group of registers. The state of the Select inputs would determine the particular register from which the data came. An alternative application is as a function generator. The device can generate two functions of three variables. This is useful for implementing highly irregular random logic.

PIN CONFIGURATION

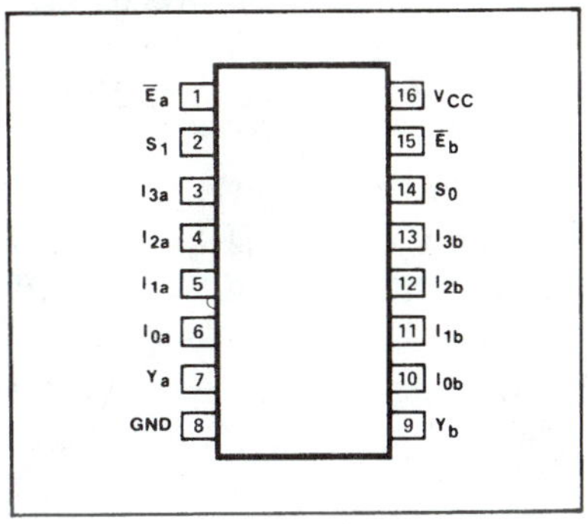

LOGIC SYMBOL

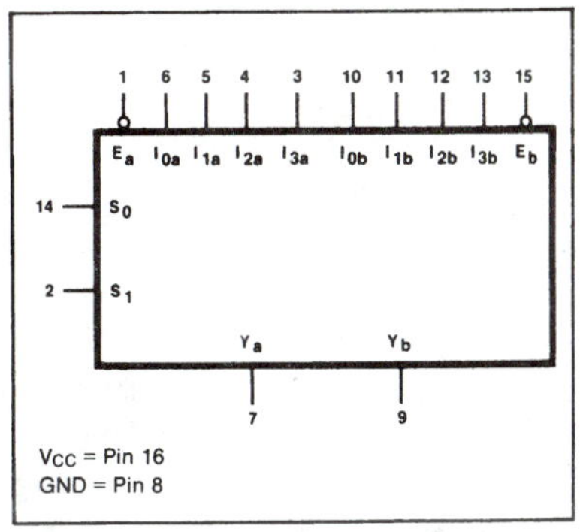

LOGIC DIAGRAM

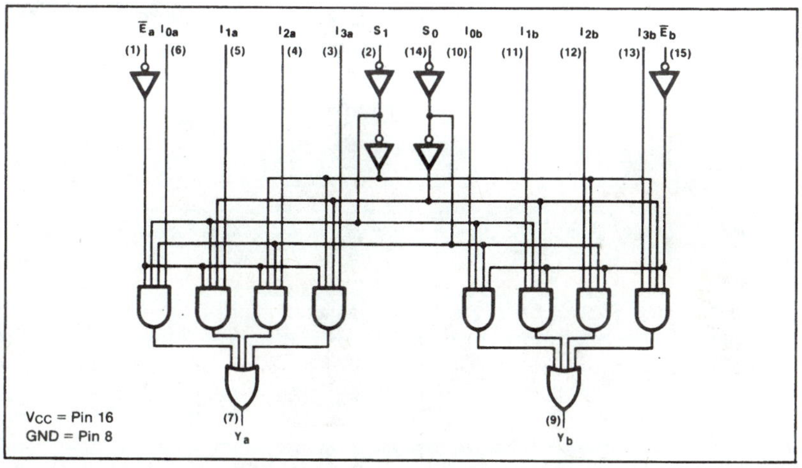

TRUTH TABLE

SELECT INPUTS		INPUTS (a or b)					OUTPUT
S_0	S_1	$\overline{E}$	I_0	I_1	I_2	I_3	Y
X	X	H	X	X	X	X	L
L	L	L	L	X	X	X	L
L	L	L	H	X	X	X	H
H	L	L	X	L	X	X	L
H	L	L	X	H	X	X	H
L	H	L	X	X	L	X	L
L	H	L	X	X	H	X	H
H	H	L	X	X	X	L	L
H	H	L	X	X	X	H	H

Figure 5.13 74LS153 Dual Four-Line Multiplexer

individual input lines (I_{0a} to I_{3a} and I_{0b} to I_{3b}) for each mux and separate output lines (Y_a and Y_b). The only lines shared in common to the two muxes are the control lines, S_0 and S_1, that determine which of the four inputs are to appear at the output. By applying the equivalent binary value to the S inputs, the corresponding I input is sent to the enabled mux's Y output. For example, let's apply 1001B to I_{3a} through I_{0a} and 1010B to I_{3b} through I_{0b}. When $\overline{E_a}$ and $\overline{E_b}$ are brought low to assert both enables, both mux sections are enabled. We can now determine what appears at the Y outputs for each of the four combinations applied to the S inputs of the 74LS153.

When both S inputs are low, inputs I_0 of both muxes appear at the Y outputs. For this example, since I_{0a} is high and I_{0b} is low, then $Y_a = 1$ and $Y_b = 0$. The next input combination of $S_1 = 0$ and $S_0 = 1$ selects the I_{1a} and I_{1b} inputs, causing $Y_a = 0$ and $Y_b = 1$. Setting $S_1 = 1$ and $S_0 = 0$ causes Y_a and Y_b both to be low by connecting the I_{2a} and I_{2b} inputs to the outputs. The last combination, setting both S inputs high, selects the I_3 inputs to appear at the output, forcing both Y outputs high.

The reverse function, the demultiplexer, is illustrated in Figure 5.14. The select inputs, S_1 and S_0, select which of the four AND gates will pass the state of the IN signal to its output. For instance, if S_1 and S_0 are both set low, then two of the three inputs to the lower AND gate are high and its output is totally dependent on the state of the IN signal. The other three AND gates are held low, because one or both of their select inputs are low. Thus we have directed the IN signal to appear at the Y_A output.

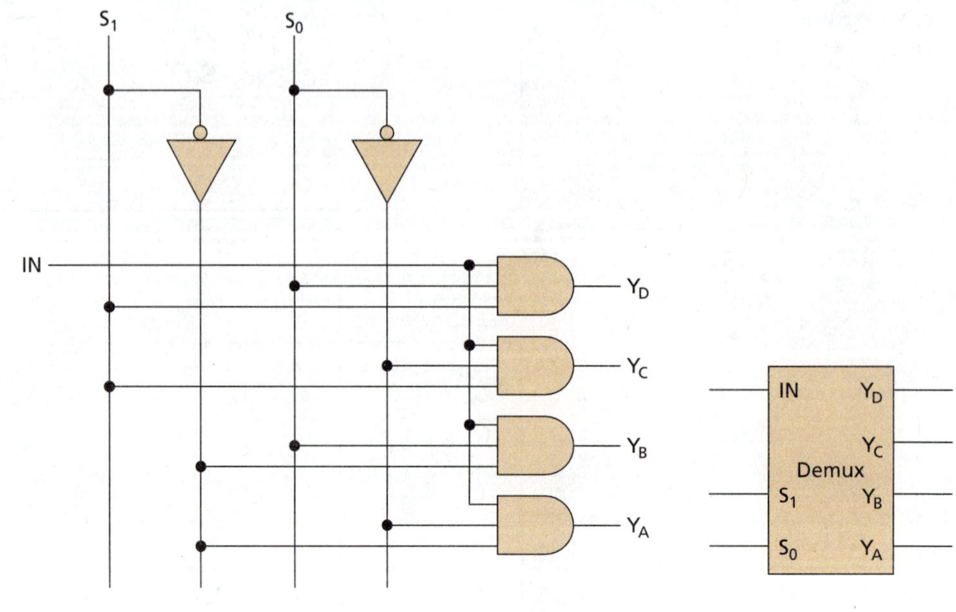

Figure 5.14 A 1-to-4-Demultiplexer

Application Note

Applications for Muxes and Demuxes

A common use of the multiplexer and demultiplexer is in digital and data communications. In some applications, many data channels are multiplexed onto a single channel in a timed sequence. They are then transported through any number of means along this single channel to a distant receiver, where they are demultiplexed and sent along separate channels to their ultimate destination. The advantage of this type of system is to reduce the central traffic path to a single channel carrier instead of numerous lines, one for each channel.

Section 5.4

Practice Problems

1. The following signals are applied to a 74LS153:
   I_{3a} to I_{0a} = 1100 I_{3b} to I_{0b} = 0110 $\overline{E_a} = 0$ $\overline{E_b} = 1$
   $S_1 = 1$ $S_0 = 1$
   What appears at both outputs and why?
2. The following signals are applied to the circuit of Figure 5.14: IN = 1, $S_1 = 1$, and $S_0 = 0$. What appears at the outputs, and why?

5.5 Controlled Registers

The parallel-load register of Figure 5.15 is controlled by the use of a 1-of-2 decoder and a two-input mux. The decoder is formed using the LOAD input, and the single inverter gate is attached to that input. The two decoded outputs are LOAD when LOAD is set high and REFRESH when LOAD is set low. These two signals are applied to each mux feeding the D inputs of the register. The parallel inputs D_0 to D_3 are fed to the AND gates of the muxes that are enabled by the LOAD signal. Q outputs of the D flip-flops are returned to the D inputs through the AND gates of the muxes when enabled by the REFRESH signal. The clock input is permanently connected to the CP inputs of each flip-flop in the register. By applying a constant CLOCK signal to the CP inputs, we eliminate the concern about noise on that line. Here's how the register works.

Start by applying data to the D_3–D_0 inputs—let's use 1010B as an example data input. Asserting the LOAD signal high turns on each AND gate connected to the D_n inputs, transferring the data 1010B, through each AND gate and OR gate to the D inputs of the flip-flops. On the next clock pulse following the enabling of the LOAD signal, the data are written into the flip-flops. Negating the LOAD signal shuts off the LOAD from the decoder and turns on the REFRESH line. This allows the outputs of each flip-flop to be returned to

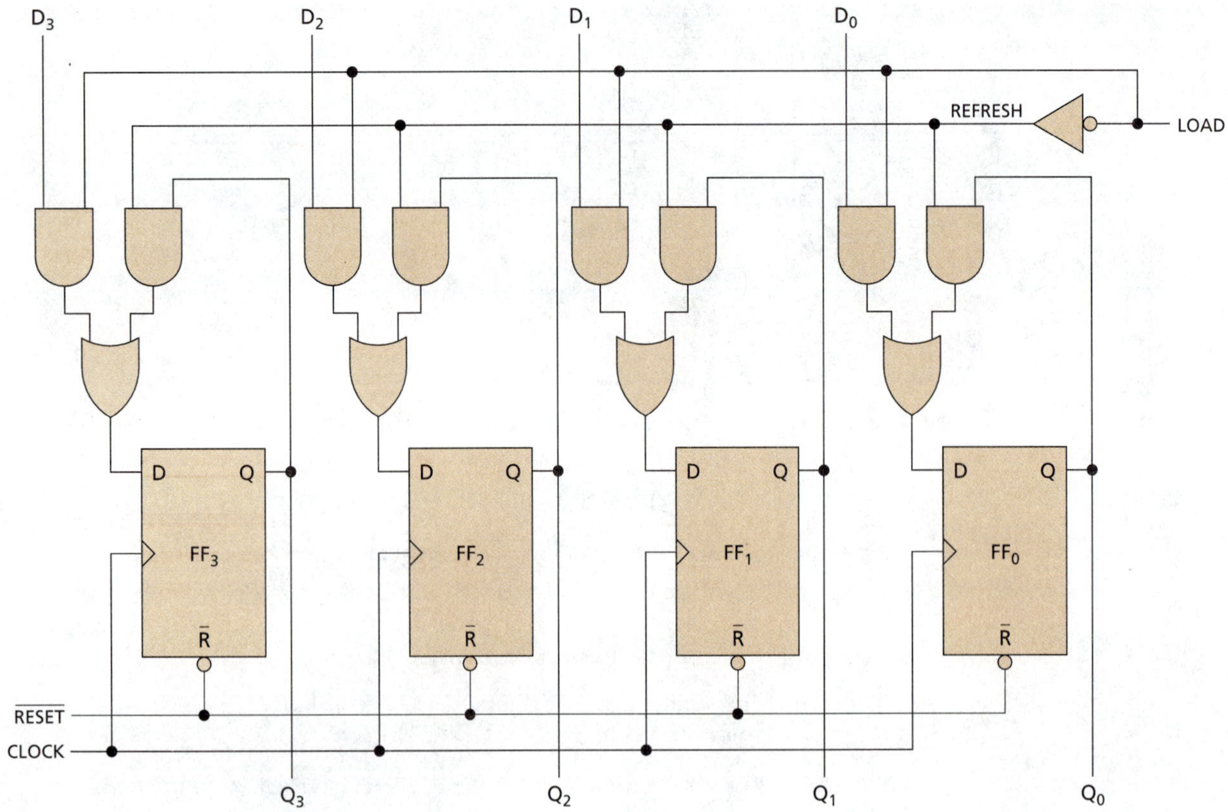

Figure **5.15** Controlled Parallel In/Out Register

Refreshing *the process of rewriting data into a flip-flop or memory to retain those data.*

each D input, thus keeping the same data in the register by rewriting or **refreshing** the data continually until LOAD is reasserted and new data are written into the register.

EXAMPLE 5.4

Explain how the data word 1100B is written into the controlled parallel register of Figure 5.15.

SOLUTION The four-bit binary word is placed onto D_3 to D_0. When LOAD is brought high, the inverter gate forces a low to the REFRESH inputs of each AND gate whose other input is connected to the individual Q output of each flip-flop. This causes each of these AND gates to output a 0 to their respective OR gate. The high on LOAD, itself, enables the LOAD AND gates on, causing each of their outputs to go to the state of each D_n line ($D_n \cdot 1 = D_n$). The OR gates, in turn, pass the outputs from the second AND gates to the D inputs of the flip-flops ($D_n + 0 = D_n$). On the next clock transition, the state of each D input is latched into each D flip-flop and appears at each Q output. Negating

LOAD after the data have been written preserves those data until the next word is loaded in.

The circuit in Figure 5.15 is the heart of the 74LS173 quad D flip-flop in Figure 5.16. The LOAD input has been changed to include two active low enabling signals $\overline{IE}_1$ and $\overline{IE}_2$, which are logically combined through an AND function to produce the LOAD signal. In this case, both input enables must be low in order for the data on the D_n input lines to be latched into the D flip-flops. A second enhancement in the 74LS173 is the addition of three-state buffers between the Q outputs of the D flip-flops and the output of Q_n lines. Three-state buffers are devices that pass digital information (1 and 0 as two of the three states of the buffers' output) when enabled and block data when disabled. In the disabled state, the output appears as if it were disconnected or floating—neither a 1 or a 0. These buffers are controlled by two output enables, $\overline{OE}_1$ and $\overline{OE}_2$, both of which must be brought low to

FEATURES

- **Edge triggered D-Type register**
- **Gated input enable for "hold" do nothing mode**
- **3-State output buffers**
- **Gated output enable control**
- **Pin compatible with the 8T10 and DM8551**

FUNCTIONAL DESCRIPTION

The "173" is a 4-Bit Parallel Load Register with clock enable control, 3-State buffered outputs and master reset. When the two Clock Enable ($\bar{E}_1$ and $\bar{E}_2$) inputs are LOW, the data on the D inputs is loaded into the register synchronously with the LOW-to-HIGH Clock (CP) transition. When one or both $\bar{E}$ inputs are HIGH one setup time before the LOW-to-HIGH clock transition, the register will retain the previous data. The Data inputs and Clock Enable inputs are fully edge-triggered and must be stable only one setup time before the LOW-to-HIGH clock transition.

MODE SELECT—FUNCTION TABLE

REGISTER OPERATING MODES	INPUTS					OUTPUTS
	MR	CP	$\bar{E}_1$	$\bar{E}_2$	D_n	Q_n (Register)
Reset (clear)	H	X	X	X	X	L
Parallel Load	L	↑	l	l	l	L
	L	↑	l	l	h	H
Hold (No change)	L	X	h	X	X	q_n
	L	X	X	h	X	q_n

3-STATE BUFFER OPERATING MODES	INPUTS			OUTPUTS
	Q_n (Register)	$\overline{OE}_1$	$\overline{OE}_2$	Q_0, Q_1, Q_2, Q_3
Read	L	L	L	L
	H	L	L	H
Disabled	X	H	X	(Z)
	X	X	H	(Z)

NOTES
H = HIGH voltage level
h = HIGH voltage level one setup time prior to the LOW-to-HIGH clock transition
L = LOW voltage level
l = LOW voltage level one setup time prior to the LOW-to-HIGH clock transition
q_n = Lower case letters indicate the state of the referenced input (or output) one setup time prior to the LOW-to-HIGH clock transition
X = Don't care
(Z) = High impedance "off" state
↑ = LOW-to-HIGH transition

The Master Reset (MR) is an active HIGh asynchronous input. When the Mr is HIGH all four flip-flops are reset (cleared) independently of any other input condition.

The 3-State output buffers are controlled by a 2-input NOR gate. When both Output Enable ($\overline{OE}_1$ and $\overline{OE}_2$) inputs are LOW, the data in the register is presented at the Q outputs. When one or both $\overline{OE}$ inputs is HIGH the outputs are forced to a high impedance "off" state. The 3-State output buffers are completely independent of the register operation; the $\overline{OE}$ transition do not affect the clock and reset operations.

Figure 5.16 74LS173 Quad-D-Type Flip-Flop (3-state)

LOGIC DIAGRAM

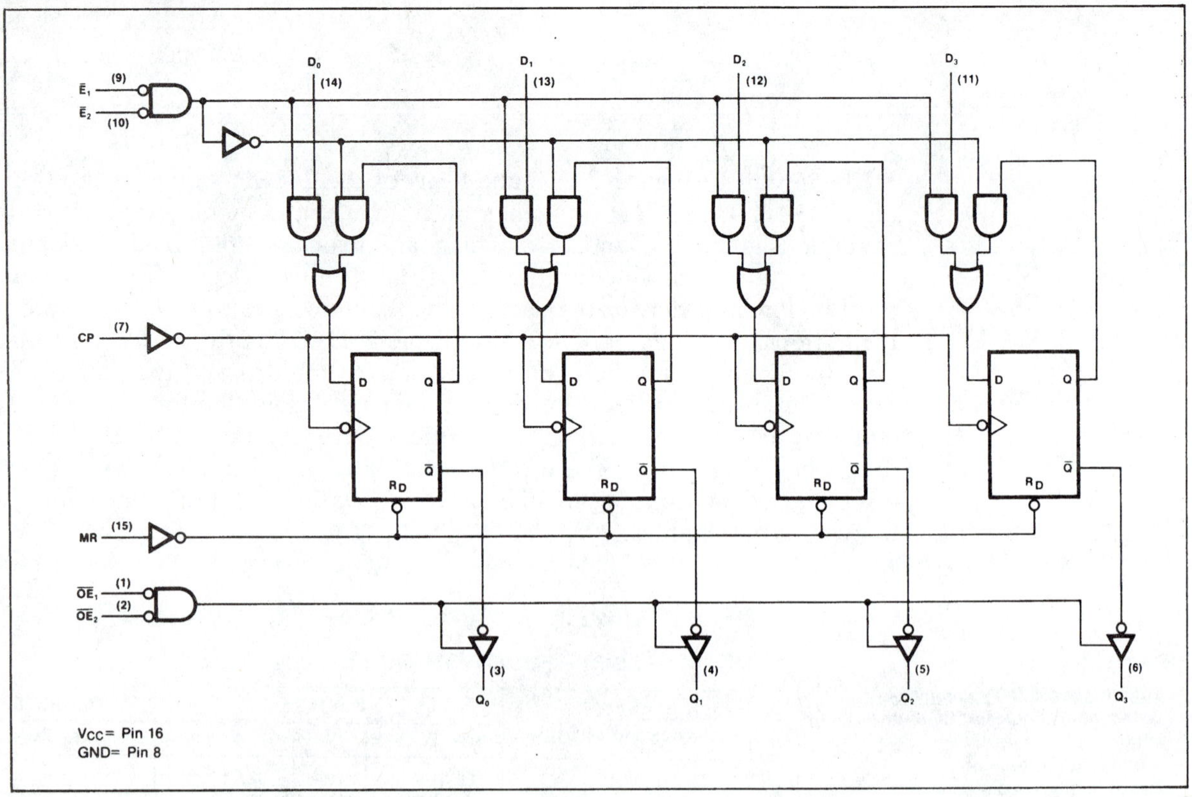

Figure **5.16** Continued

allow the outputs of the D flip-flops to appear on the Q_n lines. If either output enable is set high, the outputs of the buffers will be a high impedance, effectively disconnecting the 74LS173 from external lines connected to the Q_n outputs. The output enables let the 74LS173 be used in a bus environment. A **bus** is a set of parallel lines all of which carry similar types of digital information. For instance, a group of eight lines all transferring bits of a data word forms an eight-bit **data bus.** We'll discuss three-state buffers and buses further in later chapters.

Bus *a set of digital lines that carry the same type of binary information.*

Data bus *a set of electrical conductors that carry data information.*

Figure 5.17 shows a logical block representation for the 74LS173 quad-D flip-flop, along with a table that summarizes the operation of the chip. All four flip-flops are reset so that Q = 0 when the master reset (MR) input is asserted high. Remaining activity in the 74LS173 requires the MR line to be inactive or low. Placing data on the D_n input line and setting both input enables low causes the data to be written into the D flip-flops on the rising edge of the next CLOCK signal. Outputs may be read anytime by asserting both output enables low. Since the writing and reading of all four flip-flops are controlled at the same time, the 74LS173 is often used as a four-bit parallel in/out register.

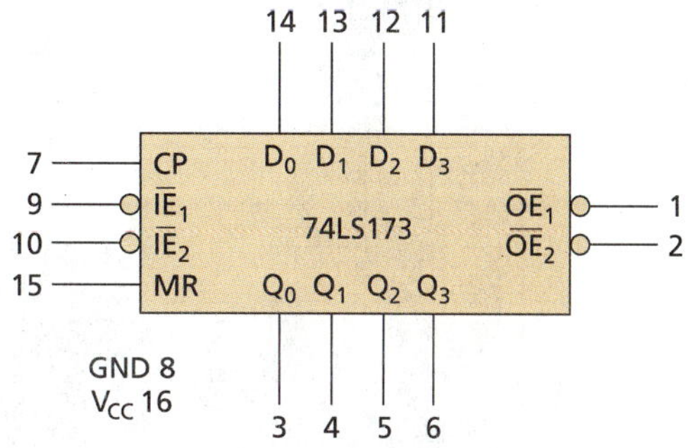

(a) Quad-D Logic Symbol

MR	$\overline{IE}_1$	$\overline{IE}_2$	$\overline{OE}_1$	$\overline{OE}_2$	D_0–D_3	CP	Q_0–Q_3
1	X	X	X	X	X	X	0 : Reset
0	X	X	X	1	X	X	High Impedance
0	X	X	1	X	X	X	High Impedance
0	X	X	0	0	X	X	Q out of D flip-flops
0	X	1	0	0	X	↑	No Change
0	1	X	0	0	X	↑	No Change
0	0	0	0	0	Data	↑	D_0–D_3 Data
0	0	0	0	0	X	Steady state	No Change

(b) Quad-D Truth Table

X = "don't care" state

Figure 5.17 74LS173 Quad-D Flip-Flop Datum

EXAMPLE 5.5

What steps are required to store the binary data 1100B into a single 74LS173 chip?

SOLUTION Place the data onto the D_n inputs. With a continuous clock applied to the CP input, bring both $\overline{IE}$ lines low. Each flip-flop will be set or reset, depending on each individual data input, on the next positive transition of the CLOCK signal. The data now stored into the register can be read anytime by bringing both output enables low.

Shift registers can be controlled in the same way that the parallel-load register is controlled. In Figure 5.18, a SHIFT control signal is used to determine when the register will shift and when it will hold the value presently entered into it, much in the same manner as used in the LOAD

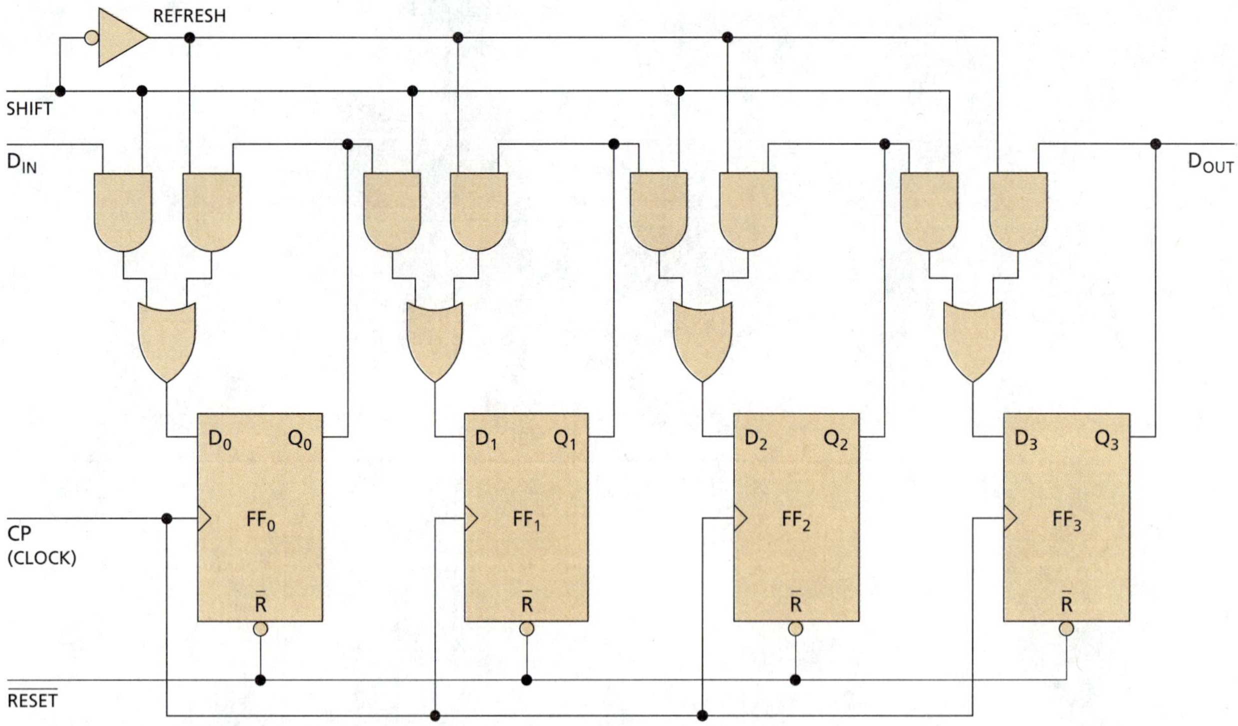

Figure 5.18 Shift-Left Register

register. Once again, the tried and true AND–OR multiplexer is put into use along with a 1-of-2 decoder (the inverter gate). D_{IN} is applied to the input of the first AND gate that is gated on with the SHIFT signal, which feeds an OR gate that derives the D input to the FF_0. Similarly, Q_0 is applied to the next AND gate that is gated on by the SHIFT input, which drives an OR gate connected to D_1 and so forth. A second REFRESH AND gate receives its input from the Q output of each flip-flop and returns that output state to the D input of each flip-flop when the SHIFT input is low. When the SHIFT signal is asserted high, the AND gates connecting the Q outputs of one flip-flop to the D inputs of the next flip-flop are gated on. The REFRESH AND gates that connect Q outputs to their own D inputs are gated off from the 0 on the REFRESH line connected to the inverter output. Thus, with SHIFT on the data are shifted left in the same manner as they were with the register in Figure 5.2, from the LSB toward the MSB. Negating SHIFT causes the AND gates connected to that signal to be gated off and turns on the REFRESH AND gates, which take the Q output of each gate and return their states to the D inputs, thereby preserving the current contents of each flip-flop.

EXAMPLE 5.6

Complete the timing diagram in Figure 5.19 for the controlled four-bit shift register of Figure 5.18.

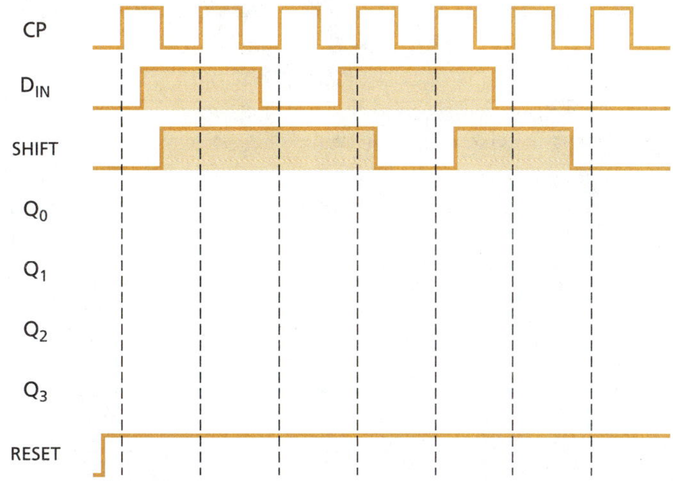

Figure 5.19 Shift Example 5.6 (problem)

SOLUTION As long as SHIFT is high, the shift register shifts data through on progressive clock cycles. When SHIFT is brought low, the shift process stops and the current contents of the register are retained.

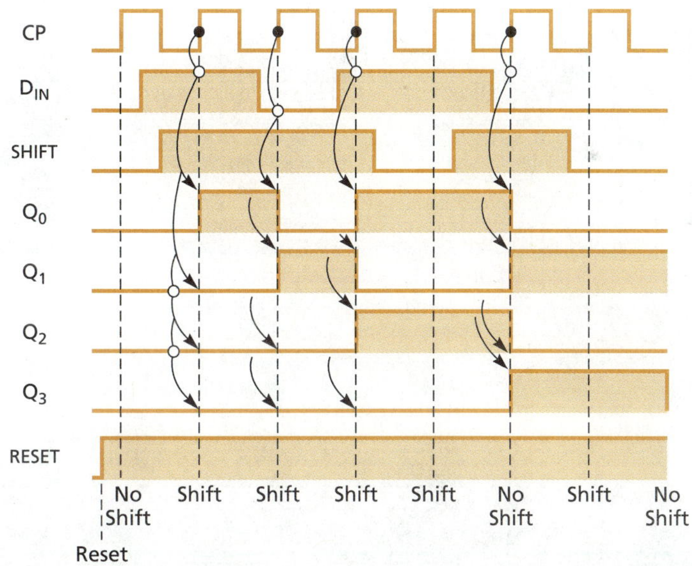

Figure 5.20 Shift Example 5.6 (solution)

The 74LS series has several register types that combine serial shift capabilities with parallel in/out functions. One such device is the 74LS95 pictured in Figure 5.21a. The control of this register is slightly different from that of the previous examples. An AND/OR mux is still used to select whether the circuit is to shift left if Q_0 is considered to be the least significant bit or shift right if Q_0 is used to hold the most significant bit, or parallel load. In addition to controlling which signal is applied to each D input, the selection to shift or load also controls which CLOCK signal is used. Separate clock inputs (CP_1 and CP_2) are available, one for the shift clock and one for the parallel-load clock. They may be different CLOCK signals, or they may be connected to the same clock source. A truth table summarizing the functions of the 74LS95 is included in Figure 5.21b.

One way to achieve a no-change condition, in which the register retains the current state, is to hold both CP inputs at a steady state. A second method for

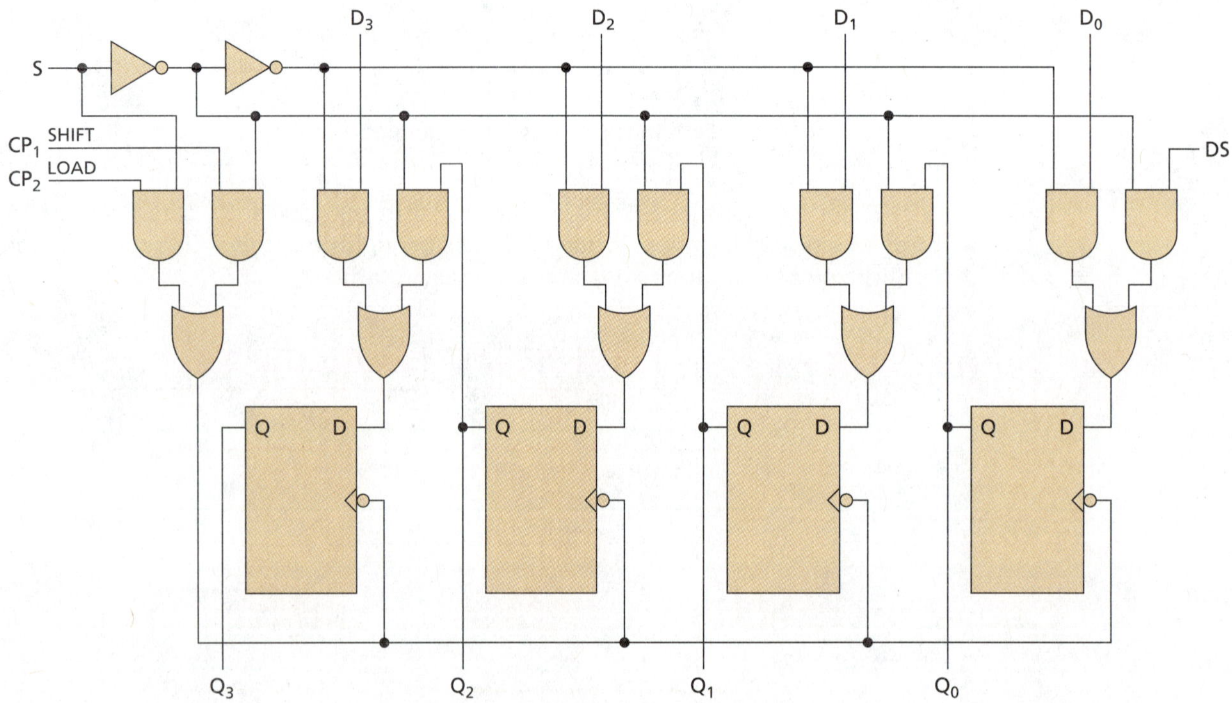

(a)

S	$\overline{CP}_1$	$\overline{CP}_2$	D_0–D_3	Q_0–Q_3
0	Steady State	X	X	No Change
0	↓	X	X	Shift
1	X	Steady State	X	No Change
1	X	↓	Data	D_0–D_3 Data
X	Steady State	Steady State	X	No Change

(b) 74LS95 Truth Table

X = "don't care" state

Figure 5.21 74LS95:Four-Bit Parallel-Load/Shift-Left Register achieving a no-change condition occurs by negating the S (shift) input and holding the parallel-load clock input (CP_2) at a steady state. Setting S high, the parallel-load function is selected by applying a CLOCK signal to CP_2. Data on D_0–D_3 are loaded, in parallel, into the register. Finally, selecting the shift function (S = 1) shifts data through the register at the rate of the CLOCK signal applied to CP_1.

EXAMPLE 5.7

Complete the timing diagram in Figure 5.22 for the 74LS95.

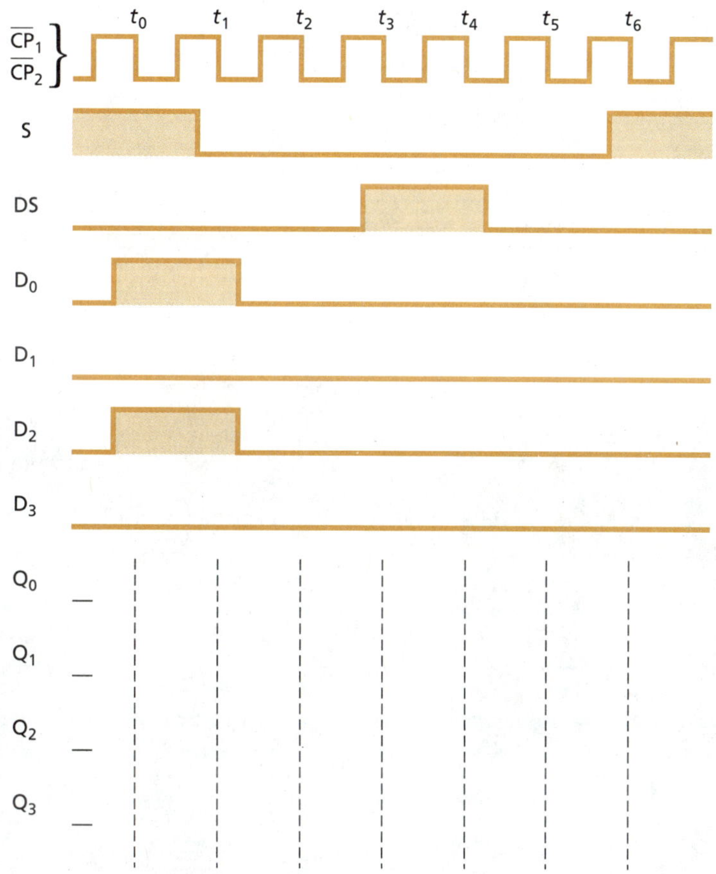

Figure 5.22 Example 5.7 (problem)

SOLUTION The timing diagram in Figure 5.22 is completed referencing the truth table in Figure 5.21. The results are shown in Figure 5.23. At t_0, S is high and clocking signals are applied to both CP inputs. This loads the data on D_0–D_3 into the register. Bringing S low before t_1 starts the shift process. The state of DS is shifted into the first flip-flop, reflected by the state of Q_0 on the timing diagram. The remaining bits shift to each succeeding flip-flop. Just before t_6, S is returned high and the register is loaded once again from the data on D_0 through D_3.

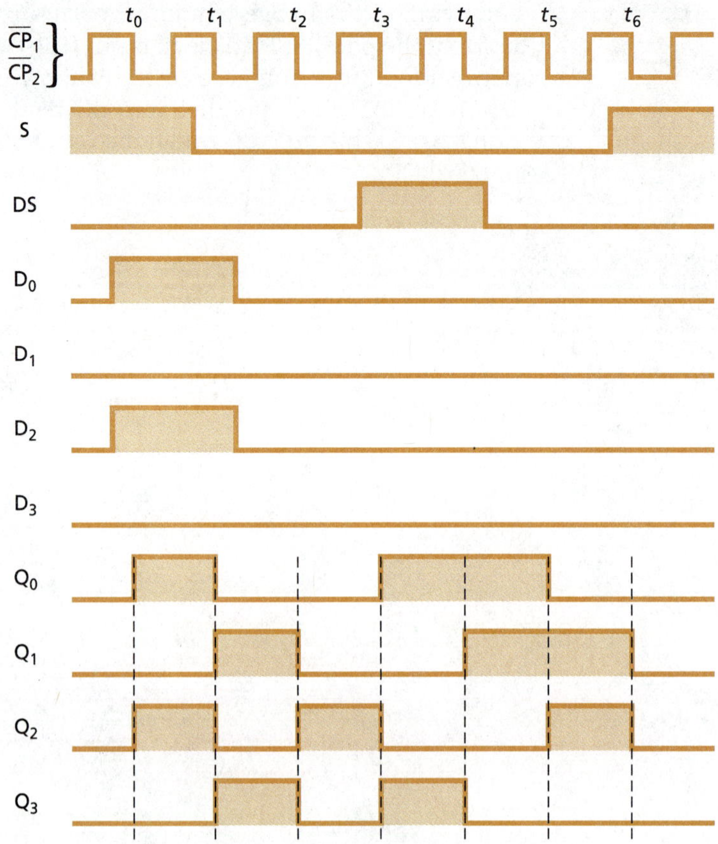

Figure 5.23 Example 5.7 (solution)

Section 5.5

Practice Problem

1. Match each register of Figure 5.24 with its correct name: "shift left," "shift right," or "parallel load."

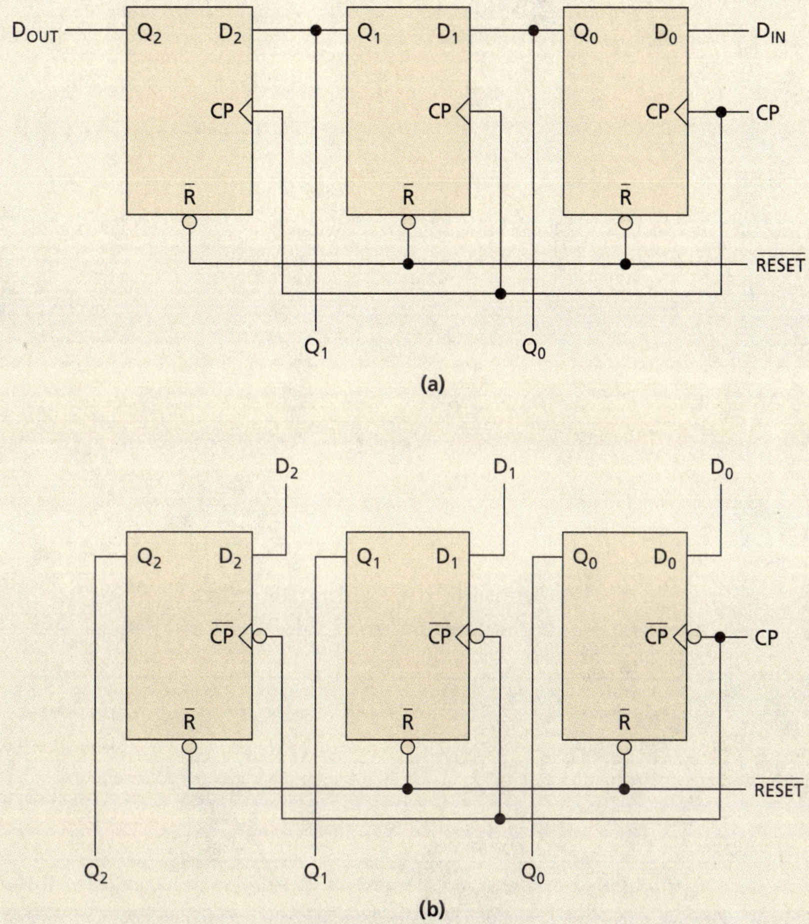

Figure 5.24 Practice Problem 5.5(1)

2. Complete the timing diagram of Figure 5.25 for the parallel-load register in Figure 5.24b.

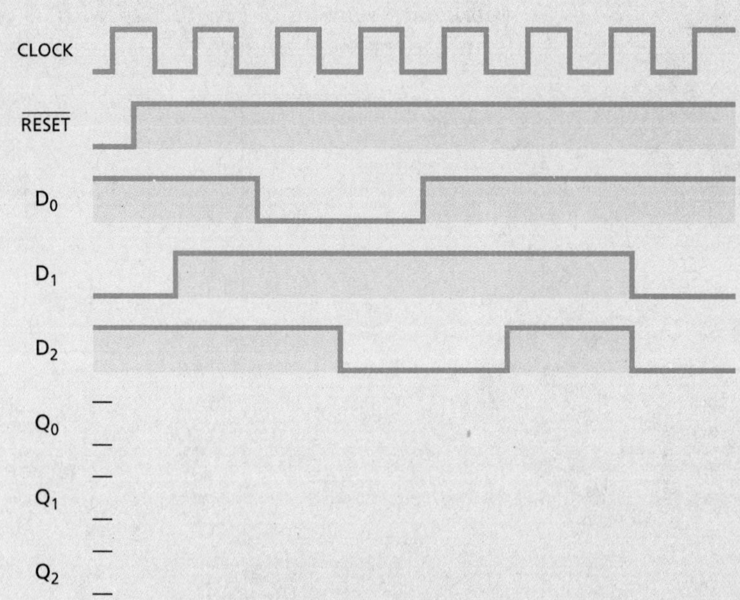

Figure 5.25 Practice Problem 5.5(2)

3. Complete the timing diagram of Figure 5.26 for the controlled shift-left register in Figure 5.24.

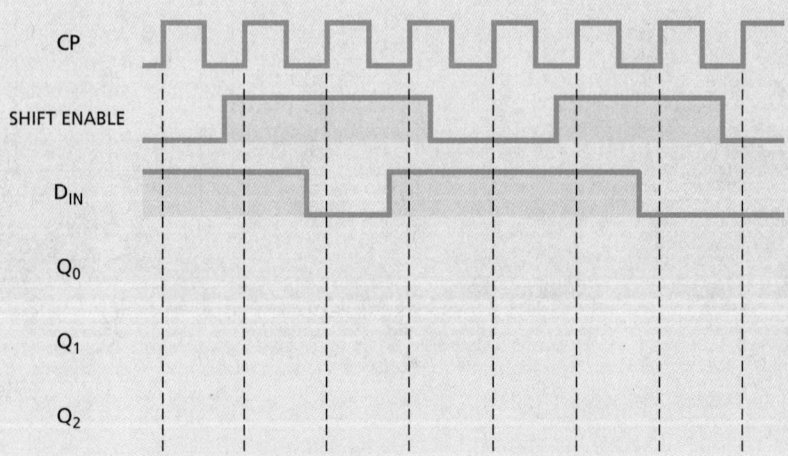

Figure 5.26 Practice Problem 5.5(3)

4. Complete the timing diagram in Figure 5.27 for the 74LS173 quad-D flip-flop.

Figure 5.27 Practice Problem 5.5(4)

5. Complete the timing diagram in Figure 5.28 for the 74LS95 four-bit register.

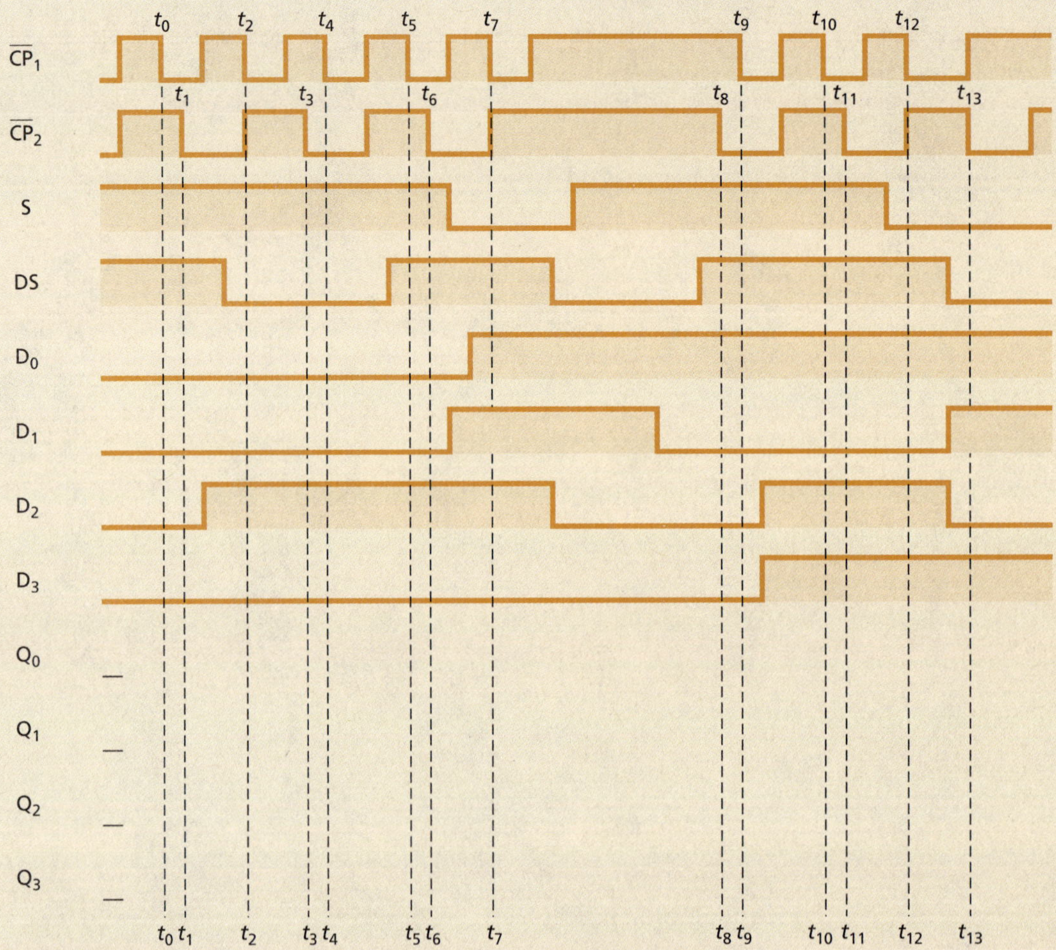

Figure 5.28 Practice Problem 5.5(5)

6. Identify each circuit in Figure 5.29, and describe what each is designed to do.

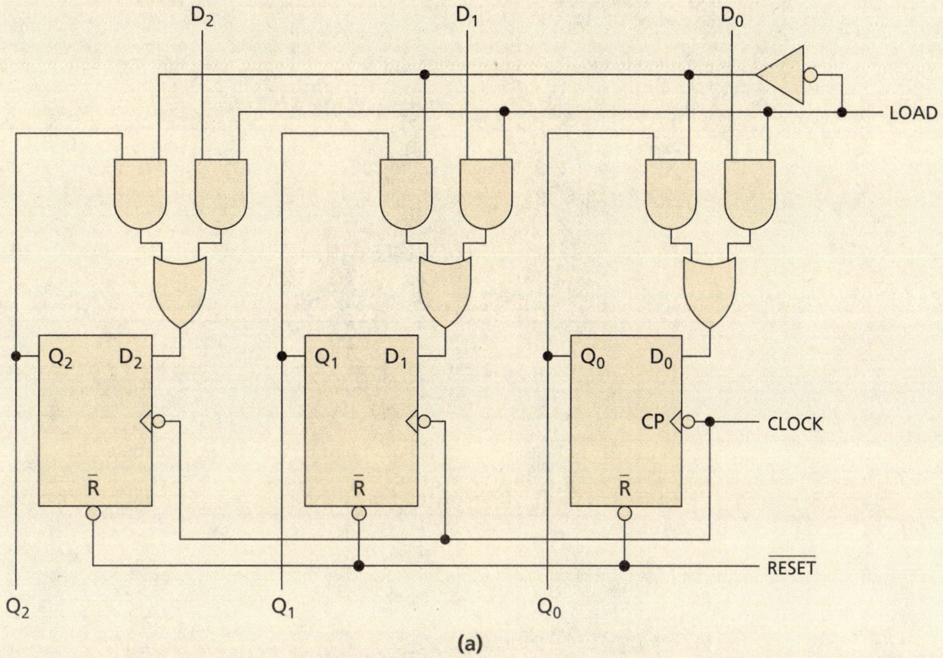

Figure 5.29 Practice Problem 5.5(6)

5.6 Solid State Memories

Registers provide a means for temporarily holding digital data used by other circuits. For example, two registers may be employed to hold the A and B words for an adder/subtracter as shown in Figure 5.30. The data in the registers are held only as long as needed to produce the sum from the adder/subtracter. To hold programs and large amounts of data for longer periods of time requires devices, called **memories**, that have a larger **capacity** than registers. *Capacity* is the term that designates the number of binary bits a circuit can handle, or in the case of registers and memories, the amount of binary data a circuit can retain. A pair of four-bit registers, used together to act as an eight-bit register, like those shown in Figure 5.30, has a capacity of eight bits. This section explores the two main types of solid state memories composed of static memory and dynamic memory devices.

Memories *devices that retain digital data.*

Capacity *the amount of data that can be held by a memory device or circuit.*

The basic unit of a memory device is its memory **cell**, which is the circuitry required to hold a single bit of binary data. The bipolar static memory cell in Figure 5.31 is a cross-coupled latch using **transistor–transistor logic (TTL)** logic family transistors. TTL is the logic family to which the 74LS series of logic devices belongs. Some of the transistors in TTL devices are made with multiple emitters. To understand how they work, look at Figure 5.32. Consider

Cell *a memory element that holds one digital bit.*

TTL (transistor–transistor logic) *a family of digital devices that use transistors as switching elements.*

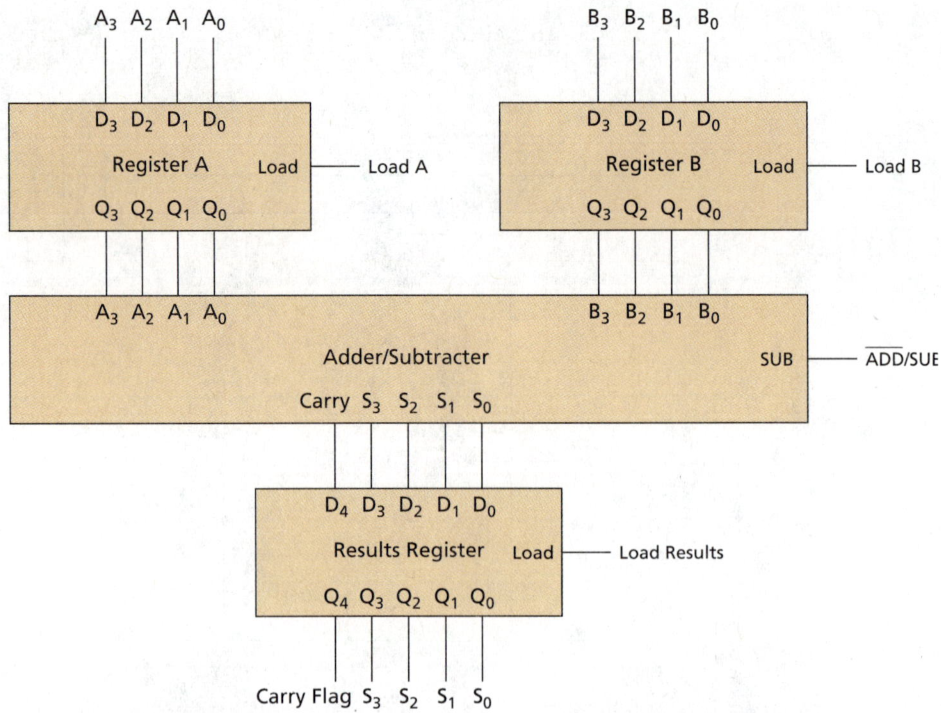

Figure 5.30 Register Application

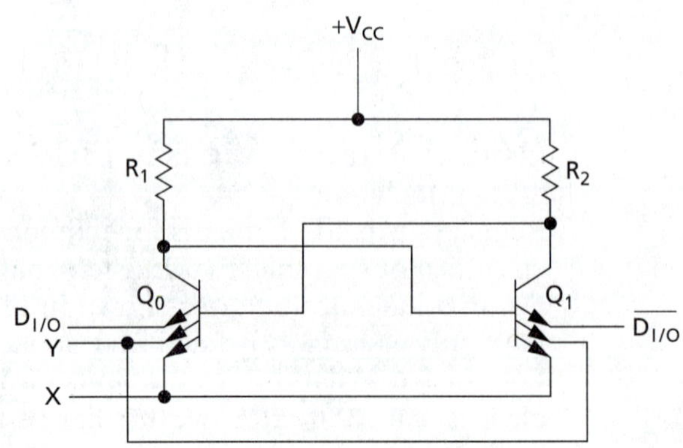

Figure 5.31 Static Bipolar Memory Cell

the emitter to base junctions of the multi-emitter transistor in Figure 5.32a as a set of diode PN junctions as shown in Figure 5.32b. Applying a logic 0 to any of the emitter inputs, as shown by the ground applied to input C, in Figure 5.32c, causes that PN junction to be forward biased. A junction voltage of approximately 0.7 V appears at the common node point for the three emitter–base junctions. It doesn't matter what the other two inputs are, the voltage at the node is 0.7 V with respect to ground. When high logic levels are applied to all three emitter inputs, as shown in Figure 5.32d, all three PN junctions are

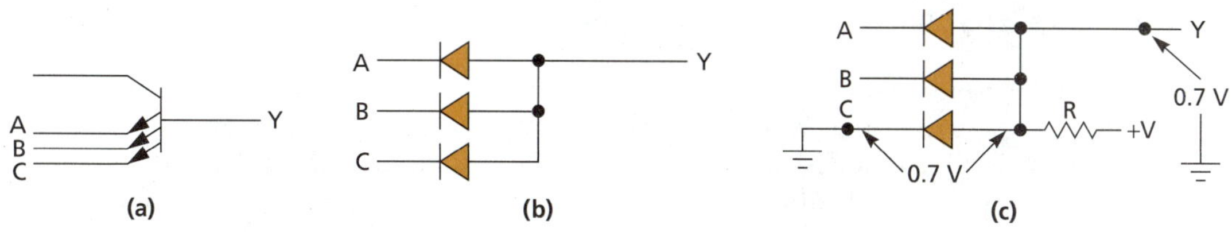

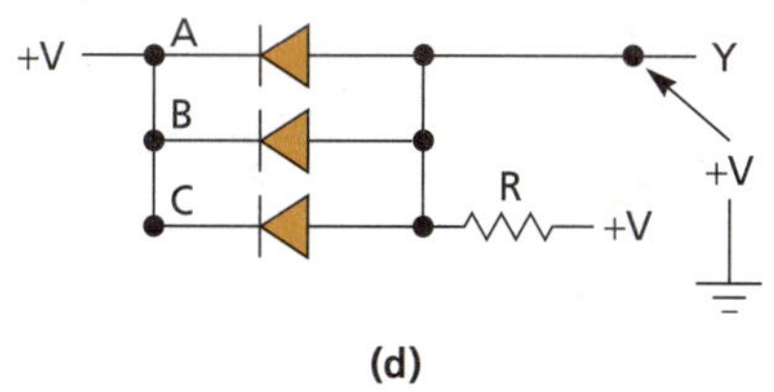

A	B	C	Y
0	0	0	0
0	0	1	0
0	1	0	0
0	1	1	0
1	0	0	0
1	0	1	0
1	1	0	0
1	1	1	1

0 = Ground 1 = +V

(e)

Figure 5.32 Multi-emitter Transistor

back biased and do not conduct. The node point senses a high voltage through the pull-up resistor, R. This action is summarized in the truth table in Figure 5.32e, which shows us that the three emitter–base junctions act like an AND logic function.

Address *a binary value that signifies the location of a data word in a memory.*

Applying 0's to either X or Y input **address** line in the memory cell of Figure 5.31 deselects the memory cell. An address is the binary code that is decoded to select one or more memory cells by asserting these address inputs. For the bipolar static memory, the active state for the address inputs is logic 1. The lows applied to the static memory cell cause their corresponding base–emitter junction to be forward biased, negating the influence of the other two emitter inputs and latching the current state of the cell as long as X and Y remain low. For example, a static cell holds a logic 1 determined by the conduction of Q_1. For the cell to hold this current information, it is not selected (X and Y = 0). Q_1 is conducting, causing its collector to be approximately 0.1 V, $V_{CE(SAT)}$ of the transistor. This low voltage is cross-coupled to Q_0's base, assuring that Q_0 is cut off. The collector of Q_0 tries to reach V_{CC}, because $V_{CE(CUTOFF)}$ makes the transistor appear open—there is no conduction through the transistor. This rising high voltage is cross-coupled back to the base of Q_1, which clamps it at 0.7 V, V_{BE} of Q_1. This positive voltage supplies the drive current to keep Q_1 conducting. Compare the similarities between this description and that of the cross-coupled NAND in Chapter 4.

To write a 1 into this memory cell, it has to be accessed and the correct binary information supplied onto its data input/output line ($D_{I/O}$). Binary address information is decoded to supply a high logic level to the X and Y emitters of the cell which is being accessed. Before this access can be completed, the correct data must be available at $D_{I/O}$. In this case, a logic 1 is applied to the third emitter of Q_0 and a logic 0 to the third emitter of Q_1. Since all the emitters of

Q_0 are high, each emitter–base junction is biased off, resulting in Q_0 being cut off. The X and Y inputs to Q_1 are also 1, keeping their respective emitter–base junctions off, but the third emitter is connected to the complement of $D_{I/O}$, making that emitter low. This, coupled with the positive drive current from Q_0's collector resistor, drives Q_1 on. The low state at Q_1's collector will assure that the base–emitter junctions of Q_0 remain off, keeping that transistor cut off. Figure 5.33 summarizes the current paths and bias voltages for storing a 1 into this cell.

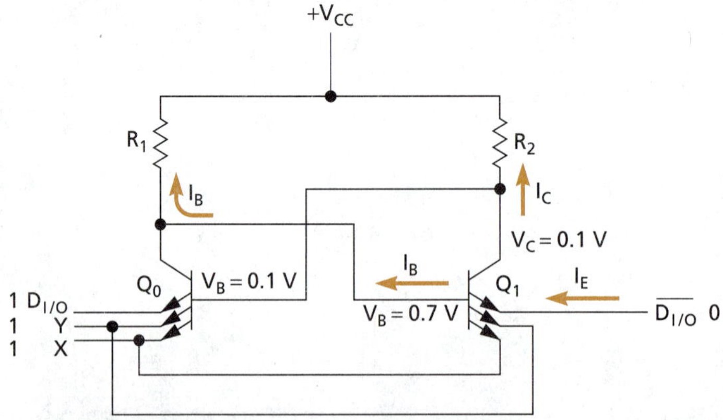

Figure 5.33 Writing a Logic 1 into a Static Cell

Write cycle *the sequence of events and signals that allows a data word to be entered into a memory location.*

Read cycle *the sequence of events that allows a device to access memory and receive data from it.*

The process of storing a logic state into a memory cell is called a write operation, and the time it takes to accomplish the storage is called a **write cycle.** To write a 0 into the static cell only requires that $D_{I/O}$ be low instead of high when the cell is accessed. In that case, Q_0 conducts and Q_1 is cut off. To get data from the cell, called a **read cycle** operation, both X and Y must be set high again to address the cell, but instead of applying a voltage to D_{IO}, a load is applied. If Q_0 is conducting because of a previously stored logic 0, current will flow through the load resistor. If Q_0 is not conducting because of a previously written 1, then no current will be flowing through the load. Therefore, no voltage across the load during a read cycle indicates the data bit in the cell was a 1. A voltage across the load resistor indicates a stored 0. What about the state of the cell after it has been read? It is unchanged, the load resistor assured that the flip-flop remained on and holding its current latched state. This is an example of a **nondestructive read,** which does not destroy the data bit in the cell when it is read. Destructive read devices, on the other hand, cause data to be lost once read, and require those data to be rewritten or refreshed at the completion of the read cycle.

Nondestructive read *data are not destroyed when they are read from memory.*

When the computer is turned off, the static memory cell loses the data it is holding. When we speak of this loss of data due to a loss of power, we refer to the memory as being **volatile.** Look at the cell in Figure 5.31 and remove V_{CC} and then reapply it. What is the state of the latch once V_{CC} is reapplied? First, once V_{CC} is removed, both transistors quit conducting, losing the currently latched data. After power is reapplied, the state of the latch belongs to the

Volatile memories *memories that lose their data when power is shut off.*

swifter of the two transistors. Whichever wins the race to turn on, determines the state of the latch. In essence, the data at turn-on are undeterminable and are often referred to as "garbage."

Static cells are organized into a matrix as shown in Figure 5.34, which shows a complete static memory structure in functional block form. Address lines A_0–A_9 feed two decoders that generate 1 of 64 X-line signals and 1 of 16 Y-line signals. Eight data input/output lines D_0–D_7 are buffered to control the direction of data—into the memory when being written and out from the memory when being read. This block also supplies the terminating load resistance required to keep the cells operating when they are being read.

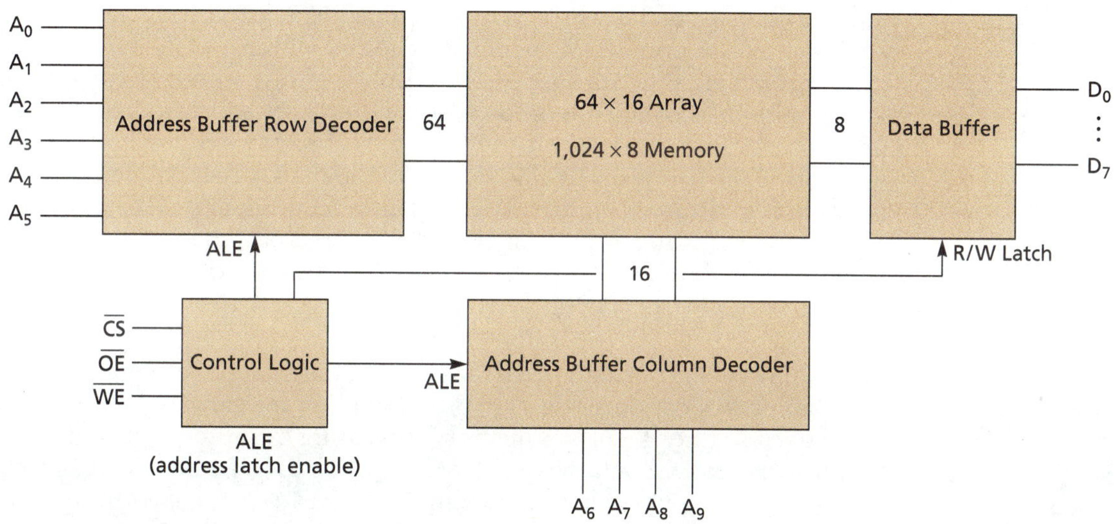

Figure 5.34 1 K × 8 Bit RAM

The cell array is organized as 64 × 16 matrix with each combination of X-line and Y-line accesses enabling eight cells at once. Data are written or read using the eight separate data lines. There only remains for the device to know when it is being accessed and whether to do a read or a write operation. One additional functional box, located in the lower left corner, is used for control logic. For this memory, there are three control inputs, chip select ($\overline{CS}$), output enable ($\overline{OE}$), and write enable ($\overline{WE}$):

Chip select ($\overline{CS}$): A memory can neither be read or written to without an active chip select. When this signal input is inactive, input data are ignored and the output data are disconnected from the data lines using a three-state buffer function.

Write enable ($\overline{WE}$): Once the chip is selected by asserting $\overline{CS}$ and properly addressed, data can be written into it by asserting $\overline{WE}$.

Output enable ($\overline{OE}$): Once the chip is selected by asserting $\overline{CS}$ and properly addressed, data can be read from the memory by asserting $\overline{OE}$.

The function of these signals is often summarized in a truth table such as the following, which uses their active low control signals:

$\overline{CS}$	$\overline{WE}$	$\overline{OE}$	DATA LINES, D_n
1	X	X	Open state
0	0	1	Write cycle
0	1	0	Read cycle
0	0	0	Undefined

The total number of bits, called a memory's capacity, tell how much digital information can be held by a single memory. This capacity is found by first determining the number of memory locations within the memory chip, and then multiplying that number by the number of bits stored at each location. The number of locations depends on the number of address lines used to define those locations. In the chip in Figure 5.34, 10 address lines, A_0–A_9, are used to generate the X-line and Y-line signals to the matrix that defines each group of eight cells within the matrix. The total number of separate locations of these eight-bit words is found by raising the base 2 to a power equal to the number (n) of address lines:

$$\text{Number of locations} = 2^n$$

The total number of memory cell locations in the memory of Figure 5.34 is 2^{10} or 1,024. Each location holds eight bits of data, so the capacity of this chip can be written as either 1,024 × 8 memory, indicating the number of locations and the bits per location; or as having a total capacity of 8,192 bits. Often the larger binary numbers, when denoting memory devices, are referred to by a shorthand notation that uses the nearest thousands as a value. For instance, the solution to this example may also be written 1 K × 8 memory or as having 8 K bits. For a quick reference, the more common shorthand references and their actual values are

2^{10} = 1,024 is 1 K
2^{11} = 2,048 is 2 K
2^{12} = 4,096 is 4 K
2^{13} = 8,192 is 8 K
2^{14} = 16,384 is 16 K
2^{15} = 32,768 is 32 K
2^{16} = 65,536 is 64 K

Why is 65,536 written 64 K instead of 65 K? The reason lies in the value of KB, which is shorthand for 1,024. Hence 64 K is 64 times 1 K, or 64 times 1,024, which is 65,536.

Address bus *a set of electrical conductors that carry address information.*

Addresses used to define the locations within the complete memory of a computer system are determined by the system's designer. A binary value is placed onto the address lines, called an **address bus,** that is used to define all memory locations within the system. These locations are found on several memory chips like those in Figure 5.34. To define more than 1,024 locations for a total memory system composed of several memory chips, we must have an address bus larger than the number of address lines required by a single memory chip. For example, an address bus used to access a number of $1{,}024 \times 8$ memories might contain 16 bits, A_0–A_{15}. Lines A_0 to A_9 are connected from the address bus to each of the memory devices directly. These inputs define memory locations within the device. The remaining address lines are decoded to generate a signal that is applied to the chip selects (CS) of each memory chip. The process of using part of the address bus to select which memory chip is accessed is called **address decoding**, and an example system configuration, using address decoding, is shown in Figure 5.35.

Address decoding *the process of decoding the information on the address bus to select a memory device.*

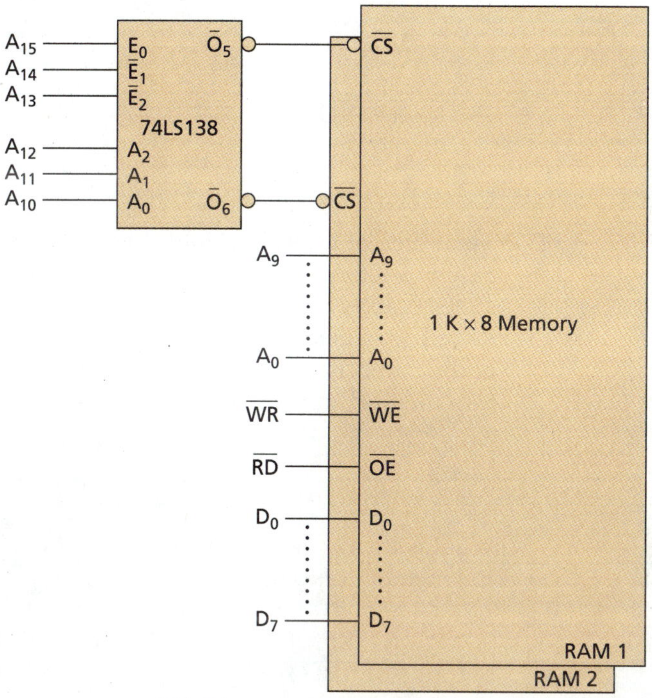

Note: A_0–A_9, $\overline{WR}$, $\overline{RD}$, and D_0–D_7 connected in parallel to RAM 2.

Figure 5.35 Address Decoding

In Figure 5.35, address lines A_{15}, A_{14}, and A_{13} are connected to the enable inputs of a 74LS138 1-of-8 decoder chip. Address lines A_{12}, A_{11}, and A_{10} supply the three binary inputs to the 74LS138 that will generate the individual outputs from the decoder. For any output to become active, A_{15} must be high and A_{14} and A_{13} both low to assert all three enables. The specific output from the 74LS138 is determined by the binary value placed on its A_2, A_1, and A_0 inputs by address lines A_{12}, A_{11}, and A_{10}, respectively. The correlation is direct, with the output number being the same as the binary value on the inputs. Every line of the address bus is terminated to an input in the circuit. Output $\overline{5}$ from the 74LS138 is connected to the $\overline{CS}$ input of the memory chip number RAM1. Given this circuit, the address values placed on the address bus that access a location in this memory chip can be determined. First the states on the address lines into the 74LS138 decoder required to enable the memory's chip select are

A_{15}	A_{14}	A_{13}	A_{12}	A_{11}	A_{10}	Address bus
E_0	$\overline{E_1}$	$\overline{E_2}$	A_2	A_1	A_0	74LS138 connections
1	0	0	1	0	1	States to assert $\overline{CS}$ of RAM1

The first location accessed in the memory occurs when all the address inputs, A_0–A_9, to the chip are low. The highest binary address that can access a location in the chip occurs when all the address inputs, A_0–A_9, are high. Any other combination of 1's and 0's on these address lines will access one of the other 1,024 locations within the device. Therefore, the lowest address than can be used to access a location in the memory is

A_{15}	A_{14}	A_{13}	A_{11}	A_{10}	A_9	A_8	A_7	A_6	A_5	A_4	A_3	A_2	A_1	A_0	or 9400H.
1	0	0	1	0	1	0	0	0	0	0	0	0	0	0	

The highest value is

A_{15}	A_{14}	A_{13}	A_{12}	A_{11}	A_{10}	A_9	A_8	A_7	A_6	A_5	A_4	A_3	A_2	A_1	A_0	or 97FFH.
1	0	0	1	0	1	1	1	1	1	1	1	1	1	1	1	

The range of address values, in hexadecimal for the circuit example of Figure 5.29 is 9400H to 97FFH. This means that placing any combination of binary states that have an equivalent hex value between these two values onto the address bus, will access a location in memory chip RAM1. To complete the access of the chip, either $\overline{OE}$, to read data from the memory, or $\overline{WE}$, to write new data into memory, must also be asserted.

EXAMPLE 5.8

What is the range of address values that allows access to a memory location in the memory chip RAM2 in Figure 5.35?

SOLUTION Since the chip select for the second memory device is connected to output 6 of the 74LS138, the lowest address value needed to access this chip is

A_{15}	A_{14}	A_{13}	A_{12}	A_{11}	A_{10}	A_9	A_8	A_7	A_6	A_5	A_4	A_3	A_2	A_1	A_0	or 9800H
1	0	0	1	1	0	0	0	0	0	0	0	0	0	0	0	

and the highest value is

A_{15}	A_{14}	A_{13}	A_{12}	A_{11}	A_{10}	A_9	A_8	A_7	A_6	A_5	A_4	A_3	A_2	A_1	A_0	or 9BFFH
1	0	0	1	1	0	1	1	1	1	1	1	1	1	1	1	

giving an address range of 9800H to 9BFFH to access a location in memory chip RAM2.

An alternate method used for address selection within a memory chip is illustrated in Figure 5.36. This is also a 1 K × 8 memory, but instead of 10 separate address input lines, there are only 5. Two new signals, row address select

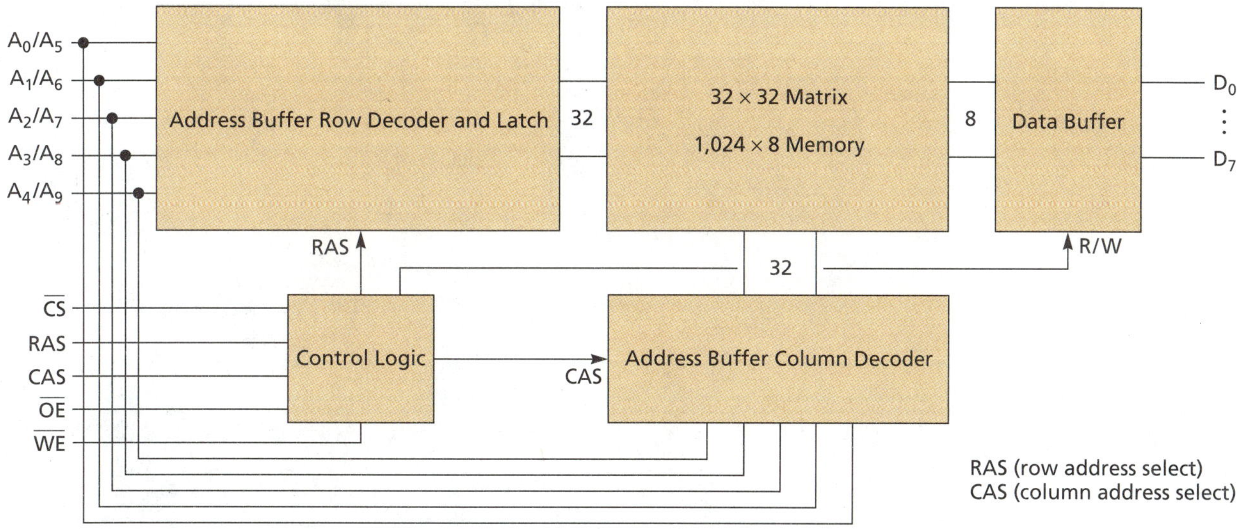

Figure 5.36 1 K × 8 Bit RAM with Row and Column Selection

RAS *(row address select) latch lower half of a memory address.*

CAS *(column address select) latch upper half of a memory address.*

(RAS) and column address select **(CAS)**, are added to the memory chip to provide a means of accessing 1,024 locations with 5 address inputs, $A_{0/5}$ through $A_{4/9}$, instead of 10 lines, A_0–A_9, used in the earlier memory chip. Each address line receives two address inputs at different times. For instance, $A_{0/5}$ is connected to A_0 first and then switched to A_5 later. When the chip is addressed, the lower five bits of the address bus, A_0–A_4, are sent to the $A_{0/5}$ to $A_{4/9}$ address lines of the chip, $\overline{CS}$ and RAS are made active and CAS inactive. This causes the binary value on the address inputs to be latched into the row address buffer and decoder within the chip. Switching memory address line inputs $A_{0/5}$ through $A_{4/9}$ from the lower address bus, A_0–A_4 to the next five address bits, A_5–A_9, and asserting CAS while $\overline{CS}$ and RAS are still active, causes the binary inputs on the address inputs to be latched into the column buffer and decoder. The outputs of the two decoders select the location in the matrix to be accessed.

Figure 5.37 is a circuit and timing diagram used to aid in the addressing requirements for this memory device. Notice that address lines A_{15}–A_{10} are connected identically to the circuit in Figure 5.35. Their purpose is, as before, to turn on the device's chip select, CS. The remaining 10 address lines are fed to a multiplexer that connects A_0–A_4 to the memory chip when the select input (S) is a 0 and connects A_5–A_9 to the same five address lines of the memory chip when S is a 1. A three-stage shift register is used to supply the RAS, CAS, and multiplexer select (MUX) signals. When the address value is valid, a signal called address enable (AEN) releases the reset hold on the shift register and allows the high input to be shifted through the register. The first clock cycle applied to the register sets the RAS signal high, latching addresses A_0–A_4 into the memory's row address buffer. The second clock shift sets MUX to change from a 0 to a 1, causing the multiplexer to change the binary value on the address inputs to the memory chip from A_0–A_4 to A_5–A_9. The third clock shift sets CAS high, latching in the second half of the address into the chip's column address latch. A half cycle later, AEN is negated, causing the register to be reset

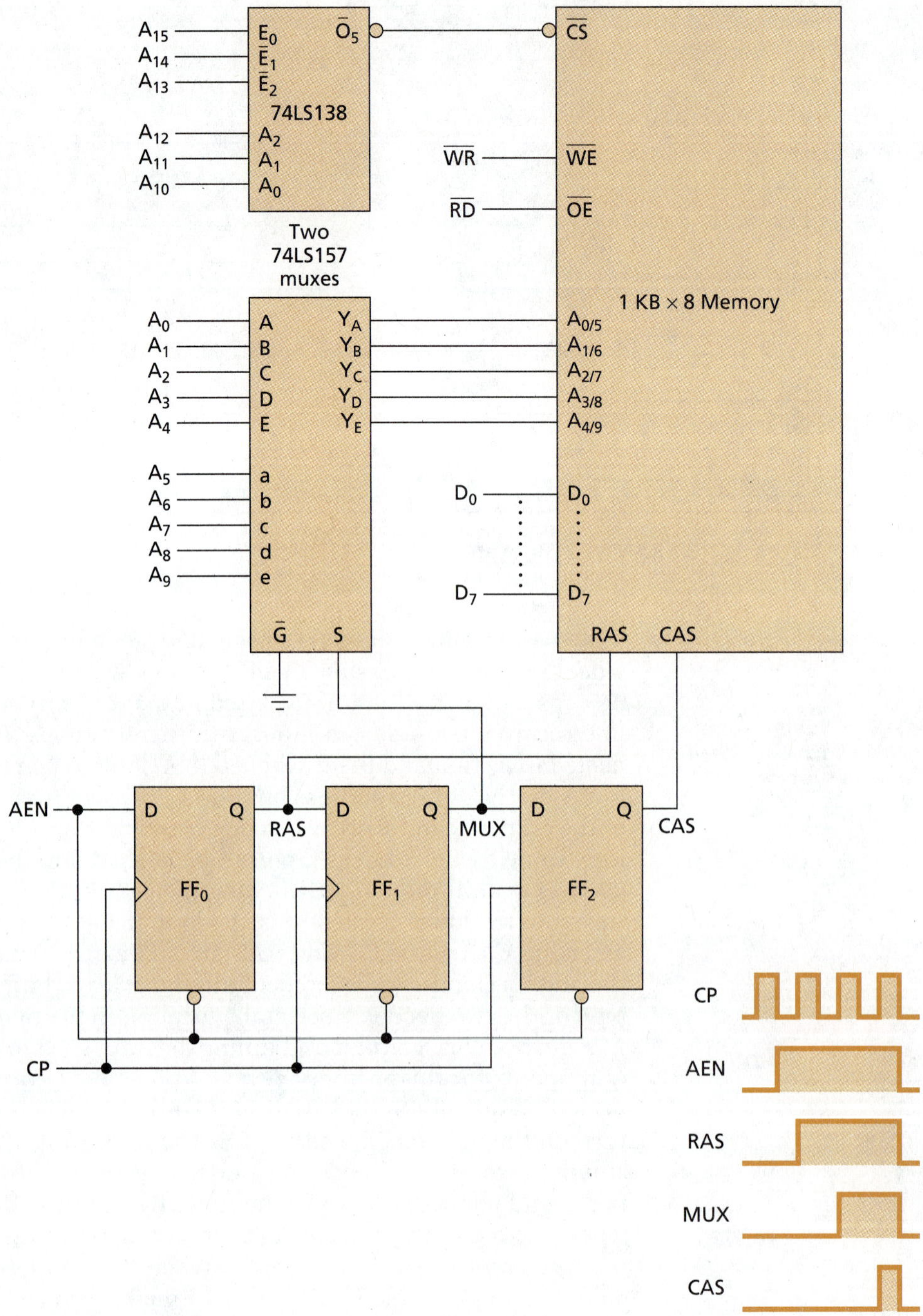

Figure 5.37 RAS/CAS Address Decoding

again, awaiting the next AEN signal. In this manner, 10 address bits are supplied to the 5 address inputs of the memory chip to select a location from among 1,024 locations.

Section 5.6

Practice Problems

1. How many address lines are there in a 4 K × 8 memory? How many data lines?
2. What is the total capacity, it bits, of a 4 K × 8 memory?

5.7 Dynamic Memories

The memories described in Section 5.6 are static bipolar memories. The basic cell for that type of memory is the transistor cross-coupled latch. A certain amount of standby current is required to retain the data in the cell. This current causes heat to be generated from the circuits, requiring some amount of physical area to dissipate the heat given off by the chip. This need for enough physical area to dissipate generated heat limits the amount of circuitry, or packing density, included in the chip. One way to reduce the power consumed by a memory device and increase its packing density is to use insulated gate field effect transistors (IGFETs) instead of bipolar transistors.

DRAM *dynamic random access memory.*

The cell of a **dynamic random access memory**, or **DRAM**, shown in Figure 5.38, is made from IGFET devices, which take advantage of their input capacitor characteristics to store binary data, rather than a flip-flop latch. Since the gate is insulated from the rest of the device, there is a capacitance (C) between gate (g) and source (s) called C_{gs}, in the range of 25 pf. This capacitance uses the gate and source material as the plates of the capacitor and the insulating material between them as the dielectric. The high impedance between the gate and source combine with the capacitance to present a comparatively long RC (resistor–capacitor) charge or discharge time for this input capacitance.

A 1 is written into the memory cell, by bringing the gate input of Q_1 high and placing a logic 1 at DATA IN. Current from source to drain of Q_1 charges the

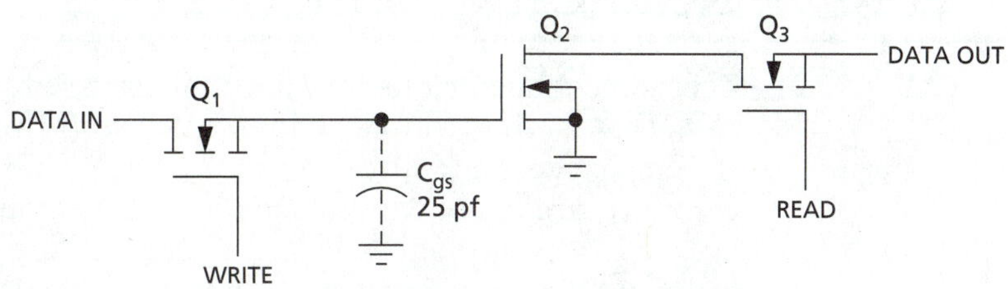

Figure 5.38 Dynamic Memory Cell

C_{gs} capacitance of Q_2. Removing the input to the gate of Q_1 leaves no ready discharge path for Q_2's C_{gs}, so it holds its charge, storing the logic 1 written into it. Q_1 and Q_3 are both cut off, since neither gate is made high. In actuality, the charge on C_{gs} will leak off through the high-impedance resistance of Q_2 gate to source. The time it takes to reduce the charge on C_{gs} is about 20 ms. Thus to retain data in this memory cell, the data must be rewritten or refreshed, before the charge on C_{gs} falls too low. To read the memory, the gate of Q_3 is turned on. If there is a charge on Q_2's C_{gs}, then current will flow from ground through Q_2 and Q_3 to the DATA OUT line. Lack of a charge on C_{gs} (indicating a logic 0) produces no current flow when Q_3 is turned on.

Although it may seem that refreshing dynamic memories adds an unwanted operation to the memory device, there are some very important advantages to using dynamic memories. First, notice that in Figure 5.38 there is no indication of a source supply to the memory cell. Therefore, very little standby current is required to maintain data in the memory. Current is required to allow the refresh circuit and address decoders to function, but that is minimal compared to static bipolar memories, which require constant current for each cell to hold its latched state. The reduced power consumption of this memory device requires a smaller physical area to dissipate heat generated by conducting devices. This translates into an increase in packing density for a given physical memory area. Basically, more memory cells can be placed into a single memory chip.

Dynamic memories are arranged in a manner similar to static memories, so that address-decoding functions operate the same as discussed in Section 5.6.

Section 5.7

Practice Problems

1. Compare static and dynamic memories. What are the advantages of one compared to the other?
2. What is the capacity, in bits, of a 32 K × 8 dynamic memory?
3. Which memory type, static or dynamic memory, is preferred usage in a notebook computer? Why?

5.8 Read-Only Memories

RAM *random access memory.*

The semiconductor memory devices discussed thus far are read/write memories. That is, they can be written to or read from at any time. They are also **random access memories** or **RAMs,** since any memory location can be accessed regardless of which location was previously accessed. By common usage and convention, read/write memories used in personal computers are called RAM chips. This term is not entirely correct, because many locations within these memories are accessed sequentially, as if that portion of the memory were a **sequential access memory** or **SAM.** However, because of this convention, we will also refer to read/write memories, in general, as RAMs. Another type of memory device is the **read-only memory** or **ROM.** These devices can be programmed (written into) only once. The data stored in them are permanent and nonvolatile, so these devices retain the data after power has been shut off.

SAM *sequential access memory.*

ROM *read-only memory.*

Figure 5.39 is used to illustrate the basic concept behind a ROM chip. Each AND gate has a unique combination of inputs configured as a 1-of-4 decoder. Each of the decoder outputs may be connected through a fuse to any number of the five OR gates available. These connections are shown in Figure 5.39 using fuse symbols. A lack of a fuse indicates a nonconnection or open condition. As an example, AND gate U_0 has its outputs connected to OR gates U_4, U_5, and U_8. When inputs A_0 and A_1 are set low, giving the address input a value of 00B, then AND gate U_0 turns on, supplying a 1 to OR gates U_4, U_5, and U_8. The output read on the five output lines, D_4 through D_0, is 10011B. Each time both address inputs are set to 00B, the output will be 10011B. The two input address lines and the five output data lines in this example can be restated as "Address location 00B holds the data 10011B" or, in hex, "Location 0H holds the data 13H."

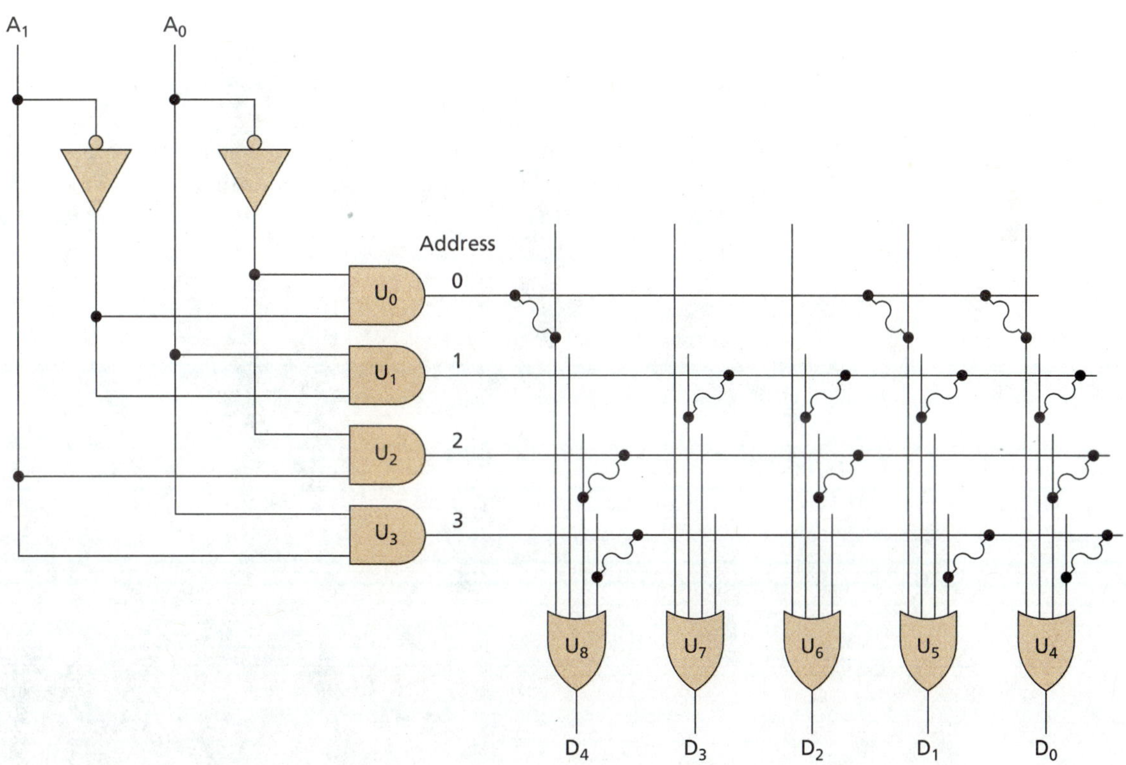

Figure 5.39 ROM Example

EXAMPLE 5.9

What is the data content of each remaining location in the ROM of Figure 5.39?

SOLUTION Following the same sequence used to uncover the data at location 0, the remaining locations hold this information in hex:

ADDRESS	DATA (HEX)
0	13
1	0F
2	15
3	13

Notice that two locations (or more, for that matter) can hold the same data. In this example, locations 0H and 3H both hold 13H.

The data in the ROM in Figure 5.39 are nonvolatile, because they depend on the blown-fuse links and not on flip-flops or IGFET devices. Data are not lost when power is removed from the memory chip. These fuses are blown using a ROM programmer and a "fuse map" detailing which ones are to be open or blown. At first the process of programming a ROM chip was left to the manufacturer of the device. Users would supply the data to be programmed into a ROM, and the IC manufacturer, for a fee, would program the chip.

ROM devices are laid out similar to read/write memory chips. The major difference is the lack of a write control input. Figure 5.40 shows a typical 4 K × 8

Application Note

Application for ROM Devices

Each time you approach a computer and fire it up, it goes into a dance and does a number of mystical operations. Among these are self-checking itself, configuring the machine to recognize the keyboard and disk drives, and operating the monitor. Many other processes are being performed before you finally get the DOS prompt or Windows screen. How did the computer know how to do all these things? Essentially, the processor inside the computer fetched an initialization program from its read-only memory area. This program not only configured the keyboard, disk, and monitor interfaces, but also accessed the disk to retrieve the DOS and Windows application programs to further set up the computer to allow you to use it. This initialization program must be present for your computer to "boot up" each time you turn it on. Therefore it must be stored in a memory that is unchangeable and nonvolatile—a basic application for ROM memory.

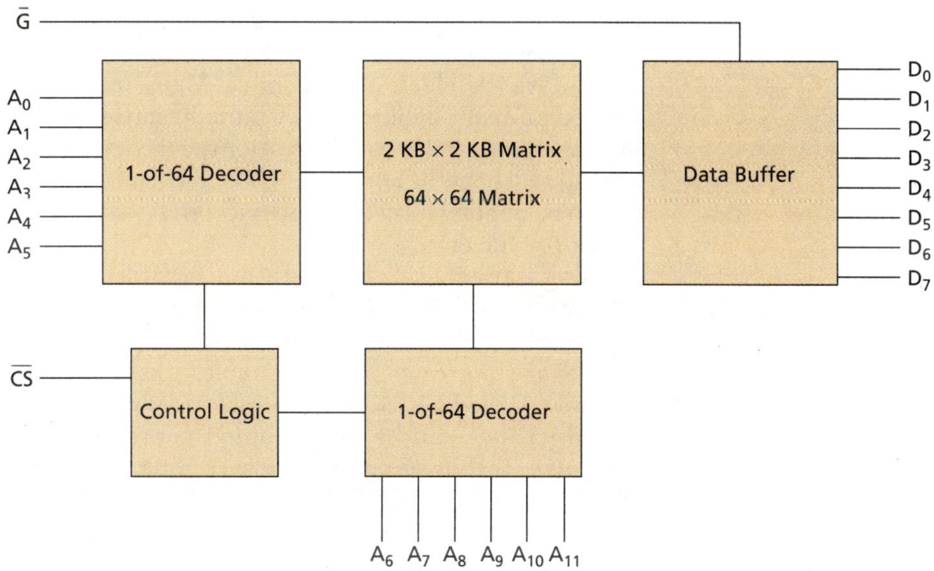

Figure 5.40 4 K × 8 Bit ROM

ROM. It has 12 address lines to select among the 4,096 memory locations, each of which holds eight bits of data. A chip select ($\overline{CS}$) is provided, as it is in a read/write memory device, for selection of the ROM chip by an address decoder. Lastly, an output enable (for some odd reason usually designated using the letter $\overline{G}$) is available for control of the output tri-state buffers between the internal memory data lines and data output lines (D_0–D_7). To retrieve the data from a ROM location, both the chip's $\overline{CS}$ and $\overline{G}$ lines must be asserted low and an address placed on the input address lines. After a short time, called access time, data appear on the data lines. **Access time** is defined as the time required between addressing a memory location and actually receiving the data from the location and is used as a chief parameter of a memory device's performance.

Access time *the time between addressing a memory and receiving the data, at that address, onto the data bus.*

One problem with ROM memories is that the data or programs contained in the device eventually become obsolete and must be replaced. The user was once forced to resubmit the new data to an IC manufacturer and pay again to have it reprogrammed into another ROM. Eventually, IC memory manufacturers developed a programmable ROM that end users could program themselves as long as they had a PROM programmer. **PROM** is the acronym for "programmable read-only memory." They are ROM-type memories that can be programmed once and read or used many times. This innovation resolved the problem of having to submit and wait for data or a program to be entered into a ROM, because the user could do the actual programming of the memory. However, it still meant that a new ROM had to be bought each time the program was to be updated, because PROMs could only be programmed once.

PROM *programmable read-only memory, which is provided by a class of memory devices that can only be read after data have been permanently stored in them.*

The next evolution in ROM technology was the erasable PROM or **EPROM.** "Fuses" were no longer links that were permanently open. Instead they became charge particle areas that could be oriented in an open or closed position, much like a switch. They were programmed in much the same way as PROMs, but

EPROM *erasable read-only memory.*

they were also erasable by using a concentrated ultraviolet light applied anywhere from 15 to 30 minutes, depending on the actual device, through a window above the electronic circuitry. Many users, afraid that long exposure to normal ambient light would eventually cause these devices to lose their memory, taped over these windows after the device was programmed. No harm, but also no real benefit. For an EPROM to become erased, a strong ultraviolet light had to be applied for a long time. Room light or sunlight by itself is not enough to destroy the data.

EEPROM *electrically erasable read-only memory.*

Using ultraviolet light to erase a PROM is a nuisance and also requires a little patience. Still, it is an improvement over the need to acquire a new PROM each time you needed to make a change in the contents of the ROM. Today, ROMS are not only programmable and erasable, but are also electrically erasable. Called **EEPROM** for "electrically erasable programmable read-only memory," they can be programmed and erased using the same PROM programmer. The erase process took as little time as the programming process, which is relatively fast. Software and burners exist that automatically erase an EEPROM before the new data are programmed into it. The user need only supply the correct disk file and request from the EEPROM burner software that the EEPROM be programmed. The software program takes over from there and does the job.

Section 5.8

Practice Problems

1. Compare a ROM with a read/write memory.
2. In general, when is the use of ROMs preferred?
3. What advantage was gained by developing PROMs? EPROMs? EEPROMS?

5.9 Memory Systems

Full address decoding *decodes all address lines to determine which memory location is accessed.*

The address-decoding scheme of Figure 5.35 used every available address line to decode the chip select inputs and location selections of the memory chips in the circuit. The term applied for this type of decoding scheme is **full address decoding.** The result of this type of decoding method is that each memory location has a unique address and no two addresses can be used to select the same location. The down side of this type of scheme is that every address line must be physically connected to either the decoder or memory chips. Some memory designs are small, with few actual memory chips used and do not require assignment of every address value of memory location. An alternate method of address decoding, called **partial address decoding,** is used to handle these smaller memory systems.

Partial address decoding *uses a portion of the address bus to determine the location of data in a memory system.*

In Figure 5.41, there are 16 address lines available, which can account for 2^{16} or 65,536 different memory address locations. However, only three memory devices are used in this memory. One is a 4 K × 8 bit ROM, and two are 4 K × 8 bit RAM chips, for a total of 12 K memory locations. Since no additional addressing is needed for this design, we need only decode enough different addresses to allow correct selection of the 12 K locations. A 74LS138 1-of-8 decoder is again employed to decode part of the address bus to assert individual chip selects. One of its enable inputs is connected to a control signal called $\overline{MEM}$, which must be asserted every time one of the memory chips is to be accessed. The two remaining enables are connected to their active state logic levels.

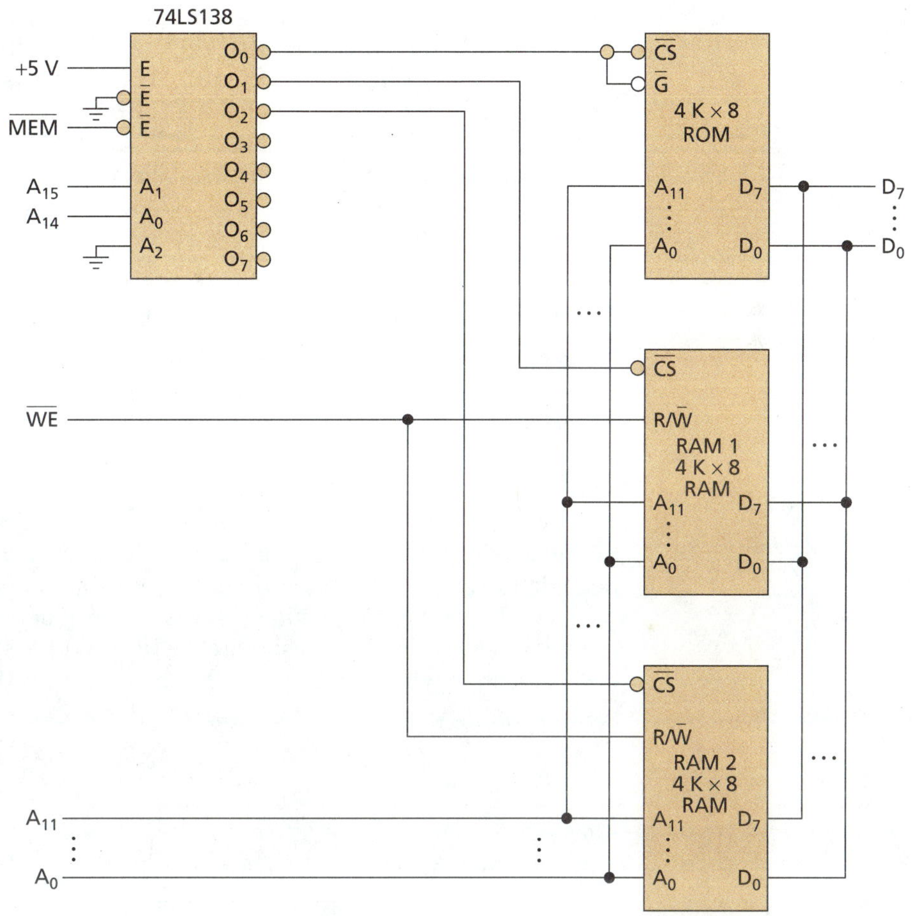

Figure 5.41 Partial Address Decoding

Address lines A_{15} and A_{14} supply two of the three binary inputs to the 74LS138 that will cause one of the eight outputs to be asserted. The third input is permanently grounded, making its logic level constantly 0. Output O_0 drives the chip select of the ROM memory, while outputs O_1 and O_2 are used to select RAM1 and RAM2 respectively. Address lines A_0–A_{11} are used to access the individual memory location in the device whose chip select has been asserted. Following the process used earlier, let's determine the address ranges for all three memory chips in the circuit. Notice that address lines A_{12} and A_{13} are not used in the circuit, so their state is a "don't care" in that it makes no difference to the operation of the circuit. Initially, we will set these "don't care" inputs to zero. Let's start with the ROM chip. To turn on its chip select, address lines A_{15} and A_{14}, must be low. This asserts output O_0 of the decoder:

A_{15}	A_{14}	A_{13}	A_{12}	A_{11}	A_{10}	A_9	A_8	A_7	A_6	A_5	A_4	A_3	A_2	A_1	A_0
0	0	0	0	x	—	—	—	—	—	—	—	—	—	—	x

When the x bits are all low, we are accessing the first location in the ROM chip. When they are high, we access the last location, making the range of address values, in hex, 0000H to 0FFFH for the ROM chip. Before we continue, let's examine the effect of our two "don't care" addresses, A_{13} and A_{12}. Each can be in a 0 state or a 1 state, meaning that there are four different combinations these two bits can have: 00B, 01B, 10B, and 11B. Does this mean that more than one address value can access a single location? Yes! Look at the first location that we designated as 0000H. Since A_{13} and A_{12} can be any one of four combinations of 1 and 0, then this location can also be accessed by addresses 1000H, 2000H, and 3000H. Is there a problem here? Not actually, if you follow these two basic guidelines when designing a partial address-decoding scheme:

1. Any number of addresses can be used to access a single memory location.
2. No address can access more than one memory location.

This means that while the first ROM location, in our example, can be accessed by *four different* addresses, each of those four addresses can *only* access that location and no others. Consider what would happen if the second guideline is violated and two different ROM locations were accessible by, say, location 2000H. The data from both locations would be mixed on the data bus, and the result would be garbage. When using partial address decoding, the designer selects which of the multiple address ranges are to be designated for each memory area. This does not mean that the other addresses are not usable, but it does reduce the confusion as to which ones to use. For instance, I—as the designer of the scheme—say to keep A_{13} and A_{12} always low and designate the ROM address range as 0000H to 0FFFH.

We continue with our decoding by including RAM1 and RAM2 memories:

RAM1:

A_{15}	A_{14}	A_{13}	A_{12}	A_{11}	A_{10}	A_9	A_8	A_7	A_6	A_5	A_4	A_3	A_2	A_1	A_0
0	1	0	0	x	—	—	—	—	—	—	—	—	—	—	x

RAM2:

A_{15}	A_{14}	A_{13}	A_{12}	A_{11}	A_{10}	A_9	A_8	A_7	A_6	A_5	A_4	A_3	A_2	A_1	A_0
1	0	0	0	x	—	—	—	—	—	—	—	—	—	—	x

Memory map *a listing of addresses and devices they access.*

Combining the ranges of all three memory chips into a single **memory map**, we can see the address designations for our system:

CHIP	START	END
ROM	0000H	0FFFH
RAM1	4000H	4FFFH
RAM2	8000H	8FFFH

A by-product of address decoding is memory expansion. In the preceding example, we had three individual 4 K × 8 bit memory devices, which we expanded to a memory system of 12 K × 8 bit. It took two additional address lines to access more than the 4 K locations of each individual chip. Suppose we wanted to make a 32 K × 8 bit memory system using these 4 K × 8 bit chips. An easy division shows that we need eight of these chips to achieve 32 K total locations. Locations within each chip are still accessed by the same 12 address inputs, A_0–A_{11}. However, to separate and enable each of the eight chips, individually, requires three additional address lines.

EXAMPLE 5.10

What changes must be made to the address-decoding circuit of Figure 5.41 to let the system access 32 KB of memory instead of 12 KB?

SOLUTION The binary input to the 74LS138 that was tied to ground (permanent zero) must be connected to either A_{13} or A_{12} instead, so that the decoder can now select one of eight different memory chips.

Memory expansion can also be used to increase data bus sizes. For instance, let's make a 4 K × 16 bit memory using 4 K × 8 bit memory devices. As shown in Figure 5.42, two 4 K × 8 bit chips are required to get a 16-bit data bus. Each of the eight data lines from the two chips are connected to a separate data line, D_0 to D_{15}. All the address inputs are tied together and returned to A_0–A_{11}. In addition, the write enable and output enable of both chips are connected together and controlled by one write and one output enable signal. To have 16 bits available at one time, *both* memory chips must be enabled at the same time

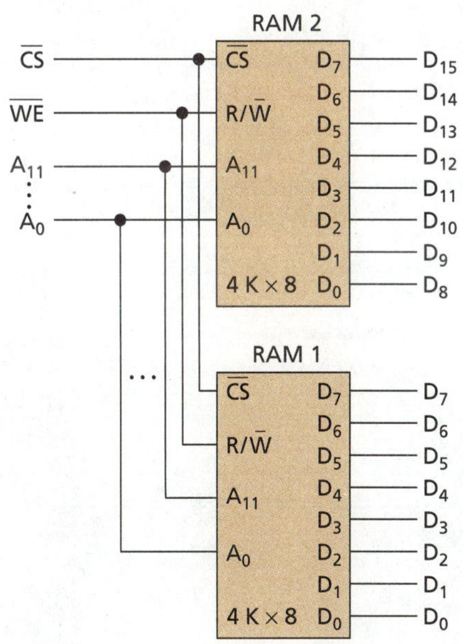

Figure 5.42 4 K × 16 Bit Memory

so that each supplies its eight-bit half of the data. As such, the chip select inputs to both devices are connected together and run from the same address decoder output.

We can mix the two memory expansion techniques to arrive at a 32 K × 16 bit memory system. The 74LS138 decoder supplies chip select signals to pairs of 4 K × 8 bit memories so that the selection by each address is for 16 bits of data. There are eight pairs of these 4 KB chips to bring the memory up to a 32 K × 16 bit size, as illustrated in Figure 5.43.

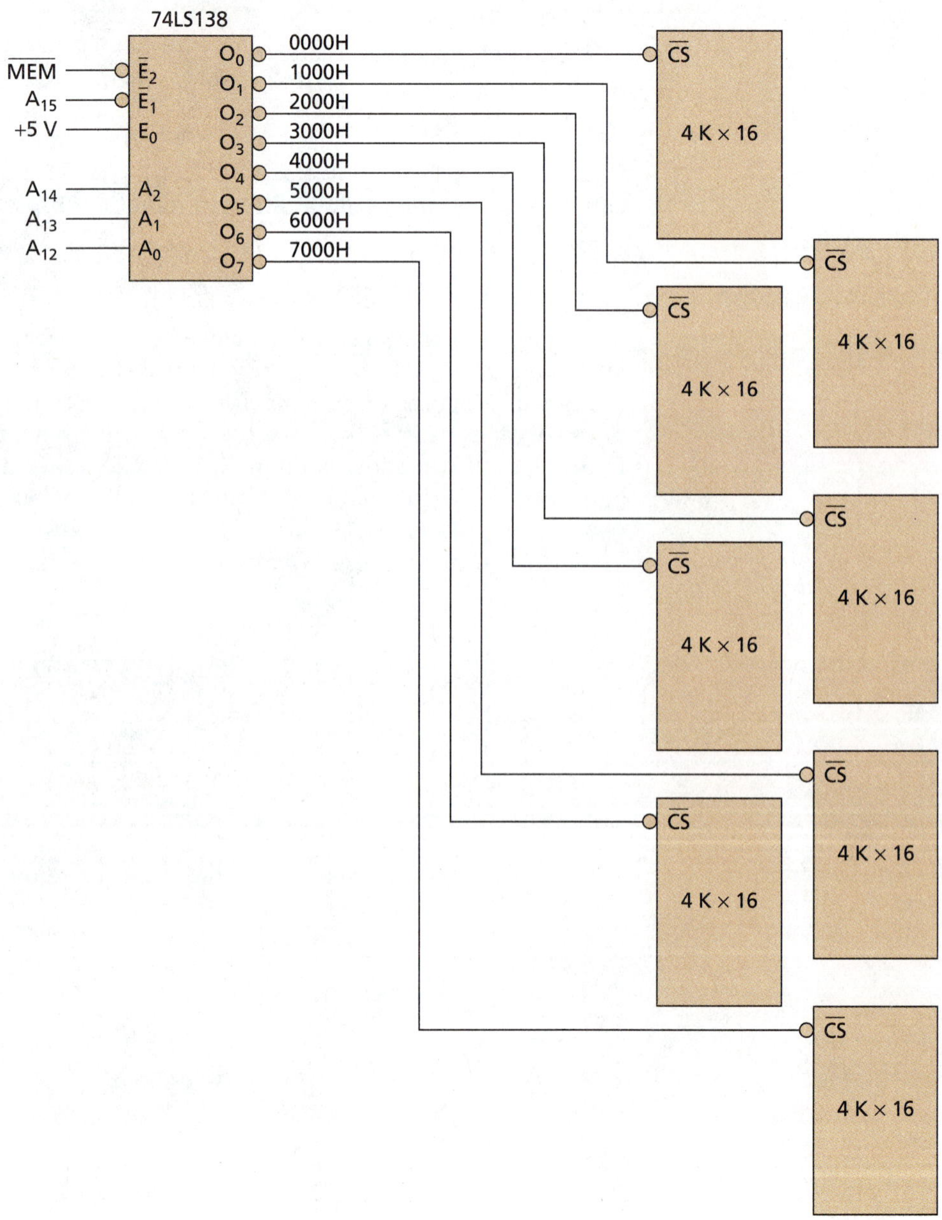

Figure 5.43 32 K × 16 Bit Memory

Section 5.9

Practice Problem

1. Assuming that "don't care" address lines are a logic 0, what is the range of address for the memory scheme in Figure 5.44?

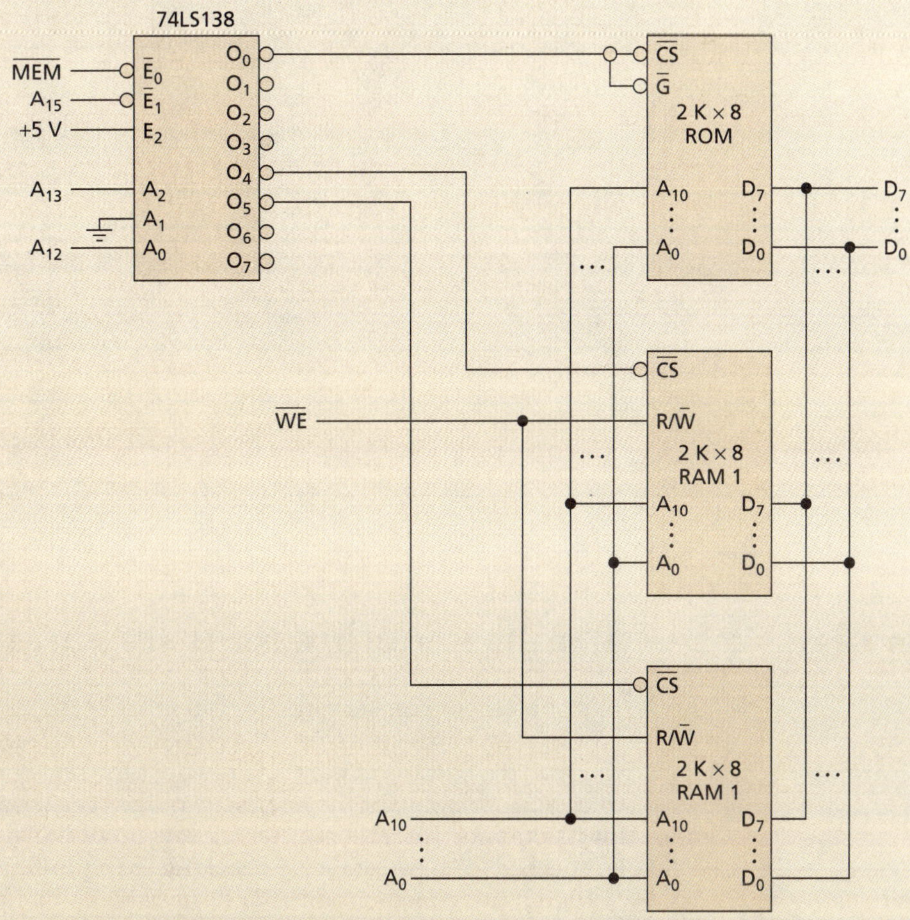

Figure 5.44 Practice Problem 5.9(1)

2. Show how the 1 K × 4 bit memory chip in Figure 5.45 is used to create a memory whose size is 4 K × 8 bit with consecutive addresses beginning at location 4C00H.

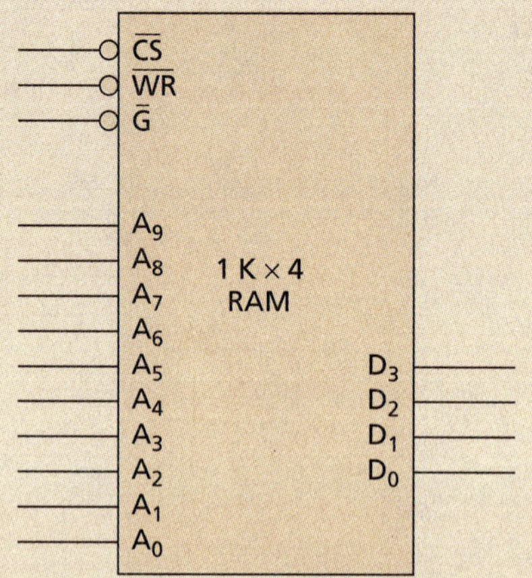

Figure 5.45 Practice Problem 5.9(2)

5.10 Troubleshooting Registers and Memories

To troubleshoot register circuits, you must begin, as always, by knowing what the circuit is supposed to do. Does it parallel-load, shift left, or shift right? Can you read the outputs in parallel, and does the register have multiple functions? Once you are aware of the operation of the register, apply its inputs and check for its outputs. Discrete registers, made up of individual flip-flop devices, can be checked for a bad stage by checking each flip-flop input and output.

__SSI__ small-scale integration; defines a group of devices that contain a small amount of logic circuitry. A quad two-input NAND chip in an example.

__MSI (medium-scale integration)__ a group of devices that logic circuits with a small amount of complexity. A quad D-flip-flop and shift register chips are examples.

__LSI (large-scale integration)__ a group of devices that contain a vast amount of logic circuitry such as a microprocessor.

Digital integrated-circuit (IC) devices come in varying sizes depending on the amount of logic circuit included in the chip. Devices such as the 74LS00 quad two-input NAND gate are considered to be **small-scale integrated chips (SSI)** because they contain a small amount of logic circuitry. Devices such as the 74LS138 1-of-8 decoder contain a good deal more circuitry than a 74LS00 and are part of a group of ICs known as **medium-scale integration (MSI)** devices. Microprocessors and associated devices are packed with an extensive amount of logic circuits and are grouped into **large-scale integration (LSI)** devices. Registers, since they contain several flip-flop stages and additional logic circuits, are MSI devices and can only be checked as a whole. In other words, if a part of a MSI register fails, the entire chip is replaced.

The two obvious areas to check in a register circuit are usually the two most forgotten areas, power and clock input. A problem is detected in a register circuit, say, an output is incorrect. How many ways can it be wrong? First, the output could be the wrong logic level, indicating an inoperative stage in the register. Second, the output could appear floating; that is, neither a logic 1 nor logic 0. When this occurs, check the V_{CC} and ground pins. If they are all right, then you probably have an open output stage. Third, the output may not ever change when the register is enabled to load or shift. Check the clock input before proceeding further. In all cases, check the control inputs, power, and clock. If they are all good, then your register chip is bad and should be replaced.

Memory devices can be fairly easy to troubleshoot or somewhat difficult. It depends largely on the circuit where they are used. Most memory devices are used in some form of computer system. Thus the control signals used to write and read memories are constantly in flux. They are turned on and off at very regular intervals. To troubleshoot these memories requires their control signals to be applied in a constant manner. This is usually done by running a debug program in the computer with the memory under test. This program may simply do a write operation to one memory location, followed by a read operation of that same location—over and over. This causes the write enable, output enable, and chip select inputs to be applied at a constant rate. These can be used to help check the memory out.

In Figure 5.46, the memory location 6200H is being constantly accessed and the data 55H written to it. The oscilloscope was used to detect that the write

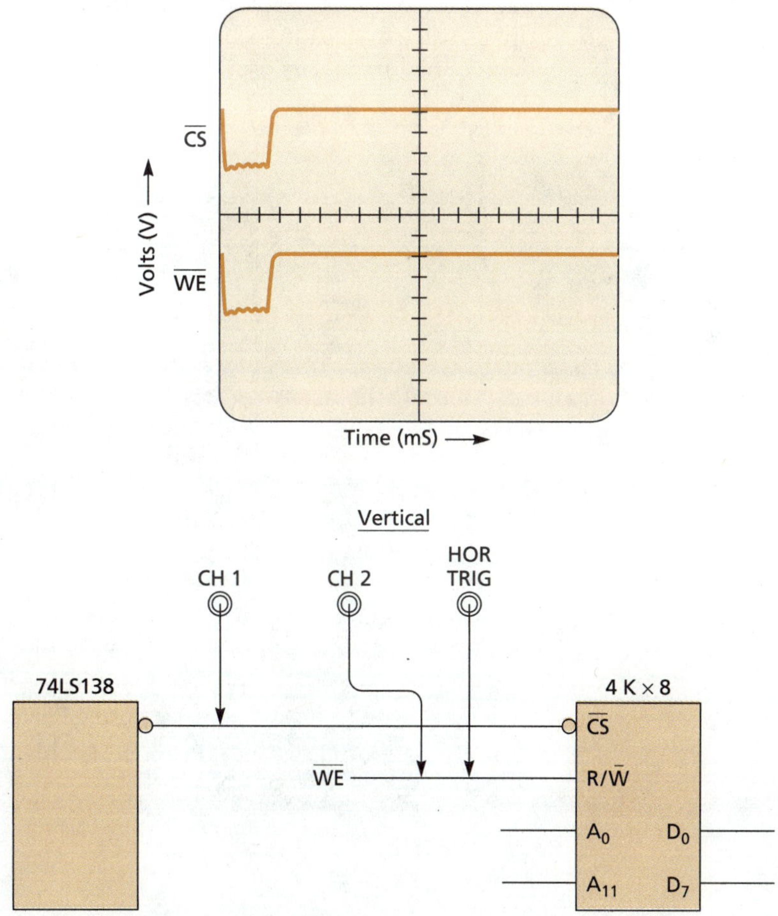

Figure 5.46 Testing Memory Signals

Section 5.10

Practice Problems

1. The eight-bit parallel-load register of Figure 5.47 was loaded with the data E3H. What is wrong with the circuit, and why?

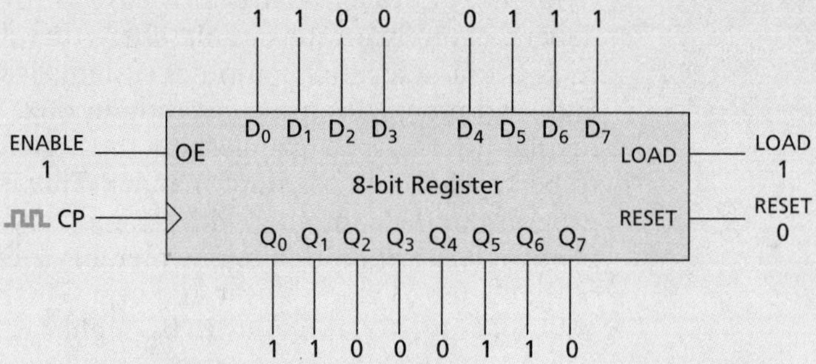

Figure 5.47 Practice Problem 5.10(1)

2. The oscilloscope pattern shown in Figure 5.48 is triggered by the $\overline{WE}$ signal when writing to location 3B25H. The chip select input is shown on the upper trace and address line A_0 on the lower trace. What problem can be determined from this oscilloscope pattern?

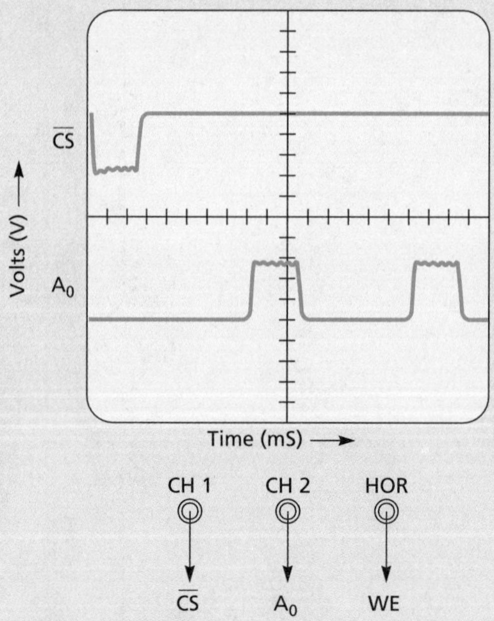

Figure 5.48 Practice Problem 5.10(2)

enable ($\overline{WE}$) signal is asserted at a regular interval. This signal is then used to trigger the horizontal sweep of the oscilloscope. This is accomplished by connecting the vertical and horizontal trigger inputs to the $\overline{WE}$ signal and setting the horizontal trigger to the external mode. Depending on the oscilloscope used, the controls are set to trigger the horizontal sweep with the falling edge of the active low $\overline{WE}$ signal. This causes the horizontal sweep to begin each time the $\overline{WE}$ is asserted. Now, whenever we look at another signal, we can see its state at the time the memory is being written. For instance, on the oscilloscope pattern in Figure 5.47, we show the chip select ($\overline{CS}$) signal going to the memory chip. You will note, by the oscilloscope pattern, that this signal goes low, as it should, at the same time as the $\overline{WE}$ signal. This means that the memory is being selected and written to.

Next, the data bus and address bus inputs to the memory can be checked for their correct state at the time the chip is being written to. Why is this important? Because, in a computer, the data bus and address bus lines are constantly changing as the computer continually accesses different memory locations. We are only concerned about their states when the computer accesses the particular memory chip in question.

If all the inputs are correct for the write cycle, then trigger the horizontal sweep with the output enable ($\overline{G}$) input. You may have to develop a simple circuit that ANDs the output enable with the chip select line to get this memory's correct output enable, because the computer, while doing only one write cycle, does many read cycles to get the instructions to write and read to this memory location. Again, check the address lines to make sure the correct location is being accessed. Lastly, check the output data lines to see if alternating ones (55H) are being read from that location. Good write signals and bad output data indicate either a bad memory chip or some problem associated with the output data lines.

5.11 Designing Memory Systems

Many factors are considered before designing a memory system. They include the size of the overall memory, types of memory devices, address allocations, required access time, and the possibility of future expansion of memory space. Almost every personal computer made today advertises its memory size and how much additional memory can be added to it. This requires that the address decoding, for those additional memory locations, be in place, even though the actual chips are not present. A system designer takes these considerations and begins by developing a memory map detailing the address allocations to be made for the system, usually starting with the ROM memory area.

Each computer has a boot-up or initialization program that runs whenever the computer is first turned on. This program sets up the myriad of devices and circuits within the computer to allow the user to have access to the system. A good portion of this initialization program resides in permanent read-only memory chips on the computer's mother board. In order to read and execute these programs, a microprocessor must first be able to address the ROM's locations. The addresses for these locations are determined by the starting address that the microprocessor sends out after it has been powered up and reset. From here it fetches the first instruction. We delve into more details of the computer's operation in later chapters. It's enough to say, at this point, that the microprocessor determines where this boot-up ROM must reside in the memory map. The amount of ROM and additional address allocations are determined by the size and content of the actual initialization program.

Once ROM has been allocated, the remaining address space left can be assigned to RAM chips. The amount of space usable is fixed by the number of address lines available from the microprocessor. For instance, an INTEL 8088 microprocessor has 20 address lines and can access up to 1 megabyte (MB) of memory while a Motorola 68000 has 24 address lines and can access a total of 16 MB of memory. Some of this memory space is dedicated to use by the initialization program, but a major portion is allocated to the user—you or me. The designer maps out these locations, as shown in a typical map, in Figure 5.49, and then precedes to develop the hardware required to properly decode the address bus and supply the write and output enables to access the memory devices. If these devices have a RAS/CAS addressing capability, then the designer must also consider the circuitry to generate the RAS, MUX, and CAS sig-

19	18	17	16	15	14	13	12	11	10	9	8	7	6	5	4	3	2	1	0	Device and Range
0	0	0	0	0	0	0	0	X ←											→ X	RAM 1 00000–00FFF
0	0	0	0	0	0	0	1	X ←											→ X	RAM 2 01000–01FFF
0	0	0	0	0	0	1	0	X ←											→ X	RAM 3 02000–02FFF
0	0	0	0	0	0	1	1	X ←											→ X	RAM 4 03000–03FFF
1	1	1	1	1	1	1	0	X ←											→ X	ROM 1 FE000–FEFFF
1	1	1	1	1	1	1	1	X ←											→ X	ROM 2 FF000–FFFFF

X = variable address lines

Figure 5.49 Memory Map

Section 5.11

Practice Problems

Make up a memory map with the following requirements:

1. Microprocessor accesses location FF0000H after reset.
2. Boot program is 12 kilobytes (KB) long.
3. Boot program uses two scratch pads in RAM, each 1 KB wide.
4. The user is allocated 640 KB of RAM space.

nals so that these devices can be properly accessed. Dynamic memory usage adds a refresh requirement to complete the system using them.

Summary

Memory devices come in many flavors—temporary short-term memory, called *registers,* read/write long-term storage, and read-only nonvolatile memories. Registers can be loaded and read either serially or in parallel. Serial data require a single transmission line, but it takes time to shift the data through the memory. Parallel loading and reading of data is quicker, but requires separate data lines for each bit of information loaded or read.

Solid state memories are used for most storage applications in computers today. These fall into two basic categories, read/write or RAM, and read-only or ROM memories. RAM memories are used when regular updating of information is required. That information is then used by an application program for computation and decision-making purposes. This information or data can be read in random or sequential order. RAMs are volatile in that the data held in them are lost when power is turned off. A permanent type of memory required when the same data are to be available at any time is the read-only memory or ROM. This memory is nonvolatile and can hold information that may be accessed randomly (such as look-up tables) or sequentially (such as an initialization program). Today's ROMs are electrically programmable and erasable, which allows them to be updated when required.

Decoders and encoders are used to translate between single signal lines and their binary equivalents. They can also be used to translate between different types of binary combinations. Multiplexers and demultiplexers play an important part in many digital circuits by directing the flow of signal traffic through various circuits. They find applications in environments where it is necessary to reduce or expand the number of signal paths in use.

Memory expansion relies on correctly decoding address inputs to enable individual chips in which a desired location resides.

Address decoding may be full or partial, depending on the size of the address bus and the amount of memory locations to be accessed within a given system. The main guideline when designing an address-decoding scheme is that while many addresses may be available to access a single location, no address is to access more than one location.

RAS and CAS inputs to a memory device allow larger-capacity memories to be accessed by fewer address inputs. Half of the address bus is latched into the memory address buffer when RAS is asserted, and the second half when CAS is asserted. A multiplexer is used to switch between the two halves of the address bus, supplying the correct half in conjunction with the proper control signal, RAS or CAS.

Registers and memory devices have different troubleshooting requirements, depending on how they are deployed in a circuit. Sometimes they are controlled with static signals and can be checked by applying required inputs and looking at the results. At other times their control inputs are constantly changing, and inputs to be examined must be synchronized to a known signal occurrence.

Designing memory systems begins by knowing what the computer system requirements are, beginning with the first address accessed by a microprocessor. ROM initialization programs begin at this address. Remaining address allocations depend on the read/write memory access needs of the initialization program and the amount of memory allocated for user application programs and data.

Glossary

access time the time between addressing a memory and receiving the data, at that address, onto the data bus.

address a binary value that signifies the location of a data word in a memory.

address bus a set of electrical conductors that carry address information.

address decoding the process of decoding the information on the address bus to select a memory device.

bus a set of digital lines that carry the same type of binary information.

capacity the amount of data that can be held by a memory device or circuit.

CAS (column address select) latch second half of an address

cell a memory element that holds one digital bit.

data bus a set of electrical conductors that carry data information.

decoder a logic circuit that translates binary codes into individual outputs.

demultiplexer (demux) a logic circuit that channels an input into one of many outputs.

DRAM dynamic random access memory.

encoder a logic circuit that translates numerous individual inputs into a binary code.

EEPROM electrically erasable read-only memory.

EPROM erasable read-only memory.

full address decoding decodes all address lines to determine which memory location is accessed.

LSI (large-scale integration) a group of devices that contain a vast amount of logic circuitry such as a microprocessor.

memories devices that retain digital data.

memory map a listing of addressees and devices they access.

MSI (medium-scale integration) a group of devices that logic circuits with a small amount of complexity. A quad D-flip-flop and shift register chips are examples.

multiplexer (mux) a logic circuit that selects an output from many inputs.

nondestructive read data are not destroyed when they are read from memory.

parallel the movement of all the bits of a binary data word at one time.

partial address decoding uses a portion of the address bus to determine the location of data in a memory system.

PROM programmable read-only memory, which is provided by a class of memory devices that can only be read after data have been permanently stored in them.

RAM random access memory.

RAS (row address select) latch the first half of an address into a memory (row location).

read cycle the sequence of events that allows a device to access memory and receive data from it.

refreshing the process of rewriting data into a flip-flop or memory to retain those data.

register a logic circuit that temporarily stores digital data.

ROM read-only memory.

SAM sequential access memory.

serial the movement of binary data one bit at a time.

shift left the serial movement of data from the least significant bit position in a register toward the most significant bit.

shift right the serial movement of data from the most significant bit position in a register toward the least significant bit.

SSI small-scale integration; defines a group of devices that contain a small amount of logic circuitry. A quad two-input NAND chip is an example.

strobe a signal of short duration that initiates an event or process.

TTL (transistor–transistor logic) a family of digital devices that use transistors as switching elements.

volatile memories memories that lose their data when power is shut off.

write cycle the sequence of events and signals that allows a data word to be entered into a memory location.

Solutions to Practice Problems

Section 5.2

1.

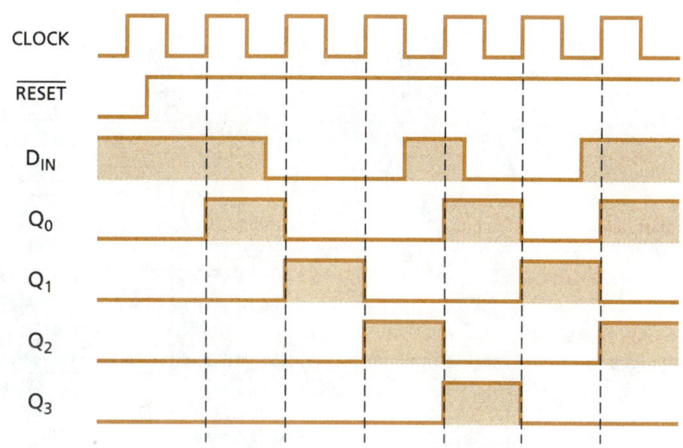

Figure 5.50 Practice Problem 5.2(1) (solution)

2.

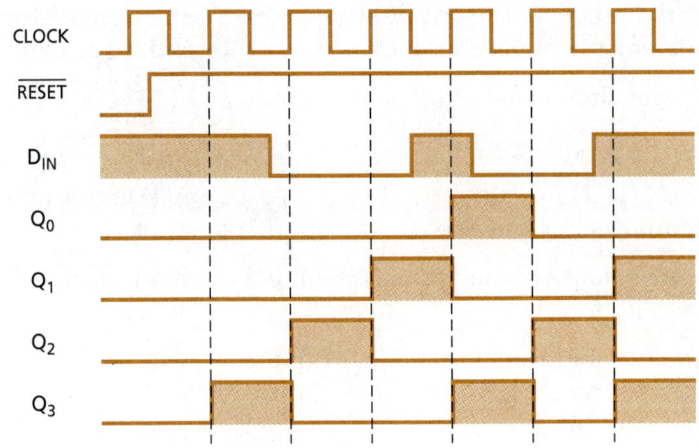

Figure 5.51 Practice Problem 5.2(2) (solution)

Section 5.3

1.

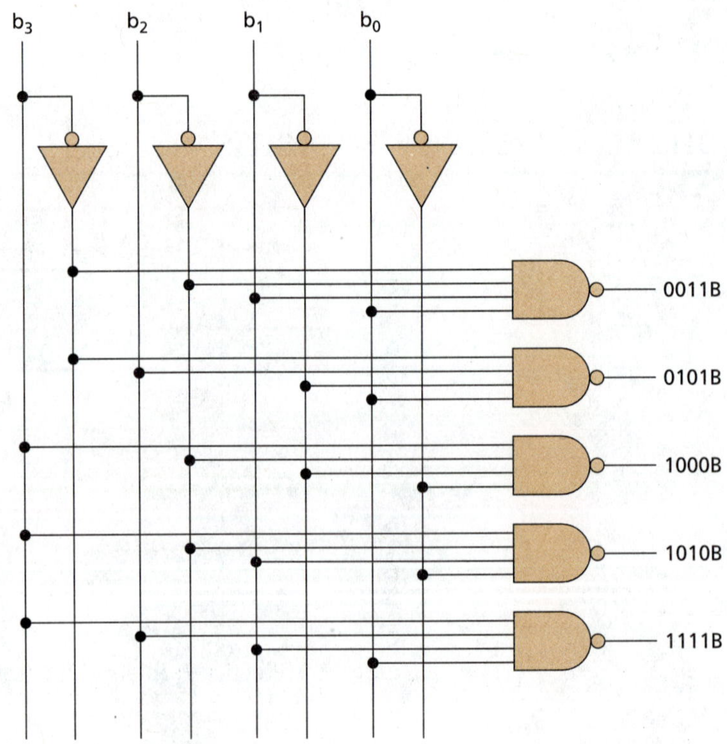

Figure 5.52 Practice Problem 5.3(1) (solution)

2. Figure 5.53

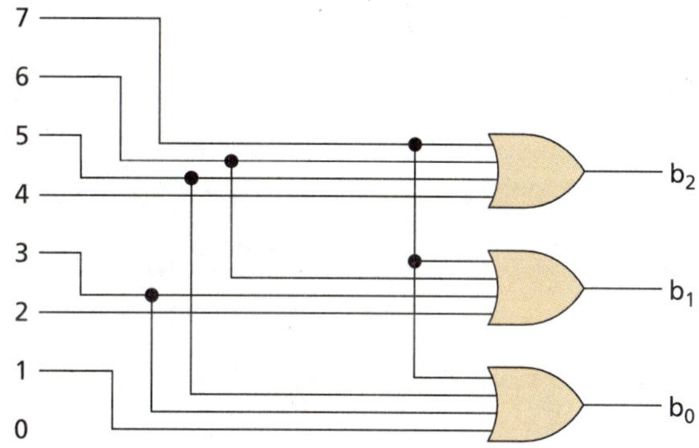

Figure 5.53 Practice Problem 5.3(2) (solution)

Section 5.4

1. Output Y_b is low because the enable for side b is negated. Output Y_a is high because it is enabled and input I_{3a} is selected by S_1 and S_0.

2. Y_c is high because it is selected and passes the logic state of IN to its output. The other outputs are all low because they are deselected.

Section 5.5

1. **a.** Shift left **b.** Parallel load **c.** Shift right **d.** Parallel load using 74LS173 **e.** Shift left using 74LS173

2.

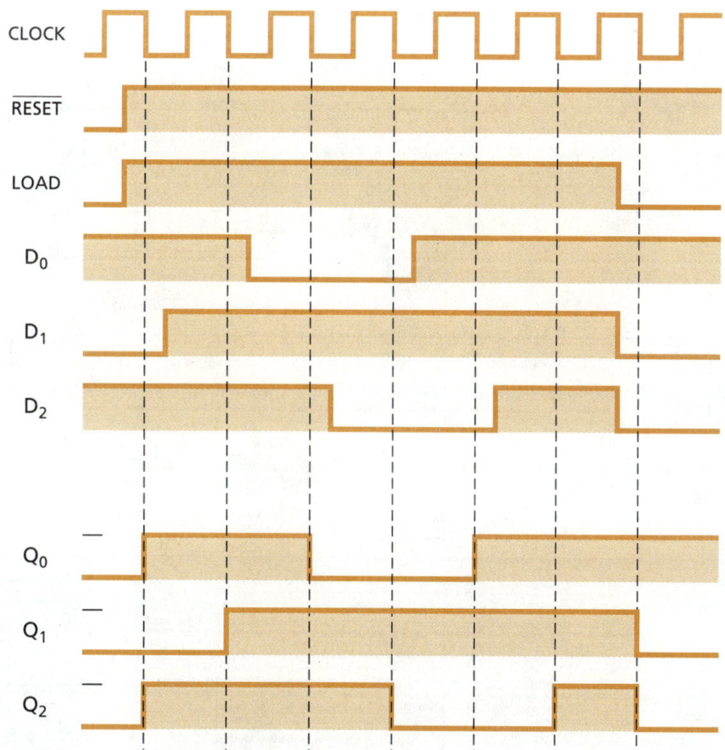

Figure 5.54 Practice Problem 5.5(2) (solution)

3.

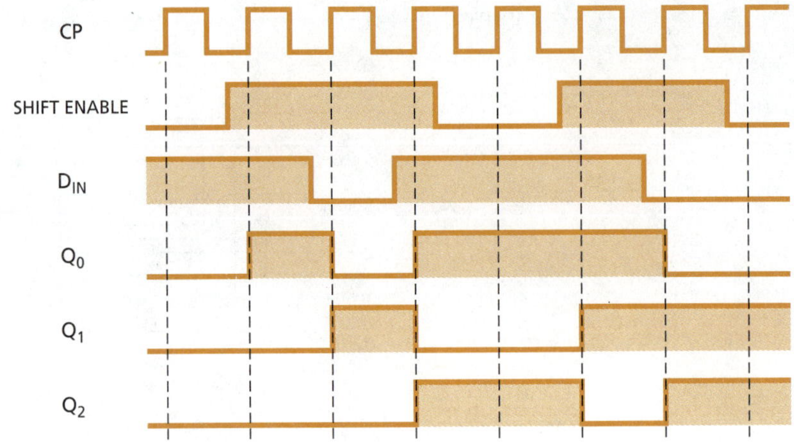

Figure 5.55 Practice Problem 5.5(3) (solution)

4.

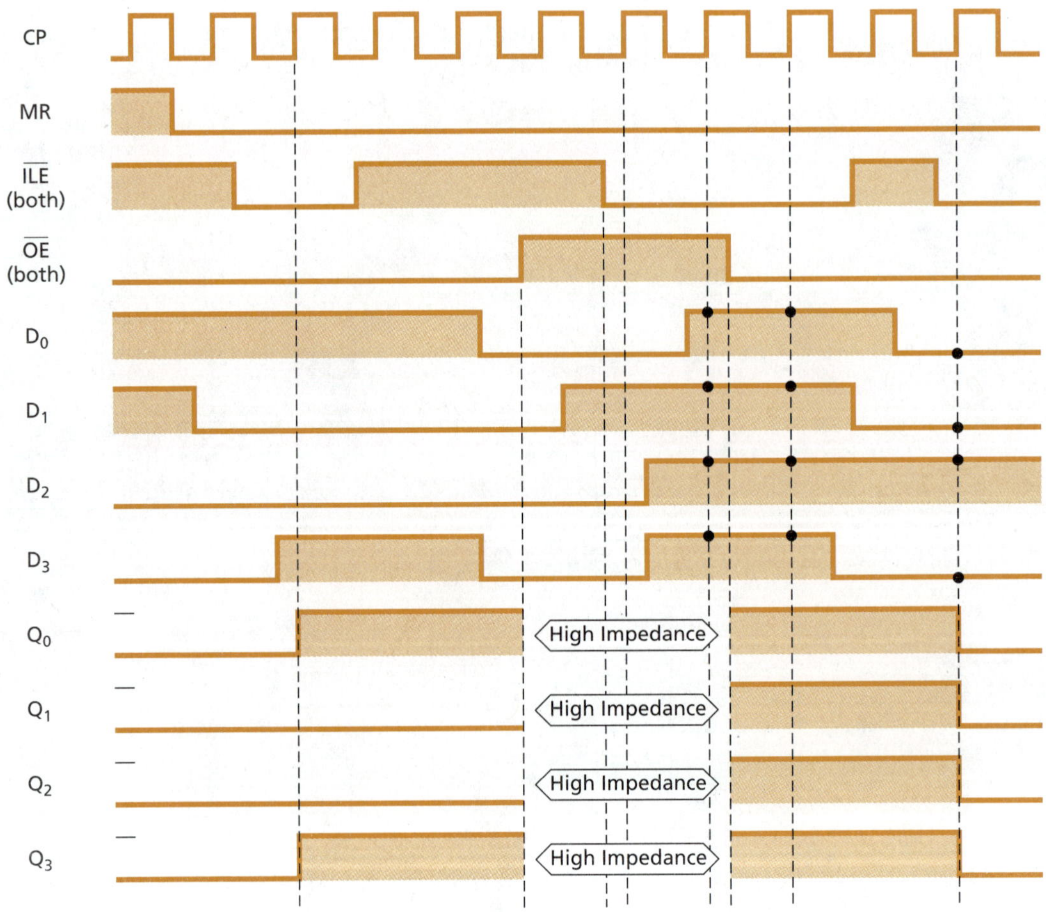

Figure 5.56 Practice Problem 5.5(4) (solution)

5.

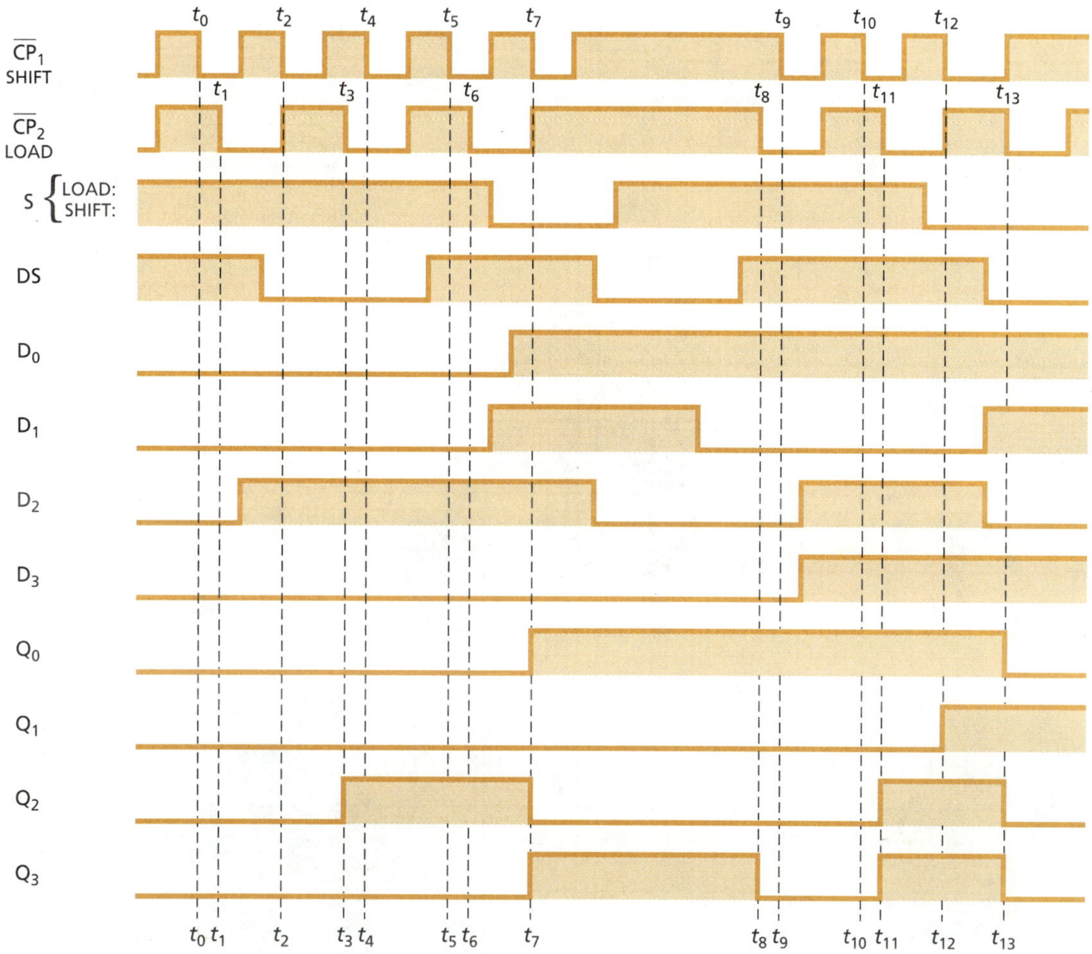

Figure 5.57 Practice Problem 5.5(5) (solution)

6. a. Parallel load **b.** Shift left **c.** Parallel load and shift left

---

Section 5.6

1. 12 address lines, 8 data lines **2.** 32 KB or 32,768 bits

---

Section 5.7

1. Dynamic memories require periodic refreshing, whereas static do not. Since dynamic memories use less power, they have a higher packing density.

2. 256 KB or 262,144 bits.

3. Dynamic memories are used because of their low power drain.

Section 5.8

1. ROM, nonvolatile, read-only memory
RAM, volatile read/write random access memory

2. ROMs are preferred whenever data or programs must be retained even when power is shut off

3. PROM, user-programmable read-only memory
EPROM, can be erased using ultraviolet light and reprogrammed
EEPROM, electrically erasable and reprogrammable

Section 5.9

1. ROM: 0000H–07FFH RAM1: 2000H–27FFH RAM2: 3000H–37FFH

2.

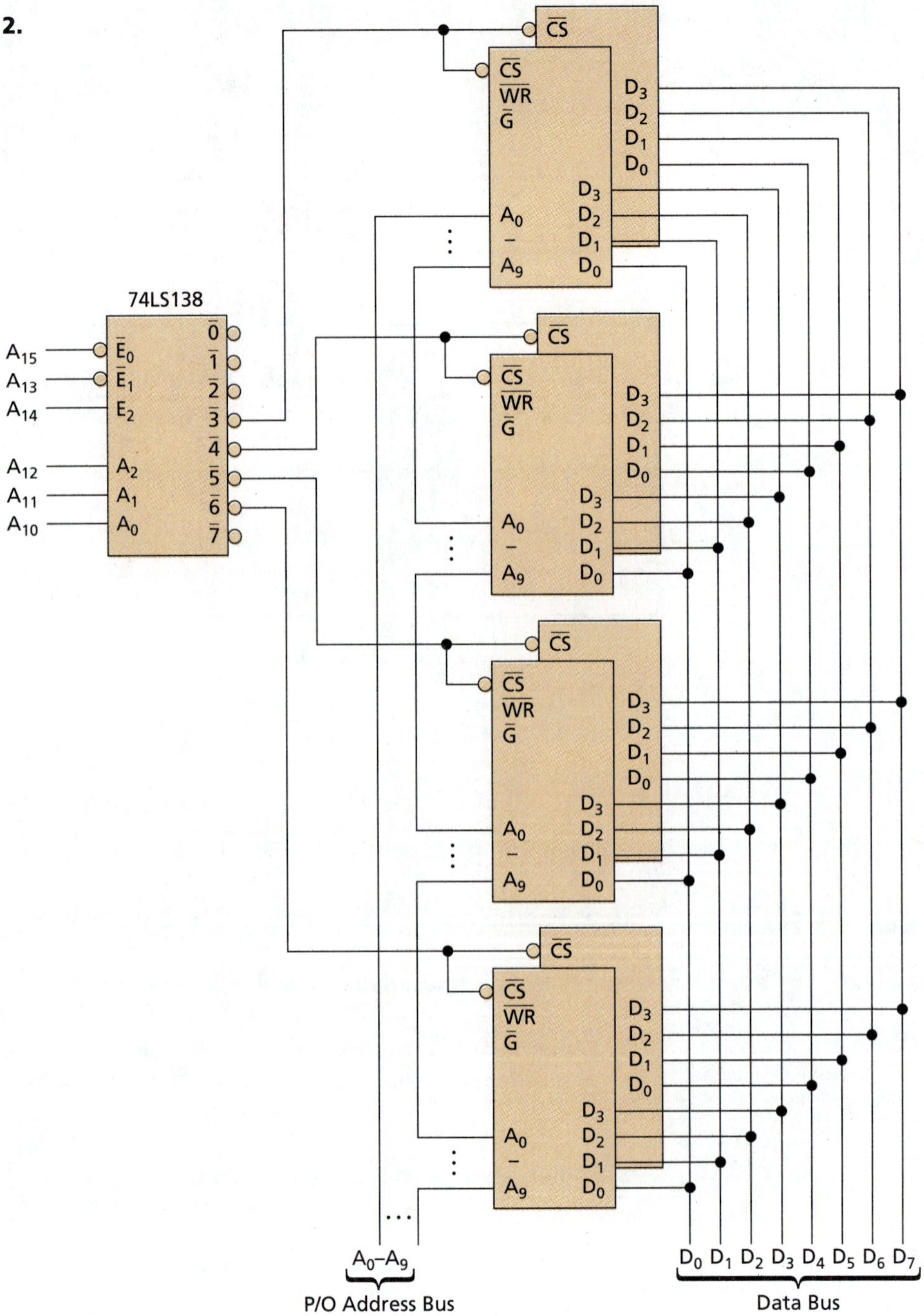

Figure 5.58 Practice Problem 5.9(2) (solution)

Section 5.10

1. The most significant bit is bad—the last flip-flop inside the chip is bad.
2. Address line A_0 is not high when the chip is being selected. Something is wrong with this input.

Section 5.11

1 and 2. System ROM (boot program): FF0000H–FF3FFFH (14 address lines: 16 KB needed for 12 KB program).

3. Boot-up RAM scratch pads: 000000H—000FFFH (2 KB RAM).
4. User RAM: 100000H–1FFFFFH (1 MB minimum for 640 KB space).

No conflicting addresses allowed, and assigned space for each section must be sufficient. Actual addresses (other than ROM) can be different.

Problems

5.1 Define the following terms:
 a. Register b. Parallel data c. Serial data

5.2 What advantage does a parallel-load register have over a serial-shift register? What advantage does the latter have over the former?

5.3 Complete the timing diagram of Figure 5.59 for the parallel-in / parallel-out register of Figure 5.60.

5.4 What is required to load a four-bit data word into a 74LS173 quad-D flip-flop register?

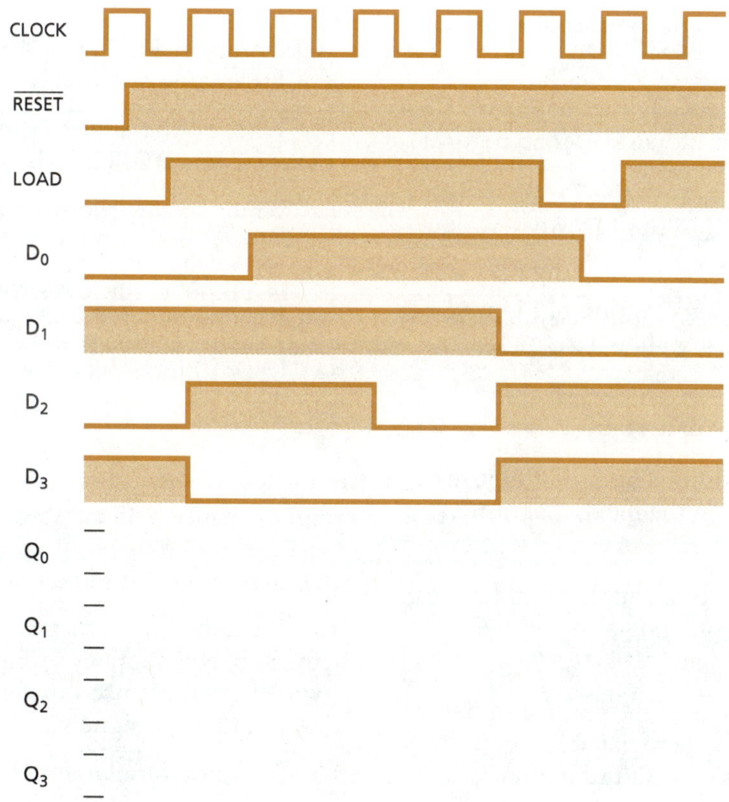

Figure 5.59 Problem 5.3

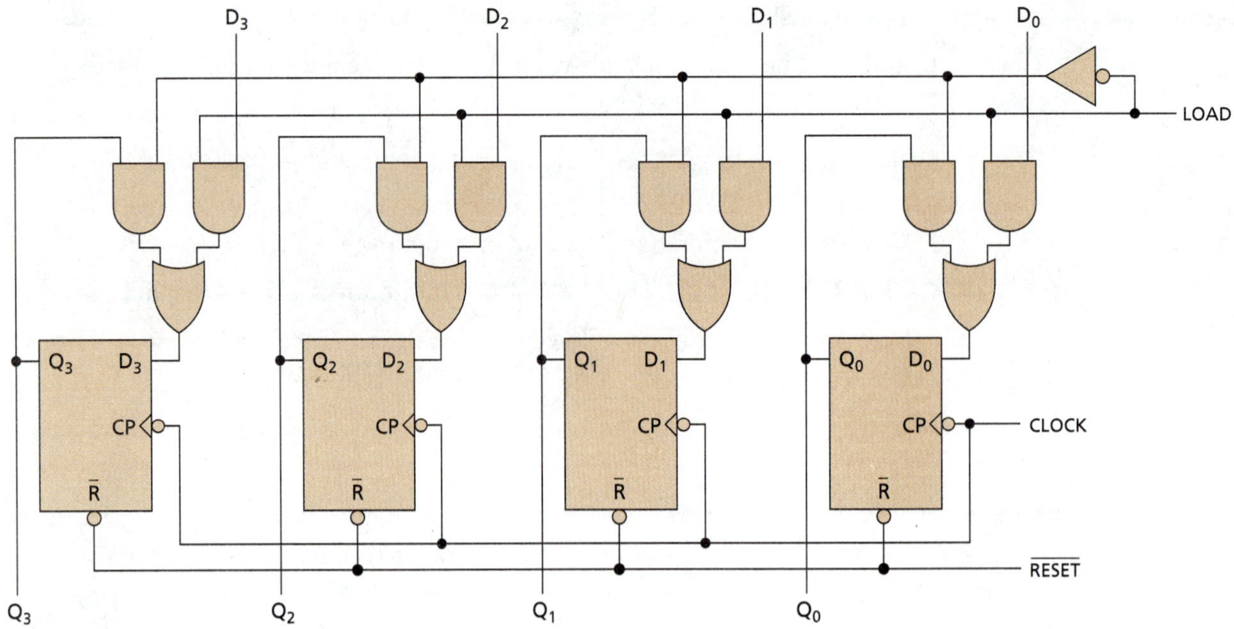

Figure 5.60 **Parallel-Load Register for Problem 5.3**

5.5 What signal(s) is (are) required to enable the 74LS173 outputs onto the output pins of the chip?

5.6 Complete the timing diagram of Figure 5.61 for the 74LS173 four-bit parallel-load register.

5.7 What types of circuits in the 74LS173 determines whether the flip-flops receive new data or refresh themselves?

5.8 Complete the timing diagram of Figure 5.62 for the shift-left register shown in the figure.

5.9 Describe the difference between shifting data left and shifting data right.

5.10 Complete the timing diagram of Figure 5.63 for the register in the block diagram of Figure 5.64.

5.11 Give an example of an application for a circuit capable of accepting data in parallel and shifting data out serially.

5.12 Describe the states of the control signals of the 74LS95 to cause the register to

a. Parallel load
b. Shift data out
c. Retain current data

5.13 Define the following memory terms:

a. Volatile
b. Access time
c. Memory cell
d. Address
e. Chip enable or select

5.14 What do the abbreviations RAM and ROM mean? Can a ROM also be a RAM? Explain.

5.15 What is a destructive read cycle? What must be done following the reading of data?

5.16 Is a bipolar static read/write memory volatile? Explain.

5.17 Does a bipolar static read/write memory employ a destructive read cycle? Explain your answer.

5.18 How is a memory word in the memory circuit of Figure 5.65 accessed and read? Describe the signals required and what goes on in the memory for it to be read.

5.19 Describe how a memory location is accessed in the memory of Figure 5.66. What type of circuit is used to generate the RAS, MUX, and CAS signals?

5.20 Compare bipolar static read/write memories and dynamic MOSFET memories. What

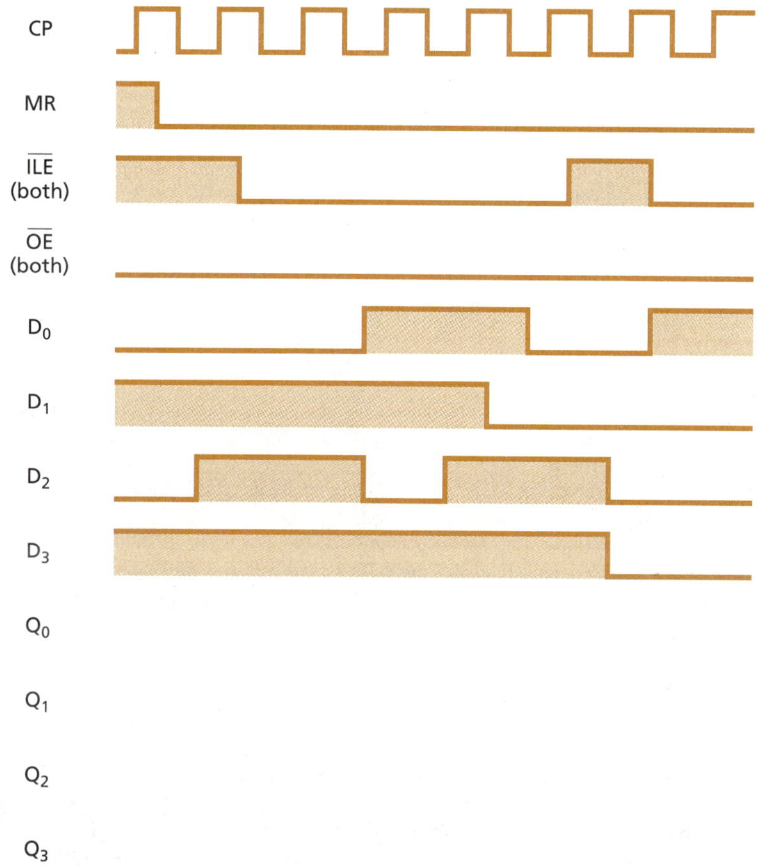

Figure 5.61 Problem 5.6

are the advantages of one compared to the other?

5.21 Why are dynamic memories refreshed?

5.22 What characteristic of a MOSFET is employed to store a logic 1 in a dynamic memory?

5.23 Why is the packing density in a dynamic memory so high?

5.24 What are the factors that govern packing density?

5.25 What is the purpose of a ROM memory?

5.26 What are the differences among a ROM, PROM, EPROM, and EEPROM?

5.27 Draw the circuit for a binary-to-hexadecimal (4-to-16) decoder.

5.28 Draw the circuit for a hexadecimal-to-binary encoder (16-to-4).

5.29 For the inputs to the 74LS138 shown in Figure 5.67, which output is asserted? What is the state of that asserted line?

5.30 Draw a circuit that uses a 74LS138 to turn on the chip selects of four different RAM chips based on this binary code:

A_4	A_3	A_2	A_1	A_0	RAM NUMBER
0	0	0	1	0	$\overline{RAM1}$
0	0	0	1	1	$\overline{RAM2}$
0	0	1	0	1	$\overline{RAM3}$
0	0	1	1	1	$\overline{RAM4}$

5.31 Explain what multiplexers and demultiplexers do.

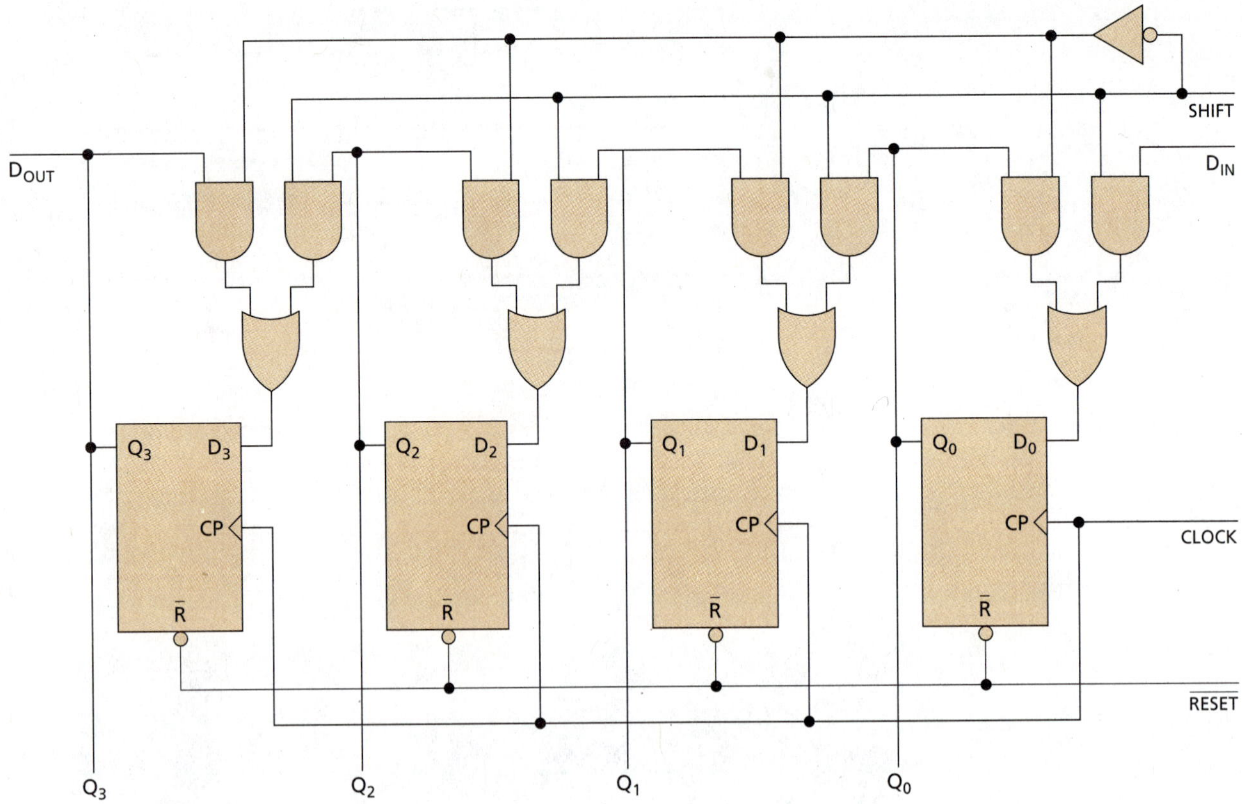

Figure 5.62 **Shift-Left Register for Problem 5.8**

5.32 Draw the logic circuit for a 3-to-1 multiplexer. It is to be enabled continuously, which eliminates the need for a separate enable input.

5.33 Draw the logic circuit for a 1-to-8 demultiplexer. This circuit is enabled continuously, eliminating the need for a separate enabling input.

5.34 What are the states of the outputs of the 74LS153 of Figure 5.68 with the inputs shown? Explain why each output is in the state you indicate.

5.35 What are the results of applying the signals shown to the circuit of Figure 5.69?

5.36 Design an address decoder circuit that enables several memory devices in accordance with the truth table shown in Figure 5.70.

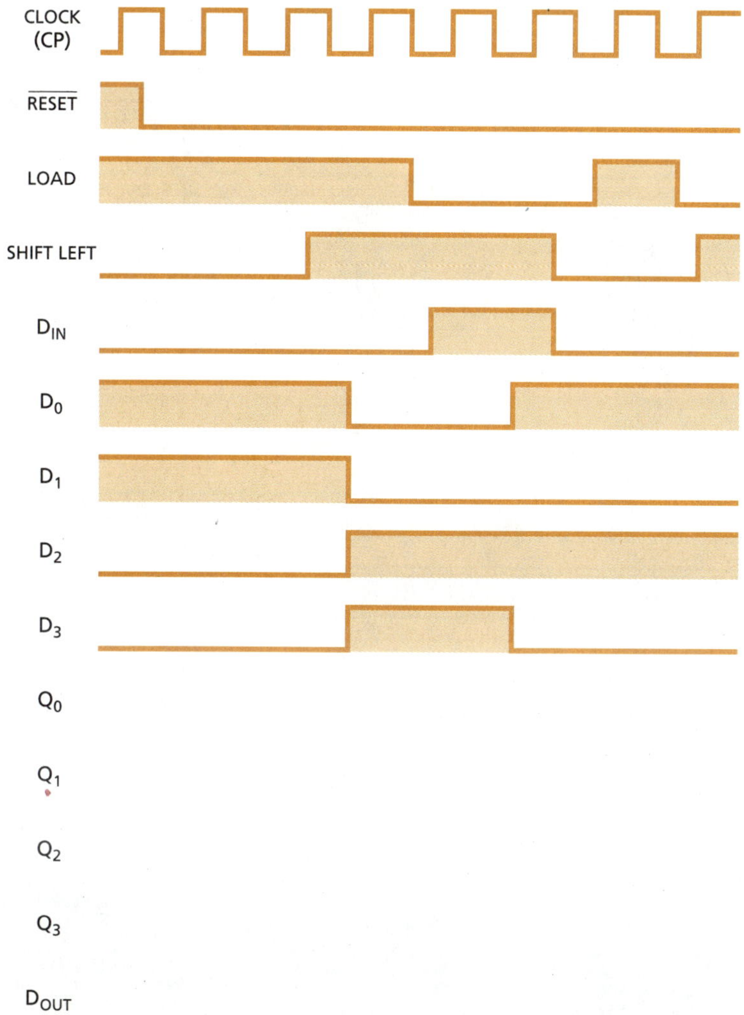

Figure 5.63 Problem 5.10

RESET	SHIFT LEFT	LOAD	CLOCK (CP)	Function
1	X	X	X	$Q_3, Q_2, Q_1, Q_0 = 0$
0	1	0	↑	Shift Left
0	0	1	↑	Parallel Load
0	1	1	X	Undefined

X = "don't care" state

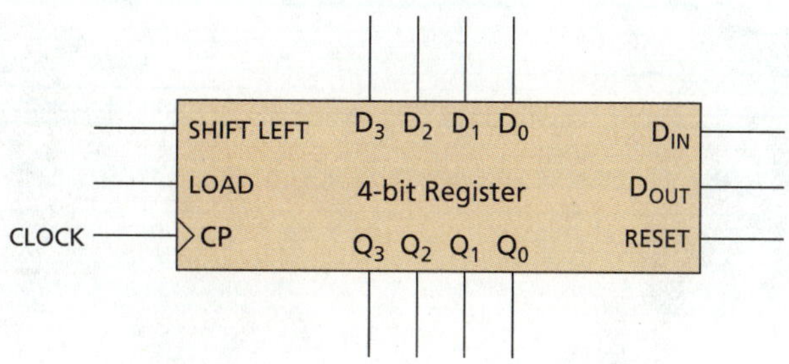

Figure 5.64 Four-Bit Register Problem 5.10

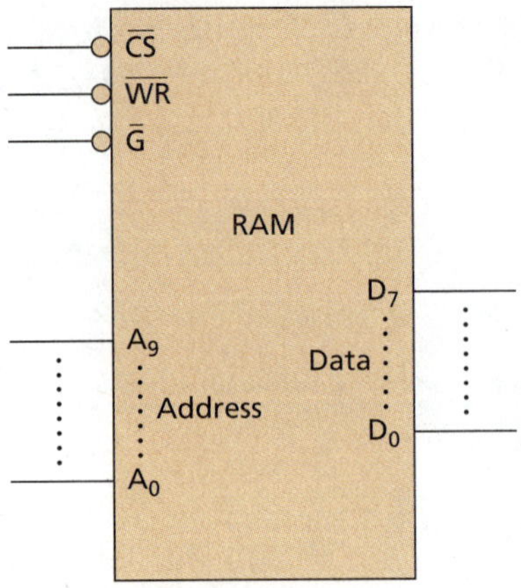

Figure 5.65 Problem 5.18

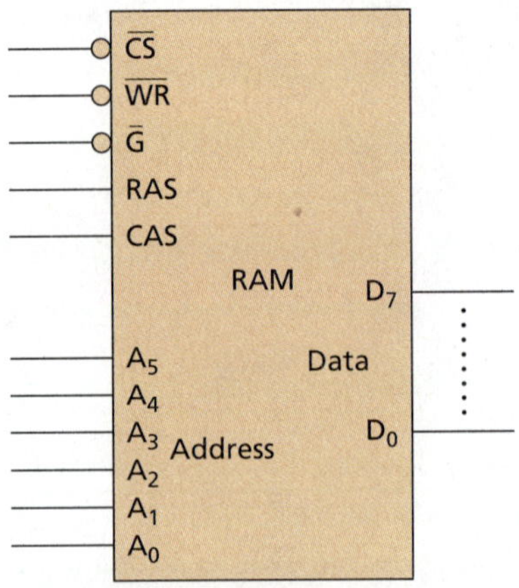

Figure 5.66 Problem 5.19

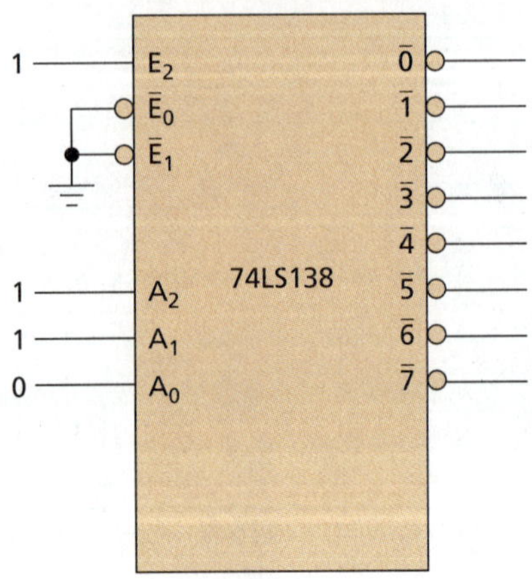

Figure 5.67 Problem 5.29

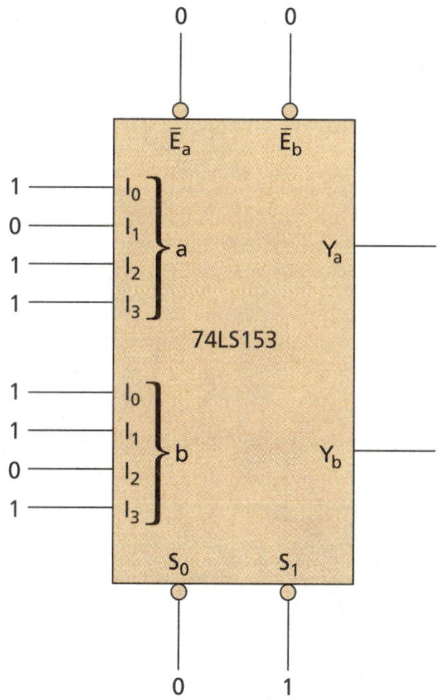

Figure 5.68 Problem 5.34

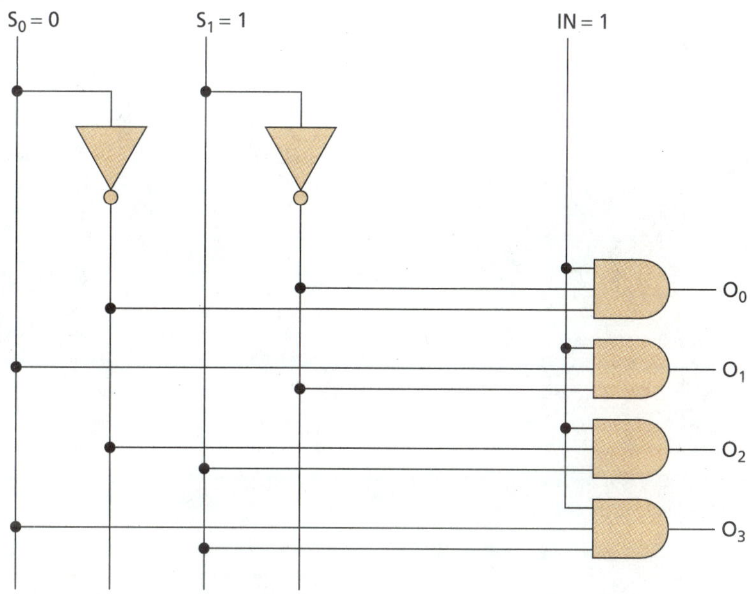

Figure 5.69 Problem 5.35

Address Range (in hex)	Signal
20000 to 23FFF	$\overline{\text{ROM}}$
24000 to 27FFF	$\overline{\text{RAM1}}$
28000 to 2BFFF	$\overline{\text{RAM2}}$
2C000 to 2C007	$\overline{\text{PNTR}}$
30000 to 30007	$\overline{\text{FDISK}}$

Figure 5.70 Problem 5.36

SIX

Programmable Logic Devices

KEY TERMS

CUPL (Universal Compiler for Programmable Logic Devices)
GAL (generic or gate array logic)
JEDEC (Joint Electronic Devices Engineering Council)
PAL (programmable array logic)
PLA (programmable logic array)
PLD (programmable logic device)
PROM (programmable read-only memory)

OBJECTIVES

After reading this chapter, the student should be able to

- Discuss the basic circuitry of a programmable logic device (PLD), including its array and output configurations.
- Describe the differences and similarities between different types of PLDs, along with some applications.
- Explain how PLD designs are created and listed in a file that is compiled by software programs such as CUPL into a JEDEC standard form, and how the JEDEC file is used to enter the design into the PLD.
- Describe various types of PLD designs as examples for the process of designing logic circuits: logic expression design, truth table design, D-register design, and state table design.
- Discuss applications for PLD designs.

OVERVIEW

This chapter discusses the theory, use, and applications of programmable logic devices (PLDs). These devices are described using the popular GAL16V8 device as a prime sample device. The CUPL universal compiler software package used for compiling PLD designs is used as an example of supporting software packages for PLD design and implementations. A student version of CUPL is supplied with the text along with copies of all .PLD files used in the text.

6.1 Introduction

PLD *programmable logic device designates a group of logic devices whose circuits and functions are programmable.*

Digital electronics and the circuitry used to implement new designs are continually evolving. The need for customized circuits led to the creation of **programmable logic devices** or **PLDs.** Originally, if a circuit implementation was designed to be included in an IC package, it would first be constructed on a prototype board layout and then tested and revised as needed. Once the circuit was operating properly, a contract was made with an IC manufacturer to implement the design in an IC package. The time required by the IC manufacturer and the expense incurred were offset by the opportunity to receive a customized logic device in an IC package. If a custom design appeared to have applications beyond the original customer use, the IC manufacturer would add that component to the company's product line. Still, it would be better, expense aside, if engineers could have the ability to produce their own custom IC device within a minimal time period. This is the first major benefit of programmable logic devices. A secondary, and not so minimal effect, was the lowering of cost to have this capability. After initial expenses for software and hardware, the only expense is the PLD devices themselves.

Programmable logic devices were developed to satisfy the increasing need to create customized logic circuits at a customer site. A logic design was developed so that it could be programmed into a PLD. Programming was achieved by first creating the sum-of-products terms for the desired logic circuit, converting the design into a plot of fuse links associated with a particular PLD, and then programming the PLD to contain the logic circuit of the original design. Using PLDs to implement custom designs is relatively inexpensive, because unprogrammed PLDs are inexpensive and the software required to implement designs in PLDs only incurs an initial expense.

6.2 The PLD Base Circuit

PLDs come in many forms, but the general basis of all these devices is the AND/OR circuit shown in Figure 6.1. Inputs are selected to be connected ("fused") to the AND gates to form the AND terms of a sum-of-products expression. Connection of the outputs of the AND gates to OR gates is also programmable, allowing selection of the OR function in a sum-of-products expression. Leaving a "fuse" intact allows an input to be connected. "Blowing" the "fuse" opens the connection, leaving the open input in a nonasserted state. The words "fuse" and "blowing" are in quotes because PLDs do not have fuses in the standard sense. If they did, a blown fuse would remain open forever, but many PLDs are reprogrammable, which means that there is a method for closing an open fuse as well as blowing one open. The inputs and their complements are brought to all the AND gates through fuses. AND gate outputs, in turn, are connected to OR gate inputs through additional fuses. A "blown fuse" at the input of the AND gate leaves that input high, whereas a "blown fuse" at the input of an OR gate leaves that input low. In other words, blown fuses leave their corresponding inputs terminated in an inactive state. Fuses that remain

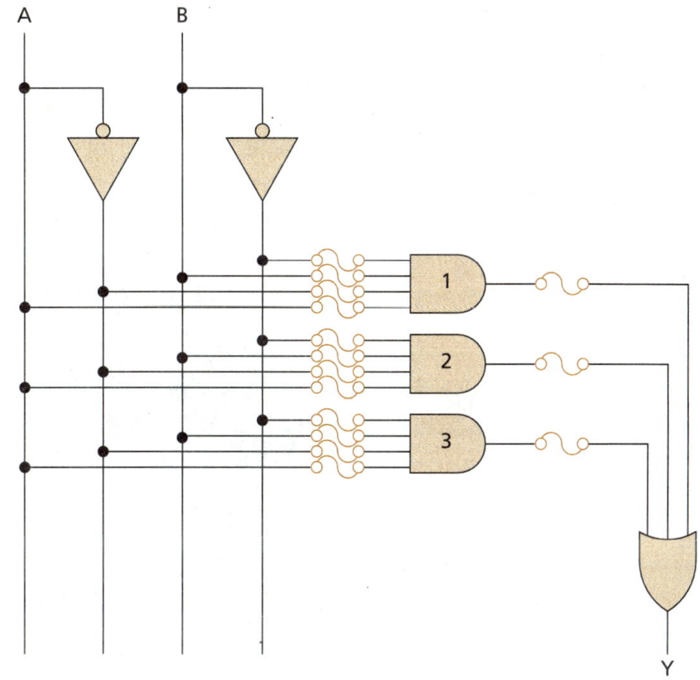

Figure 6.1 PLD AND/OR Cell

intact transfer input information to the respective gate. To illustrate the impact of the logic states resulting from fuse condition, let's implement an exclusive OR expression using the PLD example in Figure 6.1.

The sum-of-product terms expression for an exclusive OR is $\overline{A}B + A\overline{B}$. To implement it using the AND/OR circuit of Figure 6.1, the fuses connecting A and $\overline{B}$ to AND gate 1 and those connecting $\overline{A}$ and B to AND gate 2 are blown. The inputs to AND gate 3 are left alone, but the fuse on the gate's output is blown, making that input to the OR gate a logic 0, which does not affect the output of the OR gate. Blowing the fuses mentioned implements the exclusive OR logic expression, as illustrated in Figure 6.2. Unblown fuses retain their connections whether they are from the inputs to the AND gates, or from AND gate outputs to OR gate inputs. In Figure 6.2, trace the paths that are still connected, and the resulting expressions at the output will be that of the exclusive OR.

PLDs contain AND/OR circuits with many available inputs and outputs. A drawing showing all the connections among inputs and AND and OR gates would have many lines, so a simplified way to illustrate an AND/OR PLD circuit, shown in Figure 6.3, is used. Each single line to the input of a gate represents all the input lines to that gate. An X where two lines cross indicates that a connection is made from an input line to the line it crosses, whereas the lack of an X shows an open (fuse blown) connection between those two lines. A dot on two crossed lines indicates a permanent connection. The implementation of the exclusive OR discussed in the previous paragraph is illustrated in Figure 6.4.

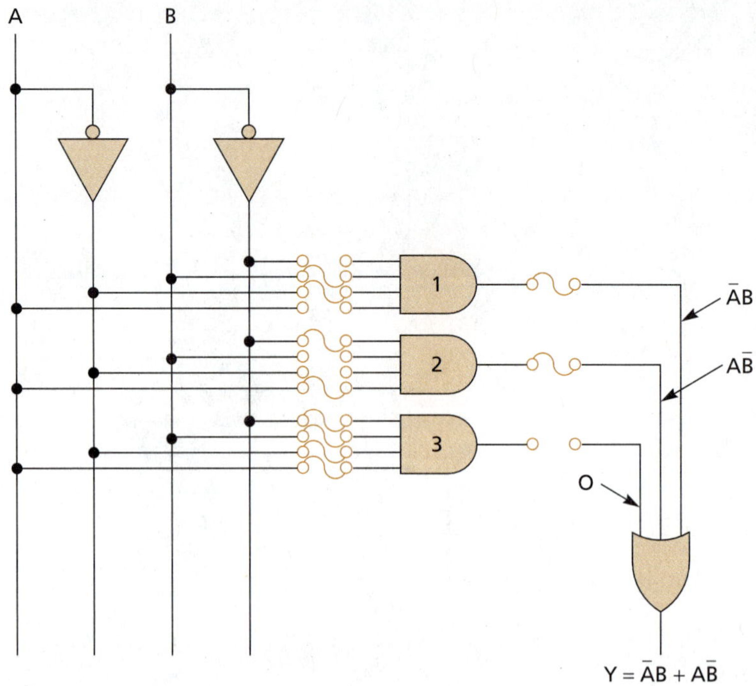

Figure 6.2 Implementing Exclusive OR

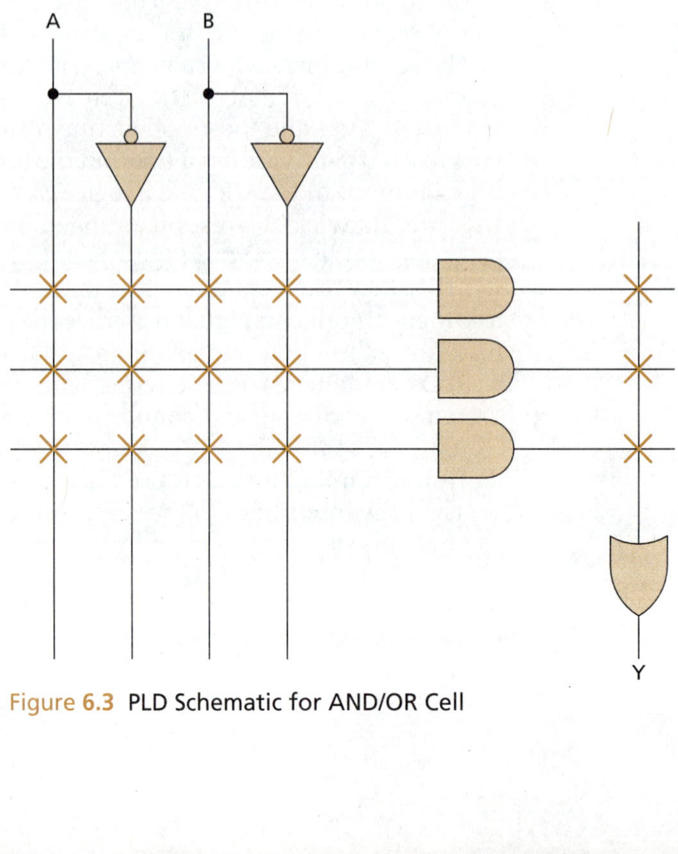

Figure 6.3 PLD Schematic for AND/OR Cell

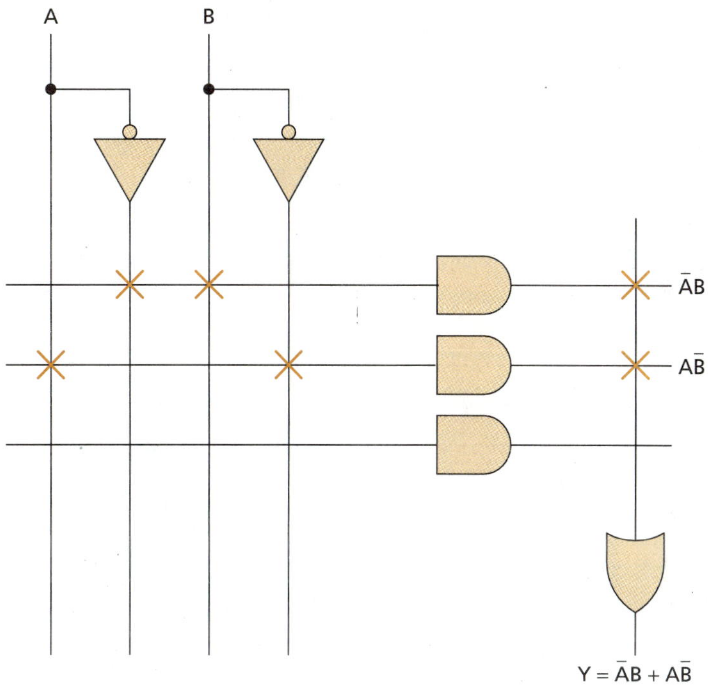

Figure 6.4 PLD Implementing of XOR

EXAMPLE 6.1

Show how the circuit in Figure 6.5 is implemented using a PLD.

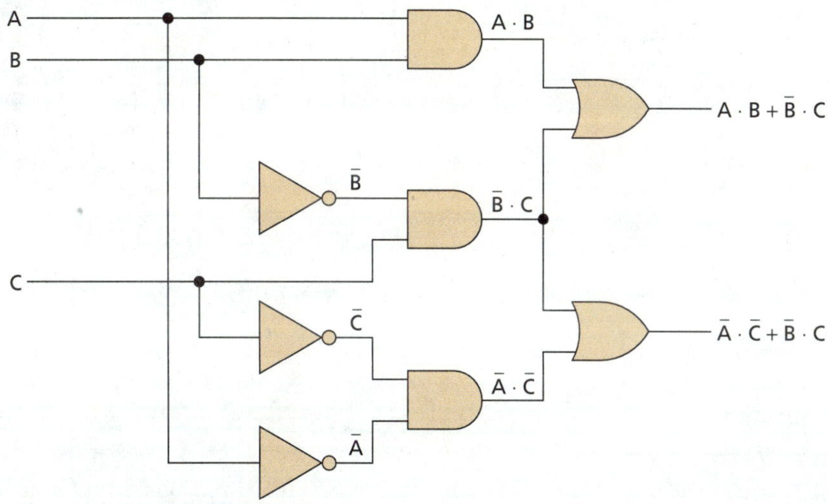

Figure 6.5 Example Logic Circuit

SOLUTION Two AND/OR circuits are needed. The implementation of the logic circuit in Figure 6.5 is represented in Figure 6.6. Actual Boolean expressions are written on the figure to verify the implementation:

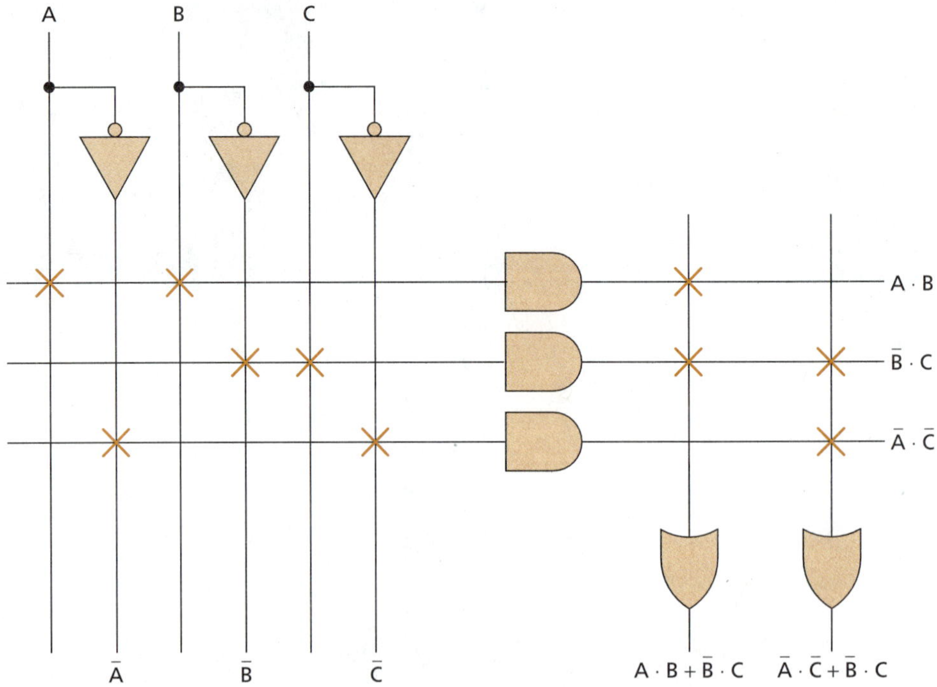

Figure **6.6** Implementing Example Logic Circuit

Section 6.2

Practice Problems

Implement the following logic expressions using PLDs. Do not forget to convert to a sum-of-products expression, first, if necessary.

1. $\overline{A}BC + \overline{A}B\overline{C} + A\overline{B}$
2. $(A + \overline{B} + C)(\overline{A} + B)$
3. $\overline{(\overline{A} + B + \overline{C})(A + \overline{B} + \overline{C})(\overline{A} + \overline{B} + \overline{C})}$

6.3 Types of PLDs

PAL *programmable array logic devices are those with programmable AND arrays only.*

PROM *programmable read-only memory devices with programmable OR arrays only.*

Illustrated in Figure 6.7 are three basic types of PLDs, which are classified by the elements in them that are programmable. **Programmable array logics** or **PALs,** shown in Figure 6.7a, are PLDs whose AND inputs are programmable and whose OR inputs are fixed. This allows the AND product terms to be selectable but leaves the OR sums fixed.

The **programmable read-only memory** or **PROM** reverses the options of a PAL. AND inputs are fixed while OR inputs are programmable. This device is

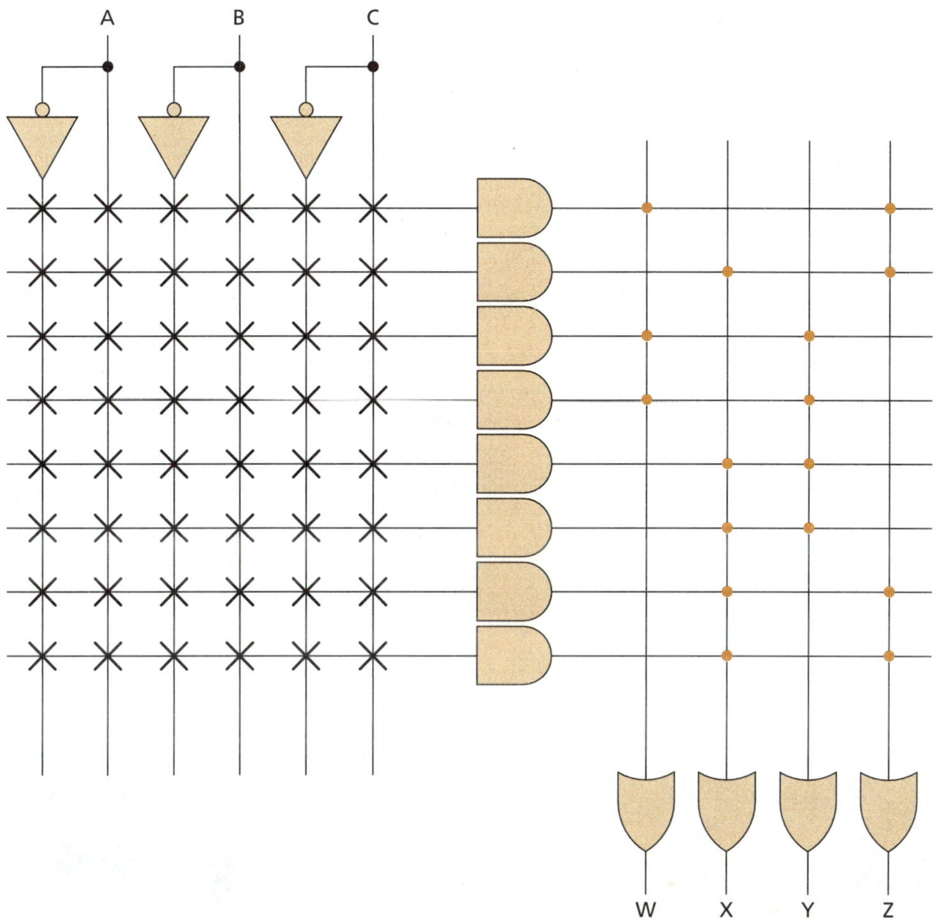

(a)

Figure 6.7 PLD Types

termed a *memory,* because it will hold specific data within its circuit. The AND inputs act as address lines to select the location within the device for a particular piece of information. That information is programmed into the OR side of the device. In Figure 6.7b, a PROM with four inputs (A, B, C, and D) is shown. Any given combination of 1's and 0's will always turn on the same AND gate. For instance, input 1001 will force AND gate 9 to a 1. All other AND gates will be 0. That 1 from AND gate 9 causes OR gates 2 and 4 to go high, while 1 and 3 remain low. Whenever 1001 is placed at the input, the output will be 0101. Change the input and you get a different output, but you will always get the same output for a given input. The device "remembers" the data (outputs at OR gates 1, 2, 3, and 4) for any given input combination on A, B, C, and D.

PLA *programmable logic array devices with programmable OR and AND arrays.*

GAL *generic or gate array logic devices are used for PLD designs for a specific PAL family of devices.*

The third type of PLD, shown in Figure 6.7c, is the most versatile and expensive type, because the **programmable logic array (PLA)** has both programmable AND and OR arrays. A fourth classification has recently emerged to describe a device that can be used in place of a number of devices within a PAL device family. It is called a **generic or gate array logic device** or, simply, **GAL.**

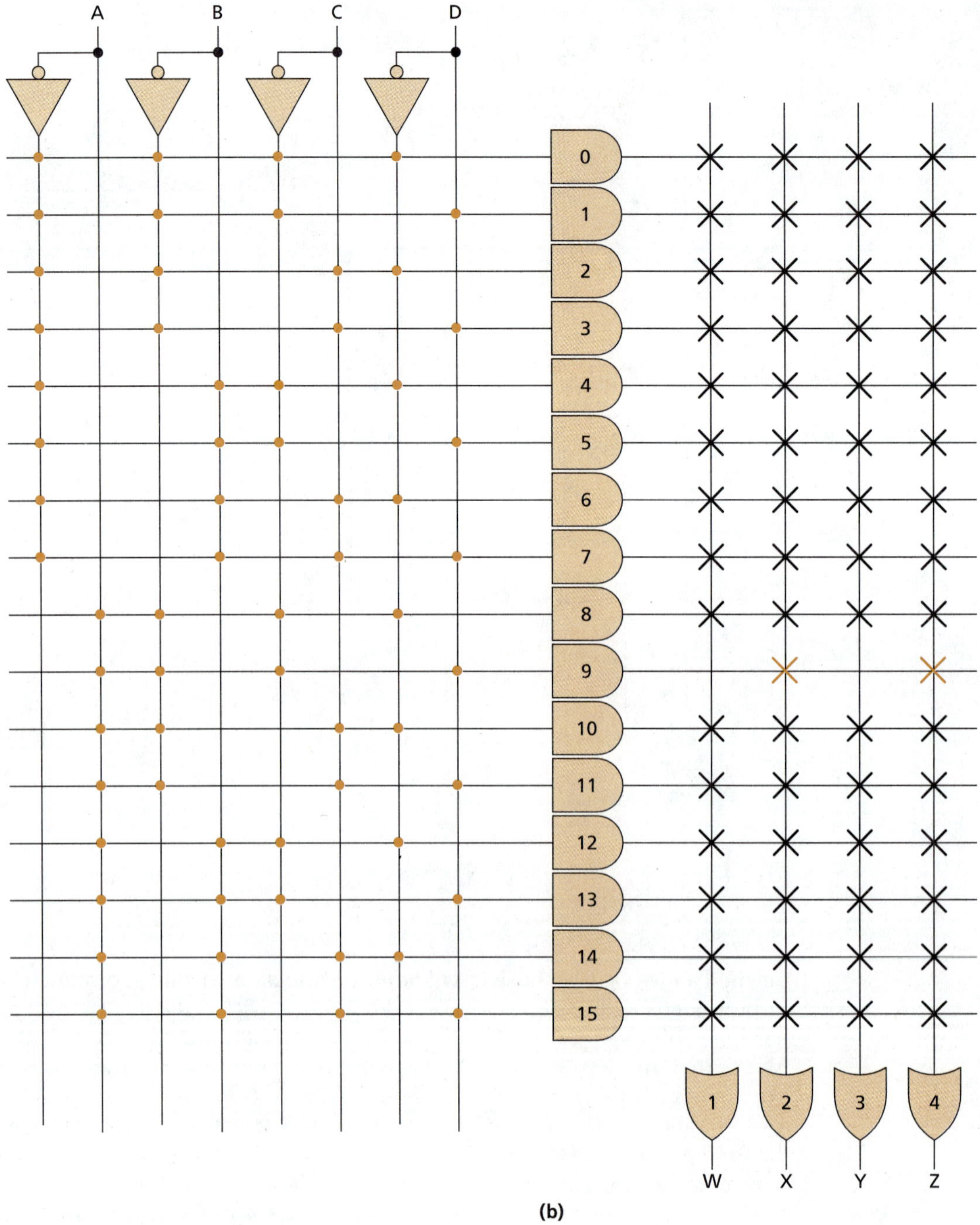

Figure 6.7 Continued

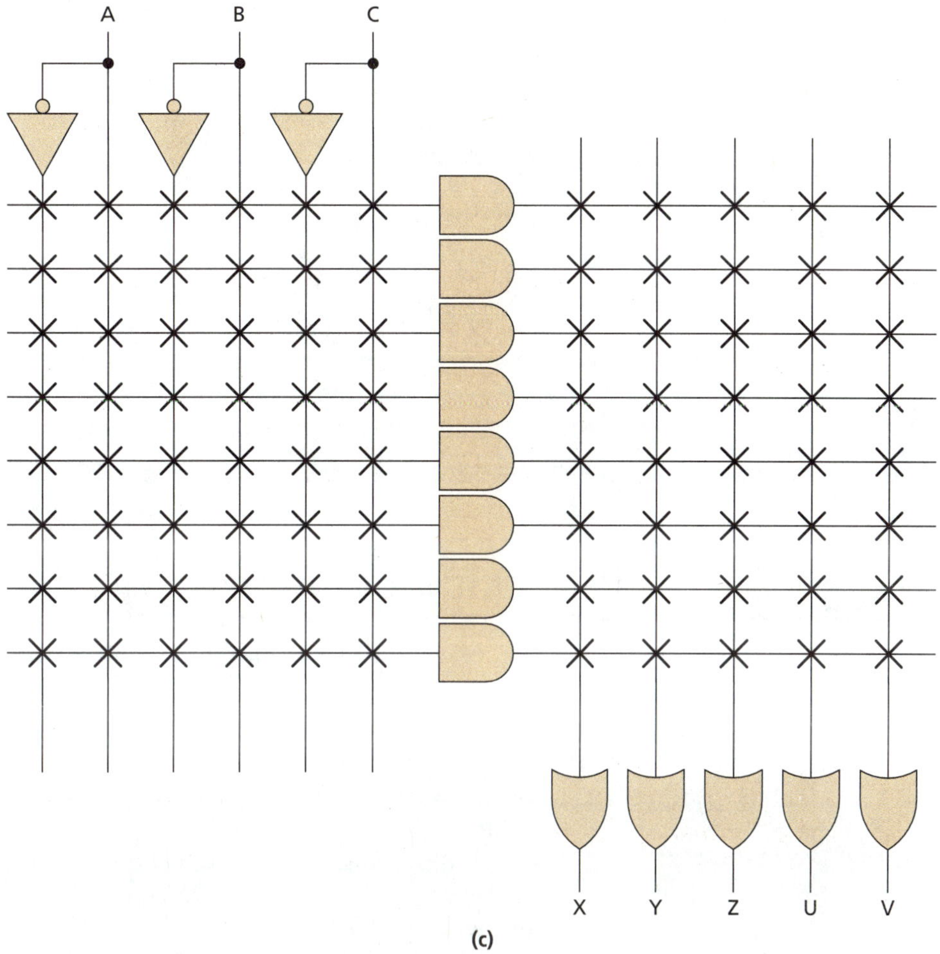

Figure 6.7 Continued

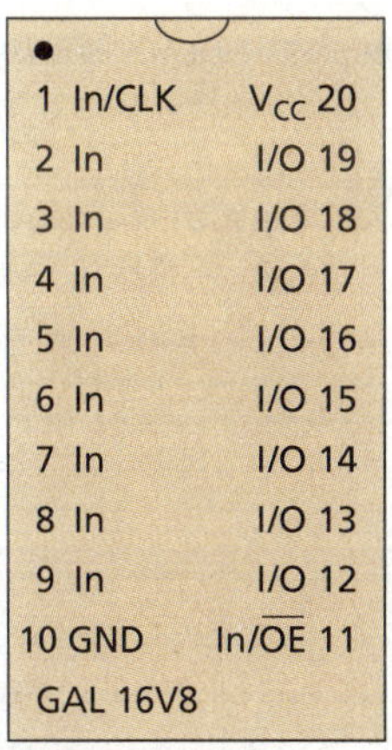

Figure 6.8 GAL16V8 Pin Diagram

The part number for a PLD briefly describes some of the characteristics of the device. It tells what type of device it is, how many inputs and outputs it has, and the type of output available with the device. A popular PLD, the GAL16V8, will be used as an example for deciphering this part number. GAL indicates this PLD is a gate array logic device with programmable AND inputs. The number 16 indicates the number of possible inputs to the device, whereas 8 indicates the total number of available outputs. This does not mean that there are 24 input and output pins, because 8 of the input pins can be programmed to be output pins instead. This is done to reduce the total number of pins in a PLD package and to give a degree of flexibility in the design and use of the device.

Figure 6.8 is the pin diagram for a GAL16V8. Note on the left that there are eight input pins and on the right eight I/O pins. I/O stands for input (I) or output (O) function for that pin. There are also V_{CC} and ground pins for power, a clock (CLK) input, and an output enable (OE) pin. The clock pin is used when the PLD design includes synchronous flip-flop register-type outputs and the output enable is used to control three-state buffer output functions. Both these functions are optional in the GAL16V8 and can be ignored if not included in a circuit design.

Section 6.3

Practice Problems

1. A sum-of-product terms expression has the same number of inputs to each AND gate. Which device—a PLA, a PAL, or a PROM—would be most suited to implement that expression, and why?
2. A complex digital circuit requires three logic circuits whose sum-of-products expressions have the same number of inputs to the OR gates, but have a differing number of AND inputs. Which PLD device type would be most appropriate to implement these expressions, and why?

6.4 PLD Output Options

The letter between the two numbers in the PLD part number (such as the V between 16 and 8 for a GAL16V8) denotes something about the output configuration for the device. Some of the more common and popular output options shown in Figure 6.9 are

1. H means *active high,* which is the normal output state; for example, $\overline{Y} = \overline{A}B + A\overline{B}$.
2. L means *active low,* which is the complement of an output; for example, $\overline{Y} = \overline{A}B + A\overline{B}$, which is the same as $Y = \overline{\overline{A}B + A\overline{B}}$ and is the complement of the active high example just given. The logic that produces the results is the same, but in the active state for an H-designated chip the output is high, whereas in the active low (L) option that same output is low.
3. R means *registered,* which is a clocked output latch. The last state of the output remains until a clock is applied to the chip.
4. V means *variable or programmable output architecture* that allows many output options to be selected. This is the type actually used in Figure 6.9, which includes several of the other output types.
5. C means *complementary output,* which makes an active high and an active low output available for the same circuit, for example, $Y = \overline{A}B + A\overline{B}$ and $\overline{Y} = \overline{A}B + A\overline{B}$. Again, the same logic yields the active high and active low results, which are complements of each other.

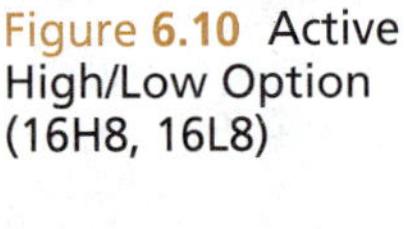

Figure 6.10 Active High/Low Option (16H8, 16L8)

Additional options that do not appear in the part number are the open collector option, three-state buffering, and feedback to an input.

To implement the choice of active high or active low, PLDs incorporate an XOR gate and use the controlled inverter function of that gate, as shown in Figure 6.10. If the fuse is left intact, the output state of the XOR will be the same as the input from the OR gate. An open fuse places a 1 at the XOR input, causing the input from the OR gate to be inverted at the output of the XOR supplying the active low or inverted output function.

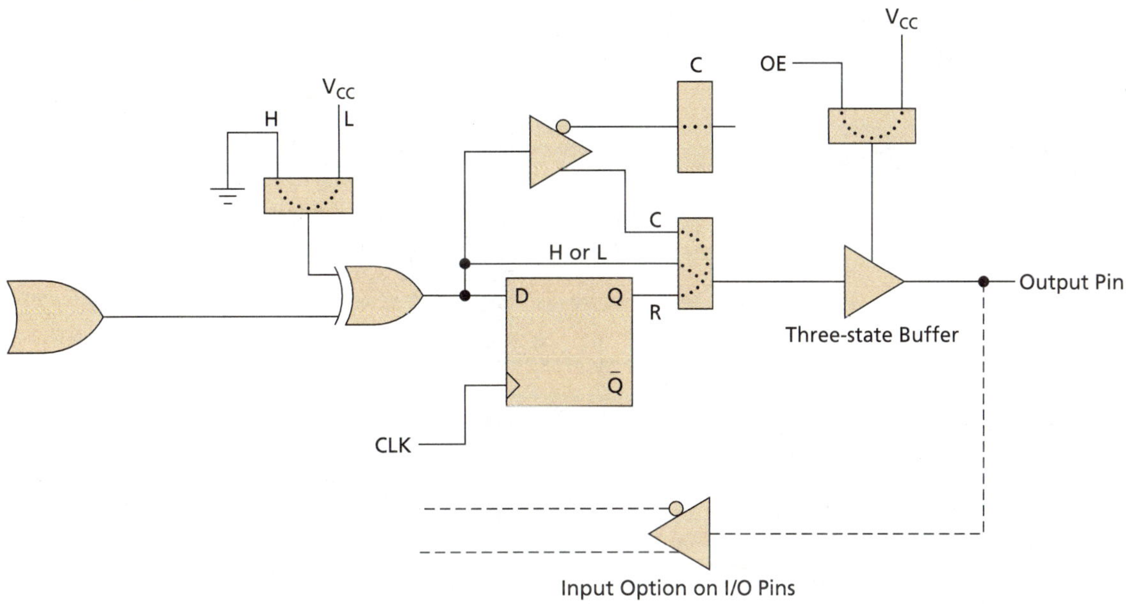

H = High or normal output
L = Low or inverted output
C = Complementary outputs—not available on common PLDs
R = Registered
CLK = Clock
OE = Output enable
V = Verstile—includes most options

Figure 6.9 Output Options for GAL16V8

Section 6.4

Practice Problems

For each of the following part numbers, give the numbers of inputs and outputs and the type of output of the device.

1. PAL22V10 **2.** GAL20R6 **3.** PAL26L12 **4.** PLA10H20

6.5 The GAL16V8 in Detail

The GAL16V8 is an inexpensive and popular PLD. It has 16 pins that are used for inputs and outputs. Of these pins, 8 can only be used for input functions, while 8 others are selectable as either an input or output function (I/O on Figure 6.11). Pin 1 may also be programmed to operate as a CLOCK signal input for synchronizing registered outputs. Pin 11 can also be used as an output enable, allowing outputs to appear as either open circuits (nothing out) or as

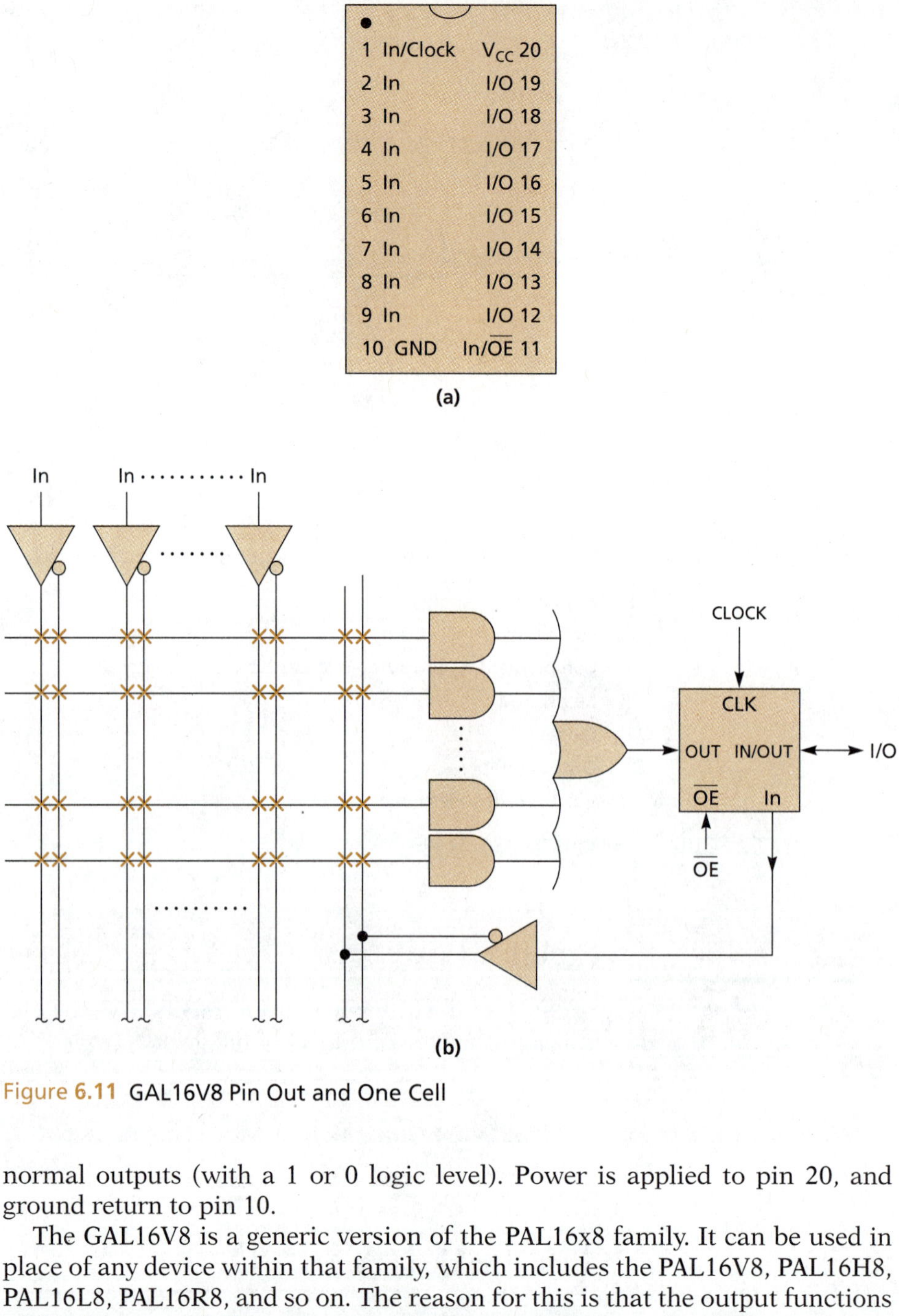

Figure 6.11 GAL16V8 Pin Out and One Cell

normal outputs (with a 1 or 0 logic level). Power is applied to pin 20, and ground return to pin 10.

The GAL16V8 is a generic version of the PAL16x8 family. It can be used in place of any device within that family, which includes the PAL16V8, PAL16H8, PAL16L8, PAL16R8, and so on. The reason for this is that the output functions of the device are fully programmable (as noted, *V* means "variable output architecture"). The selections that can be made (also shown in Figure 6.11) are active high (normal), active low (inverted), registered (latching), buffered (using the output enable), and feedback (output returned to an input).

The internal configuration of the GAL16V8 shows that the device is actually a PAL with programmable AND inputs and fixed OR connections. Figure 6.11 shows one section of the GAL16V8. Figure 6.12 shows the complete GAL16V8.

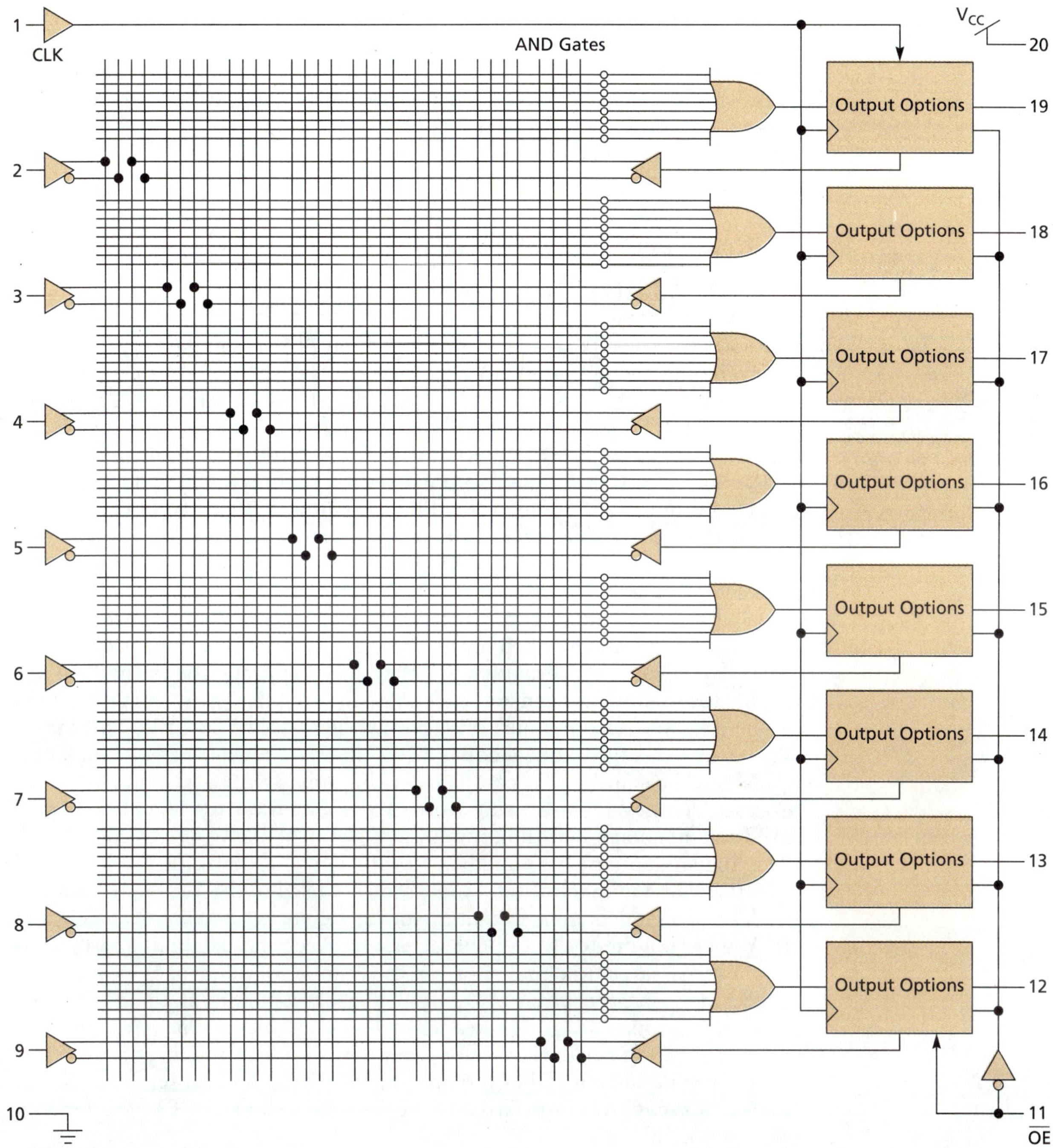

Figure 6.12 GAL16V8 Logic Diagram

The I/O select box contains the logic circuitry used to select the option used for the corresponding I/O pin, which includes an input option as well as all the output previously listed. Note that initially each input and its complement are fed through fuse (X) connections to each of the AND gates and that each AND output is permanently connected to its selected OR gate.

Placing a logic design into the GAL16V8 is a matter of "blowing" the correct set of fuses so that the connections that remain form the sum-of-products expression for the desired circuit. To do this, a logic design is compiled, using a PLD software package, to create a standard fuse map file. This file is used to direct a PLD programming device to burn the correct fuses. A popular universal compiler used today is called CUPL.

6.6 The Universal Compiler for Programmable Logic Devices

CUPL Universal Compiler for Programmable Logic devices is a compiler program that translates PLD designs into JEDEC files.

JEDEC the Joint Electronic Devices Engineering Council sets standards for PLD design and program formats.

One software package available for compiling PLD designs into a format for programming the actual device is called the **Universal Compiler for Programmable Logic Devices** or **CUPL.** CUPL takes a file written on a text editor specifically for the compiler and converts it into a **Joint Electronic Devices Engineering Council (JEDEC)** standard JC-42.1 file, with a file extension of .JED, which a PLD programming unit uses to program the PLD device. To use CUPL, you must first create a text file containing a PLD design that is to be compiled. Logic Devices, the company that distributes CUPL software, has a student version called *CUPL STARTER KIT.* It has most of the basic functions of its senior version and is more than adequate for use in and beyond the classroom/lab environment. We refer to the starter kit software simply as CUPL, because the functions described for its use work alike in the senior package.

CUPL supplies the user with a template file containing the format and syntax that the software recognizes and can easily compile. The format is included on the disk that accompanies this text, as the template text file, TMPL.PLD. When using this file, make a copy of it so that you always have a blank copy available. The template is completed using any ASCII text editor or word processor. However, when using word processors, the completed file must be saved as a nondocument file containing no control characters. A copy of that file is shown in Figure 6.13 and begins with some header information.

Notice that a number of reserved words are used in the header, followed by five X's and a semicolon. Information for the header is filled in by replacing the X's. You are not limited to five characters, but keep these data as short as possible. Brevity saves time and effort on the number of keystrokes to enter, and reduces storage space required to save each character. Each statement in CUPL ends with a semicolon (;). The header includes a name for this particular PLD design, a part number, date, revision, company, designer, and location. All these items are optional, but if they are not used, leave the X's and semicolons. This header information is transferred to the documentation (.DOC) file after compilation is completed.

The next block on the form is a comment area. Comments are recognized by the /* character at the beginning of the comment line and a */ at the end of the comment. Full line comments must have these symbols on each line. This empty box can be used to identify the circuit being designed. Below this box is a comment box used for identifying the device to be used with this design. Comments are ignored by the CUPL compiler at run time and are there for documentation only.

```
Name          XXXXX;
Partno        XXXXX;
Date          XX/XX/XX;
Revision      XX;
Designer      XXXXX;
Company       XXXXX;
Assembly      XXXXX;
Location      XXXXX;

/***********************************************************/
/*                                                         */
/*                                                         */
/*                                                         */
/***********************************************************/
/*  Allowable Target Device Types:                         */
/***********************************************************/

/**  Inputs  **/

Pin         =        ;      /*                         */
Pin         =        ;      /*                         */
Pin         =        ;      /*                         */
Pin         =        ;      /*                         */
Pin         =        ;      /*                         */
Pin         =        ;      /*                         */
Pin         =        ;      /*                         */
Pin         =        ;      /*                         */
Pin         =        ;      /*                         */
Pin         =        ;      /*                         */
Pin         =        ;      /*                         */
Pin         =        ;      /*                         */
Pin         =        ;      /*                         */
Pin         =        ;      /*                         */

/**  Outputs  **/

Pin         =        ;      /*                         */
Pin         =        ;      /*                         */
Pin         =        ;      /*                         */
Pin         =        ;      /*                         */
Pin         =        ;      /*                         */
Pin         =        ;      /*                         */
Pin         =        ;      /*                         */
Pin         =        ;      /*                         */

/** Declarations and Intermediate Variable Definitions **/


/**  Logic Equations  **/
```

Figure 6.13 TMPL.PLD

Following the device type box is a comment informing the user that the next field is used for identifying pins by variable names. An area is set aside for input pins and one for output pins. At this point, CUPL does not care which is which. CUPL will respond only to the relationship selected involving the pins. This area of the form is only used for *declaring* pin names. For instance,

PIN 1 = CONTROL;   /* assigns the name CONTROL to pin 1. */

The declarations and intermediate variable definitions area is where you insert the first level of logic functions for your design. There are three ways to enter information to be compiled by CUPL. They are *Boolean* or *logic expression, truth table,* and *state table.* Finally, the logic equations area is used to declare the final logic relationship for your design. Once more, recall that all comments are ignored by CUPL, so all these declarations, pin names, intermediate variables, and logic equations could be in a single area. The separation is for the convenience of the user and to clarify documentation.

A few notes on CUPL syntax have been already stated in the previous paragraph:

1. Semicolons follow each statement.
2. Comments begin with /* and end with */.
3. The assignment symbol is the equals (=) sign.

In all expressions, at least one space character is required to separate delimiters, although there appears to be no limit to the total number of spaces between fields. For example, assigning the variable name FROG to pin 1 can be written as

PIN 1 = FROG;   or   PIN 1   = FROG

However, it *cannot* be written as

PIN 1=FROG;

Additional CUPL syntax includes the following logic operators shown in their order of precedence:

1. Parentheses ( ) set the delimiters for a term expression; for example, (AB + C).
2. ! means logic inversion, and precedes the term; for example, !A
3. & means logic AND operator; for example, A & B
4. # means logic OR operator; for example, A # B
5. $ means exclusive OR operator; for example, A $ B

EXAMPLE 6.2

Write the CUPL language version of the following Boolean expressions:

1. $A\overline{B} + \overline{A}B$ **2.** $\overline{A}B\overline{C} + \overline{A}BC + \overline{AC}$ **3.** $A \oplus B \oplus C$

SOLUTION The first term of expression 1 is A and NOT B, which translates into CUPL language as A & !B. The second term becomes !A & B. Add the OR function between them, and the final CUPL expression is A & !B # !A & B. The order of precedence causes the NOTs to be done first, followed by the AND functions, and, finally, by the OR function. You may recognize the original logic expression as the sum-of-products form of the exclusive OR function $A \oplus B$. An optional way of writing the expression is to use CUPL's exclusive OR operator, the $: A $ B.

In expression 2, the NOTs must be watched closely. Here is its solution—check it against the original expression to follow the ! syntax:

!A & B & !C # !A & B & C # !(A & C)

Expression 3 is an exercise in using the $ XOR syntax:

A $ B $ C

Variable names used in CUPL expressions must be declared using an assignment (=) symbol. They can start with an alpha (letter), numeric (number), or the underscore character (_), and can contain up to 31 total characters; one of those characters *must* be an alpha character. Variables are *case sensitive,* meaning that the words *CAT* and *cat* are two different variable names. Variable names ending with a numeric value are considered to be index values that can be used when declaring an array of variables. CUPL recognizes a list of reserved words and symbols that cannot be used for variable names. They are included in Table 6.1.

Table 6.1 CUPL Reserved Words

APPEND	ASSEMBLY	ASSY	COMPANY	CONDITION
DATE	DEFAULT	DESIGNER	DEVICE	ELSE
FIELD	FLD	FORMAT	FUSE	IF
JUMP	LOC	LOCATION	MACRO	MIN
NAME	NODE	OUT	PARTNO	PIN
PINNODE	PRESENT	REV	REVISION	SEQUENCE
SEQUENCED	SEQUENCEJK	SEQUENCERS	SEQUENCET	TABLE
&	*	:	;	@
#	+	.	'	$
( )	[ ]	{ }	..	!
^	-	/	/* */	=
=>	'			

EXAMPLE 6.3 Identify which of the following variable names are invalid:

MOUNT 3_DOG_NIGHT SIX DAYS _123

TRUE_BLUE DATE 10/10/93 10_OCT_93

SOLUTION MOUNT and 3_DOG_NIGHTs are both valid. They contain at least one alpha character, do not use reserved words, and are within the 31 character limit. Variable SIX DAYS is invalid because it contains a space between SIX and DAYS. _123 is also invalid because it does not contain an alpha character. TRUE_BLUE is valid but DATE is not, because it is a reserved word. 10/10/93 contains a reserved symbol (/) and no alpha character. 10_OCT_93 is valid.

The equals operator (=>) is used to equate two elements as having equal values. This differs from the assignment operator (=), which is used to assign variables. Variable names can be equated to each other. Expressions and constants can be equated to one another or to variables as well. Here are some examples of how to use the equals operator:

HOW => 2C says that the variable HOW is equal to the hexadecimal 2C. Before 2C can be equated to HOW, HOW must be assigned as a valid variable name.

INPUT => OUTPUT states that the value of the variable OUTPUT is equal to the variable INPUT.

Numerical values used for any CUPL syntax can be represented in a variety of number bases. The number base in use is indicated by a preceding character surrounded by a single quote ('). The default number system used by CUPL is hexadecimal, so any value appearing without a modifying character is assumed by CUPL to be hex. The indicators for all the base number systems are

1. 'b' indicates binary, base 2
2. 'd' indicates decimal, base 10
3. 'o' indicates octal, base 8
4. 'h' indicates hexadecimal, base 16

Numerical values can be expressed as a range of values as well as singular values. To signify a range, the values are bracketed with [ ] and have two dots (. .) between them. An example range of decimal values is 'd'[0 . . 9].

EXAMPLE 6.4

What is the CUPL syntax for the following numeric values?

Octal 457
Hex range from 0 to FF
Decimal 1,234
Binary range of BCD values from 0 to 9

SOLUTION

'o'457 for the octal number
'h'[00 . . FF] or [00 . . FF] for the hex range
'd'1234 for the decimal value
'b'[0000 . . 1001] for the binary BCD range of values

Section 6.6

Practice Problems

1. Identify which of the following CUPL variable names are invalid:
   a. abort_2 b. Test 5 c. 1993 d. _Out e. FIELD

2. Translate each of the logic expressions into CUPL syntax:
   a. $\overline{\overline{A}BD} + \overline{CD} + A\overline{B}$ b. $(\overline{A} + BC)(\overline{BCD + AD})$
   c. $(\overline{\overline{A} + B})(C\overline{D}) + A\overline{B}D$ d. $ABC + \overline{AC} + BD$

6.7 Entering a PLD Design into a CUPL Template File

A number of methods are available for entering a logic design that is to be compiled by CUPL. The first method uses a logic expression format to describe a logic circuit. CUPL is friendly enough to allow the designer to enter a design with minimal effort. For example, look at the circuit in Figure 6.14. This is essentially a "do nothing" circuit, being used as an example to illustrate the steps for using CUPL. PLD programmers need a sum-of-products expression placed in a JEDEC file, but the designer need not develop this form. Instead, he or she lets the CUPL compiler do that job.

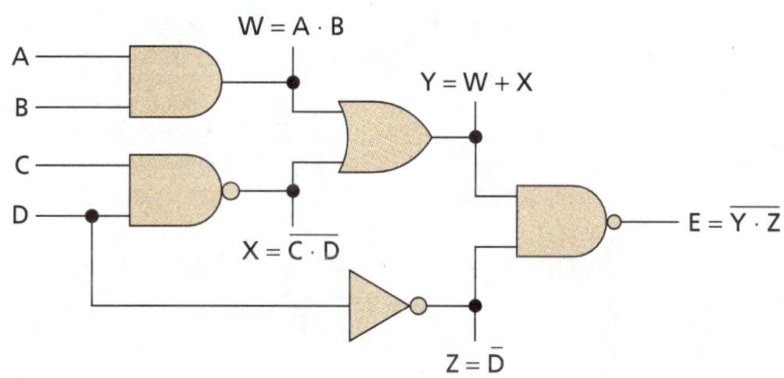

Figure 6.14 CUPL Example Circuit

The first task is to determine what pins are required and to assign a variable name to each of them. In the case of Figure 6.14, the names A, B, C, D, and E are used for input and output variables. The PLD that is used to hold this design is a GAL16V8. Referring to the pin diagram in Figure 6.11, we select pin 2 to be input variable A to avoid using the CLK input pin 1, pin 3 to be input variable B, 4 to be C, and 5 to be D. On the output side, we use pin 18 for output variable E.

A copy of the template file is made using the DOS copy command

COPY TMPL.PLD EXAMPLE.PLD

and then a text editor, like DOS Editor, Q Edit, or a word processor, such as Wordstar, Word, or Word Perfect, in non-document mode, is used to fill the template in. As an option, you can use the EZ Edit editor included with the CUPL software. The first area to be modified is the header area, where we will enter a name for this design (EXAMPLE.PLD), part number (G16V8), designer's name (MILLER), date (10/14/93), and location (PHOENIX). Following that information are the pin assignments. Make sure to delete unused pin assignment statements, to avoid having the compiler give errors for unassigned pins. At this point EXAMPLE.PLD looks like Figure 6.15.

```
        Listing of File: D:CUPLFILE/EXAMP1.PLD Page: 1

Name         Examp1.pld;
Partno       16V8;
Date         10/14/93;
Revision     XX;
Designer     Miller;
Company      XXXXX;
Assembly     XXXXX;
Location     XXXXX;

/*************************************************************/
/*                                                           */
/*    Example Circuit for CUPL Compilation                   */
/*                                                           */
/*************************************************************/
/*  Allowable Target Device Types: GAL16V8                   */
/*************************************************************/

/**  Inputs  **/

Pin 2 = A;      /*                                */
Pin 3 = B;      /*                                */
Pin 4 = C;      /*                                */
Pin 5 = D;      /*                                */

/**  Outputs  **/

Pin 18 = E;     /*                                */

/** Declarations and Intermediate Variable Definitions **/


/**  Logic Equations  **/
```

Figure 6.15 EXAMP1.PLD

The next step is to develop the logic expression for the circuit. To simplify the process, begin by assigning logic expressions to each of the intermediate points in the circuit, (W, X, Y, and Z), instead of trying to determine the expression for the entire circuit at one time:

$$W = A \cdot B$$

$$X = \overline{C \cdot D}$$

$$Y = W + X$$

$$Z = \overline{D}$$

These are entered, using CUPL syntax, into the intermediate variable area on the template:

W = A & B;

X = !(C & D);

Y = W # X;

Z = !D;

Lastly, the final expression for E is written, in CUPL form, based on the intermediate inputs to the final NAND gate. This is entered into the logic expression area on the template as

E = !(Y & Z);

Notice that using intermediate variables minimizes the job of creating and reducing logic expressions, decreasing possible errors on the part of the designer. The completed file, EXAMPLE.PLD, is shown in Figure 6.16.

```
Name        Examp1.pld;
Partno      16V8;
Date        10/14/93;
Revision    XX;
Designer    Miller;
Company     XXXXX;
Assembly    XXXXX;
Location    XXXXX;

/************************************************************/
/*                                                          */
/* Example Circuit for CUPL Compilation                     */
/*                                                          */
/************************************************************/
/*  Allowable Target Device Types: GAL16V8                  */
/************************************************************/
```

Figure 6.16 Completed Template EXAMP1.PLD

```
/**  Inputs  **/

Pin 2 = A;      /*                                 */
Pin 3 = B;      /*                                 */
Pin 4 = C;      /*                                 */
Pin 5 = D;      /*                                 */

/**  Outputs  **/

Pin 18 = E;      /*                                 */

/** Declarations and Intermediate Variable Definitions **/

W = A & B;
X = !(C & D);
Y = W # X;
Z = !D;

/**  Logic Equations  **/

E = !(Y & Z);
```

Figure 6.16 Continued

Section 6.7

Practice Problems

1. Complete a CUPL template for the following logic expression:

   $$\text{Today} = \text{MON} \cdot \overline{\text{TUE}} \cdot \text{WED} + \overline{\text{THUR} \cdot \text{FRI}} + \overline{\overline{\text{WKEND}} \cdot \overline{\text{VACATION}}}$$

2. Complete a CUPL template for the following logic expression:

   $$X = A \cdot \overline{\text{NO}} \cdot \text{TODO} + B \cdot \overline{\text{YES}} \cdot \text{TODO} + \overline{A} \cdot \overline{B} \cdot \text{REST}$$

6.8 Compiling a .PLD File Using CUPL

Install the CUPL starter software by following the directions in Appendix B. Included with the CUPL software package is a menu file called MCUPL.EXE. Invoking this file by typing MCUPL at the DOS prompt brings up a Logo screen, which can be removed by pressing the return key. This brings up the *main menu screen* shown in Figure 6.17. The main menu presents the basic operating commands in the main menu area of the screen. To select one of these commands move the highlight bar, using the up and down arrow keys or mouse, until the desired command is highlighted, and then press the return key or click the left mouse button. Before doing this, a brief description of the command appears in a *message center area* to the right of the main menu area. Since the default

```
◀MAIN MENU▶                              ◀Message Center▶

<E>dit Design File
<C>ompile CUPL File
<L>ook at DOC File
View Expanded <M>acro (MX) File              Create or Edit
<R>eview Error LST File                  a design (.PLD) file
<J>EDEC File Editor
Ed<I>t Simulation Input File
<S>imulate CUPL File
<V>iew Simulation Results
<D>evice Selection
<B>uild/View Device Library              ◀Operator Input▶
<P>ALASM to CUPL
<H>elp (CUPL Quick Reference)
<T>utorial for PLD's
Design E<X>amples
<U>ser's Editor
<A>dditional Product Information
S<Y>stem Information
<Q>uit

       ◀Device = none▶
              CUPL/Starter Kit v4.2a - 10/93
```

Figure 6.17 CUPL Starter Main Menu:

```
◀Device Selection Guide▶                 ◀Message Center▶

<1> P16L8
<2> P16R4
<3> P16R6
<4> P16R8                                  Press F1 for a list
<5> P20L8                                of device information
<6> P20L10
<7> P20R4
<8> P20R6
<9> P20R8
<A> P22V10
<B> G16V8                                ◀Operator Input▶
<C> G16V8A
<D> G20V8
<E> G20V8A
<U>nselect Device

<R>eturn to main menu                    >C:\CUPL\960SAMPL.PLD

       ◀Device = none▶
              CUPL/Starter Kit v4.2a - 10/93
```

Figure 6.18 Device Selection Menu

selection highlighted is *EZ Edit,* the command description describes this command as involving a text editor.

At the bottom of the main menu is a note indicating that the device currently selected is "None." This is because there is no default device selection made by the CUPL program. Below the message center area is an *operator input box.* If an operator input is required or is an option, a > prompt and blinking cursor will appear in this area. The first command that should be selected, when compiling a .PLD file, is to select the device you wish to compile a JEDEC file for. This is done by moving the highlight bar, using the down arrow key, to the *device selection option* and pressing the return key.

The next screen is the device selection guide shown in Figure 6.18, which allows you to select a device type by highlighting it and pressing return or clicking with the mouse. Highlight GAL16V8 or GAL16V8A, and press return or click the left mouse button. This changes the message at the bottom of the main menu screen to Device = G16V8, and returns you to that screen. The next command needed to compile the .PLD file is COMPILE CUPL FILE. After this command is highlighted and the return key pressed, the directory screen of Figure 6.19 appears.

The *directory area* lists any of the .PLD files that are currently on the default drive. If any of these files are to be compiled, move the highlight bar to that file and press return. If another file on a different drive is to be compiled, type the drive, path, and filename in the *operator input box* next to the > prompt and press return. After you press the return key, the *CUPL compiler options menu* in Figure 6.20 is brought up.

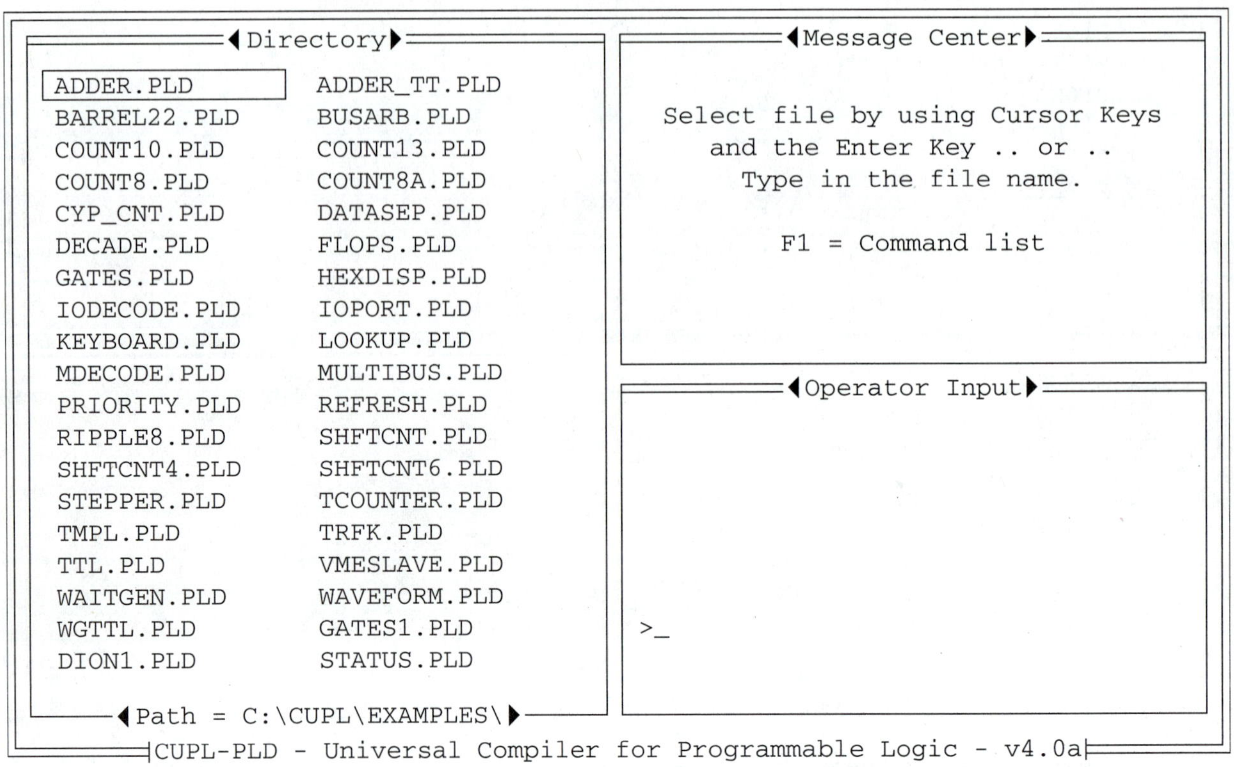

Figure 6.19 .PLD File Directory

```
═══◄CUPL Compiler►═══                    ═══◄Message Center►═══
  <J>EDEC download format
  <H>EX download format
  <I> HL download format
  <N>ame of JEDEC file = PLD file           Produces a JEDEC download
  <A>bsolute file                                  format file
  <L>isting file
  <E>xpanded macro definition file
  <X>panded product terms
  <F>use plot/chip diagram
  <B>erkeley PLA file                    ───◄Press F5 when finished►───
  <C>reate PALASM file                   ═══◄Operator Input►═══
  <U>se a specified library
  <S>imulate after compilation
  <W>aveform simulation output
  <O>ne-hot-bit state machine
  <K> Optimization methods
  <Z> Output file formats                >C:\CUPL\960SAMPL.PLD
  <M>inimization Levels
  <R> More Options

──────◄Flags = m1►──────
═══════════╡CUPL/Starter Kit v4.2a - 10/93╞═══════════
```

Figure 6.20 Compiler Options

The CUPL compiler options menu lists the options you want the compiler to perform during the process. Options are selected and deselected by moving the highlight to the desired item and pressing return. An asterisk appears next to a selected option. The options selected on the CUPL compiler menu (Figure 6.21) for our example are

1. JEDEC download file—this creates the .JED file
2. Name of JEDEC file = PLD file—keeps filename the same
3. Listing the file—lists .PLD file with any error messages
4. Expanded product terms—shows the SOP terms in the .DOC file
5. Fuse plot/chip diagram—includes the fuse plots and chip layout diagrams in the .DOC file

The message between the message center and operator input boxes says to press key F5 when you are finished making option selections. Pressing F5 begins the compiling process. On the screen, CUPL shows the progress being made and how long each section of the compiling process takes, as shown in Figure 6.22.

If the compilation is successful, no error messages will be displayed and you will be asked to press any key to continue. Pressing any key returns you to the main menu screen. On your target disk will be the additional files:

1. <filename>.JED Fuse plot file
2. <filename>.LST Listing file
3. <filename>.DOC Documentation file

```
CUPL Compiler                                  Message Center

* <J>EDEC download format
  <H>EX download format
  <I> HL download format
* <N>ame of JEDEC file = PLD file          Produces a JEDEC download
  <A>bsolute file                                format file
* <L>isting file
  <E>xpanded macro definition file
* <X>panded product terms
* <F>use plot/chip diagram
  <B>erkeley PLA file                        Press F5 when finished
  <C>reate PALASM file                           Operator Input
  <U>se a specified library
  <S>imulate after compilation
  <W>aveform simulation output
  <O>ne-hot-bit state machine
  <K> Optimization methods
  <Z> Output file formats
  <M>inimization Levels                    >
  <R> More Options

            Flags = jnlxfm1
                      CUPL/Starter Kit v4.2a - 10/93
```

Figure 6.21 Selected Options

```
CUPLPLD: Universal Compiler for Programmable Logic
Version 4.0a   Serial# MD-40A-9015
Copyright (C)  1983,1990 Logical Devices, Inc.
Licensed to DEVRY

cuplx
time: 4 secs
cupla
time: 2 secs
cuplb
time: 3 secs
cuplm
time: 1 secs
cuplc
time: 6 secs
total time: 16 secs
Press any key to continue
_
```

Figure 6.22 Compiler Messages

Figure 6.23 is a condensed copy of the documentation file for the example circuit that was designed in the previous section and compiled using CUPL as described in this section. It was condensed by removing some of the fuse plot pages that showed the unused sections of the GAL16V8.

A detailed header starts the documentation file that identifies the device and the version of CUPL used to compile it. Next is a copy of the expanded product

```
*************************************************************
                         EXAMPLE.PLD
*************************************************************

CUPL          4.0a Serial# MD-40A-9015
Device        g16v8s  Library DLIB-h-81-9
Created       Wed Oct 13 11:23:25 1993
Name          EXAMPLE.PLD
Partno        16V8
Revision      XX
Date          10/14/93
Designer      Miller
Company       XXXXX
Assembly      XXXXX
Location      Phoenix

============================================================
                   Expanded Product Terms
============================================================

E =>
    A & B
  # !C
  # !D

W =>
    A & B

X =>
    !C
  # !D

Y =>
    A & B
  # !C
  # !D

Z =>
    !D
```

Figure 6.23 .DOC File Example

```
==================================================================
                           Symbol Table
==================================================================

Pin Variable                              Pterms    Max     Min
Pol   Name       Ext    Pin     Type       Used    Pterms   Level

    A                    1       V          -        -        -
    B                    2       V          -        -        -
    C                    3       V          -        -        -
    D                    4       V          -        -        -
    E                    18      V          3        8        1
    W                    0       I          1        -        -
    X                    0       I          2        -        -
    Y                    0       I          3        -        -
    Z                    0       I          1        -        -

LEGEND   F : field      D : default variable       M : extended node
         N : node       I : intermediate variable  T : function
         V : variable   X : extended variable      U : undefined

                  ==================================================
                                                    Fuse Plot
                  ==================================================

                  Syn      02192 - Ac0     02193 x

                  Pin #19   02048   Pol x   02120   Acl -
                   00000 xxxxxxxxxxxxxxxxxxxxxxxxxxxxxxxx
                   00032 xxxxxxxxxxxxxxxxxxxxxxxxxxxxxxxx
                   00064 xxxxxxxxxxxxxxxxxxxxxxxxxxxxxxxx
                   00096 xxxxxxxxxxxxxxxxxxxxxxxxxxxxxxxx
                   00128 xxxxxxxxxxxxxxxxxxxxxxxxxxxxxxxx
                   00160 xxxxxxxxxxxxxxxxxxxxxxxxxxxxxxxx
                   00192 xxxxxxxxxxxxxxxxxxxxxxxxxxxxxxxx
                   00224 xxxxxxxxxxxxxxxxxxxxxxxxxxxxxxxx
                  Pin #18   02049   Pol x   02121   Acl x
                   00256 x-x-----------------------------
                   00288 -----x--------------------------
                   00320 ---------x----------------------
                   00352 xxxxxxxxxxxxxxxxxxxxxxxxxxxxxxxx
                   00384 xxxxxxxxxxxxxxxxxxxxxxxxxxxxxxxx
                   00416 xxxxxxxxxxxxxxxxxxxxxxxxxxxxxxxx
                   00448 xxxxxxxxxxxxxxxxxxxxxxxxxxxxxxxx
                   00480 xxxxxxxxxxxxxxxxxxxxxxxxxxxxxxxx

                  LEGEND    X : fuse not blown
                            - : fuse blown
```

Figure 6.23 Continued

```
============================================================
                       Chip Diagram
============================================================

               ______________________
              |                      |
              |  EXAMPLE.PLD         |
         x---|1                 20|---x Vcc
       A x---|2                 19|---x
       B x---|3                 18|---x E
       C x---|4                 17|---x
       D x---|5                 16|---x
         x---|6                 15|---x
         x---|7                 14|---x
         x---|8                 13|---x
         x---|9                 12|---x
     GND x---|10                11|---x
              |______________________|
```

Figure 6.23 Concluded

terms in CUPL syntax, followed by a symbol table for declared variables. Pin names are associated with their pin numbers and the type of the variable. In our example, all the pin name functions were variable—that is, either a 1 or 0 could be applied to them, because they are not at a fixed input or output level. The intermediate terms are also listed and shown to be intermediate. The total number of product terms for intermediate and output variables are also charted on this symbol table.

The fuse plot follows next and would normally represent a plot of all output pins, but only pin 19 (unused) and pin 18 (used) are illustrated here. An *X* means a connection is made, and a dash (–) means an open "fuse." The last part of the documentation file is the chip diagram, which is necessary to show someone which pins are used.

One other CUPL command that is useful at this point is the *edit design entry command* of the main menu. Highlighting this command and pressing enter asks for a file to be edited. In the operator input area, type in the drive, path, and filename, and then press the return key. This brings up EZ Edit with the desired text file in the work area, as shown in Figure 6.24.

On the bottom of the editor screen is a message line that contains the filename PRESS F1 FOR HELP, and a box containing the current line and page numbers. Pressing F1 calls up a help window, shown in Figure 6.25, that lists the control keys used with this editor.

The last line of the help screen says to press the ALT-X key to save the edited file or quit the editor. Pressing ALT-X displays the *save/quit window* on the screen as seen in Figure 6.26. Highlight the choice, and press return. The file will be saved into the same file originally called up. The editor can be used to create a new text file, in which case invoking Save will ask for a filename. Selecting Quit will exit the editor without asking if you want to save the file,

```
 0                          EZ Edit 2.00
Name          ex1.pld;
Partno        16V8;
Date          01/05/93;
Revision      01;
Designer      Miller;
Company       DeVry;
Assembly      StdAl;
Location      Phoenix;

/**************************************************/
/*                                                */
/*          Example Circuit for !X                */
/*                                                */
/**************************************************/
/*  Allowable Target Device Types: 16V8           */
/**************************************************/

/**  Inputs  **/

Pin 2 = A;
Pin 3 = B;
Pin 4 = C;
Pin 5 = D;
  <a:\ex1.pld             >      Press F1 for Help  <  1  2  >
```

Figure 6.24 EZ Edit with File Loaded

```
 0                               EZ Edit 2.00
Name          ex1.pld;
Partno        16V8;
Date     ----------------------------------- Help ------------------------------
Revision
Designer   Up Arrow       : Cursor up          | Home       : Beginning of line
Company    Down Arrow     : Cursor down        | End        : End of line
Assembly   Left Arrow     : Cursor left        | PgUp       : Page up
Location   Right Arrow    : Cursor right       | PgDn       : Page down
         ------------------------------------------------------------------------
/*******   Ctrl-Home      : Top of screen      | F7         : Border color
/*         Ctrl-End       : Bottom of screen   | Shift-F7   : Border bkgnd
/*         Ctrl-PgUp      : Top of file        | F8         : Text color
/*         Ctrl-PgDn      : Bottom of file     | Shift-F8   : Text bkgnd
/*******   Ctrl-Bkspace   : Deletes a line     |
/*  Allo   Backspace      : Deletes a char     |
/*******   F2             : Switch Window      |
           Alt-X          : Save/Quit          |

/**  Inp ------------------------- Press any key to continue ----------------------

Pin 2 = A;
Pin 3 = B;
Pin 4 = C;
Pin 5 = D;
  <a:\ex1.pld             >      Press F1 for HELP            <  1  2  >
```

Figure 6.25 EZ Edit Help Menu

```
0                                          EZ Edit 2.00
N    ┌───────┐    ex1.pld;
P    │ Save  │    16V8;
D    │ Quit  │    01/05/93;
R    └───────┘    01;
Designer          Miller;
Company           DeVry;
Assembly          StdAl;
Location          Phoenix;

/*****************************************************/
/*                                                   */
/*          Example Circuit for !X                   */
/*                                                   */
/*****************************************************/
/*  Allowable Target Device Types: 16V8              */
/*****************************************************/

/**  Inputs  **/

Pin 2 = A;
Pin 3 = B;
Pin 4 = C;
Pin 5 = D;
  <a:\ex1.pld            >        Press F1 for HELP  <  1  2  >
```

Figure 6.26 EZ Edit Save/Quit

so make sure you have saved a file before quitting the editor. Otherwise it will be lost.

Section 6.8

Practice Problem

List the step-by-step process, using menu-driven CUPL software, to compile EXAMPLE.PLD, located in drive A: path LOGIC, into the following three files: EXAMPLE.JED, EXAMPLE.DOC, and EXAMPLE.LST. The device to be used is a GAL16V8.

6.9 CUPL Design Using Variable Ranges

Variable names that use subscripts can be declared and used as a range of variables. For instance, the address bus of a memory chip has the designations A_0 through A_{11}. These are declared in CUPL, following the range syntax, as

PIN [2 . . 13] = [A0 . . A11];

This statement assigns each pin in sequential order to a variable name A followed by an index number, so that PIN 2 is assigned A0, PIN 3 assigned A1, and so forth. To show an effective application of range variables, we will create the template file to compile the MUX circuit design of Figure 6.27 in a single GAL16V8. The complete template file is

```
Name      MUXMEM.PLD;
Partno    GAL16V8;
Date      05/05/94;
Revision  XX;
Designer  Miller;
Company   XXXXX;
Assembly  XXXXX;
Location  Phoenix;

/*******************************************************/
/*                                                     */
/*     Multiplexer for Figure 6.27                     */
/*                                                     */
/*******************************************************/
/*   Allowable Target Device Types: GAL16V8            */
/*******************************************************/

/**  Inputs  **/

Pin [2..9]   = [A0..A7];       /* Address Inputs */
Pin [12..13] = [A8..A9];       /* Address Inputs */
Pin 14 = S;                    /* Select Input   */

/**  Outputs  **/

Pin [19..15] = [MA0..MA4]; /* Outputs to Memory */

/**  Logic Equations  **/

[MA0..MA4] = [A0..A4] & !S # [A5..A9] & S;
```

Figure 6.27 MUX Design Using Ranges

The format used in this template makes use of the range syntax for variables with common designations. Pin [2 . . 9] = [A0 . . A7]; states that pins 2 through 9 are assigned the names A0 through A7 sequentially. That is, pin 2 is A0, pin 3 is A1, pin 4 is A2, and so forth. The range format is also used in the equation defining the relationship between the outputs, MA0 through MA4, address inputs A0 through A9, and the select control input S. The CUPL statement

[MA0 . . MA4] = [A0 . . A4] & !S # [A5 . . A9] & S;

says that output MA0 is equal to input A0 when S is low and to A5 when S is high. Similarly, MA1, the next output, equals A1 when S is low and A6 when S is high, which are the next two inputs in each range.

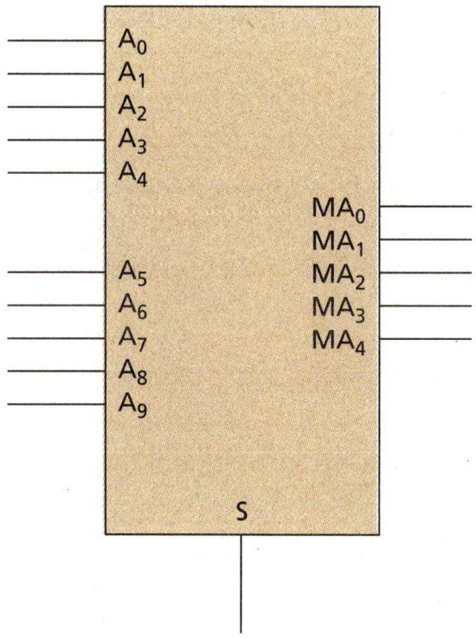

Figure 6.27 Concluded

6.10 Truth Table Design

CUPL allows three different methods, which can be compiled from a .PLD file to a .JED fuse file. The first, design by logic expression, has been used to introduce CUPL. A second method, using a state table design, is explored later in the chapter. A third method that CUPL recognizes for implementing a PLD design uses truth tables. To start using this method requires a truth table that describes the desired logic circuit. In explaining how to set up the truth table for CUPL, we will use the truth table example in Figure 6.28.

The truth table in Figure 6.28 has four inputs labeled A, B, C, and D. Input D is designated as the least significant bit (LSB) when considering bits A, B, C,

A	B	C	D	T	W	X	Y	Z
0	0	0	0	0	0	0	0	1
0	0	0	1	0	0	0	1	0
0	0	1	0	0	0	1	0	0
0	0	1	1	0	1	0	0	0
0	1	0	0	1	0	0	0	0
0	1	0	1	1	0	0	0	1
0	1	1	0	1	0	0	1	0
0	1	1	1	1	0	1	0	0
1	0	0	0	1	1	0	0	0
1	0	0	1	1	1	0	0	1
1	0	1	0	1	1	0	1	0
1	0	1	1	1	1	1	0	0
1	1	0	0	1	1	1	0	1
1	1	0	1	1	1	1	1	0
1	1	1	0	1	1	1	1	1
1	1	1	1	0	0	0	0	0

Figure 6.28 Truth Table Example

and D as a four-bit binary word. Five outputs—T, W, X, Y, and Z—are generated by the 16 combinations of the four inputs. Again, output Z is considered to be the LSB of the five-bit output word TWXYZ. To translate this table into CUPL syntax, you must begin by declaring a field for the inputs and a field for the outputs. After the fields have been declared, a truth table is declared defining the relationship between each combination of the input field and the desired results of the output field. The reserved word FIELD is used to declare a field and has the following syntax:

FIELD <field name> = [element, element, element . . . element];

with the least significant bit located as the rightmost element. In our example, the input field is declared as

FIELD INPUT = [A, B, C, D];

and the output field is

FIELD OUTPUT = [T, W, X, Y, Z];

Once the fields are declared, then the table itself can be formed using this syntax:

TABLE <field name> => <field name> {<table listing>}

The table may be listed in a number of different ways. The most common is to list the inputs and outputs as they appear on the table. These values can be represented as binary, octal, decimal, or hexadecimal numbers. CUPL defaults to hex if no other system designation is used. The first line of the table in Figure 6.28 has all inputs low (a hex 0) and only output Z high. Because there are four input bits, they can be represented by their hex equivalent. As there are five output bits, we will retain them in their binary form. The truth table is declared, and the first line of the table is translated into CUPL syntax as

TABLE INPUT => OUTPUT {

0 => 'b'00001;

This says that an input on A, B, C, and D of 'h'0 or 'b'0000 yields an output on T, W, X, Y, and Z of 'b'00001. Similarly, the second line is written as 1 => 'b'00010; note the use of the equality (=>) symbol as opposed to the more familiar assignment (=) operator. The completed CUPL table is

```
TABLE INPUT => OUTPUT {
0 => 'b'00001;   1 => 'b'00010;   2 => 'b'00100;
3 => 'b'01000;   4 => 'b'10000;   5 => 'b'10001;
6 => 'b'10010;   7 => 'b'10100;   8 => 'b'11000;
9 => 'b'11001;   A => 'b'11010;   B => 'b'11100;
C => 'b'11101;   D => 'b'11110;   E => 'b'11111;
F => 'b'00000;   }
```

To complete the template input, pins and output pins are declared before field declarations. The completed template for this example is shown in Figure 6.29.

```
Name        fig6_29;
Partno      16V8;
Date        11/02/93;
Revision    XX;
Designer    Miller;
Company     XXXXX;
Assembly    XXXXX;
Location    Phoenix;

/**********************************************************/
/*                                                        */
/*     Figure 6.29: Truth Table Example                   */
/*                                                        */
/**********************************************************/
/*  Allowable Target Device Types: 16V8                   */
/**********************************************************/

/**  Inputs  **/

Pin 2 = A;       /*                                */
Pin 3 = B;       /*                                */
Pin 4 = C;       /*                                */
Pin 5 = D;       /*                                */

/**  Outputs  **/

Pin 18 = T;      /*                                */
Pin 17 = W;      /*                                */
Pin 16 = X;      /*                                */
Pin 15 = Y;      /*                                */
Pin 14 = Z;      /*                                */

/** Declarations and Intermediate Variable Definitions **/

FIELD INPUT = [A, B, C, D];
FIELD OUTPUT = [T, W, X, Y, Z];

/**  Logic Equations  **/

TABLE INPUT => OUTPUT {

0 => 'b'00001; 1 => 'b'00010; 2 => 'b'00100; 3 => 'b'01000;
4 => 'b'10000; 5 => 'b'10001; 6 => 'b'10010; 7 => 'b'10100;
8 => 'b'11000; 9 => 'b'11001; A => 'b'11010; B => 'b'11100;
C => 'b'11101; D => 'b'11110; E => 'b'11111; F => 'b'00000;
}
```

Figure 6.29 Truth Table Example

EXAMPLE 6.5

The following logic expression and truth table in Figure 6.30 represents two logic circuits that are to be implemented using a single GAL16V8. Complete the CUPL template file required for the design to be compiled.

$$F = \overline{A} \cdot \overline{B} + \overline{C \cdot D} \oplus E$$

A	B	C	Y
0	0	0	1
0	0	1	1
0	1	0	0
0	1	1	1
1	0	0	0
1	0	1	0
1	1	0	1
1	1	1	1

Figure **6.30** Example 6.5 Truth Table

SOLUTION As with any template, fill out the heading with as much information as you need. Next declare the pin assignments for input and output pins. We will need inputs for A, B, C, D, and E and two output pins, F and Y. Then translate the logic expression into CUPL syntax:

F = !A & !B # !(C & D) $ E;

There are two fields to be declared for the truth table:

FIELD INS = [A, B, C]; FIELD OUTS = [Y];

Lastly, put the table together:

TABLE INS => OUTS {

'o'0 => 'b'1; 'o'1 => 'b'1; 'o'2 => 'b'0;

'o'3 => 'b'1; 'o'4 => 'b'0; 'o'5 => 'b'0;

'o'6 => 'b'1; 'o'7 => 'b'1; }

All the pieces put together are shown in the template of Figure 6.31.

```
Listing of File: D:\CUPLFILE\LOGTAB.PLD Page: 1

Name        LOGTAB.PLD;
Partno      16V8;
Date        11/13/95;
Revision    XX;
Designer    Miller;
Company     XXXXX;
Assembly    XXXXX;
Location    Phoenix;

/*******************************************************************/
/*                                                                 */
/* Figure 6.31 Truth Table and Logic Equation Design Example       */
/*                                                                 */
/*******************************************************************/
/*  Allowable Target Device Types: GAL16V8                         */
/*******************************************************************/

/**  Inputs  **/

Pin 2 = A;      /*                                  */
Pin 3 = B;      /*                                  */
Pin 4 = C;      /*                                  */
Pin 5 = D;      /*                                  */
Pin 6 = E;      /*                                  */

/**  Outputs  **/

Pin 18 = F;      /*                                  */
Pin 17 = Y;      /*                                  */

/** Declarations and Intermediate Variable Definitions **/

FIELD INPUT = [A, B, C];
FIELD OUTPUT = [Y];

/**  Logic Equations  **/

F = !A & !B # !(C & D) $ E;

TABLE INPUT => OUTPUT {
'o'0 = 'b'1;     'o'1 = 'b'1;     'o'2 = 'b'0;     'o'3 = 'b'1;
'o'4 = 'b'0;     'o'5 = 'b'0;     'o'6 = 'b'1;     'o'7 = 'b'1;
}
```

Figure 6.31 Truth Table and Logic Equation Design Example

Application Note

PLD Decoder Application: Display Driver

PLD devices have been used to implement decoders and encoders. This is somewhat natural, since the basic cell of a PLD is the AND/OR matrix. The easiest method to implement decoders and encoders is to use the truth table design and equate each input with the decoded or encoded output required. Decoders and encoders have been customized to fit specialized needs. One such need is a hexadecimal-to-seven-segment LED decoder. A seven-segment light-emitting diode (LED) matrix is shown in Figure 6.32. There are seven LEDs arranged so that when the correct voltage is applied to them, they light to display a character. For instance, applying a ground to the common ground and a forward-bias voltage to inputs a, b, and c causes only those three segments to light, showing the character 7. A decoder that translates BCD codes to the correct seven-segment codes to display the characters 0 to 9 are readily available and fairly inexpensive. However, there are very few hexadecimal-to-seven-segment decoders available. Here is where a PLD comes in handy. The first step is to set down a truth table relating the four binary inputs for hexadecimal values. Next determine which of the seven segments need to be lit to display each character. Set these up as the seven outputs of the truth table. Complete the PLD template in Figure 6.33, and compile it into a .JED file.

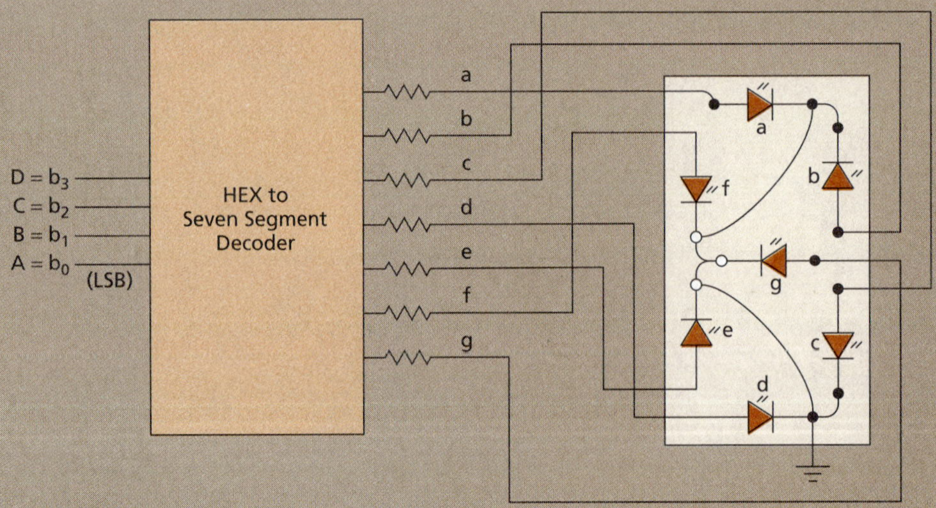

Figure 6.32

```
                Listing of File: A:\HEXLED.PLD Page: 1

Name        hexled;
Partno      16V8;
Date        06/01/95;
Revision    02;
Designer    Miller;
Company     XXXXX;
Assembly    XXXXX;
Location    Phoenix;

/**********************************************************/
/*                                                        */
/*   Figure 6.33: Hex to 7 Segment Decoder                */
/*                                                        */
/**********************************************************/
/*  Allowable Target Device Types: GAL16V8                */
/**********************************************************/

/**  Inputs  **/

Pin 2 = A;
Pin 3 = B;
Pin 4 = C;
Pin 5 = D;

/**  Outputs  **/

Pin 12 = a;
Pin 13 = b;
Pin 14 = c;
Pin 15 = d;
Pin 16 = e;
Pin 17 = f;
Pin 18 = g;

/** Declarations and Intermediate Variable Definitions **/

FIELD input = [D, C, B, A];
FIELD output = [a, b, c, d, e, f, g];

/**  Logic Equations  **/

TABLE input => output {
        0 => 7E; 1 => 30; 2 => 6D; 3 => 79;
        4 => 33; 5 => 5B; 6 => 5F; 7 => 70;
        8 => 7F; 9 => 73; A => 77; B => 1F;
        c => 4E; D => 3D; E => 4F; F => 47;
        }
```

Figure 6.33 Hex- to Seven-Segment Decoder

Section 6.10

Practice Problems

1. Complete the CUPL template for the truth table of Figure 6.34.

PAT	DON	TOM	JERRY	CAT	MOUSE	DOG
0	0	0	0	0	0	1
0	0	0	1	1	0	0
0	0	1	0	1	0	0
0	0	1	1	0	1	0
0	1	0	0	0	1	1
0	1	0	1	0	1	1
0	1	1	0	1	1	0
0	1	1	1	1	1	0
1	0	0	0	1	0	1
1	0	0	1	0	1	1
1	0	1	0	1	1	0
1	0	1	1	0	1	1
1	1	0	0	1	0	1
1	1	0	1	1	0	0
1	1	1	0	1	0	1
1	1	1	1	0	1	0

Figure **6.34**

2. Complete the CUPL template for the logic expression and truth table of Figure 6.35.

$$\text{THIS} = (\overline{\text{FAR}} + \text{NEAR} + \overline{\text{TIME} \cdot \text{COST}}) \cdot \text{MOST}$$

FAR	NEAR	TIME	THAT	THEM
0	0	0	0	0
0	0	1	0	1
0	1	0	1	0
0	1	1	1	1
1	0	0	1	0
1	0	1	0	1
1	1	0	0	0
1	1	1	1	1

Figure **6.35**

6.11 Implementing a Full Adder on a GAL16V8

As with many digital logic circuits, the full adder can be implemented using a programmable logic device (PLD) such as a GAL16V8. The full adder is created, first using logic expressions for the SUM and CARRY and then using the full-adder truth table developed earlier.

The first step as in all PLD designs is to create the text file that will be compiled into the JEDEC file required to program the device. For CUPL, this requires completing a copy of the TMPL.PLD file for the full-adder circuit. For the PLT burner, the text file to use is the PLTMP.PLD file. A copy of TMPL.PLD is made using the DOS command

A:>COPY TMPL.PLD ADDER.PLD

A text editor is then used to fill in the ADDER.PLD template as shown in Figure 6.36.

```
Name          Adder;
Partno        16V8;
Date          05/26/93;
Revision      XX;
Designer      Miller;
Company       XXXXX;
Assembly      XXXXX;
Location      XXXXX;

/************************************************************/
/*                                                          */
/*    Full adder using Logic Expressions                    */
/*                                                          */
/************************************************************/
/*  Allowable Target Device Types: GAL16V8                  */
/************************************************************/

/**   Inputs   **/

Pin 2 = A;      /*  Input Number A   */
Pin 3 = B;      /*  Input Number B   */
Pin 4 = CIN;    /*  Carry In         */

/**   Outputs   **/

Pin 12 = SUM;
Pin 13 = CARRY;

/**   Logic Equations   **/
SUM = A $ B $ CIN;
CARRY = A & B # A & CIN # B & CIN;
```

Figure 6.36 Full Adder Using Logic Expression

Match the CUPL logic equations to the logic expressions developed earlier, in Chapter 3, for the full adder. Notice how the CUPL syntax relates to standard logic notation. Following the header information, input pins A, B, and C and SUM and CARRY output pins are declared. There are no intermediate equations—only two final expressions:

```
SUM = A $ B $ C;        /* $ - exclusive OR function */
CARRY = A & B # A & C # B & C;      /* & - AND   # - OR */
```

EXAMPLE 6.6

Explain how the CUPL syntax relates to the actual logic expressions.

SOLUTION Besides using the correct syntax, attention must also be paid to precedence of operation. In the SUM equation, there is no difficulty, because the exclusive OR function is the only one used. However when we encounter the CARRY expression, there are both AND and OR functions. The order of precedence dictates that the AND functions are performed before the OR functions. This expression could also be written

```
CARRY = (A & B) # (A & C) # (B & C);
```

In this case, the parentheses serve only to clarify the appearance of the expression. Both forms produce the same result.

Once the ADDER.PLD file is completed and saved, it is time to compile it into JEDEC and document files using the CUPL programs. The compiled JEDEC file (.JED) can now be taken to any PLD burner capable of programming a GAL16V8 from the .JED file. A PLD chip is placed in the burner, and the burner software is run.

Creating the full adder using the truth table begins by translating that table into CUPL format. This is done by declaring the input and output fields of the truth table:

```
FIELD NUMIN = [C, B, A];
FIELD RESULTS = [SUM, CARRY];
```

These declarations create fields with names NUMIN and RESULTS. NUMIN is a three-bit field composed of inputs A, B, and C, and RESULTS is a two-bit field using SUM and CARRY as bit elements. Commas are used to separate elements of the fields. The least significant bit of each field is the rightmost element (A, in the case of the NUMIN field). Once the fields are declared, the truth table can be created:

TABLE NUMIN => RESULTS

{ 'b'000 => 'b'00; 'b'001 => 'b'10; 'b'010 => 'b'10;

'b'011 => 'b'01; 'b'100 => 'b'10; 'b'101 => 'b'01;

'b'110 => 'b'01; 'b'111 => 'b'11;

Put together, the truth table form for the full adder is shown in Figure 6.37.

```
Name         Adder;
Partno       16V8;
Date         05/26/93;
Revision     XX;
Designer     Miller;
Company      XXXXX;
Assembly     XXXXX;
Location     XXXXX;

/*********************************************************/
/*                                                       */
/*   Full Adder Using the Truth Table Method             */
/*                                                       */
/*********************************************************/
/*  Allowable Target Device Types: GAL16V8                */
/*********************************************************/

/**  Inputs  **/

Pin 2 = A;      /*  Number A  */
Pin 3 = B;      /*  Number B  */
Pin 4 = CIN;    /*  Carry In  */

/**  Outputs  **/

Pin 12 = SUM;
Pin 13 = CARRY;

/**  Logic Equations  **/

FIELD NUMIN = [CIN, B, A];
FIELD RESULTS = [SUM, CARRY];

TABLE NUMIN => RESULTS
{ 'b'000 => 'b'00;  'b'001 => 'b'10;  'b'010 => 'b'10;
  'b'011 => 'b'01;  'b'100 => 'b'10;  'b'101 => 'b'01;
  'b'110 => 'b'01;  'b'111 => 'b'11;
}
```

Figure 6.37 Full-Adder Truth Table

Section 6.11

Practice Problems

1. Create the CUPL template for a full subtracter, using logic expressions.
2. Create the CUPL template for a full subtracter, using the truth table method.

6.12 Four-Bit Adder PLD Circuits

When I discussed what can or cannot be done with PLDs, I pointed out that one must pay close attention to a particular PLD's pin complement and the internal resources of the device. Paying attention to the pin designations as inputs and outputs is fairly straightforward. You must select input pins to handle input signals, and output pins for output signals. A less obvious need is being aware of internal limitations. When you first look at a diagram of the AND/OR blocks of a PLD, you may become slightly overwhelmed by the amount of logic shown on the diagram. With all that logic available, you may feel you should be able to make any logic circuit with a PLD device as long as you have enough input and output pins. Wrong! In this section, it will become clear why, as we try to compile our four-bit adder design.

Our four-bit adder requires four inputs for number A, four for number B, and a ninth input pin for carry in. Output pins required are four for sum, and one for the carry flag. These are nicely accommodated by the GAL16V8, which has 11 inputs available after the five outputs are designated.

As done previously, the first step is to create the PLD template for a four-bit parallel adder. This requires the declaration of four A number inputs, four B number inputs, four SUM outputs, a carry in (CIN), and a carry out (COUT). For this design, we will use logic expressions for each sum (S_n and carry out (C_n). The completed template is shown in Figure 6.38.

Note that there are enough input and output pins to accommodate the four-bit adder circuit. Next compile the PLD file using CUPL. A compilation begins, and the screen shows the ongoing results:

```
cuplx
time: 7 secs
cupla
time: 5 secs
cuplb
time: 42 secs
cuplm
time: 1 secs
cuplc
[000600] excessive number of product terms: S3
[000600] excessive number of product terms: S2
[000600] excessive number of product terms: S1
[000600] excessive number of product terms: COUT
```

```
Name         4BitAdd;
Partno       16V8;
Date         06/03/93;
Revision     XX;
Designer     Miller;
Company      XXXXX;
Assembly     XXXXX;
Location     XXXXX;

/*********************************************************/
/*                                                       */
/*      Four-Bit Parallel Adder                          */
/*                                                       */
/*********************************************************/
/*  Allowable Target Device Types: GAL16V8               */
/*********************************************************/

/**  Inputs  **/

Pin 2 = A0;     /*  A Number LSB                  */
Pin 3 = A1;     /*                                */
Pin 4 = A2;     /*                                */
Pin 5 = A3;     /*  A Number MSB                  */
Pin 6 = B0;     /*  B Number LSB                  */
Pin 7 = B1;     /*                                */
Pin 8 = B2;     /*                                */
Pin 14 = B3;    /*  B Number MSB                  */
Pin 13 = CIN;   /*  Carry In                      */

/**  Outputs  **/

Pin 19 = S3;    /*  Sum MSB                       */
Pin 18 = S2;    /*                                */
Pin 17 = S1;    /*                                */
Pin 16 = S0;    /*  Sum LSB                       */
Pin 15 = COUT;  /*  Carry Out                     */

/** Declarations and Intermediate Variable Definitions **/
C0 = A0 & B0 # A0 & CIN # B0 & CIN;  /* Intermediate */
C1 = A1 & B1 # A1 & C0 # B1 & C0;    /* Carries      */
C2 = A2 & B2 # A2 & C1 # B2 & C1;

/**  Logic Equations  **/
S0 = A0 $ B0 $ CIN;
S1 = A1 $ B1 $ C0;
S2 = A2 $ B2 $ C1;
S3 = A3 $ B3 $ C2;
COUT = A3 & B3 # A3 & C2 # B3 & C2;
```

Figure 6.38 Four-Bit Adder

time: 7 secs
total time: 62 secs
CUPL errors encountered!
Press any key to continue

What happened? The error messages indicate that the sum-of-products expressions for S1, S2, S3, and COUT were too large for the resources of the GAL16V8 that the file was being compiled for. How is this possible with such a simple, straightforward design? The answer comes to light when you examine the documentation file of Figure 6.39 for this design.

Why did the sum-of-products terms get so large? Think about our circuit for a moment. The carry out from the sum of the two least significant bits is $A_0B_0 + B_0(CIN) + A_0(CIN)$. This becomes the expression used as the carry input to the next full adder in the four-bit adder. The sum of the next adder (S_1) is A_1 plus B_1 plus C_0 where C_0 is expressed as the carry out from the first adder. To put it another way,

$$S_1 = [A_0B_0 + A_0(CIN) + B_0(CIN)(A_1\overline{B}_1 + \overline{A}_1\overline{B}_1)$$
$$+ [\overline{A_0B_0 + A_0(CIN) + B_0(CIN)}](A_1\overline{B}_1 + \overline{A}_1B_1)$$

Factor that into the second adder's carry out, and use that as the next adder's carry in. The results are shown in the documentation file. What happened here is that our design exceeded the *internal* resources of the PLD. There just aren't enough sections in the AND/OR arrays to accommodate the large sum-of-products terms.

Does this mean that we cannot implement the four-bit adder using PLDs? Of course not, it just means that what is needed is a new approach or perhaps a different PLD. One approach is to reduce the number of product terms that the GAL16V8 must handle. To do this, we will make the interconnections between full adder carry outputs and carry inputs using external connections instead of internal ones. In this solution, start by creating four full adders in the PLD design, and bring the carry outs and carry ins of each adder to a pin. When the design is completed, connect the carry functions externally, as shown in Figure 6.40. This will change our requirements for input and output pin allocations to

Four number A inputs
Four number B inputs
Four carry inputs
Four carry outputs
Four sum outputs

This gives a total of 20 inputs and outputs, which brings up a second problem—the GAL16V8 does not have a total of 20 inputs and outputs. It only has a total of 16. That means we need to use a different PLD chip or to use more than one GAL16V8 to bring about the final solution. The chosen solution was to stay with the GAL16V8 and use two devices: one holds three full adders and the other the additional fourth full adder. The choice was made because the remaining resources of the second PLD chip can be used to implement a four-bit two's complement adder/subtracter or a four-bit BCD adder circuit. The completed templates for both PLDs are shown in the template of Figure 6.41.

```
*************************************************************
                          4BitAdd
*************************************************************

CUPL         4.0a Serial# MD-40A-9015
Device       g16v8s Library DLIB-h-891-9
Created      Thu Jun 03 21:15:49 1993
Name         4BitAdd
Partno       16V8
Revision     XX
Date         06/03/93
Designer     Miller
Company      XXXXX
Assembly     XXXXX
Location     XXXXX

=============================================================
                   Expanded Product Terms
=============================================================

C0 =>
    A0 & B0
  # A0 & CIN
  # B0 & CIN

C1 =>
    A1 & B1
  # A0 & A1 & B0
  # A1 & B0 & CIN
  # A0 & A1 & CIN
  # A0 & B0 & B1
  # B0 & B1 & CIN
  # A0 & B1 & CIN

C2 =>
    A2 & B2
  # A1 & A2 & B1
  # A0 & A2 & B0 & B1
  # A0 & A2 & B1 & CIN
  # A2 & B0 & B1 & CIN
  # A0 & A1 & A2 & B0
  # A0 & A1 & A2 & CIN
  # A1 & A2 & B0 & CIN
  # A1 & B1 & B2
  # A0 & B0 & B1 & B2
  # A0 & B1 & B2 & CIN
  # B0 & B1 & B2 & CIN
  # A0 & A1 & B0 & B2
  # A0 & A2 & B2 & CIN
  # A1 & B0 & B2 & CIN
```

Figure 6.39 Four-Bit Adder DOC File

```
COUT =>
      A3 & B3
    # A2 & A3 & B2
    # A1 & A3 & B1 & B2
    # A0 & A3 & B0 & B1 & B2
    # A0 & A3 & B1 & B2 & CIN
    # A3 & B0 & B1 & B2 & CIN
    # A0 & A1 & A3 & B0 & B2
    # A1 & A3 & B0 & B2 & CIN
    # A0 & A1 & A3 & B2 & CIN
    # A1 & A2 & A3 & B1
    # A0 & A2 & A3 & B0 & B1
    # A0 & A2 & A3 & B1 & CIN
    # A2 & A3 & B0 & B1 & CIN
    # A0 & A1 & A2 & A3 & B0
    # A1 & A2 & A3 & B0 & CIN
    # A0 & A1 & A2 & A3 & CIN
    # A2 & B2 & B3
    # A1 & B1 & B2 & B3
    # A0 & B0 & B1 & B2 & B3
    # A0 & B1 & B2 & B3 & CIN
    # B0 & B1 & B2 & B3 & CIN
    # A0 & A1 & B0 & B2 & B3
    # A1 & B0 & B2 & B3 & CIN
    # A0 & A1 & B2 & B3 & CIN
    # A1 & A2 & B1 & B3
    # A0 & A2 & B0 & B1 & B3
    # A0 & A2 & B1 & B3 & CIN
    # A2 & B0 & B1 & B3 & CIN
    # A0 & A1 & A2 & B0 & B3
    # A1 & A2 & B0 & B3 & CIN
    # A0 & A1 & A2 & B3 & CIN

S0 =>
      A0 & !B0 & !CIN
    # A0 & B0 & CIN
    # !A0 & !B0 & CIN
    # !A0 & B0 & !CIN

S1 =>
      !A0 & A1 & !B0 & !B1
    # !A0 & A1 & !B1 & !CIN
    # A1 & !B0 & !B1 & !CIN
    # A0 & A1 & B0 & B1
    # A0 & A1 & B1 & CIN
    # A1 & B0 & B1 & CIN
    # A0 & !A1 & B0 & !B1
    # A0 & !A1 & !B1 & CIN
    # !A1 & B0 & !B1 & CIN
    # !A0 & !A1 & !B0 & B1
    # !A0 & !A1 & B1 & !CIN
    # !A1 & !B0 & B1 & !CIN
```

Figure 6.39 Continued

```
S2 =>
     !A1 & A2 & !B1 & !B2
   # !A0 & !A1 & A2 & !B0 & !B2
   # !A0 & A2 & !B0 & !B1 & !B2
   # !A0 & !A1 & A2 & !B2 & !CIN
   # !A0 & A2 & !B1 & !B2 & !CIN
   # !A1 & A2 & !B0 & !B2 & !CIN
   # A2 & !B0 & !B1 & !B2 & !CIN
   # A1 & A2 & B1 & B2
   # A0 & A2 & B0 & B1 & B2
   # A0 & A2 & B1 & B2 & CIN
   # A2 & B0 & B2 & B2 & CIN
   # A0 & A1 & A2 & B0 & B2
   # A1 & A2 & B0 & B2 & CIN
   # A0 & A1 & A2 & B2 & CIN
   # A1 & !A2 & B1 & !B2
   # A0 & !A2 & B0 & B1 & !B2
   # A0 & !A2 & B1 & !B2 & CIN
   # !A2 & B0 & B1 & !B2 & CIN
   # A0 & A1 & !A2 & B0 & !B2
   # A1 & !A2 & B0 & !B2 & CIN
   # A0 & A1 & !A2 & !B2 & CIN
   # !A1 & !A2 & !B1 & B2
   # !A0 & !A1 & !A2 & !B0 & B2
   # !A0 & !A2 & !B0 & !B1 & B2
   # !A0 & !A1 & !A2 & B2 & !CIN
   # !A0 & !A2 & !B1 & B2 & !CIN
   # !A1 & !A2 & !B0 & B2 & !CIN
   # !A2 & !B0 & !B1 & B2 & !CIN

S3 =>
     !A2 & A3 & !B2 & !B3
   # !A1 & !A2 & A3 & !B1 & !B3
   # !A1 & A3 & !B1 & !B2 & !B3
   # !A0 & !A1 & !A2 & A3 & !B0 & !B3
   # !A0 & !A1 & A3 & !B0 & !B2 & !B3
   # !A0 & !A2 & A3 & !B0 & !B1 & !B3
   # !A0 & A3 & !B0 & !B1 & !B2 & !B3
   # !A0 & !A1 & !A2 & A3 & !B3 & !CIN
   # !A0 & !A1 & A3 & !B2 & !B3 & !CIN
   # !A0 & !A2 & A3 & !B1 & !B3 & !CIN
   # !A0 & A3 & !B1 & !B2 & !B3 & !CIN
   # !A1 & !A2 & A3 & !B0 & !B3 & !CIN
   # !A1 & A3 & !B0 & !B2 & !B3 & !CIN
   # !A2 & A3 & !B0 & !B1 & !B3 & !CIN
   # A3 & !B0 & !B1 & !B2 & !B3 & !CIN
   # A2 & A3 & B2 & B3
```

Figure 6.39 Continued

```
# A1 & A3 & B1 & B2 & B3
# A0 & A3 & B0 & B1 & B2 & B3
# A0 & A3 & B1 & B2 & B3 & CIN
# A3 & B0 & B1 & B2 & B3 & CIN
# A0 & A1 & A3 & B0 & B2 & B3
# A1 & A3 & B0 & B2 & B3 & CIN
# A0 & A1 & A3 & B2 & B3 & CIN
# A1 & A2 & A3 & B1 & B3
# A0 & A2 & A3 & B0 & B1 & B3
# A0 & A2 & A3 & B1 & B3 & CIN
# A2 & A3 & B0 & B1 & B3 & CIN
# A0 & A1 & A2 & A3 & B0 & B3
# A0 & A1 & A2 & A3 & B3 & CIN
# A1 & A2 & A3 & B0 & B3 & CIN
# A2 & !A3 & B2 & !B3
# A1 & !A3 & B1 & B2 & !B3
# A0 & !A3 & B0 & B1 & B2 & !B3
# A0 & !A3 & B1 & B2 & !B3 & CIN
# !A3 & B0 & B1 & B2 & !B3 & CIN
# A0 & A1 & !A3 & B0 & B2 & !B3
# A1 & !A3 & B0 & B2 & !B3 & CIN
# A0 & A1 & !A3 & B2 & !B3 & CIN
# A1 & A2 & !A3 & B1 & !B3
# A0 & A2 & !A3 & B0 & B1 & !B3
# A0 & A2 & !A3 & B1 & !B3 & CIN
# A2 & !A3 & B0 & B1 & !B3 & CIN
# A0 & A1 & A2 & !A3 & B0 & !B3
# A0 & A1 & A2 & !A3 & !B3 & CIN
# A1 & A2 & !A3 & B0 & !B3 & CIN
# !A2 & !A3 & !B2 & B3
# !A1 & !A2 & !A3 & !B1 & B3
# !A1 & !A3 & !B1 & !B2 & B3
# !A0 & !A1 & !A2 & !A3 & !B0 & B3
# !A0 & !A1 & !A3 & !B0 & !B2 & B3
# !A0 & !A2 & !A3 & !B0 & !B1 & B3
# !A0 & !A3 & !B0 & !B1 & !B2 & B3
# !A0 & !A1 & !A2 & !A3 & B3 & !CIN
# !A0 & !A1 & !A3 & !B2 & B3 & !CIN
# !A0 & !A2 & !A3 & !B1 & B3 & !CIN
# !A0 & !A3 & !B1 & !B2 & B3 & !CIN
# !A1 & !A2 & !A3 & !B0 & B3 & !CIN
# !A1 & !A3 & !B0 & !B2 & B3 & !CIN
# !A2 & !A3 & !B0 & !B1 & B3 & !CIN
# !A3 & !B0 & !B1 & !B2 & B3 & !CIN
```

Figure 6.39 Concluded

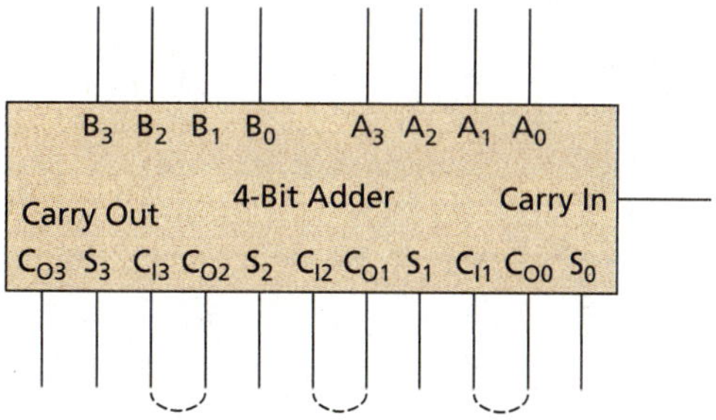

Figure 6.40 Four-Bit Parallel Adder PLD Solution

```
Name          3BitAdd;
Partno        16V8;
Date          10/28/93;
Revision      XX;
Designer      Miller;
Company       XXXXX;
Assembly      XXXXX;
Location      Phoenix;

/**********************************************************/
/*                                                        */
/*    Three-Bit Parallel Adder                            */
/*                                                        */
/**********************************************************/
/*  Allowable Target Device Types: 16V8                   */
/**********************************************************/

/**   Inputs   **/

Pin 1 = A0;        /*                              */
Pin 2 = B0;        /*                              */
Pin 3 = CIN;       /*                              */
Pin 4 = A1;        /*                              */
Pin 5 = B1;        /*                              */
Pin 6 = C1;        /*                              */
Pin 7 = A2;        /*                              */
Pin 8 = B2;        /*                              */
Pin 9 = C2;        /*                              */

/**   Outputs   **/

Pin 19 = S0;       /*                              */
Pin 18 = C0;       /*                              */
Pin 17 = S1;       /*                              */
Pin 16 = C01;      /*                              */
Pin 15 = S2;       /*                              */
Pin 14 = C02;      /*                              */
```

Figure 6.41 Two PLD Four-Bit Adder

```
/**  Logic Equations  **/

S0 = A0 $ B0 $ CIN;
S1 = A1 $ B1 $ C1;
S2 = A2 $ B2 $ C2;
C0 = A0 & B0 # A0 & CIN # B0 & CIN;
C01 = A1 & B1 # A1 & C1 # B1 & C1;
C02 = A2 & B2 # A2 & C2 # B2 & C2;


Name         Full_Adder;
Partno       16V8;
Date         05/26/93;
Revision     XX;
Designer     Miller;
Company      XXXXX;
Assembly     XXXXX;
Location     XXXXX;

/***********************************************************/
/*                                                         */
/*  Full Adder to Be Used with Three-Bit Parallel Adder    */
/*                                                         */
/***********************************************************/
/*  Allowable Target Device Types: GAL16V8                 */
/***********************************************************/

/**  Inputs  **/

Pin 1 = A3;      /*  Input Number A  */
Pin 2 = B3;      /*  Input Number B  */
Pin 3 = CIN;     /*  Carry In        */

/**  Outputs  **/

Pin 18 = COUT;
Pin 19 = S3;

/**  Logic Equations  **/

S3 = A3 $ B3 $ CIN;
COUT = A3 & B3 # A3 & CIN # B3 & CIN;
```

Figure 6.41 Continued

When these two designs were compiled, the GAL16V8 resources were more than sufficient to hold the completed design. The interconnections for the two chips are shown in Figure 6.42.

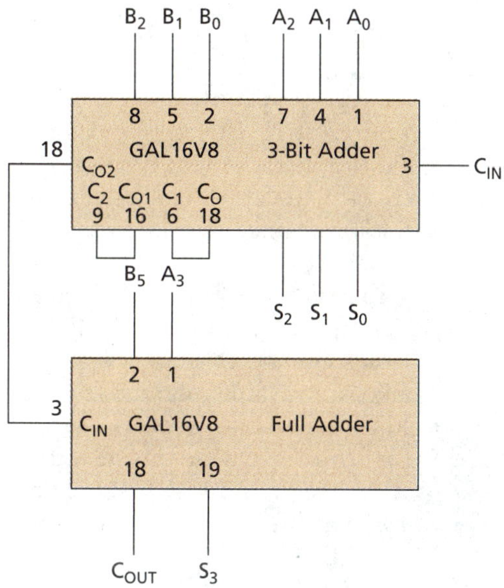

Figure 6.42 Four-Bit Adder Using Two 16 V8 PLDs

Section 6.12

Practice Problem

A GAL22V10 is another PLD whose pin diagram is shown in Figure 6.43. It has the pin function resources necessary to support a four-bit adder. However, external connections are still required between the COUT and CIN pins of each full adder. Create the CUPL template for the four-bit adder using the GAL22V10.

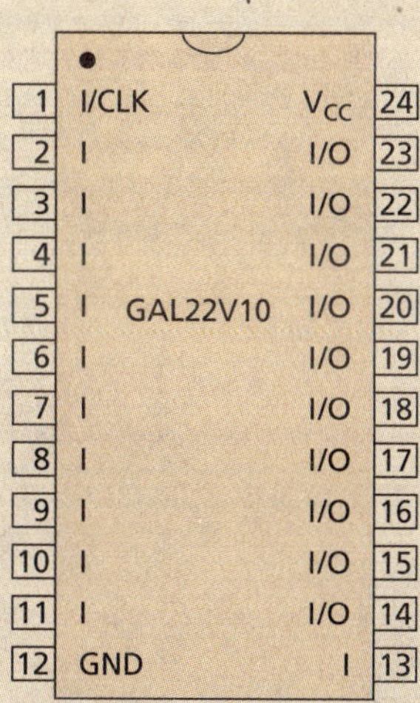

Figure 6.43 GAL22V10 Pin Diagram

6.13 Designing Counters Using CUPL

The registered outputs of a GAL16V8 are actually D-type flip-flops forcing any counter or register design using the GAL16V8 to adhere to the requirements and functions of that type of flip-flop. The state table of Figure 6.44 describes a two-bit synchronous up counter with an asynchronous RESET signal. From this state table, logic expressions can be derived for the D inputs to D-type flip-flops used in the GAL16V8.

To begin the counter PLD design, declare the CLK input pin as variable clock and one input pin as RESET. Two output pins are declared using variable names Q0 and Q1. Logic expressions for D_0 and D_1 are derived from the state table in the same manner as used with earlier counter designs done in the text. Q0 changes state on every clock pulse, making the first flip-flop act as a toggle flip-flop. To achieve this, using a D flip-flop, connect the $\overline{Q}$ to the D input or in CUPL syntax, D0 = !Q0. Q1 changes state whenever Q0 is high, which, restated, says that the D input of Q1 is the opposite state of its Q output when Q0 is set. The CUPL expression for this function uses the exclusive OR operator: D1 = Q0 $ Q1.

To translate CUPL expressions to include registered outputs, in this example, requires the use of an output modifier to be used with the logic expression. The syntax for these modifiers is

<output>.<modifier> = <expression>

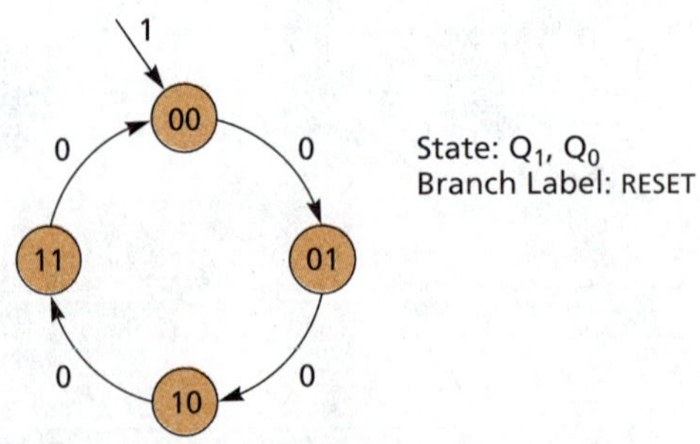

(a) State Diagram

RESET	Q_{1N}	Q_{0N}	Q_{1N+1}	Q_{0N+1}
1	X	X	0	0
0	0	0	0	1
0	0	1	1	0
0	1	0	1	1
0	1	1	0	0

X = "don't care" states.

(b) State Table

Figure **6.44** State Table for Two-Stage Synchronous Up Counter

Some commonly used modifiers are

.D	D flip-flop
.J	J input for a J–K flip-flop
.K	K input for a J–K flip-flop
.OE	three-state buffered output

Writing the expressions without the modifiers yields the logic circuit required to drive the D inputs:

D0 = !Q0;

D1 = Q0 $ Q1;

but these will not make the outputs registered and will require connections to external flip-flops. For our example, to include the flip-flop function to the PLD design, change the expressions to use the D flip-flop modifier:

Q0.D = !Q0;

Q1.D = Q0 $ Q1;

These expressions now state: Yield the D flip-flop output (Q) using the expressions (!Q0 and Q0 $ Q1) as the D inputs for each flip-flop. This set of expressions works, because Q0 and Q1 have been declared as output pin functions and the D modifier causes the CUPL compiler to select the D flip-flop registered output of the GAL16V8. The complete .PLD template file listing is shown in Figure 6.45.

Using logic expressions to design synchronous counters is fine for low mod numbers such as the mod-4 two-bit counter just developed. However, when the mod number of the counter increases significantly, developing these logic expressions becomes increasingly difficult. CUPL provides an alternate method for synchronous counter design that translates a state table directly to CUPL syntax and form. This method requires using the additional CUPL functions $DEFINE, FIELD, and SEQUENCE.

$DEFINE is used to equate a variable name (var) with a constant value or expression. The basic syntax for the $DEFINE statement is

$DEFINE <var name> <value>

Important: A ; is *not* included at the end of this statement! As an example, we can define the variable PIE as a constant with the value of 314 by writing this DEFINE statement:

$DEFINE PIE 314

Now the term PIE can be used anywhere the constant value 3.14 is required.

The FIELD statement is the same one used for truth tables and has the syntax

FIELD <field name> = [var, var, var . . .];

Lastly, the SEQUENCE statement defines the counter sequence as it appears on the state table, and includes both **PRESENT** and **NEXT** statements. Asynchronous inputs are added by making the NEXT statement conditional. First, an unconditional NEXT statement has the syntax

PRESENT <state_var> NEXT <state_var>

```
Name        2STGUP.PLD;
Partno      GAL16V8;
Date        03/16/94;
Revision    XX;
Designer    Miller;
Company     XXXXX;
Assembly    XXXXX;
Location    PHOENIX;

/*************************************************************/
/*                                                           */
/*    TWO-STAGE BINARY SYNCHRONOUS UP COUNTER                */
/*                                                           */
/*************************************************************/
/*  Allowable Target Device Types: 16V8                      */
/*************************************************************/

/**  Inputs  **/

Pin 1 = CLOCK;       /* CLOCKING SIGNAL INPUT */
Pin 2 = RESET;

/**  Outputs  **/

Pin 19 = Q0;       /* Q OUTPUT OF FLIP FLOP 0 */
Pin 18 = Q1;       /* Q OUTPUT OF FLIP FLOP 1 */

/**  Logic Equations  **/

Q0.D = !RESET & !Q0;      /* MAKES FF0 A TOGGLE FLIP FLOP */
Q1.D = !RESET & (Q1 $ Q0);
```

Figure 6.45 Two-Stage Up Counter .PLD File

where state_var is a declared state of the counter. and a conditional NEXT is written as

PRESENT <state_var> IF <var or expression> NEXT <state_var>;

Conditional NEXT statements may be extended to several conditions:

PRESENT <state_var> IF <var or expression> NEXT <state_var>;
IF <var or expression> NEXT <state_var>;

A DEFAULT statement is available that places the counter into a specified state if none of the conditions of the IF statements are met. The syntax for DEFAULT is

DEFAULT NEXT <state_var>;

All these are put together within the SEQUENCE format:

```
SEQUENCE <field name> {
        PRESENT – NEXT statements;
        DEFAULT if used;
        }
```

To illustrate how the state table format works, let's construct the CUPL file for the mod-4 counter of Figure 6.44, using the state design method. Input and output pins are declared as they are in any CUPL template design. To start the state design, first define the four counter states using the $DEFINE statement:

```
$DEFINE S0 0
$DEFINE S1 1
$DEFINE S2 2
$DEFINE S3 3
```

Next, declare the counter state field, which defines the order of the bits that would normally be used in the state circles in a state diagram or on a state table:

```
FIELD ORDER = [Q1,Q0];
```

Take these pieces and use them to create the counter sequence adding an active high RESET asynchronous input as a condition for each NEXT statement. The default state is 00 when RESET is set high:

```
SEQUENCE ORDER {
        PRESENT S0   IF !RESET NEXT S1;
        PRESENT S1   IF !RESET NEXT S2;
        PRESENT S2   IF !RESET NEXT S3;
        PRESENT S3   IF !RESET NEXT S0;
        DEFAULT                NEXT S0;
        }
```

The sequence says that the counter will transition from the present state to the next state, indicated by declared constants S0 to S3, if RESET is inactive. The only default condition is an active reset, which forces the counter into the S0 state. S0 is defined as 0H, which is 00B in this case, indicating that both flip-flops are reset. The completed CUPL file is shown in Figure 6.46.

```
Name        STATE.PLD;
Partno      GAL16V8;
Date        03/16/94;
Revision    XX;
Designer    MILLER;
Company     XXXXX;
Assembly    XXXXX;
Location    PHOENIX;
```

Figure 6.46 CUPL State Table Design

```
/*************************************************************/
/*                                                           */
/*  Two-Stage Binary Synchronous Up Counter State Design    */
/*                                                           */
/*************************************************************/
/*  Allowable Target Device Types: 16V8                      */
/*************************************************************/

/**  Inputs  **/

Pin 1 = CLOCK;        /* CLOCKING SIGNAL INPUT */
Pin 2 = RESET;

/**  Outputs  **/

Pin 19 = Q0;
Pin 18 = Q1;

/**  Logic Equations  **/

$DEFINE S0 0
$DEFINE S1 1
$DEFINE S2 2
$DEFINE S3 3

FIELD ORDER = [Q1, Q0];

SEQUENCE ORDER {
        PRESENT S0   IF !RESET NEXT S1;
        PRESENT S1   IF !RESET NEXT S2;
        PRESENT S2   IF !RESET NEXT S3;
        PRESENT S3   IF !RESET NEXT S0;
        DEFAULT                NEXT S0;
        }
```

Figure **6.46** Continued

EXAMPLE 6.7

Create a CUPL design for a mod-4 counter with both RESET and ENABLE input controls.

SOLUTION A third input, ENABLE, is added to the in declarations. The only other change to the previous design is that each PRESENT transition now has two conditions instead of one. If ENABLE is asserted, then the counter counts, and if it is not asserted, the counter retains its current state. In both cases, RESET must be inactive. Since each condition within the present statements accounts for both states of the ENABLE signal when RESET is inactive, the default will represent what occurs when RESET is asserted. The sequence now is

```
SEQUENCE ORDER {
        PRESENT S0   IF ENABLE & !RESET NEXT S1;
                     IF !ENABLE & !RESET NEXT S0;
        PRESENT S1   IF ENABLE & !RESET NEXT S2;
                     IF !ENABLE & !RESET NEXT S1;
        PRESENT S2   IF ENABLE & !RESET NEXT S3;
                     IF !ENABLE & !RESET NEXT S2;
        PRESENT S3   IF ENABLE & !RESET NEXT S0;
                     IF !ENABLE & !RESET NEXT S3;
        DEFAULT                         NEXT S0;
        }
```

The completed CUPL file for this design is shown in Figure 6.47. This completed PLD design, when compiled and programmed into a PLD, operates the same as an equivalent discrete two-bit counter circuit, but the major difference is that the entire PLD design is incorporated into a single chip, whereas the discrete design requires flip-flop, AND, OR, and XOR logic gate ICs.

```
Name         STATE_n2.PLD;
Partno       GAL16V8;
Date         03/16/94;
Revision     XX;
Designer     MILLER;
Company      XXXXX;
Assembly     XXXXX;
Location     PHOENIX;

/************************************************************/
/*                                                          */
/*  Two-Stage Binary Synchronous Up Counter State Design   */
/*                  with enable and reset                   */
/************************************************************/
/*  Allowable Target Device Types: 16V8                     */
/************************************************************/

/**  Inputs  **/

Pin 1 = CLOCK;        /* CLOCKING SIGNAL INPUT */
Pin 2 = RESET;
Pin 3 = ENABLE;

/**  Outputs  **/

Pin 19 = Q0;
Pin 18 = Q1;
```

Figure 6.47 Mod-4 State Design with Enable

```
/**  Logic Equations  **/

$DEFINE S0 0
$DEFINE S1 1
$DEFINE S2 2
$DEFINE S3 3

FIELD ORDER = [Q1, Q0];

SEQUENCE ORDER {
        PRESENT S0  IF ENABLE & !RESET NEXT S1;
                    IF !ENABLE & !RESET NEXT S0;
        PRESENT S1  IF ENABLE & !RESET NEXT S2;
                    IF !ENABLE & !RESET NEXT S1;
        PRESENT S2  IF ENABLE & !RESET NEXT S3;
                    IF !ENABLE & !RESET NEXT S2;
        PRESENT S3  IF ENABLE & !RESET NEXT S0;
                    IF !ENABLE & !RESET NEXT S3;
        DEFAULT                        NEXT S0;
        }
```

Figure 6.47 Continued

Section 6.13

Practice Problems

1. Create a CUPL design for a free-running synchronous mod-8 counter with RESET.
2. Create a CUPL file for a synchronous mod-5 counter with RESET and ENABLE inputs. The sequence is to count from 0H to 4H.

Summary

Programmable logic devices offer an affordable way to implement custom logic design. Their accessibility and ease of use contribute to their continued success. Several good development software packages exist to aid in using PLDs. One popular package is the Universal Compiler for Programmable Logic Devices or CUPL. As a menu-driven PLD compiler, CUPL is easy to use with a large library of PLD devices. Any PLD programming device that recognizes the JEDEC file form can be used with the .JED files compiled by CUPL.

A specific format used by CUPL is summarized in a TMPL.PLD template file supplied with the CUPL software. This file accepts logic designs in logic expressions using CUPL syntax operators; truth table form, using field and table syntaxes, or state diagram form, using a sequence based on current and next state conditions. Once correctly completed, the .PLD file is compiled into several different files based on selected options. These files are the JEDEC (.JED) fuse plot file, and a .DOC documentation file. CUPL software also includes an editor to assist in making on-the-spot corrections to the .PLD text file.

Individual PLD manufacturers supply compiler and programmer software without cost. Care must be taken when applying these tools for education purposes, because they may be too cumbersome to use or may not be applicable for inexpensive PLD devices.

PLDs come in three basic formats, based on which arrays are fixed and which are programmable. Those forms are PLA, programmable logic arrays; PAL, programmable array logic; and PROM, programmable read-only memory. Outputs of PLD devices take a number of forms, including buffered, complemented, registered, and three-state buffered. A universal output form is the variable type, which allows programmable selection of the output format.

A generic form of PLD, the GAL, can be substituted for any device within a specified PAL family. A commonly used PLD, the GAL16V6, has a total of 16 input and output pins, including 8 fixed input types and 8 selectable as inputs or outputs. A clock input is also available for registered output operation. A further option in a GAL16V8 is an output enable for three-state control of the chip's output functions.

The process of designing a PLD application, compiling it into a JEDEC file, and burning the design into a PLD is relatively easy using CUPL software. Other PLD compilers are available from PLD manufacturers for little or no cost, but they are usually more difficult to use or apply only to that manufacturer's devices.

Glossary

CUPL Universal Compiler for Programmable Logic devices is a compiler program that translates PLD designs into JEDEC files.

GAL generic or gate array logic devices are used for PLD designs for a specific PAL family of devices.

JEDEC the Joint Electronic Devices Engineering Council sets standards for PLD design and program formats.

PAL programmable array logic devices are those with programmable AND arrays only.

PLA programmable logic array devices with programmable OR and AND arrays.

PLD programmable logic device designates a group of logic devices whose circuits and functions are programmable.

PROM programmable read-only memory devices with programmable OR arrays only.

Solutions to Practice Problems

Section 6.2

1.

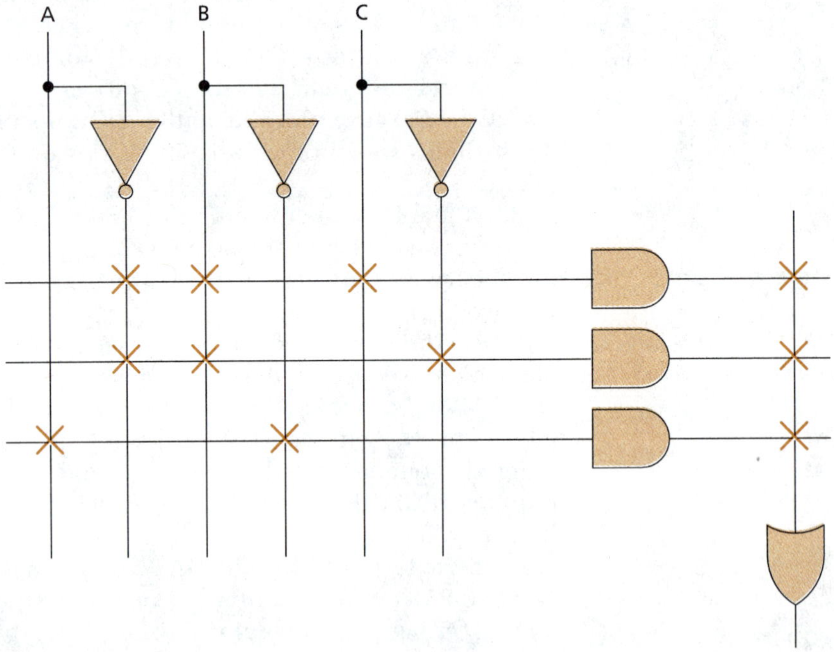

Figure 6.48 Practice Problem 6.2(1) (solution)

2.

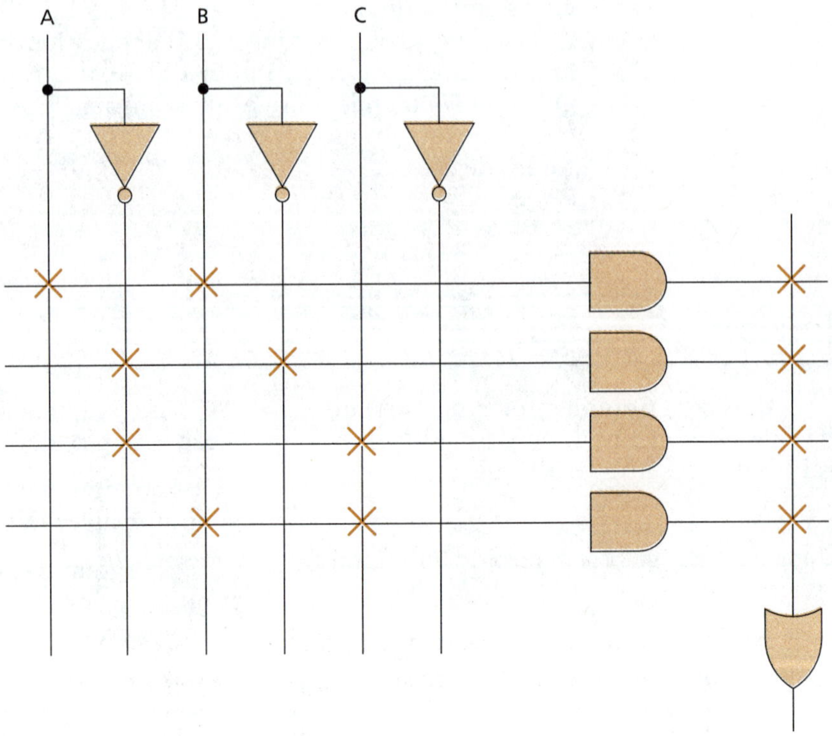

Figure 6.49 Practice Problem 6.2(2) (solution)

3.

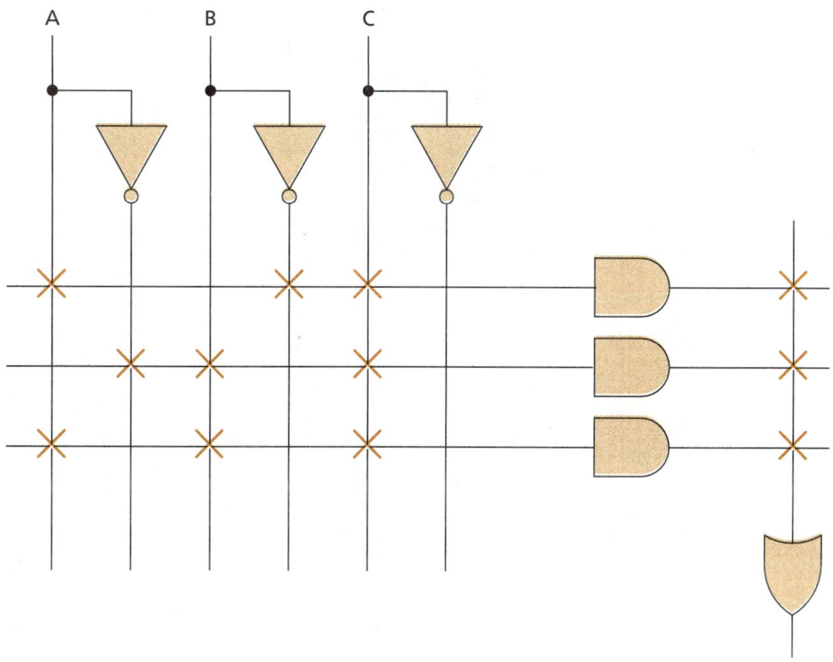

Figure 6.50 Practice Problem 6.2(3) (solution)

Section 6.3

1. PROM, because of the fixed AND array
2. PAL, because of the fixed OR array

Section 6.4

1. 22 in, 10 out, variable output
2. 20 in, 6 out, registered output
3. 26 in, 12 out, active low (complemented) output
4. 10 in, 20 out, active high (normal) output

Section 6.6

1. **a.** valid
   b. Not valid—contains space
   c. Not valid—no alpha character
   d. Valid
   e. Not valid—reserved word
2. **a.** !A & B & D # ! (C & D) # A & !B
   b. (!A # B & C) & !(B & C & D # A & D)
   c. ! (!A # !B) & (C & !D) # A & !B & D
   d. A & B & C # !(A & C) # B & D

Section 6.7

1.

```
Name         Today.Pld;
Partno       16V8;
Date         10/15/93;
Revision     XX;
Designer     Miller;
Company      XXXXX;
Assembly     XXXXX;
Location     Phoenix;

/*******************************************************************/
/*                                                                 */
/*    Answer to Practice Problem 6.7(1)                            */
/*                                                                 */
/*******************************************************************/
/*  Allowable Target Device Types: GAL16V8                         */
/*******************************************************************/

/**  Inputs  **/

Pin 2          = MON      ;        /*                              */
Pin 3          = TUES     ;        /*                              */
Pin 4          = WED      ;        /*                              */
Pin 5          = THUR     ;        /*                              */
Pin 6          = FRI      ;        /*                              */
Pin 7          = WKEND    ;        /*                              */
Pin 8          = VACATION ;        /*                              */

/**  Outputs  **/

Pin 18         = TODAY    ;        /*                              */

/**  Logic Equations  **/

TODAY = MON & !TUES & WED # !(THUR & FRI) # !(!WKEND & !VACATION);
```

Figure 6.51 Practice Problem 6.7(1) (solution)

2.

```
Name          X_ANSWER;
Partno        16V8;
Date          10/15/93;
Revision      XX;
Designer      Miller;
Company       XXXXX;
Assembly      XXXXX;
Location      Phoenix;

/***********************************************************/
/*                                                         */
/*   Answer to Practice Problem 6.7(2)                     */
/*                                                         */
/***********************************************************/
/*  Allowable Target Device Types: GAL16V8                 */
/***********************************************************/

/**  Inputs  **/

Pin 2        = A      ;       /*                         */
Pin 3        = B      ;       /*                         */
Pin 4        = NO     ;       /*                         */
Pin 5        = YES    ;       /*                         */
Pin 6        = TODO   ;       /*                         */
Pin 7        = REST   ;       /*                         */

/**  Outputs  **/

Pin 18       = X      ;       /*                         */

/**  Logic Equations  **/

X = A & !NO & TODO # B & !YES & TODO # !A & !B & REST;
```

Figure 6.52 Practice Problem 6.7(2) (solution)

Section 6.8

Select Device . . GAL . . 16V8
Compile . . A:\LOGIC\EXAMPLE.PLD
Compile Options . . create JEDEC file . . JEDEC file same name as .PLD . . List file . . expanded product terms . . fuse plot/pin diagram
Press F5 . . press any key when compilation is finished

Section 6.10

1.

```
Name         PP6 11_1;
Partno       16V8;
Date         11/02/93;
Revision     XX;
Designer     Miller;
Company      XXXXX;
Assembly     XXXXX;
Location     Phoenix;

/*************************************************************/
/*                                                           */
/*   Solution to Practice Problem 6.10(1)                    */
/*                                                           */
/*************************************************************/
/*  Allowable Target Device Types: 16V8                      */
/*************************************************************/

/**  Inputs  **/

Pin 2 = PAT;       /*                               */
Pin 3 = DON;       /*                               */
Pin 4 = TOM;       /*                               */
Pin 5 = JERRY;     /*                               */

/** Outputs **/

Pin 18 = CAT;      /*                               */
Pin 17 = MOUSE;    /*                               */
Pin 16 = DOG;      /*                               */

/** Declarations and Intermediate Variable Definitions **/

FIELD GUYS = [ PAT, DON, TOM, JERRY ];
FIELD PETS = [ CAT, MOUSE, DOG ];

/**  Logic Equations  **/

TABLE GUYS => PETS {

0 => 'b'001;    1 => 'b'100;    2 => 'b'100;    3 => 'b'010;
4 => 'b'011;    5 => 'b'011;    6 => 'b'110;    7 => 'b'110;
8 => 'b'101;    9 => 'b'011;    A => 'b'110;    B => 'b'011;
C => 'b'101;    D => 'b'100;    E => 'b'101;    F => 'b'010;
}
```

Figure 6.53 Solution to Practice Problem 6.10(1) (solution)

2.

```
Name          PP6_11_2;
Partno        16V8;
Date          11/02/93;
Revision      XX;
Designer      Miller;
Company       XXXXX;
Assembly      XXXXX;
Location      Phoenix;

/**********************************************************/
/*                                                        */
/*   Answer to Practice Problem 6.10(2)                   */
/*                                                        */
/**********************************************************/
/*  Allowable Target Device Types: 16V8                   */
/**********************************************************/

/**  Inputs  **/

Pin 2 = FAR;        /*                                */
Pin 3 = NEAR;       /*                                */
Pin 4 = TIME;       /*                                */
Pin 5 = COST;       /*                                */
Pin 6 = MOST;       /*                                */

/**  Outputs  **/

Pin 18 = THIS;      /*                                */
Pin 17 = THAT;      /*                                */
Pin 16 = THEM;      /*                                */

/** Declarations and Intermediate Variable Definitions **/

FIELD LENGTH = [ FAR, NEAR, TIME ];
FIELD CHOICE = [ THAT, THEM ];

/**  Logic Equations  **/

THIS = MOST & (!FAR # NEAR # !( TIME & COST ));

TABLE LENGTH => CHOICE {

'b'000 => 'b'00;    'b'001 => 'b'01;    'b'010 => 'b'10;
'b'011 => 'b'11;    'b'100 => 'b'10;    'b'101 => 'b'01;
'b'110 => 'b'00;    'b'111 => 'b'11      }
```

Figure 6.54 Practice Problem 6.10(2) (solution)

Section 6.11

1.

```
        Listing of File: D:\CUPL\FULLSUB.PLD Page: 1

Name        PP6.12#1;
Partno      16V8;
Date        06/03/95;
Revision    XX;
Designer    Miller;
Company     XXXXX;
Assembly    XXXXX;
Location    Phoenix;

/************************************************************/
/*                                                          */
/*  Full Subtractor Solution to Practice Problem 6.11(1)    */
/*                                                          */
/************************************************************/
/*  Allowable Target Device Types: GAL16V8                  */
/************************************************************/

/**  Inputs  **/

Pin 2 = A;
Pin 3 = B;
Pin 4 BW_IN;

/**  Outputs  **/

Pin 18 = DIFF;
Pin 17 = BW_OUT;

/**  Logic Equations  **/

DIFF = A $ B $ BW_IN;    /* difference = A - B - borrow in */
BW_OUT = !A & B # !A & BW_IN # B & BW_IN;   /* borrow out */
```

Figure 6.55 Practice Problem 6.11(1) (solution)

2.

```
          Listing of File: D:\CUPL\SUBTT.PLD Page: 1

Name         PP6.12#2;
Partno       16V8;
Date         06/03/95;
Revision     XX;
Designer     Miller;
Company      XXXXX;
Assembly     XXXXX;
Location     Phoenix;

/*******************************************************/
/*                                                     */
/*                                                     */
/*   Full Subtractor Truth Table Method                */
/*******************************************************/
/*  Allowable Target Device Types: GAL16V8              */
/*******************************************************/

/**  Inputs  **/

Pin 2 = A;
Pin 3 = B;
Pin 4 = BW_IN;

/**  Outputs  **/

Pin 18 = DIFF;
Pin 17 = BW_OUT;

/** Declarations and Intermediate Variable Definitions **/

FIELD SUBIN = [ BW_IN, A, B ];
FIELD SUBOUT = [ DIFF, BW_OUT ];

/**  Logic Equations  **/

TABLE SUBIN => SUBOUT {
        0 => 'b'00;  1 => 'b'11;  2 => 'b'10;  3 => 'b'00;
        4 => 'b'11;  5 => 'b'01;  6 => 'b'00;  7 => 'b'11;
        }
```

Figure 6.56 Practice Problem 6.11(2) (solution)

Section 6.12

```
Name        4BitGd;
Partno      22V10;
Date        06/05/93;
Revision    XX;
Designer    Miller;
Company     XXXXX;
Assembly    XXXXX;
Location    XXXXX;

/************************************************************/
/*                                                          */
/*   Four-Bit Parallel Adder                                */
/*                                                          */
/************************************************************/
/*  Allowable Target Device Types:                          */
/************************************************************/

/**  Inputs  **/

Pin 2 = A0;        /* A number least significant bit */
Pin 3 = A1;        /*                                */
Pin 4 = A2;        /*                                */
Pin 5 = A3;        /*                                */
Pin 6 = B0;        /* B number least significant bit */
Pin 7 = B1;        /*                                */
Pin 8 = B2;        /*                                */
Pin 9 = B3;        /*                                */
Pin 10 = C0;       /* Carry inputs                   */
Pin 11 = C1;
Pin 13 = C2;
Pin 14 = C3;

/**  Outputs  **/

Pin 15 = S0;      /* SUM least significant bit       */
Pin 16 = S1;      /*                                 */
Pin 17 = S2;      /*                                 */
Pin 18 = S3;      /*                                 */
Pin 19 = CO1;     /* Carry outputs                   */
Pin 20 = CO2;
Pin 21 = CO3;
Pin 22 = COUT;

/**  Logic Equations  **/
S0 = A0 $ B0 $ C0;
S1 = A1 $ B1 $ C1;
S2 = A2 $ B2 $ C2;
S3 = A3 $ B3 $ C3;
CO1 = A0 & B0 # A0 & C0 # B0 & C0;
CO2 = A1 & B1 # A1 & C1 # B1 & C1;
CO3 = A2 & B2 # A2 & C2 # B2 & C2;
COUT = A3 & B3 # A3 & C3 # B3 & C3;
```

Figure 6.57 Practice Problem 6.12 (solution)

Section 6.13

1.

```
Listing of File: D:\CUPL\MOD8.PLD

Name         MOD8.PLD;
Partno       16V8;
Date         08/21/94;
Revision     XX;
Designer     MILLER;
Company      XXXXX;
Assembly     XXXXX;
Location     PHOENIX;
/************************************************************/
/*                                                          */
/*    Mod-8 Up Counter                                      */
/*                                                          */
/************************************************************/
/*  Allowable Target Device Types: GAL16V8                  */
/************************************************************/

/** Inputs **/

Pin 2 = ENABLE;
Pin 3 = RESET;
Pin 1 = CLOCK;

/** Outputs **/

Pin 18 = Q0;
Pin 17 = Q1;
Pin 16 = Q2;

$DEFINE S0 0
$DEFINE S1 1
$DEFINE S2 2
$DEFINE S3 3
$DEFINE S4 4
$DEFINE S5 5
$DEFINE S6 6
$DEFINE S7 7

FIELD ORDER = [Q2, Q1, Q0];
```

Figure 6.58 Practice Problem 6.13(1) (solution)

```
SEQUENCE ORDER {
    PRESENT SO IF !RESET & ENABLE NEXT S1;
               IF !RESET & !ENABLE NEXT S0;
    PRESENT S1 IF !RESET & ENABLE NEXT S2;
               IF !RESET & !ENABLE NEXT S1;
    PRESENT S2 IF !RESET & ENABLE NEXT S3;
               IF !RESET & !ENABLE NEXT S2;
    PRESENT S3 IF !RESET & ENABLE NEXT S4;
               IF !RESET & !ENABLE NEXT S3;
    PRESENT S4 IF !RESET & ENABLE NEXT S5;
               IF !RESET & ENABLE NEXT S4;
    PRESENT S5 IF !RESET & ENABLE NEXT S6;
               IF !RESET & !ENABLE NEXT S5;
    PRESENT S6 IF !RESET & ENABLE NEXT S7;
               IF !RESET & !ENABLE NEXT S6;
    Listing of File: D:\CUPL\MOD8.PLD

PRESENT S7 IF !RESET & ENABLE NEXT S0;
           IF !RESET & !ENABLE NEXT S7;
DEFAULT                         NEXT S0;
}
```

Figure 6.58 Continued

2. Listing of File: D:\CUPL\MOD5.PLD Page: 1 of 1

```
Name        MOD5.PLD;
Partno      16V8;
Date        08/21/94;
Revision    XX;
Designer    MILLER;
Company     XXXXX;
Assembly    XXXXX;
Location    PHOENIX;

/************************************************************/
/*                                                          */
/*   Mod-5 Up Counter                                       */
/*                                                          */
/************************************************************/
/*  Allowable Target Device Types: GAL16V8                  */
/************************************************************/

/**  Inputs  **/

Pin 2 = ENABLE;
Pin 3 = RESET;
Pin 1 = CLOCK;

/**  Outputs  **/
```

Figure 6.59 Practice Problem 6.13(2) (solution)

```
Pin 18 = Q0;
Pin 17 = Q1;
Pin 16 = Q2;

$DEFINE S0 0
$DEFINE S1 1
$DEFINE S2 2
$DEFINE S3 3
$DEFINE S4 4

FIELD ORDER = [ Q2, Q1, Q0 ];

SEQUENCE ORDER {
         PRESENT S0 IF !RESET & ENABLE NEXT S1;
                    IF !RESET & !ENABLE NEXT S0;
         PRESENT S1 IF !RESET & ENABLE NEXT S2;
                    IF !RESET & !ENABLE NEXT S1;
         PRESENT S2 IF !RESET & ENABLE NEXT S3;
                    IF !RESET & !ENABLE NEXT S2;
         PRESENT S3 IF !RESET & ENABLE NEXT S4;
                    IF !RESET & !ENABLE NEXT S3;
         PRESENT S4 IF !RESET & ENABLE NEXT S0;
                    IF !RESET & !ENABLE NEXT S4;
         DEFAULT                        NEXT S0;
         }
```

Figure 6.59 Continued

Problems

6.1 What type of logic circuit forms the core of a basic PLD cell?

6.2 What form do logic expressions take before a JEDEC file can be created?

6.3 What are the different types of PLD devices, and what makes each type unique?

6.4 For each of the following applications, select which PLD device type—PAL, PLA, or PROM—is best suited for use.

a. A look-up table whose data are found at sequential locations in the device.

b. A logic circuit with numerous outputs. Some of these outputs are asserted by similar input combinations. The number of input variables for each combination varies.

c. An alarm is to be triggered by several different source combinations. One combination, for instance, monitors five doorway entries, while another input combination monitors three windows in a particular room.

d. A combination lock that generates an open signal if the correct four inputs are applied and an alarm signal if any other combination of four inputs are applied.

6.5 Redraw the following logic circuits, using PLD symbology:

a.

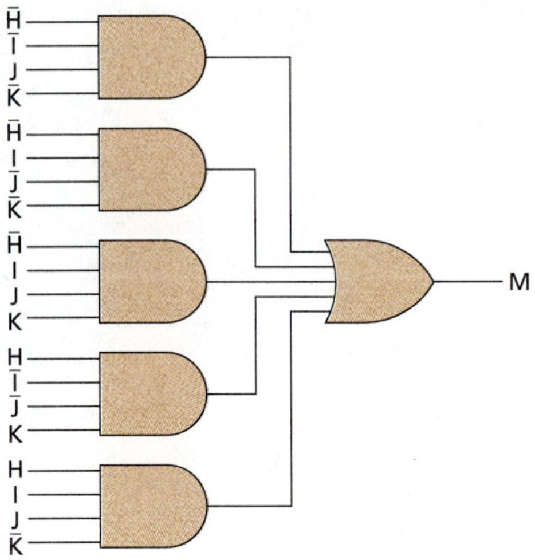

Figure 6.60 Problem 6.5a

b.

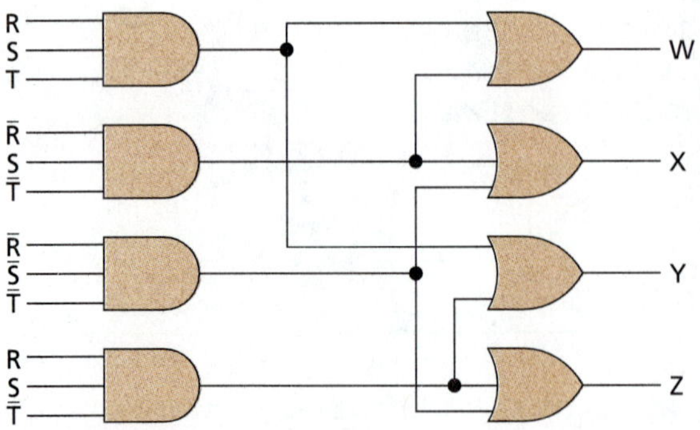

Figure 6.61 Problem 6.5b

c.

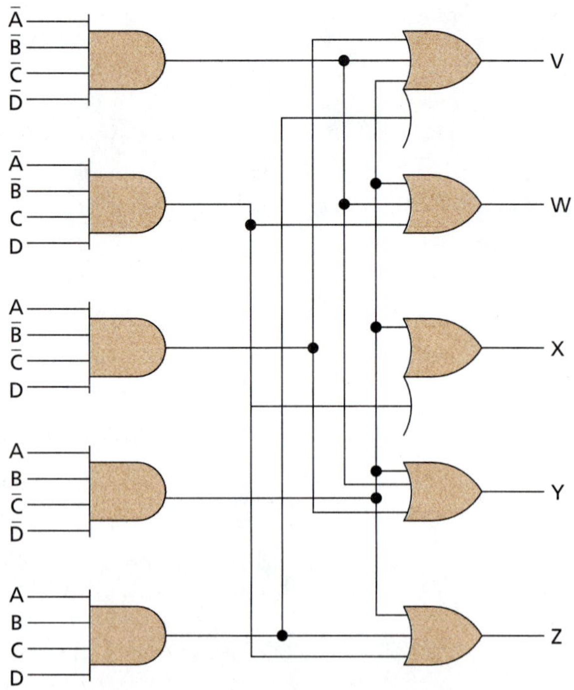

Figure 6.62 Problem 6.3c

6.6 Redraw the following logic circuits using PLD symbology:

a.

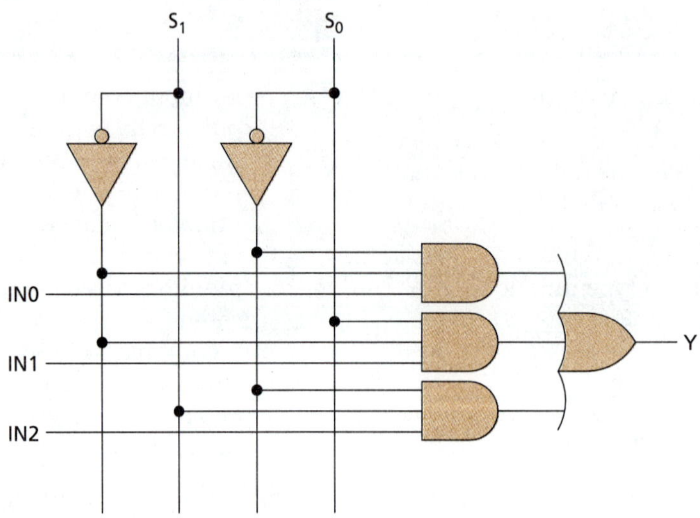

Figure 6.63 Problem 6.6a

b.

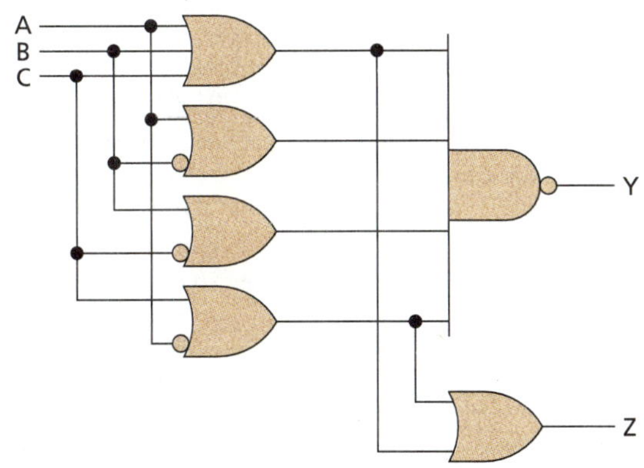

Figure 6.64 Problem 6.6b

c.

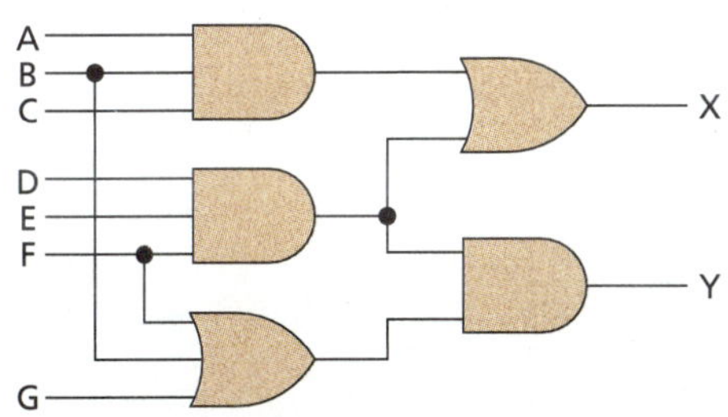

Figure 6.65 Problem 6.6c

6.7 Draw PLD logic diagrams for each of these SOP expressions:

a. $A \cdot \overline{B} \cdot \overline{C} + \overline{A} \cdot \overline{B} \cdot C + \overline{A} \cdot B \cdot \overline{C}$

b. $C \cdot D + A \cdot \overline{B} \cdot D + B \cdot \overline{C} \cdot \overline{D}$

c. $E \cdot T \cdot C + B \cdot \overline{E} \cdot \overline{T} + \overline{B} \cdot C$

d. $S \cdot T \cdot U \cdot V + \overline{T} \cdot \overline{V} + \overline{S} \cdot \overline{U} \cdot \overline{V}$

6.8 Draw PLD logic diagrams for each of these SOP expressions:

a. $\overline{A} \cdot \overline{B} \cdot C + \overline{A} \cdot B \cdot \overline{C} + A \cdot \overline{B} \cdot C$

b. $\overline{C} \cdot D + A \cdot \overline{B} \cdot \overline{D} + B \cdot C \cdot D$

c. $E \cdot \overline{T} \cdot \overline{C} + \overline{B} \cdot E \cdot T + B \cdot C$

d. $S \cdot \overline{T} \cdot \overline{U} \cdot V + \overline{T} \cdot \overline{V} + S \cdot U \cdot V$

6.9 For each part number given, how many input and output pins are available, and what is the indicated output type?

a. PAL24H10 **b.** PLA8V12 **c.** GAL16R8
d. GAL22L10 **e.** PAL24V8

6.10 Translate each of the following logic expressions into CUPL syntax:

a. $A \cdot B(\overline{C + D}) + \overline{B} \cdot C$

b. $A + \overline{C} + B \cdot (\overline{C} + \overline{A \cdot D})$

c. $S \cdot T + \overline{X \cdot R} + S \cdot \overline{T} \cdot \overline{X}$

d. $(\overline{F + G})(\overline{E \cdot G} + F \cdot \overline{D}) + \overline{F \cdot G \cdot H}$

e. $(A \oplus B \oplus C) + \overline{A} \cdot \overline{B} \cdot \overline{C}$

f. $\overline{H \cdot O \cdot P} + \overline{T \cdot O} + \overline{I \cdot T}$

6.11 What are the file extensions used by and generated by the CUPL software? Define what each file type is.

6.12 Write the correct syntax for declaring input pin 6 as variable HOT and output variable pin 18 as COOL_IT.

6.13 Declare the CUPL fields, and create the table declarations for the following truth tables:

a.

A_1	A_2	A_3	B_5	B_3	B_1	B_0
0	0	0	0	0	0	0
0	0	1	0	1	0	1
0	1	0	1	0	1	0
0	1	1	1	1	0	0
1	0	0	0	0	1	1
1	0	1	1	0	0	1
1	1	0	0	1	1	0
1	1	1	1	1	1	1

Figure 6.66 Problem 6.13a

b.

DO	IT	TO	DAY	AND	BE	DONE
0	0	0	0	0	1	0
0	0	0	1	0	1	1
0	0	1	0	0	0	1
0	0	1	1	1	0	1
0	1	0	0	1	0	0
0	1	0	1	1	0	0
0	1	1	0	0	1	1
0	1	1	1	0	1	0
1	0	0	0	0	0	1
1	0	0	1	0	1	1
1	0	1	0	0	0	1
1	0	1	1	0	1	0
1	1	0	0	1	1	1
1	1	0	1	1	1	1
1	1	1	0	1	0	1
1	1	1	1	1	0	0

Figure 6.67 Problem 6.13b

6.14 Declare the CUPL fields, and create the table declarations for the following truth tables:

a.

TIME	IS	NOW	GO	FOR	IT
0	0	0	0	0	0
0	0	1	1	0	0
0	1	0	1	0	1
0	1	1	0	1	0
1	0	0	1	0	0
1	0	1	0	1	1
1	1	0	0	1	0
1	1	1	1	1	1

Figure **6.68** Problem 6.14a

b.

A_0	A_1	A_2	A_3	D_0	D_1	D_2	D_3
0	0	0	0	0	0	0	0
0	0	0	1	0	1	1	0
0	0	1	0	0	1	1	1
0	0	1	1	1	1	1	1
0	1	0	0	1	1	0	0
0	1	0	1	0	1	1	0
0	1	1	0	1	1	1	0
0	1	1	1	1	1	1	1
1	0	0	0	1	0	0	0
1	0	0	1	1	0	0	1
1	0	1	0	1	1	0	0
1	0	1	1	0	0	0	1
1	1	0	0	0	0	1	0
1	1	0	1	1	0	1	1
1	1	1	0	1	0	0	1
1	1	1	1	1	1	1	1

Figure **6.69** Problem 6.14b

6.15 Fill in a CUPL template using the following criteria: Select a file name, and use your name, date, and location for header information. Use a GAL16V8 device. The logic expressions are
$\text{RUN} = \text{GO} + \overline{\text{STOP}} \cdot \overline{\text{OOPS}} + \text{HELP}$
$\text{DONE} = \text{GO} \cdot \overline{\text{HELP}} + (\overline{\text{STOP} \oplus \text{OOPS}})$

6.16 Complete the CUPL template for the truth table of Figure 6.70 and the logic circuit in Figure 6.71, both using a single GAL16V8. Select a filename, and use your name, date, and location for header information.

6.17 Determine which of the variable names are valid CUPL variables. For those that are invalid, state why they are invalid:
a. 3_STORY **b.** all_gone **c.** Top This
d. 1234_GO **e.** TABLE **f.** 1993
g. BE_DONE **h.** BAD?

6.18 Which of the following variable names are valid CUPL variables? For those that are invalid, state why they are invalid:
a. FORCE_ONE **b.** 'h'45CB **c.** $DOLLAR
d. GOOD_1 **e.** MAKE (IT) **f.** ABC[1 . . 5]
g. small_LTR **h.** !GREAT!

6.19 What is the first step after inserting your data disk and calling up the CUPL program when compiling a CUPL.PLD file into a .JED file?

6.20 If there is a syntax error during the CUPL compile procedure, how does the user know it and where can the user find the error and which line it is on?

6.21 Complete the CUPL templates for a 4-to-16 decoder using two GAL16V8's.

6.22 Create a CUPL TMPL file for an address decoder that generates the following signals: ENRAM, ENROM, RAS, CAS, and MUX. The inputs are address lines A12–A15, MEMR, and MEMW. The outputs are asserted in response to these criteria:

ENRAM for addresses with an upper nibble (A15–A12) of 1000B to 1100B.

EN	A	B	C	D	P	L	DE
1	0	0	0	0	0	1	0
1	0	0	0	1	0	1	0
1	0	0	1	0	1	1	0
1	0	0	1	1	1	1	0
1	0	1	0	0	0	0	0
1	0	1	0	1	1	0	1
1	0	1	1	0	0	0	1
1	0	1	1	1	0	0	0
1	1	0	0	0	0	1	0
1	1	0	0	1	0	1	0
1	1	0	1	0	1	0	1
1	1	0	1	1	1	0	1
1	1	1	0	0	1	1	1
1	1	1	0	1	1	0	1
1	1	1	1	0	1	1	0
1	1	1	1	1	0	0	1
0	X	X	X	X	0	0	0

X = "don't care" states.

Figure 6.70 Problem 6.16

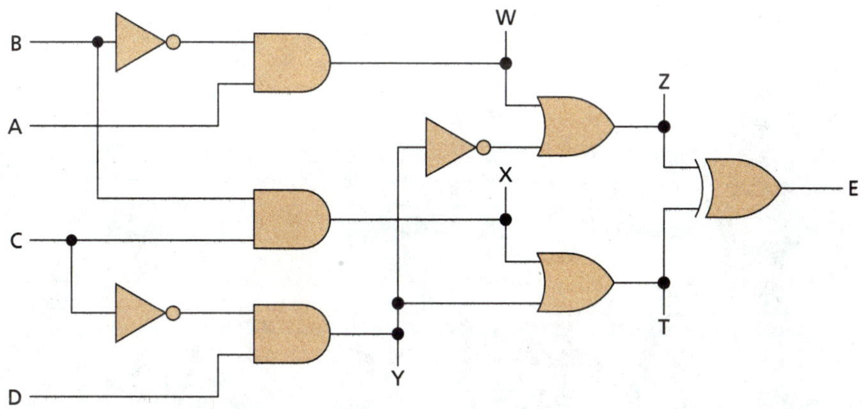

Figure 6.71 Problem 6.16

ENROM for addresses with an upper nibble of 1111B.
RAS whenever MEMR or MEMW are asserted.
MUX one clock cycle after RAS is asserted.
CAS one clock cycle after MUX is asserted.
RAS, MUX, and CAS are reset to 0 when MEMR and MEMW are both negated.

seven

Converting Between Digital and Analog Systems

KEY TERMS

accuracy
ADC
average conversion time
counter or ramp ADC
DAC
dual-slope ADC
end of conversion (EOC)
end of process (EOP)
flash or simultaneous ADC
full-scale output
maximum conversion time
monotonicity curve
percent of resolution
precision
resolution
scaling op amp
start conversion (SC)
step
successive approximations ADC
up/down ramp or continuous ADC
voltage resolution
weighted resistor ladder DAC

OBJECTIVES

After completing this chapter, the student should be able to

- Describe a resistor ladder network that uses a set of resistors that uses current sources for digit–weight products.
- Describe an R/2R resistor ladder that uses only two different resistor values, one which is twice the other, to manage the equivalent digit–weight products.
- Describe a flash analog-to-digital converter (DAC) made up of a set of op amps that alter saturation levels as the analog input is changed. The op amp outputs are used as a binary representation of the analog input.
- Explain the operation of several types of counter ADCs, including a ramp converter, with comparisons to illustrate advantages and disadvantages of all types.
- Explain the operation of successive approximations and why it is the ADC of choice.
- Describe the instrumentation ADC used in multimeters and other digital test equipment.
- Outline methods for troubleshooting and designing DAC and ADC circuits.

OVERVIEW

This chapter takes the student through circuits that convert between digital values and analog quantities. First the conversion of digital to analog systems is explored, using two types of resistor ladder network converters. The benefits and disadvantages of each are explained, along with design criteria for both types of converters.

After a description of digital-to-analog resistor networks, analog-to-digital conversion is discussed. Four basic types of converters are shown, with comparisons of benefits and applications. Data sheets for medium-scale integration (MSI) chip versions of analog-to-digital and digital-to-analog converters, including applications, are described and analyzed.

7.1 Introduction

The human experience with electronic devices and equipment is based on a combination of digital-and-analog circuitry and applications. From a simple on/off switch to a compact disc player, both analog and digital electronic circuits abound. Audio playback was, for the longest time, in the prime domain of analog electronics. Today's compact disc (CD) players convert reflected laser light to digital signals and then into the analog sounds we enjoy listening to. Microwave ovens have digital controllers that govern time and microwave energy with precision and dependability. Even telephone conversations experience changes from analog voice to digital signals and back to voice at the receiving end. To integrate analog signals and digital sequences into coherent units, we need a group of circuits to perform conversions between both analog and digital types of electronic signals.

7.2 Digital-to-Analog Conversion

A straightforward method of converting a digital quantity into an equivalent analog value applies the principle of weighted digits. Each binary position has its own unique numerical weight, which can be represented by an electrical quantity such as voltage or current. The least significant bit (LSB) has a weight of 1, which can be 1 volt (V), 1 milliamp (ma), or 1 ohm (Ω). The next position has twice the weight, with the position following doubling the weight value once more, and so forth. To convert a binary number into an equivalent analog quantity, add the weighted electrical quantities for each bit position that is a logic 1. For instance, the quantity 1001B, with 1 ma representing the least significant bit, would produce a total current of

$$1 \times 1\text{ ma} + 0 \times 2\text{ ma} + 0 \times 4\text{ ma} + 1 \times 8\text{ ma} = 1\text{ ma} + 8\text{ ma} = 9\text{ ma}$$

Simply put, the binary value 1001B would have an equivalent analog current value of 9 ma.

EXAMPLE 7.1

What is the analog equivalent to the binary quantity 1001101B? The least significant bit produces a current value of 2.5 ma.

SOLUTION Add up the current weight values for each logic 1 in the binary number

$$2.5\text{ ma} + 10\text{ ma} + 20\text{ ma} + 160\text{ ma} = 192.5\text{ ma}$$

So 1001101B with an LSB of 2.5 ma produces a total current of 192.5 ma.

Weighted resistor ladder DAC *a digital-to-analog resistor network that electrically sums the weights of each bit position to yield the analog equivalent of a digital input.*

All that is needed now is a circuit that will electrically generate the sum of the current weight values to produce an analog equivalent value for each given binary input. Such a circuit, called a **weighted resistor ladder,** is shown in Figure 7.1. Binary number inputs are placed at b_0, the least significant bit (LSB) through b_5, the most significant bit (MSB), and the analog result is measured at V_A across load resistor R_L. Applying a logic 1 voltage level to resistors R_0 through R_5 generates the currents for each bit position. R_0, connected to the LSB, has the highest value to produce the lowest current quantity. R_1, used to generate the current weight value for the next binary position, must be half the value of R_0 to produce twice the weighted current. R_2 is half the value of R_1, to yield four times the current through the LSB resistor, R_0. This continues for R_4 and R_5. All the resistors in the ladder network are tied to a common node point that terminates in load resistor R_L. This resistor has a value of ten times R_0 so that it will have minimal effect on the conversion process. Each resistor's current value represents the weight of each given bit position. That weight doubles in proportion to the current through each resistor, as the position progresses from the LSB to the MSB position with total current being summed through R_L. The voltage across R_L will vary according to the total sum of the currents into the common node point from each resistor in the network, which is determined by the actual binary value placed at the inputs.

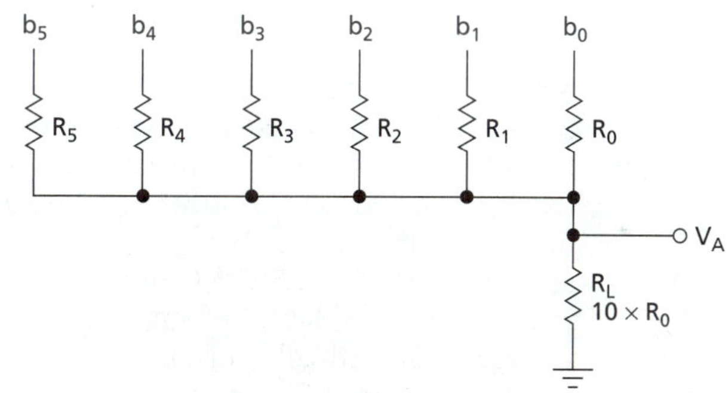

Figure 7.1 Resistor Ladder DAC

Consider a resistor network that is designed to allow 1 μa of current through R_0 when a logic 1 is placed at input b_0. To see how the conversion process works, let's determine the voltage at V_A for each of the following binary number inputs to the network, given an R_L that has a value of 250 kΩ:

1. 100110B **2.** 001101B **3.** 000001B **4.** 111111B

The currents through each resistor value double from one bit position to the next, so the current weight values for each bit are $b_0 = 1$ μa, $b_1 = 2$ μa, $b_2 = 4$ μa, $b_3 = 8$ μa, $b_4 = 16$ μa, and $b_5 = 32$ μa. To find the voltage at V_A, add up the weight current values for each input that is a logic 1. Logic 0 inputs have a current value of 0 μa regardless of which position receives that logic 0 as an input. Multiply the total current by the load resistor value of 250 kΩ to get the equivalent voltage out for each binary input:

1. 100110B = 32 μa + 4 μa + 2 μa = 38 μa   38 μa × 250 kΩ = 9.5 V
2. 001101B = 8 μa + 4 μa + 1 μa = 13 μa   13 μa × 250 kΩ = 3.25 V
3. 000001B = 1 μa × 250 kΩ = 0.25 V
4. 111111B = 32 μa + 16 μa + 8 μa + 4 μa + 2 μa + 1 μa = 63 μa
   63 μa × 250 kΩ = 15.75 V

The currents produced in each resistor of the resistor ladder are dependent on the voltage used to represent a logic 1 level and the actual values of each resistor. Inputs b_0 to b_5, combined with the ladder resistors, form current sources that feed node point V_A. To fully analyze an actual resistor ladder network requires the application of a direct current network analysis theorem such as nodal analysis or superposition. One concept for exploring multisource DC networks was developed by a scientist named Millman. Millman stated that the DC voltage at a node point in a circuit containing multiple sources can be calculated by dividing the sum of the current sources feeding that node by the sum of the conductances measured from that node to ground. Applying Millman's theory to the resistor ladder circuit of Figure 7.1, the voltage at V_A translates to

$$V_A = \frac{\dfrac{V_1 \cdot b_0}{R_0} + \dfrac{V_1 \cdot b_1}{R_0/2} + \dfrac{V_1 \cdot b_2}{R_0/4} + \dfrac{V_1 \cdot b_3}{R_0/8} + \dfrac{V_1 \cdot b_4}{R_0/16} + \dfrac{V_1 \cdot b_5}{R_0/32}}{\dfrac{1}{R_0} + \dfrac{1}{R_0/2} + \dfrac{1}{R_0/4} + \dfrac{1}{R_0/8} + \dfrac{1}{R_0/16} + \dfrac{1}{R_0/32}}$$

where V_1 is the voltage level for a logic 1. The current source for each branch is the logic 1 voltage times the bit value, b_n (0 or 1) divided by the series resistance ($R_0/2n$), while conductance is the reciprocal of each branch resistance ($1/R_0/2n$). Resistor R_L is omitted from the equation for two reasons. The first is that its selected value is 10 times the largest value (R_0) in the network, which makes its effect negligible, giving an error of less than 0.1 percent. Its current source (ground) is $0/R_L = 0$, and its acceptance is $1/R_0/10$, which introduces a minor error into the calculation. More importantly, V_A is actually a thevenized voltage taken with all loads, including R_L removed from node V_A. As we will see shortly, this is the actual case in the digital-to-analog final circuit, where R_L is removed and replaced with a scaling op amp.

Multiplying top and bottom of the fraction by R_0 and simplifying each fraction (that is, $V_1 \cdot b_5/1/32 = 32 \cdot V_1 \cdot b_5$ yields the equation

$$V_A = \frac{V_1 \cdot b_0 + 2 \cdot V_1 \cdot b_1 + 4 \cdot V_1 \cdot b_2 + 8 \cdot V_1 \cdot b_3 + 16 \cdot V_1 \cdot b_4 + 32\, V_1 \cdot b_5}{1 + 2 + 4 + 8 + 16 + 32}$$

Factor out V_1 from the numerator, and add the values in the denominator:

$$V_A = \frac{V_1(b_0 + 2 \cdot b_1 + 4 \cdot b_2 + 8 \cdot b_3 + 16 \cdot b_4 + 32 \cdot b_5)}{63}$$

The final format for the equation for V_A results from recognizing that

1. In the denominator, 63 represents $2^6 - 1$ for the six-bit input ladder network in Figure 7.1. In general, the denominator term is $2n - 1$, where n is the number of binary inputs.
2. Binary bits b_0 through b_5 will be 0 or 1 logic levels, and the sum of the products of the weights, multiplied by the bit values, represents the deci-

mal value for the input binary number, abbreviated BIN for binary input. V_1 represents the actual voltage value used to represent a logic 1 level, using 0 V to represent a logic 0.

Substituting the generalized quantities from steps 1 and 2, the formula for the analog voltage, V_A, takes on the final form shown in Equation 7.1:

$$V_A = \frac{BIN \cdot V_1}{2^n - 1} \tag{7.1}$$

Notice that the analog voltage is independent of actual resistor values as long as these values retain a ratio of 2 to 1 for each successive bit position.

EXAMPLE 7.2

Make a chart showing the relationship of the analog out compared to binary input for a four-bit resistor ladder DAC using a logic 1 voltage level of 5 V.

SOLUTION For the numerator of equation 7.1, $2^4 - 1 = 15$. Substitute 5 V for V_1 in the equation. Make a chart showing the computed V_A for each B_{IN} combination:

BINARY INPUT	B_{IN}	V_A
$b_4 \ldots b_0$	Decimal	Voltage
0000	0	0
0001	1	0.333
0010	2	0.667
0011	3	1.0
0100	4	1.33
0101	5	1.67
0110	6	2.0
0111	7	2.33
1000	8	2.67
1001	9	3.0
1010	10	3.33
1011	11	3.67
1100	12	4.0
1101	13	4.33
1110	14	4.67
1111	15	5.0

There are two observations that we can make, based on Example 7.2. First, when the maximum binary input is applied (1111B in this case), V_A takes a value equivalent to the voltage of a logic 1 level (V_1). A maximum binary input produces a full-scale analog output voltage, so we can say that, for the resistor ladder, full-scale voltage is the same as that for a logic 1 level.

A second observation from the chart in Example 7.2 is that there is a correlation between the binary input (B_{IN}) and the analog output. They are related by a factor of one-third. That is, the decimal equivalent value is exactly one-third of the voltage output. It is often desirable to make the output voltage take on a more direct relationship with the decimal equivalent value of the binary input. This is accomplished by using a **scaling op amp,** as shown in Figure 7.2. Adding the scaling amplifier to the weighted resistor ladder is the final step in completing the weighted ladder **digital-to-analog converter** or **DAC** circuit.

Scaling op amp *a feedback amplifier used as a voltage multiplier or scaler.*

DAC *digital-to-analog converter.*

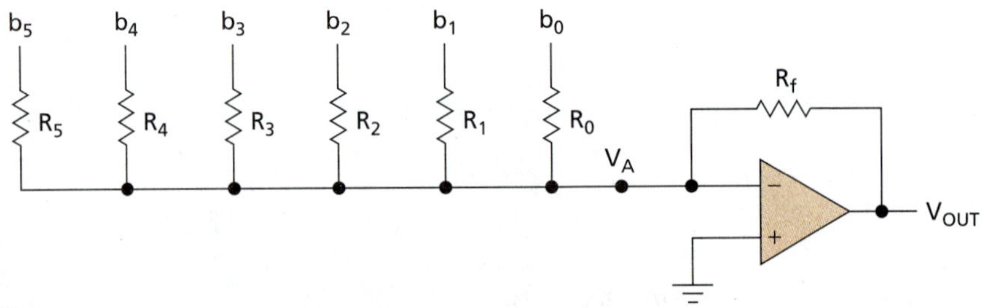

Figure 7.2 Resistor Ladder DAC with Scaling Amplifier

A scaling op amp circuit is, basically, a negative feedback amplifier whose voltage gain (A_v) is the ratio of output voltage (V_{OUT}) divided by input voltage (V_{IN}). The gain of a negative or inverting feedback amplifier can also be computed by dividing the feedback resistor (R_f) connected between the negative or inverting input and the op amp output, by the input resistance (R_{IN}). The ratio is preceded by a minus sign to indicate the inverting aspect of the inverting feedback amplifier. These relationships are summarized by the following equations:

$$A_V = -\frac{V_{OUT}}{V_{IN}} = -\frac{R_f}{R_{IN}}$$

The input to the op amp is the resistor network, and the scaling factor (the gain of the amplifier) is set by the value of feedback resistor, R_f. Notice the lack of a load resistor, R_L, in this circuit, because we are now using the op amp input to terminate the ladder network output. Another aspect of inverting feedback op amp amplifiers is that both inputs to the op amp measure the same voltage level. In this case, since the positive input is returned directly to ground or 0 V, the voltage measured at the negative input will also be 0 V. This is called a *virtual ground,* because the negative input is not physically tied to ground even though the voltage measures 0 V. This virtual ground acts as a terminating reference for the network and feedback resistor, R_f. Since the negative input is not actually at ground, the sum of the currents at node V_A flow through resistor R_f, causing a voltage to be developed across R_f referenced to the 0 V virtual ground point. This voltage, measured at V_{OUT} of the op amp, is the analog voltage of V_A scaled by the gain factor of the op amp and equivalent to the current through R_f times the value of R_f.

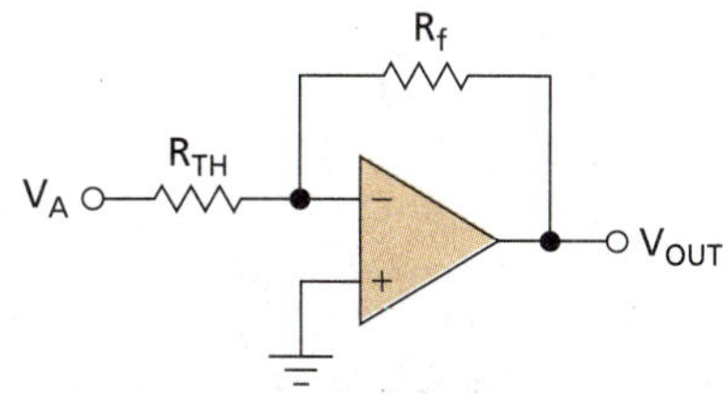

Figure 7.3 Thevenized Model of Resistor Ladder DAC

To analyze a negative feedback amplifier, we need an input voltage (V_A), input resistance (R_{IN}), feedback resistance (R_f), and an output voltage (V_{OUT}), as shown in the model of Figure 7.3. The relationship of these values in the feedback amplifier is

$$\frac{V_{OUT}}{V_A} = -\frac{R_f}{R_{IN}} \tag{7.2}$$

V_A is the calculated analog voltage derived from Equation 7.1. V_{OUT} is the desired final output, which leaves the resistor values to be determined. The negative feedback model shown in Figure 7.3 is actually the Thevenin equivalent for the scaled resistor ladder DAC. To thevenize the circuit, remove the load at node point V_A, which removes the virtual ground and allows the thevenized voltage, V_A, to be measured. The load removed is the op amp scaler itself. R_{IN} is the thevenized resistance found by connecting all the binary inputs of the ladder to ground and measuring the resistance at node V_A to ground. Since this places all the resistors in parallel, R_{IN} or R_{TH} can be calculated by adding the reciprocals of each resistor in the network and then inverting that sum. This results from the parallel configuration that the network takes when all sources are brought to ground and the resistance measured from the node (V_A) to ground:

$$\frac{1}{R_{TH}} = \frac{1}{R_0} + \frac{1}{R_0/2} + \frac{1}{R_0/4} + \frac{1}{R_0/8} + \frac{1}{R_0/16} + \frac{1}{R_0/32}$$

Multiply top and bottom by 32:

$$\frac{1}{R_{TH}} = \frac{32}{R_0} + \frac{16}{R_0} + \frac{8}{R_0} + \frac{4}{R_0} + \frac{2}{R_0} + \frac{1}{R_0}$$

Add the fractions:

$$\frac{1}{R_{TH}} = \frac{63}{R_0}$$

Invert both sides of the equation, and R_{TH} for the circuit in Figure 7.2 is $R_0/63$. Notice that 63 is the same as $2^n - 1$, so a more generalized equation for R_{TH} is

$$R_{TH} = R_0/(2^n - 1) \tag{7.3}$$

Substituting R_{TH} for R_{IN} and rearranging Equation 7.2, we can derive the equation for scaling feedback resistor R_f:

$$R_f = V_{OUT} \cdot R_{TH}/V_A \tag{7.4}$$

EXAMPLE 7.4

Find the feedback resistor required to scale the circuit of Figure 7.2 so that V_{OUT} for a B_{IN} of 100110B is 3.8 V. R_0 has a value of 24 kΩ and V_1 is 7 V, as shown in Figure 7.2.

SOLUTION First calculate R_{TH}, using Equation 7.3:

$$R_{TH} = 24\ \text{k}\Omega/63 = 381\ \Omega$$

In decimal, B_{IN} = 100110B is 38. Using Equation 7.1, we find V_A for a decimal B_{IN} of 38 to be 4.22 V.

Next plug in the appropriate values into Equation 7.4 and solve for the feedback resistor:

$$R_f = 3.8\ \text{V} \cdot 381\ \Omega/4.22\ \text{V} = 343\ \Omega$$

Section 7.2

Practice Problems

An eight-bit DAC has an R_0 value of 270 kΩ and a logic 1 voltage level of 10 V. Answer the following questions based on this circuit:

1. What is the thevenized V_A voltage for a binary input of 00110110B?
2. What is the thevenized input resistance, R_{TH}?
3. Calculate a scaling feedback resistance that yields a V_{OUT} of 25.5 V when B_{IN} is 11111111B. Assume the op amp is capable of reaching that voltage.

7.3 Digital-to-Analog Converter Specifications

Resolution *a single step change in an ADC or DAC's output.*

Step *the analog voltage change of a DAC or ADC associated with a one-bit change in a digital quantity.*

Voltage resolution *the voltage value of one step change of an ADC or DAC output.*

Percent of resolution *the percentage of the whole that one step size is.*

Full-scale output *the maximum analog value a DAC circuit obtains when the maximum binary value is applied.*

There are a number of specifications and parameters that determine the performance of a digital-to-analog converter. Foremost of these specifications is a DAC's **resolution.** Each voltage change at the analog output (V_A) created by a successive change in the binary input is called a **step.** If V_A changes from 1.23 V to 1.33 V with a binary change from 0100110B to 0100111B, then the step size of that DAC is 1.33 V − 1.23 V or 0.1 V. This can be stated as a resolution of 0.1 V per step or per bit. If the binary change is greater, then the analog voltage change can be determined by multiplying the difference in binary values by the step size. For a step size of 0.1 V, a binary change from 0111001B to 0111111B, being a binary change of 110B or 6, would cause V_A to change by 6 × 0.1 V or 0.6 V. The step size is the **voltage resolution** of a DAC. The **voltage resolution** or step size is the smallest change in the analog output of a DAC.

The **percent of resolution** of a DAC is the ratio of the step size to the maximum analog output (called the **full-scale output**). It is computed using the equation

$$\text{Percentage of resolution} = \frac{(100\%)(\text{step voltage})}{(\text{full-scale voltage})} \tag{7.5}$$

The greater the number of binary input bits, the smaller the step size, so resolution can also be stated in terms of the number of binary inputs. The total number of steps between 0 and full scale is $2^n - 1$, where n is the number of binary inputs. Stated otherwise, the maximum analog voltage occurs with a full-scale binary input, which is represented by the value of $2^n - 1$. As a result, the percent of resolution can also be found by dividing 100 percent by $2^n - 1$. Summarizing, the equations to find resolution for a DAC are

$$\text{Voltage resolution} = (\text{full-scale voltage})/(2^n - 1) \tag{7.6}$$

$$\text{Percent of resolution} = (100\%)/(2^n - 1) \tag{7.7}$$

EXAMPLE 7.5

What is the resolution of a 10-bit DAC with a full-scale voltage of 12 V?

SOLUTION The voltage resolution is found by using Equation 7.6:

$$V_{RES} = 12\text{ V}/(2^{10} - 1) = 11.73\text{ mV per step}$$

and the percent of resolution computed using Equation 7.7:

$$\%RES = 100\%/(2^{10} - 1) = 0.098\%$$

This says that a step size of 11.73 mV is 0.098% of the maximum value the DAC output can have.

__Accuracy__ how close to the actual analog value the output of DAC is.

__Precision__ the repeatability of the output of a DAC or ADC for the same input quantity.

The better the resolution, the smaller the analog change for each binary change. By dividing the full-scale voltage into a large number of steps, the tracking between binary values and analog output is improved. However, resolution does not guarantee **accuracy**. **Accuracy** is defined as how close the expected analog voltage is to the actual analog output. A DAC with a 0.098 percent resolution can still have a 3 percent accuracy if each translation between a digital quantity and the analog voltage creates a 3 percent error between expected and actual analog values.

A DAC can have good resolution and accuracy, and poor **precision**. Another word for **precision** is *repeatability*. Precision asks, "Does the analog output always have the exact same value each time a given binary input is applied, or does it vary slightly (possible due to environmental or circuit aging factors)?"

To get a feel for these parameters, let's examine the resistor ladder DAC of Figure 7.2 in terms of resolution, accuracy, and precision. We will start with accuracy as the easiest to determine. The accuracy of the analog voltage out is dependent on the accuracy of the logic 1 and 0 voltage levels and the values of the resistors in the network. A resistor that is inaccurate will produce an error in the current it supplies to the summing node. The same is true for a logic level that is in error. If a logic 1 is supposed to be 5 V, then it must always be 5 V if an accurate analog voltage is to be expected.

Next let's look at precision. Precision resistors and good op amps tend to be very stable devices over a fair range of temperatures. Even aging does not seem to cause major problems with these components. As such, it is safe to say that the resistor ladder is a fairly precise circuit. It will return the same set of values for a given set of binary inputs each time they are applied.

Now let's look at resolution. Recall that the greater number of bits, the smaller the percentage of resolution and the better the tracking correlation. What happens to the resistor values as you try to increase resolution? Each time another bit is added, the new resistor value must be halved. That means that the resistor values between the first and last bits are going to get further apart. At some point, this difference becomes too much and causes errors. Either the smaller values are influenced by lead resistance or the extremely large values become susceptible to noise. In either case, there is a limit to the practical resolution that a resistor ladder can be used for. Another consideration is the expense to make some of the custom precision resistor values required to complete the resistor ladder network.

A method for dynamically testing a DAC is to apply the outputs of a free-running binary up counter to the binary inputs of the DAC and observe the resulting analog output with an oscilloscope. If all is working correctly, the output should appear as a series of voltage steps as shown in Figure 7.4. Each step represents a unit binary change, therefore each should have the same voltage change from the preceding step. This "staircase" voltage is called a **monotonicity curve.** One can tell at a glance if there is a major problem with the DAC.

Monotonicity curve *an output signal generated by a DAC whose binary input is changing in an up count sequence. It is used to test the performance of the DAC.*

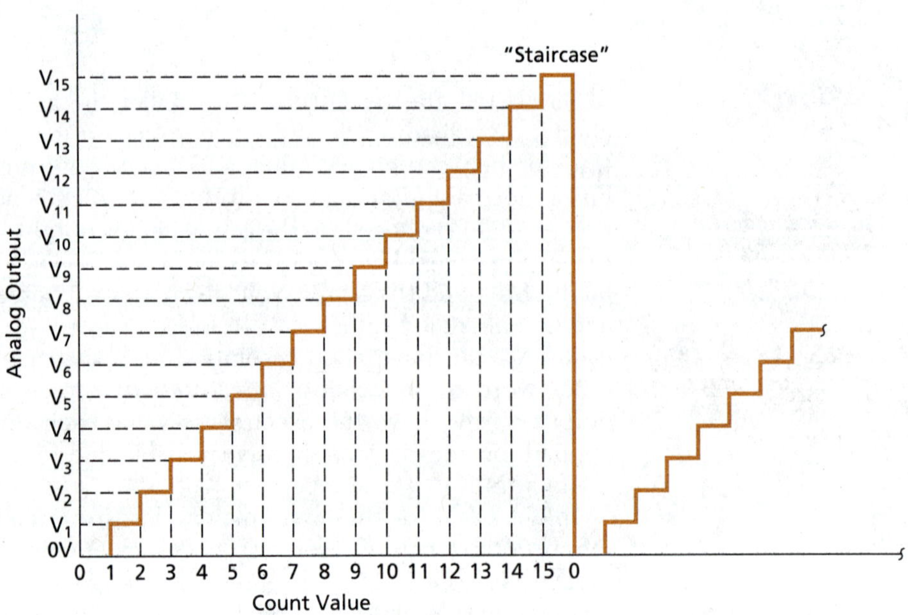

Figure 7.4 Four-Bit Monotonicity Curve

For instance, if b_2 was in error—always low, for example—then the curve would reflect this problem, as shown in Figure 7.5.

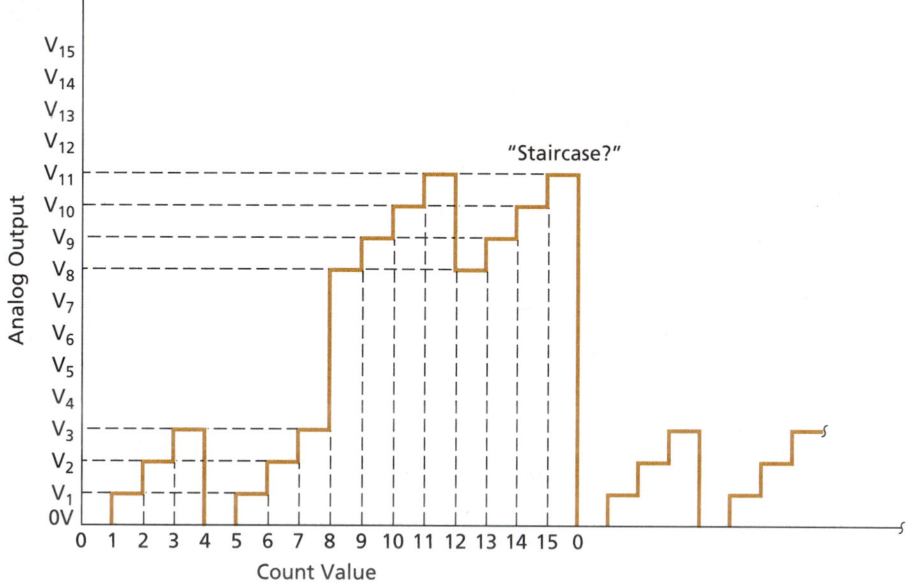

Figure 7.5 Four-Bit Monotonicity Curve with b_2 Bad

Notice that the first few counts of the counter do appear to step correctly. But when the counter reaches 00100B, instead of showing the next step the output drops back to 0 and the staircase appears to restart. However, when the count reaches 01000B, the output jumps from "3" to "8." This indicates that the problem with b_2 is that it is constantly low.

Other information that can be gathered from the monotonicity curve includes accuracy and precision. Measurements can be made on each step (or groups of steps) to check how accurate the output is for each binary input. Precision or repeatability can be observed by changing the oscilloscope's time base so that more than one full staircase can be observed. Consecutive monotonicity curves can then be checked to see how well the output repeats the given sequence.

Section 7.3

Practice Problems

1. Calculate the voltage resolution and percent of resolution for an eight-bit DAC with a full-scale voltage of 26 V.
2. Draw a monotonicity curve for a four-bit DAC with a full-scale voltage of 15 V. How much is the step voltage?

7.4 The R/2R Resistor Ladder

The weighted resistor ladder, discussed in the preceding section, has the advantage that it has a straightforward, binary weight sum conversion. The down side is twofold. First, the number of different precision resistor values means the converter is expensive to make. Second, the extreme resistor values between the least and most significant bits of the converter limit the practical resolution of the DAC. To overcome both these disadvantages, the R/2R ladder of Figure 7.6a was developed. It consists of resistors with only two different values, one twice the value of the other—hence the name R/2R. The currents in each of the bit position legs are summed through the resistor network rather than at a node point. The correct relationship of binary input to analog output is proven by developing an equation for V_A, as was done with the weighted resistor ladder.

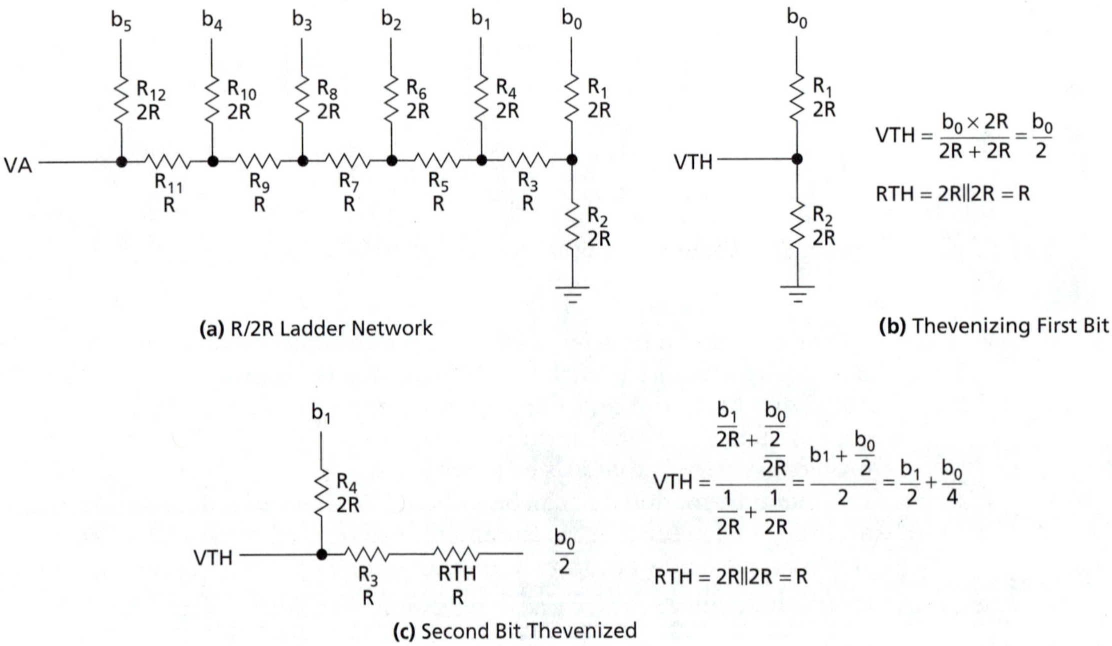

Figure 7.6 R/2R Ladder

The process is begun by thevenizing the least significant bit first, making the rest of the ladder a load at the junction of R_1 and R_2. The resulting V_{TH}, shown in Figure 7.6b, is the voltage at b_0 divided by 2. R_{TH} for b_0 is 2R in parallel with 2R or simply the value of R. The Thevenin equivalent of b_0 replaces the least significant leg, and the next bit position is thevenized, as shown in Figure 7.6c. Millman's theorem is used here to find V_{TH} for this thevenized network. The sum of the current sources is

$$\frac{b_1}{2R} + \frac{b_0/2}{R + R}$$

This sum is divided by the sum of conductances, 1/2R + 1/2R:

$$V_{TH} = \frac{\dfrac{b_1}{2R} + \dfrac{b_0/2}{2R}}{1/2R + 1/2R}$$

Multiplying the top and bottom by 2R yields

$$V_{TH} = \frac{b_1 + b_0/2}{2} = b_1/2 + b_0/4$$

The process is repeated for the remaining bit positions of the ladder network, as shown in Figure 7.7. The final V_{TH}, which is the V_A being sought, is the Thevenin equivalent of b_5 and the V_{TH} of b_4:

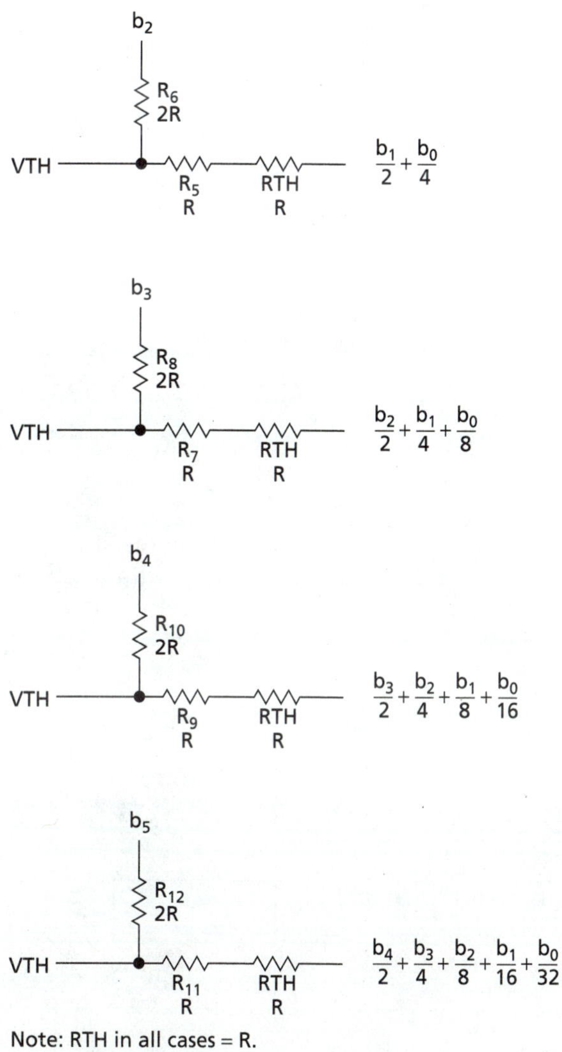

Figure 7.7 Thevenizing R/2R Ladder

$$V_A = \frac{\dfrac{b_5}{2R} + \dfrac{b_4/2 + b_3/4 + b_2/8 + b_1/16 + b_0/32}{R + R}}{1/2R + 1/2R}$$

$$V_A = b_5/2 + b_4/4 + b_3/8 + b_2/16 + b_1/32 + b_0/64$$

Convert each fraction to a common denominator and then summing the results:

$$V_A = \frac{32b_5 + 16b_4 + 8b_3 + 4b_2 + 2b_1 + b_0}{64}$$

The bit values are actually the weighted bit values of each position (1 or 0 times the voltage of a logic 1 level = $B_{IN} \cdot V_1$) and the denominator is equivalent to 2^n, where n is the number of bits in the ladder network. The equation for V_A can be generalized to

$$V_A = \frac{B_{IN} \cdot V_1}{2^n} \tag{7.8}$$

EXAMPLE 7.7

What is the analog voltage for a 10-bit R/2R resistor ladder that uses a logic 1 voltage of 3.5 V for the following input binary values: 1100010101B and 0001100111B?

SOLUTION The B_{IN} decimal equivalents for the two binary numbers are 789 and 103. V_A for these inputs using Equation 7.8 are

$$V_A = (789)(3.5\text{ V})/(1024) = 2.7\text{ V}$$

$$V_A = (103)(3.5\text{ V})/(1024) = 0.352\text{ V}$$

The R/2R ladder, like the resistor ladder, requires a scaling op amp so that the analog voltage out (V_{OUT}) can be adjusted to a useful range of values. Figure 7.8

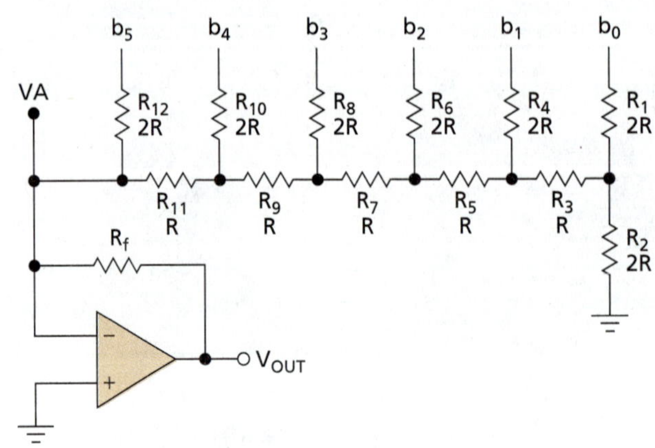

(a) R/2R Ladder DAC with Scaling Amplifier

Figure 7.8 R/2R Ladder DAC

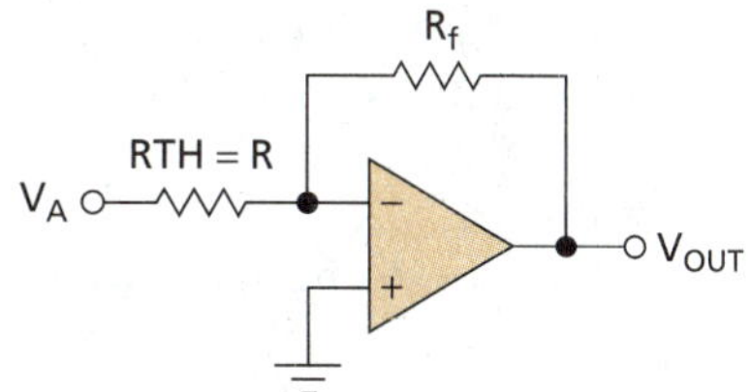

(b) Thevenin Equivalent Circuit

Figure **7.8** Continued

shows the complete circuit for the R/2R DAC and the thevenized equivalent circuit for calculating the scaling feedback resistor, R_f. In the final thevenized equivalent circuit of the R/2R ladder network, the series resistance, R_{IN}, is equal to the value of resistor R.

EXAMPLE 7.6

For the six-bit DAC of Figure 7.8, what is the value of R_f required to set V_{OUT} to 6.3 V at full scale using a logic 1 voltage level of 3.5 V? The resistor value, R, is 1 kΩ. What is the resolution of this DAC?

SOLUTION Start by using Equation 7.8 to find full-scale V_A:

$$V_A = (63)(3.5\text{ V})/(64) = 3.45\text{ V}$$

Next use Equation 7.3 to solve for R_f, keeping in mind that R_{IN} or R_{TH} is equal to the value of R or 1 K

$$R_f = (6.3\text{ V})(1\text{ k}\Omega)/(3.45\text{ V}) = 1.83\text{ k}\Omega$$

The resolution for the DAC is found by using Equations 7.6 and 7.7:

$$V_{RES} = 3.45/63 = 0.055\text{ V at }V_A$$

$$V_{RES} = 6.3/63 = 0.1\text{ V at }V_{OUT}$$

$$\%RES = 100\%/63 = 1.59\%$$

The following chart summarizes the two resistor ladder networks in terms of design criteria, for purpose of comparison:

PARAMETER	WEIGHTED RESISTOR	R/2R
V_A	$B_{IN} \cdot V_1/(2^n - 1)$	$B_{IN} \cdot V_1/2^n$
R_{TH}	$R/(2^n - 1)$	R
R_f	$V_{OUT} \cdot R_{TH}/V_A$	$V_{OUT} \cdot R_{TH}/V_A$
%RES	$100\%/(2^n - 1)$	$100\%/(2^n - 1)$

The other major differences between the two ladder networks are the resistor values used and the consequences of using each type of network. The weighted resistor ladder employs many different precision values, whereas the R/2R uses only two precision values. This limits the practical resolution of the resistor ladder substantially more than the R/2R ladder, due to the extreme values of resistors required for the resistor ladder. The accuracy of both ladder DACs depends on the accuracy of the resistor values and on the voltage source for the logic 1 voltage level.

Application Note

Digital-to-Analog Conversion Application

A digital CD player converts light reflected from the CD surface into a sequence of digital bits. These bits are then converted into analog signals to produce the music or computer programs that we use. One reason why CD digital music is better than that of older vinyl analog records is the lack of surface noise generated by a needle used to pick up the sounds from records. Also, additional noise can be eliminated from digitized music by removing some of the lower bits where noise usually resides. These bits have very little effect on the actual music, because the desired analog equivalence of the digitized audio comes from the higher-order bits of the digital sequences. DACs are used to convert the digits created by sensing the reflected light into analog voltage magnitudes of the replicated audio signal.

Section 7.4

Practice Problems

A seven-bit R/2R uses a logic 1 voltage level of 6 V and an R value of 1.5 kΩ. Compute the following values:

1. Full-scale voltage
2. Voltage resolution
3. R_f for a full-scale V_{OUT} of 7.8 V
4. Percent of resolution

7.5 A DAC on a Single MSI Chip, National Semiconductor's DAC0830

R/2R ladder DACs are incorporated into medium-scale integrated (MSI) chips, which include additional circuitry to allow the chip to be used in a microprocessor-controlled system. Such a device, shown in Figure 7.9, is a DAC0830 made by National Semiconductor. An eight-bit binary value is

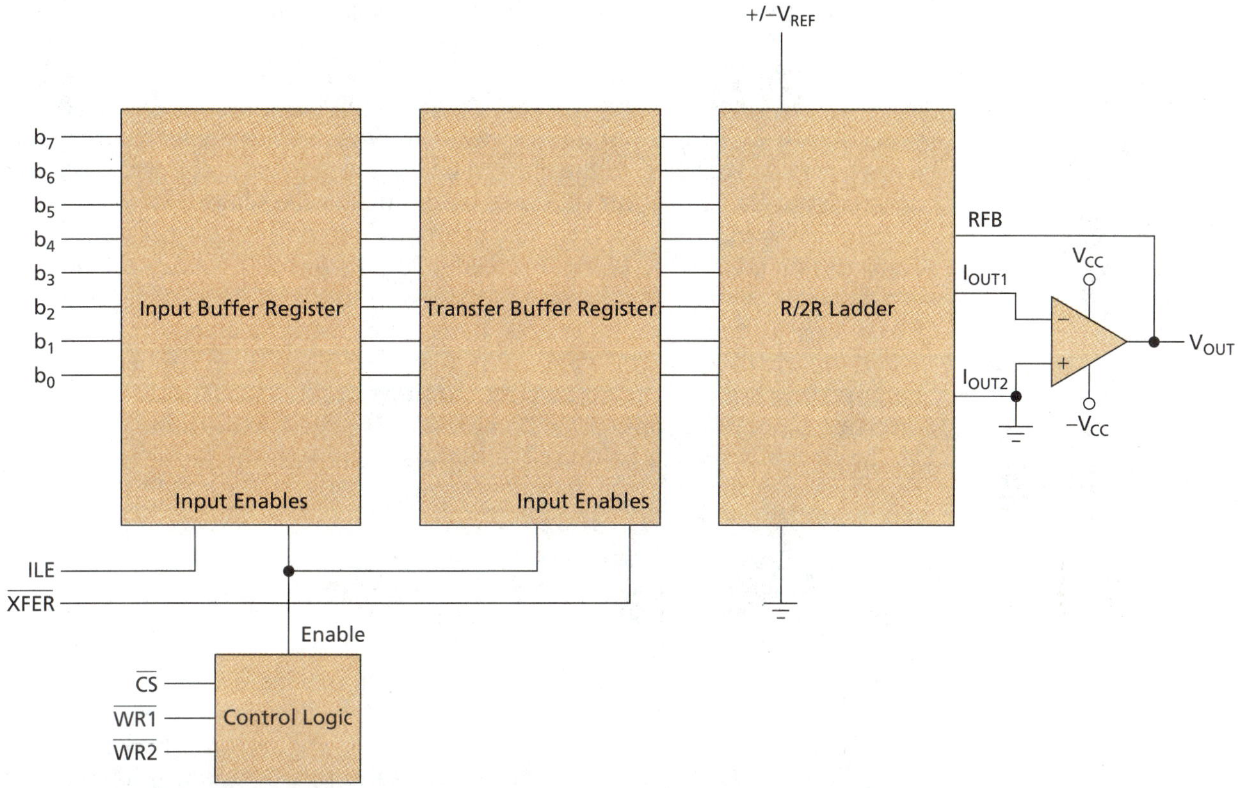

Figure 7.9 DAC0830 Digital-to-Analog Converter

applied to inputs DI_0 to DI_7, which are connected internally in the chip to an input buffer register. This register is loaded with the data on the input lines when the following signals are asserted:

1. $\overline{CS}$ (chip select), in combination with ILE enables the WR1 function to load the input register.
2. $\overline{WR_1}$ (write enable) is used to synchronize write signals from a microprocessor when entering data into the DAC's input register.
3. ILE (input load enable) latches the binary data into the input buffer register.

The use of the chip select and write enable signals are based on microprocessor functions that we will discuss in larger detail later in the text. Suffice it to say, for purposes of using the chip by itself, all three of these inputs must be turned on in order for the data on the input lines to be stored into the input register.

A second buffer register, called the *transfer register,* accepts and latches the data from the input register when these two signals are asserted, along with the three just described:

4. $\overline{WR_2}$ is a second write control input that enables data to be written into the transfer register.
5. $\overline{XFER}$ is the transfer control signal used to latch the data into the transfer register.

The outputs from the transfer register are connected to the binary inputs of an eight-bit R/2R register used to convert the binary information into an analog voltage at I_{OUT}. The maximum analog output at I_{OUT1} with respect to I_{OUT2} is selected by the value applied to $\pm V_{REF}$ input, which has an allowable maximum value of ±10 V. The I_{OUT} outputs supply the inputs to an external op amp whose output may be returned to the chip to take advantage of an internal feedback resistor, R_{FB}. Using this internal resistor maintains the scaling factor of 1. The scaling factor can be made less than 1 by inserting a resistor between the op amp output and the R_{FB}. Using this feedback resistor sets the scale factor of the op amp to 1, so that the maximum analog voltage from the op amp is equivalent to V_{REF}. Different scaling is accomplished by using an external feedback resistor for the op amp inverting amplifier. I_{OUT2} should be connected to the noninverting input of the op amp and to the ground. I_{OUT1} is connected to the inverting input of the op amp to complete the scaled DAC circuit.

The device is said to be double buffered when both the input and transfer registers are controlled by external signals applied to ILE and $\overline{XFER}$ inputs. In this state, the input data applied to $D_0 - D_7$ must first be latched into the input register by toggling ILE high and then low. These data then must be moved to the transfer buffer controlled by $\overline{XFER}$ before they can be converted by the R/2R network. Single buffering is achieved by asserting $\overline{XFER}$ low continually and using ILE to control when data are allowed into the input buffer of the chip. Once latched into the input register, the data then are immediately passed through the transfer buffer to the ladder network. Finally, a flow-through condition can be achieved by asserting $\overline{CS}$, $\overline{WR}_1$, $\overline{WR}_2$, $\overline{XFER}$, and ILE continually. When this is done, binary data applied to D_0 through D_7 are sent immediately through both the input and transfer buffer registers to the R/2R ladder and converted to an analog voltage.

Application Note

DAC0830

A compact disc has digital audio recorded on it by using a laser to "burn" a binary image into it. When a second laser is used to "read" the disc, its light is reflected back to a sensor. The reflected light is converted to form a serial stream of digital information that must be changed to an analog signal so that we may enjoy the sounds recorded on the disc. A converter such as the DAC0830 is used because the conversion process can be synchronized with the information coming off the disc. Each eight bits of serial information is converted to parallel, using a serial-in, parallel-out register, and then sent to the DAC. The DAC must only convert the information sent to it by the serial-to-parallel register. When the register is finished shifting in eight bits and is ready to transfer these eight bits to the DAC, it supplies a signal to enable the ILE input of the DAC configured to operate in a single buffered operation. This controls when the DAC does its conversion, so that the audio reproduced is synchronized with digital information read from the disc.

Section 7.5

Practice Problems

1. What is the percent of resolution for the DAC0830?
2. What is the voltage resolution of the DAC0830 if a maximum V_{REF} is used?
3. Draw the monotonicity curve for a DAC0830 with a V_{REF} of 7.5 V.
4. What type of buffering is used to allow the creation of the monotonicity curve?

7.6 Analog-to-Digital Conversion

__Flash__ or __simultaneous ADC__ a high speed analog to digital converter.

The process of converting analog input signals into digital words is more complex than translating digital words to analog levels. Either the circuit will be more complex, or the circuit will require something in addition after the conversion is completed. The later requirement is represented in the first analog-to-digital (ADC) circuit we will discuss, the **flash** or **simultaneous** converter.

The flash ADC uses op amps configured as comparators that compare the input analog voltage to a reference voltage developed by using a voltage divider connected between the maximum V_{REF} and ground. The four-bit example, shown in Figure 7.10, uses a resistor voltage divider between a 5 V reference voltage and ground. Each resistor value is 1 kΩ, making the voltage drops

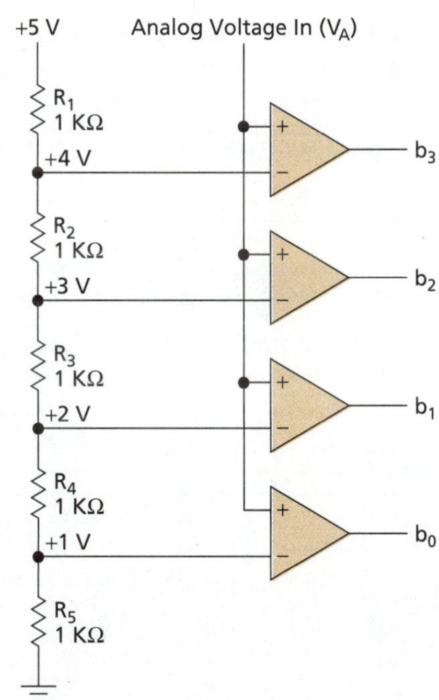

Figure 7.10 Flash ADC

across each resistor even (1 V in this case). That results in a 1 V voltage at the inverting input of op amp connected to b_0, 2 V into the next op amp, 3 V into the next one, and 4 V into the top op amp. The noninverting inputs of all the comparators are connected to the analog input voltage. If the analog voltage in is 0 V, then the voltage at the inverting input of all four comparators is more positive than the 0 V at the noninverting inputs. This causes all the op amps to saturate negatively to $-V_{CC}$. Increasing the analog voltage input to just above 1 V causes b_0 to switch to positive saturation of $+V_{CC}$, since the noninverting input is now more positive than the inverting one. As the analog voltage is increased further, each op amp output in turn changes from negative saturation to positive saturation. Selecting $-V_{CC}$ to represent a logic 0 and $+V_{CC}$, a logic 1, the circuit operation can be summarized with the following truth table:

V_{analog} In	b_3	b_2	b_1	b_0
0 V	0	0	0	0
1 V	0	0	0	1
2 V	0	0	1	1
3 V	0	1	1	1
4 V	1	1	1	1

The conversion time for the flash ADC is limited only by the propagation time of the op amps switching from one saturation level to another. The outputs can track analog voltages that increase or decrease.

There are two main advantages of the flash converter. The first one is resolution. To generate a number of binary bits from the flash converter, many op amps are required, specifically $2^n - 1$ op amps. The second drawback is that the binary output does not change in a binary counting sequence. Considering U_0 as the least significant bit of the converter output, the numeric sequence of the ADC in Figure 7.10 is 0, 1, 3, 7, and 15—not a good sequence for the 0 V, 1 V, 2 V, 3 V, and 4 V inputs. Some form of binary decoding or encoding is required to produce a more meaningful binary sequence.

EXAMPLE 7.8

Create the truth table template for a CUPL design to convert the existing outputs of the flash converter in Figure 7.8 to a sequence that tracks the voltage inputs.

SOLUTION The truth table for the conversion is

INPUTS				OUTPUTS		
U_3	U_2	U_1	U_0	b_2	b_1	b_0
0	0	0	0	0	0	0
0	0	0	1	0	0	1
0	0	1	1	0	1	0
0	1	1	1	0	1	1
1	1	1	1	1	0	0

This is a good example for an application of a PLD where a custom logic design is required. A decoder/encoder does not exist that produces the preceding conversion. However, it is a comparatively simple task to set up a PLD design using a GAL16V8 to do the job.

The header portion of the template is filled out as in previous examples. Next the input and out pins are declared, using range variables:

PIN [2 . . 5] = [U0 . . U3];

PIN [19 . . 17] = [b0 . . b2];

Then the truth table fields are declared:

FIELD ADC = [U3, U2, U1, U0];

FIELD BINOUT = [b2, b1, b0];

And, finally, the truth table is declared:

TABLE ADC => BINOUT {

0 => 0;   1 => 1;   3 => 2;   7 = 3;   F = 4;

}

***Counter** or **ramp ADC** the first of the counter-type analog-to-digital converters.*

***Start conversion (SC)** a signal used to begin an analog-to-digital conversion sequence.*

***End of conversion (EOC)** signal that indicates when an ADC has completed the conversion of an analog signal to its digital equivalent.*

Example 7.8 shows a comparatively easy way to resolve one of the FLASH ADC's disadvantages. However, the other one—resolution—remains, and can only be improved by changing to a different type of converter. The first step in improving the ADC operation is the **counter** or **ramp** ADC pictured in Figure 7.11.

This time a single comparator is used to compare the analog input voltage, V_A, and a reference voltage, V_{REF}, produced by a R/2R DAC. The output of the comparator is used to control when a binary up counter will count. The output of the counter is both the inputs to the reference DAC and the binary word out. The circuit works in this fashion: The counter is initially held at a reset condition by negating the **start conversion (SC)** input, keeping it from counting and holding its output to 00000000B. V_{REF} for this state is 0 V, the output of the DAC, which converts the binary 00000000B input to an equivalent analog voltage out. Next an analog voltage is applied to V_A. Immediately the comparator saturates positive, since this input is greater than 0 V, asserting the count enable of the counter. Once the reset is released by setting SC high, the counter begins to count, with V_{REF} increasing as the binary counter value increases. Eventually the voltage at V_{REF} becomes greater than V_A, causing the comparator to saturate negative, disabling the count enable and asserting the **end of conversion (EOC).** The counter stops with the equivalent binary value of V_A held on the output digital lines, D_0–D_7. Many ramp converters include a register connected to the counter binary outputs. This register is reloaded when EOC is asserted, thus holding the binary output until the next conversion is completed. Such a register is included in Figure 7.11.

The ramp ADC resolves the problem of resolution. Consider the converters in Figures 7.10 and 7.11. For the flash converter to have the same resolution as the ramp converter, we require 255 comparators, a 256-resistor voltage divider, and an encoder to convert the 255 comparator outputs to eight binary bits. This

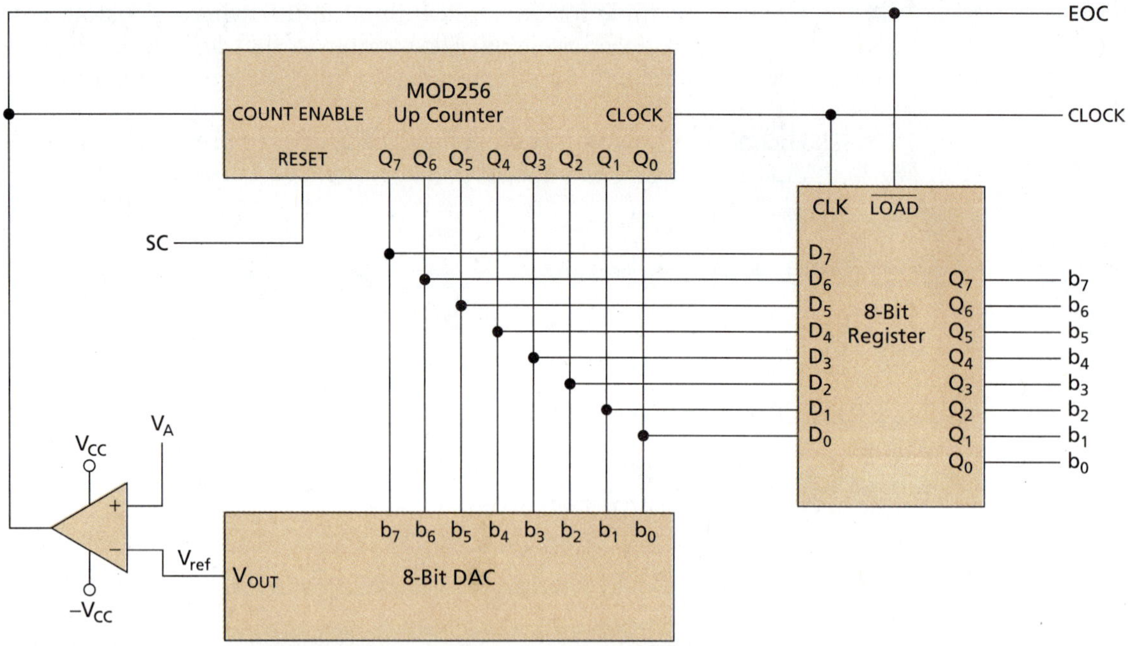

Figure **7.11** Ramp or Counter ADC

amount of circuitry is excessive when compared to the ramp converter with the same resolution. The ramp converter will require a single comparator, an eight-bit counter, an eight-bit register and an eight-bit DAC. Doubling the resolution to 16 bits requires double the amount of circuitry for the flash converter, but only requires larger counters, registers, and DACs for the ramp converter. Given this advantage of the ramp over the flash, what are the drawbacks of the ramp converter?

Maximum conversion time *the time it takes for an ADC circuit to produce a digital equivalent output for a change from minimum to maximum analog input voltage.*

Maximum time of conversion is a major specification for an ADC. It measures how long it takes a converter to produce a binary output once an analog input is applied. The flash converter is only limited by the propagation time of the op amp and encoder circuit, thus the reason for its name. In comparison, the main drawback of the ramp converter is its time of conversion. Each time a conversion is needed, the circuit must be reset and allowed to count up to the analog input voltage. The maximum conversion time required by an ADC is the time needed to convert an analog voltage into a full-scale binary out ($2^n - 1$ or 11111111B in the circuit of Figure 7.11). It takes one clock cycle for each count of the counter, so the maximum conversion time (t_{max}) for the ramp converter is

$$t_{max} = (2^n - 1)/f_{clock} \quad (7.9)$$

EXAMPLE 7.9

Compare the maximum conversion times for an eight-bit flash ADC using propagation times of 500 nS and for an eight-bit ramp ADC with a 100 KHz clock frequency.

SOLUTION Assuming a propagation time of 500 nS for the op amps and another 500 nS for the encoder, all conversions using the flash converter would take 1,000 nS to complete.

Selecting a counter frequency of 100 KHz for the ramp ADC yields a maximum conversion time of

$$t_{max} = 255/100 \text{ KHz} = 2.55 \text{ mS}$$

for the ramp to have the same conversion time as the flash converter, in this example, requires a clock with a frequency of

$$f_{clock} = 255/1{,}000 \text{ nS} = 255 \text{ MHz}$$

Section 7.6

Practice Problems

1. Compare the flash ADC to the ramp ADC.
2. What is the maximum conversion time for a 10-bit ramp ADC operating with a clock frequency of 125 KHz?

7.7 Additional Counter-Type ADCs

***Up/down ramp** or **continuous ADC** an analog-to-digital converter used to overcome the lack of tracking of the analog input present in the ramp converter.*

To overcome the inability of the ramp ADC to track dropping analog voltages, the **up/down ramp** or **continuous ADC**, as shown in Figure 7.12, was developed. An up/down binary counter replaces the binary up counter of the ramp converter. Instead of controlling when the counter will count, the output of the comparator now controls which direction the counter counts. Once again, the DAC analog output is used as a reference to compare the current state of the counter to the analog input voltage at V_A. If V_A is larger than V_{REF}, then the comparator is saturated positive and the counter counts up. Once V_{REF} rises above V_A, the counter reverses direction and counts down. Because of this type of control, the digital output will "track" V_A as it varies up and down. In this way, a continuous ADC can deliver a constant varying binary output for, as an example, an audio input. This overcomes the need to reset the counter each time a new conversion is desired. Instead, the current binary output is sensed and used as needed.

A slight disadvantage of the continuous converter is seen when a steady state analog voltage is applied to V_A. Because the counter does not stop counting, the binary output will fluctuate 1 bit around a steady state voltage. Begin by allowing V_{REF} to finally increase above V_A. This causes the comparator to saturate negative and the counter count down, one count below V_A. Now V_{REF} is again below V_A, causing the comparator to saturate positive and forcing the counter to count one count above V_A. This causes the comparator to saturate negative again and the counter to count down. This fluctuation will continue as long as V_A is held steady.

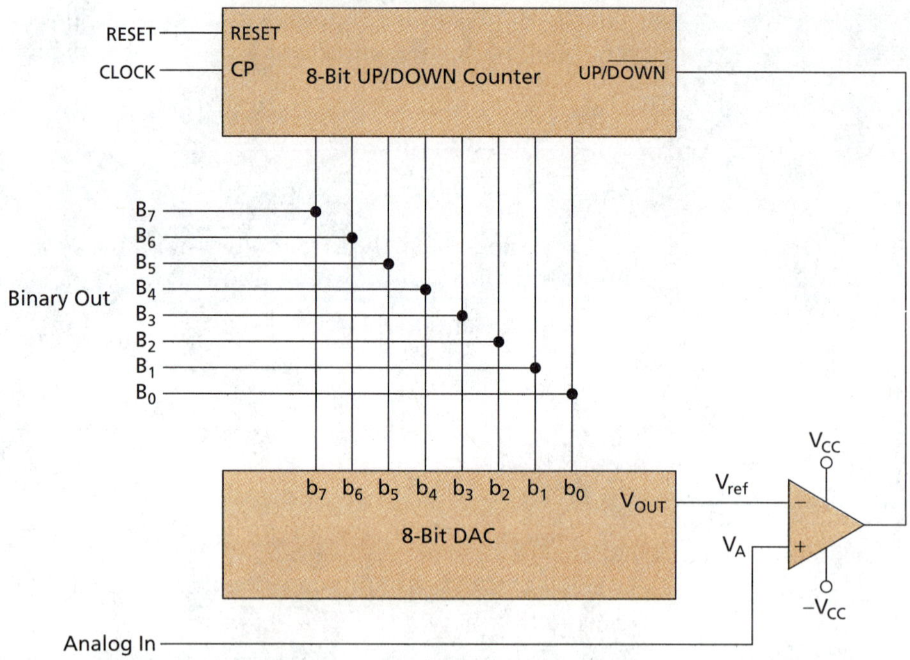

Figure 7.12 Continuous ADC

Maximum conversion time for a continuous ADC is the same for a ramp ADC, which is $2^n - 1$ divided by the frequency of the clock input to the up/down counter. This may seem strange, because the tracking capability usually allows for quicker conversion times. However, the maximum conversion time is a measure of worst-case conditions, and it is possible for an analog signal to suddenly change from the lowest level to the highest within the sampling time of the converter. When this occurs, the converter takes the maximum time to complete the conversion of the peak voltage to its full-scale digital equivalent. However, the overall performance of the continuous converter is better than the ramp converter, and to compensate for the disparity an additional timing specification, known as the **average conversion time** (t_{ave}), is used as a separate measure of performance. This "specification" is, simply, the maximum conversion time divided by 2.

Average conversion time *the time it takes an up/down ramp analog to digital converter to produce the digital value.*

EXAMPLE 7.9

What is the maximum and average conversion time for a 12-bit continuous ramp using a 250 KHz clock?

SOLUTION Equation 7.9 is used to find the maximum conversion time. The results are then divided by 2 for the average conversion time:

$$t_{max} = (2^n - 1)/f_{clock} = 4095/250 \text{ KHz} = 16.38 \text{ mS}$$

$$t_{ave} = t_{max}/2 = 16.38 \text{ mS}/2 = 8.19 \text{ mS}$$

Successive approximations ADC *an analog-to-digital conversion that works by taking numerous approximations of the analog input until those approximations produce the exact digital equivalent output.*

The most commonly used ADC in medium-scale integrated circuits is the **successive approximations ADC** of Figure 7.13. This circuit overcomes both the problem of maximum conversion time and the fluctuation aspect of the continuous converter while maintaining the resolution advantage gained by the ramp converter. A performance improvement in conversion time is realized by the process used by this converter to accomplish the change from analog voltage to a digital value. The converter operates by literally guessing at what the binary value will be and then successively adjusting that approximation until the actual binary equivalent of the analog input voltage is achieved.

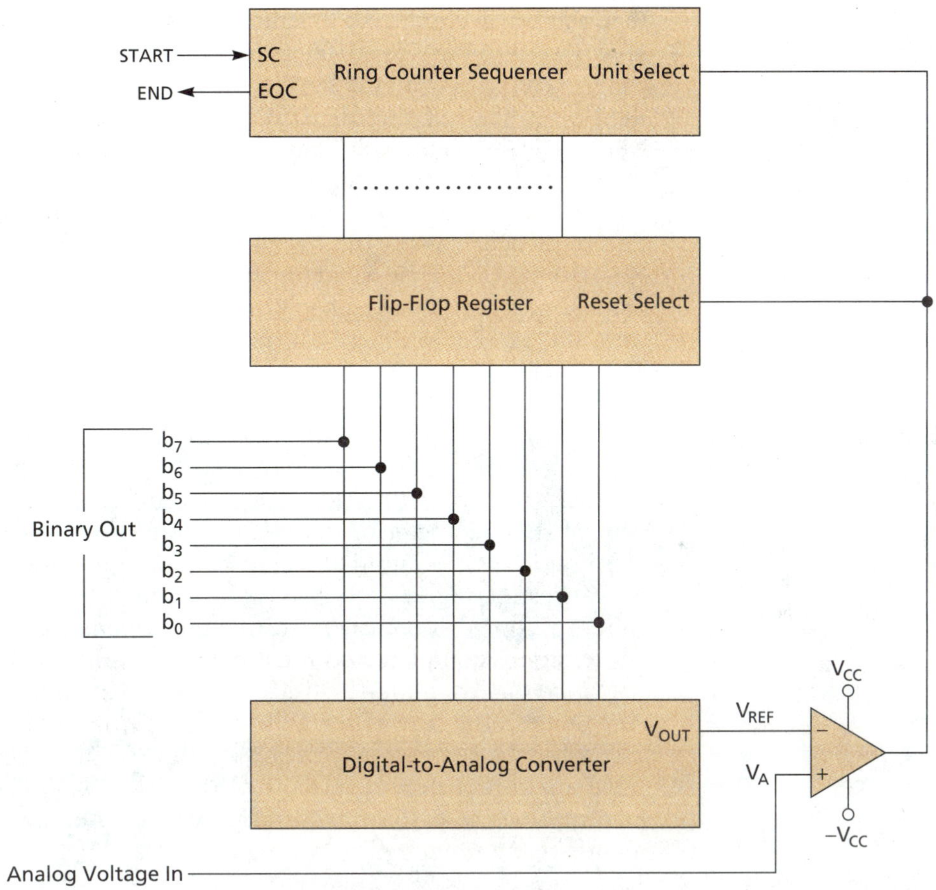

Figure 7.13 Successive Approximations ADC

The eight-bit binary output in Figure 7.13 is taken from a set of flip-flops arranged as an eight-bit register. Each of these flip-flops are set, one at a time in successive turn, starting with the most significant bit flip-flop, by a ring counter sequencer. The resulting binary output is fed into a DAC, whose output analog voltage (V_{REF}) is compared to the actual analog input voltage (V_A). If V_{REF} is larger than V_A, then the flip-flop just set is reset and the sequencer moves on to the next flip-flop. If V_{REF} is smaller than V_A, then the flip-flop just set remains set and the sequencer moves on to the next flip-flop. Once the sequencer moves to the next flip-flop, the state of the last flip-flop remains unchanged until a new conversion is begun. The process is repeated until all the flip-flops have been set, compared, and their required state selected. The binary value in the flip-flop register after all eight have been set and tested is equivalent to the V_A input voltage.

To see how this process works, let's first set up an example conversion. We shall assume that the DAC, used in the successive approximation ADC, is scaled to produce 1 mV per step. Once this is established, we can see how the sequence of an eight-bit successive approximation ADC works to convert an analog voltage of 100 mV to a binary value of 01100100B. First the most significant flip-flop is set. The DAC outputs 128 mV, the equivalent of 10000000B. This is larger than the 100 mV input, so this flip-flop is reset and the next one is set. The DAC now outputs 64 mV for 01000000B. This flip-flop will remain set, because the 64 mV reference from the DAC is lower than the 100 mV input at V_A. The third flip-flop is now set, and the DAC rises to 96 mV, equivalent to the 01100000B from the eight-bit register. This flip-flop also remains set, and the next one is set by the sequencer. This increases the DAC to 112 mV, which is greater than the V_A of 100 mV, causing this last flip-flop to be reset, dropping V_{REF} back to 96 mV. The sequencer causes the next flip-flop to set, bringing V_{REF} up to 104 mV (01101000B), which is, again, more than V_A, forcing this flip-flop to again reset, and the next one becomes set. This brings V_{REF} to 100 mV (01100100B), where we want it to be, but there are still two more flip-flops to be set and sampled. The next reading at V_{REF} goes to 102 mV, causing the next to last flip-flop to be reset. The same happens to the least significant flip-flop, which caused V_{REF} to be 101 mV, again, higher than V_A. After all the flip-flops were set and tested, the resulting binary value in the register is 01100100B for the 100 mV input.

Successive approximation ADCs are equipped with a start conversion (SC) control signal to determine when the conversion should begin and an end-of-conversion (EOC) output signal to indicate when the conversion is completed. SC causes the registers to clear and begins the sequencer at the most significant flip-flop position. EOC informs the system, using the converter, when the conversion is complete and the binary data are valid and can be used.

Notice that each flip-flop must be set and sampled before the conversion is complete. In the case of the eight-bit ADC, this means that it took eight clock cycles for the sequencer to sequence through all eight flip-flops. It takes this same amount of time regardless of the value of V_A. Thus for a successive approximations ADC, the maximum conversion—or any conversion time, for that matter—is summarized by this equation:

$$t_{max} = n/f_{clock} \tag{7.10}$$

where n is the number of binary bits at the output (also the number of flip-flops in the circuit).

EXAMPLE 7.10

Compare the maximum and average conversion times of a 12-bit continuous ADC to a 12-bit successive approximations ADC, both operating with a clock frequency of 250 KHz.

SOLUTION The times for the continuous converter were computed in Example 7.9 as t_{max} = 16.38 mS and t_{ave} = 8.19 mS. For the successive approximations converter, both times have the same value and are found by using Equation 7.10:

$$t_{max} = t_{ave} = 12/250\text{ KHz} = 48\ \mu\text{S}$$

As can be seen by comparing Equations 7.8 and 7.10 or looking at Example 7.10, the successive approximations ADC is considerably faster than the ramp and continuous converters based on maximum and average conversion time. There is no bit fluctuation when converting a steady state analog value, because the use of the SC control signal determines when the conversion takes place. A single conversion is made each time SC is triggered until EOC goes active. These signals are controlled and detected by a microprocessor-based computer system, which can initiate starts and detect EOCs well within the successive approximations time frame.

The following table summarizes and compares the benefits and down sides of each of the ADC converters:

CONVERTER TYPE	RESOLUTION	SPEED	CIRCUIT	NOTES
Flash	Poor	Fastest	Simplest	Needs encoder
Ramp or counter	Good	Slowest	Basic	No tracking
Up/down or continuous	Good	Moderate	Basic	Fluctuation
Successive approximations	Good	Very good	Complex	Use SC and EOC

Section 7.7

Practice Problems

1. What is the percent of resolution and conversion time for a 10-bit successive approximations ADC operating with a clock frequency of 1.75 MHz?
2. Compare the resolution and conversion times of a 10-bit ramp converter and a 10-bit successive approximations converter. Both converters operate with a 1 MHz clock frequency.

7.8 Successive Approximation in a Single Chip, National Semiconductor's ADC0808

National Semiconductor makes a number of ADCs, among which is the ADC0808, an eight-bit successive approximations converter shown in Figure 7.14. The chip is equipped with an eight-channel analog multiplexer that allows

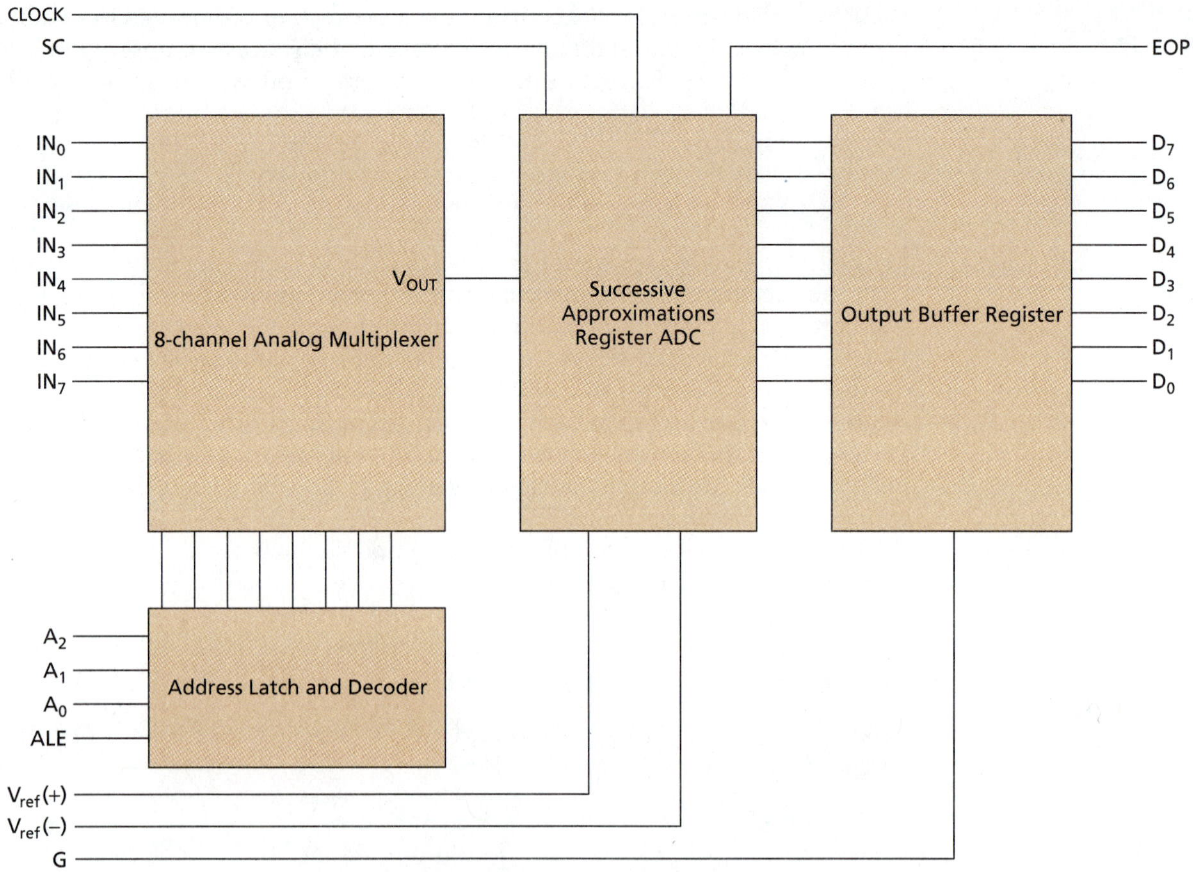

Figure 7.14 ADC0808 Analog-to-Digital Converter

one of eight analog inputs to be sent to the ADC converter circuit. Address lines A_0, A_1, and A_2 select which of the analog inputs, IN_0 to IN_7, is used to supply the analog voltage to the ADC. The address selection must be latched into a 1-of-8 decoder buffer using control input ALE (address latch enable). This process merely chooses one of the analog channels, it does not start the conversion. A separate signal, called SC (start conversion) is used to initiate the conversion sequence of the successive approximations converter. The time it takes the converter to complete its conversion depends on the frequency of the CLOCK input. As with all successive approximations converters, the total conversion time is the number of bits (in this case, 8) divided by the CLOCK frequency. When the conversion is complete, an **end-of-process (EOP)** line is asserted, and the final results are latched into an output register. There is a three-state buffer between the output register and binary output lines D_0 to D_7. This buffer passes data out when the output enable (G) signal is asserted.

End of process (EOP) *same as end of conversion (EOC).*

The process a ADC0808 goes through to convert an analog voltage applied to IN_4 to a digital value at D_0 to D_7, starts by applying the signal to that input and setting the inputs A_2–A_0 to 100B. Line A_2 represents the most significant bit of a three-bit binary number applied to these inputs. To select an analog input, the equivalent binary number of that input is used as an address, such that 100B will select IN_4. ALE is toggled to latch the address into the decoder register. Now that channel 4 is selected, SC is toggled high and back low again to start the conversion process. After eight clock cycles, the converter completes the conversion, latches the results in the output register, and asserts EOP. Asserting input G causes the output register's contents to be placed on D_0 to D_7.

The reference voltages ($V_{REF(+)}$ and $V_{REF(-)}$) determine the maximum and minimum analog voltages that are to be applied to any one of the analog inputs. The voltage resolution of each step is computed by taking the difference of the two reference voltages and dividing that difference by $2^n - 1$. The rate of conversion is solely dependent on the clock frequency and is calculated by the same method described for successive approximation converters in Section 7.7; that is, $t_{conv} = 8/f_{clock}$.

EXAMPLE 7.11

What is the percentage of resolution for the ADC0808? What is the voltage resolution for a ADC0808 with a $V_{REF(+)} = 10$ V and $V_{REF(-)} = 0$ V? What is the conversion time for a ADC0808 using a clock frequency of 340 KHz?

SOLUTION The percent of resolution is calculated the same as with all DACs and ADCs:

$$\text{Percent resolution} = 100\%/(2^n - 1) = 100\%/255 = 0.39\%$$

The voltage resolution is found by dividing the voltage reference by $2^n - 1$:

$$V_{RES} = (V_{REF(+)} - V_{REF(-)})/255 = 10\text{ V}/255 = 39.2\text{ mV}$$

The maximum conversion time is found by multiplying the time of one clock cycle, which is the reciprocal of the clock frequency, by 8, the number of bits in the converter:

$$t_{conv} = \text{number of bits } (n) \times \text{time period of one cycle } (tp)$$

$$\text{since } tp = 1/\text{clock frequency } (f_{CLOCK} \text{ then } t_{conv} = n/f_{CLOCK})$$

$$t_{conv} = 8/340\text{ kHz} = 23.53\ \mu\text{S}$$

Besides controlling the conversion using SC and EOP, the ADC0808 can also be operated in a continuous or flow-through mode by connecting SC and EOP together. An OR gate, as shown in Figure 7.15, is used in between the two lines to provide a means of starting the first conversion. Initially, a channel is selected as before and its address latched into the decoder buffer using the ALE signal. The analog voltage is applied to the channel input, and the START signal is toggled. This initiates the first conversion. At the end of the first

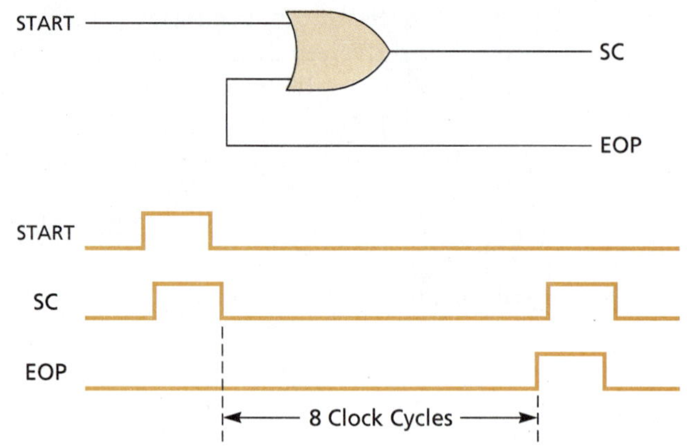

Figure 7.15 Continuous Conversion Gate

conversion, the ADC0808 asserts EOP high for one clock cycle. On the trailing edge of the EOP pulse, the SC is signaled to begin the next conversion. The process continues as long as START is kept low. If START is brought high, then the input to SC is always high and fails to trigger the conversion process.

Application Note

Application for the ADC0808 Converter

Back to compact discs. The process for recording data onto a CD is to take the audio signal, convert it to digital data, serialize the digital data, and change the binary bits to laser pulses that burn the digital music onto the disc. The ADC0808 is well equipped to handle the process of converting the continually changing audio signal to continually changing eight-bit digital words. The rate at which the audio signal is sampled is the conversion rate of the ADC, which in turn depends on the frequency of the clock input. A microprocessor can be used to coordinate the samples supplied by the ADC and apply them to the laser driver.

Section 7.8

Practice Problems

Compute the following parameters for a ADC0808 operating with a $V_{REF(+)}$ of 6 V and $V_{REF(-)}$ of 0 V and a clock frequency of 275 KHz.

1. Percentage of conversion
2. Voltage resolution
3. Conversion time

7.9 Instrumentation ADC

Dual-slope ADC *analog-to-digital converter used in instrumentation.*

Many digital voltmeters use another type of ADC where the time of conversion is not as important as noise rejection. This type of converter, shown in Figure 7.16, is called a **dual-slope** analog-to-digital converter. The circuit works by starting the counter at 0 and applying an analog voltage to V_{IN}. Capacitor C, the feedback path of integrating op amp U_1 charges for a fixed time using the

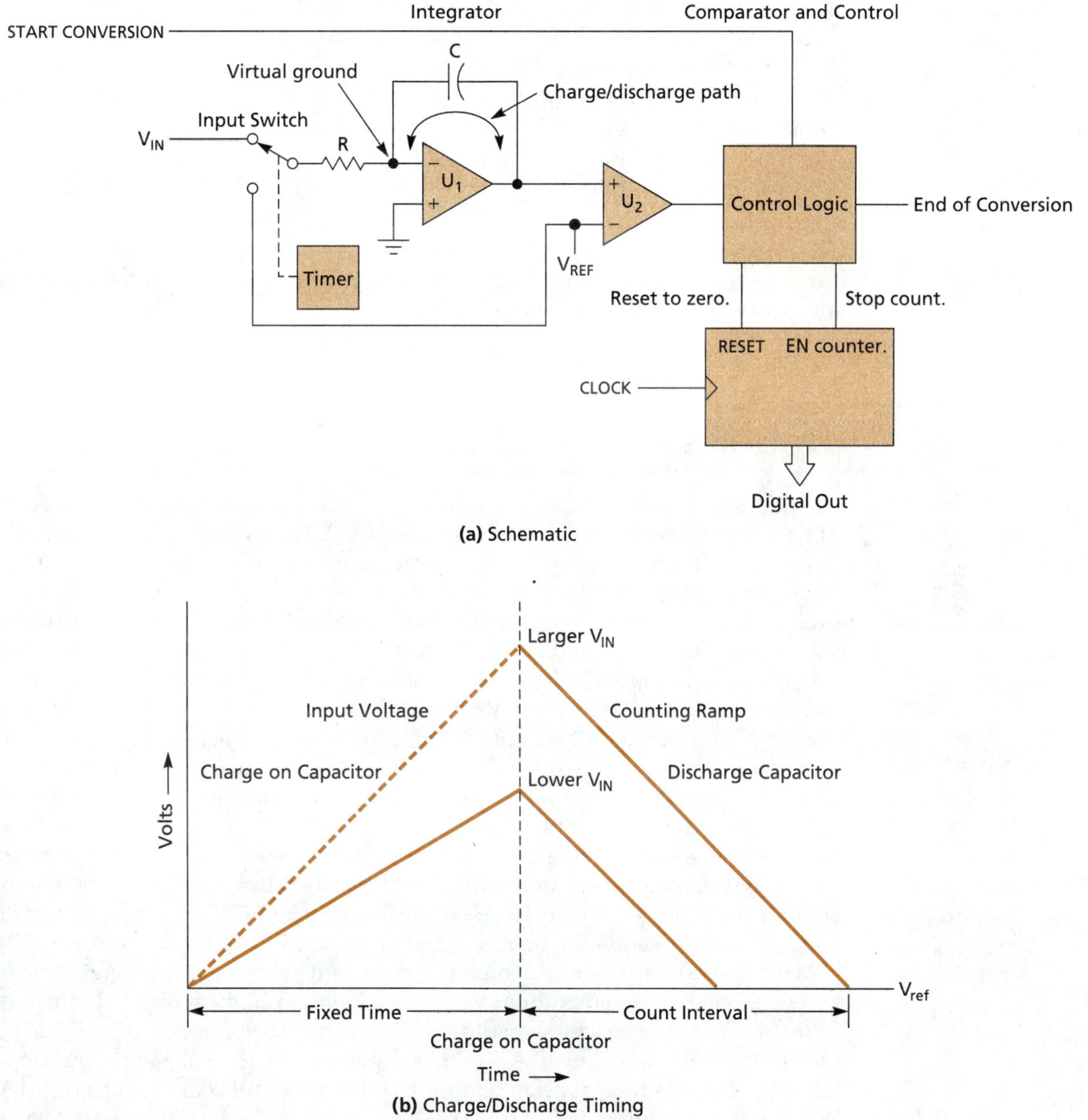

Figure 7.16 Dual-Slope Integrating ADC

virtual ground at the inverting (−) input of op amp U_1 as a reference. The total charge on the capacitor for this fixed time depends solely on the value of V_{IN}. This charge is referred to the "up slope" of the dual-slope reference.

After the fixed time is done, the input switch disconnects the input of U_1 from V_{IN} to a reference voltage, V_{REF}, which also supplies an input to U_1. At this time, the counter is allowed to count. Capacitor C starts to discharge, which is the "down slope" of the dual-slope reference, and U_1 op amp's output begins to fall. When U_1 reaches V_{REF}, it causes the second op amp, U_2, to switch and stop the counter, which now holds the binary equivalent of V_{IN}. The larger V_{IN}, the more capacitor C charges and the longer it takes to discharge down to V_{REF}. The longer the discharge time, the higher the counter counts. There is a direct proportion to the voltage at V_{IN} and the final count value from the counter.

Maximum conversion time for the dual-slope converter is twice that of a ramp counter type, or

$$\text{Max conversion time} = \frac{2(2^n - 1)}{\text{clock frequency}}$$

The accuracy of the dual-slope converter depends on the accuracy of V_{REF}. Noise reduction occurs because V_{IN} supplies the charge voltage to an op amp integrator, which rejects noise as part of the integration function.

7.10 Troubleshooting DACs

As with all previous circuits, the first step in troubleshooting DACs is to determine what the problem is, and the handiest tool for that job is the monotonicity curve shown in Figure 7.17. To generate the curve in Figure 7.17, a free-running mod-16 binary up counter was used to drive the four-bit DAC. An oscilloscope is used to monitor the output of the DAC. A correct output is shown in Figure 7.17a, indicating 15 steps from 0 V to full-scale voltage. Since this is showing a good curve, you will notice a step change for each binary change. The voltage at any point can be checked for accuracy, and by changing the time base of the horizontal sweep, precision or repeatability can be observed. Figure 7.17b illustrates the appearance of the monotonicity curve when a bit is bad; in this case, input bit b_2 is open, adding nothing to the analog voltage out. Notice that the first few steps are all right, since the lower two bits of the DAC are good. The analog voltage out representing the count from 0000B to 0011B tracks correctly, indicating that the two lower bits are good. When the counter increments to 0100B, the output falls back to 0 V, indicating that b_2 is still low when it should be high.

With or without the monotonicity curve, DAC problems are diagnosed by knowing what an analog output value should be for a given digital input. The difference between an actual reading and the expected voltage can be used to indicate which bit is bad by dividing that difference by the voltage resolution of the DAC. For instance, take an eight-bit DAC with a full-scale voltage of 5.1 V. When a binary input of 01011100B (92) is applied to the DAC, the output reads

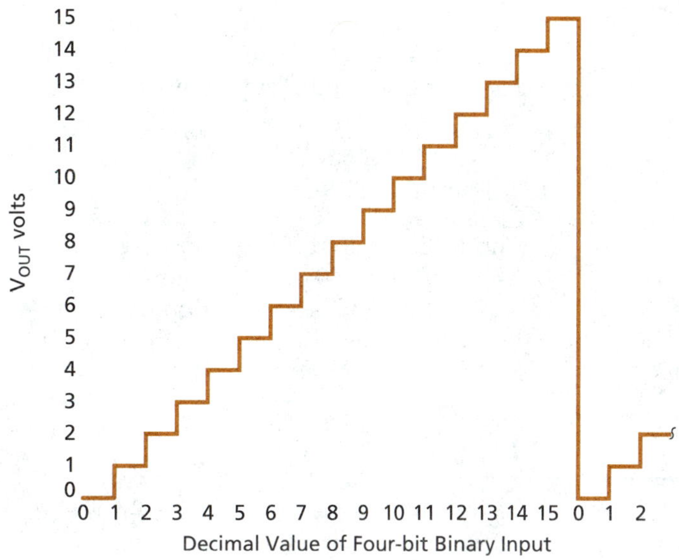

(a) Good Monotonicity Curve

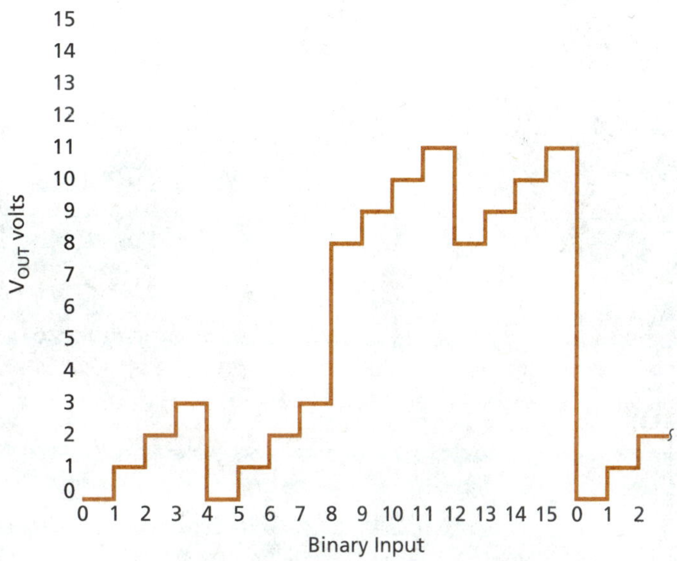

(b) Monotonicity Curve with b_2 Open (0)

Figure 7.17 Monotonicity Curves

1.68 V. We need to determine if this is a good or bad reading, and if it is bad, which bit is at fault. The first step is to determine what the analog voltage out should be. This is found by multiplying the voltage resolution by the count value:

$$V_{RES} = V_{\text{full scale}} \div 2^n - 1 = 5.1\text{ V}/255 = 0.02\text{ V per bit}$$

$$V_{OUT} = V_{RES} \times \text{count} = 0.02\text{ V/bit} \times 92 = 1.84\text{ V}$$

We see that 1.68 V is incorrect because it should be 1.84 V, given the indicated binary input. To find which bit is bad, find the difference between the two analog quantities and divide it by the voltage resolution:

$$V_{OUT\ expected} - V_{OUT\ actual} = 1.84\ V - 1.68\ V = 0.16\ V$$

$$\text{Bad bit} = \text{difference} \div V_{RES} = 0.16\ V/0.02\ V\ \text{per bit} = 8\ \text{or}\ 00001000B$$

This indicates that bit b_3 is missing or always low. Once the incorrect bit is established, the circuit must be examined further to find out why the bit is missing or low. There could be an open input to the DAC or a short from b_3 to ground. At least by using the method just described, the faulty bit is located and can be corrected.

Section 7.10

Practice Problems

1. Which bit is bad, as indicated by the monotonicity curve in Figure 7.18 for the DAC to produce that curve?

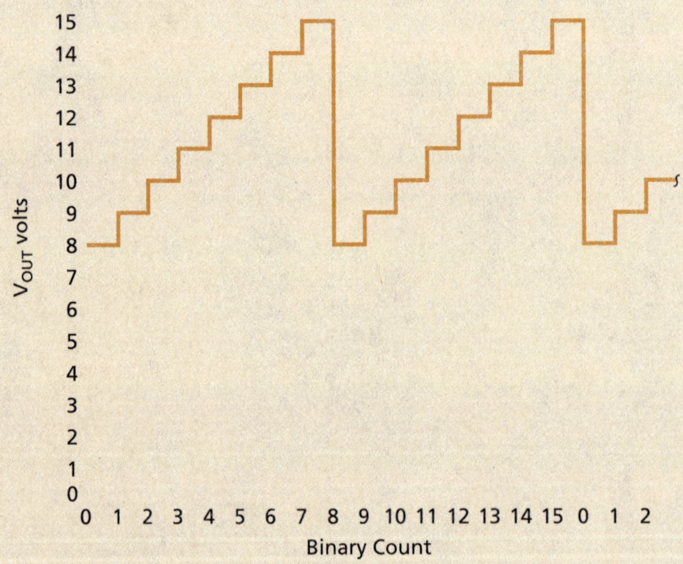

Figure 7.18 Practice Problem 7.10(1)

2. The analog voltage measured at V_{OUT} of a 10-bit DAC with an input of 0011011110B and a full-scale voltage of 24 V is approximately 5.969 V. Which bit is bad? Is the bad bit a logic 0 or a logic 1?

7.11 Troubleshooting ADCs

Troubleshooting ADCs is a bit more complex than troubleshooting DACs, because there are more component parts to the ADC circuits. Basic ADCs, other than the flash converter, include some form of counter/register, a DAC, and a comparator. When an ADC proves to be failing, it will be necessary to isolate the failed circuit element of the DAC before attempting to determine which component is bad. The output of the comparator will determine what the counter/register is supposed to be doing at any given time. If the counter/register is not responding correctly to the stimulus from the comparator, then the counter/register is the area to explore. If the counter is putting out the correct response, then maybe the DAC is not converting the digital input to the correct analog output. Lastly, because this is a closed loop, the comparator may not be switching correctly, given the reference voltage from the DAC and the analog input voltage.

Much of the discussion in this section and the previous one on troubleshooting DACs becomes moot when using medium-scale integrated (MSI) DACs and ADCs. In general, since much of the converting hardware is in a single chip, if the inputs to that chip are correct, but the output isn't, then the chip is bad and the circuit is easily repaired by replacing the failed IC. Troubleshooting circuits incorporating these chips is primarily a task of isolating the chip that is not doing its job rather than of determining what is wrong with the individual internal components of that chip.

7.12 Designing DACs

Many of the design criteria for a DAC are based on the requirements for the circuit. They include

1. Number of bits for the DAC or its resolution
2. Range of analog voltages to determine full-scale voltage
3. Required accuracy
4. Voltage of a logic 0 and a logic 1

Taking accuracy first, the greater the degree of accuracy required, the more precise the resistors need to be and the more reliable a voltage source representing a logic 1 must be. An additional consideration, which involves both the resistors of the ladder and the source voltage, is the amount of power the DAC itself consumes. Lower values of resistance result in a more stable DAC, but draw a large current, taxing the power source. Too-large resistor values do not draw the current, but are more susceptible to noise and temperature influences. A happy medium must be selected for a reliably operating DAC circuit.

Let's design a DAC with a resolution of 0.196 percent and a full-scale analog output voltage of 5 V. We will make a logic 0 equal to 0 V and logic 1 to 12 V. The DAC is to have $+/-3$ percent accuracy, so the resistors used in the ladder network must be at least 1 percent resistors. Our first step is to determine how many bits our DAC is to have. We find this from the resolution figure of 0.196 percent. This figure of merit is equivalent to 100 percent divided by the total number of steps in the DAC conversion from 0 to full scale. We can find this number by dividing 100 percent by 0.196 percent.

Number of steps = 100%/0.196% = 510.2 or 511

Next, the number of bits in the DAC is determined by the number of bits required to yield 511 total steps:

$$2^n - 1 = 511 \qquad n = 9$$

We will use an R/2R ladder to achieve a DAC with a nine-bit resolution, and select 5 kΩ and 10 kΩ resistor values for this R/2R ladder. These values are low enough for reasonable stability of the DAC without taxing the power source too heavily. A nine-bit DAC with a logic 1 voltage of 12 V produces a full-scale voltage of

$$V_A = B_{IN} \times V_{logic\ 1} \div 2^n = 511 \times 12/512 = 11.98\ V$$

The desired V_{OUT} at full scale is 5 V, so we need to scale the output using an op amp inverting amplifier. V_{IN}, at full scale, is a V_A value of 11.98 V, and V_{OUT} is 5 V. R_{IN} is R or 5 kΩ, and R_f is the scaling feedback resistor value we need to find:

$$R_f = R_{IN} \cdot V_{OUT}/V_{IN} = 5\ k\Omega \times 5/11.98\ V = 2.09\ k\Omega$$

The completed circuit is shown in Figure 7.19.

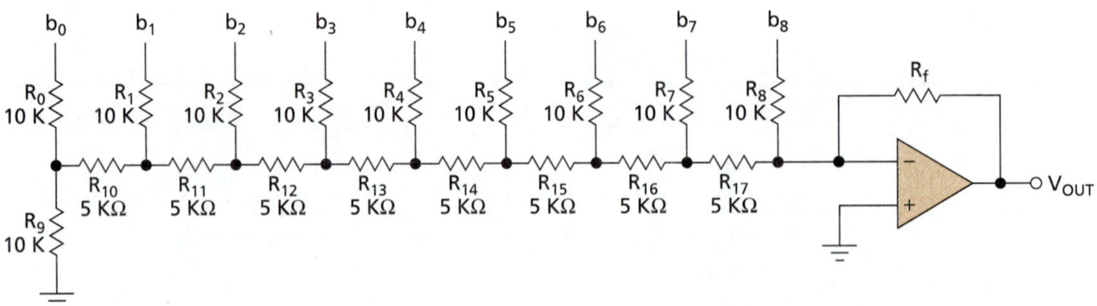

Figure 7.19 R/2R Ladder Design

Section 7.12

Practice Problem

Design an R/2R ladder DAC using an R value of 5 kΩ with a resolution of 0.7874 percent, a 5 V logic 1 level. The scaling op amp will yield a full-scale output voltage of 25 V.

Summary

To interface analog signals to a computer for purposes of evaluation, control, or other processing, an analog-to-digital converting circuit is required. To convert digital results to analog values requires a digital-to-analog conversion, which is performed by using either the weighted resistor or R/2R ladder network buffered by a scaling operational amplifier configured as an inverting feedback amplifier. The weighted resistor ladder network has the advantage of simplicity and direct translation between current values and weighted binary values. The R/2R network is more frequently used because the weighted resistor ladder has resolution limitations related to the differing resistor values in the ladder.

Several types of ADC circuits can be used to convert analog signals to their digital equivalents. If speed is of utmost importance, then a flash or simultaneous converter is used. This type of converter has limited resolution, because an operational amplifier is required for each digital bit. In addition, some form of encoding is required if weighted digital output values are required to represent the analog inputs. The ramp- or counter-type ADC overcomes the resolution and encoding disadvantages of the flash converter at the sacrifice of speed. Ramp converters require a signal to initiate conversion by resetting a counter to allow the ADC to count up to the equivalent digital value for each analog voltage sampled. The up/down ramp or continuous converter overcomes the need to reset the ADC and can be used to track analog changes. Unfortunately, this converter produces a continually fluctuating change of one bit for a steady state analog input.

The most commonly used ADC is the successive approximations converter. This converter produces a digital output by approximating the equivalent digital quantity a number of times until the approximation matches the actual value. The time of conversion is always the same regardless of the analog input. This converter combines the resolution advantage gained by the ramp converter and the tracking ability of the up/down ramp converter with a marked improvement of conversion time when compared to the other counter-type converters.

Medium-scale integration (MSI) ICs contain basic DAC and ADC circuits, usually in combination with control logic and buffer registers. An example of each type was presented in this chapter. These MSI chips are designed to interface and operate with microprocessor-controlled computer systems.

Dual-slope ADC circuits are used in digital meters, where noise rejection and accuracy are more crucial than the time of conversion. Dual-slope converters work by using the charge and discharge times of an integrator capacitor to control the time a counter counts. The level of the input voltage determines the degree to which the capacitor charges, which governs the time it takes the capacitor to discharge, allowing the counter to count to a binary value equivalent to the input analog voltage. The maximum time of conversion for the dual-slope converter is twice the time of a ramp-type converter.

Troubleshooting converter circuits requires the technician to be aware of the expected results for a given input to each circuit. DAC troubleshooting is aided by the use of a monotonicity curve to give a visual presentation of the DAC's performance. ADC circuit troubleshooting requires the isolation of circuit elements with the ADC counter circuit. When either conversion type is done using an MSI chip, the process of troubleshooting an ADC chip involves checking that the inputs are all correct before labeling the chip as defective.

Designing converting circuits requires knowledge of what is expected of the circuit and how accurate, precise, and stable the circuit is to be. Power consumption needs are weighed against long-term circuit stability in determining the range of resistor values used in the DAC ladder networks.

Glossary

accuracy how close to the actual analog value the output of a DAC is.

ADC analog-to-digital converter.

average conversion time the time it takes an up/down ramp analog to digital converter to produce the digital value.

counter or **ramp ADC** the first of the counter-type analog-to-digital converters.

DAC digital-to-analog converter.

dual-slope ADC analog-to-digital converter used in instrumentation.

end of conversion (EOC) signal that indicates when an ADC has completed the conversion of an analog signal to its digital equivalent.

end of process (EOP) same as end of conversion (EOC).

flash or **simultaneous ADC** a high speed analog to digital converter.

full-scale output the maximum analog value a DAC circuit obtains when the maximum binary value is applied.

maximum conversion time the time it takes for an ADC circuit to produce a digital equivalent output for a change from minimum to maximum analog input voltage.

monotonicity curve an output signal generated by a DAC whose binary input is changing in an up count sequence. It is used to test the performance of the DAC.

percent of resolution the percentage of a whole that one step size is.

precision the repeatability of the output of a DAC or ADC for the same input quantity.

resolution a single step change in an ADC or DAC's output.

scaling op amp a feedback amplifier used as a voltage multiplier or scaler.

start conversion (SC) a signal used to begin an analog-to-digital conversion sequence.

step the analog voltage change of a DAC or ADC associated with a one-bit change in a digital quantity.

successive approximations ADC an analog-to-digital conversion that works by taking numerous approximations of the analog input until those approximations produce the exact digital equivalent output.

up/down ramp or **continuous ADC** an analog-to-digital converter used to overcome the lack of tracking of the analog input present in the ramp converter.

voltage resolution the voltage value of one step change of an ADC or DAC output.

weighted resistor ladder DAC a digital-to-analog resistor network that electrically sums the weights of each bit position to yield the analog equivalent of a digital input.

Solutions to Practice Problems

Section 7.2

1. $V_A = 2.12$ V **2.** $R_{TH} = 1.06$ kΩ **3.** $R_f = 2.7$ kΩ

Section 7.3

1. $V_{RES} = 0.102$ V per step
%RES = .39%

2. $V_{STEP} = 1.0$ V

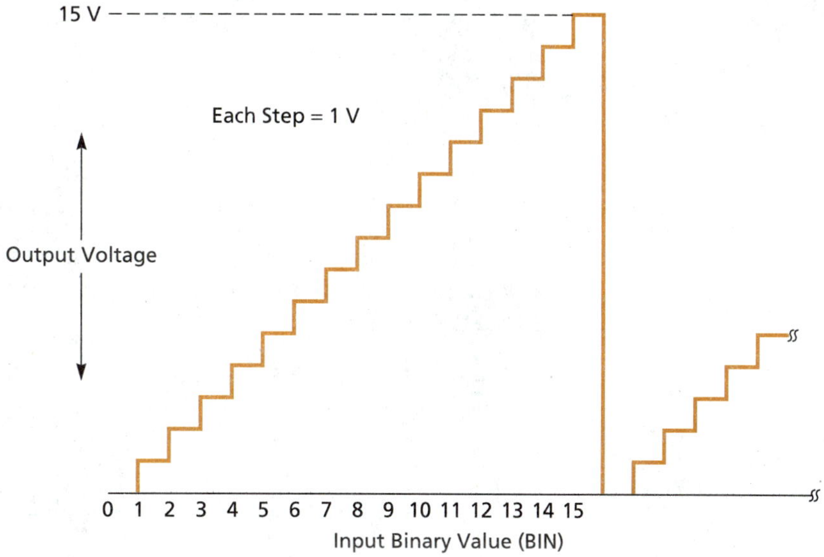

Figure 7.20 Practice Problem 7.3(2) (Solution)

Section 7.4

1. a. $V_{\text{full scale}} = 5.953\text{ V}$ **b.** $V_{RES} = 0.047\text{ V per step}$
c. $R_f = 1.965\text{ k}\Omega$ **d.** $\%RES = 0.7874\%$

Section 7.5

1. 0.39% **2.** Maximum $V_{RES} = 0.039\text{ V per step}$

3.

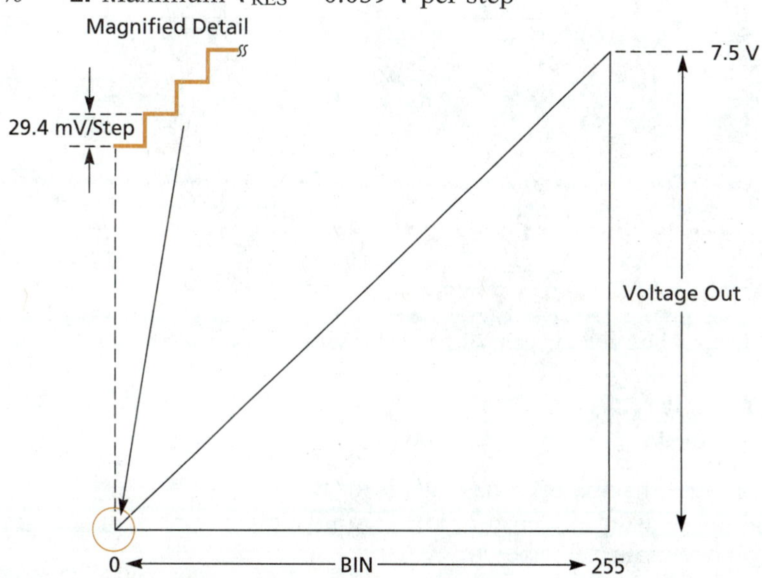

Figure 7.21 Practice Problem 7.5(3) (Solution)

4. Flow-through

Section 7.6

1. A flash converter is faster, but a ramp converter needs less circuitry for the same resolution. Flash converters can track increasing or decreasing analog voltages. Ramp converters require a start-of-conversion input to restart at a binary value of zero in order to track analog voltages correctly.

2. Time of conversion = 8.184 mS

Section 7.7

1. Percent of resolution = 0.0978%
Conversion time = 5.7143 μS

2. Maximum conversion time for 10-bit ramp ADC = $(2^{10} - 1)$/1 MHz = 1023 μS maximum conversion time for 10-bit successive approximations ADC = 10/1 MHz = 10 μS.

Both converters have the same resolution = 100%/$(2^{10} - 1)$ = 0.098%

Section 7.8

1. a. 0.392% **b.** 0.02353 V **c.** 29.1 μS

Section 7.10

1. Bit b_3 is always high

2. $V_{RES} = 24/(2^{10} - 1) = 0.0235$ V
$V_{OUT} = (222)(0.0235) = 5.208$ V

Bad bit = (5.959 − 5.208)/0.0235 = 32, which shows bit b_5 always high.

Section 7.12

Design is a seven-bit R/2R ladder using 5 kΩ and 10 kΩ resistors and a feedback resistor value of 25.2 kΩ.

Problems

7.1 Define the following digital to analog (DAC) terms:
a. Resolution **b.** Accuracy **c.** Precision

7.2 What is the relationship of the following DAC parameters to a DAC circuit? How do they differ from each other?
a. Resolution **b.** Accuracy **c.** Precision

7.3 Compute the analog voltage at V_A for the weighted resistor ladder in Figure 7.22 given the binary input data shown.

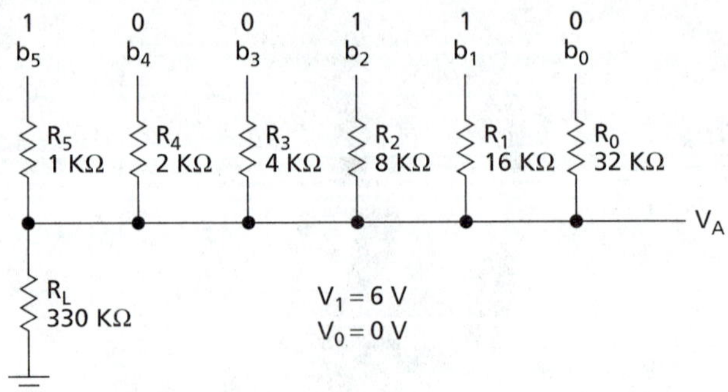

Figure 7.22 Problem 7.3

7.4 Compute the analog voltage at V_A for the weighted resistor ladder in Figure 7.23 given the binary input data shown.

7.5 The op amp scaler of Figure 7.24 is added to the weighted resistor ladder network in Figure 7.22. Compute the value of R_f required to cause V_{OUT} to have a voltage resolution of 0.1 V for each step.

7.6 The op amp scaler of Figure 7.24 is added to the weighted resistor ladder network in Figure 7.23. Compute the value of R_f required

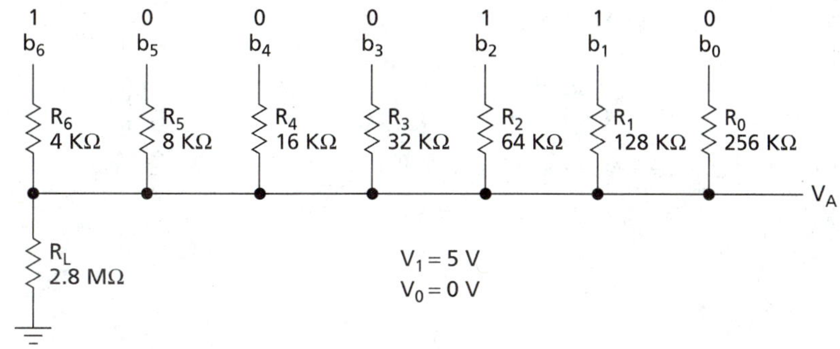

Figure 7.23 Problem 7.4

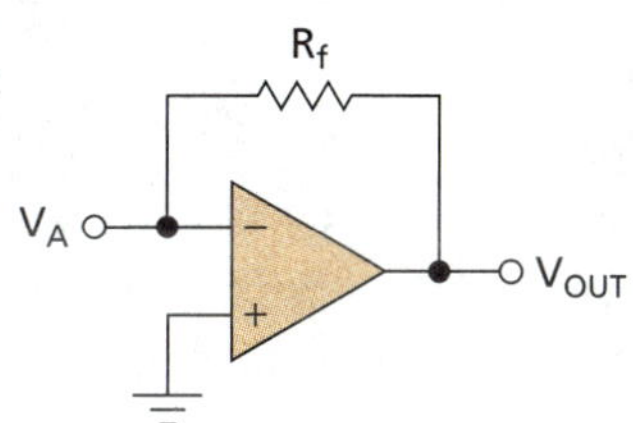

Figure 7.24 Problems 7.5 and 7.6

to cause V_{OUT} to have a voltage resolution of 0.2 V for each step.

7.7 What is the percent of resolution of a seven-bit DAC?

7.8 What is the percent of resolution of a 16-bit DAC?

7.9 What is the voltage of resolution of a seven-bit DAC with a full-scale voltage level of 3.75 V?

7.10 What is the voltage resolution of a 16-bit DAC with a full-scale voltage level of 6.56 V?

7.11 Compare a weighted resistor ladder network DAC to an R/2R network DAC. Which is generally preferable to use? Why?

7.12 What are the advantages of an R/2R ladder network compared to a weighted resistor ladder network?

7.13 What is the full-scale voltage at V_A for the R/2R network of Figure 7.25? What value of R_f is needed to make that maximum output 7.7 V?

7.14 What is the full-scale voltage at V_A for the R/2R network of Figure 7.26? What value of R_f is needed to make that maximum output 15 V?

7.15 What is a monotonicity curve? How is it generated?

7.16 What DAC parameters can be tested using a monotonicity curve?

7.17 For the DAC0830, what signals are asserted to put the chip in a double-buffered operating mode?

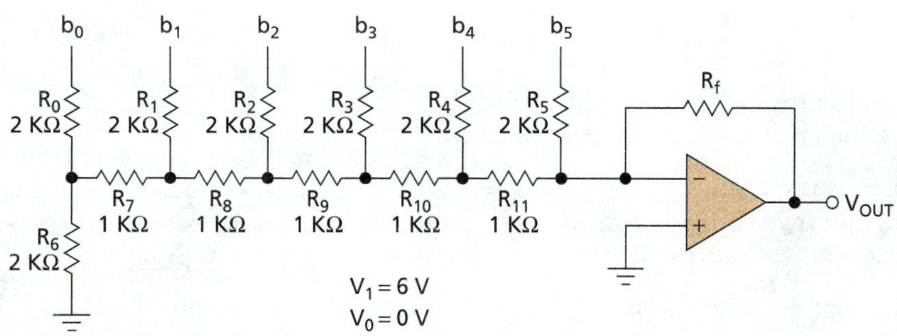

Figure 7.25 Problem 7.13

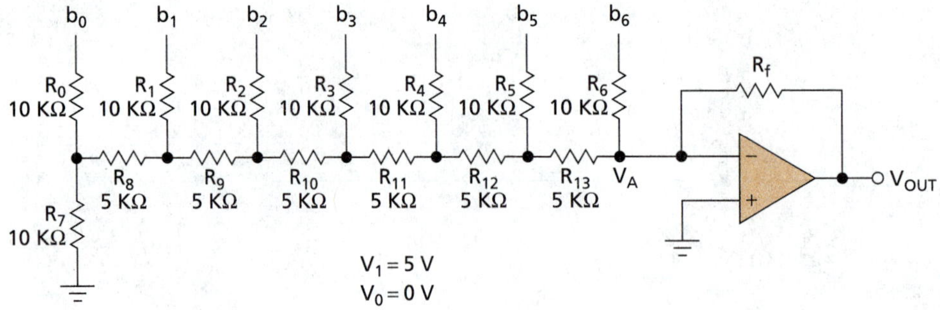

Figure 7.26 Problem 7.14

7.18 What signals are asserted to place the DAC0830 into a single-buffered operation? Flow-through operations?

7.19 What is the percent of resolution for a DAC0830?

7.20 What is the voltage resolution of a DAC0830 using a V_{REF} of 8 V?

7.21 Which ladder network is used in a DAC0830?

7.22 What determines the accuracy of DAC0830?

7.23 Define these analog-to-digital (ADC) terms:
a. Maximum conversion time
b. Average conversion time

7.24 Which of the ADC converters uses average conversion time as a figure of merit? Why?

7.25 What are the advantages and disadvantages of a flash analog-to-digital converter (ADC)?

7.26 Which disadvantage of a flash converter was the ramp ADC designed to overcome?

7.27 What is the maximum conversion time of a 15-bit ramp ADC using a 330 KHz clock signal?

7.28 What is the maximum conversion time of a 16-bit ramp ADC using a 1.44 MHz clock signal?

7.29 What is the disadvantage of the ramp converter that the up/down ramp or continuous converter was designed to overcome?

7.30 What is the maximum and average conversion times for a 10-bit continuous converter that uses a 1.8 MHz clock signal?

7.31 Describe how a successive approximations converter changes an analog voltage into a digital word.

7.32 What improvement in the performance of counter-type ADCs is provided by a successive approximations converter?

7.33 What is the maximum conversion time of a 14-bit successive approximations ADC connected to a 575 KHz clock?

7.34 What is the maximum conversion time of a 16-bit successive approximations converter using a 1.44 MHz clock signal?

7.35 Compare the conversion times of a successive approximations ADC and a ramp ADC, both using the same clock signal and having the same resolution.

7.36 Which type of ADC is used in the ADC0808 chip?

7.37 Describe how one of the analog inputs is selected to be converted to digital using an ADC0808.

7.38 What does the EOP signal indicate? How is it used to allow the ADC0808 to continuously convert an analog input?

7.39 List the improvement each ADC achieved when compared to the circuit that preceded it. The progression is flash, ramp, up/down ramp, and successive approximations.

7.40 Give the ADC type that matches each of these descriptions:
a. Fastest counter type
b. Fluctuates one bit with steady state input

c. Requires an encoder for regular digital outputs
d. Must be reset before each conversion
e. Has the poorest resolution
f. Has the fastest conversion time

7.41 What is the main benefit of a dual-slope analog-to-digital converter?

7.42. What is the primary application for a dual-slope ADC?

7.43 Compare the maximum conversion time of a dual-slope ADC to both a counter ramp ADC and a successive approximations converter.

eight

The Binary Architecture Basic Electronic (BABE) Computer

KEY TERMS

accumulator or A-register
address bus
address buffer
ALU (arithmetic logic unit)
assembler program
assembly or source program
BABE (binary architecture basic electronic) computer
buffer or B-register
conditional jump
control logic
data bus
execute cycle
fixed-length instruction set
flag
input/output or I/O interface
input/output or I/O or peripheral device
instruction decoder
instruction pointer
instruction register
labels
loop
machine or object program
mnemonic
opcode
operand
program
temporary register
variable-length instruction set

OBJECTIVES

After completing this chapter, the student should be able to

- Discuss the BABE computer from a functional point, beginning with its programming model, detailing the resources available to programmers for the BABE computer.
- Present and discuss the BABE computer instruction set.
- Explain how instructions are performed using the BABE computer, starting with the instruction opcode and operand fetch cycles.
- Discuss the execute cycles of various BABE instructions.
- Illustrate assembly language programming principles using the BABE instruction set and computer.
- Show how the hardware or logic circuits for the BABE computer support the functional diagram of the system.
- Describe applications of bus concepts found in the BABE's data, address, and control bus operations.

OVERVIEW

The purpose of this chapter is to integrate a number of logic circuits into a basic digital computer system. This computer architecture is used to illustrate how a computer fetches and interprets program instructions. Concepts of machine coding and assembly-level programming are introduced, along with program applications that illustrate the power of even this very basic computer system.

8.1 Introduction

The operation of a digital computer relies on the movement of binary information from one circuit in the system to another. Logic circuits at the destination locations use the information, generating a result that depends on the nature of the information and the purpose for its use. The three basic types of binary information in a digital computer are addresses, data, and instructions. Addresses define locations for the storage and retrieval of data in memory and input and output devices. Data are actual binary information, which can be numbers, character codes, or any other defined information. Instruction codes define the operations that a computer is to perform. The deeper meaning and use of all three types of binary information and how computers manage them are explored using a computer model called the **binary architecture basic electronic computer** or **BABE computer.**

BABE (binary architecture basic electronic) computer *a generic computer system used to illustrate the functions and operation of a microprocessor based computer system. It is upwardly compatible with the INTEL 80x86 processor line.*

8.2 Functional Blocks of the BABE Computer

The block diagram of the BABE computer is shown in Figure 8.1. In the center of the diagram are two main bus structures, the address and data buses. A **bus** is a group of wires or electrical paths that pass similar types of digital information. The **address bus** for the BABE computer has 16 lines, each holding one of 16 address bits, individually denoted consecutively as A_0 to A_{15}. Binary information defining memory, input device, and output device locations are carried on the address bus. The **data bus** is a group of eight lines, labeled D_0 to D_7, that are responsible for passing digital data from one unit to another within the computer system. Each of the units connected to the buses is equipped with control signals that determine which unit's output is placing information on the buses and which unit is expected to receive that information. As an example, to move a byte of data from the buffer (B) register to the accumulator (A) register, the output of the B-register must be enabled and the A-register must have an active load signal. Control signals on the various functional blocks that are labeled starting with an L are input load enables. For instance, LACC stands for *load accumulator.* When active, this signal causes the data on the data bus to be loaded into the A-register. Outputs are enabled onto the data bus with signals beginning with an E. Using the A-register again, the EACC signal, when asserted, places the contents of the accumulator onto the data bus.

Address bus *a group of signal paths that carry binary address information.*

Data bus *a group of signal paths that carry binary data information.*

EXAMPLE 8.1

Using Figure 8.1 as a reference, which signals are asserted to transfer data between the following units?

1. Temporary register to the instruction pointer
2. A-register to the I/O interface
3. ALU to the accumulator

SOLUTION The units that supply the data require their E lines to be asserted. Units that are to accept the data require their "L" line to be turned on. For each of the examples shown, the signals to be asserted are

1. ET enables the temporary register output, and LIP loads the instruction pointer.
2. EACC enables the A-register, and LIO loads the I/O interface.
3. EALU places ALU contents onto the data bus, and LACC loads those data into the A-register (accumulator).

Opcode *shorthand notation for operation code, which is the binary code for a program instruction.*

Operand *defines the location or value of binary data.*

A program is a list of instructions that are to be performed (executed) consecutively. The codes for these instructions, called **opcodes,** for *operation codes,* and corresponding **operands** are stored in memory at consecutive locations. Operands are either memory locations of data or the actual data themselves

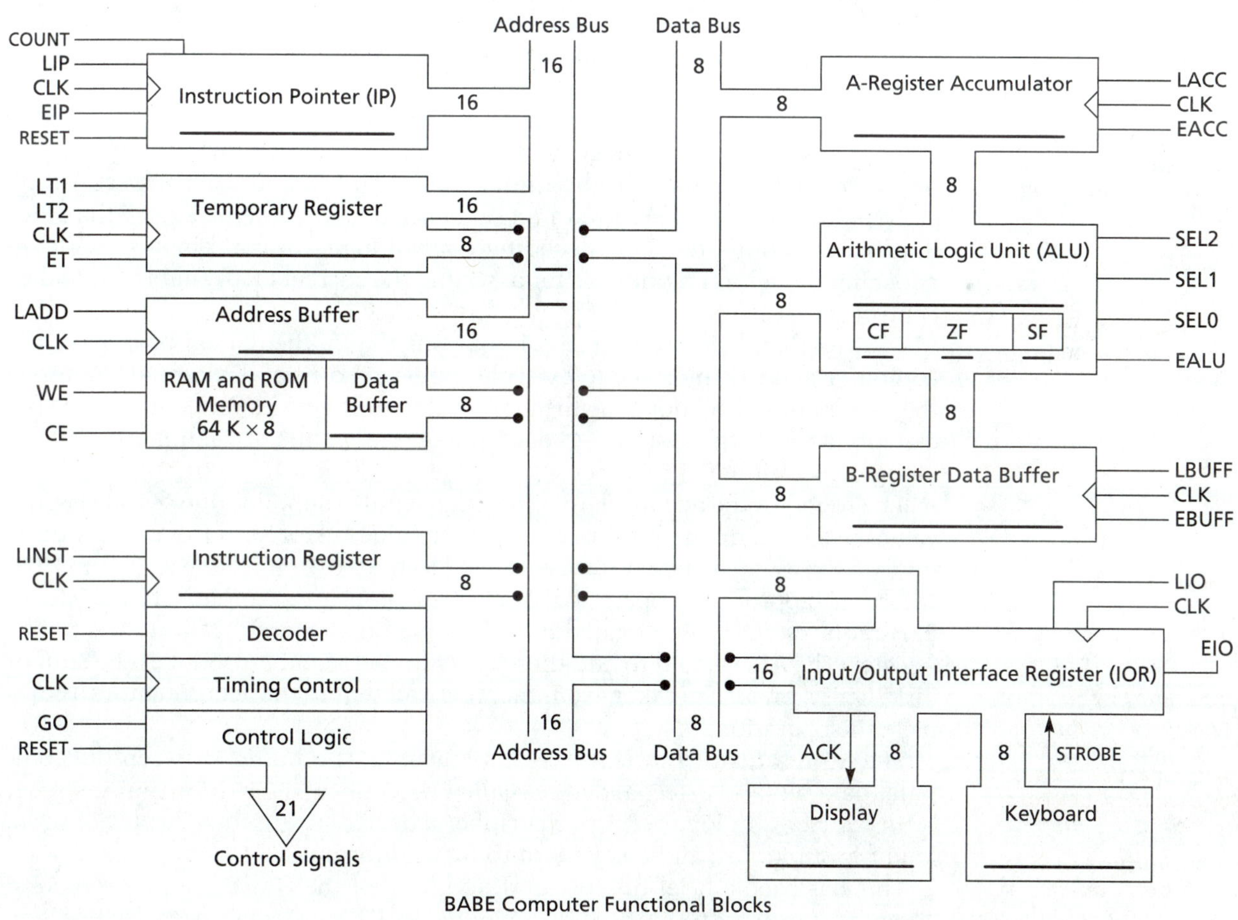

Figure 8.1 BABE Computer Functional Blocks

Instruction pointer *a binary up counter in a processor that keeps track of the next instruction to be fetched and executed.*

Instruction register *a register in a computer that stores the opcode fetched from memory.*

Address buffer *parallel-load register used to store address information that is applied to the computer's address bus.*

Temporary register *a register that holds a value that will be used very quickly by another section of the computer.*

Instruction decoder *computer circuit that decodes the binary bits of the opcode. The decoded outputs are used by the control logic to generate the signals that direct the computer's operation.*

Control logic *the part of the computer that generates the control signals that direct the computer's operation.*

Accumulator *or* ***A-register*** *a register that is used to hold the results of arithmetic and logic operations.*

Buffer *or* ***B-register*** *a section of the BABE computer that holds data usually used by the ALU.*

ALU (arithmetic logic unit) *the functional block of a computer that performs arithmetic and logic operations.*

Input/output *or* ***I/O interface*** *a buffer register that holds data from the computer that are to be used by peripheral devices connected to the interface.*

Input/output *or* ***I/O*** *or* ***peripheral device*** *device connected to the external computer buses, usually through some form of interface.*

that instructions use. By storing opcodes and operands in consecutive order in memory, the execution of a program can flow as a list of instructions. Later in the chapter, we will see how this occurs, but for now the first functional unit illustrated in Figure 8.1, in the upper left corner, is responsible for keeping track of the orderly execution of instructions. The **instruction pointer (IP)** contains the address of the memory location of the next opcode or operand to be fetched, enabling an instruction to be retrieved, decoded, and executed. The contents of the IP are used to access memory. The information at the location pointed to by IP are transferred to the **instruction register** if it is an opcode, or to the **address buffer** if the data constitute an operand address. Address operands are first loaded into the **temporary register** one byte at a time and then transferred from the temporary register to the address buffer as a 16-bit address.

The address buffer receives address information from the address bus and uses that information to access a location in the computer's **memory.** The memory is composed of two main types of memory devices, read/write random access memory (RAM), and read-only memory (ROM). The contents of ROM are usually initialization programs or routines that allow the system to be configured for use. RAM memory contains current user programs and data. RAM memory is written into by asserting the chip enable (CE) and write enable (WE) signals. RAM and ROM are read by asserting the CE signal by itself, which enables the outputs of a data buffer to pass the memory data to the data bus.

The instruction register accepts binary opcodes, which are decoded in the **instruction decoder.** Decoded signals are combined with phase sequence signals from a ring counter in the timing control block and further decoded by **control logic** to generate the correct sequence of control signals to all the other units of the computer. This decoding control logic, in combination with sequencing signals, is responsible for asserting the correct load, enable, and other control signals.

The righthand side of Figure 8.1 contains the mathematical portion of the computer and the interface to external input and output devices. At the top is the **A-register,** also known as the **accumulator.** It gets its name from the special job it has in the system—to accumulate the results of each mathematical or logic operation. For instance, on completion of an ADD instruction, the sum will be found in the accumulator. This register also supplies one of the original values used in a math operation. The other value is stored in the **B-register** or **buffer register.** The two numbers are fed to the **arithmetic logic unit (ALU),** where the appropriate operation is performed. As an example, adding 5CH to 22H requires 5CH to be loaded into the accumulator and 22H into the buffer register. The ALU is told to add these two numbers, and it produces the sum of 7EH. This sum is then taken from the ALU and sent to the accumulator to complete the operation.

The **input/output interface register** is used as a buffer between the computer data bus and external devices called **peripherals, input/output devices,** or **I/O devices.** In Figure 8.1 two peripheral devices are used, a display of some kind for output use and a keyboard to input information.

This has been a brief functional description of the BABE computer blocks. The next section introduces a portion of the BABE's instruction set. Using a few basic instructions, a step-by-step description of program execution can show how the units of the BABE computer function.

Section 8.2

Practice Problems

1. List the control signals that are required to perform the following data transfers:
   a. Load the B-register from memory.
   b. Load the address buffer from the instruction pointer.
   c. Load the accumulator from the keyboard.
2. List the sequence of control signals required to move 5CH and 22H from memory, and to load those data into the accumulator and buffer register, and then to take the sum from the ALU and place it into the accumulator.

8.3 Instruction Set Basics for the BABE Computer

Program *a list of successive instructions to be executed by a computer.*

Machine *or* ***object program*** *program written using the actual binary codes for a computer's instructions and operands.*

Assembly *or* ***source program*** *semiliterate programs that use a computer's instruction set. These programs are a single level above machine or object programs.*

Mnemonic *shorthand code for an instruction that is used when writing source or assembly code programs.*

Labels *used in an assembly program to identify an address location or constant value.*

Assembler program *software that converts a source program into object code.*

A **program** is a series of instructions that indicate operations for a computer to perform. In a digital computer, these instructions are translated into binary codes called operation codes or opcodes. Binary data used by these instructions are stored in memory locations called operands. A complete instruction is made of its opcode and appropriate operand. A program written using these binary **machine codes** is called an **object program.** Today's sophisticated microprocessors have object programs based on opcodes and operands that are too involved to code by hand. Instead, a slightly higher level of instruction format is supplied for these processors. Programs written at this level are called **assembly** or **source programs** and are made up of short abbreviated descriptions of instruction opcodes called **mnemonics** (the *M* is silent). Operands are represented by hexadecimal or decimal numbers or **labels** defining those numbers. **Assembler programs** are used to convert the source programs to executable object code.

This section looks at a partial list of the instructions used by the BABE computer. These instructions are used to develop source and object programs.

The first group of instructions is used to move binary data from one section of the BABE computer to another. Figure 8.1 is reproduced as needed to illustrate the activity each instruction performs. Each instruction for the BABE computer has an eight-bit opcode, the upper four bits defining the operation and the lower four bits defining the type of operand the instruction uses. The general syntax format for these instructions is

```
<8-bit opcode> <destination operand>, <source operand>
```

Whenever the **< >** are used, it will mean that something is to be written in place of this general indicator. For instance, in the preceding instruction format, **<8-bit opcode>** means that an actual opcode mnemonic is placed in this field. In the general format, the opcode is followed by a space, the destination operand, a comma, and then the source operand. As an example, to move data from a location in memory to the A-register, the instruction mnemonic is

```
MOV A,[<mem>]
```

where **<mem>** stands for a memory address. The square brackets, **[ ],** are used to enclose the address value. Whenever these brackets are used, it means that the value inside of them is used to point to a location in memory.

EXAMPLE 8.2

Write the instruction to move an eight-bit data word from location 1200H to the A-register.

SOLUTION Replace <mem> in the syntax for the MOV A,[<mem>] instruction with 1200H:

```
MOV A,[1200H]
```

Assemblers have various ways to indicate hexadecimal values. The one in Example 8.2 follows the form used throughout this textbook. Instruction assemblers written for INTEL devices use either a trailing H or leading 0 characters to indicate hex values. Motorola assemblers use a leading $. Omitting these characters causes the assemblers to interpret the values as decimal numbers. For example, 01234, 1234H, and $1234 are three ways to write the hexadecimal value 1234, depending on the assembler used. Similarly, 01234 or 1234 can be the decimal number 1,234 depending on the assembler in use. It is crucial that the user be very familiar with the assembler syntax that is presently in use. This text will continue to use the trailing H form for hexadecimal values.

An assembler written for the BABE computer would translate the mnemonic instruction form into a machine (or object) code form. The object code for the **MOV A,[<mem>]** instruction is **01H.**The **0** portion of the opcode defines the operation as a data move. The **1** denotes the A-register as the destination and indicates the requirement for a memory address operand. The instruction decoder in Figure 8.1 interprets this object code and determines that a memory address operand follows. In addition, since addresses are 16 bits long, the decoder also determines that there will be two bytes of operand information. Table 8.1 lists some of the MOV instructions and their object codes.

Table 8.1 BABE MOV Instructions

MNEMONIC	OPCODE
MOV A,[<mem>]	01
MOV B,[<mem>]	02
MOV A,<immed>	03
MOV B,<immed>	04
MOV [<mem>],A	05
MOV [<mem>],B	06

The abbreviation **immed** stands for immediate data and means that the operand value is used immediately as data instead of seeking data from a location in memory. **MOV A,5CH** moves the hexadecimal number 5CH into the A-register. **MOV A,[<mem>]** and **MOV B,[<mem>]** loads registers A and B respectively from the memory location specified as the **<mem>** operand address. **MOV [<mem>],A** and **MOV [<mem>],B** cause the contents of the A- and B-registers to be stored at memory locations specified by the **<mem>** operand addresses.

Figure 8.2 is a copy of the BABE computer showing which signals are active to move data from location 1200H into the A-register. For this illustration, the data at location 1200H are A4H. In the next section, we will see how the op-code was fetched and where the 1200H in address decoder came from. For now, we are just concerned with observing how data are moved from memory, once addressed, and stored in the A-register. Chip enable, CE, is asserted to enable the memory's data buffer contents (A4H for this example) onto the data bus, while LACC loads the A-register with data A4H from the data bus.

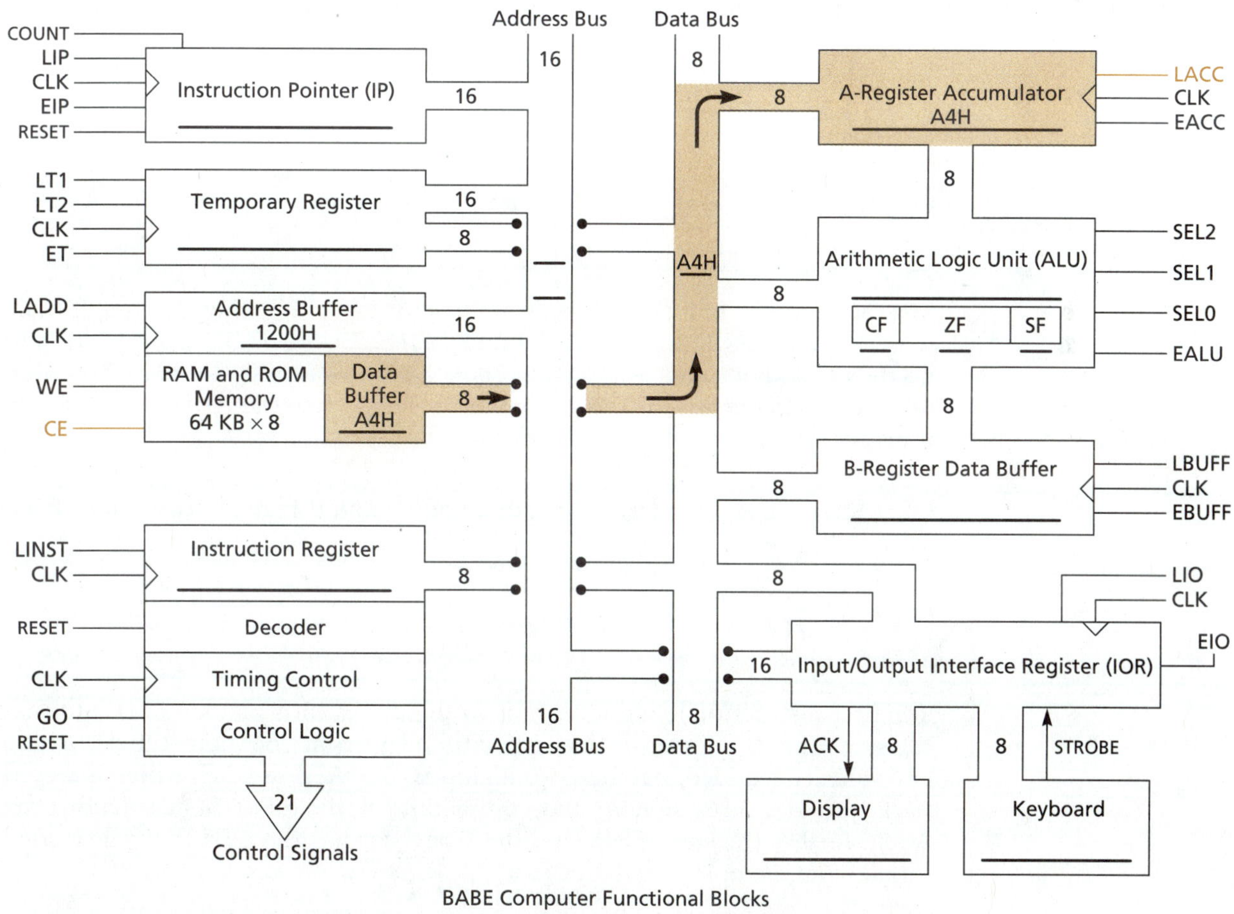

Figure 8.2 Data Flow Example

Table 8.2 ADD and SUB Instructions

MNEMONIC	OPCODE
ADD [<mem>]	11
SUB [<mem>]	12
ADD <immed>	13
SUB <immed>	14
ADD B	15
SUB B	16

The next group of instructions is the add and subtract arithmetic group. Their opcodes all have a 1H in the upper nibble, just as all the MOV instructions have a 0H. A partial list of the add and subtract instructions is listed in Table 8.2. The first thing you notice for this group of instructions is that there is only one operand. This is because in the BABE computer, the A-register, whose special job is to perform as an accumulator, is always the destination register for arithmetic instructions. That is, the sum or difference will be found in the A-register at the completion of each of these instructions. The operands for each indicate the source data to be added to the A-register. This means that a known value must be in the A-register before these instructions can be properly executed.

EXAMPLE 8.3

List the sequence of instructions to add two numbers together that are initially stored in memory locations 1200H and 1201H. Use the ADD B instruction to do the actual addition. If location 1200H has 5CH and 1201H has A2H, what are the final contents of the A- and B-registers after the ADD B instruction is executed?

SOLUTION The sequence requires that both A- and B-registers are loaded from the data in memory before the ADD B instruction can be used:

```
MOV A,[1200H]        ; 5CH -> A-register
MOV B,[1201H]        ; A2H -> B-register
ADD B               ; A + B -> A-register
```

The first instruction loads A-register from 1200H, thus placing 5CH into that register. The B-register is loaded with A2H from location 1201H. These two values are added together, producing 5CH + A2H = FEH, which is stored back into the A-register. At the completion of the ADD B instruction, the A-register has the sum, FEH, and the B-register has the A2H that was loaded into it from memory.

Section 8.3

Practice Problems

1. List the instruction sequence that subtracts the value 6CH from the data at location 4400H and stores the difference at location 128CH.
2. List an alternate sequence to accomplish the same results as in Practice Problem 1.

8.4 The Fetch Cycle

There are two types of fetch cycles: an opcode fetch and an operand fetch. To understand the need for these cycles, consider a program stored in memory. The binary opcodes and operands are placed in succeeding locations; for example

LOCATION	OPCODE OR OPERAND
1000H	01H
1001H	54H
1002H	20H

This sequence starts with the opcode for a **MOV A,[2054H]** instruction, which is 01H, stored at location 1000H. Next follows 54H, the low byte of the 16-bit operand address at location 1001H and finally, 20H, the upper byte of the operand address at location 1002H. For this instruction to be performed or executed, it must first be fetched (read) from memory. The opcode must be stored (written) into the instruction register and the operand fetched and stored into the address buffer of the memory so that the data at memory location 2054H can be accessed and moved to the A-register.

The job of the instruction pointer (IP), in the BABE computer, is to keep track of where the computer is in the program by "pointing" to the memory location that has the next program byte to be fetched. In this example, the instruction pointer starts with the value 1000H, the address of the first byte of the instruction. Because this is the first byte, the computer logic knows that the computer is to start an opcode fetch cycle.

The first phase of an opcode fetch, shown in Figure 8.3, begins by enabling the output of the instruction pointer onto the address bus. In our example, this places the value 1000H onto the address bus. The LADD control signal is asserted, to cause the address buffer to be loaded from the address bus with the 1000H value from the instruction pointer. This can be summarized in shorthand as **IP —> ADD BUFF.**

The address in the address buffer accesses a location in the memory that contains the opcode of the instruction to be executed. On the second phase of the fetch cycle, as shown in Figure 8.4, memory is accessed by asserting the CE input. This places the contents of location 1000H, which is 01H, onto the data

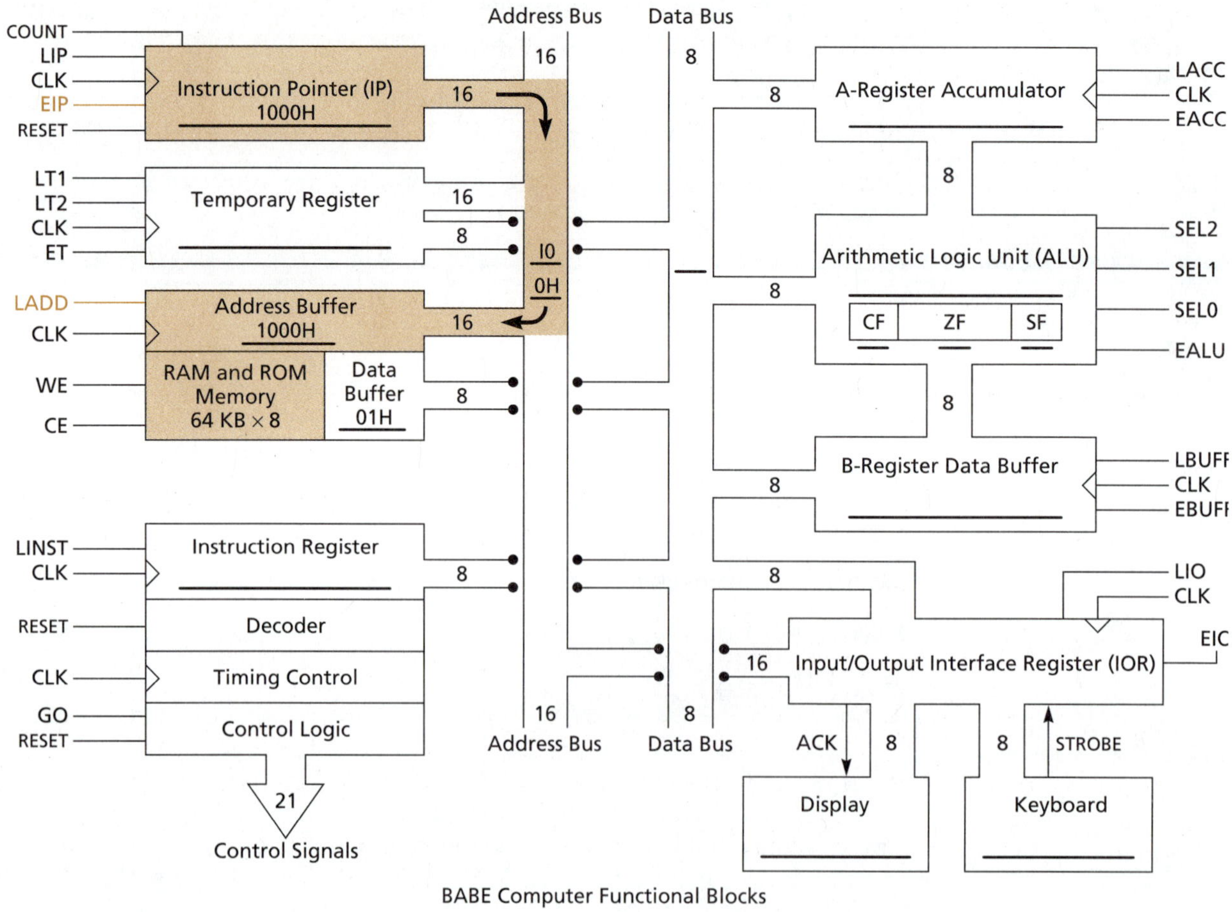

Figure 8.3 Opcode Fetch Cycle Phase 1

bus. Signal LINST is also inserted, causing the instruction register to be loaded with opcode 01H from the data bus.

The third and final phase of an opcode fetch cycle enables the instruction pointer to count for one clock cycle, causing its contents to be increased by 1, also called incrementing by 1, so that it now points to the next location in the program (1001H) as shown in Figure 8.5.

The instruction decoder and control logic decode the 01H opcode and determine that an operand fetch from memory is required to follow the opcode cycle. It begins the same way as an opcode fetch cycle by enabling the contents of IP onto the address bus and loading them into the address buffer as shown in Figure 8.6. In the second phase of the operand fetch cycle, memory is read as it was in the second phase of the opcode fetch cycle, but this time the data read are sent to the lower byte of the temporary address register instead of the instruction register as shown in Figure 8.7. The last phase of the operand fetch cycle is the same as the last phase of the opcode fetch cycle, which is to increment IP so that it points to the next byte in the sequence, as shown in Figure 8.8.

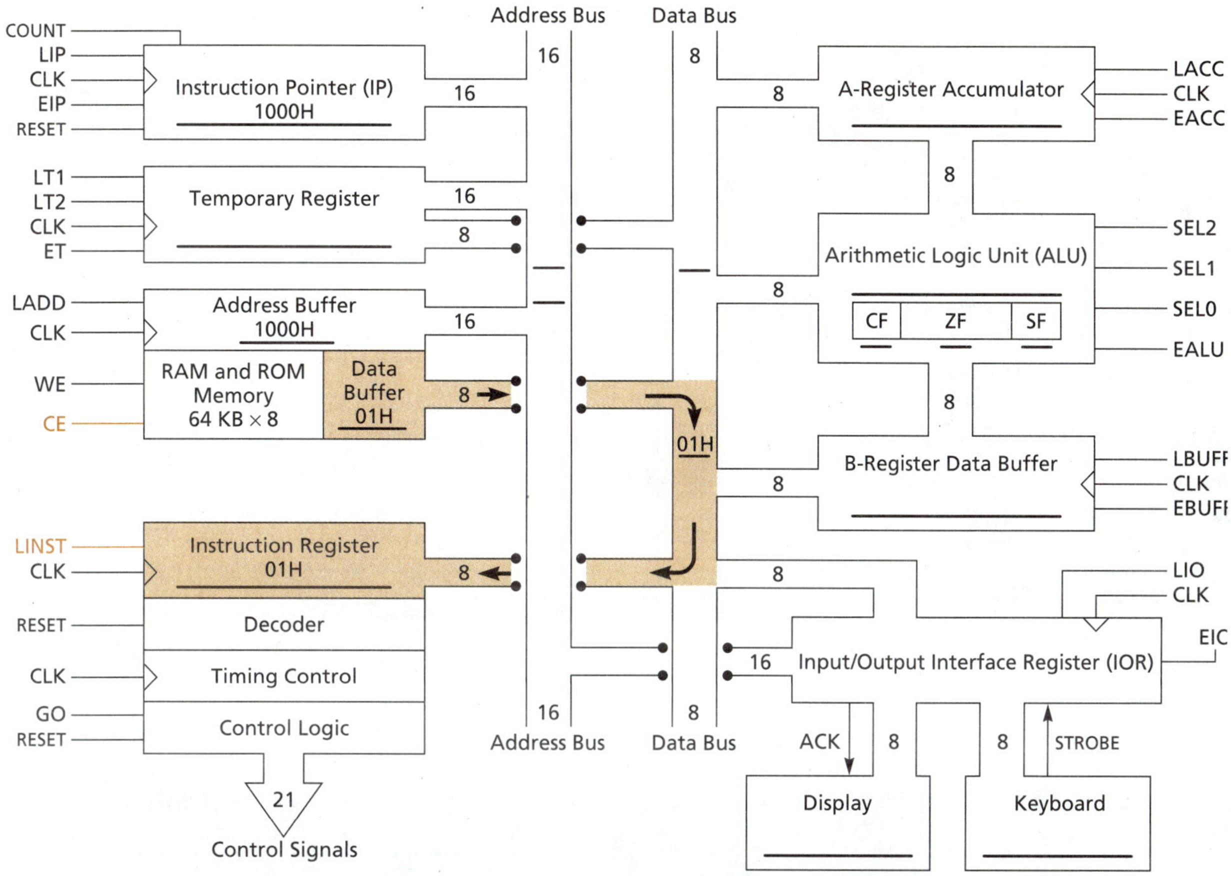

BABE Computer Functional Blocks

Figure 8.4 Opcode Fetch Cycle Phase 2

The instruction decoder also determines that this is a two-byte operand, and a second operand fetch cycle is required to get the second byte of the address. The computer repeats all three phases of the operand fetch cycle, this time fetching the upper byte of the address from location 1002H and storing it in the upper byte of the temporary register. The complete operand address now resides in the temporary address register and is used by the execute cycle of the instruction to point to the memory location that holds the operand data word.

Every instruction has an opcode fetch cycle, but some instructions, such as **ADD B,** do not require an operand fetch, since both source and destination elements—the A- and B-registers, in this case—are defined in the opcode. Other instructions, such as **MOV A,<immed>,** require a single operand fetch. The single byte operand, for this instruction, is the 8-bit data byte that is moved immediately into the A-register. Remaining instruction types, such as **MOV A,[<mem>],** require two 8-bit operands to define a complete address location of the data that are used by the instruction. Two operand fetch cycles are necessary to complete the fetch of the 16-bit address.

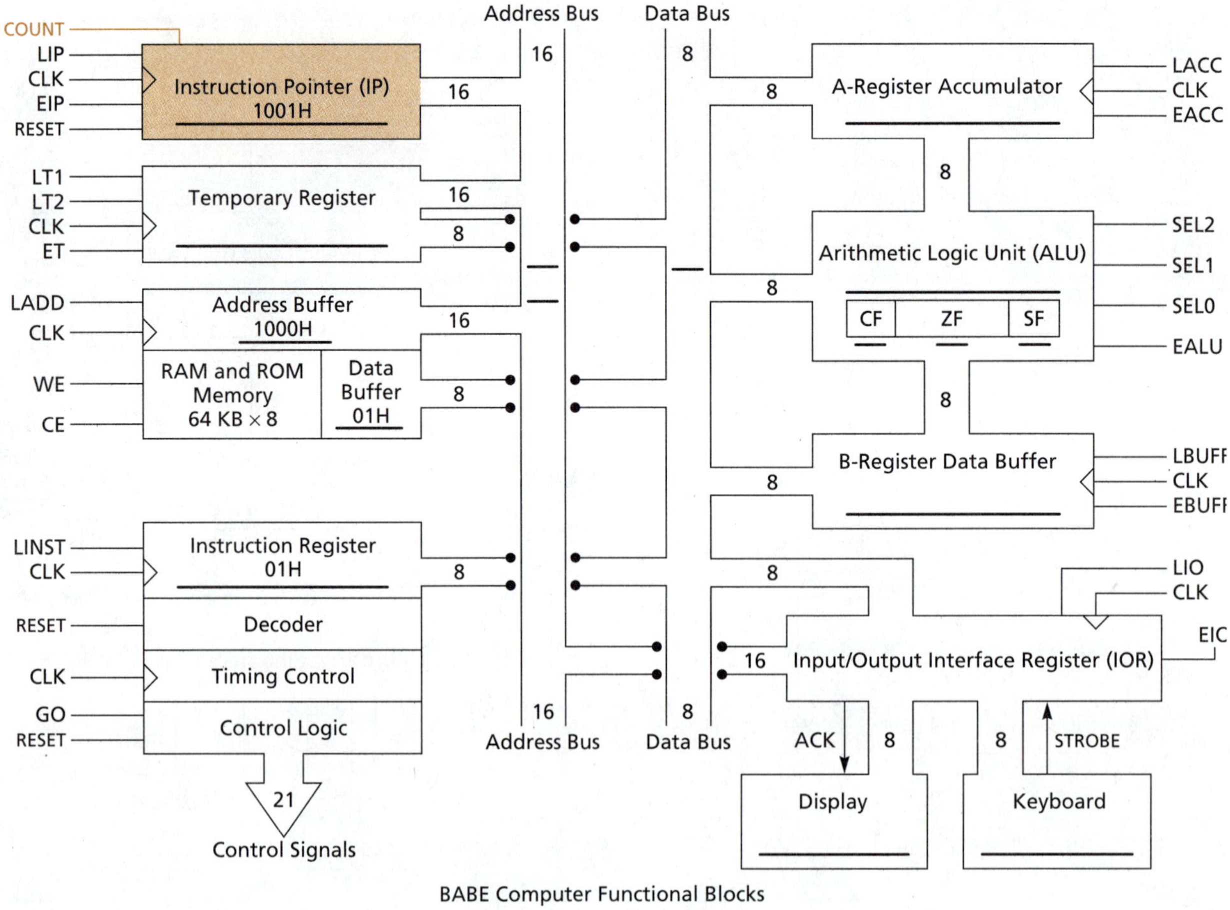

Figure 8.5 Opcode Fetch Cycle Phase 3

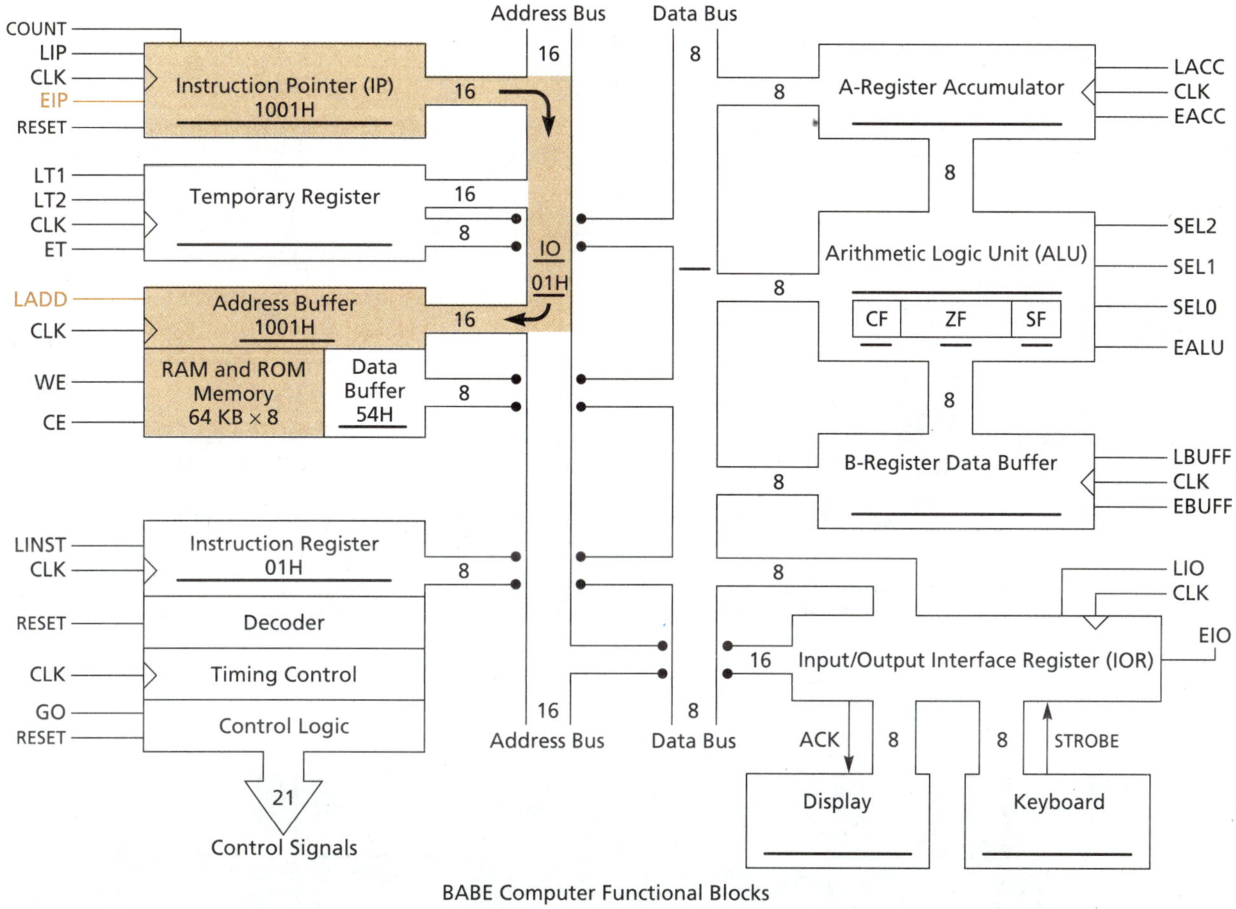

Figure 8.6 Operand Fetch Cycle Phase 1

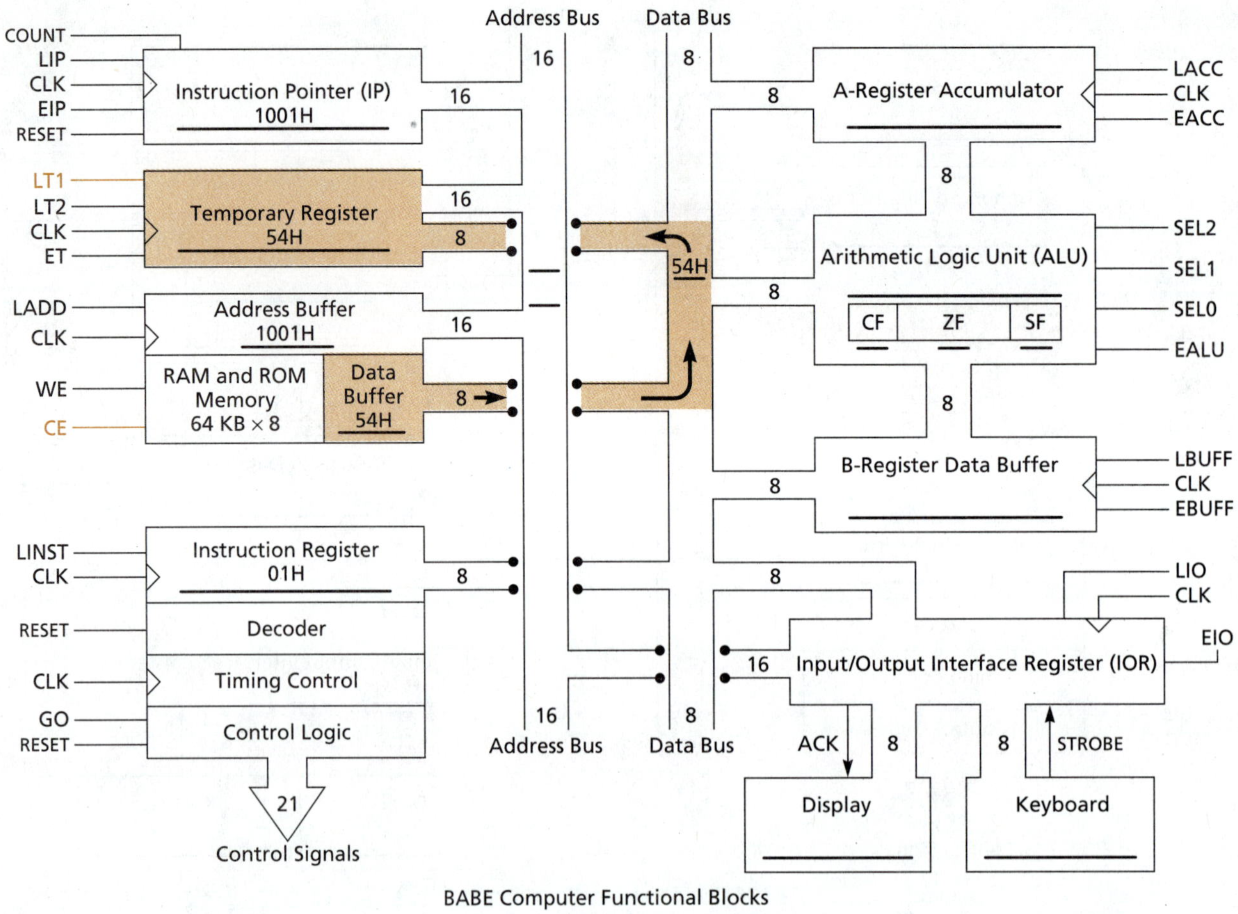

Figure 8.7 Operand Fetch Cycle Phase 2

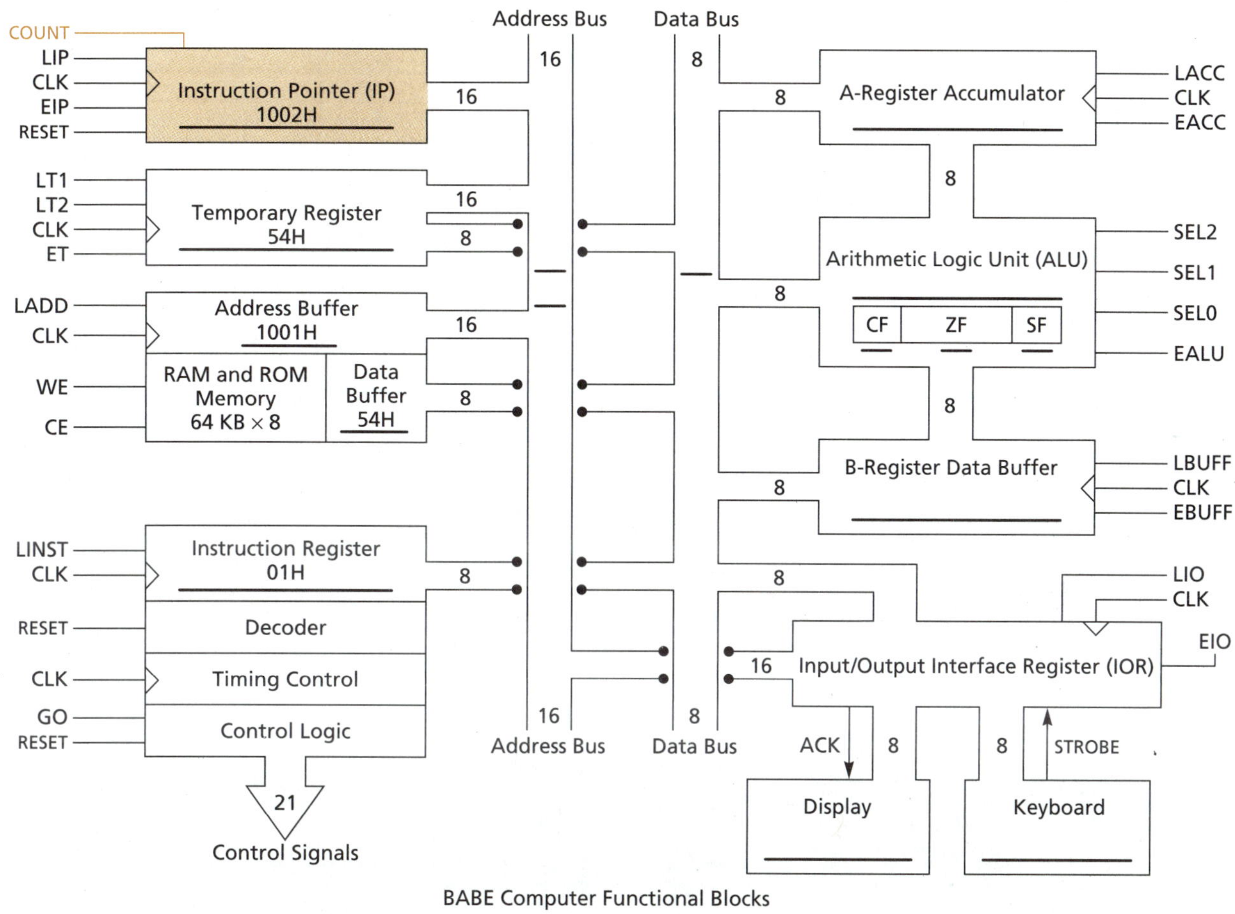

Figure 8.8 Operand Fetch Cycle Phase 3

Section 8.4

Practice Problems

1. Determine the number of operand fetch cycles required by all of the instructions in Tables 8.1 and 8.2.
2. Write the sequence of events for the opcode and operand fetch cycles of a **MOV B,[3312H]** instruction stored in memory, starting at location 82CCH. What are the contents of the following registers after the opcode and operand fetch cycles are complete?
   a. IP b. Instruction register c. Temporary register
   d. Address buffer e. Memory

8.5 The Execute Cycle

Execute cycle *the sequence of events that causes an instruction operation to be performed (executed).*

To complete an instruction, it must be *executed,* which does not mean "wiped out" or "killed," but rather that the operation defined by the instruction must be *performed.* Whereas opcode and operand fetch cycles follow the same sequence for each instruction requiring them, the **execute cycle** is different for each instruction, because that cycle depends on what the instruction is intended to do. Let's start by completing the **MOV A,[A254H]** instruction in the preceding section. To do so, we'll assume that location A254H has the data 7CH in it.

In the first phase of the execute cycle of the **MOV A,[A254H]** instruction, the contents of the temporary register are transferred to the address buffer by asserting ET and LADD as shown in Figure 8.9, allowing the operand location to be accessed. Figure 8.10 shows the second and last phase of the execute cycle for the **MOV A,[A254H]** instruction. The memory data buffer is enabled when CE is asserted, placing the data, 7CH, from location A254H, onto the data bus. The data are then loaded into the A-register by asserting the LACC control

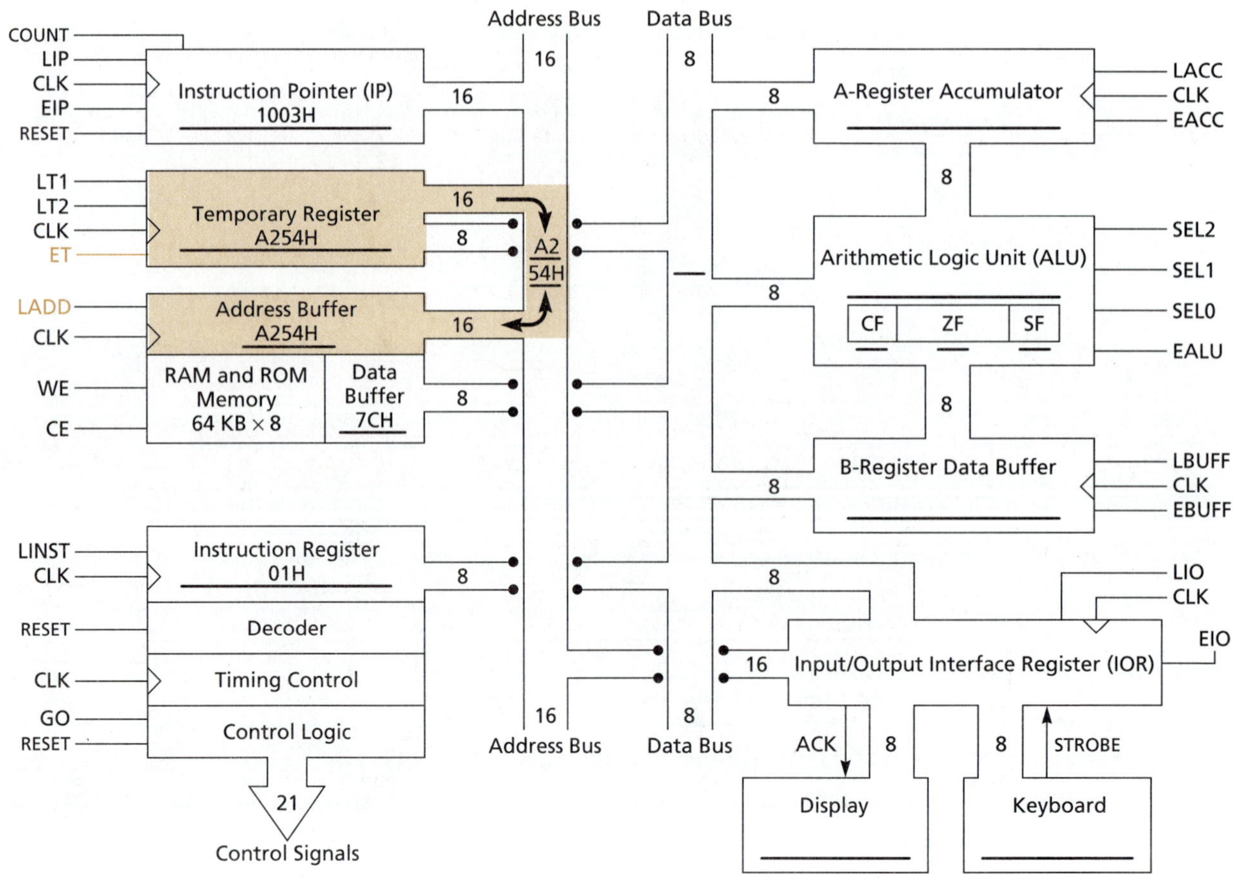

BABE Computer Functional Blocks

Figure 8.9 **MOV A,[<mem>] Execute Cycle Phase 1**

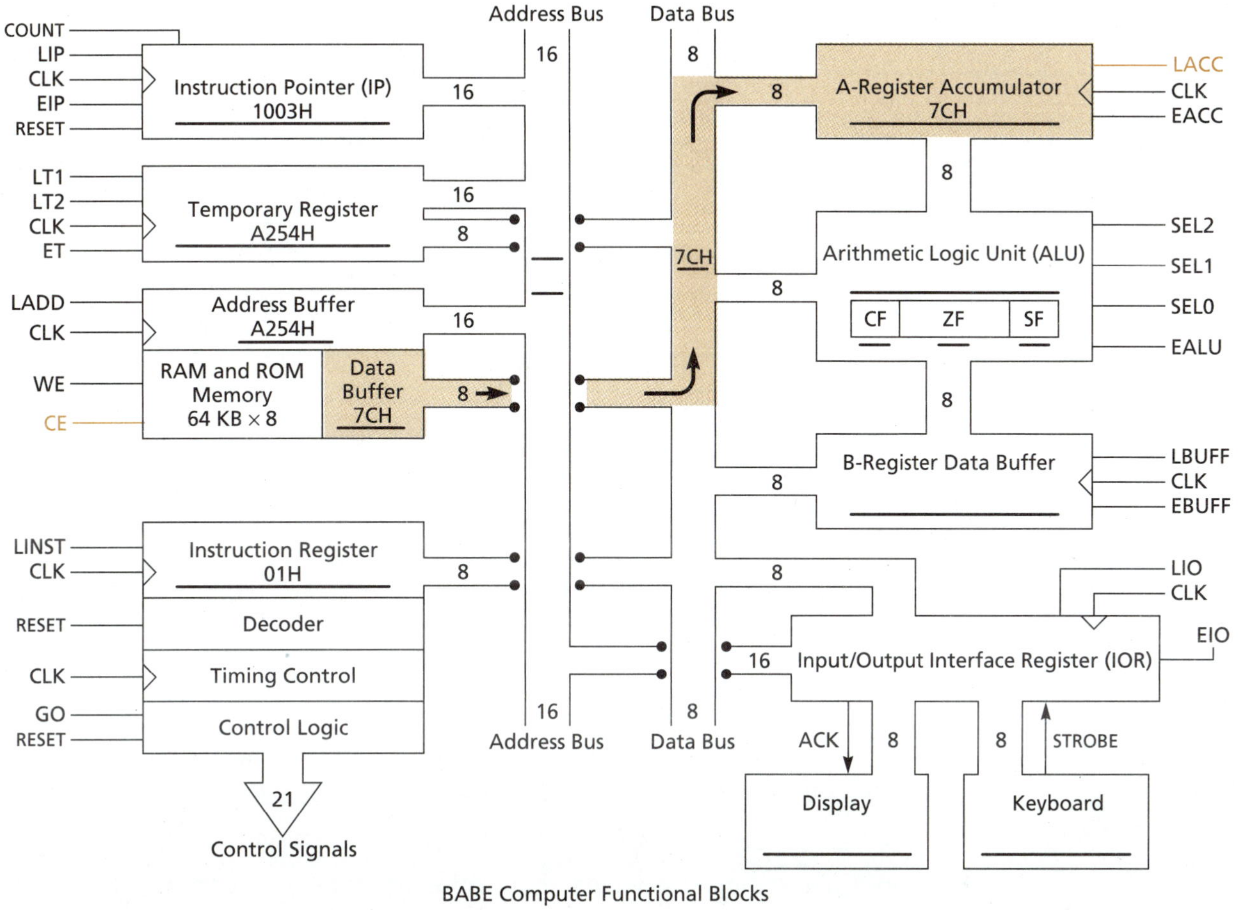

Figure 8.10 **MOV A,[<mem>]** Execute Cycle Phase 2

signal. The instruction is now completed, and the computer will automatically begin the fetch cycle for the next instruction. Table 8.3 summarizes the sequence of fetching and executing the **MOV A,[A254H]** instruction: To contrast execute cycle sequence for different instruction types, we next consider the sequence used to execute an **ADD B** instruction. It is assumed that the A- and B-registers contain data previously fetched from memory. The **ADD B** opcode is fetched from the program list in memory, as before, and loaded into the instruction register using the same opcode fetch sequence:

```
IP —> ADD BUFF
MEM —> INST REG
Increment IP
```

When the instruction decoder and the control logic analyze this opcode, they determine that no operand is required and direct the computer to start the instruction's execute cycle. Since numbers are already in the A- and B-registers, their sum appears in the arithmetic logic unit (ALU). Think of this section of the ALU as a two's complement adder/subtracter, and recall that once numbers

Table 8.3 MOV A,[<mem>] Sequence

ACTIVITY	MEANING
IP -> ADD BUFF	;Begin opcode fetch.
MEM -> INST REG	;Opcode to instruction ;register.
Increment IP	;End opcode fetch.
IP -> ADD BUFF	;Begin operand fetch.
MEM -> TEMPORARY LOW	
Increment IP	
IP -> ADD BUFF	
MEM -> TEMPORARY HIGH	
Increment IP	;Complete operand fetch.
TEMP -> ADD BUFF	;Begin execute cycle.
MEM -> A-REGISTER	;Complete instruction.

are placed on the A and B inputs of that circuit, the sum appears at its outputs. The execute cycle of an ADD B instruction is only required to take the sum output of the ALU and place it in the A-register. To do this, the correct input logic levels, which are 000B, are set on SEL_2, SEL_1, and SEL_0 inputs of the ALU to select the add operation. Control signal EALU is asserted to enable the sum outputs of the ALU onto the data bus. LACC is asserted, causing the A-register to be loaded from the data bus with the output word from the ALU. The complete execute cycle for the ADD B instruction is summarized as

```
ALU (SEL2, SEL1, SEL0 = 000B) -> A-Register
```

Since there are fewer steps in the sequence of the **ADD B** instruction, it is completed in less time than the **MOV A,[A254H]** instruction.

EXAMPLE 8.4

Compare the sequence required to add the contents of two memory locations using the **ADD [<mem>]** and **ADD B** instructions.

SOLUTION To use the **ADD [<mem>]** instruction, the A-register must first be loaded with data, whereas to use the **ADD B** instruction, both A- and B-registers must first be loaded with data. The instruction sequences to perform both additions are

ADD [<mem>]	ADD B
MOV A,[<mem>]	MOV A, [<mem>]
ADD [<mem>]	MOV B,[<mem>]
	ADD B

The sequence of events to execute each set of instructions is

FIRST INSTRUCTION OPCODE FETCH CYCLES:

ADD [<mem>]	**ADD B**
IP —> ADD BUFF	IP —> ADD BUFF
MEM —> INST REG	MEM —> INST REG
INC IP	INC IP
FIRST OPERAND FETCH CYCLES:	
IP —> ADD BUFF	IP —> ADD BUFF
MEM —> TEMP LOW	MEM —> TEMP LOW
INC IP	INC IP
IP —> ADD BUFF	IP —> ADD BUFF
MEM —> TEMP HIGH	MEM —> TEMP HIGH
INC IP	INC IP
FIRST EXECUTE CYCLES:	
TEMP —> ADD BUFF	TEMP —> ADD BUFF
MEM —> A-Register	MEM —> A-Register
SECOND INSTRUCTION FETCH CYCLES:	
IP —> ADD BUFF	IP —> ADD BUFF
MEM —> INST REG	MEM —> INST REG
INC IP	INC IP
SECOND OPERAND FETCH CYCLES:	
IP —> ADD BUFF	IP —> ADD BUFF
MEM —> TEMP LOW	MEM —> TEMP LOW
INC IP	INC IP
IP —> ADD BUFF	IP —> ADD BUFF
MEM —> TEMP HIGH	MEM —> TEMP HIGH
INC IP	INC IP
SECOND EXECUTE CYCLES:	
TEMP —> ADD BUFF	TEMP —> ADD BUFF
MEM —> B-Register	MEM —> B-Register
SEL_2, SEL_1, SEL_0 = 000B	
ALU —> A-Register	
THIRD OPCODE FETCH CYCLE:	
None	IP —> ADD BUFF
	MEM —> INST REG
	INC IP
THIRD EXECUTE CYCLE:	
None	SEL_2, SEL_1, SEL_0 = 000B
	ALU —> A-Register

Section 8.5

Practice Problems

1. Write out the sequence of events required to subtract the data in location 1234H from the data in location 1210H. The states of SEL_2, SEL_1, and SEL_0, for a subtract operation, are 001B.
2. If each step in the process of executing an instruction takes one clock cycle, compare the total time it takes to fetch and execute both sequences in Example 8.4 for a system using a 5 MHz clock.

8.6 The Remaining BABE Computer Instructions

Table 8.4 is the complete instruction set for the BABE computer shown with instructions organized in operational groups, specified by the binary code of the upper nibble of each opcode. The first two groups, the MOV, ADD, and SUB instructions were previously discussed in Section 8.5. The logic group, each with an upper nibble of 2H, is made up of shift, decrement, and increment instructions. The shift instructions operate on the A-register only and cause the binary bits in that register to shift one position, left or right, depending on the direction indicated by the shift instruction. Each time a shift instruction is fetched and executed, the bits in the A register are shifted one binary place. For example, if the A-register contains 6CH or 01101100B before a shift left (**SHL**), it will contain D8H or 1101100B after the **SHL** is done. The **SHL** instruction causes

Table 8.4 BABE Instruction Set

GROUP	MNEMONIC	OPCODE	FLAGS
Data Move (0X)	MOV A, [<mem>]	01	None
	MOV B, [<mem>]	02	None
	MOV A, <immed>	03	None
	MOV B, <immed>	04	None
	MOV [<mem>],A	05	None
	MOV [<mem>],B	06	None
Add (1X) and Sub (4X)	ADD [<mem>]	11	All
	SUB [<mem>]	41	All
	ADD <immed>	13	All
	SUB <immed>	43	All
	ADD B	17	All
	SUB B	47	All
Logic (2X)	SHL	28	All
	SHR	29	All
	DEC A	2A	ZF
	DEC B	2C	ZF
	INC A	2D	ZF
	INC B	2E	ZF
I/O (3X & 5X)	IN	30	None
	OUT	5F	None
Program Flow (X1)	JMP <mem>	81	None
	JZ <mem>	91	ZF = 1
	JNZ <mem>	A1	ZF = 0
	JM <mem>	B1	SF = 1
	JP <mem>	C1	SF = 0
	JC <mem>	D1	CF = 1
	JNC <mem>	E1	CF = 0
Stop (6)	HALT	6F	None

the binary bits to be shifted left once, toward the most significant bit (MSB) and a 0 is shifted into the least significant bit (LSB). Therefore, 0110 1100B shifted left once becomes 1101 1000B or D8H after it has been shifted left once. Shifting 6CH right one bit using the **SHR** instruction produces this result:

```
6CH = 0110 1100B <SHR> 0011 0110B = 36H
```

Shifting a binary word left once has the effect of doubling the binary value, whereas shifting right effects a divide-by-2 result. In the examples used in the previous paragraph, for the **SHL** instruction 6CH or $108 \times 2 = 216$ or D8H, and for the **SHR** instruction $6CH \div 2 = 36H$.

Increment (**INC**) and decrement (**DEC**) instructions increase or decrease a register by 1. If the A-register has 3BH and an **INC A** instruction is executed, the A-register contents will be incremented to 3CH. Similarly, a **DEC A** instruction causes the results to reduce by one or to 3AH using the same initial sample data.

EXAMPLE 8.5

Using the same original data in the A- and B-registers for each instruction, determine the contents of each register following the execution of these instructions:

Original data: A-register = 4AH B-register = 57H
Instructions: **INC A SHL DEC B SHR**

SOLUTION **INC A** causes the A-register to increment by 1, to 4BH.
SHL shifts the A-register left once, to 94H.
DEC B decreases B-register by one, to 56H.
SHR shifts the A-register right once, to 25H.

Two instructions, **IN** and **OUT,** pass data back and forth between peripheral devices connected to the input/output interface register (IOR) and the computer's central processing unit (CPU). The **IN** instruction takes data from the keyboard, through IOR, and stores them into the A-register. The **OUT** instruction transfer data from the A-register to IOR, which is then passed onto the display unit.

A final group of instructions alters the flow of program execution. Up to this point, one instruction followed another sequentially and the instruction pointer (IP) easily kept track of the instruction flow, since it was incremented each time a fetch cycle was performed. The jump group of instructions is designed to alter the contents of the IP so that the next instruction is fetched from a new memory location. The purpose behind the use of the jump group of instruction is to cause a new set of instructions to be executed. The decision to do this may be unconditional, meaning it must be done without question, or conditionally, requiring some test to be made before the change of flow or jump is performed. If the test fails, the jump is not taken and the program continues normal execution. To understand how this works, consider an unconditional jump instruction sequence stored at location 1040H.

1040H 81H
1041H 78H
1042H 03H

The instruction is **JMP 0378H** and says to jump or "GOTO" location 0378H immediately. The sequence begins by fetching the opcode from location 1040H, which is initially pointed to by the contents of the instruction pointer. This moves opcode 81H into the instruction register and increments the IP to 1041H. The instruction decoder interprets the code 81H as an unconditional jump instruction that requires two operand fetch cycles to get the address to jump to. As the data for each operand are fetched, they are stored into the temporary register and the IP is incremented to 1043H. Figure 8.11 illustrates the last phase of the operand fetch cycle for the JMP instruction.

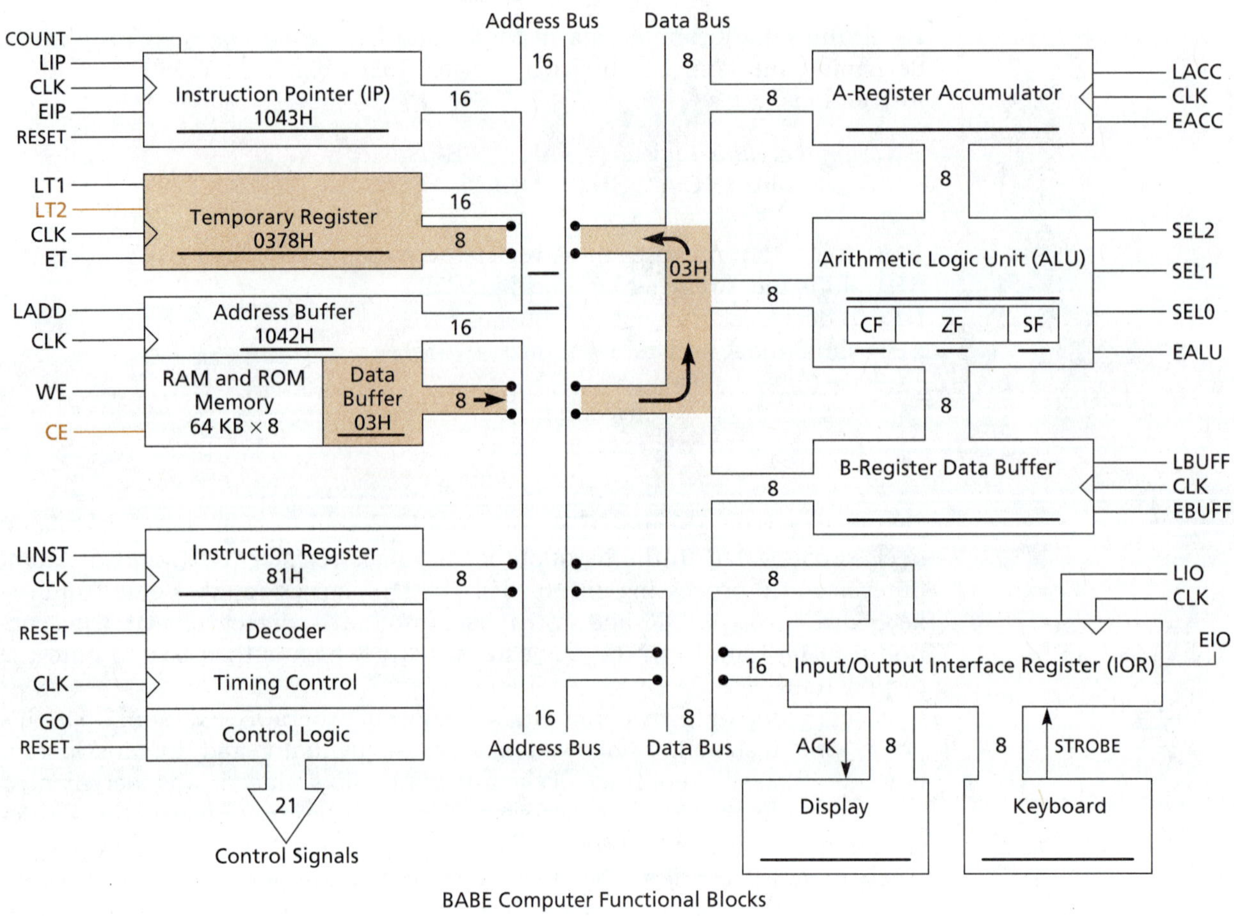

Figure 8.11 **JMP** Operand Fetch Cycle 2

The execute cycle of the jump instruction, shown in Figure 8.12, enables the temporary register outputs and loads the IP with 0378H. On the next opcode fetch cycle, the IP uses this new value to fetch the next instruction form location 0378H instead of 1043H. A change in program flow has been made. Additional jump instructions on Table 8.4 indicate change of program flow only if a particular condition or test is met. These jump instructions are called **conditional jumps.**

Conditional jump *an instruction that causes a change in flow of a program if a specific condition is met.*

Flag *flip-flop that reflects a status condition of an operation's result.*

Logic within the arithmetic logic unit tests the results of specific instructions and sets or resets flip-flops, called condition **flags,** that indicate the conditions of these results as follows:

ZF Zero flag: when set indicates the results are zero and when reset indicates the results are nonzero.

CF Carry flag: indicates if a carry resulting from addition, or a borrow during subtraction, has occurred. Lack of a carry or borrow resets the carry flag.

SF Sign flag: a set state indicates the results are a negative value; reset indicates a positive value.

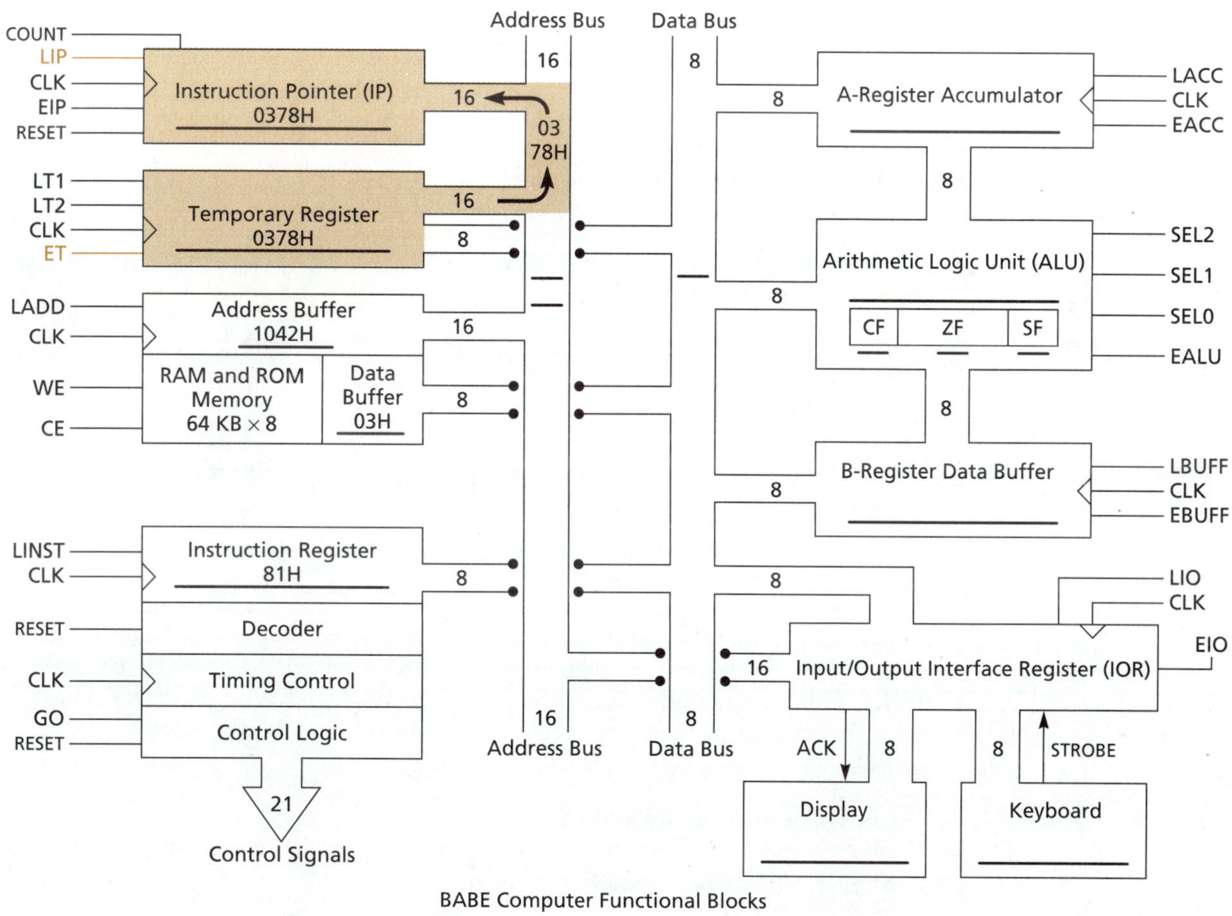

Figure 8.12 **JMP** Execute Cycle

On Table 8.4, the rightmost column indicates which flags are set or reset by each instruction. Notice that the MOV, stop, and I/O groups neither have an effect on the state of the flags nor does their operation depend on the state of the flags. In contrast, all the flags are set or reset by the result of executing one of the arithmetic group instructions. For instance, adding 1DH to E3H, using one of the ADD instructions, results in an answer of 100H, which is too big to fit into the eight-bit A-register. Consequently, the lower eight bits, 00H, are placed into the A-register and the upper 1H sets **CF** to a 1, indicating that a carry or overflow beyond the eight-bit capacity of the A-register resulted from the addition. Since all eight bits of the sum that remain in the A-register are low or 00H, the **ZF** will also be set to indicate the zero content of the A-register. The most significant bit of the A-register contents is a zero, signifying that the value in the A-register is positive, which causes **SF** to be reset.

EXAMPLE 8.6

What are the states of the flags for each of the following arithmetic operations?

78H + 3CH   9BH + E2H   CDH − 3BH   6BH − 99H

SOLUTION In this first case, 78H + 3CH = B4H. The states of the flags are

CF = 0, indicating no carry
ZF = 0, indicating nonzero
SF = 1, indicating a negative number (MSB = 1)

In the second example, the sum is 9BH + E2H = 17DH, which causes the flags to take these states:

CF = 1, showing the carry beyond 8 bits
ZF = 0, indicating nonzero results
SF = 0, indicating a positive number (MSB = 0)

In the third case, the difference is CDH − 3BH = 91H and the flags are

CF = 0, since there was no borrow
ZF = 0, indicating nonzero results
SF = 1, indicating a negative number

In the last case, 99H is larger than 6BH, so a borrow is required to perform the subtraction. Imagine the borrow as a 1 included with the 6BH, to make it 16BH. This value cannot be held by an eight-bit register, so the carry flag (CF) is set to indicate the need to borrow. The difference is 16BH − 99H = D2H. To summarize the flag states for this last case,

CF = 1, indicating a need to borrow
ZF = 0, indicating nonzero results
SF = 1, signifying a negative number

For the jump group, the righthand column in Table 8.4 indicates the conditional state of the flag required to cause the jump instruction to change program flow. If the condition is not met then the next instruction is fetched and executed as if the jump instruction were not included in the program. The conditional jump instructions and their definitions are

JZ jump if zero (ZF = 1)
JNZ jump if not zero (ZF = 0)
JC jump if carry (CF = 1)
JNC jump if no carry (CF = 0)
JP jump if positive (SF = 0)
JM jump if minus (SF = 1)

The conditional jump instructions are used by programmers to allow the computer to make decisions based on the result of previous instructions. As an example, we shall write a program sequence that subtracts two numbers, and if the results are positive, store the results in memory. If the results are negative, send the results to the display. The program sequence to perform the function that directs the computer to decide where to send the results of the subtraction is

```
            MOV A,<mem-1>          * Get one number.
            MOV B,<mem-2>          * Get second number.
            SUB B                  * Do the subtraction.
            JM  <mem-3>            * Is the result negative?
            MOV <mem-4>,A          * Store positive value in
                                     memory.
<mem-5>          .
                 .                 * Other instructions.
                 .
<mem-3>     OUT                    * Output negative value.
            JMP <mem-5>            * Enter program to continue.

note: <mem-number> represents an actual memory address.
```

The first three instructions fetch the data and do the subtraction. The JM instruction asks the question, "Are the results negative?" by jumping if the sign flag, SF, is set. If it is, "GOTO" or jump to location mem-3. If SF is not set, continue the program to the next instruction (**MOV <mem-4>,A**). After this MOV instruction, other program instructions may be present, for additional program functions, starting at mem-5. If the jump is taken, the OUT instruction is fetched and executed to send the negative results to the display, followed by an unconditional jump instruction that returns the program flow to those additional instructions starting at mem-5.

Most microprocessor devices have their flags arranged into a status, flag, or condition code register. This register appears on a microprocessor block diagram as a separate block. The BABE computer uses so few flags that they are included as an internal three-bit register within the arithmetic logic unit (ALU) instead of being shown as a separate register.

Section 8.6

Practice Problems

1. What are the contents of the A-register and the B-register after each of the following instructions is executed? Use original data 5CH for each instruction:
   a. **SHR** b. **ADD D2H** c. **INC A** d. **SHL**

2. Original contents of the registers and memory are

   A = B4H B = 09H 1430H = 9CH

   What are the contents of the A-register after each of these instructions are executed? Use the original data for each example.
   a. **MOV A,[1430H]** b. **ADD [1430H]** c. **SUB B** d. **DEC A**

3. Initially, all three flags are reset to 0 before the execution of any instruction in Practice Problem 2. What is the state of the flags after each instruction in Practice Problem 2 is executed?

8.7 Writing Source Code Programs

There are many approaches to constructing a program. Writing programs at the source code (mnemonic) level presents an interesting twist to program writing. Upper-level languages such as BASIC, PASCAL, or C contain instructions and operations that cause several actions to take place. They are also formalized into a format that lends itself to programming in structured or block form. Source code programs, in contrast, tend to be free form, although they can be somewhat structured. The limitations imposed by the abbreviated instruction set used by the BABE computer precludes the use of structured programming. The intention here is to present a correlation between a program and its execution using the BABE computer architecture.

In the previous section, we saw how a program segment was used to illustrate the use of a conditional jump instruction. This segment subtracted two numbers and directed the difference to memory if it were positive and to the display if it were negative. For the first complete program that we will do, the computer is directed to subtract two numbers and if the result is positive, the result is sent to the display. If the result is negative, then perform the two's complement on the result and sent that to the display. The first step is to organize the flow of the program. This may be done graphically using any number of standard (or nonstandard) flow charts. Basically, a block is drawn for each major function the computer is to perform. The blocks are connected to show the flow from one block to another until the program is finished. This is used to help organize your thoughts about how the program is expected to complete a given task. We would start with a block to indicate the fetching of the data.

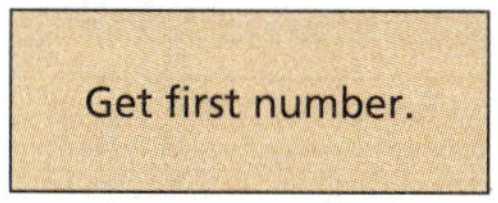

The next block indicates the subtraction process and is connected to the first block by a straight line.

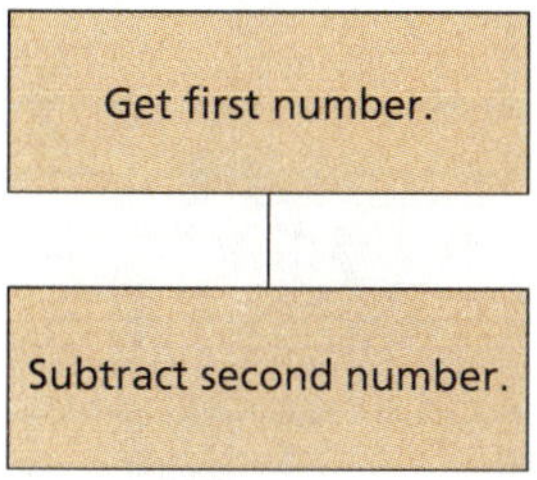

In order to continue with what the program needs to do, a decision must be made. If the difference is positive, send the results to the display and, if negative, take the two's complement and send that to the display. The question asked at this point is either "Is the number positive?" or "Is the number negative?" Either one will work, and neither in this case is preferred. So we will select the negative question. The question box is a diamond and flows from the preceding block.

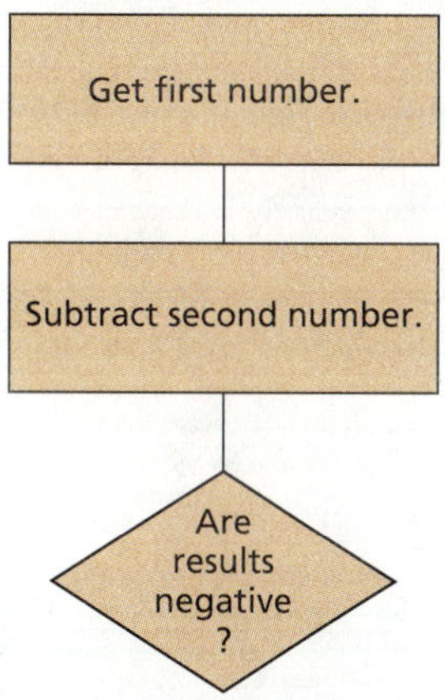

There are two possible answers: Yes, the results are negative, or no, the results are positive. If the answer is no, then the program direction continues as it was flowing, down in the flow chart. If the answer is yes, then the program flow changes direction. These flows are indicated by lines from the diamond.

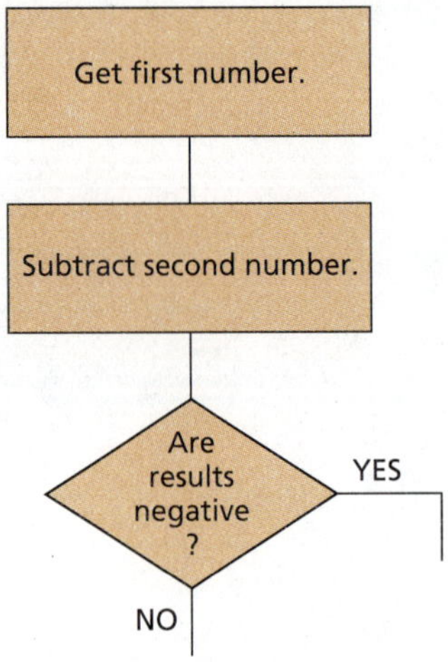

Following the "no" program flow, the next process is to send the positive value to the display. Input and output functions are shown in a rectangle that has leaned over (parallelogram):

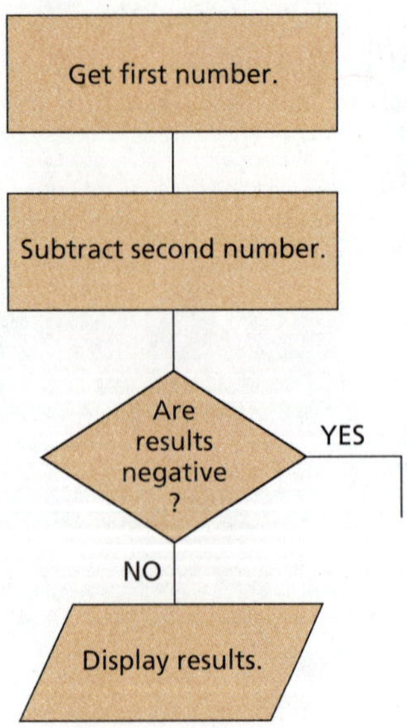

And then stop the program:

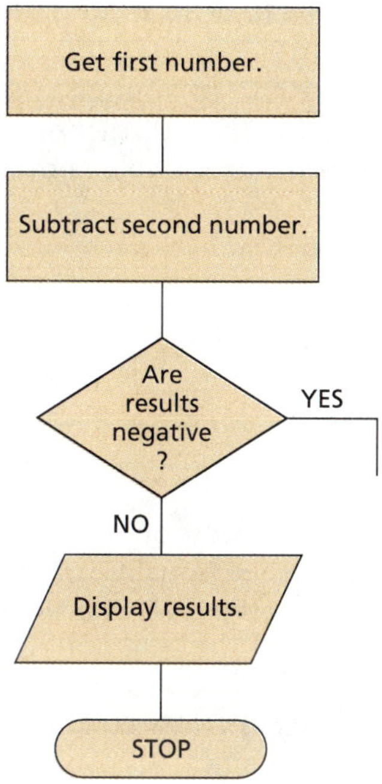

What remains is to chart the "yes" program flow. This requires the number to be two's-complemented and then displayed. The complement process will take one block, after which we will return to the main flow to the display block.

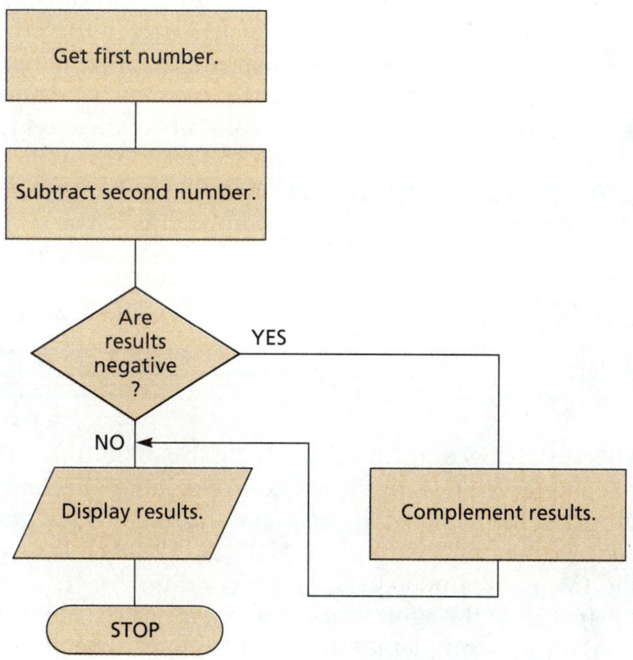

The next step is to translate the flow diagram into an actual program using the instruction set of the BABE computer. For purposes of clarification, let's assign memory location 1200H to contain the first number and location 1201H, the other number for the subtraction function. The first block fetches the first number and translates into a move from memory to the A-register instruction:

```
MOV A,[1200H]
```

The second block indicates the performance of the subtract instruction following the MOV instruction:

```
MOV A,[1200H]
SUB [1201H]
```

Now that the subtraction is done, a decision must be made regarding the sign of the difference. This is accomplished by using a conditional jump instruction. From Table 8.4, we determine that the SUB instruction sets or resets the sign flag (SF) according to the results of the subtraction process. In our example, the question being asked is "Are the results negative?" In terms of the sign flag, this is the same as asking, "Is SF set?" The conditional jump instruction that poses this question is **JM <mem>** since, according to Table 8.4, the jump is taken if the SF is set. For purposes of illustration, let's use location 2400H as the one jumped to if the number is negative. Adding the JM instruction to the program increases the number of instructions to three:

```
MOV A,[1200H]
SUB [1201H]
JM  2400H
```

If the question is answered no, indicating that the number is positive, then the jump is not taken and the next instruction in the program is fetched and executed. According to our flow diagram, that instruction causes the results to be displayed. The instruction that will transfer the results to the display unit is an OUT instruction that moves the contents of the A register to the output portion of the I/O interface. This operation is followed by the instruction HALT, which stops execution of the program. Extending the program to include the OUT and HALT instructions makes it now appear as

```
MOV A,[1200H]
SUB [1201H]
JM  2400H
OUT
HALT
```

This is the flow of the program from top to bottom if the jump instruction does not change the flow of the program execution. For purposes of illustration, this program list is placed in memory starting at location 2300H. That is, the opcode of the first instruction, **MOV A,[1200H],** is stored at location 2300H. If the JM instruction experiences a condition of "true"—that is, the SF is set by the results of the subtraction being a negative value—then the alternate operation of two's-complementing the results is to be performed. The set of instructions to do this starts at location 2400H, where the JM instruction directed the instruction pointer to find them. A number is complemented by first subtracting it from FFH and then adding 1H. Currently, at this point in the program, the difference is in the A-register resulting from the subtraction operation. In order to subtract it from FFH, that number has to be placed into the B-register and FFH into the A-register. This is done by moving the contents of the A-register to memory, loading the A-register with FFH, and then doing the subtraction using the memory location where the original difference was just stored. Finally, 1H is added to the difference of FFH and the original negative number. The instructions to do all this are

```
MOV [1202H],A   *Save the contents of A-register.
MOV A,FFH       *Load the A-register with FFH.
SUB [1202H]     *Subtract the negative number from the FFH.
ADD 1H          *Complete the two's complement.
```

The only thing remaining to be done is to display the results and stop the program execution. Since there already are OUT and HALT instructions, we will use an unconditional jump to return to the main program at this point. The difficulty is to know what address to use with the JMP instruction. This can be determined by examining the main program flow and seeing what address is used to hold the opcode for the OUT instruction. To do this, we must first translate the source code into machine code starting at location 2300H. It is not actually necessary to translate each opcode into hex, because the only thing we are interested in is the number of locations used up until the opcode for the OUT instruction. We do need to convert enough of the source listing to object listing to determine the OUT opcode location.

We start with the first instruction whose opcode resides at location 2300H. It has two operand bytes that hold the address for the source of the MOV instruction for a total of three bytes. The two operand bytes occupy the next two memory locations in the program list, 2301H and 2302H, leaving the next location 2303H to hold the opcode byte of the SUB instruction. The program can be rewritten showing the locations for each instruction's opcode. The number of bytes required to represent each instruction and their operands are used to determine the actual locations of all the instruction opcodes:

```
2300H MOV A,[1200H]    * 3-byte instruction
2303H SUB [1201H]      * 3-byte
2306H JM  2400H        * 3-byte
2309H OUT              * 1-byte
230AH HALT             * 1-byte
```

Since each of the first three instructions has two byte operands, they each use three locations (one for the opcode byte and two for the two byte operands). Notice that the OUT instruction has no operand and requires the use of a single memory location. To complete the negative branch portion of the program, we now have an address for the OUT instruction's opcode, which is 2309H. This becomes the operand value for the unconditional jump needed to cause program execution to return to the correct place, the OUT instruction. The completed negative branch program is

```
2400H MOV [1202H],A     * Save the contents of A-register.
2403H MOV A,FFH         * Load the A-register with FFH.
2405H SUB [1202H]       * Subtract the negative number from
                          FFH.
2408H ADD 1H            * Complete the two's complement.
240AH JMP 2309H         * Return to OUT instruction.
```

Also included in the negative branch portion of the program listing are the addresses for each of the opcodes, starting with 2400H for the MOV instruction. Note that instructions that use immediate data only require two bytes—one for the opcode and one for the byte of data.

A second example will help to illustrate the process a little more. This program multiplies the two numbers in locations 1200H and 1201H and stores the results in locations 1202H and 1203H. The reason for requiring two locations to store the product is that when two bytes are multiplied, their product is a 16-bit or word-size value that requires two 8-bit memory locations to hold its value. The first part of the problem is to figure out how to multiply two numbers when there is no multiply instruction. The answer is using successive additions. If you add the number 6 to itself five times, the sum is the same as the product of 6×5.

The flow diagram for the program is shown in Figure 8.13. It starts by setting the accumulator and location 1202H to zero, since the beginning product before the program is run is nothing. To the zero in the accumulator, add the number at location 1200H (first number). Check to see if a carry was set by this addition. Of course, adding anything to 0 will reset the CF, because there will be no carry. Since there isn't a carry, decrease the second number in location 1201H by 1, to signify that the first number was added once.

As long as we are not multiplying by 1, there will be a nonzero result from decreasing the second number by 1. This causes the second question "Is the second number nonzero?" to be answered yes and returns program flow to the ADD instruction, adding the first number to the accumulator a second time. This time, a carry may or may not occur. If it does occur and sets CF, then the first jump is taken and location 1202H is increased by 1 to accumulate the number of carry occurrences.

The second number is decreased again, since another addition has been done. As long as the second number is not yet zero, the process will continue. When the second number reaches zero, the last process is performed. This stores the accumulated result from the A-register into memory at 1203H and stops the program. Locations 1202H, which holds the accumulated carries from the additions, and 1203H, which now has the accumulator final results, contain the 16-bit product of the two 8-bit numbers. To satisfy yourself that this works, take any two numbers and run them through the program flow diagram. The results in 1202H and 1203H will be the product of those two numbers.

When a program segment repeats itself a number of times, it is said to be in a **loop**. When the condition is satisfied that causes the program to stop repeating, it is said to "fall out of the loop." There is nothing standard or sacred about the looping terminology, but it is common practice. For the multiplication program, as long as the second number is not zero, the program repeats staying in the loop. When the second number finally reaches zero, program execution falls out of the loop and continues with the next instruction.

Loop *a set of instructions that are repeatedly executed a given number of times.*

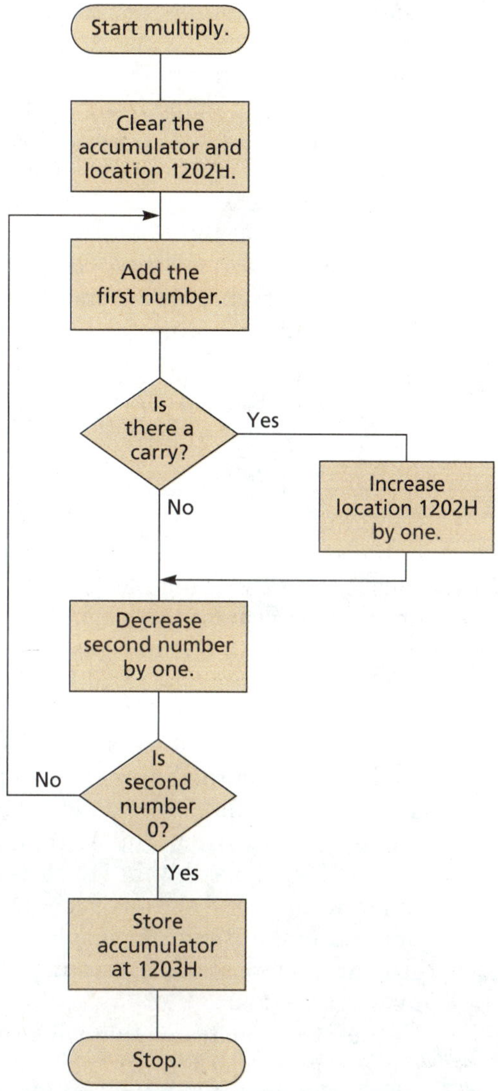

Figure 8.13 Multiply Flow Chart

In some respects, the flow diagram looks deceptively easy, but, due to limitations in the BABE instruction set it is going to require a lot of data manipulation to get each process to work. Consider the first process block, which says to clear the accumulator, which will hold the lower byte of the product, and memory location 1202H, which will hold the accumulated carry results that form the upper byte of the product. What is desired is to set both the A-register and location 1202H to 0. This is accomplished by moving 0 into the A-register and then storing the contents of the A-register (now 0) into memory location 1202H:

```
MOV A,OH
MOV [1202H],A
```

Adding the first number to the accumulator is straightforward, as is the use of the JC (jump if carry) instruction. The only difficulty is not knowing what exact address to jump to if CF is set. One way to handle this problem is to temporarily assign a name to the memory location, which will be replaced by the actual address when it is known. A name that identifies an address location is called *label* and can be used anywhere in a program in place of the actual address value. We will use a label, called MSB for most significant byte, to define the location that the JC instruction jumps to. A second label, called SUM, defines the location for the opcode of the ADD instruction, that is used by the other jump instruction operand to allow return of program execution to the ADD instruction within the loop:

```
    MOV A,OH
    MOV [1202H],A
SUM ADD A,[1200H]
    JC MSB
```

Continuing with the main program execution flow, resulting from the answer to the JC instruction of "the carry flag is not set," we encounter a block that calls for decrementing the second number, which is held in location 1201H. In order to decrement the second number, the program will fetch that number from location 1201H and place it into the B-register, then decrement the B-register, and return the results back to location 1201H. Following this process, a test is made to check if the second number is zero or not to determine if there are more additions to perform or not. The reason that the number has to be fetched and returned to memory is that the B-register is used by the ADD instruction. Recall that the adder circuit in the ALU block sums the numbers in the A- and B-register. The ADD [1200H] instruction fetches the data from location 1200H, places those data into the B-register, and then the sum is transferred to the A-register. As such, leaving the decremented result of the second number in the B-register would cause that value to be lost when the ADD instruction is repeated. As a result, that value is saved back to memory before the loop is repeated.

The ZF can be tested following the move that returned the count value from the B-register back to memory, because MOV instructions do not alter the states of the flags. In this case, the flags reflect the result of decrementing the contents of the B-register. Our program list continues on with the decrementing of the second number:

```
    MOV A,OH
    MOV [1202H],A
SUM ADD A,[1200H]
    JC MSB
DWN MOV B,[1201H]
    DEC B
    MOV [1201H],B
    JNZ SUM
```

The zero flag (ZF) is tested to see if the results of decrementing the number at location 1201H is zero or not. If it is not zero, execution is returned to the ADD instruction at location SUM. If it is zero, then the program falls out of the loop, stores the accumulated results in the A-register at location 1203H, and stops the program:

```
     MOV A,OH
     MOV [1202H],A
SUM  ADD A,[1200H]
     JC MSB
DWN  MOV B,[1201H]
     DEC B
     MOV [1201H],B
     JNZ SUM
     MOV [1203H],A
     HALT
```

The only part of the program remaining to be written is the carry branch. Once again, the B-register will be used, so the contents of the A-register are not disturbed. The same difficulty arises that developed when decrementing the number at 1201H. First the contents of location 1202H, used to hold the accumulated carries, are to be moved into the B-register, the B-register incremented once, and the results stored back into location 1202H. Once this is accomplished, program flow is returned to location DWN using an unconditional jump instruction. DWN identifies the location of the **MOV B,[1201H]** opcode and is used to return the execution of the program, through a **JMP DWN** instruction, back to checking if more additions are to be performed:

```
MSB  MOV B,[1202H]
     INC B
     MOV [1202H],B
     JMP DWN
```

Putting all the sections together and correctly commenting the program produces this source code:

```
* This program multiplies one number with another
* and stores the product in two successive locations.
*
* Initialize accumulator, and carry location to 0.
*
      MOV A,OH
      MOV [1202H],A
*
* Begin multiply routine, using successive additions.
*
SUM   ADD A,[1200H]
      JC MSB                 * check for carry
*
* Get second number, decrement, and check for 0.
*
DWN   MOV B,[1201H]
      DEC B
      MOV [1201H],B
      JNZ SUM
```

```
*
* Multiplication complete, store lower byte.
*
      MOV [1203H],A
      HALT
*
* Accumulate carries from addition as product MSB.
*
MSB   MOV B,[1202H]
      INC B
      MOV   [1202H],B
      JMP DWN                * Return to sum loop.
```

Full-line comments in a source program are usually preceded by a delimiter character, in this case, an asterisk (*). These comments are used to briefly describe the function of a block of code. The comment field on an instruction line, as used with the JC MSB and JMP DWN instructions are used to clarify a specific instruction's use. A source program should be commented sufficiently for someone to be able to understand what the program is doing. On the other hand, it should not be so overcommented that it becomes cumbersome to read.

Section 8.7

Practice Problems

1. Create a flow chart for a program that inputs two numbers from the keyboard and stores them in two locations. The program then adds the two numbers, stores the results in memory, and sends the results to the display.
2. Write the source code for Practice Problem 1, using the labels NUMB1, NUMB2, and SUM to define memory locations for data use. No jump instructions are required for this program.

8.8 BABE Computer Hardware

The actual circuitry that comes together in the BABE computer is created from numerous digital circuits that have been discussed in previous chapters throughout this text. The circuit for the first functional block, the instruction pointer, shown in Figure 8.14, uses the 74LS193 binary counter. This chip is a four-bit parallel-load ripple up/down counter. Four 193s are connected in cascade by wiring terminal count up (C_0) of one 193 to the up clock input of the next counter. The address bus of the computer is connected to the load inputs and the outputs of two 74LS244 three-state buffer chips. The starting count of the counter can be loaded into the 193s by placing that value on the address bus and asserting LIP, which turns on the load input of the counters. EIP signal must remain inactive when this is done. The counter counts up when the signal labeled COUNT is set high and passes the CLK to the UP clock input of the first

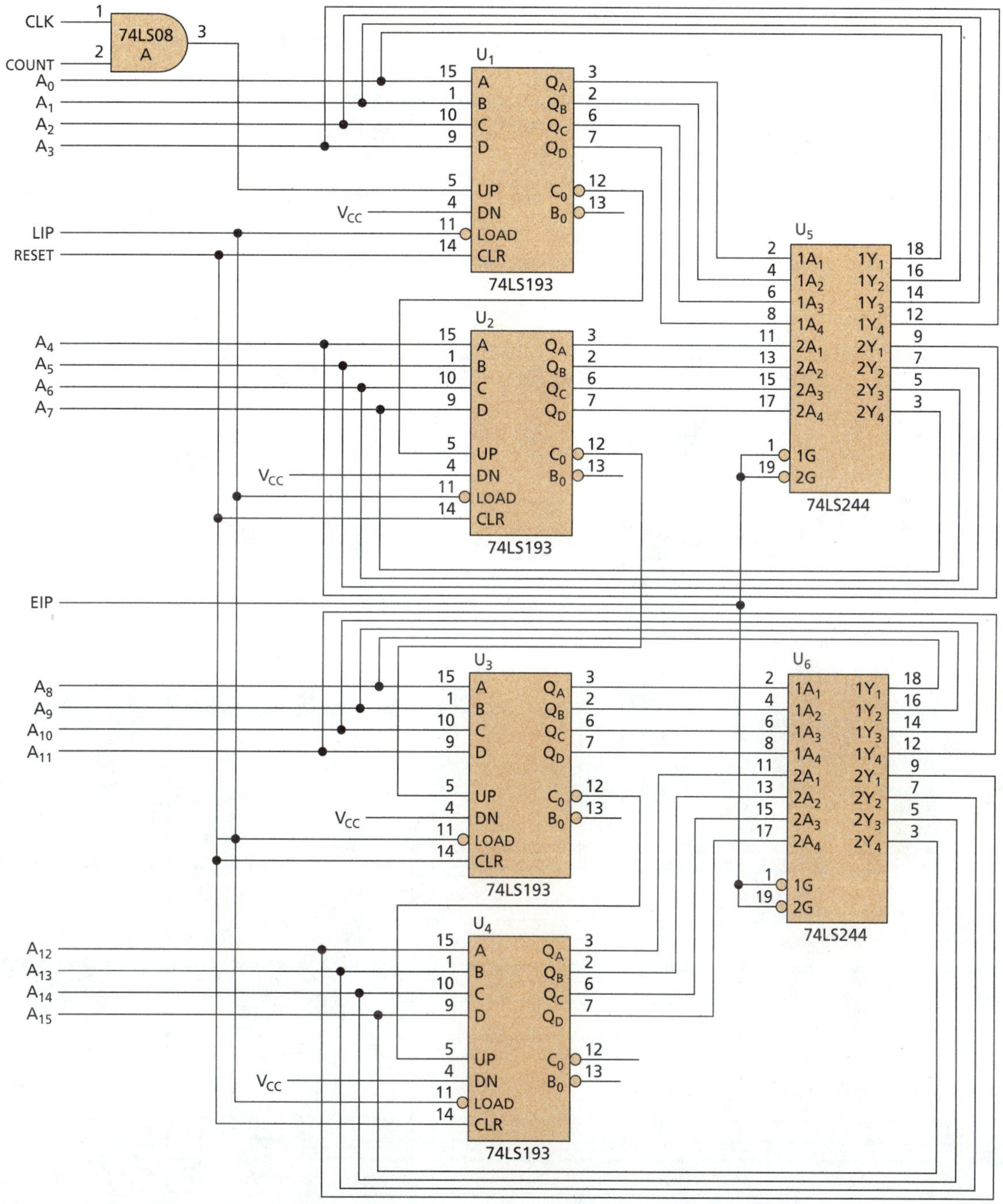

Figure 8.14 Instruction Pointer (IP)

74LS193. The instruction pointer is cleared to 0000H by using the RESET input signal. Instruction pointer outputs are placed onto the address bus when EIP is asserted and LIP is negated. The EIP signal enables the outputs of the three-state buffer chips (74LS244), which are connected to the address bus.

The temporary register is used to temporarily store both 8-bit halves of a 16-bit address operand before forming a full 16-bit address, which is then transferred to the address buffer or instruction pointer. It is also used to temporarily hold immediate data fetched as an instruction operand. The circuit, shown in Figure 8.15, has four 74LS173 quad-D flip-flops that are used as two 8-bit parallel-load regis-

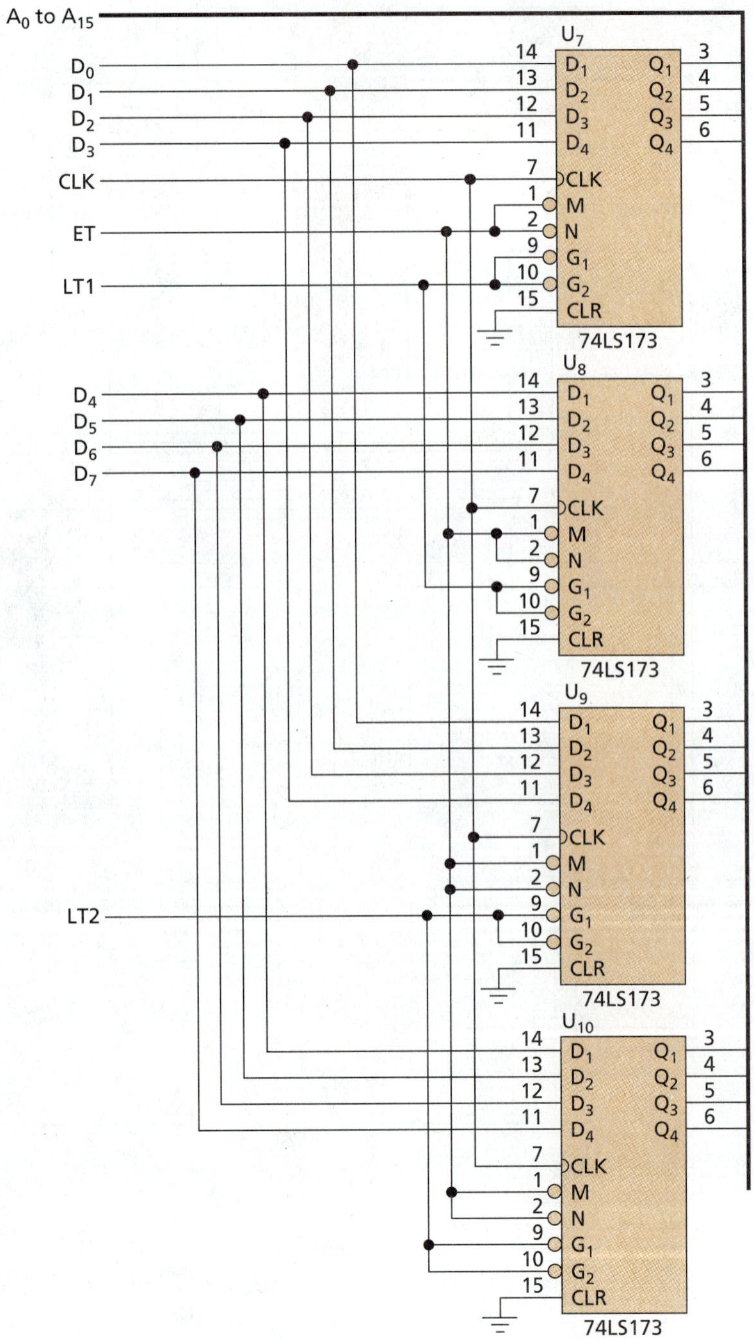

Figure 8.15 Temporary Register

ters, one 8-bit register for each byte of an address operand. One pair of 173s are loaded when LT1 is made low, and the second pair of 173s are loaded using LT2. All 16 outputs of the 74LS173s are enabled at the same time when ET is asserted.

The memory unit, Figure 8.16, has a 64K × 8 RAM memory, which internally may encompass more than one actual memory chip. Address selection

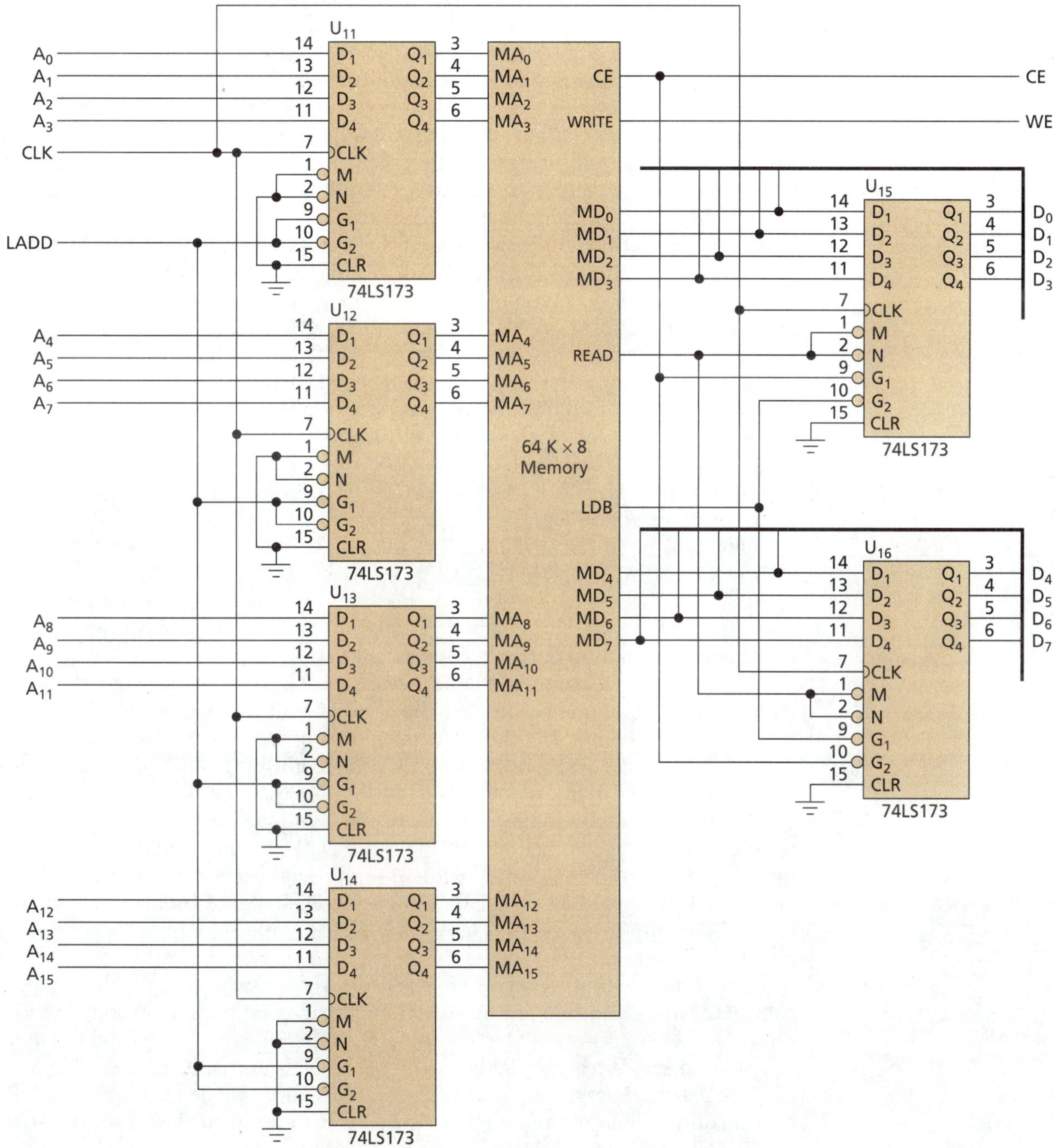

Figure 8.16 Memory Limit

of a location within this memory comes from the outputs of the address buffer whose inputs are connected to the address bus. The address buffer is another set of four 74LS173 quad-D flip-flops, which are loaded by applying the LADD signal. The outputs of the 173s are always enabled, passing the contents of the D flip-flops immediately to the RAM address inputs (MA_0–MA_{15}). RAM memory data outputs (MD_0–MD_7) supply inputs to two more 74LS173 quad-D flip-flop chips used as a data buffer. The input enables for these two chips are on constantly, so that data from the memory are immediately loaded into them. They make up the eight-bit data buffer, whose output is returned to the data bus when CE is asserted. The load data buffer signal, LDB, is on as long as the memory is powered, causing the data buffers to be constantly updated with data from the memory. Data are written into the memory directly when CE and WE are both asserted turning on the write function in the RAM memory. Asserting CE by itself enables the memory's output and turns on the READ input to the data buffers, passing memory data from the RAM chips through the 173 buffers to the data bus. When WE and CE are asserted together, READ is disabled, removing the outputs of 173 data buffer from the data bus.

The instruction register and decoder of Figure 8.17 uses two more 74LS173 quad-D flip-flops and two 74LS154 hex (1 of 16) decoders. The outputs of the 173s are always enabled, whereas the inputs are asserted using the LINST signal. This eight-bit register is cleared to 00 by the RESET signal. An instruction's opcode is stored into the instruction register in phase 2 of the opcode fetch cycle. The upper nibble defines the operation and is decoded by decoder U19. The outputs of that decoder are labeled MOV for opcode 0, ADD/SUB for opcode 1, LOGIC (2), I/O (3), JUMP (A), and HALT (E). Compare these codes to the instruction table (Table 8.4). The lower nibble is decoded in U18 and shows the operand type for the instruction. The outputs of the lower nibble are labeled MEMA (1), MEMB (2), IMMEDA (3), IMMEDB (4), STORA (5), STORB (6), and HALT (E). These codes do not correspond, by name, to all instructions, but these are the lines that will go low by decoding the binary code of the lower nibble of each instruction on Table 8.4.

The A-register (accumulator) and B-register (buffer) use the same identical circuit shown in Figure 8.18. On the figure, dual control signals are used to show control lines for the A-register on top and B-register on bottom. Two 74LS173 quad-D flip-flops are used as an eight-bit parallel-in/parallel-out register. Inputs, which are loaded using the LACC or LBUFF control signal, are connected directly to the data bus of the computer. Outputs go directly to the data inputs of the arithmetic logic unit (ALU) and to a 74LS244 octal three-state buffer. The outputs of the 244 are enabled back to the data bus, using control signal EACC or EBUFF. These registers are used to supply numerical data inputs to the ALU and, through the three-state buffer, data to the data bus.

The arithmetic unit, Figure 8.19, uses an eight-bit two's complement adder/subtracter to do addition and subtraction, an eight-bit parallel-in/shift-left-or-right/parallel-out register to handle the SHL and SHR instructions and another eight-bit adder/subtracter that is used to add 1 (INC) or subtract one (DEC) to the A-register or B-register. SEL_2, SEL_1, and SEL_0 are decoded using a 74LS138 to determine which of the functions the ALU is to perform: ADD (000B), SUB (001B), SHL (010B), SHR (011B), INC (100B and 110B), or DEC (101B and

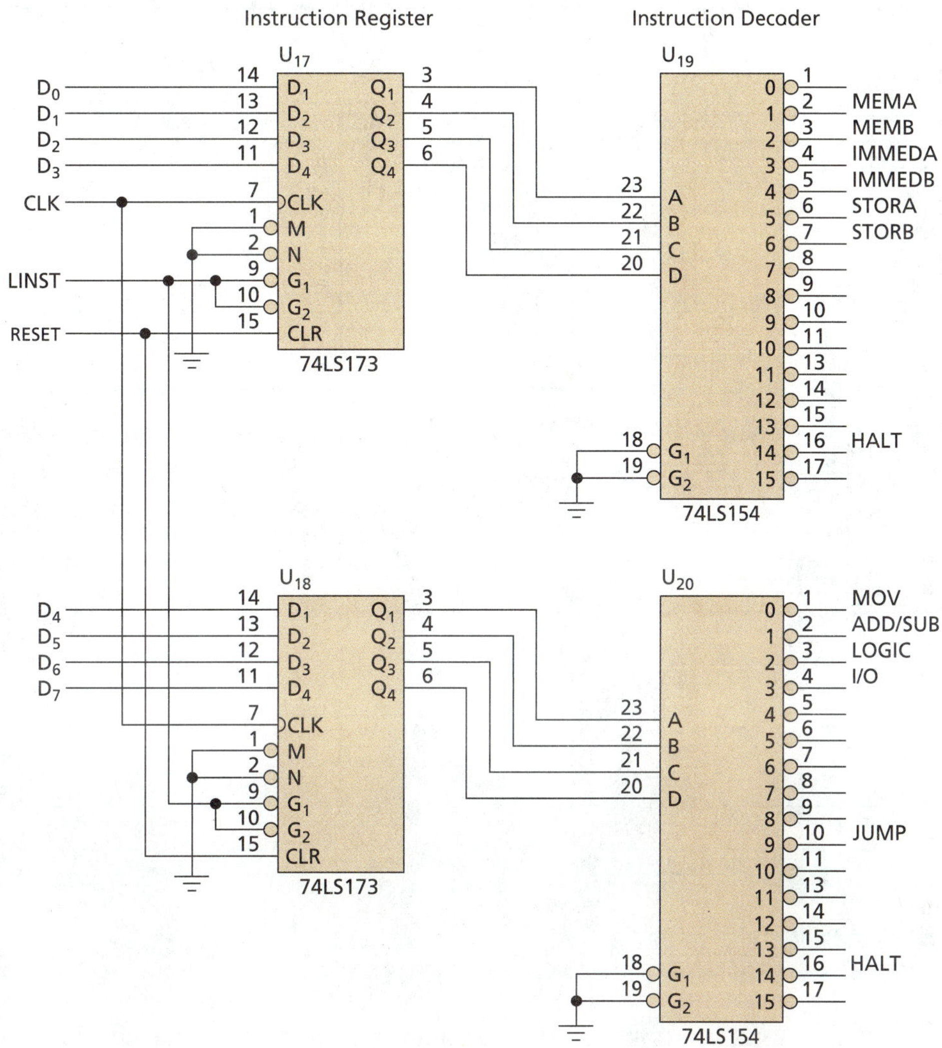

Figure 8.17 Instruction Register

111B). The NAND logic develops separate increment and decrement control signals and selects between the A-register and the B-register. Outputs from the three ALU units are fed through an internal eight-bit bus to a 74LS244 three-state buffer, where they are returned to the computer's data bus. Notice that the inputs to this circuit are supplied from the A-register and the B-register and not the data bus.

Figure 8.20 is the I/O (input/output) interface circuit. Here again, 74LS173s are used to form one input buffer register (U31 and U32), and one output buffer register (U33 and U34). When a key is pressed at the keyboard, the key code is presented to the input of the input buffer and loaded in using the STROBE signal. These data can be accessed by the computer by asserting the EIO signal. Output data to the display are first sent and loaded into the output buffer

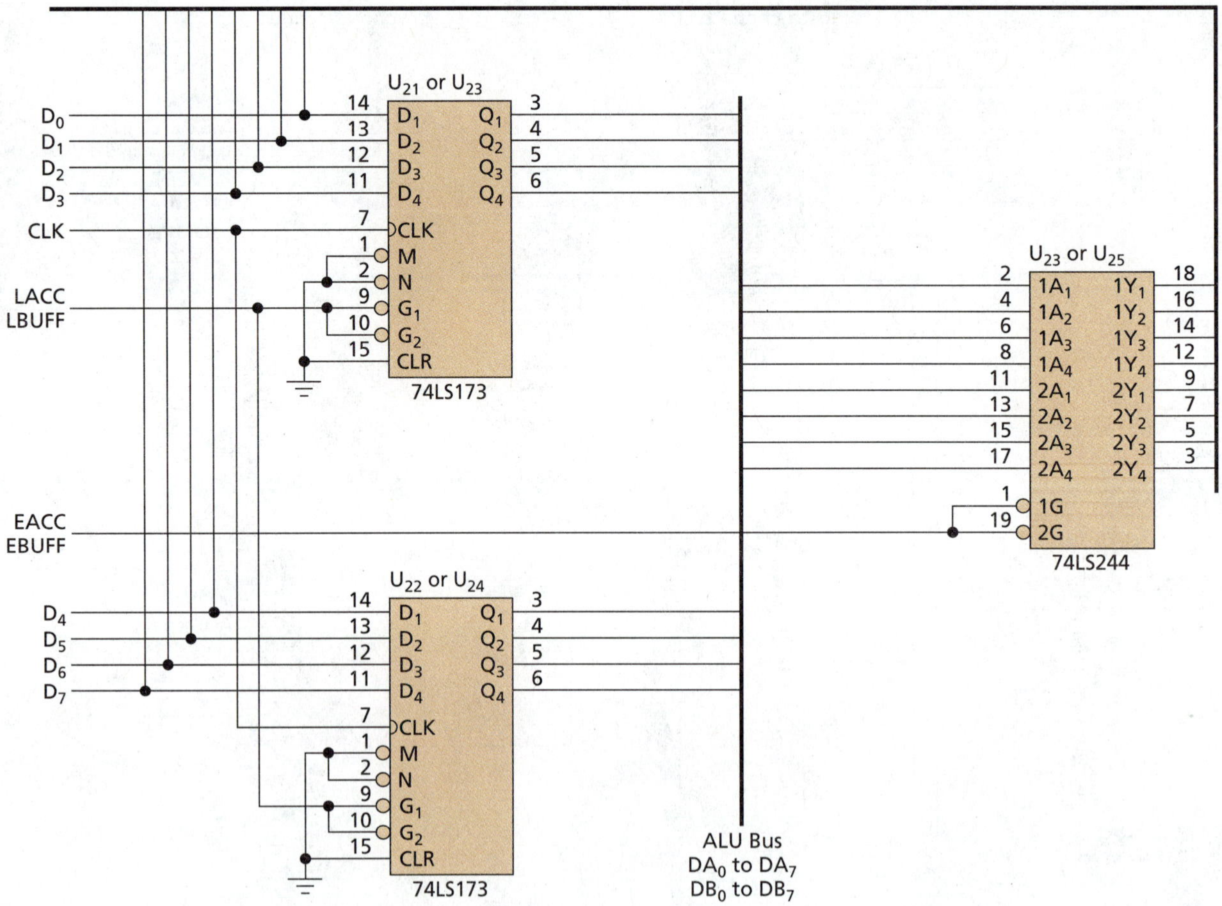

Figure 8.18 A- and B-Registers

using the LIO signal. The data are transferred to the display when ACK (acknowledge) is turned on by the display system.

The phase timing control for the BABE computer fetch and execute cycles comes from a mod-6 ring counter, shown in Figure 8.21. The counter is preset to the first phase, T_0, of an instruction cycle, by asserting reset or decoding the HALT instruction. The three methods to preset the counter, using logic in U41, is to complete an opcode fetch cycle that has an operand, complete the first of two operand fetch cycles, or complete an execute cycle. When the fetch cycle is complete, the decode logic in the instruction decoder determines if the instruction requires an operand. If an operand fetch is to come next, the ring counter is returned to T_0, the first phase, to start the operand fetch. If a second operand fetch follows the first one, the ring counter is again returned to phase T_0 to fetch the second operand. If there is no operand to be fetched or if only one operand fetch is required, when those cycles are completed, the ring counter continues with the execute cycle, which always begins with phase T_3. That is because phase T_0, T_1, and T_2 make up the three phases of an opcode or operand fetch cycle. When the execute cycle is finished, which might be at

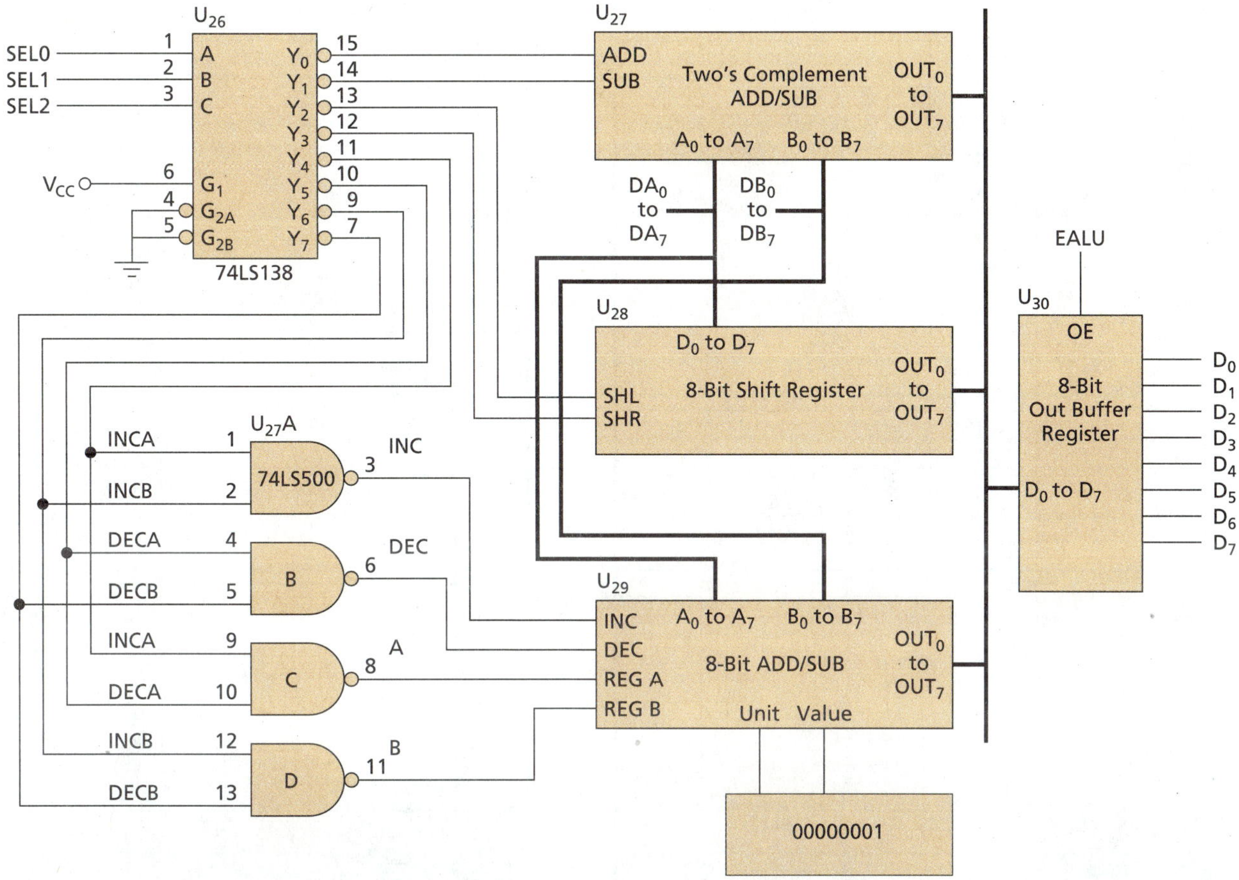

Figure 8.19 Arithmetic Logic Unit (ALU)

phase T_3, T_4, or T_5, depending on the instruction being executed, the ring counter is returned to the phase T_0 to begin the next opcode fetch cycle. Instructions that require different numbers of clock cycles to complete the fetch and execute cycles are called **variable-length instructions**, whereas computers that have the same number of cycles for each instruction are called **fixed-length instruction** computers. Fixed-length instructions each use the maximum cycles required to complete any one instruction, for all instructions. This uses a number of "do-nothing phases" for many instructions. Computers based on fixed-length instructions require simpler logic to implement the timing, whereas variable-length instructions execute in less time.

Variable-length instruction set *instructions that take only the number of clock cycles required to complete their execution.*

Fixed-length instruction set *instructions that take the same number of clock cycles to be completed.*

The control logic block, Figure 8.22, has a large number of logic circuits whose inputs are the decoded instructions from the instruction decoder and T_0 to T_5 phase outputs from the timing ring counter. These inputs are used to assert the correct output control signals at the correct time. For instance, the T_0 phase of every instruction is the first phase of an opcode fetch cycle. T_0 is also the first phase of an operand fetch. Both of these do the same thing—take the contents of the instruction pointer and send them to the address buffer. To do this, EIP (enable instruction pointer) and LADD (load address buffer) are

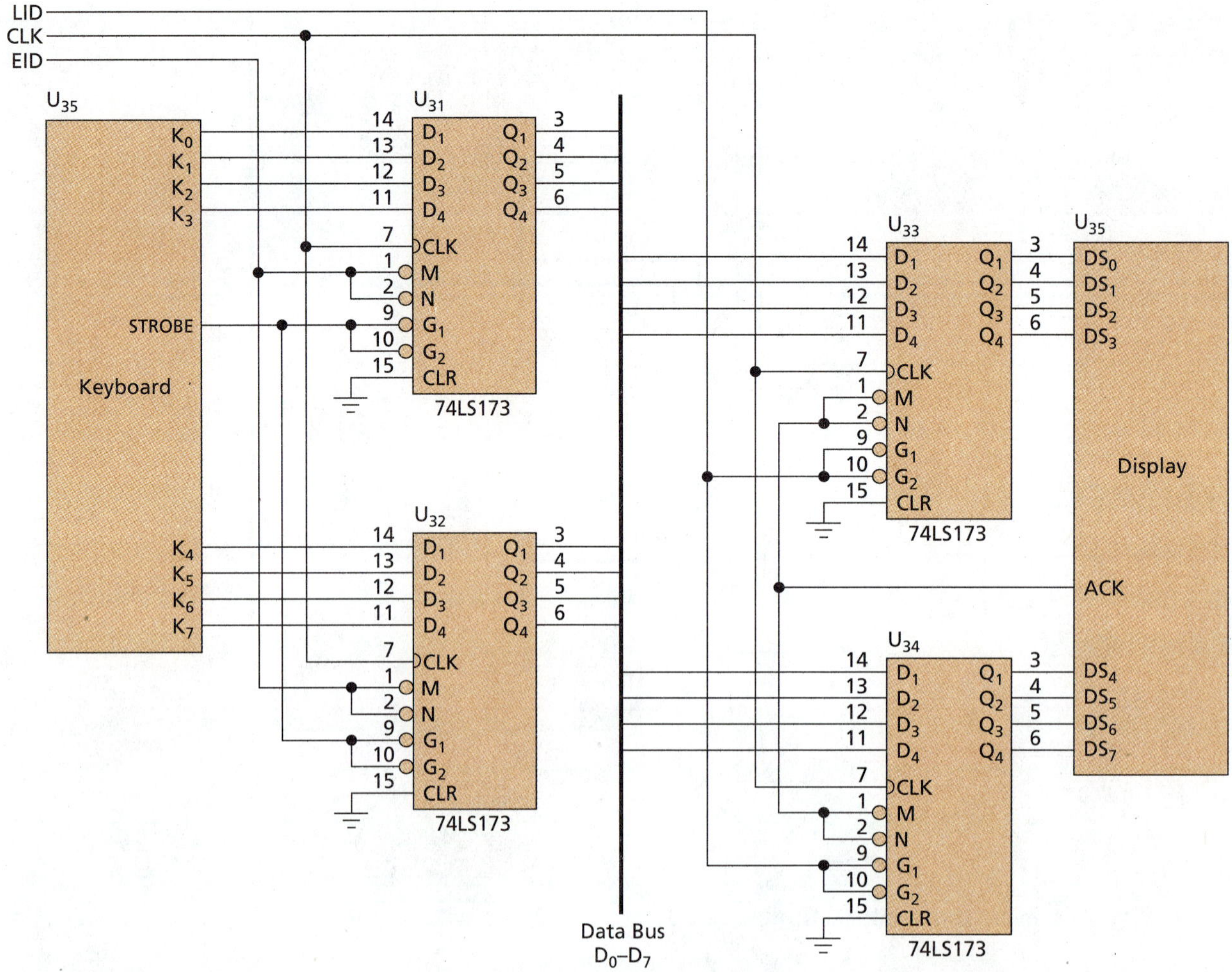

Figure 8.20 I/O Interface

turned on. All remaining control outputs are left inactive. The control logic in this block must respond to a high on T_0 by asserting EIP and LADD. Table 8.5 summarizes each instruction with respect to T_0–T_5, the decoded inputs and control signal outputs asserted at each phase of the instruction. The logic in the block, based on Table 8.5, is implemented using several GALs.

EXAMPLE 8.7

Use Table 8.5 to describe the phase-by-phase sequence of **ADD [1200H]** and **SUB 4CH** instructions. Location 1200H holds the data 67H, and the A-register initially contains CEH for both instructions. IP contents for the opcode of each instruction are 4420H.

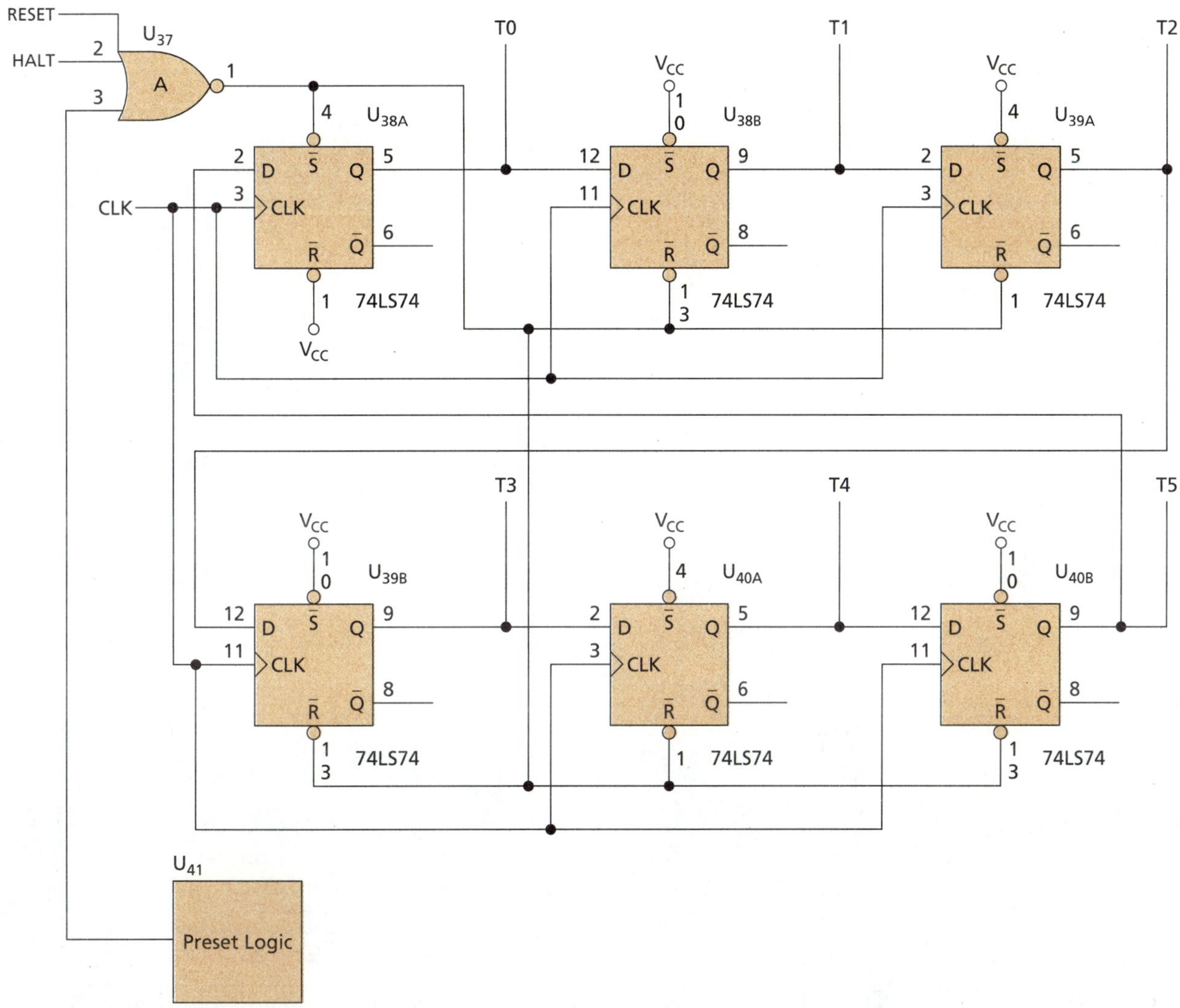

Figure 8.21 Timing Control

SOLUTION The first instruction has the general syntax **ADD [<mem>]** and begins, as all instructions do, with an opcode fetch cycle:

```
phase T0: EIP, LADD      ; 4420H -> address buffer
      T1: CE, LINST      ; 11H -> instruction register
      T2: COUNT          ;4421H -> IP
```

Next follow two operand fetch cycles:

```
T0: EIP, LADD            ; 4421H -> address buffer
T1: CE, LT1              ; 00H -> temporary register low
T2: COUNT                ; 4422H -> IP
```

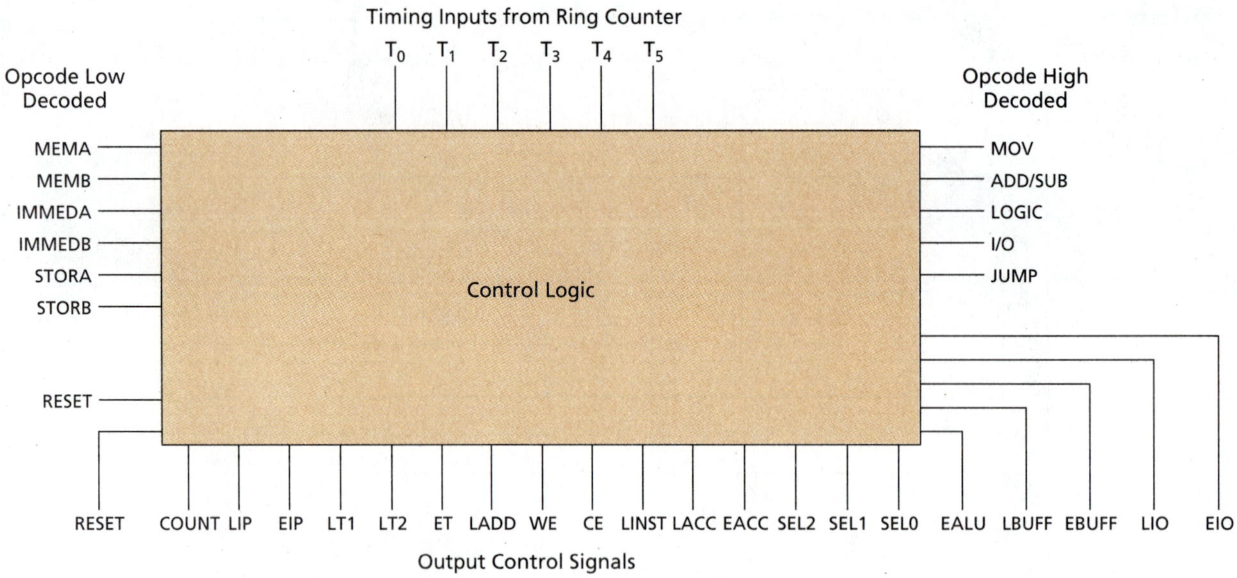

Figure 8.22 Control Block

```
T0: EIP, LADD                ; 4422H -> address buffer
T1: CE, LT2                  ; 12H -> temporary register high
T2: COUNT                    ; 4423 -> IP
```

Finally, the execute cycle:

```
T3: ET, LADD                 ; 1200H -> address buffer
T4: CE, LBUFF                ; 67H -> B-register
T5: EALU, LACC, S2-S0 = 000B       ; CEH + 67H = 135H
    ; 35H -> A-register; CF = 1; SF and ZF = 0
```

The second instruction starts out the same way, by fetching the opcode. However, there is only one byte of operand with this immediate-mode instruction:

```
T0: EIP, LADD                ; 44200H -> address buffer
T1: CE, LINST                ; 14H -> instruction register
T2: COUNT                    : 4421H -> IP
T0: EIP, LADD                : 4421H -> address buffer
T1: CE, LT1                  : 4CH -> temporary register low
T2: COUNT                    : 4422H -> IP
```

The execute cycle of the subtract immediate instruction is:

```
T3: ET, LBUFF                : 4CH -> B-register
T4: EALU, LACC, S2-S0 = 001B       ; CEH - 4CH = 82H
     82H -> A-register. SF = 1, CF and ZF = 0
```

Table 8.5 Control Logic

OPCODE FETCH: T_0: EIP, LADD
T_1: CE, LINST
T_2: COUNT, FF

FIRST OPERAND FETCH:
T_0: EIP, LADD
T_1: CE, LT1
T_2: COUNT, TF

THIRD OPERAND FETCH:
T_0: EIP, LADD
T_1: CE, LT2
T_2: COUNT

EXECUTE CYCLE:

INSTRUCTION	T_3	T_4	T_5
MOV A, [<mem>]	ET, LADD	CE, LACC	TRESET
MOV B, [<mem>]	ET, LADD	CE, LBUFF	TRESET
MOV A, <immed>	ET, LACC	TRESET	
MOV B, <immed>	ET, LBUFF	TRESET	
MOV [<mem>],A	ET, LADD	CE, WE, EACC	TRESET
MOV [<mem>],B	ET, LADD	CE, WE, EBUFF	TRESET
ADD [<mem>]	ET, LADD	CE, LBUFF	EALU, LACC SEL2–0 = 000
SUB [<mem>]	ET, LADD	CE, LBUFF	EALU, LACC SEL2–0 = 001
ADD <immed>	ET, LBUFF	EALU, LACC SEL2–0 = 000	TRESET
SUB <immed>	ET, LBUFF	EALU, LACC SEL2–0 = 001	TRESET
ADD B	EALU, LACC SEL2–0 = 000	TRESET	
SUB B	EALU, LACC SEL2–0 = 001	TRESET	
SHL	EALU, LACC SEL2–0 = 010	TRESET	
SHR	EALU, LACC SEL2–0 = 011	TRESET	
INC A	EALU, LACC SEL2–0 = 100	TRESET	
DEC A	EALU, LACC SEL2–0 = 101	TRESET	
INC B	EALU, LACC SEL2–0 = 110	TRESET	
DEC B	EALU, LACC SEL2–0 = 111	TRESET	
Jcc, JMP	EIP, LIP IF TRUE TRESET IF FALSE	TRESET	
IN	EIO, LACC	TRESET	
OUT	EACC, LIO	TRESET	
HALT	TRESET		

EXAMPLE 8.8

If each phase requires one clock cycle, compare the time of execution, in clock cycles between the ADD and SUB instructions of Example 8.7.

SOLUTION The total number of clock cycles for the ADD instruction is 3 for opcode fetch, 6 for the two operand fetches, and 3 for the execute cycle—total, 12 clock cycles. The SUB instruction required the same 3 opcode fetch clock cycles, but needed only 3 for the single operand and 2 for the execute cycle—total, 8 clock cycles. Therefore, the immediate-mode instruction executed faster than the memory reference instruction.

Section 8.8

Practice Problems

1. Describe the activity the BABE computer performs when executing these two instructions stored in memory, starting at location A48CH. Use two sample data entries, and explain the difference in the response of the JC instruction. First sample data are A-register = 5AH and B-register = 13H. The second sample data are A-register = 5AH and B-register = B4H:

```
SUB B
JC 1234H
```

   a. First describe using the block diagrams.
   b. Then describe each phase for both instructions.

2. How does an accumulator (the A-register) differ from any other register (B-register, for instance) in function?

3. Using a 2.5 MHz clock, calculate the time it takes to execute these instructions (assuming each phase takes one clock cycle):
   a. **MOV A,[3400H]** b. **INC A** c. **JM 62C0H**
   d. **SUB [7788H]** e. **OUT** f. **HALT**

Summary

The BABE computer is a prototype 8-bit computer. It has an 8-bit data bus and a 16-bit address bus. Functionally, the blocks that make up the computer are

1. Instruction pointer, which keeps track of program execution
2. Memory, including an address and data buffer
3. Instruction register and decoder to interpret instruction opcodes
4. Control sequencer and logic, which generates the control signals for each phase of each instruction
5. A-register, which serves as the computer's accumulator
6. B-register, which supplies the second number needed for the computer's arithmetic operations

7. Arithmetic logic unit, which does the adding, subtracting, shifting, incrementing, and decrementing of data
8. Input/output interface that receives data from a keyboard and outputs data to a display

The BABE computer has a short instruction set that includes instructions to move data, add and subtract numbers, shift, increment and decrement numbers, test numbers (conditional jumps) and import and export data. The instructions are executed using a variable instruction length that causes programs to be run in minimal time.

The basic functions of the BABE computer are the same as those of commercial 8-, 16-, and 32-bit CISC (complex instruction set computers) systems. The basic format, block functions, and operations are patterned after the INTEL Corporation of 80x86 microprocessor computers.

Glossary

accumulator or **A-register** a register that is used to hold the results of arithmetic and logic operations.

address bus a group of signal paths that carry binary address information.

address buffer parallel-load register used to store address information that is applied to the computer's address bus.

ALU (arithmetic logic unit) the functional block of a computer that performs arithmetic and logic operations.

assembler program software that converts a source program into object code.

assembly or **source program** semiliterate programs that use a computer's instruction set. These programs are a single level above machine or object programs.

BABE (binary architecture basic electronic) computer a generic computer system used to illustrate the functions and operation of a microprocessor based computer system. It is upwardly compatible with the INTEL 80x86 processor line.

buffer or **B-register** a section of the BABE computer that holds data usually used by the ALU.

conditional jump an instruction that causes a change in flow of a program if a specific condition is met.

control logic the part of the computer that generates the control signals that direct the computer's operation.

data bus a group of signal paths that carry binary data information.

execute cycle the sequence of events that causes an instruction operation to be performed (executed).

fixed-length instruction set instructions that take the same number of clock cycles to be completed.

flag flip-flop that reflects a status condition of an operation's result.

input/output or **I/O interface** a buffer register that holds data from the computer that are to be used by peripheral devices connected to the interface.

input/output or **I/O** or **peripheral device** device connected to the external computer buses, usually through some form of interface.

instruction decoder computer circuit that decodes the binary bits of the opcode. The decoded outputs are used by the control logic to generate the signals that direct the computer's operation.

instruction pointer a binary up counter in a processor that keeps track of the next instruction to be fetched and executed.

instruction register a register in a computer that stores the opcode fetched from memory.

labels used in an assembly program to identify an address location or constant value.

loop a set of instructions that are repeatedly executed a given number of times.

machine or **object program** program written using the actual binary codes for a computer's instructions and operands.

mnemonic shorthand code for an instruction that is used when writing source or assembly code programs.

opcode shorthand notation for *operation code,* which is the binary code for a program instruction.

operand defines the location or value of binary data.

program a list of successive instructions to be executed by a computer.

temporary register a register that holds a value that will be used very quickly by another section of the computer.

variable-length instruction set instructions that take only the number of clock cycles required to complete their execution.

Solutions to Practice Problems

Section 8.2

1. a. LBUFF, CE **b.** LADD, EIP **c.** LACC, EIO

2. CE, LACC CE, LBUFF EALU, LACC

Section 8.3

1. **MOV A,[4400H]**
SUB 6CH
MOV [128CH],A

2. **MOV A,[4400h]**
MOV B,6CH
SUB B
MOV [128CH],A

Section 8.4

1. **MOV A,<immed>, MOV B,<immed>, ADD <immed>,** and **SUB <immed>** require a one-byte data operand fetch cycle. **MOV A,[<mem>], MOV B,[<mem>], MOV [<mem>], A, MOV [<mem>], B, ADD [<mem>],** and **SUB [<mem>]** require two operand fetch cycles. **ADD B** and **SUB B** do not require an operand.

2. 82CCH —> address buffer
02H —> instruction register
IP incremented to 82CDH
82CDH —> address buffer
12H —> temporary register low
IP incremented to 82CEH

82CEH —> address buffer
33H —> temporary register high
IP incremented to 82CFH

a. 82CFH **b.** 02H **c.** 3312H **d.** 82CEH **e.** 33H

Section 8.5

1. IP —> address buffer
01H —> instruction register
increment IP
IP —> address buffer
10H —> temporary register low
increment IP
IP —> address buffer
12H —> temporary register high
increment IP
1210H —> address buffer
contents of 1210H —> A-register
IP —> address buffer
41H —> instruction register
increment IP
IP —> address buffer
34H —> temporary register low
increment IP
IP —> address buffer
12H —> temporary register high
increment IP
1234H —> address buffer
contents of 1234H —> B-register
SEL_2, SEL_1, SEL_0 = 001B; ALU —> A-register

2. 4.6μs for **ADD [<mem>]** sequence
5.2μs for **ADD B** sequence

Section 8.6

1. a. 2EH **b.** 2EH **c.** 5DH **d.** B8H

2. a. A-register = 9CH
B-register = 09H
b. A-register = 50H
B-register = 9CH
c. A-register = ABH
B-register = 09H
d. A-register = B3H
B-register = 09H

3. a. ZF = 0, CF = 0, SF = 0
b. ZF = 0, CF = 1, SF = 0
c. ZF = 0, CF = 0, SF = 1
d. ZF = 0, CF = 0, SF = 0

Section 8.7

1.

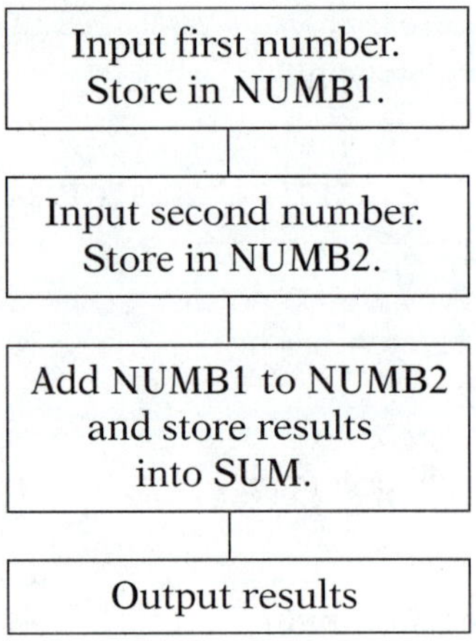

2.
```
IN
MOV [NUMB1],A
IN
MOV [NUMB2],A
ADD [NUMB1]
MOV [SUM],A
OUT
HALT
```

Section 8.8

1. **a.** ALU set to subtract (SEL inputs), difference sent to A-register.

   1234H loaded into temporary register.

   If CF = 1, 1234H moved from temporary register to IP execute next instruction at location 1234H.

   If CF = 0, fetch the next instruction from memory following JC instruction.

   b.
```
                A48CH —> address buffer         ; EIP, LADD
                47H —> instruction register     ; CE, LINST
                increment IP                    ; COUNT
                difference —> A-register        ; SEL = 001B; EALU, LACC
                A48DH —> address buffer         ; EIP, LADD
                DIH —> instruction register     ; CE, LINST
                increment IP                    ; COUNT
                A48EH —> address buffer         ; EIP, LADD
                34H —> temporary low            ; CE, LT1
                increment IP                    ; COUNT
                A48FH —> address buffer         ; EIP, LADD
                12H —> temporary high           ; CE, LT2
                increment IP                    ; COUNT
Sample data 5AH − 13H makes CF = 0              ; no borrow
                fetch next opcode from A490H
Sample data 5AH − B4H makes CF = 1              ; borrow required
                1234H —> IP                     ; ET, LIP
                fetch next opcode from 1234H
```

2. An accumulator receives the results of arithmetic operations.

3. a. 4.4 μS **b.** 1.6 μS **c.** 4.0 μS **d.** 4.8 μS **3.** 1.6 μS
f. 1.2 μS

Problems

8.1 List the units of the BABE computer that have three-state buffers on their outputs. *Hint:* These outputs are connected back to the address or data bus.

8.2 List the BABE units that have outputs that directly connect to another unit without three-state buffers.

8.3 What is the purpose of the instruction pointer?

8.4 How does the instruction pointer keep track of program execution?

8.5 What control signal allows the instruction pointer to count?

8.6 Which control signal enables instruction pointer contents to be placed onto the address bus?

8.7 Which control signal allows the instruction pointer to be loaded from the address bus?

8.8 What is the purpose of loading the instruction pointer with a new address? Give two reasons.

8.9 At what point in a JMP instruction cycle is the program counter loaded with a new address?

8.10 At what point in an instruction cycle sequence is the instruction pointer incremented? Why is this done?

8.11 What is unique about an accumulator?

8.12 What are the outputs of the accumulator connected to? Why are these connections made?

8.13 Which instructions cause data to be loaded into the BABE accumulator?

8.14 Which instructions read the accumulator? Where is the data moved to by these instructions?

8.15 Which BABE units supply inputs to the ALU for purposes of addition and subtraction?

8.16 List the control signals that move data from the ALU to the accumulator to complete an ADD instruction.

8.17 What does an opcode define?

8.18 What does an operand define? How many different types of operands are usable with the BABE computer instruction set? What are they, and how do they differ from each other?

8.19 Describe the opcode fetch cycle of the BABE computer. Which signals are asserted for each phase of the opcode fetch cycle?

8.20 Is the opcode fetch cycle different for any one of the BABE's instructions? Explain your answer.

8.21 How does an operand fetch cycle differ from an opcode fetch cycle?

8.22 Describe the steps that the BABE computer does to complete an opcode fetch and an operand fetch.

8.23 Which of the BABE's instructions reference memory locations to fetch or store data?

8.24 Which of the BABE's instructions do not require an operand? Why do they lack an operand?

8.25 Describe the complete operation of a **INC A** instruction, giving the control signals that are asserted for each phase and the action these control signals cause to happen.

8.26 Describe the complete operation of a **ADD B** instruction, giving the control signals that are asserted for each phase and the action these control signals cause to happen.

8.27 What are the steps required for data to be read from a memory location and sent to the buffer register?

8.28 What are the steps required to store the data in the buffer register into a memory location?

8.29 What is the name for the decoder used as the instruction decoder?

8.30 List the sets of inputs sent to the control unit to produce the synchronized control signals for the rest of the BABE computer.

8.31 What is the mod number of the state ring counter in the control unit?

8.32 Which units of the BABE computer are directly affected by the RESET signal? Why are the results of asserting the RESET signal?

8.33 What is the purpose of the IN and OUT interface?

8.34 Which instructions affect the IN and OUT registers? What other register is involved in data transfers to the OUT and from the IN registers?

8.35 How long does it take to execute the following instructions in a BABE computer with a 1 MHz clock:

a. **MOV [3340H],A** b. **ADD B** c. **INC A**
d. **HALT** e. **JM 1200H** f. **SUB 34H**

8.36 How long does it take to execute the following instructions in a BABE computer with a 2.5 MHz clock:

a. **MOV A,[BC49H]** b. **DEC B** c. **SUB B**
d. **OUT** e. **JC 3CC4H** f. **ADD 5FH**

8.37 Describe the process the BABE computer goes through to fetch and execute a **JMP 5C06H** instruction.

8.38 Describe the process the BABE computer goes through to fetch and execute a **JNZ ABCDH** instruction.

8.39 What are the conditions of the CF, ZF, and SF as a result of each of the following additions?

a. 34H + 77H b. 80H + 80H
c. C3H + DDH d. 6CH + A3H

8.40 What are the conditions of the CF, ZF, and SF as a result of each of the following subtractions?

a. 55H − 55H b. 4DH − 8EH
c. D3H − AFH d. 12H − C7H

8.41 Write a CUPL PLD file for a pair of GAL16V8 devices to hold the upper and lower nibbles of the instruction decoder circuit of Figure 8.17.

8.42. Write a CUPL PLD file for a GAL16V8 to hold the ALU select decoder of Figure 8.19. This GAL includes the 74LS138 and 74LS00 circuits.

8.43 Write a source program that adds two numbers that are currently stored in two memory locations called **NUMB1** and **NUMB2.** If the sum produces a carry, shift the A-register left and display the results. If there is no carry, shift the A-register right and display the results.

8.44 Write a source program that inputs a single exam grade number (two BCD numbers between 00 and 99) from the keyboard and outputs the equivalent exam letter grade in ASCII code to the display unit. "A" = 41H, "B" = 42H, "C" = 43H, "D" = 44H, and "F" = 46H.

8.45 Translate the source program of Problem 8.43 into object code stored beginning at location 1400H.

8.46 Translate the source program of Problem 8.44 into object code stored in memory beginning at location 4AC0H.

8.47 Write a source program that inputs a weight value, compares it with two limit values stored at locations TOP and BOTTOM, and outputs the code values FFH, 00H, and 3CH, stored in locations MUCH, LESS, and OKEH respectively. If the weight value exceeds the value in TOP, send the code from MUCH. If the weight falls below the value in BOTTOM, send the code in LESS. If the weight lies between the upper and lower limits in TOP and BOTTOM, send the code from OKEH to the OUT register.

8.48 Write a source program that inputs the ASCII code for a hexadecimal number and then sends the actual hexadecimal value to the output register. Do not create a lookup table of ASCII and hexadecimal values.

8.49 Write a source program that inputs the number of stages in a binary counter and then sends the counter's mod number, in hex, to the OUT register. *Hint:* The mod number is equal to 2^n, where n is the number of stages in the counter. Also, each time a number is shifted left (toward the MSB), it is doubled.

8.50 Write a source program that inputs two hexadecimal numbers and sends the hex product of those two numbers to the OUT register. The largest input numbers are FFH, and the product of FFH and FFH is FE01H. The computer will wait for any nonzero input after the lower byte of the sum is sent out before sending out the upper byte.

nine

Logic Families, Parameters, and Specifications

KEY TERMS

absolute maximum ratings
bus contention
CMOS (complementary metal oxide semiconductor)
Darlington output
discrete logic gate
DTL (diode–transister logic)
ECL (emitter-coupled logic)
Fanout (FO)
FAST TTL
HCMOS
high-impedance or open state of an output
hold time (t_h)
IGFET (insulated gate field effect transistor)
I_{OH}, I_{IH}, I_{OL}, and I_{IL}
LSI (large-scale integration)
MSI (medium-scale integration)
NMOS (N-channel MOS)
noise margin
nominal
open-collector output
PMOS (P-channel MOS)
power dissipation (PD)
propagation time or delay
RTL (resistor–transistor logic)
Schottky diode
set-up time or t_s
SSI (small-scale integration)
three-state buffer output
totem pole output
TTL (transistor–transistor logic)
V_{OH}, V_{OL}, V_{IH}, and V_{IL}
wired AND and wired OR

OBJECTIVES

After completing this chapter, the student should be able to

- Briefly describe the history of logic families.
- Explain what is included in the TTL family of devices and what the differences are among its members.
- Describe the specifications for a family of logic devices, using the TTL family.
- Discuss different types of TTL output configurations, including the totem pole, open-collector, and three-state buffered types, and show applications for each.
- On complete TTL data sheet, point out each of the device parameters and specifications.
- Describe the MOS family of devices, used when current drain is a large consideration, and compare it to the TTL family.
- Describe methods and considerations for interfacing different family type devices in a single circuit, and explain which are possible and what must be dealt with when mixing device types.

OVERVIEW

In this chapter, we explore the parameters, specifications, and operation of logic gates of different technologies. Each technology is classified as a logic family. For instance, the group of logic devices manufactured using transistor–transistor logic (TTL) are all members of the TTL family. This includes variations and enhancements made to the basic technology, such as methods to increase operating speed and reduce power consumption. Sample data sheets are used to aid the student in understanding and using them when working with various devices. Consideration is given to device specifications and parameters to assist in design and troubleshooting problems.

This chapter presumes that the student has a basic working knowledge of diodes, bipolar transistors, field effect transistors (FETs), and circuit analysis. This knowledge need not be in depth but should include basic concepts such as barrier potential, Ohm's law, diode and transistor operating ranges, and FET on and off conditions.

9.1 Introduction

The first attempts at creating digital logic circuits involved using combinations of vacuum tubes and mechanical relays to perform logic functions. These logic circuits were large and consumed large quantities of power. An early improvement to benefit the digital field was the invention of the bipolar junction transistor, which greatly reduced both the size and power consumption of logic circuits. Still, to create the simplest of digital systems using discrete transistor circuits required a fair amount of physical space and reasonably large power sources. The march toward smaller devices from which to develop digital logic systems took a quantum leap with the advent of the integrated circuit (IC), which allowed many device circuits to be integrated onto a comparatively small physical area. The first commercially successful IC contained a combination of resistor and transistor elements arranged to operate as logic gates.

9.2 Early Technologies

RTL (resistor–transistor logic) *a logic technology that preceded the development of TTL.*

DTL (diode–transister logic) *a logic technology that preceded the development of TTL.*

The first practical integrated logic family of devices was called **resistor–transistor logic** or **RTL.** Resistors and transistors were combined to form AND, OR, NAND, NOR, inverter, and other logic gates and devices. An inverter gate example is shown in Figure 9.1. This logic family did not survive long because of problems with heat dissipation and output loading. The second generation of logic devices belonged to a family called **diode–transistor logic** or **DTL.** They used a diode matrix to perform the logic function, followed by a transistor amplifier, as shown in the NAND gate example in Figure 9.2. To illustrate the gate's operation, let's develop the truth table for it and verify that this is a NAND gate.

Applying 0 logic level (0 V) to both inputs A and B causes diodes D_1 and D_2 to be forward biased toward V_{CC} through R_1. The barrier voltage of the diodes places 0.7 V at the anode of diode D_3. This voltage is not sufficient to forward bias D_3 and the base-to-emitter junction of the transistor. R_3 supplies a ground

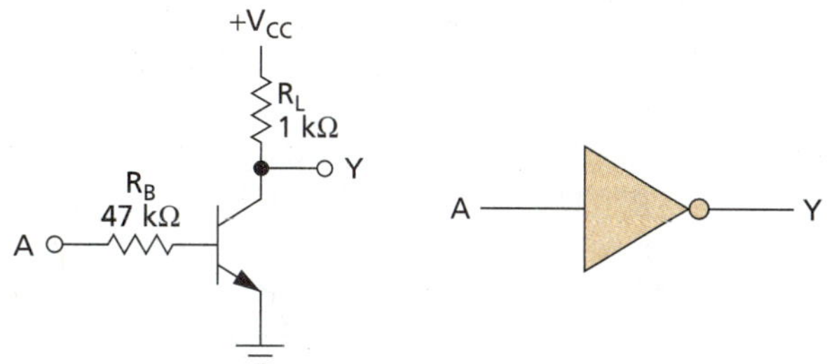

Figure 9.1 RTL Inverter

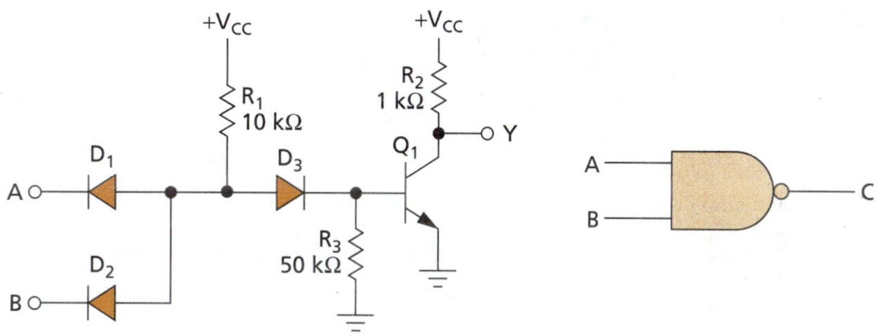

Figure 9.2 DTL NAND Gate

potential to the base of Q_1, assuring that it is not conducting. As a result, output Y sees source current through collector resistor R_2 to V_{CC} and high potential for a logic 1 output. The first line on the truth table is

A	B	Y
0	0	1

Maintaining a logic 0 on input A and changing B to a logic 1 state by applying V_{CC} to input B sets the input state for the next truth table entry. Diode D_1 remains forward biased, but diode D_2 is now reverse biased and does not conduct. However, the forward conduction of D_1 keeps the anode of D_3 at 0.7 V, and Q_1 remains cut off. The output at Y is again a high logic level. Reversing the inputs to A and B now turns on diode D_2 and shuts off diode D_1. The conducting diode again holds D_3 and Q_1 nonconducting. The truth table to this point is

A	B	Y
0	0	1
0	1	1
1	0	1

The last combination of inputs occurs when inputs A and B are both brought high. D_1 and D_2 are both cut off, and the anode of D_3 now sees a positive potential at V_{CC} through R_1, while its cathode sees a low potential at ground through R_3. D_3 is forward biased and begins to conduct through both R_1 and R_3. The voltage drop across R_3 reaches 0.7 V rapidly and forward biases the base to emitter junction of Q_1. The transistor quickly saturates, producing sink current at output Y and a voltage potential of 0.1 V, the $V_{CE(SAT)}$ of the transistor, appears at Y as a logic 0 state. This completes the truth table for the DTL NAND gate:

A	B	Y
0	0	1
0	1	1
1	0	1
1	1	0

This type of analysis will be used again to determine the logic function of other logic family devices. You may wonder why we take the effort to analyze the internal workings of an integrated circuit. Essentially, you cannot open one up and repair a faulty transistor within the chip. If it is bad, toss it out and replace it with another chip! The reason for the analysis is based on several key points. First, to understand how a circuit is supposed to react, it would be helpful to understand its operation. Secondly, isn't it about time you stopped wondering what *really* goes on inside that little chip? Then there comes the time when you have to replace a faulty chip or need to design a logic circuit. Functionally, the operation of a specific logic circuit is the same regardless of the family it belongs to. A NAND gate is a NAND gate is a NAND gate—that sort of notion. So won't any NAND gate do when you need to replace a defective one? As long as the pin out of the replacement chip is the same and the functions are the same, why not? We shall see why not as we progress through this chapter. Also when designing with logic chips, you cannot arbitrarily select logic devices without paying attention to their logic family and the consequences of using a particular device.

9.3 Transistor–Transistor Logic Family of Devices

TTL (transistor–transistor logic) *a transistor-based logic technology.*

One of the more popular integrated logic families in use today is **transistor–transistor logic** or **TTL** technology, which uses bipolar transistors to do the logic and amplifying functions. Figure 9.3 shows a TTL NAND gate, which uses multiple emitter–base junctions in the input transistor, Q_1. These emitter–base junctions actually perform the AND logic function, and the remaining circuit elements accomplish the inversion required by the NAND logic function. The collector of Q_1 is connected to the base of Q_2, which is called the *phase splitter*

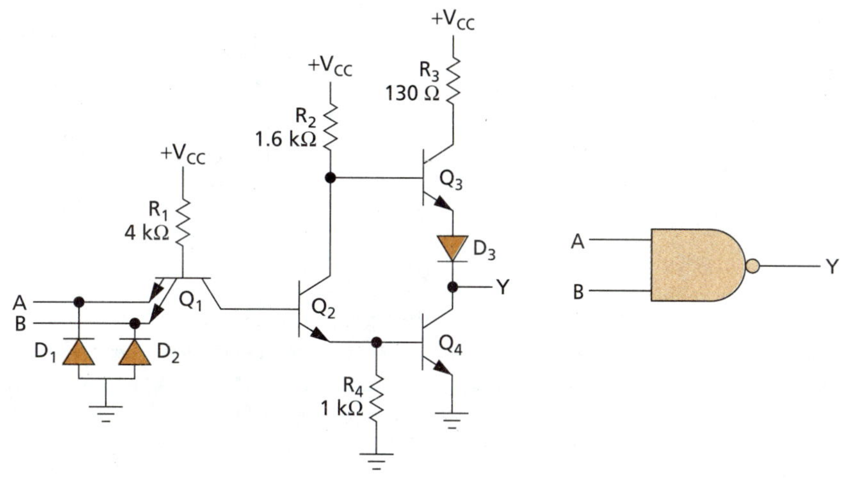

Figure 9.3 TTL NAND Gate

transistor. The emitter output of Q_2 uses this part of the circuit as an emitter follower whose output is in phase with the base input, to drive Q_4. The output at Q_2's collector—using the same circuit, but now as a common emitter amplifier—supplies an output 180° out of phase with the base input, to drive the base of Q_3. The output circuit, called a **totem pole output,** consists of Q_3 on top (close to the source voltage supply, V_{CC}), and Q_4 on the bottom (to sink the output toward the ground).

Totem pole output *a logic output configuration that provides both source and sink currents for logic 1 and logic 0 output states.*

The multiple emitter–base junctions of Q_1 form a P–N junction circuit that operates in the same manner as the diode circuit in the DTL NAND in the previous section. When a logic 0 is applied to any of the emitter inputs, that junction becomes forward biased and causes Q_1 to conduct, dropping its collector potential close to 0 V. This low voltage cuts off Q_2, which acts as the collector load for Q_1. With Q_2 off, there is no drive current to Q_4, so it in turn cuts off, preventing the output from seeing ground current. Another consequence of the off state of Q_2 is that its collector voltage tries to go to V_{CC}. As a result, the base potential to Q_3 rises until Q_3 goes into conduction, supplying source current and a logic 1 voltage potential at the output of the device. This circuit operates the same as long as one of the emitters is brought low.

The only combination without a low on any emitter is when they are both set to a logic 1. This causes Q_1 to cut off, because none of its emitter–base junctions are forward biased. The base at Q_2 sees a positively driven current through the collector–base circuit of the nonconducting Q_1. Q_2 turns on, driving Q_4's base, turning on that transistor, resulting in a low V_{CE} voltage of around 0.1 V, which is representative logic 0 state at the output. Since Q_2 is now conducting, its collector approaches a low potential, cutting off output transistor Q_3, preventing it from allowing current to the output from V_{CC}. Figure 9.4 summarizes the logic inputs, conducting transistors, and the resulting logic output as presented on a truth table. From this truth table, we can see that this gate does, indeed, perform a NAND logic function.

Diodes D_1 and D_2 serve a dual purpose. First they discharge any static electricity that may be built up when handling the chip outside of a circuit. Second,

Inputs		Transistor States				Output
A	B	Q_1	Q_2	Q_3	Q_4	Y
0	0	On	Off	On	Off	1 Source
0	1	On	Off	On	Off	1 Source
1	0	On	Off	On	Off	1 Source
1	1	Off	On	Off	On	0 Sink

Figure 9.4 TTL NAND Table

the diodes remove any negative switching spikes, which occur when a high level switches to a low, at the inputs of the device. Diode D_3 prevents both output transistors from being turned on at the same time. It forces the base voltage of Q_3 to be at least 1.4 V before both Q_3 and D_3 turn on. The following list summarizes the parts of a TTL NAND gate and their functions:

Q_1 Input transistor that supplies the AND logic function
Q_2 Phase splitter, supplies in-phase drive to Q_4 and out-of-phase drive to Q_3
Q_3 Output transistor that supplies current to V_{CC}
Q_4 Output transistor that sinks current from ground
D_1 and D_2 Static charge protection
D_3 Switching diode, assures that Q_3 is cut off when Q_4 is conducting

The NOR gate shown in Figure 9.5 uses the same conceptual ideas for operation within the device as used for the NAND gate. The major differences are

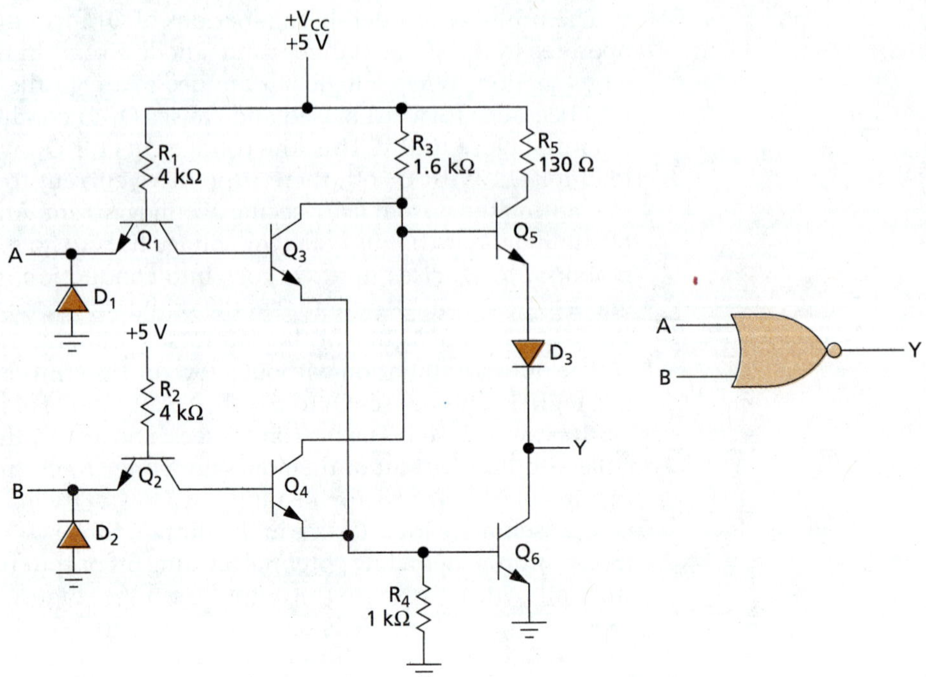

Figure 9.5 TTL NOR Gate

that there are two separate input transistors (Q_1 and Q_2) instead of one, and two parallel phase splitters (Q_3 and Q_4). The input transistors operate the same as the NAND Q_1. When one of the inputs are brought low, that input transistor conducts, cutting off its corresponding phase splitter (Q_3 or Q_4). When one of the inputs is brought high, its input transistor cuts off, driving its corresponding phase splitter on. Both phase splitters have their collectors and emitters connected together, so that the collector potential of one is in parallel with the other. If either transistor is conducting, the collector voltage of the pair will be low, in turn forcing Q_5 to be cut off and driving Q_6 on, through the emitter current of the conducting phase splitter transistor, setting the output to a logic low. In order for the collector voltage of the parallel phase splitters to be high, both input transistors must be conducting, which is achieved by applying a logic 0 to each one's input emitter. The activity in the TTL NOR gate is summarized in the truth table of Figure 9.6.

Inputs		Transistor States						Output
A	B	Q_1	Q_2	Q_3	Q_4	Q_5	Q_6	Y
0	0	On	On	Off	Off	On	Off	1 Source
0	1	On	Off	Off	On	Off	On	0 Sink
1	0	Off	On	On	Off	Off	On	0 Sink
1	1	Off	Off	On	On	Off	On	0 Sink

Figure 9.6 TTL NOR Truth Table

9.4 TTL Specifications

There are many specifications and parameters associated with a logic device. In this section, we will try to make some sense out of a good many of them and why they are important. A little later on in the chapter, actual data sheets are used to show where these parameters and specifications are listed.

Fanout (FO) *the number of similar devices that can be driven by a single logic gate's output.*

I_{OH}, I_{IH}, I_{OL}, and I_{IL} *input and output current characteristics of a logic device.*

One of the foremost specifications of a logic gate is its **fanout** or **FO.** This specification is derived by dividing a gate's output current parameter by its input current parameter. The purpose is to determine how many inputs a single output can drive and still maintain a correct logic level. There are two separate output current parameters and two separate input ones. $\mathbf{I_{OH}}$ specifies the output current available from a gate when its output is a logic 1 or high. $\mathbf{I_{OL}}$ lists the output current available when the output is at a logic 0 or low level. Input current specifications denote the input current drawn when a logic 0 ($\mathbf{I_{IL}}$) or a logic 1 ($\mathbf{I_{IH}}$) is applied to the input of the gate. Fanout is calculated by dividing output current by input current for the same logic state. A gate's fanout is the lower of the fanouts calculated for each logic level by using these equations:

$$FO_H = I_{OH}/U_{IH} \qquad FO_L = I_{OL}/I_{IL} \tag{9.1}$$

EXAMPLE 9.1

What is the fanout of a NAND gate with the following parameters?

$I_{OH} = 12$ ma $\quad I_{OL} = 15$ ma $\quad I_{IH} = 0.2$ ma $\quad I_{IL} = 1.2$ ma

SOLUTION Calculate fanout for each level, and then select the lower value. Keep in mind that you cannot drive a partial gate, so always round your result down.

$FO_H = 12$ ma/0.2 ma $= 60 \quad FO_L = 15$ ma/1.2 ma $= 12.5$

This gate's fanout is 12. That is, the output of the gate can drive 12 of the same type of gate and maintain correct logic levels at that output.

A second important parameter is a gate's power dissipation. This also has more than one value. There is a standby dissipation caused when the circuit is not switching, and there is an operating specification. Since the operating power dissipation will be greater due to the activity within the device's circuits, it will be the value of concern. When deciding on the use of a particular logic family in a logic design, accumulative power dissipation will dictate the size of the power source used to run the circuit. Even more crucial, given the number of portable computers and other electronic devices, is the drain caused by logic circuits on the batteries that make these units portable in the first place. We shall compare logic family power considerations with respect to many of today's applications a bit later in the chapter.

Portable electronic units have come about as the ability to squeeze more and more circuit functions into smaller IC packages has improved. The term for the amount of circuitry found inside of an IC is **packing density.** Many factors determine the actual number of transistors and the like that can be placed into a single chip. One factor is the aforementioned power dissipation. The more a chip dissipates power, the more heat it must also dissipate and the hotter it gets. Of large concern to microprocessor manufacturers as they continue to increase the amount of electronics included in a single chip that are operated at faster clock speeds. Because of the increase in circuits and higher operating speeds, more heat is required to be dissipated away from the chip. If this heat is not quickly removed from the body of the IC, it could easily consume itself by burning out a number of internal circuit elements.

Another factor controlling packing density is the actual physical size of the individual circuit components. Look back at the TTL NAND and NOR gates of Figures 9.1 and 9.3. A fair amount of discrete components—the transistors, diodes, and resistors—is required to make a complete gate. The number of these components limits the number of gates that can be included in a single IC.

V_{OH}, V_{OL}, V_{IH}, and V_{IL} logic device input and output voltage levels for logic 1 and logic 0.

Output and input voltage limits for logic levels have many important aspects. The most important requirement is that they specify what is acceptable for a given logic level in a particular gate. For instance, $V_{OH(min)}$ determines the lowest voltage sensed at a gate's output for a logic 1 or high level. $V_{OL(max)}$ shows the highest logic 0 or low level acceptable from this gate. Equally important are similar input voltage specifications that specify what the limits of an input can sense as reliable logic level inputs. They are **$V_{IH(min)}$** and **$V_{IL(max)}$**, and their values will differ from their output counterparts. As an example, a $V_{OH(min)}$ might be 3.8 V, whereas an input specification for the same gate might be 3.6 V. The disparity between these input and output values stems from the possibility that an output high voltage might experience some degradation before it reaches an input to another gate. The difference between the two specifications is this degradation and is called **noise margin**.

Noise margin *the difference between an output logic state voltage level and an input logic state voltage level.*

Noise margin is determined by finding the difference between the allowable extremes for output and input logic levels. The equations for calculating noise margin are

$$NM_H = V_{OH(min)} - V_{IH(min)} \qquad NM_L = V_{IL(max)} - V_{OL(max)} \tag{9.2}$$

Noise margin can be illustrated graphically, as shown in Figure 9.7. Looking at this figure, you will note that there is an area between the high-level minimums and low-level maximums. This is a virtual no-man's land for logic devices. Voltages within this range are neither a logic 1 or a logic 0. There is no way to predict what the output of a gate would be when the input is within this range.

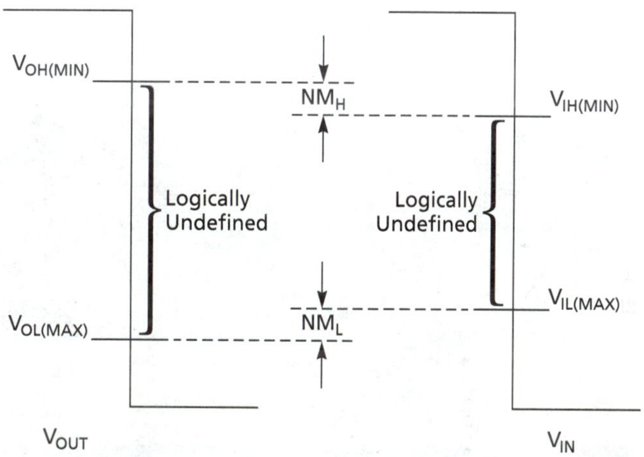

Figure 9.7 Noise Margin

Section 9.4

Practice Problem

Using the following specifications for a TTL NOR gate, determine its fanout and noise margins.

$V_{OH(min)} = 3.95\text{ V} \qquad V_{OL(max)} = 0.75\text{ V}$

$V_{IH(min)} = 3.77\text{ V} \qquad V_{IL(max)} = 0.86\text{ V}$

$I_{OH} = 17\text{ ma} \qquad I_{OL} = 15\text{ ma} \qquad I_{IH} = 0.46\text{ ma} \qquad I_{IL} = 0.98\text{ ma}$

9.5 Output Configurations of TTL Gates

There are several different forms of output drive circuits for TTL gates. The type used in the NAND and NOR gates of Figures 9.3 and 9.5 is commonly referred to as the *totem pole* output because of the "stacking" of the output transistors and switching diode. It consists of a sink and source transistor separated by the diode. Recall that the purpose of the diode is to assure that the source transistor is not conducting when the sink transistor is turned on.

Darlington output *output of a logic circuit that provides high-speed switching of heavily loaded outputs.*

Another form of TTL output is the **Darlington output** of Figure 9.8. The Darlington transistor pair, Q_3 and Q_4, supply a higher drive current capability and faster switching time than the totem pole output. In addition, the base-to-emitter potential of the two transistors replaces the need for the diode that was in the totem pole configuration. Q_5 operates as Q_4 did in the totem pole, to supply sink current for a 0 logic level.

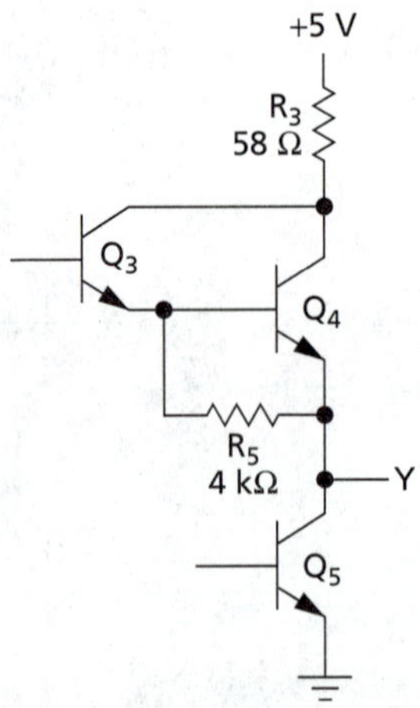

Figure 9.8 Darlington Output

Both the totem pole and Darlington outputs cannot be directly connected to other outputs. Try to imagine connecting two NAND outputs together that have totem pole output forms. Now, consider what happens when one output drives toward 0 logic sinking current through its Q_4. At the same time, the other NAND gate tries to drives its output toward V_{CC} through its Q_3. The 0 logic would dominate, but the excessive current driving through both output transistors may well destroy one or the other. To connect the outputs together, a third type of logic gate output, called an open-collector output, is required to AND them together.

Open-collector output *logic devices whose output transistors or FETs do not have a collector or drain pull-up resistor.*

Two types of TTL outputs are available to overcome the problem of connecting outputs directly to one another. The first, pictured in Figure 9.9, is called the **open-collector output**. The upper transistor(s) and applicable diode are removed from the totem pole or Darlington configurations, leaving the sink output transistor without any connection to its collector except the output pin. This allows gate outputs to be connected together as shown, using the two

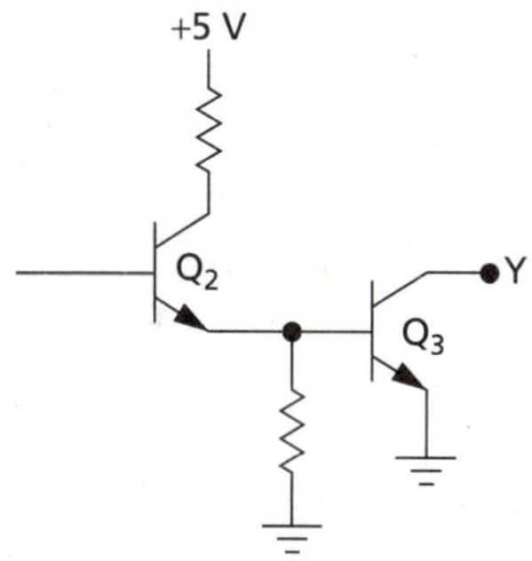

Figure 9.9 Open-Collector Output

open-collector AND gates, in Figure 9.10. The outputs of the AND gates are connected together in a wired AND configuration, which is indicated by the dotted AND gate in Figure 9.10. Any number of outputs can be brought to the same point. A pull-up collector resistor (labeled R_L for load resistor in Figure 9.10) is required to supply a collector current path for the output transistors connected to point Y.

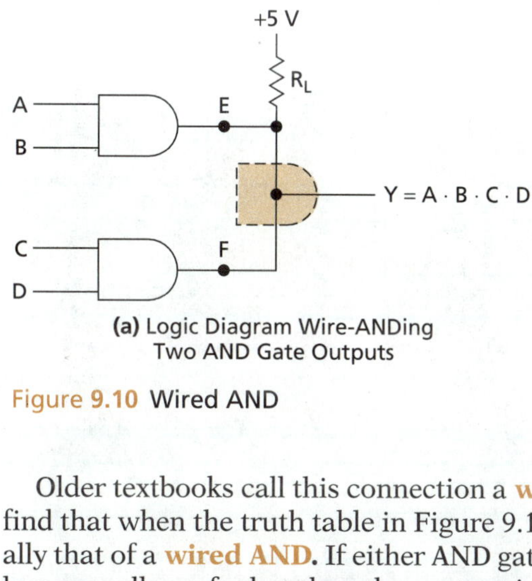

(a) Logic Diagram Wire-ANDing Two AND Gate Outputs

E	F	Y
0	0	0
0	1	0
1	0	0
1	1	1

(b) Truth Table of Wired AND Portion

Figure 9.10 Wired AND

***Wired AND** and **wired OR** simulated logic function that occurs when open collector outputs are connected together.*

Older textbooks call this connection a **wired OR** configuration, but you will find that when the truth table in Figure 9.10 is developed, the function is actually that of a **wired AND.** If either AND gate's output drives low, point Y will be low regardless of what the other output is trying to do. The reason for this is that the low is produced from the conduction of the sink transistor to ground. The opposite or high state occurs when the sink transistor is cut off. There is no source transistor. The logic high current is derived through the load resistor when all the output transistors are not conducting.

The first three lines of the truth table, using E and F as inputs to the wired AND and Y as its output, shows one of the sink transistors conducting by the presence of a 0 at E or F or both. This holds point Y to a 0. The last line shows both transistors cut off (logic 1 state on E and F) allowing point Y to go high through resistor R_L. The truth table is that of an AND gate, so why has this connection been called a wired OR? Recall from Chapter 2 on Boolean algebra and

DeMorgan relationships that AND and OR gates can be expressed as opposite functions. This table can be interpreted as an AND gate: Output Y goes high only when E AND F both go high or $Y = E \cdot F$. It can also be interpreted as an OR function: Output Y goes low if E OR F goes low or $\overline{Y} = \overline{E} + \overline{F}$. The preference of designation may come from the application of the wired configuration at the time of its use.

EXAMPLE 9.2

Draw a wired AND circuit that implements the expression $Y = \overline{AB} \cdot \overline{CD} \cdot \overline{EF} \cdot \overline{GH}$.

SOLUTION Use four open collector NAND gates, and connect all their outputs together with a pull-up resistor as shown in Figure 9.11.

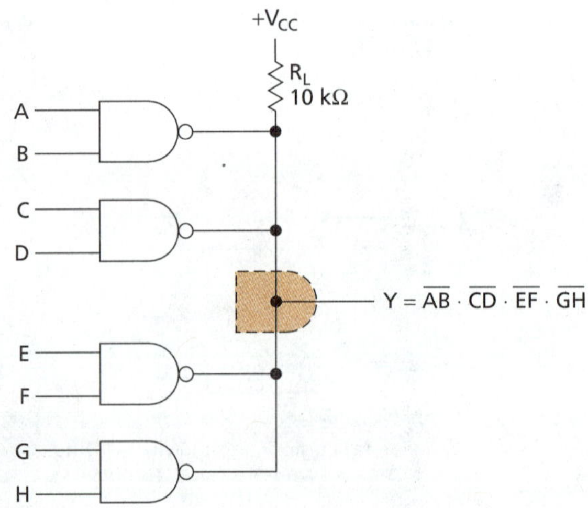

Figure 9.11 Wired AND Solution

The second output configuration that allows outputs to be directly connected together performs a different function from that of the open collector gate. This additional output configuration is called a **three-state buffer output** and is incorporated within the structure of the device. Figure 9.12 shows a NAND gate with a three-state buffered output. Besides the two logic inputs and one logic output, there is an additional control input. This input is usually designated as an ENABLE signal and when asserted, enables the logic output state to appear at the logic output of the gate. When the ENABLE signal is negated, the gate's output is effectively disconnected from the logic output pin. This output now appears to be "floating," because it is no longer terminated with the gate's output. Because of this, this state is often referred to as

Three-state buffer output *a logic output type that allows the output to be "disconnected" from its output pin. Places the output in an open or high-impedance state.*

***High-impedance** or **open state of an output** a condition in which the logic output of a device appears to be disconnected from its output pin.*

the **open state** of the output. Open states are characterized as having a high impedance and very low current, if any at all. As a result, this state is also called the **high-impedance state** of the three-state buffer. Unfortunately, too many users incorrectly refer to this state as the three-state condition of the output. Keep in mind that "three-state" actually means three states, and the device could hardly be in three states at once. The three states are an output logic level 0, output logic level 1, and the open or high-impedance state. The symbol for the three-state NAND and a truth table summarizing the functions of the gate are included in Figure 9.12.

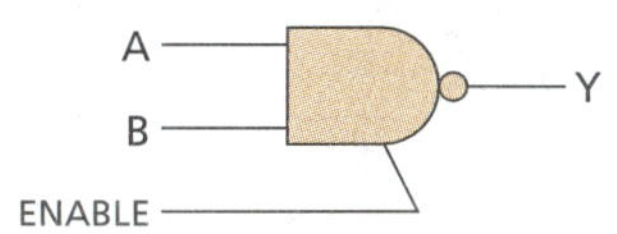

(a) Symbol

ENABLE	A	B	Y
0	X	X	Open
1	0	0	1
1	0	1	1
1	1	0	1
1	1	1	0

X = don't care state

(b) Truth Table

Figure **9.12** Three-State NAND

Three-state buffers often come in their own IC to allow any logic device's output to be connected through a three-state function. The logic symbol for a three-state buffer and its truth table are shown in Figure 9.13. The symbol for this device is a buffer (or amplifier) triangle with three leads, an input, output, and an enable control input. The enable input is often shown coming in at an angle, but that configuration is not universally adapted.

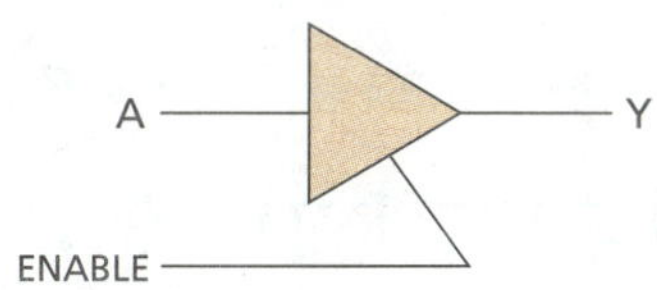

(a) Symbol

ENABLE	A	Y	Three States
0	X	Open	State 1
1	0	0	State 2
1	1	1	State 3

X = don't care state

(b) Truth Table

Figure **9.13** Three-State Buffer

Application Note

Three-State Application Note

***Bus contention** a condition that occurs when two or more devices try to place their outputs on the same bus at the same time.*

Three-state buffering is used to allow connection of many outputs to a common electrical path. Control of which output is placed onto the path is achieved through the enable inputs. These electrical paths are called *buses.* Essentially, a bus is a set of parallel conductors that carry similar information. For instance, a binary number of eight bits can be carried on a data bus of eight wires. Each conductor would carry one bit of the numerical data.

Bus contention *a condition that occurs when two or more devices try to place their outputs on the same bus at the same time.*

The output that is currently on the bus is a function of a process called **bus contention,** because the outputs literally contend for use of the bus. Only one set of outputs can be placed on the bus at one time. Otherwise, multiple outputs would mix and produce a result that did not look like any one output. Contention needs to be resolved to determine which output uses the bus at any given time. This is controlled by other signals in the digital circuit that control the states of the output enable signals.

One frequent use of three-state buffering and buses occurs in the personal computer that has many devices and circuits interconnected through three major bus systems. One bus system transfers data throughout the system and is called *data bus.* A second bus holds address information that identifies the electrical location of devices and circuits. This bus is the *address bus.* The last bus generally doesn't use three-state buffers, but controls them instead and is called the *control bus.*

Section 9.5

Practice Problems

1. Draw a logic circuit that wire-ANDs 4 two-input AND gates and complete the wired AND truth table.
2. Design a circuit that connects four outputs to a common output line through three-state buffers. Use a 1-of-4 decoder to control which output is selected to be on the line based on the binary selection made to the decoder.

9.6 Other TTL Device Types

Schottky diode *low-impedance, low-barrier voltage diodes added to TTL transistors to decrease logic gate propagation time.*

Propagation time (t_p or delay *the time period between applying an input signal and receiving an output change.*

Variations of the TTL type of integrated circuit were developed to improve operating speed and/or reduce power consumption. One variation uses a device called a **Schottky diode** to reduce the switching time of a transistor from saturation (logic 0 state) to cutoff (logic 1 state). The delay in performing the change of state is known as the **propagation time** or **delay** of the transistor and of the logic gate. More specifically, the propagation time delay is the time between applying an input change and sensing a change in output. The junction potential of a Schottky diode is approximately 0.4 V, which is less than that of a regular PN junction potential of a 0.7 V. When a transistor is turned on full (saturated), the normally back-biased PN junction from base to collector becomes forward biased. There is a 0.7 V potential from base to emitter and 0.1 V potential from collector to emitter, resulting in a 0.6 V potential between the base and collector. This causes the collector-to-base junction to be forward biased and conduct. Causing the transistor to cease conducting from the saturated condition requires time to bring the base to collector barrier voltage down from the 0.6 V and cause it to become reverse biased again. Placing a Schottky diode in parallel with the base-to-collector junction, as shown in Figure 9.14, prevents the base-to-collector junction from reaching a full 0.6 V potential. Instead, when the transistor saturates, the Schottky diode becomes forward biased and clamps the base-to-collector junction at 0.4 V. As the transistor switches to cutoff, the base-to-collector barrier has to drop 0.4 V instead of the 0.6 V it would have dropped without the Schottky diode. As a result, the transistor switches considerably faster. The symbol for the transistor with the Schottky diode in place, commonly referred to as a Schottky transistor, is shown in Figure 9.14. Typical propagation delay for Schottky TTL is 4.5 to 5.0 nS, which is three to four times faster than regular TTL, which is typically 15 to 22 nS.

(a) Schottky Diode in Place **(b)** Schottky Transistor Symbol

Figure **9.14** Schottky Transistor

Regular TTL part numbers are designated as 74xx where xx is some number from 00 on up. For instance, 7400 is the part number for a quad two-input TTL NAND IC. This means that inside a 7400 integrated circuit (IC) there are four (quad) separate NAND gates, each with two inputs. Schottky TTL devices are similarly designated with an S following the 74. A quad two-input Schottky TTL NAND IC is designated 74S00.

An extended variation on the Schottky diode is one of the more popular series of TTL devices known as the low-power Schottky or LS series. Their designation shows an LS instead of an S following the 74. For instance, the quad two-input LS NAND is designated 74LS00. Low power is achieved by reducing the current flow of the gates by increasing the value of a number of resistors. Compare the schematic of the 74S00 NAND gate of Figure 9.15 with the 74LS00 NAND gate of Figure 9.16. Notice that base, emitter, and collector resistor values in the LS device have been increased. Overall, the LS device uses less power and generates less heat than either the Schottky or regular TTL device. The supply current required for each type of gate is 8–12 ma for TTL, 16–36 ma for Schottky TTL and 1.6–4.4 ma for LS devices. Increasing the resistor values also increases the time it takes to reduce barrier potentials of saturated transistors and diodes affecting the gate's propagation time. Propagation time of the LS series is typically between 10 and 15 nS, almost twice that of the Schottky device, but still faster than the regular TTL gates.

FAST TTL *the latest form of TTL logic, whose devices operate faster than any TTL device that have preceded it.*

TTL technology continues to be improved in speed of operation and power consumption. The next TTL series is called **FAST TTL** and is designated by an F following the 74. For instance, the NAND gate used earlier, in FAST technology, is designated 74F00. The FAST family retains a comparatively lower power drain with supply currents between 2.8 and 10.2 ma while decreasing propagation time compared to LS devices. Typical propagation delays range from 4.3 to 5.0 nS.

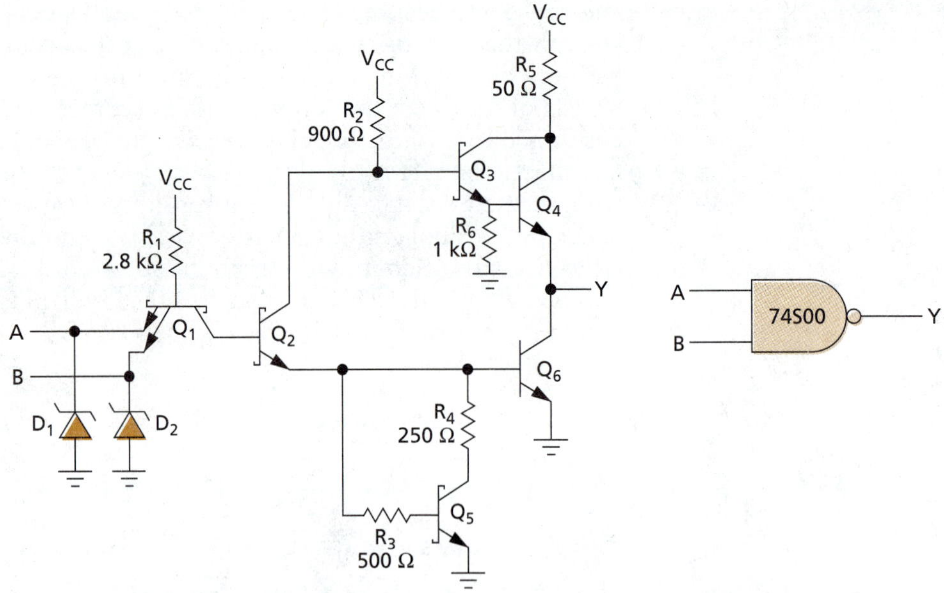

Figure 9.15 Schottky 74S00 NAND Gate

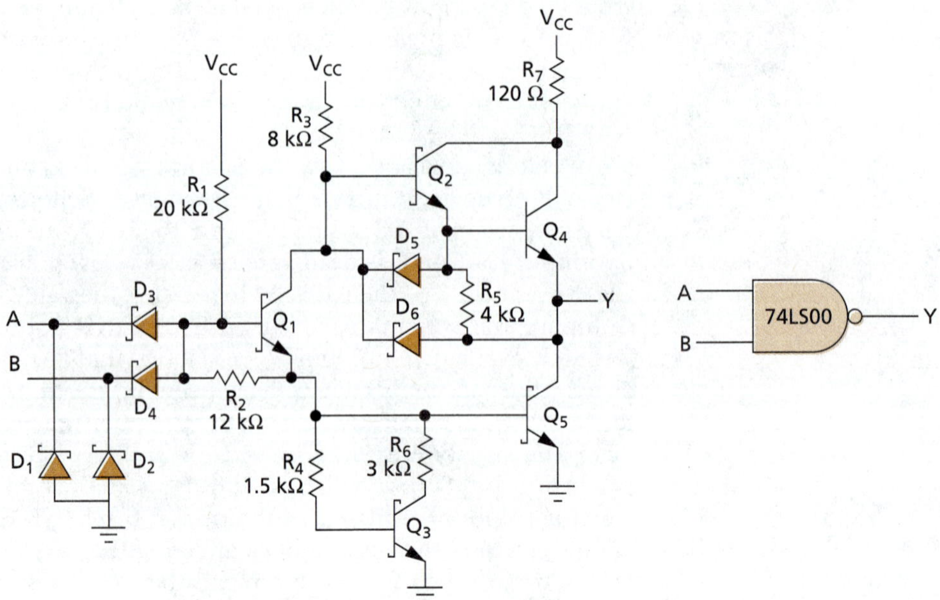

Figure 9.16 74LS00 Low-Power Schottky NAND Gate

It should be noted that most of the TTL device types discussed also come in a military version designated as 54xx instead of 74xx. These devices are essentially the same as the commercial 74 series, but are subjected to tighter specifications and quality testing.

From a logic standpoint, similar TTL gates operate the same; a NAND gate is a NAND gate, but from a design standpoint propagation delay and power consumption are two crucial considerations. With microprocessor clock speeds exceeding 150 MHz in today's personal computers, supporting logic devices must be made to also operate at higher rates. The other consideration that has always been present, but becomes increasingly important, is noise. Devices switching around or below 10 nS are affected by rapid noise spikes on power source lines as well as input and output circuit connections. Noise is a consideration in determining what to do with open inputs to gates and unused gates within a single IC.

9.7 TTL Data Sheets

Specifications and additional information for logic devices are found in data manuals. General information for a family of logic devices is usually presented first, along with a listing of available devices for that family. Each device has a data sheet that presents specific details about that particular device. For instance, Figure 9.17 details the data sheet for a 74x00 quad two-input NAND gate integrated circuit. A title containing the part numbers applicable to the data sheet is accompanied by a pin out diagram of the device. If the device is a bit more complex, a brief functional description of that device would also be included. Such a description is not needed for the 74x00. Notice by the title that this data sheet includes specifications for the three common types of TTL devices, the 7400, 74S00, and 74LS00 NAND gates.

Absolute maximum ratings *parameters that can be applied to a device without destroying it.*

Nominal-normal operating *such as voltage.*

One of the first set of specifications given in a data sheet is the **absolute maximum ratings** chart. The values indicated in this chart are those that can be applied to the device without damaging it. *They are not operating values.* As an example, the maximum value for V_{CC} is shown as 7.0 V. This means that a voltage of that potential can be applied to the V_{CC} power pin of the chip without destroying it. However, the **nominal** or **normal operating** source voltage applied to V_{CC} is 5.0 V.

The fanout table following the absolute maximum ratings chart gives the input and output current values of each TTL type, for one of the four NAND gate sections. From these numbers, fanout can be calculated using Equation 9.1, described earlier in this chapter. Direct current (DC) characteristics for the 74X00 series NAND gate yield the supply-operating current (I_{CC}) required for the device when all four gate outputs are in a high or low state. Power dissipation can be estimated from these values by multiplying these currents by the V_{CC} source voltage. The alternating current (AC) characteristics represent the operating specifications for the logic gates, including propagation delay times in terms of transition time from logic high to low and low to high states.

Power dissipation (PD) *the amount of power consumed by a logic device and dissipated as heat.*

Another parameter of concern is **power dissipation (PD)**, which is a measure of the power consumed by the operation of a device. It can be calculated by multiplying the source voltage (V_{CC}) by the current used by the device (I_{CC}). Power dissipation translates into heat generated by the operation of the device. The higher the power used, the more heat it generates and, as a consequence, the more physical area required to dissipate the heat away from the device.

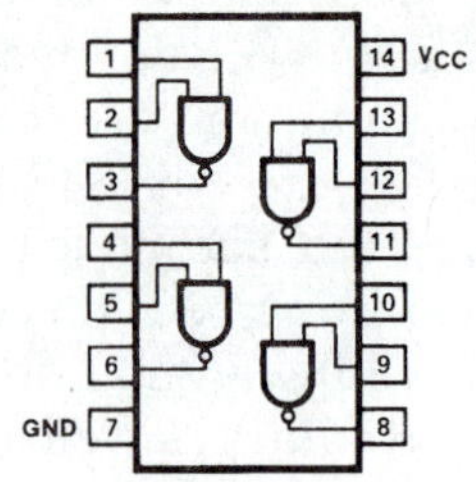

Figure A

ABSOLUTE MAXIMUM RATINGS

PARAMETER	54/74, 8200, 9300	54H/74H	54S/74S, 82S00	54LS/74LS
V_{CC} Supply voltage, continuous (Note a)	7.0V	7.0V	7.0V	7.0V
Input voltage, continuous (Notes a & b)	−0.5V to +5.5V	−0.5V to +5.5V	−0.5V to +5.5V	−0.5V to 15V(b)
Input current, continuous	−30mA to +5mA	−15mA to +5mA	−30mA to 5mA	−30mA to +1mA
Voltage applied to HIGH outputs (Note a)	−0.5V to V_{CC}	−0.5V to 7.0V	−0.5V to 7.0V	−0.5V to V_{CC}.
Voltage applied to "off" Open Collector outputs (Notes a & c)	−0.5V to 7.0V	−0.5V to 7.0V	−0.5V to 7.0V	−0.5V to 7.0V
Current into LOW standard output, continuous	30mA	40mA	40mA	15mA
Current into LOW Buffer output, continuous	80mA	100mA	100mA	50mA
Operating free air temperature range	−55°C to +125°C			
Storage temperature range	−65°C to +150°C			

NOTES

a. Voltages are referenced to device ground terminal

b. The following LS device inputs are limited to 5.5V input breakdown: All inputs of LS181, Clock inputs of LS90, LS92, LS93, LS196, LS197, LS290, LS293, LS390, LS393 and LS490.

c. Some open collector devices are specially processed to handle higher output voltages of from 15V to 30V. The Absolute Maximum voltage for these devices is 10% over the specified V_{OUT} test condition.

INPUT AND OUTPUT LOADING AND FAN-OUT TABLE (See Note a)

PINS		54/74	54H/74H	54S/74S	54LS/74LS
Inputs	I_{IH} (μA)	40		100	20
	I_{IL} (mA)	-1.6		-4.0	-0.4
Outputs	I_{OH} (μA)	250		250	250
	I_{OL} (mA)	48		60	12/24 [a]

DC CHARACTERISTICS OVER OPERATING TEMPERATURE RANGE (See Note b)

PARAMETER		TEST CONDITIONS	54/74		54H/74H		54S/74S		54LS/74LS		UNIT
			Min	Max	Min	Max	Min	Max	Min	Max	
I_{CCH}	Supply current	V_{CC} = Max, V_{IN} = 0		8.5				36		2.0	mA
I_{CCL}	Supply current	V_{CC} = Max, $V_{IN} \geq 4.5$		54				80		12	mA

Figure 9.17 74x00 Data Sheet and Appendix A

EXAMPLE 9.3

What are the fanout, power dissipation, and propagation delays for a 74LS00 NAND gate?

SOLUTION The fanout is derived by dividing the output current by the input current using Equation 9.1. From the data sheet, for the 74LS00,

$$I_{OH} = 400\ \mu a \quad I_{OL} = 8\ ma \quad I_{IH} = 20\ \mu a \quad I_{IL} = 360\ \mu a$$

and the fanout equals

$$FO_H = 400\ \mu a/20\ \mu a = 20 \qquad FO_L = 8\ mA/360\ \mu a = 22$$

The fanout for the 74LS00 is 20, the lower fanout value. Power dissipated by the 74LS00 is computed by multiplying the supply current by the supply voltage. For the 74LS00, $I_{CCH} = 1.6$ ma and $I_{CCL} = 4.4$ ma. The maximum power dissipation derives from using the higher current multiplied by the power source voltage value:

$$\text{Power dissipation} = 4.4\ ma \times 5\ V = 22\ mW$$

Propagation delays are read directly from the AC characteristics table and are 15 nS for both high- to low- and low- to high-level transition directions of the 74LS00.

The remaining specifications are specifically for the 74LS00 quad two-input NAND gate and come from the general characteristics for low-power Schottky devices presented in the front of the data manual and included in Figure 9.17 under the heading of *Low Power Schottky (54LS/74LS) Family Characteristics.* Presented in this table are the logic 1 and logic 0 input and output voltage levels and currents for different output configurations such as open collector (I_{OH}) and three-state buffer off current (I_{OZ}). Shown last in Figure 9.17 is a fanout table giving the number of inputs of each type of TTL type that can be driven by each TTL type. For instance, the 74LSxx can drive a maximum of twenty of its own 74LSxx devices, or five 74xx devices, or four 74Sxx devices.

EXAMPLE 9.4

What is the noise margin for a 74LS00?

SOLUTION From the data sheet, you find the output and input voltage levels from the logic 1 and 0 operating states and then, using Equation 9.2, take the difference to find the noise margin:

$$NM_H = V_{OH} - V_{IH} = 2.7\ V - 2.0\ V = 0.7\ V$$

$$NM_L = V_{IL} - V_{OL} = 0.8\ V - 0.35\ V = 0.45\ V$$

A comparison of specifications for TTL devices, using the 74x00 NAND gate, is summarized in Table 9.1. The specifications included on the table are input and output currents for both logic 1 and logic 0 levels, maximum input and minimum output voltage levels for logic 0 and logic 1, power dissipation (PD), fanout, and noise margin. Propagation times are listed as the time for the output to perform a low to high transition (tp_{LH}) and high to low transition (tp_{HL}). As expected, the gate with the shortest propagation time is the FAST TTL, followed by the Schottky type (74S00), and the one with the lowest power dissipation is the LS variety (74LS00).

EXAMPLE 9.5

From Table 9.1, what logic level would you apply to the inputs of unused NAND gates within each device type?

Table 9.1 Logic Families Comparison Chart

SPECIFICATION	7400	74H00	74S00	74LS00	FAST
I_{IH} (μa)	40	50	50	20	100
I_{IL} (ma)	1.6	2.0	2.0	0.36	0.6
I_{OH} (μa)	400	500	1000	400	1000
I_{OL} (ma)	16	20	20	8	20
I_{CC} or I_{DD} (mA)	12	40	36	4.4	10.2
t_{PLH} (nS)	22	10	4.5	15	5.0
t_{PHL} (nS)	15	10	5.0	15	4.3
V_{IH} (volts)	2.0	2.0	2.0	2.0	2.0
V_{IL} (volts)	0.8	0.8	0.8	0.8	0.8
V_{OL} (volts)	0.4	0.4	0.5	0.5	0.5
V_{OH} (volts)	2.4	2.4	2.7	2.7	2.5
PD (mW)	60	200	180	40	100
Fanout	10	10	10	20	10
NM high (volts)	0.4	0.4	0.7	0.7	0.5
NM low (volts)	0.4	0.4	0.3	0.3	0.3

SOLUTION The choice of logic level to be applied to unused gates is determined by the current drawn for a particular output state. Connecting both inputs of the NAND gate to a high forces the output to a low state, drawing a current of 40 μa (I_{IH}) + 16 ma (I_{OL}) for a total of 16.04 ma. Tying the inputs to a logic 0 forces the output to a logic 1 and causes I_{IL} + I_{OH} to be 1.6ma + 400 μa or a total of 2.0 ma to be drawn. Since tying the unused inputs low causes less current to be drawn (2.0 ma versus 16.04 ma), then it is advisable to connect unused gate inputs to ground.

Section 9.7

Practice Problems

1. Compare the fanouts for 7400, 74S00, and 74LS00 when each drives only the same type of devices. First compute the fanouts using the current values from the fanout table, and then verify the fanouts using the fanout chart at the end of the data sheet.
2. Compare the propagation delays of the 7400, 74S00, and 74LS00.
3. For the following electronics equipment, which TTL type would you use for the devices within it? Why would you select that type?
   a. Fifty MHz desk-top personal computer
   b. Portable cassette player
   c. Thirty MHz lap-top computer
   d. Portable digital multimeter

9.8 Terminating Unused Inputs

Leaving an input open to a TTL gate is *logically* the same as applying a logic 1 level to that input. The input transistor of a NAND gate does not conduct when the emitter is disconnected from a terminating point or signal, which is similar to applying a logic 1 to the emitter input of that input transistor. This causes the output of the NAND gate to go to a logic 0 when signals on the other NAND inputs are a high logic level. Essentially, all the inputs look like logic 1's, causing the NAND output to go to a 0. Setting any of the other inputs low forces the NAND output to go high. In effect, the open input to the NAND gate has no *logical* effect on the output, which responds to other inputs applied to the gate.

EXAMPLE 9.6

What is the logical expression for the output of the AND gate in Figure 9.18?

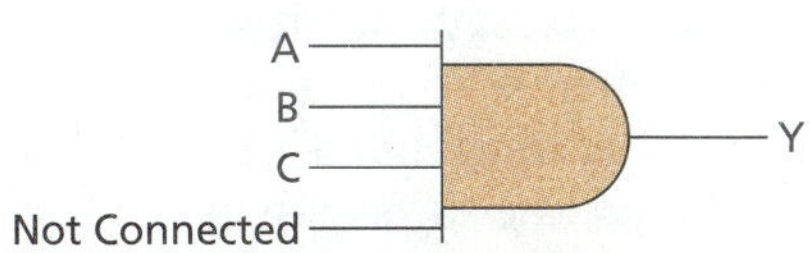

Figure 9.18 Example 9.6

SOLUTION The open input can either be ignored or considered to be in a logic 1 state. The Boolean expression is

$$Y = A \cdot B \cdot C \quad \text{or} \quad Y = A \cdot B \cdot C \cdot 1 = A \cdot B \cdot C$$

Leaving an input open to a TTL NOR gate is logically disastrous. The input, again, acts like a logic 1, and any time a logic 1 is applied to the input of a NOR gate, the output is forced low, regardless of the activity on the remaining inputs.

So then it's OK to leave an unused TTL NAND input unterminated, but you should always terminate an unused TTL NOR input with a 0 logic level, right? *Wrong!* The NOR consideration is correct, but *never* leave any input unterminated. The main reason is noise. Any open input is susceptible to induced noise spikes from nearby circuit paths and other sources and causes the gate to function erratically. This affects the logic circuits the gate is driving and can generate additional electrical noise within the device itself that translates into induced noise to other gates within the same device. Unused gates within an IC chip should also be terminated because of the possibility of noise as well as fluctuations in supply current and heat dissipation caused by the erratic behavior of the unused gate. You can also keep the power dissipation at a lower level by selecting the correct termination for the unused gate.

EXAMPLE 9.7

Which logic level is used to terminate unused inputs to 74xx series NAND and NOR gates?

SOLUTION The logic state desirable to apply an unused gate is determined by the inactive state of the gate's input. An unused input to a NAND gate is set to a high logic state, and an unused input to a NOR gate is terminated in a low state. These terminations do not affect the logic function of the gate with respect to the remaining inputs. The termination to an inactive state does reduce noise influences on the operation of the gate.

9.9 Device Parameters

Table 9.2 lists the symbols for various logic device parameters and specifications, some of which have been previously discussed. Output and input voltages specify the minimum and maximum output driving voltages and input sense voltages for the device that is expected to retain each specified logic state. For instance, V_{OH} specified for a 7400 NAND gate is 2.4 V. This is a minimum voltage level that a TTL gate can output and still be recognized as a logic 1. The V_{OL} for the same gate is 0.4 V. This is a maximum voltage that can be generated and still be considered a logic 0. We can also determine the "no man's" operating range in which the logic level measured is neither a logic 1 or 0. The difference between these two voltages—that is, the range from 0.4 to 2.4 V—is the range of the undefined output logic level.

On the input side, V_{IH} is the minimum level that can be applied to a TTL input as a logic 1 and V_{IL} is the maximum 0 logic level voltage. The range of voltages between these two levels (0.8 to 2.0 V for the 7400 gate) is an undefined input logic voltage. The difference between the voltage from a gate's output and the acceptable input voltage at the same logic level is known as the noise mar-

Table 9.2 Glossary of Specification Sheet Symbols and Terms

VOLTAGES	
V_{CC}	Collector or source supply voltage
V_{IH}	Input high logic level
V_{IL}	Input low logic level
V_{OH}	Output high voltage
V_{OL}	Output low voltage
NM_H	High-level noise margin
NM_L	Low-level noise margin
VOLTAGES	
I_{CC}	Supply current
I_{IL}	Input low logic-level current
I_{IH}	Input high logic-level current
I_{OL}	Output logic low current
I_{OH}	Output high current
I_{OS}	Output short circuit current
FI	Fan in
FO	Fanout
TIMING PARAMETERS	
C_I	Input capacitance
f_{MAX}	Maximum operating frequency
t_F	Fall time
t_R	Rise time
t_{PHL}	Propagation time high-to-low transition
t_{PLH}	Propagation delay low-to-high transition
t_H	Hold time
t_S	Setup time
t_W or PW	Pulse width time
DC	Duty cycle

gin. In effect, noise in this area is acceptable and will not affect the logic outcome from the driven gate. Noise margin is the absolute difference between the input voltage and the output logic voltage for the same logic state, as detailed in Equation 9.2. For the 7400 NAND gate, the noise margins are

$$NM_H = |V_{OH} - V_{IH}| = 0.4 \text{ V}$$
$$NM_L = |V_{OL} - V_{IL}| = 0.4 \text{ V}$$

The output current specification is the maximum current value that a device supplies while still maintaining a correct logic 1 voltage. Any additional load causes more current to be drawn and the logic 1 level to decrease below the minimum V_{OH} specification. The effect of this current value is to limit the number of similar gates that a single logic device can drive. This parameter is called the fanout, which has been discussed earlier in the chapter and expressed in Equation 9.1. Timing parameters include the already familiar propagation delay, which is defined as the transition time between applying an input change and getting a response on the output. Additional time loss occurs because of rise (t_{PLH}) and fall (t_{PHL}) times of these level changes. Additional timing considerations involve the maximum operating frequency and input capacitance. Capacitance, although not a timing parameter, has a lot of effect on device timing. This capacitance limits the rise and fall times of input signals and in conjunction with other internal capacities, determines the propagation times and maximum operating frequency of a device.

An illustration of some of the timing parameters is shown in Figure 9.19. Rise time (t_R) is the time it takes for a signal to rise from 10 to 90 percent of the maximum voltage level of a signal. Fall time (t_F) is the time it takes to do the reverse, fall from 90 to 10 percent of a signal's peak voltage. Propagation time, tp, has been defined as the time between a signal reaching a logic level at the input and getting a change of state at the output. Minimum logic 1 voltage levels and maximum logic 0 voltage levels occur during the rise and fall times of the signals applied to a gate's input. Both of the propagation times, high-to-low and low-to-high transitions, for a NAND gate are shown in Figure 9.19. The maximum operating frequency is derived by adding together a gate's high-to-low (t_{PHL}) and low-to-high (t_{PLH}) propagation times to yield the minimum pulse width (PW) of a signal that can be applied to the gate with enough time for the gate to successfully switch between logic levels. The reciprocal of this time is the maximum operating frequency for the gate. This process is illustrated in Figure 9.20.

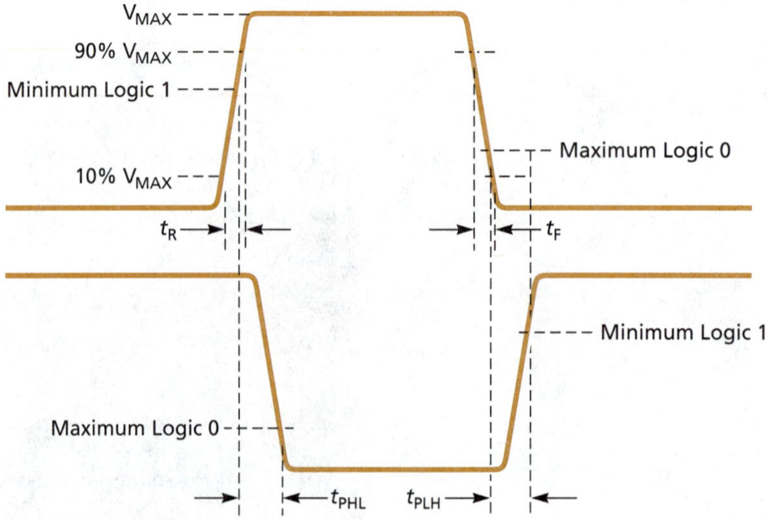

Figure 9.19 Rise (t_R), Fall (t_F), and Propagation (t_p) Times

The maximum operating frequency of the 74LS00 NAND gate is computed using the 15 nS propagation specified in the data sheet for the gate:

$$1/f_{MAX} = t_{PHL} + t_{PLH} = 15 \text{ nS} + 15 \text{ nS} = 30 \text{ nS}$$

$$f_{MAX} = 1/\text{total time} = 1/30 \text{ nS} = 33.3 \text{ MHz}$$

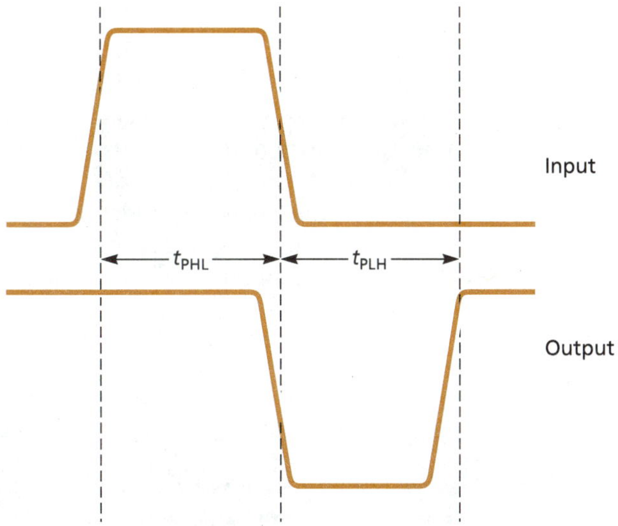

$$f_{MAX} = \frac{1}{t_{PLH} + t_{PHL}}$$

Figure 9.20 Determining f_{MAX}

Duty cycle is not generally considered a parameter or specification for a logic device. However, it can have an effect on a device's operation by determining how long a logic signal remains high and how long it remains low. Duty cycle is found by dividing the logic 1 state time of a repetitive signal by the total time of a single cycle of that signal multiplied by 100 percent. In Figure 9.21, the

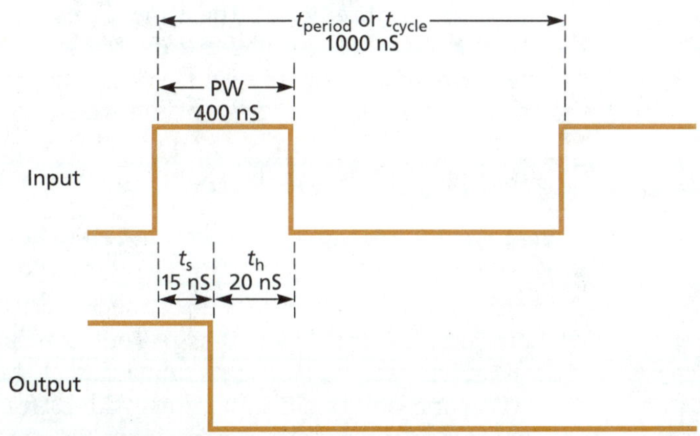

Note: Numbers are used as example values.

t_s Setup time

t_h Hold time

PW Pulse width

$$\text{Duty cycle (DC)} = \frac{PW}{t_{period}} \times 100\%$$

Figure 9.21 Timing Considerations

Set-up time or t_s *the minimum time an input must be applied before a synchronizing signal is applied.*

Hold time (t_h) *the minimum time an input must be retained to allow a circuit or device to respond to it.*

duty cycle of the input waveform is 400 nS/1,000 nS × 100 = 40 percent. In other words, the signal is high 40 percent of the time. **Set-up time** (t_s) is the minimum time that a signal must be applied to an input to achieve a desired output result. **Hold time** (t_h) is the minimum time an input signal must remain in place to assure that the desired results have enough time to occur. In Figure 9.21, the input must be present a minimum of 15 nS and must remain 20 nS after the output has begun to change to assure that the change will take place. Removing the input signal too soon may cause the output to return to its original state or go into an undefined result.

Section 9.9

Practice Problems

1. Show how the following parameters were derived for the 74LS00 NAND gate using Table 9.2:
   a. Fanout
   b. NM_H and NM_L
2. What are the maximum operating frequencies of the 74S00, 74LS00, and 74F00 NAND gates, using Table 9.2?

9.10 Emitter-Coupled Logic (ECL)

ECL (emitter-coupled logic) transistor-based logic technology having nonsaturated switching transistors.

Schottky diodes were added to TTL transistors to speed up the device's switching time, by reducing the time transistors needed to come out of saturation. Imagine, then, if the transistors never went into saturation (or cut off, for that matter) in the first place, but switched instead between two points within the operating area of the transistor's characteristic curve. The switch between the two levels of conduction would be a lot faster than forcing the transistor to switch between full conduction and cutoff. This is the concept behind **emitter-coupled logic** or **ECL** logic devices. Propagation times for ECL gates are around 0.8 nS, which is still five to six times faster then the FAST TTL technology listed in Table 9.1. One tradeoff is the high power dissipation required to keep constant currents conducting through the ECL circuitry. Typical power dissipations can run as high as 250 mW per IC. Other disadvantages of ECL gates include (1) the need for two power sources and (2) using odd operating voltage values for the two logic states, because they are not ground (0 V) and V_{CC}, as they are when using TTL. Since these odd voltages must be maintained fairly close between output drivers and inputs to gates, noise margins are tighter, meaning that ECL devices are less tolerant of noise spikes. In addition, maintaining these logic levels restricts the number of loads that can be placed on ECL outputs, keeping their fanouts very low. However, if speed is the name of the game, then you cannot top ECL devices.

Section 9.10

Practice Problem

For the following electronic equipment, which might use ECL devices. Why would you not use ECL for the others?

1. Cable TV channel box
2. A 150 MHz computer workstation control station
3. A 100 MHz lap-top computer
4. CD stereo player

9.11 The MOS Family of Devices

Speed of operation is one consideration when selecting a logic family to use. Another consideration is power dissipation and size. These both have a direct consequence on the production of portable computer equipment that run off of battery-type power sources. Obviously, the less power used by devices, the longer the battery powering them lasts. It is also desirable to keep portable equipment as small and lightweight as possible. Thus the size of components that make up these portable units must be small and contain as much electronic circuitry as possible. One restriction on the size of an integrated chip is the surface area required to dissipate the heat generated by the operation of the chip. When power dissipation is lowered in a logic circuit, less heat is generated and more logic circuits can be included within a given area of an IC chip. The term associated with the amount of circuitry that can be included within a single integrated circuit is packing density.

IGFET (insulated gate field effect transistor) *the switching device used for MOS technologies.*

NMOS (N-channel metal oxide semiconductor) *N-channel FET-based logic technology.*

One benefit of **metal oxide semiconductor (MOS)** technology is that it reduces the problem of power dissipation and increases packing density in an integrated circuit. Logic circuits derived from this technology use **insulated gate field effect transistors (IGFETs)** to perform the logic and amplifying functions of the transistors in TTL logic devices. A NAND gate made using **NMOS**, (which stands for N-channel MOS) IGFETs is shown in Figure 9.22. Immediately notice the lower component count in this gate compared to the 74LS00 TTL NAND gate in Figure 9.3. FETs Q_1 and Q_2 supply the AND and inverting functions, while Q_3 is used as a drain resistance to V_{CC}. When either input A or B is 0, its corresponding FET is cut off and not conducting. The output is returned high through the source current from Q_3. The only time output Y goes low is when a 1 is applied to both A and B inputs. These high inputs turn on Q_1 and Q_2, allowing sink current from ground to drive output Y to a 0 logic state. A truth table for this gate shows output Y high whenever an input is low and low only when A AND B are both high, satisfying the definition of a NAND gate.

The NMOS version of the NOR gate, shown in Figure 9.23, has Q_1 and Q_2 connected with drain and source in parallel. This means that if either FET conducts, output Y sees source current and a low logic level. Output Y will be high only when both inputs are low and both FETs are nonconducting. This describes the

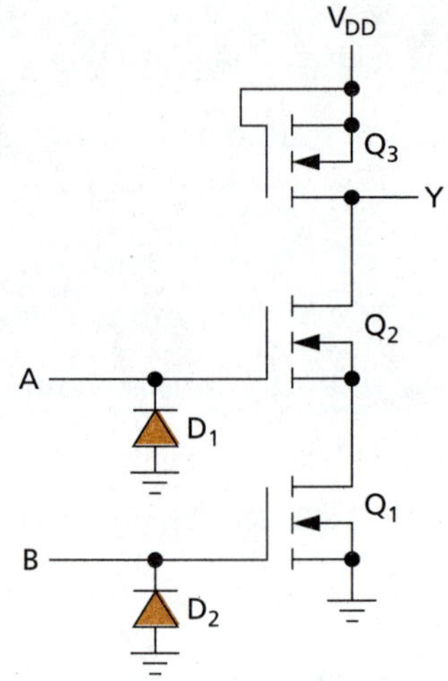

Figure 9.22 NMOS NAND Gate

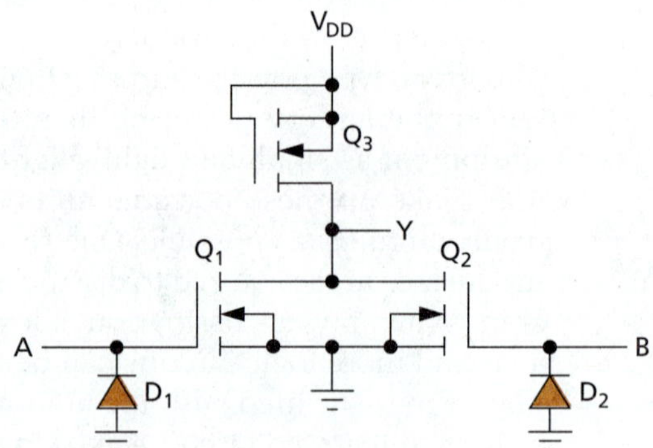

Figure 9.23 NMOS NOR Gate

PMOS (P-channel metal oxide semiconductor) *P-channel FET-based logic technology.*

CMOS (complementary metal oxide semiconductor *combined N-channel and P-channel FET-based logic technology.*

basic NOR gate whose output is low if input A OR input B is high. NMOS and similar devices created using **PMOS** (P-channel FETs) draw less current and dissipate less power than TTL devices. Further power reduction is achieved using complementary N-channel and P-channel FETs in a single gate. This family of devices is known as **complementary MOS** or **CMOS** technology.

When NMOS outputs sink current to ground, they also drive current through drain FET Q_3 to V_{CC}. The CMOS inverter in Figure 9.24, however, only sinks or sources current to the output and not through a drain resistor. This reduces power consumption and lowers heat that would have been dissipated through the drain resistance. The inverter of Figure 9.24 has an N-channel (Q_1) and P-channel (Q_2) FET connected between ground and source voltage, V_{CC}. Input A is applied to the gates of both FETs, and the output is taken between the drain connections. A high logic level at input A causes the N-channel FET to conduct and the P-channel FET to cut off. Output Y receives only the sink current

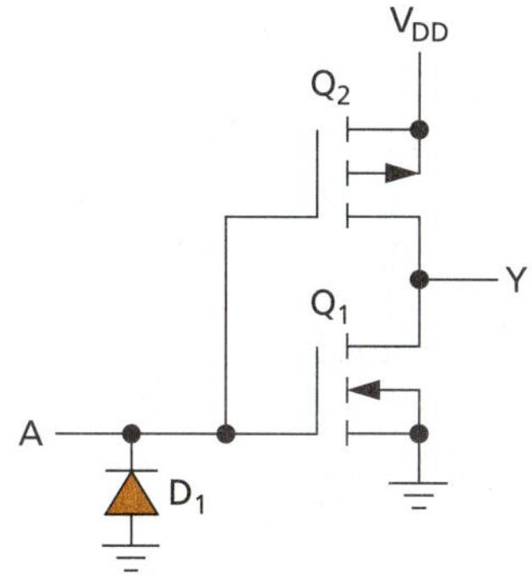

Figure 9.24 CMOS Inverter Gate

through Q_1 and a logic level of 0. A 0 applied to input A cuts off the N-channel FET and turns on the P-channel FET, allowing only source current to reach input B and forcing that output to a high state. Thus this gate performs the invert function while limiting the total current that flows in the gate to the current drawn by the load applied at output Y, at any given time.

Typical operating power source current for a CMOS gate is 1.0 μa compared to 4.4 ma for the low-power Schottky (74LSxx) device. Although the reduced power consumption is a large advantage of MOS logic devices, there are also significant disadvantages. Insulated gate FETs have a very high input impedance—in the neighborhood of 10,000 MΩ or better. This translates to a low input current that is usually less than 1 μa. This is good, but the effect of insulating the gate from the drain and source is to create a capacitance between the gate and those elements. This capacitance must be charged before a FET starts to conduct. This increases the propagation time when compared to TTL devices. A CMOS gate has propagation times of around 125 nS compared to 10 nS for TTL and 5.0 nS for Schottky TTL.

The high-impedance input also presents a handling problem that TTL gates do not experience. MOS devices are highly susceptible to static charges, which do not build up on the lower input impedance of TTL devices. A static charge can build up rapidly through improper handling until it reaches a voltage that causes the gate to the source to arc over and destroy the FET. Antistatic mats and wrists straps are used to reduce the risk of static charge buildup while handling MOS devices. Another aid in reducing the risk is the addition of reverse-biased diodes at the device gate inputs (D_1 in Figure 9.24). These are the same diodes used in TTL circuits to eliminate negative switching spikes. For MOS, they serve the more important job of reducing static charge problems.

Since MOS inputs have high impedance, it does not require much input fluctuation to cause the circuit to change. An unterminated input will appear to "drift" or "float," depending on the amount of electrical noise that exists in the surrounding environment. At one time, this floating input may appear as a logic 1 and at other times as a logic 0. Because of the high input impedance, this drift will be highly unstable, changing from one state to another at its own whim. *Always terminate an unused MOS input regardless of the level or absence of electrical noise.*

CMOS designations are similar to TTL except that a C appears after the 74 or 54 designation. Figure 9.25 is an example of a 74C00 NAND gate. A 0 on either input causes at least one of the cascaded N-channel FETs to be cut off, preventing sink current from reaching output Y. The input that is brought low turns on its corresponding P-channel FET to supply source current and a high to output Y. When both inputs are high, both N-channel FETs conduct sinking output Y and both P-channel FETs are cut off, preventing any source current from flowing. In essence, when input A AND input B are high, output Y is low, defining the NAND function.

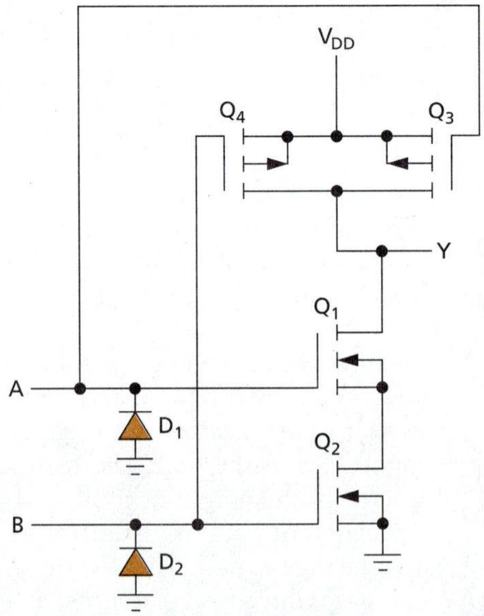

Figure 9.25 CMOS NAND Gate

The NOR gate of Figure 9.26 has its N-channel FETs in parallel and its P-channel FETs cascaded. A high input to either A or B cuts off the source current through Q_3 and Q_4 and supplies sink current to output Y through either Q_1 or Q_2 or both. Inputs of 0 on A and B turn on both P-channel FETs and cut off both N-channel FETs. This is the only time source current is present at output Y. In essence, output Y is brought low if either input A OR input B is set high, defining the NOR function.

The range of supply voltages for CMOS devices is +5 to +15 V. These devices operate faster at higher supply voltages than at the lower ones, so +12 to +15 V is the more commonly used supply voltage. This operating voltage is one reason that CMOS and TTL devices are incompatible with each other. The different propagation times are another reason that it is difficult to use both types of devices in the same circuit, but the real limiting factor is the currents required and supplied by each device type. Output low current, I_{OL}, for CMOS devices is 0.44 ma, while input low current, I_{IL}, for a 74LSxx TTL input is 0.36 ma. This means that a CMOS output can only drive a single TTL input. More about interfacing logic families later. The point now is that some of these problems have been resolved with a faster version of CMOS called **high-speed CMOS** or

HCMOS *high-speed CMOS technology.*

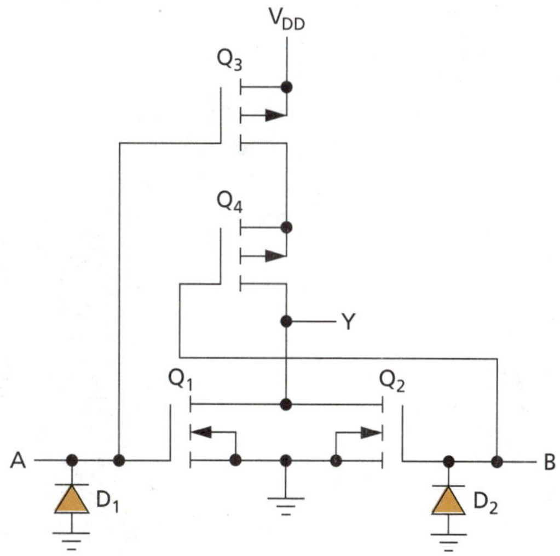

Figure 9.26 CMOS NOR Gate

HCMOS, designated as 74HCxx or 74Hxx. Some IC manufacturers also call this version of CMOS, FAST CMOS, not to be confused with FAST TTL.

HCMOS devices operate with a 5 V source voltage for voltage compatibility with TTL devices. Propagation delays have been reduced to 15 nS, which is comparable with low-power Schottky (LS) series TTL operating speeds. Output current for a low logic level has been increased to 4.0 ma so that fanouts when driving 74LSxx TTL devices are closer to 10 than 1. Table 9.3 summarizes the specifications for TTL and MOS devices.

Table 9.3 Logic Families Comparison Chart

SPECIFICATION	7400	74H00	74S00	74LS00	FAST	74HC00	74C00	ECL
I_{IH} (μa)	40	50	50	20	100	1.0	0.3	265
I_{IL} (ma)	1.6	2.0	2.0	0.36	0.6	1 μa	0.3 μa	0.5 μa
I_{OH} (μa)	400	500	1,000	400	1,000	4,000	2,100	50 ma
I_{OL} (ma)	16	20	20	8	20	4.0	0.44	50
I_{CC} or I_{DD} (ma)	12	40	36	4.4	10.2	1.0 μa	1.0 μa	26
tp_{LH} (nS)	22	10	4.5	15	5.0	15	125	0.8
tp_{HL} (nS)	15	10	5.0	15	4.3	15	125	0.8
V_{IH} (volts)	2.0	2.0	2.0	2.0	2.0	3.15	3.5	−1.13
V_{IL} (volts)	0.8	0.8	0.8	0.8	0.8	1.35	1.5	−1.48
V_{OL} (volts)	0.4	0.4	0.5	0.5	0.5	0.1	0.05	−1.63
V_{OH} (volts)	2.4	2.4	2.7	2.7	2.5	4.4	4.95	−0.98
PD (mW)	60	200	180	40	100	5 μW	5 μW	250
Fanout	10	10	10	20	10	4,000	1,466	189
NM high (volts)	0.4	0.4	0.7	0.7	0.5	1.25	1.45	0.15
NM low (volts)	0.4	0.4	0.3	0.3	0.3	1.25	1.45	0.15

Section 9.11

Practice Problems

1. For each application, select the type of device that would most likely be used and give a reason for your selection.
   a. Notebook computer
   b. Large-screen TV remote control receiver logic
   c. Pentium desk-top computer
   d. Automated production emergency cutoff circuit
2. Which device family has the following?
   a. The fastest switching speed
   b. Lowest power dissipation
   c. Highest fanout
   d. Least tolerant noise margin

9.12 Interfacing Different Logic Families

Mixing various types of TTL devices in the same logic circuit does not present too many problems. They all run at a supply voltage of 5 V, and most combinations have a sufficient fanout. One exception to this general rule is using low-power devices to drive other types of TTL devices. The nature of the lower power consumption is accomplished by reducing currents throughout the device. Lower output currents cause significant decreases in fanout. Looking back on the 74xx00 data sheet, the last chart on that sheet detailed fanout relationships among the various TTL devices. It shows a fanout of 5 when an LS type tries to drive standard TTL (74xx) inputs, which is about half the fanout normally present between like members of the TTL family of devices.

Other considerations when mixing TTL types is the propagation delay differences from one type to another. Because some gates operate faster than others, it is important to avoid mixing device types in circuits that have critical timing considerations. Signals arriving at points in the circuit at incorrect times may develop unexpected results.

Greater caution must be taken when mixing TTL, MOS, and ECL devices within the same circuits. For all the speed it presents, ECL is the most difficult to interface with other logic families. Not only are its power supplies at different voltages when compared to TTL or CMOS, but logic-level voltages are also different. To adapt ECL to TTL, for instance, requires voltage-level shifters to change ECL logic levels to TTL and back. These shifters also buffer ECL outputs from TTL inputs in an effort to match the differing input and output loading characteristics and impedances.

Many aspects must be taken into consideration when interfacing TTL or FAST TTL with CMOS or HCMOS. The older CMOS, which operated more efficiently at a supply voltage (and a high logic level) between 12 V and 15 V, requires a voltage-level shifter to change the 12 V CMOS level to TTL logic 1 voltage of 5 V and back. One method used to achieve the translation is to use logic devices with open collector (for TTL) and open drain (for CMOS) outputs. This way, as shown in Figure 9.27, pull-up resistors connected to the required level can be used to effect the change. When the open-collector TTL output switches to a logic 1 level, the output is raised to the 12 V sensed by the CMOS input. Similarly, an open-drain CMOS output raises to 5 V through its pull-up resistor when switching to a logic 1 level.

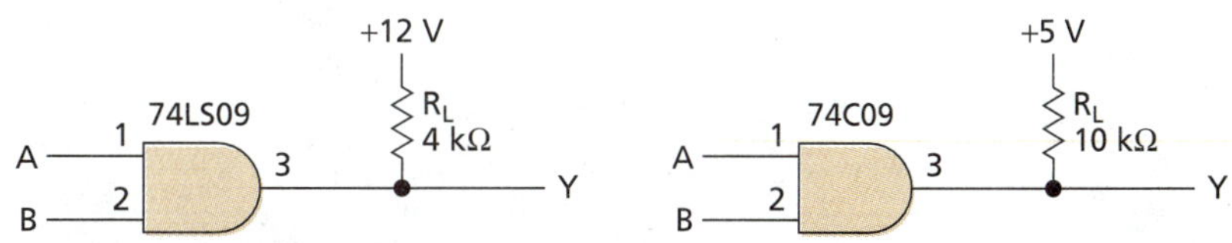

(a) TTL to CMOS-Level Shifter Using Open-Collector Output

(b) CMOS to TTL-Level Shifter Using Open-Drain Output

Figure **9.27** Logic Level Shifting

Another consideration when mixing TTL- and CMOS-type devices is fanout. Output currents for TTL are considerably higher than those for CMOS. The same can be said for input currents. When a TTL output is required to drive CMOS inputs, there is virtually no current loading problem. The high TTL output current easily drives hundreds of CMOS inputs. However, when the reverse is applied, trouble abounds. From Table 9.3, the output current for a low state from a CMOS device is shown to be 0.44 ma and the input current for a standard TTL device is at 1.6 ma. The fanout for this combination is 0.44 ma/1.6 ma or rounded to 0. Try a 74LS00 device whose input current at a low state is 0.36 ma. Now the fanout is 0.44 ma/0.36 ma or 1, which brings us to a commonly used term for non-TTL devices: *TTL compatible.* when you see this term, interpret it to mean that the device

1. Has logic output levels of 0 and 5 V
2. Has sufficient output current to drive *one* TTL load

Be careful not to make the mistake of thinking that a TTL-compatible device operates as fast as a TTL gate or has the same fanout as a normal TTL device.

Speed is still a consideration. The HCMOS, the high-speed version of the CMOS devices, operates at TTL logic levels and has a propagation delay time approaching 74LSxx types—approximately 15 nS. Mixing 74HCxx with 74LSxx devices works well as long as you do not ignore the fanout limitation when using the HCMOS to drive LS TTL. Standard CMOS runs considerably slower than most TTL devices and is inappropriate for circuits with critical timing considerations.

These are not the only aspects to consider when mixing logic technologies, but these are the more obvious considerations. Always consult data manuals and data sheets when trying to figure out how well such mixing will work within a given application.

Section 9.12

Practice Problems

1. Which of these mixed families requires level shifters?
   a. HCMOS/LS TTL b. 74xx/74LSxx c. ECL/CMOS d. 74Cxx/74LSxx

2. In Figure 9.28, which output goes low first when C is brought high? Why? Inputs A and B are previously set high.

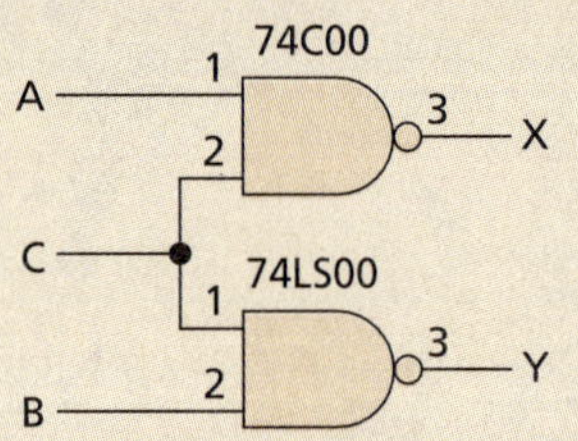

Figure 9.28 Practice Problem 9.12(2)

3. Give as many reasons as you can why HCMOS and LS TTL would not be used together in a notebook computer.

9.13 Medium-Scale Integration Data Sheet

SSI (small-scale integration) *a class of logic devices that have minimal circuitry. Examples include quad two-input AND and OR gate IC's.*

Discrete logic gate *a single logic gate usually found in small groups in SSI devices.*

MSI (medium-scale integration) *a class of logic devices that falls between discrete (SSI) and LSI devices.*

The data sheet for the 74x00 quad two-input NAND is an example of a specification for a **small-scale integration** or **SSI** device. These devices have minimal logic functions and are often referred to as **discrete logic gates**. They include many versions of the AND, OR, NAND, NOR, XOR, inverter, buffer logic gates. The next level above SSI is **medium-scale integration** or **MSI** and includes logic circuits that are a bit more complex. These include four-bit parallel adders, counters, registers, decoders, encoders, multiplexers, demultiplexers, and other logic circuits.

An example of a MSI device is shown in the data sheet for the 74LS190 BCD decade and 74LS191 four-bit binary counters (Figure 9.29). Since this device is more complex, an introductory description of its functions is included on the first page of the data sheet, along with a logic symbol and connection (pin) diagram. At the bottom of this page is a state diagram describing both counters' sequences. The description indicates that both counters are presettable, so that the starting count value can be selected.

A logic diagram is provided for the counter, which is followed by an in-depth description of their functions. It is not my intent at this time to discuss the operation of these counters, but instead to familiarize the reader with the form and content of their data sheet. It is important for a user to fully understand what the device that is to be used is all about. The detailed description of the counter's operation is an excellent place to begin.

The remaining sections of the data sheet detail the operating and electrical specifications for the counters. Many specifications are similar to those found

DESCRIPTION

The "191" is a presettable 4-Bit Binary Up/Down Counter with state changes of the counter synchronous with the LOW-to-HIGH transition of the Clock Pulse input.

The circuit features an asynchronous Parallel Load ($\overline{PL}$) input which overrides counting and loads the data present on the D_n inputs into the flip-flops. Synchronous expansion in a multistage counter is made possible by a Count Enable ($\overline{CE}$) input. The count up or count down mode is determined by an Up/Down ($\overline{U}$/D) input. A variety of methods for generating carry/borrow signals in multistage counter application is made possible by Terminal Count (TC) and Ripple Clock ($\overline{RC}$) outputs.

FEATURES

- **Synchronous, reversible 4-bit binary counting**
- **Asynchronous parallel load capability**
- **Count Enable control for synchronous expansion**
- **Single Up/Down control input**

LOGIC SYMBOL

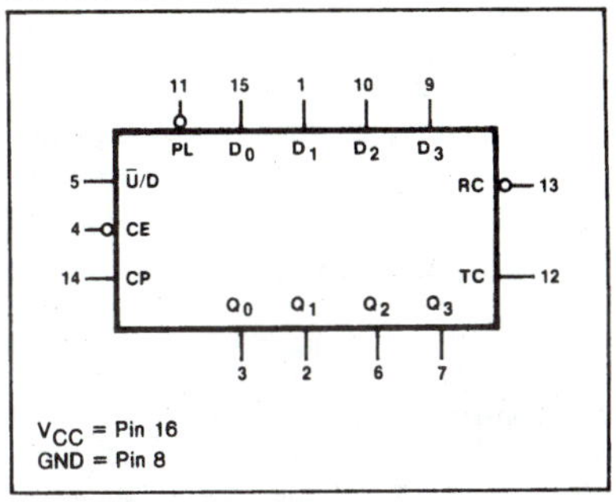

PIN CONFIGURATION

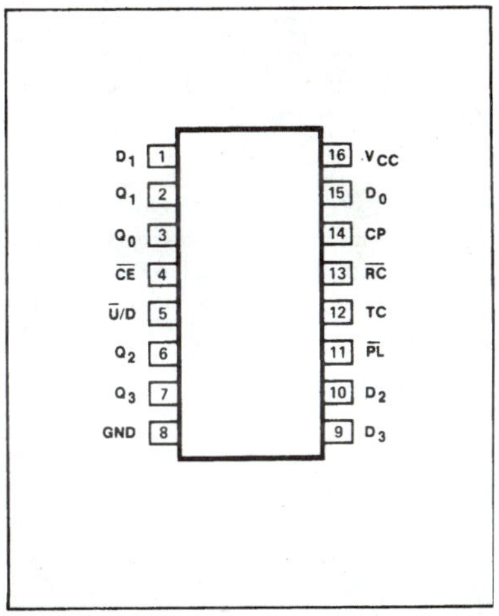

INPUT AND OUTPUT LOADING AND FAN-OUT TABLE[(a)]

PINS	DESCRIPTION		54/74	54S/74S	54LS/74LS
CP	Clock Pulse (active HIGH going edge) input	I_{IH} (μA) I_{IL} (mA)	40 −1.6		20 −0.4
$\overline{CE}$	Clock Enable (active LOW) input	I_{IH} (μA) I_{IL} (mA)	120 −4.8		60 −1.2
$\overline{U}$/D	Up/Down count control input	I_{IH} (μA) I_{IL} (mA)	40 −1.6		20 −0.4
$\overline{PL}$	Parallel Load (active LOW) input	I_{IH} (μA) I_{IL} (mA)	40 −1.6		20 −0.4
D_n	Parallel Data inputs	I_{IH} (μA) I_{IL} (mA)	40 −1.6		20 −0.4
Q_n	Counter outputs	I_{OH} (μA) I_{OL} (mA)	−800 16		−400 4/8[(a)]
TC	Terminal Count output	I_{OH} (μA) I_{OL} (mA)	−800 16		−400 4/8[(a)]
$\overline{RC}$	Ripple Clock (active LOW pulse) output	I_{OH} (μA) I_{OL} (mA)	−800 16		−400 4/8[(a)]

NOTE
a. The slashed numbers indicate different parametric values for Military/Commercial temperature ranges respectively.

Figure 9.29 74LS191 Data Sheet

FUNCTIONAL DESCRIPTION

The "191" is an asynchronously presettable Up/Down 4-Bit Binary Counter. It contains four master/slave flip-flops with internal gating and steering logic to provide asynchronous preset and synchronous count-up and count-down operation.

Asynchronous parallel load capability permits the counter to be preset to any desired number. Information present on the parallel Data inputs (D_0-D_3) is loaded into the counter and appears on the outputs when the Parallel Load ($\overline{PL}$) input is LOW. As indicated in the Mode Select Table, this operation overrides the counting function.

Counting is inhibited by a HIGH level on the Count Enable ($\overline{CE}$) input. When $\overline{CE}$ is LOW, internal state changes are initiated synchronously by the LOW-to-HIGH transition of the Clock input. The Up/Down ($\overline{U}/D$) input signal determines the direction of counting as indicated in the Mode Select Table. The $\overline{CE}$ input may go LOW when the clock is in either state, however, the LOW-to-HIGH $\overline{CE}$ transition must occur only while the Clock is HIGH. Also, the $\overline{U}/D$ input should be changed only when either $\overline{CE}$ or CP is HIGH.

MODE SELECT-FUNCTION TABLE

OPERATING MODE	INPUTS					OUTPUTS
	$\overline{PL}$	$\overline{U}/D$	$\overline{CE}$	CP	D_n	Q_n
Parallel load	L	X	X	X	L	L
	L	X	X	X	H	H
Count up	H	L	I	↑	X	count up
Count down	H	H	I	↑	X	count down
Hold "do nothing"	H	X	H	X	X	no change

TC AND $\overline{RC}$ TRUTH TABLE

INPUTS			TERMINAL COUNT STATE				OUTPUTS	
$\overline{U}/D$	$\overline{CE}$	CP	Q_0	Q_1	Q_2	Q_3	TC	$\overline{RC}$
H	X	X	H	H	H	H	L	H
L	H	X	H	H	H	H	H	H
L	L	⊔	H	H	H	H	H	⊔
L	X	X	L	L	L	L	L	H
H	H	X	L	L	L	L	H	H
H	L	⊔	L	L	L	L	H	⊔

H = HIGH voltage level steady state.
L = LOW voltage level steady state.
I = LOW voltage level one setup time prior to the LOW-to-HIGH clock transition
X = Don't care.
↑ = LOW-to-HIGH clock transition.
⊔ = LOW pulse.

LOGIC DIAGRAM

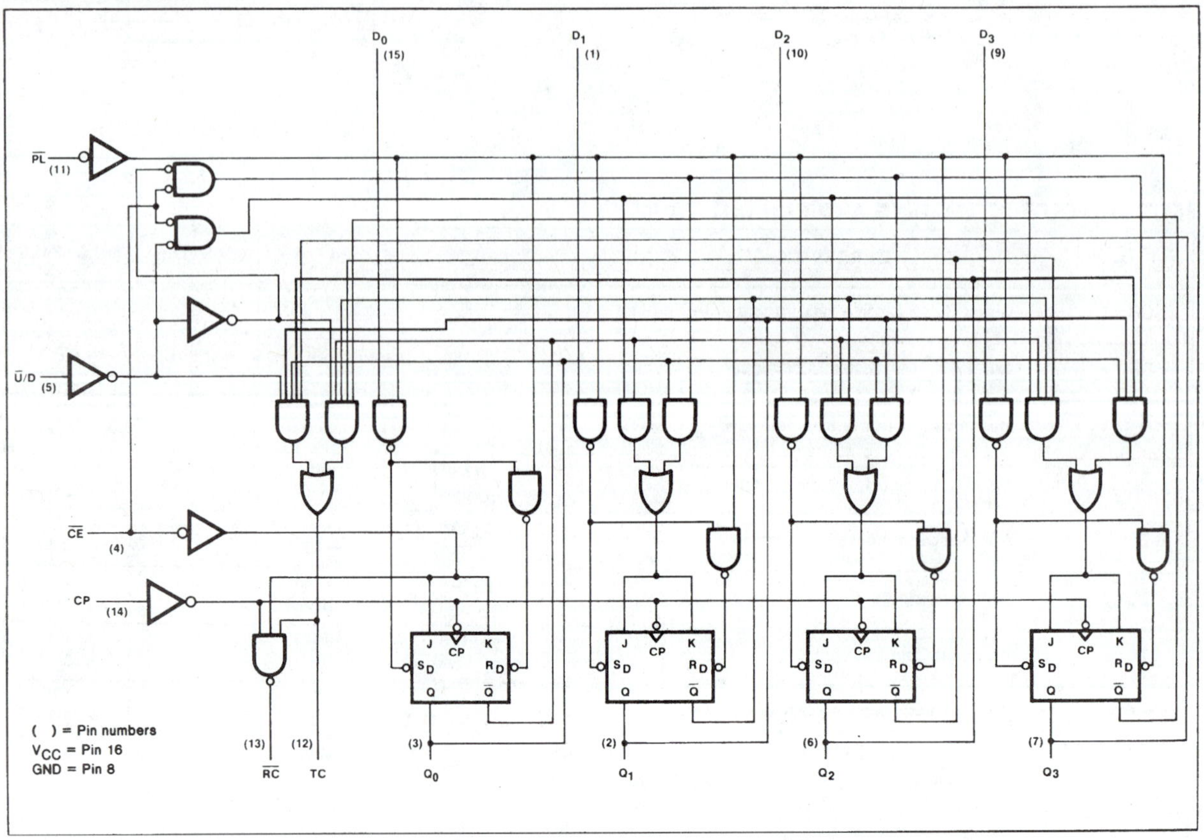

Figure 9.29 Continued

Overflow/underflow indications are provided by two types of outputs, the Terminal Count (TC) and Ripple Clock ($\overline{RC}$). The Terminal Count (TC) output is normally LOW and goes HIGH when a circuit reaches zero in the count-down mode or reaches "15" in the count-up mode. The TC output will remain HIGH until a state change occurs, either by counting or presetting, or until $\overline{U}$/D is changed. Do not use the TC output as a clock signal because it is subject to decoding spikes.

The TC signal is used internally to enable the Ripple Clock ($\overline{RC}$) output. When TC is HIGH and $\overline{CE}$ is LOW, the Ripple Clock follows the Clock Pulse (CP) delayed by two gate delays. The $\overline{RC}$ output essentially duplicates the LOW clock pulse width, although delayed in time by two gate delays. This feature simplifies the design of multistage counters as indicated in Figures A and B. In Figure A, each $\overline{RC}$ output is used as the clock input for the next higher stage. When the clock source has a limited drive capability this configuration is particularly advantageous, since the clock source drives only the first stage. It is only necessary to inhibit the first stage to prevent counting in all stages, since a HIGH signal on $\overline{CE}$ inhibits the $\overline{RC}$ output pulse as indicated in the Mode Select Table. The timing skew between state changes in the first and last stages is represented by the cumulative delay of the clock as it ripples through the preceding stages. This is a disadvantage of the configuration in some applications.

Figure B shows a method of causing state changes to occur simultaneously in all stages. The $\overline{RC}$ outputs propagate the carry/borrow signals in ripple fashion and all clock inputs are driven in parallel. The LOW state duration of the clock in this configuration must be long enough to allow the negative-going edge of the carry/borrow signal to ripple through to the last stage before the clock goes HIGH. Since the $\overline{RC}$ output of any package goes HIGH shortly after its CP input goes HIGH, there is no such restriction on the HIGH state duration of the clock.

In Figure C the configuration shown avoids ripple delays and their associated restrictions. Combining the TC signals from all the preceding stages forms the $\overline{CE}$ input signal for a given stage. An enable signal must be included in each carry gate in order to inhibit counting. The TC output of a given stage is not affected by its own $\overline{CE}$ therefore the simple inhibit scheme of Figure A and B does not apply.

N-STAGE COUNTER USING RIPPLE CLOCK

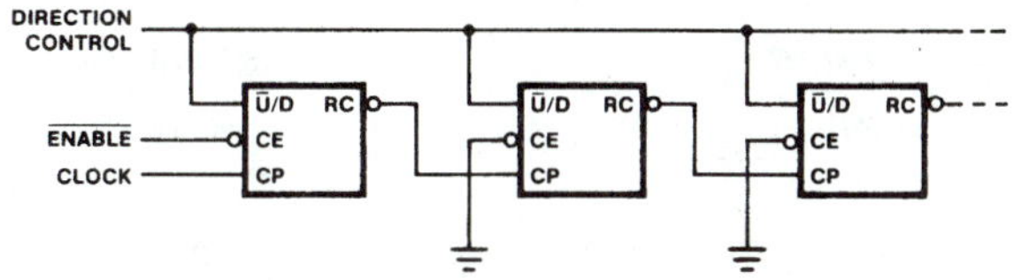

Figure A

SYNCHRONOUS N-STAGE COUNTER USING RIPPLE CARRY/BORROW

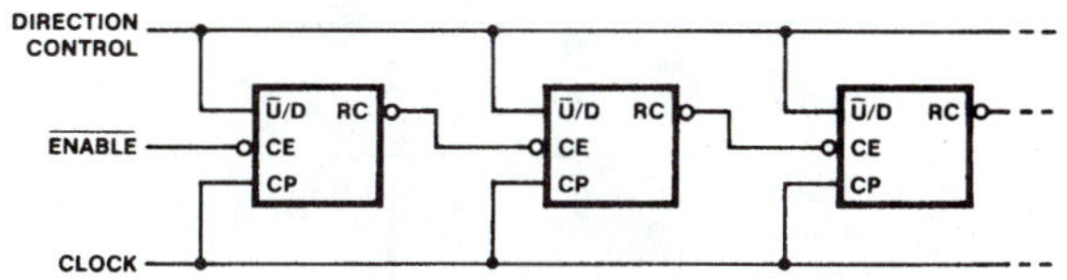

Figure B

SYNCHRONOUS N-STAGE COUNTER WITH PARALLEL GATED CARRY/BORROW

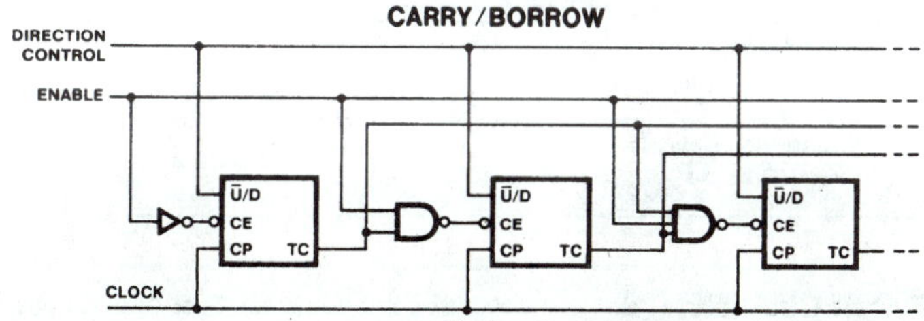

Figure C

DC CHARACTERISTICS OVER OPERATING TEMPERATURE RANGE[b]

PARAMETER		TEST CONDITIONS		54/74 Min	54/74 Max	54S/74S Min	54S/74S Max	54LS/74LS Min	54LS/74LS Max	UNIT
I_{OS}	Output short circuit current	V_{CC} = Max	Mil	−20	−65			−15	−100	mA
			Com	−18	−65			−15	−100	mA
I_{CC}	Supply current	V_{CC} = Max	Mil		99				35	mA
			Com		105				35	mA

NOTE

b. For family dc characteristics, see inside front cover for 54/74 and 54H/74H, and see inside back cover for 54S/74S and 54LS/74LS specifications.

Figure 9.29 Continued

AC CHARACTERISTICS: T_A=25°C (See Section 4 for Test Circuits and Conditions)

	PARAMETER	TEST CONDITIONS	54/74		54S/74S		54LS/74LS		UNIT
			C_L = 15pF, R_L = 400Ω				C_L = 15pF, R_L = 2kΩ		
			Min	Max	Min	Max	Min	Max	
f_{MAX}	Maximum input count frequency	Figure 1	20				20		MHz
t_{PLH} t_{PHL}	Propagation delay, Clock to Q outputs	Figure 1		24 36				24 36	ns ns
t_{PLH} t_{PHL}	Propagation delay, Clock to $\overline{RC}$ output	Figure 2		20 24				20 24	ns ns
t_{PLH} t_{PHL}	Propagation delay, Clock to TC output	Figure 1		42 52				42 52	ns ns
t_{PLH} t_{PHL}	Propagation delay $\overline{U}$/D to $\overline{RC}$ output	Figure 7		45 45				45 45	ns ns
t_{PLH} t_{PHL}	Propagation delay $\overline{U}$/D to TC output	Figure 7		33 33				33 33	ns ns
t_{PLH} t_{PHL}	Propagation delay Data to Q outputs	Figure 3		22 50				32 40	ns ns
t_{PLH} t_{PHL}	Propagation delay $\overline{PL}$ to any output	Figure 4		33 50				33 50	ns ns
t_{PLH} t_{PHL}	Propagation delay $\overline{CE}$ to $\overline{RC}$ output	Figure 2		33 33				33 33	ns ns

AC SETUP REQUIREMENTS: T_A = 25°C (See Section 4 for Test Circuits and Conditions)

	PARAMETER	TEST CONDITIONS	54/74		54S/74S		54LS/74LS		UNIT
			Min	Max	Min	Max	Min	Max	
t_W	CP pulse width	Figure 1	25				25		ns
t_W	$\overline{PL}$ pulse width	Figure 5	35				35		ns
t_s	Setup time Data to $\overline{PL}$	Figure 6	20				20		ns
t_h	Hold time Data to $\overline{PL}$	Figure 6	0				0		ns
t_{rec}	Recovery time $\overline{PL}$ to CP	Figure 5	40				40		ns
$t_s(L)$	Setup time LOW $\overline{CE}$ to clock	Figure 8	40				40		ns
$t_h(L)$	Hold time LOW $\overline{CE}$ to Clock	Figure 8	0				0		ns

Figure 9.29 Continued

AC WAVEFORMS

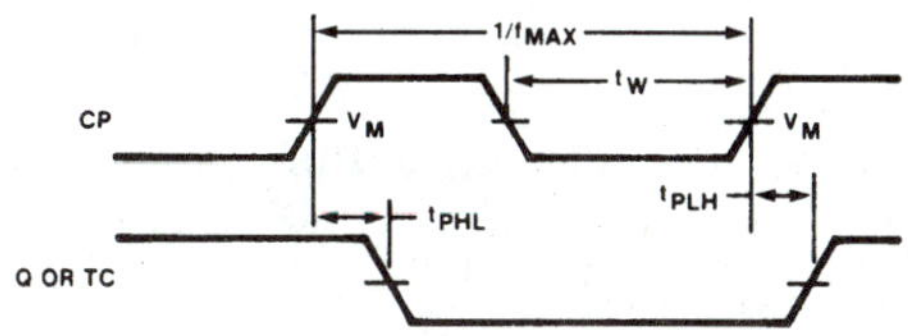

V_M = 1.5V for 54/74 and 54S/74S; V_M = 1.3V for 54LS/74LS.

Figure 1

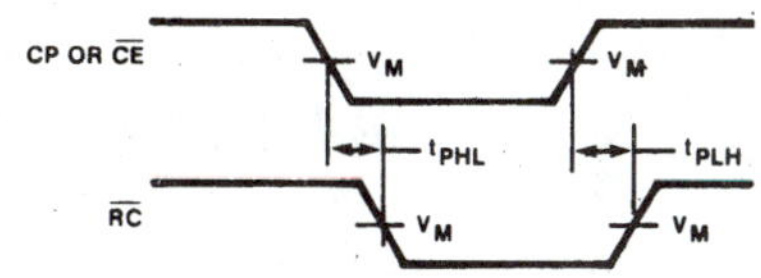

V_M = 1.5V for 54/74 and 54S/74S; V_M = 1.3V for 54LS/74LS.

Figure 2

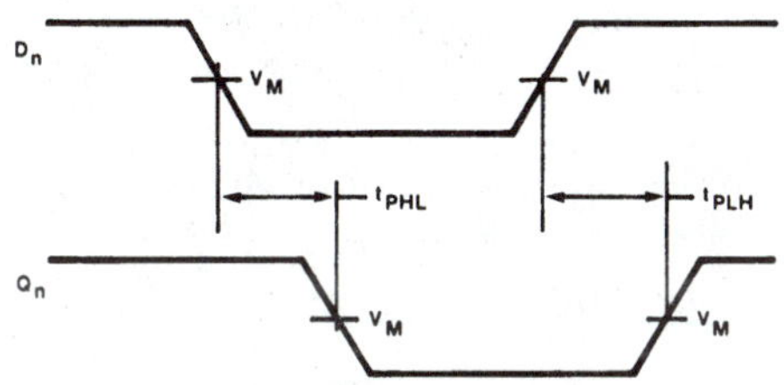

V_M = 1.5V for 54/74 and 54S/74S; V_M = 1.3V for 54LS/74LS.

Figure 3

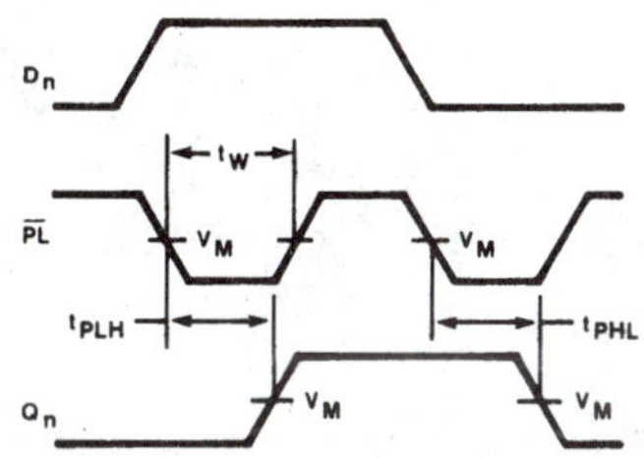

V_M = 1.5V for 54/74 and 54S/74S; V_M = 1.3V for 54LS/74LS.

Figure 4

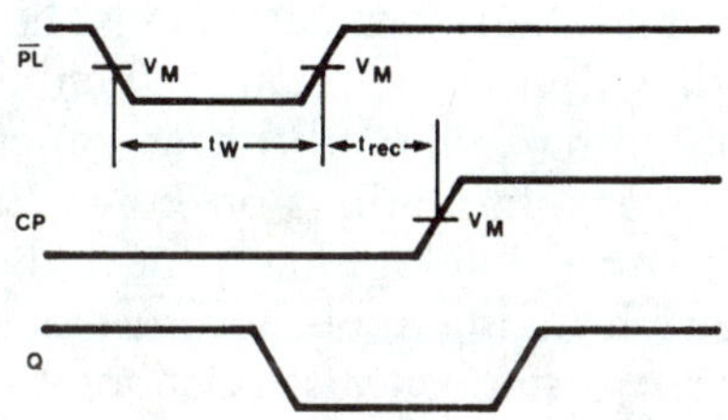

V_M = 1.5V for 54/74 and 54S/74S; V_M = 1.3V for 54LS/74LS.

Figure 5

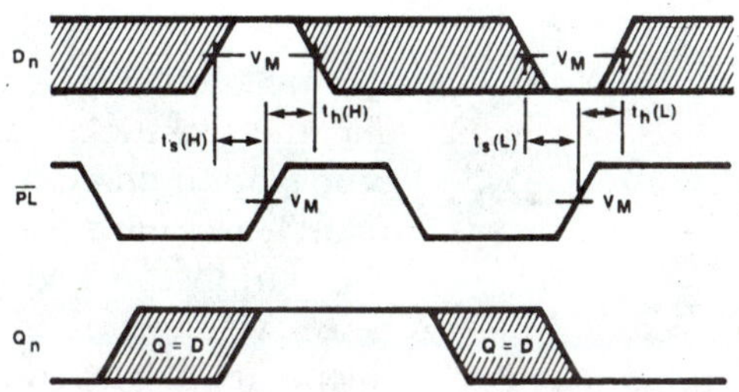

V_M = 1.5V for 54/74 and 54S/74S; V_M = 1.3V for 54LS/74LS.
The shaded areas indicate when the input is permitted to change for predictable output performance.

Figure 6

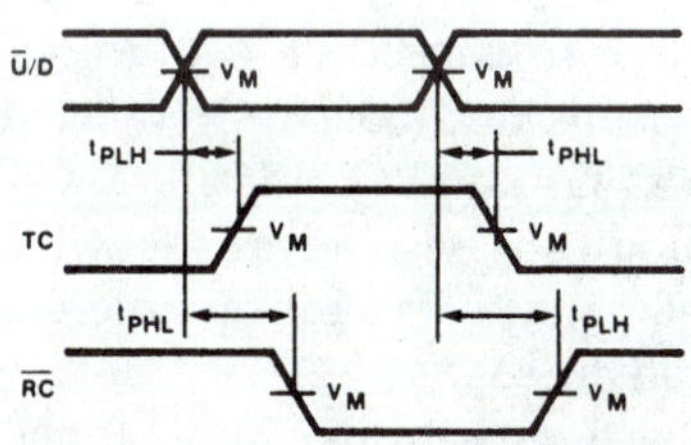

V_M = 1.5V for 54/74 and 54S/74S; V_M = 1.3V for 54LS/74LS.

Figure 7

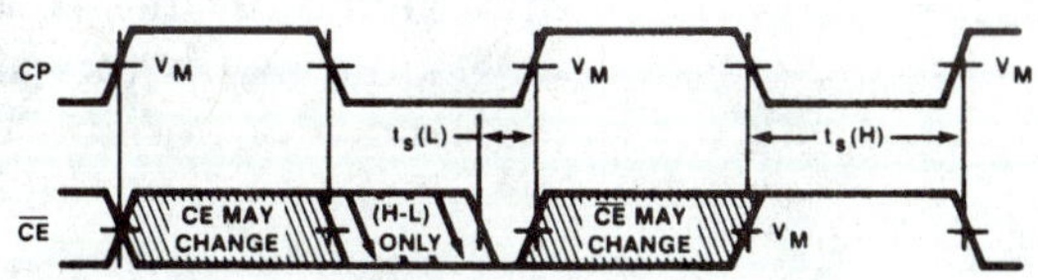

V_M = 1.5V for 54/74 and 54S/74S; V_M = 1.3V for 54LS/74LS.
The shaded areas indicate when the input is permitted to change for predictable output performance.

Figure 8

Figure 9.29 Concluded

in the SSI data sheet, including output and input currents and voltages and propagation times. This particular data sheet includes some basic applications circuits followed by timing diagrams that graphically represent the device's AC characteristics.

Data sheets for the SSI and MSI chips used throughout the text are included in Appendix A following this chapter. Data books containing an entire family of devices are available from the manufacturers of those devices. These data books are essentially reference material to be used when designing or troubleshooting digital circuits. The bare minimum of information from these data books gives the pin out and operating functions of the device you may be concerned with.

LSI (large-scale integration) *a class of logic devices that contains complex circuits and numerous functions.*

A last note about data sheets and books. We have dealt with SSI and MSI logic devices in this chapter reflecting those introduced in the text. A further level of integration is called **large-scale integration** or **LSI**. At this level, we find microprocessors and other complex devices. Individual device data sheets may be books in themselves because of the many functions these devices have. These data sheets or books are usually your first and foremost reference for any device you may encounter. It is, therefore, important that you understand how to use this information.

Summary

Having knowledge about logic devices and their parameters aids the designer and the technician as they explore digital circuits. Factors such as fanout, propagation delay, and power dissipation have direct bearings on circuit design. These parameters set restrictions and limitations on how a particular logic family can be used. Each type of logic technology has its advantages, with newer versions and newer technologies bringing about improvements or refinements of various specifications. For instance, MOS technology was advanced to reduce power and heat dissipation and increase packing density of integrated circuits. The price for this improvement over TTL devices is longer propagation times. Further refinement of MOS technology brought about the HCMOS family, which reduced propagation times to nearly match those of low-power Schottky TTL devices. Similarly, FAST TTL technology was developed to further reduce propagation time while also yielding a slight improvement in power dissipation when compared to older TTL technologies.

Requirements for faster operating speeds and reduced power dissipation, especially for portable equipment such as notebook computers, will force still further advances in logic family technologies. These advances will continue in the integrated circuit area, and it will be a major job for technicians and technologists alike to keep pace with these advances.

Understanding data sheets and books is of utmost importance in understanding how a logic device works and what its limitations and specifications are. It does not matter whether the device is SSI, MSI, or LSI—data sheets and books are usually the first available information about them. SSI, by the nature of their small size and few functions, have the simplest and most straightforward data references. As logic circuits get more complex, so do their data sheets and books. Logic devices such as counters and registers are included in MSI-size chips. Their data sheets usually include a description of their operation and accompanying truth table. Microprocessors and other multifunctional logic devices are included in the LSI class of devices. Information about them fills a small-size data book rather than a data sheet. Included in these data books are functional descriptions, block diagrams, electrical characteristics, and application notes giving examples of their operation and use.

Glossary

absolute maximum ratings parameters that can be applied to a device without destroying it.

bus contention a condition that occurs when two or more devices try to place their outputs on the same bus at the same time.

CMOS (complementary metal oxide semiconductor) combined N-channel and P-channel FET-based logic technology.

Darlington output output of a logic circuit that provides high-speed switching of heavily loaded outputs.

discrete logic gate a single logic gate usually found in small groups in SSI devices.

DTL (diode–transister logic) a logic technology that preceded the development of TTL.

ECL (emitter-coupled logic) transistor-based logic technology having nonsaturated switching transistors.

Fanout (FO) the number of similar devices that can be driven by a single logic gate's output.

FAST TTL the latest form of TTL logic, whose devices operate faster than any TTL device that have preceded it.

HCMOS high-speed CMOS technology.

high-impedance or **open state of an output** a condition in which the logic output of a device appears to be disconnected from its output pin.

hold time(t_h) the minimum time an input must be retained to allow a circuit or device to respond to it.

IGFET (insulated gate field effect transistor) the switching device used for MOS technologies.

I_{OH}, I_{IH}, I_{OL}, and I_{IL} input and output current characteristics of a logic device.

LSI (large-scale integration) a class of logic devices that contains complex circuits and numerous functions.

MSI (medium-scale integration) a class of logic devices that falls between discrete (SSI) and LSI devices.

NMOS (N-channel metal oxide semiconductor) N-channel FET-based logic technology.

noise margin the difference between an output logic state voltage level and an input logic state voltage level.

nominal-normal operating such as voltage.

open collector output logic devices whose output transistors or FETs do not have a collector or drain pull-up resistor.

PMOS (P-channel metal oxide semiconductor) P-channel FET-based logic technology.

power dissipation (PD) the amount of power consumed by a logic device and dissipated as heat.

propagation time (t_p) or delay the time period between applying an input signal and receiving an output change.

RTL (resistor–transistor logic) a logic technology that preceded the development of TTL.

Schottky diode low-impedance, low-barrier voltage diodes added to TTL transistors to decrease logic gate propagation time.

set-up time or t_s the minimum time an input must be applied before a synchronizing signal is applied.

SSI (small-scale integration) a class of logic devices that have minimal circuitry. Examples include quad two-input AND and OR gate IC's.

three-state buffer output a logic output type that allows the output to be "disconnected" from its output pin. Places the output in an open or high-impedance state.

totem pole output a logic output configuration that provides both source and sink currents for logic 1 and logic 0 output states.

TTL (transistor–transistor logic) a transistor-based logic technology.

V_{OH}, V_{OL}, V_{IH}, and V_{IL} logic device input and output voltage levels for logic 1 and logic 0.

wired AND and wired OR simulated logic function that occurs when open collector outputs are connected together.

Solutions to Practice Problems

Section 9.4

FO_H = 17 ma/0.46 ma = 36 FO_L = 15 ma/0.98 ma = 15
Fanout is the lower of the two calculations, or 15.
NM_H = 3.95 V − 3.77 V = 0.18 V NM_L = 0.86 V − 0.75 V = 0.11 V

Section 9.5

1.

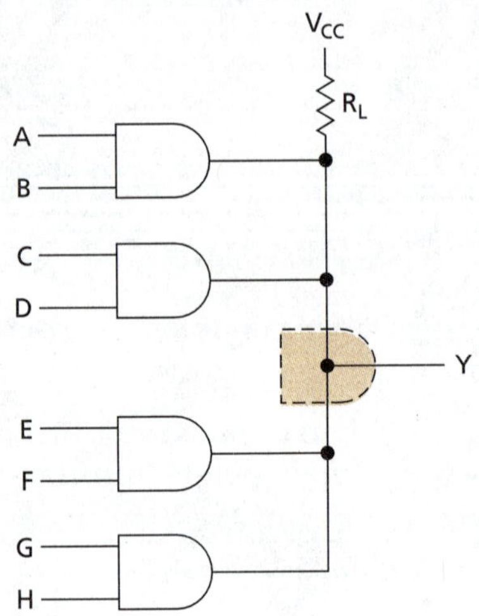

Figure 9.30 Practice Problem 9.5(1) (solution)

A B C D E F G H	Y
← Any Input Low →	0
← All Inputs High →	1

Figure 9.31 Truth Table for Practice Problem 9.5(1)

2.

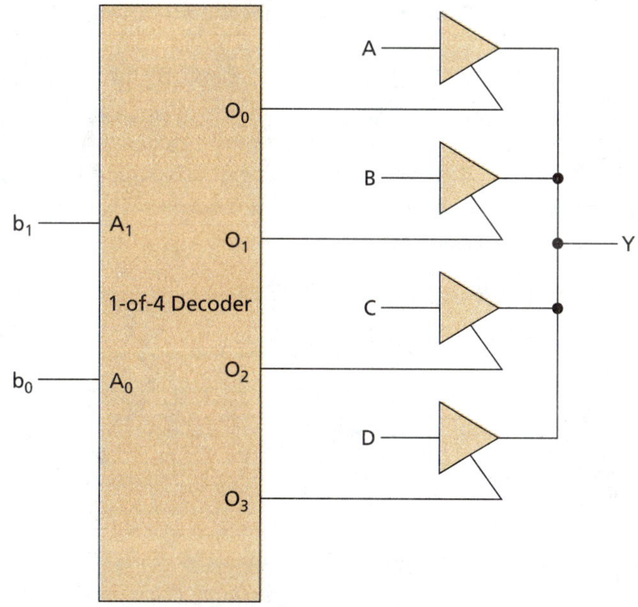

Figure 9.32 Practice Problem 9.5(2) (solution)

Section 9.7

1. Fanout 7400: 400 μa/40 μa = 10 and 16 ma/1.6 ma = 10; FO = 10
74S00: 1 ma/50 μa = 20 and 20 ma/2.0 ma = 10; FO = 10
74LS00: 400 μa/20 μa = 20 and 8 ma/.36 ma = 22; FO = 20
From the table: all fanouts concur.

2. Propagation times read from the data tables:

	7400	74S00	74LS00
t_{pHL}	15 nS	5 nS	15 nS
t_{pLH}	22 nS	4.5 nS	15 nS

3. **a.** 74Fxx—speed
b. 74LSxx—lowest power consumption
c. 74LSxx—good speed, low power
d. 74LSxx—good speed, low power

Section 9.9

1. **a.** Fanout = I_O/I_I 400 μa/20 μa = 20 8 ma/3.6 ma = 22; FO = 20
b. $NM_L = V_{OL} - V_{OL} = 0.8 - 0.5 = 0.3$ V
$NM_H = V_{OH} - V_{IH} = 2.7 - 2.0 = 0.7$ V

2. 74S00: 4.5 nS + 5.0 nS = 9.5 nS 1/9.5 nS = 105 MHz
74LS00: 2 × 15 nS = 30 nS 1/30 nS = 33.33 MHz
74F00: 5.0 nS + 4.3 nS = 9.3 nS 1/9.3 nS = 108 MHz

Section 9.10

1. Do not use ECL because: there is no need for high speed, would prefer single voltage source.

2. Use ECL for speed-critical parts of the work station.

3. Do not use ECL, because high power consumption of ECL reduces battery charge life.

4. Do not use ECL because speed is not required for CD system.

Section 9.11

1. a. HCMOS—low power consumption, good speed
   b. Any TTL for better speed
   c. ECL or FAST TTL—speed
   d. ECL—quickest

2. a. ECL
   b. 74Cxx
   c. 74HCxx
   d. ECL

Section 9.12

1. Answers c and d

2. Y, because 74LS00 has lower propagation delay time than 74C00

3. 74LSxx consumes too much power; poor fanout when 74HCxx drives 74LSxx

Problems

9.1 Compare the diode–transistor logic (DTL) technology to transistor–transistor logic (TTL) technology. What are the advantages of TTL compared to DTL?

9.2 What is the significance of the fanout specification to the logic circuit designer?

9.3 What important information is conveyed by the noise margin specification for a logic gate?

9.4 Relate power dissipation to packing density for an integrated circuit (IC) chip.

9.5 Answer the following questions about the gate in Figure 9.33:
- **a.** What type of logic function does the gate perform?
- **b.** What are the noise margins of the gate?
- **c.** What functions do each of the individual transistors perform in the gate circuit?
- **d.** What is the fanout of this gate?

9.6 Answer the following questions about the gate in Figure 9.34:
- **a.** What type of logic function does the gate perform?
- **b.** What are the noise margins of the gate?
- **c.** What functions do each of the individual transistors perform in the gate circuit?
- **d.** What is the fanout of this gate?

9.7 Describe three different types of output configurations used in TTL logic devices. What is the use for each output type?

9.8 Draw three OR gates whose outputs are joined in a wired AND configuration. Describe how the wired AND operates as an AND logic function.

9.9 Describe how the wired AND circuit in Problem 9.8 can be considered as a wired OR logic function.

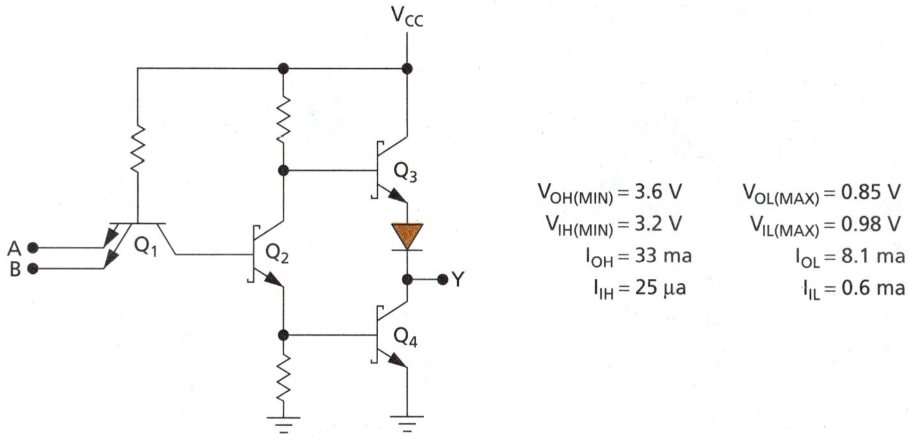

Figure 9.33 Problem 9.5

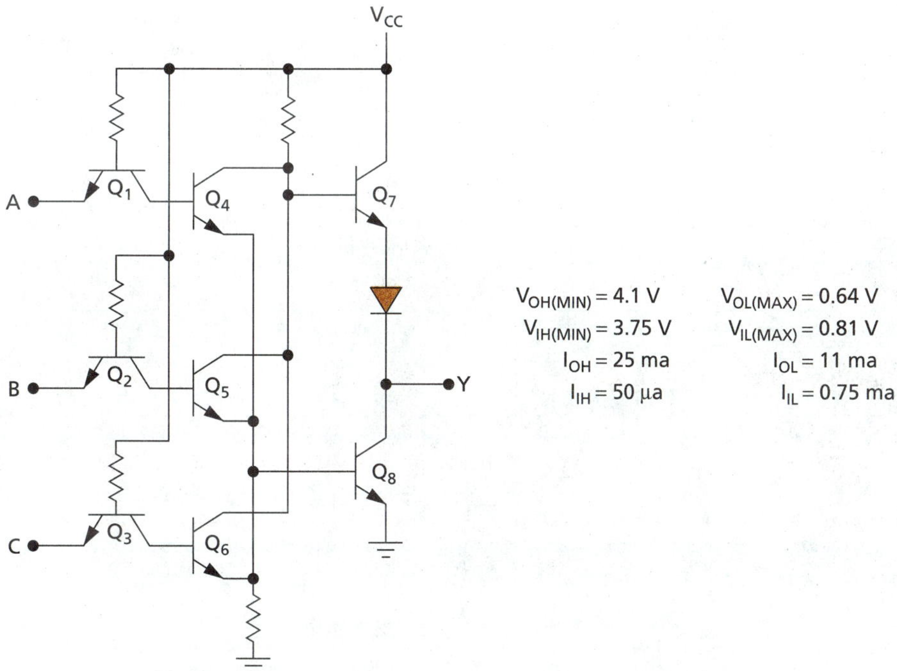

Figure 9.34 Problem 9.6

9.10 Using the data sheet for the 74LS00, determine the following specifications:

a. I_{OH} and I_{OL}

b. Fanout

c. Power dissipation

d. Propagation delay (high to low and low to high)

e. Noise margins

9.11 Compare the following devices. Which has the lowest propagation time? lowest power dissipation? highest fanout? 7400, 74S00, 74F00, and 74LS00

9.12 Give two everyday applications that would most likely use 74LSxx devices rather than 74xx devices. Why would the LS chips be preferred for these applications?

9.13 What does an absolute maximum V_{CC} of 7.5 V actually mean in reference to an IC chip?

9.14 What is the main advantage for using emitter-coupled logic devices (ECL)? Why does ECL have this advantage?

9.15 What are ECL's drawbacks, and why are they so bad?

9.16 Answer the following questions for the N-MOS gate in Figure 9.35:

a. What is the logic expression for the gate's output?

b. What is the gate's fanout?

c. Determine the gate's noise margins.

d. What function is being performed by Q_2?

e. What is the purpose of the diode in the gate circuit?

9.17 Give the advantages of MOS devices over TTL devices.

9.18 What is the main disadvantage of MOS logic gates compared to TTL? What causes this disadvantage to be present?

9.19 Give two reasons why MOS devices are preferred over TTL in notebook computers.

9.20 For the CMOS gate of Figure 9.36, answer the following questions:

a. What is the gate's fanout?

b. What is the gate's logic function?

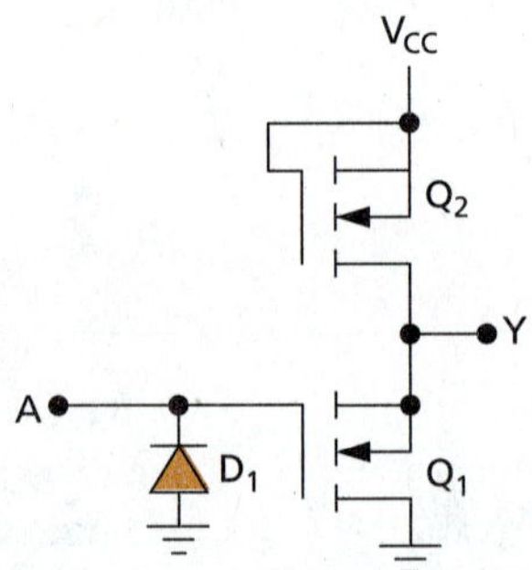

$V_{OH(MIN)} = 11.21$ V	$V_{OL(MAX)} = 0.18$ V
$V_{IH(MIN)} = 11.0$ V	$V_{IL(MAX)} = 0.24$ V
$I_{OH} = 112$ ma	$I_{OL} = 112$ µa
$I_{IH} = 2.2$ µa	$I_{IL} = 2.2$ µa

Figure 9.35 Problem 9.16

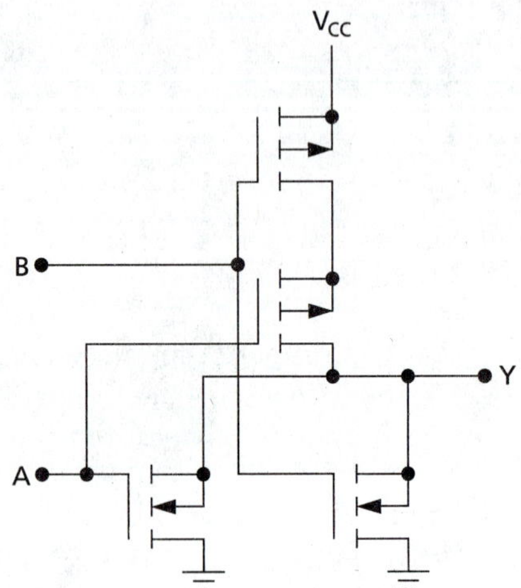

$I_{OH} = 500$ µa	$I_{OL} = 500$ µa
$I_{IH} = 7.3$ µa	$I_{IL} = 7.3$ µa

Figure 9.36 Problem 9.20

9.21 Compare the difference in operation between a TTL NAND gate and a CMOS NAND gate when an input is left open.

9.22 Why is it unadvisable to leave an unused input to any gate unterminated?

9.23 Using the specification sheet for a 74C00, determine the following characteristics for a device using a V_{DD} of 15 V:

a. Fanout
b. Power dissipation
c. Propagation delay (high to low and low to high)
d. Noise margin

9.24 When mixing logic family devices in a single logic circuit, what are some of the considerations to take into account? Give reasons for each of those considerations.

9.25 What does it mean for a CMOS device to be TTL compatible?

9.26 Compare CMOS specifications to HCMOS specifications. In what areas are the parameters for HCMOS an improvement over CMOS?

Appendixes

Appendix A

Data Sheets

1. 74xxx TTL Family of Devices General Specifications

Absolute Maximum Ratings

Exceeding these specifications could damage the chip.

1. V_{CC} = +7V
2. $V_{IN(min)}$ = − 0.5 V $V_{IN(max)}$ = +5.5 V
3. $I_{IN(min)}$ = − 30 ma $V_{IN(max)}$ = +5 ma
4. Operating temperature = −55°C to +125°C

Nominal Operating Characteristics

Normal operating specifications using V_{CC} = +5 V and ground return of 0 V.

Specification	74xx	74LS	Units
$V_{IH(min)}$	2.0	2.0	Volts
$V_{IL(max)}$	0.8	0.7	Volts
$V_{OH(min)}$	2.4	2.5	Volts
$V_{OL(max)}$	0.4	0.5	Volts
$I_{OH(max)}$	250	250	μA
$I_{OL(max)}$	16	20	mA
$I_{OZ(max)}$	40	50	μA
Fanout	10	20	Unit loads

Symbol Definitions

V	voltage	IH	input high
I	current	IL	input low
OH	output high	OZ	high-impedance output
OL	output low		

2. 74xxx Family Data Sheets

74LS00
NAND

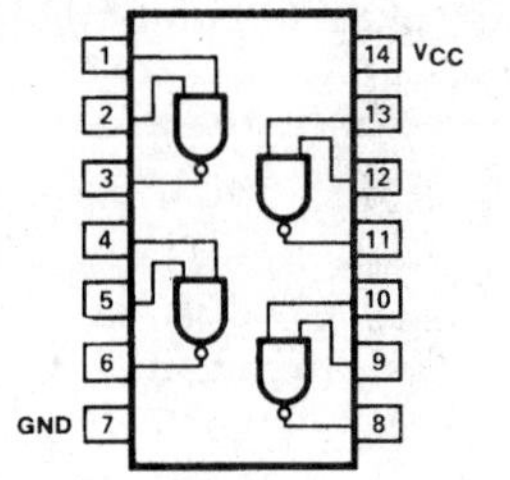

Figure A

ABSOLUTE MAXIMUM RATINGS

PARAMETER	54/74, 8200, 9300	54H/74H	54S/74S, 82S00	54LS/74LS
V_{CC} Supply voltage, continuous (Note a)	7.0V	7.0V	7.0V	7.0V
Input voltage, continuous (Notes a & b)	−0.5V to +5.5V	−0.5V to +5.5V	−0.5V to +5.5V	−0.5V to 15V(b)
Input current, continuous	−30mA to +5mA	−15mA to +5mA	−30mA to 5mA	−30mA to +1mA
Voltage applied to HIGH outputs (Note a)	−0.5V to V_{CC}	−0.5V to 7.0V	−0.5V to 7.0V	−0.5V to V_{CC}.
Voltage applied to "off" Open Collector outputs (Notes a & c)	−0.5V to 7.0V	−0.5V to 7.0V	−0.5V to 7.0V	−0.5V to 7.0V
Current into LOW standard output, continuous	30mA	40mA	40mA	15mA
Current into LOW Buffer output, continuous	80mA	100mA	100mA	50mA
Operating free air temperature range	−55°C to +125°C			
Storage temperature range	−65°C to +150°C			

NOTES

a. Voltages are referenced to device ground terminal

b. The following LS device inputs are limited to 5.5V input breakdown: All inputs of LS181, Clock inputs of LS90, LS92, LS93, LS196, LS197, LS290, LS293, LS390, LS393 and LS490.

c. Some open collector devices are specially processed to handle higher output voltages of from 15V to 30V. The Absolute Maximum voltage for these devices is 10% over the specified V_{OUT} test condition.

INPUT AND OUTPUT LOADING AND FAN-OUT TABLE (See Note a)

PINS		54/74	54H/74H	54S/74S	54LS/74LS
Inputs	I_{IH} (μA)	40		100	20
	I_{IL} (mA)	-1.6		-4.0	-0.4
Outputs	I_{OH} (μA)	250		250	250
	I_{OL} (mA)	48		60	12/24 (a)

DC CHARACTERISTICS OVER OPERATING TEMPERATURE RANGE (See Note b)

PARAMETER		TEST CONDITIONS	54/74		54H/74H		54S/74S		54LS/74LS		UNIT
			Min	Max	Min	Max	Min	Max	Min	Max	
I_{CCH}	Supply current	V_{CC} = Max, V_{IN} = 0		8.5				36		2.0	mA
I_{CCL}	Supply current	V_{CC} = Max, $V_{IN} \geq 4.5$		54				80		12	mA

2. 74xxx Family Data Sheets

AC CHARACTERISTICS T_A = 25°C (See Section 4 for Waveforms and Conditions.)

PARAMETER		TEST CONDITIONS	54/74 C_L = 15 pF R_L = 400 Ω Min	Max	54H/74H C_L = 25 pF R_L = 280 Ω Min	Max	54S/74S C_L = 15 pF R_L = 280 Ω Min	Max	54LS/74LS C_L = 15 pF R_L = 2k Ω Min	Max	UNIT
t_{PLH}	Propagation delay	Waveform 1		22		10		4.5		15	ns
t_{PHL}	Propagation delay	Waveform 1		15		10		5.0		15	ns

LOW POWER SCHOTTKY (54LS/74LS) FAMILY CHARACTERISTICS

DC CHARACTERISTICS OVER OPERATING TEMPERATURES (Unless otherwise specified on Data Sheet)

PARAMETER		TEST CONDITIONS		LIMITS Min	Typ[1]	Max	UNIT
V_{IH}	Input HIGH voltage	Guaranteed input HIGH voltage for all inputs		2.0			V
V_{IL}	Input LOW voltage	Guaranteed input LOW voltage for inputs	Mil			0.7	V
			Com			0.8	V
V_{CD}	Input clamp diode voltage	V_{CC} = Min, I_{IN} = −18mA			−0.65	−1.5	V
V_{OL}	Output LOW voltage	V_{CC} = Min, I_{OL} = 4.0mA			0.25	0.4	V
		V_{CC} = Min, I_{OL} = 8.0mA (Com. only)			0.35	0.5	V
V_{OH}	Output HIGH voltage	V_{CC} = Min, I_{OH} = −400μA	Mil	2.5	3.4		V
			Com	2.7	3.4		V
I_{OH}	Output HIGH current (open collector)	V_{CC} = Min, V_{OUT} = 5.5V				100	μA
I_{OZH}	Output "off" current HIGH (3-state)	V_{CC} = Max, V_{OUT} = 2.4V, $V_{\overline{OE}}$ = 2.0V				20	μA
I_{OZL}	Output "off" current LOW (3-state)	V_{CC} = Max, V_{OUT} = 0.4V, $V_{\overline{OE}}$ = 2.0V				−20	μA
I_{IH}	Input HIGH current[2]	V_{CC} = Max, V_{IN} = 2.7V				20	μA
I_I	Input HIGH current at Max input voltage	V_{CC} = Max, V_{IN} = 10V[3]				0.1	mA
I_{IL}	Input LOW current[2]	V_{CC} = Max, V_{IN} = 0.4V				−0.4	mA
I_{OS}	Output short circuit current	V_{CC} = Max, V_{OUT} = 0V		−15		−100	mA

DRIVING DEVICE	NUMBER OF LOADS DRIVEN 74LS	74	74H	74L	74S	8200 AND 9300	82S00
74LS	20	5	4	40	4	5	20
*74LS Buffers	60	15	12	120	12	15	60
74	40	10	8	80	8	10	40
74 Buffers	60	30	24	120	24	30	120
74H	50	12	10	100	10	12	50
74H Buffers	75	37	30	150	30	37	150
74L	9	2	1	20	1	2	9
74S	50	12	10	100	10	12	50
74S Buffers	150	37	30	150	30	37	150
8200 & 9300	40	10	8	80	8	10	40
82S00	50	12	10	100	10	12	50

*The 74LS Buffers include 3-State outputs except LS253 & LS670

2. 74xxx Family Data Sheets

74LS02
NOR

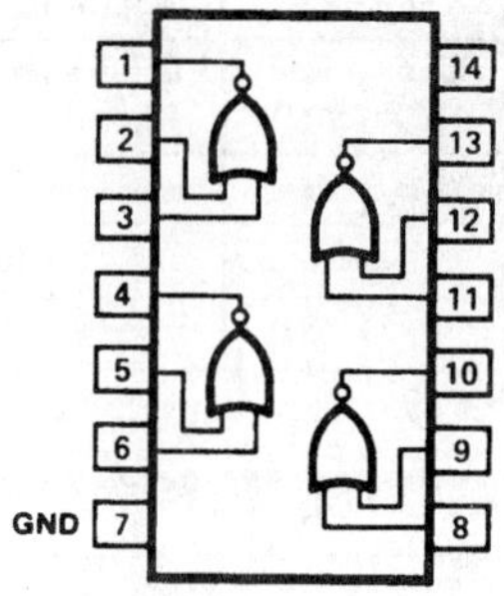

74LS04
INVERTER

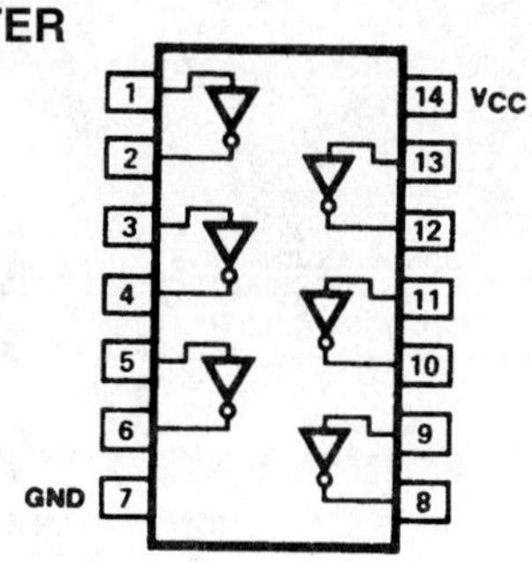

74LS08
AND

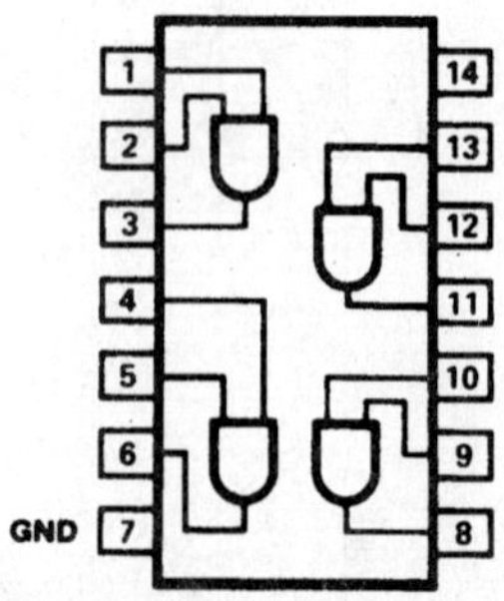

74LS32
OR

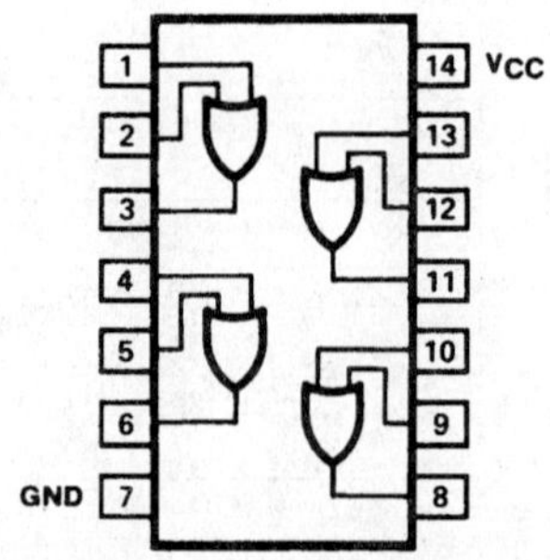

74LS244
THREE STATE BUFFER

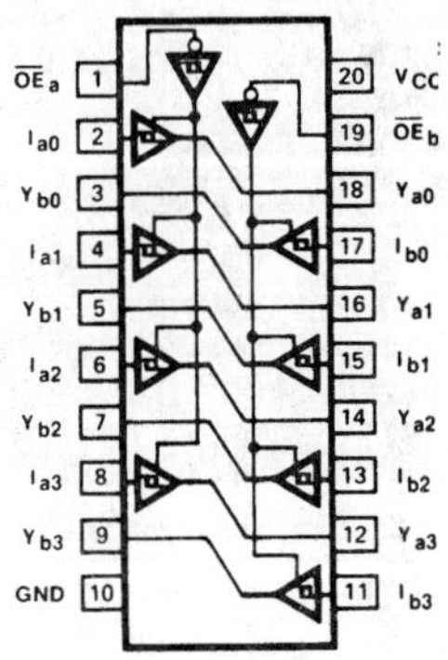

74LS244
TRUTH TABLE

INPUTS				OUTPUTS	
$\overline{OE}_a$	I_a	$\overline{OE}_b$	I_b	Y_a	Y_b
L	L	L	L	L	L
L	H	L	H	H	H
H	X	H	X	(Z)	(Z)

H = HIGH voltage level
L = LOW voltage level
X = Don't care
(Z) = High impedance (off) state

2. 74xxx Family Data Sheets

74LS10
NAND

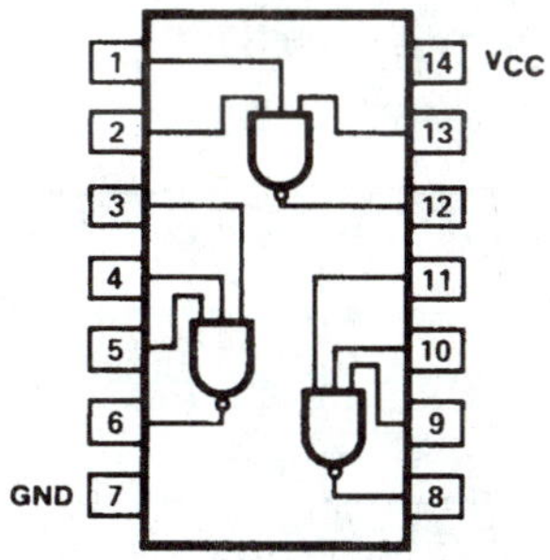

74LS20
NAND

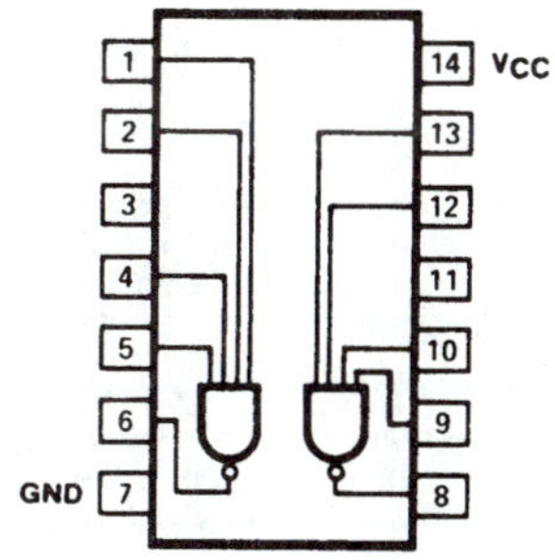

74LS30
NAND

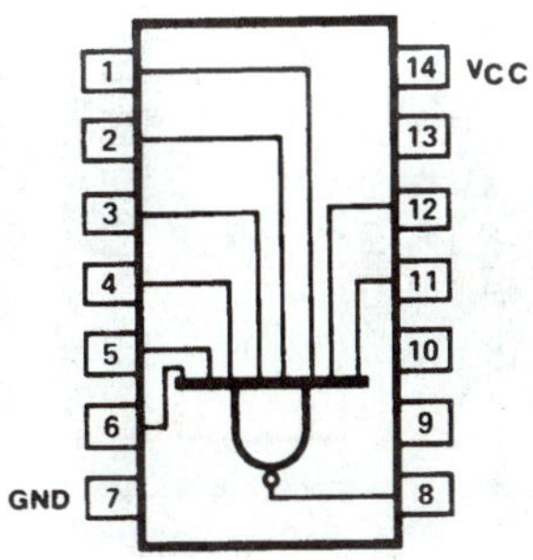

74LS86
XOR

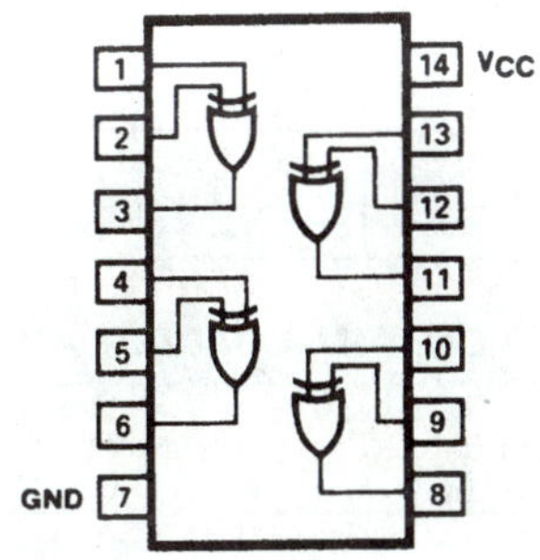

TRUTH TABLE

INPUTS		OUTPUT
A	B	Y
L	L	L
L	H	H
H	L	H
H	H	L

L = LOW voltage level
H = HIGH voltage level

2. 74xxx Family Data Sheets

74LS74
D-FLIP FLOP

DESCRIPTION

The "74" is a Dual Positive Edge-Triggered D-Type Flip-Flop featuring individual data, clock, set and reset inputs; also complementary Q and $\overline{Q}$ outputs.

Set ($\overline{S}_D$) and Reset ($\overline{R}_D$) are asynchronous active LOW inputs and operate independently of the clock input. Information on the data (D) input is transferred to the Q output on the LOW-to-HIGH transition of the clock pulse. The D inputs must be stable one setup time prior to the LOW-to-HIGH clock transition for predictable operation. Although the clock input is level sensitive, the positive transition of the clock pulse between the 0.8V and 2.0V levels should be equal to or less than the clock to output delay time for reliable operation.

LOGIC SYMBOL

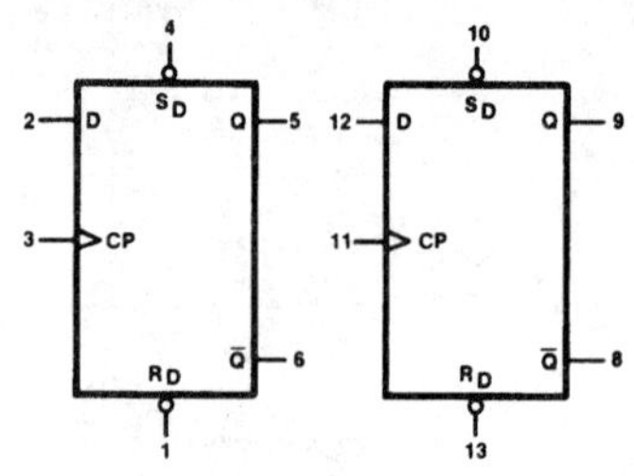

Pin numbers for pin configuration A

ORDERING CODE (See Section 9 for further Package and Ordering Information.)

PACKAGES	PIN CONF.	COMMERCIAL RANGES V_{CC} = 5V ± 5%; T_A = 0°C to +70°C		MILITARY RANGES V_{CC} = 5V ± 10%; T_A = -55°C to +125°C	
Plastic DIP	Fig. A Fig. A	N7474N N74S74N	N74H74N N74LS74N		
Ceramic DIP	Fig. A Fig. A	N7474F N74S74F	N74H74F N74LS74F	S5474F S54S74F	S54H74F S54LS74F
Flatpak	Fig. B Fig. A			S5474W S54S74W	S54H74W S54LS74W

PIN CONFIGURATION

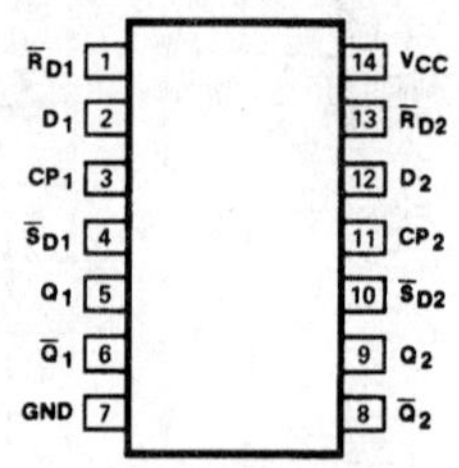

Figure A

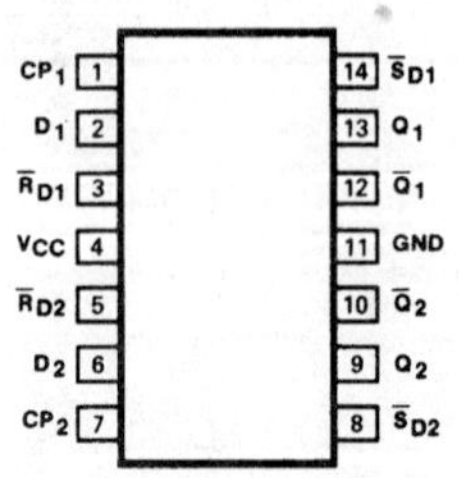

Figure B

INPUT AND OUTPUT LOADING AND FAN-OUT TABLE [a]

PINS		54/74	54H/74H	54S/74S	54LS/74LS
D Data Input	I_{IH} (μA) I_{IL} (mA)	40 -1.6	50 -2.0	50 -2.0	20 -0.36
CP Clock Input	I_{IH} (μA) I_{IL} (mA)	80 -3.2	100 -4.0	100 -4.0	40 -0.8
$\overline{S}_D$ Set Input	I_{IH} (μA) I_{IL} (mA)	80 -1.6	100 -2.0	100 -4.0	40 -0.8
$\overline{R}_D$ Reset Input	I_{IH} (μA) I_{IL} (mA)	120 -3.2	150 -4.0	150 -6.0	60 -1.15
Q & $\overline{Q}$ Output	I_{OH} (μA) I_{OL} (mA)	-400 16	-500 20	-1000 20	-400 4/8 (a)

DC CHARACTERISTICS OVER OPERATING TEMPERATURE RANGE [b]

PARAMETER		TEST CONDITIONS		54/74		54H/74H		54S/74S		54LS/74LS		UNIT
				Min	Max	Min	Max	Min	Max	Min	Max	
I_{CC}	Supply current	V_{CC} = Max	Mil		30		42		50		8.0	mA
		V_{CP} = 0V	Com		30		50		50		8.0	mA

NOTES

a. The slashed numbers indicate different parametric values for Military/Commercial temperature ranges respectively.

b. For family dc characteristics, see 54/74, 54H/74H, 54S/74S, and 54LS/74LS specification.

2. 74xxx Family Data Sheets

LOGIC DIAGRAM

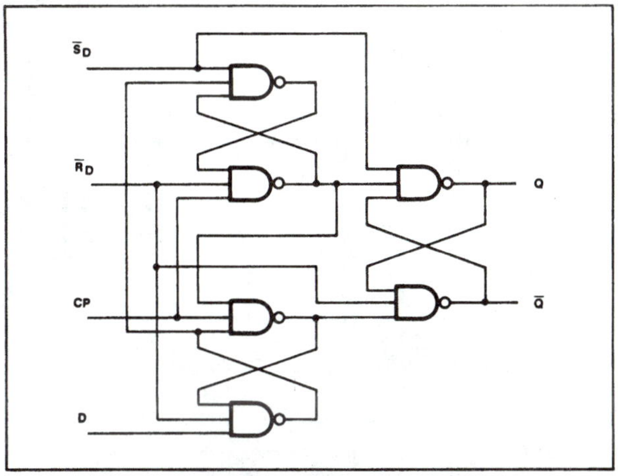

MODE SELECT—TRUTH TABLE

OPERATING MODE	INPUTS				OUTPUTS	
	$\overline{S}_D$	$\overline{R}_D$	CP	D	Q	$\overline{Q}$
Asynchronous Set	L	H	X	X	H	L
Asynchronous Reset (Clear)	H	L	X	X	L	H
Undetermined (c)	L	L	X	X	H	H
Load "1" (Set)	H	H	↑	h	H	L
Load "0" (Reset)	H	H	↑	l	L	H

H = HIGH voltage level steady state.
h = HIGH voltage level one setup time prior to the LOW to HIGH clock transition.
L = LOW voltage level steady state.
l = LOW voltage level one setup time prior to the LOW to HIGH clock transition.
X = Don't care.

AC CHARACTERISTICS $T_A = 25°C$ (See Section 4 for Waveforms and Conditions.)

PARAMETER		TEST CONDITIONS	54/74		54H/74H		54S/74S		54LS/74LS		UNIT
			$C_L = 15\,pF$ $R_L = 400\,\Omega$		$C_L = 25\,pF$ $R_L = 280\,\Omega$		$C_L = 15\,pF$ $R_L = 280\,\Omega$		$C_L = 15\,pF$ $R_L = 2k\,\Omega$		
			Min	Max	Min	Max	Min	Max	Min	Max	
f_{Max}	Maximum clock frequency	Waveform 3	15		35		75		25		MHz
t_{PLH}	Propagation delay	Waveform 3		25		15		9.0		25	ns
t_{PHL}	Clock to output	Waveform 3		40		20		9.0		40	ns
t_{PLH}	Propagation delay	Waveform 5		25		20		6.0		25	ns
t_{PHL}	Set or reset to output	Waveform 5 CP = HIGH		40		30		13.5		40	ns
t_{PHL}	Set or reset to output	Waveform 5 CP = LOW		40		30		8.0		40	ns

AC SET UP REQUIREMENTS: $T_A = 25°C$ (See Section 4 for Waveforms and Conditions.)

PARAMETER		TEST CONDITIONS	54/74		54H/74H		54S/74S		54LS/74LS		UNIT
			Min	Max	Min	Max	Min	Max	Min	Max	
$t_w(H)$	Clock pulse width (HIGH)	Waveform 3	30		15		6.0		25		ns
$t_w(L)$	Clock pulse width (LOW)	Waveform 3	37		13.5		7.3		15		ns
$t_w(L)$	Set or reset pulse width (LOW)	Waveform 5	30		25		7.0		25		ns
$t_s(H)$	Setup time (HIGH) data to clock	Waveform 3	20		10		3.0		25		ns
$t_s(L)$	Setup time (LOW) data to clock	Waveform 3	20		15		3.0		20		ns
t_h	Hold time data to clock	Waveform 3	5.0		5.0		2.0		5.0		ns

NOTE

c. Both outputs will be High while both $\overline{S}_D$ and $\overline{R}_D$ are LOW, but the output states are unpredictable if $\overline{S}_D$ and $\overline{R}_D$ go HIGH simultaneously.

2. 74xxx Family Data Sheets

54/7476
54H/74H76
54LS/74LS76

DESCRIPTION

The "76" is a Dual JK Flip-Flop with individual J, K, Clock, Set and Reset inputs. The 7476 and 74H76 are positive pulse triggered flip-flops. JK information is loaded into the master while the Clock is HIGH and transferred to the slave on the HIGH-to-LOW Clock transition. The J and K inputs must be stable while the Clock is HIGH for conventional operation.

The 74LS76 is a negative edge triggered flip-flop. The J and K inputs must be stable only one setup time prior to the HIGH-to-LOW Clock transition.

The Set ($\overline{S}_D$) and Reset ($\overline{R}_D$) are asynchronous active LOW inputs. When LOW, they override the Clock and data inputs forcing the outputs to the steady state levels as shown in the Truth Table.

LOGIC SYMBOL

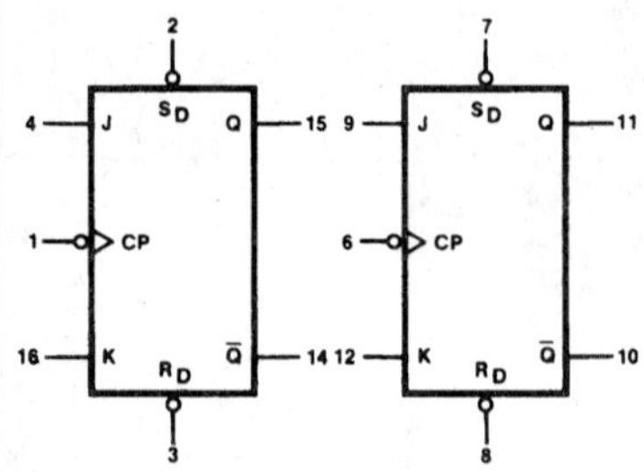

V_{CC} = Pin 5
GND = Pin 13

ORDERING CODE (See Section 9 for further Package and Ordering Information)

PACKAGES	PIN CONF.	COMMERCIAL RANGES $V_{CC} = 5V \pm 5\%$; $T_A = 0°C$ to $+70°C$	MILITARY RANGES $V_{CC} = 5V \pm 10\%$; $T_A = -55°C$ to $+125°C$
Plastic DIP	Fig A Fig A	N7476N • N74H76N N74LS76N	
Ceramic DIP	Fig A Fig A	N7476F • N74H76F N74LS76F	S5476F • S54H76F S54LS76F
Flatpak	Fig A Fig A		S5476W • S54H76W S54LS76W

PIN CONFIGURATION

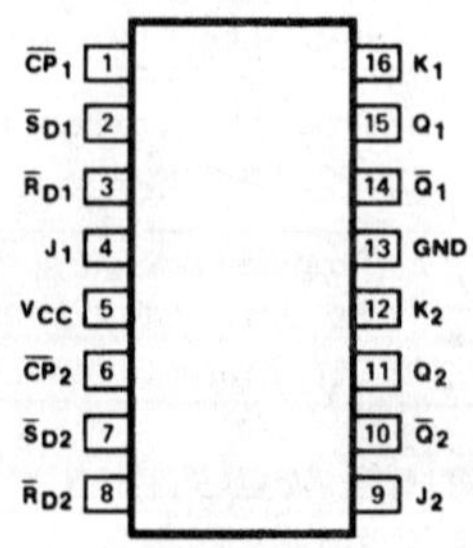

Figure A

INPUT AND OUTPUT LOADING AND FAN-OUT TABLE [(a)]

PINS			54/74	54H/74H	54S/74S	54LS/74LS
$\overline{CP}$	Clock input	I_{IH} (μA) I_{IL} (mA)	80 -3.2	50 -2.0		80 -0.8
$\overline{R}_D$	Reset input	I_{IH} (μA) I_{IL} (mA)	80 -3.2	100 -4.0		60 -0.8
$\overline{S}_D$	Set input	I_{IH} (μA) I_{IL} (mA)	80 -3.2	100 -4.0		60 -0.8
JK	Data inputs	I_{IH} (μA) I_{IL} (mA)	40 -1.6	50 -2.0		20 -0.4
Q & $\overline{Q}$ Outputs		I_{OH} (μA) I_{OL} (mA)	-400 16	-500 20		-400 4/8 [(a)]

DC CHARACTERISTICS OVER OPERATING TEMPERATURE RANGE [(b)]

PARAMETER		TEST CONDITIONS	54/74		54H/74H		54S/74S		54LS/74LS		UNIT
			Min	Max	Min	Max	Min	Max	Min	Max	
I_{CC}	Supply current	V_{CC} = Max, V_{CP} = 0V		40		50				8.0	mA

NOTES

a. The slashed numbers indicate different parametric values for Military/Commercial temperature ranges respectively.

b. For family dc characteristics, see 54/74 and 54H/74H and see 54S/74S and 54LS/74LS specification.

2. 74xxx Family Data Sheets

LOGIC DIAGRAM

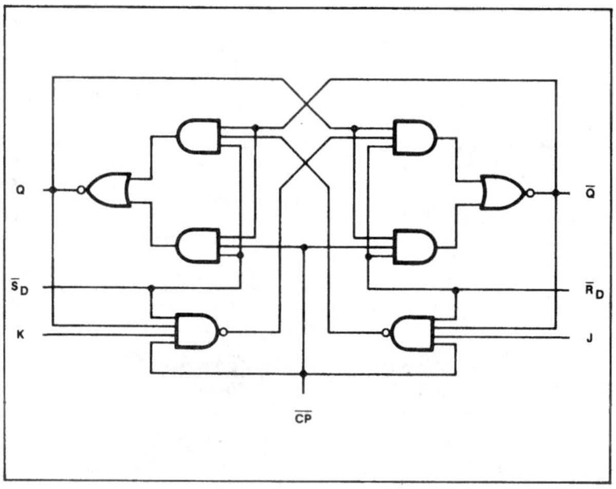

MODE SELECT—TRUTH TABLE

OPERATING MODE	INPUTS					OUTPUTS	
	$\bar{S}_D$	$\bar{R}_D$	$\overline{CP}$ (d)	J	K	Q	$\bar{Q}$
Asynchronous Set	L	H	X	X	X	H	L
Asynchronous Reset (Clear)	H	L	X	X	X	L	H
Undetermined (c)	L	L	X	X	X	H	H
Toggle	H	H	⎍	h	h	$\bar{q}$	q
Load "0" (Reset)	H	H	⎍	l	h	L	H
Load "1" (Set)	H	H	⎍	h	l	H	L
Hold "no change"	H	H	⎍	l	l	q	$\bar{q}$

H = HIGH voltage level steady state.
L = LOW voltage level steady state.
h = HIGH voltage level one setup time prior to the HIGH-to-LOW Clock transition. (e)
l = LOW voltage level one setup time prior to the HIGH-to-LOW Clock transition. (e)
X = Don't care.
q = Lower case letters indicate the state of the referenced output prior to the HIGH to LOW Clock transition.
⎍ = Positive Clock pulse.

AC CHARACTERISTICS $T_A = 25°C$ (See Section 4 for Waveforms and Conditions)

	PARAMETER	TEST CONDITIONS	54/74		54H/74H		54S/74S		54LS/74LS		UNIT
			C_L = 15 pF R_L = 400 Ω		C_L = 25 pF R_L = 280 Ω				C_L = 15pF R_L = 2kΩ		
			Min	Max	Min	Max	Min	Max	Min	Max	
f_{MAX}	Maximum Clock frequency	Waveform 4	15		25				30		MHz
t_{PLH} t_{PHL}	Propagation delay Clock to Output	Waveform 4		25 40		21 27				20 30	ns ns
t_{PLH} t_{PHL}	Propagation delay $\bar{S}_D$ or $\bar{R}_D$ to Output	Waveform 5		25 40		13 24				20 30	ns ns

AC SETUP REQUIREMENTS $T_A = 25°C$ (See Section 4 for Waveforms and Conditions)

	PARAMETER	TEST CONDITIONS	54/74		54H/74H		54S/74S		54LS/74LS		UNIT
			Min	Max	Min	Max	Min	Max	Min	Max	
$t_w(H)$	Clock pulse width (HIGH)	Waveform 4	20		12				20		ns
$t_w(L)$	Clock pulse width (LOW)	Waveform 4	47		28				13		ns
$t_w(L)$	Set or Reset pulse width (LOW)	Waveform 5	25		16				25		ns
t_s	Setup time J or K to Clock	Waveform 4	(e)		(e)				20		ns
t_h	Hold time J or K to Clock	Waveform 4	0		0				0		ns

NOTES

c. Both outputs will be HIGH while both $\bar{S}_D$ and $\bar{R}_D$ are LOW, but the output states are unpredictable if $\bar{S}_D$ and $\bar{R}_D$ go HIGH simultaneously.

d. The 74LS76 is edge triggered. Data must be stable one setup time prior to the negative edge of the Clock for predictable operation.

e. The J and K inputs of the 7476 and 74H76 must be stable while the Clock is HIGH for conventional operation.

2. 74xxx Family Data Sheets

54/7483
54LS/74LS83A

DESCRIPTION

The "83" is a high speed 4-Bit Binary Full Adder with internal carry lookahead. It accepts two 4-bin binary words (A_1-A_4, B_1-B_4) and a Carry input (C_{IN}). The sum of the two 4-bit words is combined with the carry input and presented at the four Sum outputs (Σ_1-Σ_4) and the Carry output (C_{OUT}). It operates with either HIGH or LOW operands (positive or negative logic).

FEATURES

- **High speed 4-bit binary addition**
- **Cascadeable in 4-bit increments**
- **LS83A has fast internal carry lookahead**
- **See "283" for corner power pin version**

LOGIC SYMBOL

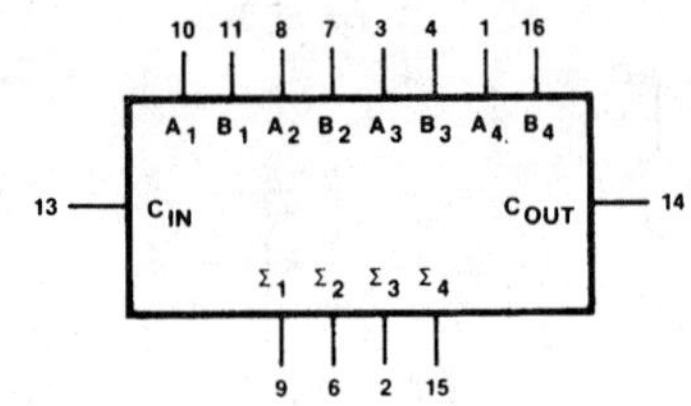

V_{CC} = Pin 5
GND = Pin 12

PIN CONFIGURATION

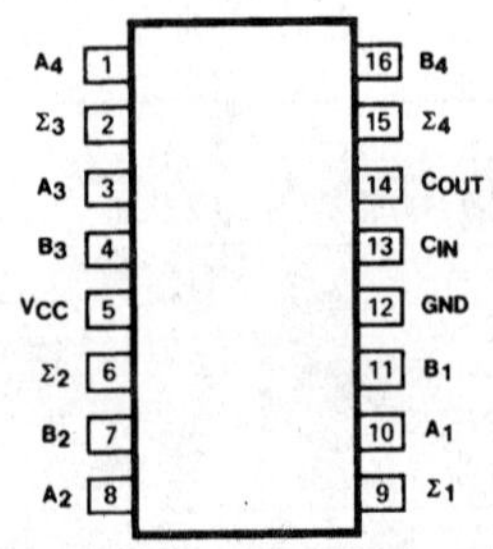

ORDERING CODE (See Section 9 for further Package and Ordering Information)

PACKAGES	COMMERCIAL RANGES V_{CC}=5V±5%; T_A=0°C to +70°C	MILITARY RANGES V_{CC}=5V±10%; T_A=−55°C to +125°C
Plastic DIP	N7483N • N74LS83AN	
Ceramic DIP	N7483F • N74LS83AF	S5483F • S54LS83AF
Flatpak		S5483W • S54LS83AW

INPUT AND OUTPUT LOADING AND FAN-OUT TABLE[a]

PINS	DESCRIPTION		54/74	54S/74S	54LS/74LS
A_1, B_1, A_3, B_3	Operand inputs	I_{IH} (μA) I_{IL} (mA)	80 −3.2		40 −0.8
A_2, B_2, A_4, B_4.	Operand inputs	I_{IH} (μA) I_{IL} (mA)	40 −1.6		40 −0.8
C_{IN}	Carry input	I_{IH} (μA) I_{IL} (mA)	80 −3.2		20 −0.4
Σ_1-Σ_4	Sum outputs	I_{OH} (μA) I_{OL} (mA)	−800 16		−400 4/8[a]
C_{OUT}	Carry output	I_{OH}(μA) I_{OL} (mA)	−400 8.0		−400 4/8[a]

NOTE
a. The slashed numbers indicate different parametric values for Military/Commercial temperature ranges respectively.

2. 74xxx Family Data Sheets

FUNCTIONAL DESCRIPTION

The "83" adds two 4-bit binary words (A_n plus B_n) plus the incoming carry. The binary sum appears on the Sum outputs (Σ_1-Σ_4) and the outgoing carry (C_{OUT}) according to the equation:

$$C_{IN}+(A_1+B_1)+2(A_2+B_2)+4(A_3+B_3)+8(A_4+B_4)=\Sigma_1+2\Sigma_2+4\Sigma_3+8\Sigma_4+16C_{OUT}$$

Where (+) = plus

Due to the symmetry of the binary add function the "83" can be used with either all active HIGH operands (positive logic) or with all active LOW operands (negative logic)-see Table "A". With active HIGH inputs, C_{IN} cannot be left open; it must be held LOW when no "carry in" is intended. Interchanging inputs of equal weight does not affect the operation, thus C_{IN}, A_1, B_1, can arbitrarily be assigned to pins 10, 11, 13, etc.

PINS	C_{IN}	A_1	A_2	A_3	A_4	B_1	B_2	B_3	B_4	Σ_1	Σ_2	Σ_3	Σ_4	C_{OUT}	
Logic Levels	L	L	H	L	H	H	L	L	H	H	H	L	L	H	
Active HIGH	0	0	1	0	1	1	0	0	1	1	1	0	0	1	(10+9=19)
Active LOW	1	1	0	1	0	0	1	1	0	0	0	1	1	0	(carry+5+6=12)

Table "A"

LOGIC DIAGRAM

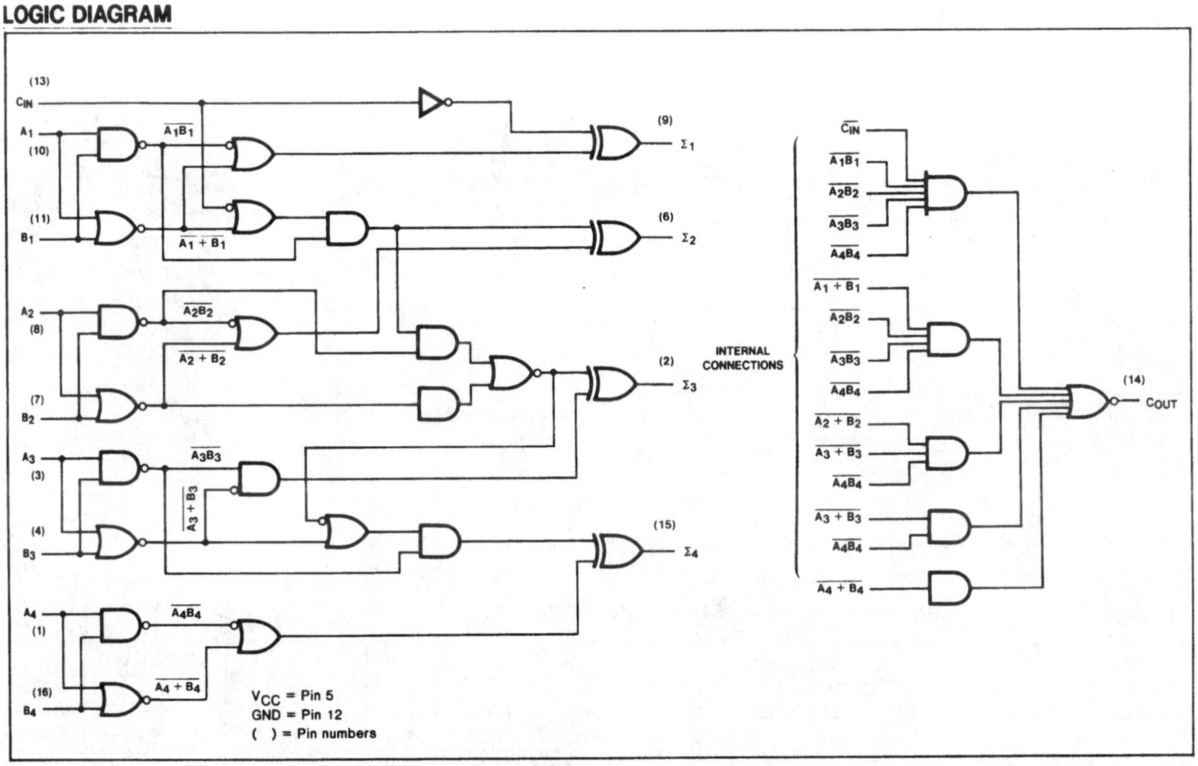

2. 74xxx Family Data Sheets

LOGIC DIAGRAM

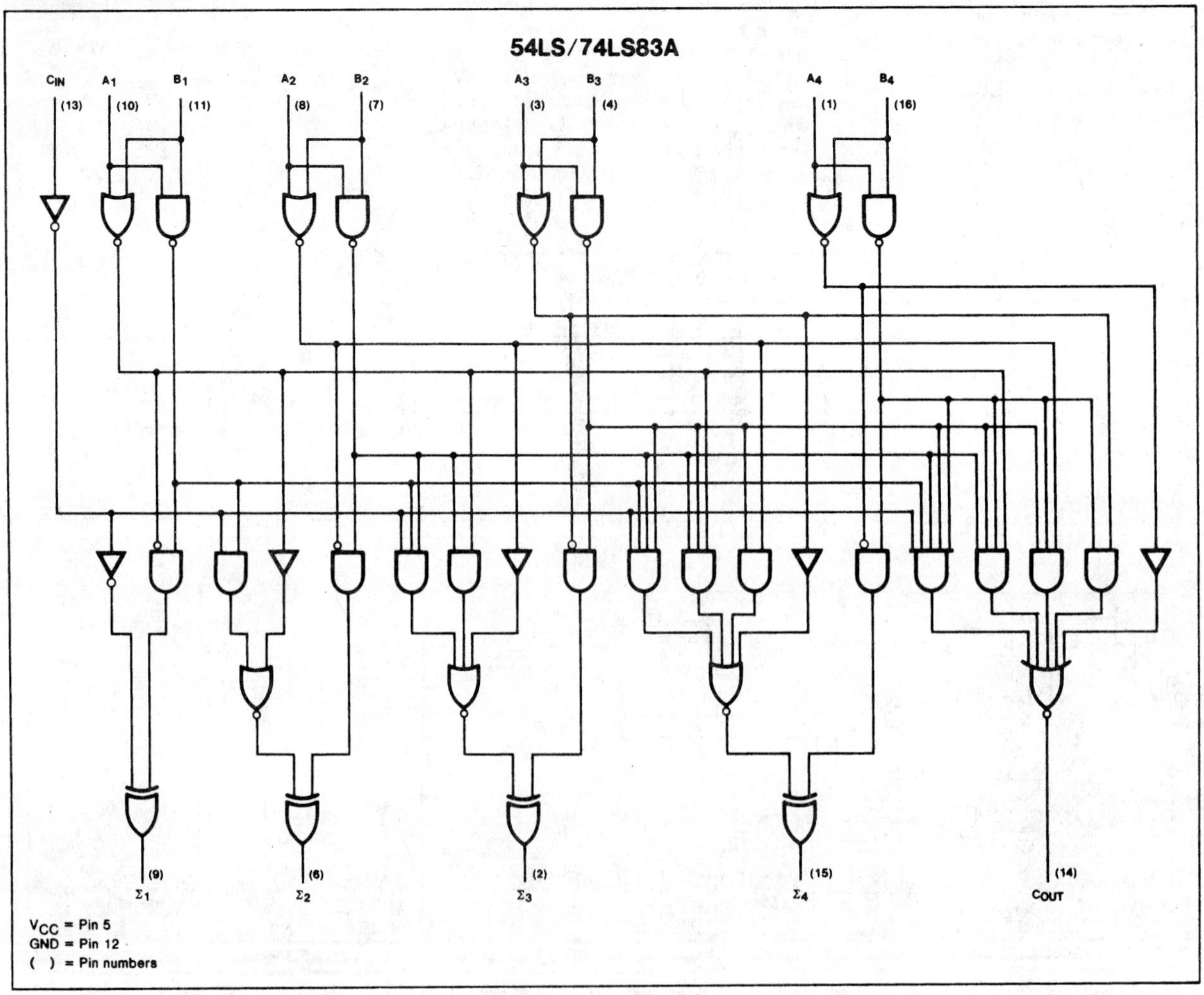

DC CHARACTERISTICS OVER OPERATING TEMPERATURE RANGE[b]

PARAMETER		TEST CONDITIONS		54/74		54S/74S		54LS/74LS		UNIT
				Min	Max	Min	Max	Min	Max	
I_{OS}	Short circuit current for C_{OUT} only	V_{CC} = Max	Mil	−20	−70			−15	−100	mA
			Com	−18	−70			−15	−100	mA
I_{CC}	Supply Current	V_{CC} = Max	V_{IN} = 0V		79				39	mA
			V_{IN} = 4.5V		79				34	mA

NOTE

b. For family dc characteristics, see 54/74, 54H/74H, 54S/74S, and 54LS/74LS specifications.

2. 74xxx Family Data Sheets

54/7493
54LS/74LS93

LOGIC SYMBOL

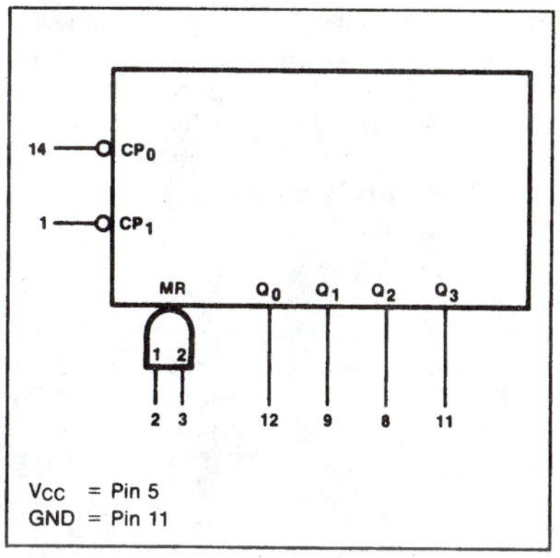

DESCRIPTION

The "93" is a high-speed 4-bit ripple type counter divided into two sections. The counter has a divide-by-two section and a divide-by-eight section which are triggered by HIGH-to-LOW transitions on the clock inputs. Either section can be used separately or tied together (Q_0 to $\overline{CP}_1$) to form a modulo-16 counter. The counter has a 2-input gated Master Reset (Clear).

PIN CONFIGURATION

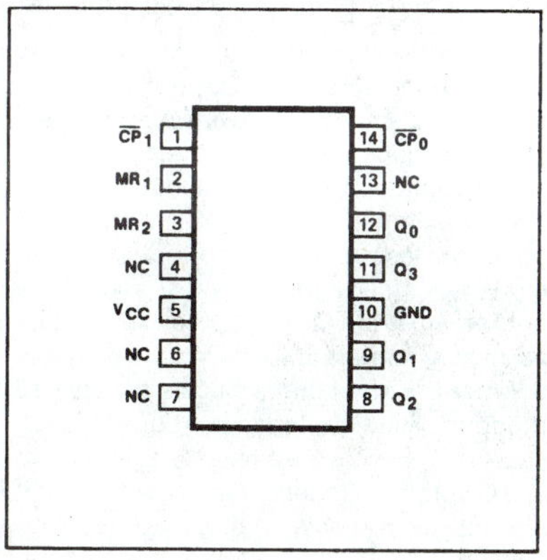

ORDERING CODE (See Section 9 for further Package and Ordering Information)

PACKAGES	COMMERCIAL RANGES V_{CC} = 5V ± 5%; T_A = 0°C to 70°C	MILITARY RANGES V_{CC} = 5V ± 10%; T_A = -55°C to +125°C
Plastic DIP	N7493N • N74LS93N	
Ceramic DIP	N7493F • N74LS93F	S5493F • S54LS93F
Flatpak		S5493W • S54LS93W

INPUT AND OUTPUT LOADING AND FAN-OUT TABLE [a]

PINS	DESCRIPTION		54/74	54S/74S	54LS/74LS
$\overline{CP}_0$	Clock (Active LOW going edge) input (to ÷2 Section)	I_{IH} (μA) I_{IL} (mA)	80 -3.2		40 -2.4
$\overline{CP}_1$	Clock (Active LOW going edge) input (to ÷8 Section)	I_{IH} (μA) I_{IL} (mA)	80 -3.2		40 -1.6
MR_1, MR_2	Master Reset (Clear) inputs	I_{IH} (μA) I_{IL} (mA)	40 -1.6		20 -0.4
Q_0	Output from ÷2 section	I_{OH} (μA) I_{OL} (mA)	-800 16		-400 4/8[a]
Q_1, Q_2, Q_3	Outputs from ÷8 section	I_{OH} (μA) I_{OL} (mA)	-800 16		-400 4/8[a]

NOTE

a. The slashed numbers indicate different parametric values for Military/Commercial temperature ranges respectively.

2. 74xxx Family Data Sheets

FUNCTIONAL DESCRIPTION

The "93" is a 4-bit ripple type Binary counter. The device consists of four master-slave flip-flops internally connected to provide a divide-by-two section and a divide-by-eight section. Each section has a separate clock input to initiate state changes of the counter on the HIGH-to-LOW clock transition. State changes of the Q outputs do not occur simultaneously because of internal ripple delays. Therefore, decoded output signals are subject to decoding spikes and should not be used for clocks or strobes. The Q_0 output is designed and specified to drive the rated fan-out plus the $\overline{CP}_1$ input of the device.

A gated AND asynchronous Master Reset (MR_1 MR_2) is provided which overrides both clocks and resets (clears) all the flip-flops.

Since the output from the divide-by-two section is not internally connected to the succeeding stages, the device may be operated in various counting modes. In a 4-bit ripple counter the output Q_0 must be connected externally to input $\overline{CP}_1$. The input count pulses are applied to input $\overline{CP}_0$. Simultaneous divisions of 2, 4, 8 and 16 are performed at the Q_0, Q_1, Q_2 and Q_3 outputs as shown in the truth table. As a 3-bit ripple counter the input count pulses are applied to input $\overline{CP}_1$. Simultaneous frequency divisions of 2, 4 and 8 are available at the Q_1, Q_2 and Q_3 outputs. Independent use of the first flip-flop is available if the reset function coincides with reset of the 3-bit ripple-through counter.

MODE SELECTION

RESET INPUTS		OUTPUTS			
MR_1	MR_2	Q_0	Q_1	Q_2	Q_3
H	H	L	L	L	L
L	H	Count			
H	L	Count			
L	L	Count			

H = HIGH voltage level
L = LOW voltage level
X = Don't care

TRUTH TABLE

COUNT	OUTPUT			
	Q_0	Q_1	Q_2	Q_3
0	L	L	L	L
1	H	L	L	L
2	L	H	L	L
3	H	H	L	L
4	L	L	H	L
5	H	L	H	L
6	L	H	H	L
7	H	H	H	L
8	L	L	L	H
9	H	L	L	H
10	L	H	L	H
11	H	H	L	H
12	L	L	H	H
13	H	L	H	H
14	L	H	H	H
15	H	H	H	H

NOTE: Output Q_0 connected to Input $\overline{CP}_1$.

LOGIC DIAGRAM

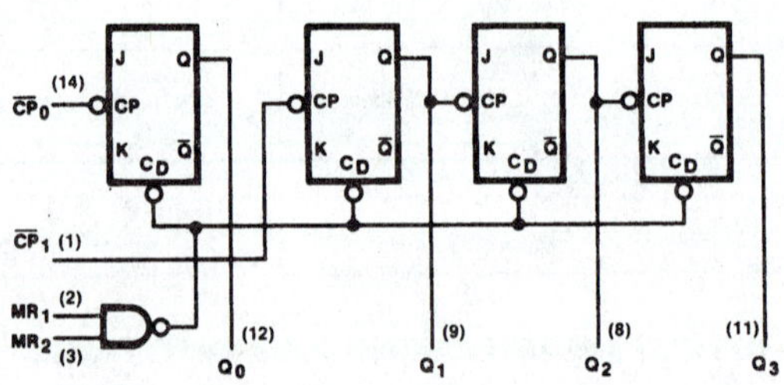

( ) = Pin Numbers
V_{CC} = Pin 5
GND = Pin 10

DC CHARACTERISTICS OVER OPERATING TEMPERATURE RANGE [b]

PARAMETER		TEST CONDITIONS		54/74		54S/74S		54LS/74LS		UNIT
				Min	Max	Min	Max	Min	Max	
I_{CC}	Supply current	V_{CC} = Max	Mil		46				15	mA
			Com		53				15	mA

NOTE

b. For family dc characteristics, see 54/74, 54H/74H, 54S/74S, and 54LS/74LS specifications.

2. 74xxx Family Data Sheets

AC CHARACTERISTICS $T_A = 25°C$ **(See Section 4 for Test Circuits and Conditions)**

PARAMETER		TEST CONDITIONS	54/74 $C_L = 15\,pF$, $R_L = 400\,\Omega$		54S/74S		54LS/74LS $C_L = 15\,pF$, $R_L = 2k\,\Omega$		UNIT
			Min	Max	Min	Max	Min	Max	
f_{MAX}	$\overline{CP}_0$ Input count frequency	Figure 1	10				32		MHz
f_{MAX}	$\overline{CP}_1$ Input count frequency	Figure 1	10				16		MHz
t_{PLH} t_{PHL}	Propagation delay $\overline{CP}_0$ input to Q_0 output	Figure 1						16 18	ns ns
t_{PLH} t_{PHL}	Propagation delay $\overline{CP}_1$ input to Q_1 output	Figure 1						16 21	ns ns
t_{PLH} t_{PHL}	Propagation delay $\overline{CP}_1$ input to Q_2 output	Figure 1						32 35	ns ns
t_{PLH} t_{PHL}	Propagation delay $\overline{CP}_1$ input to Q_3 output	Figure 1						51 51	ns ns
t_{PLH} t_{PHL}	Propagation delay $\overline{CP}_0$ input to Q_3 output	Figure 1		135 135				70 70	ns ns
t_{PHL}	MR Input to any output	Figure 2						40	ns

AC SETUP REQUIREMENTS $T_A = 25°C$ **(See Section 4 for Test Circuits and Conditions)**

PARAMETER		TEST CONDITIONS	54/74		54S/74S		54LS/74LS		UNIT
			Min	Max	Min	Max	Min	Max	
t_W	$\overline{CP}_0$ Pulse width	Figure 1	50				15		ns
t_W	$\overline{CP}_1$ Pulse width	Figure 1	50				30		ns
t_W	MR Pulse width	Figure 2	50				15		ns
t_{rec}	Recovery time MR to $\overline{CP}$	Figure 2					25		ns

AC WAVEFORMS

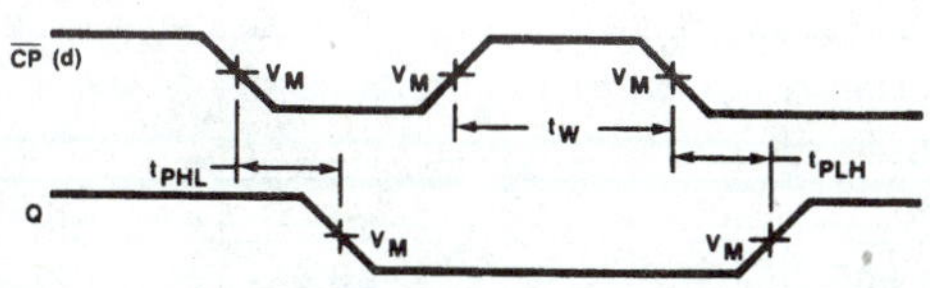

$V_M = 1.5V$ for 54/74 and 54S/74S; $V_M = 1.3V$ for 54LS/74LS.
The number of clock pulses required between the t_{PLH} and t_{PHL} measurements can be determined from the appropriate Truth Table.

Figure 1

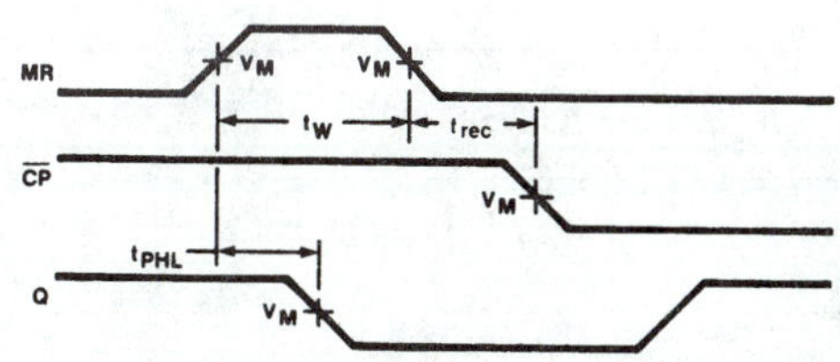

$V_M = 1.5V$ for 54/74 and 54S/74S; $V_M = 1.3V$ for 54LS/74LS.

Figure 2

2. 74xxx Family Data Sheets

54S/74S138
54LS/74LS138

DESCRIPTION

The "138" is a HIGH speed 1-of-8 Decoder/Demultiplexer. The "138" is ideal for HIGH speed bipolar memory chip select address decoding. The multiple input enables allow parallel expansion to a 1-of-24 decoder using only three "138" devices; or to a 1-of-32 decoder using four "138" devices and one inverter.

FEATURES

- **Demultiplexing capability**
- **Multiple input enable for easy expansion**
- **Ideal for memory chip select decoding**
- **Direct replacement for Intel 3205**

LOGIC SYMBOL

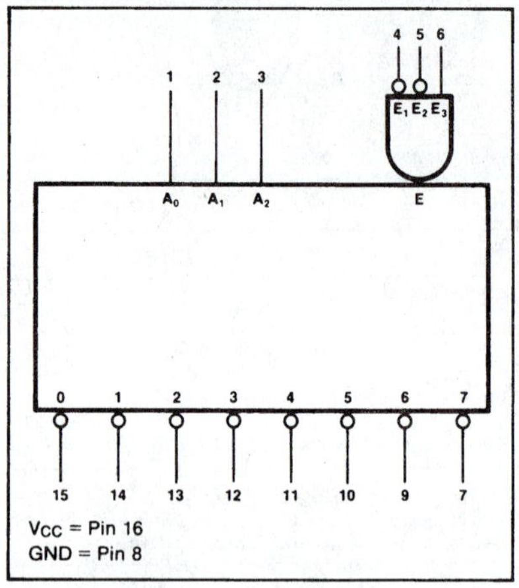

PIN CONFIGURATION

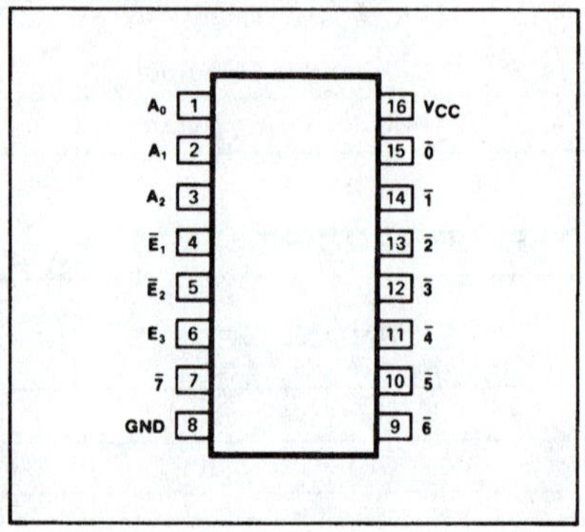

ORDERING CODE (See Section 9 for further Package and Ordering Information)

PACKAGES	COMMERCIAL RANGES $V_{CC} = 5V \pm 5\%$; $T_A = 0°C$ to $70°C$	MILITARY RANGES $V_{CC} = 5V \pm 10\%$; $T_A = -55°C$ to $+125°C$
Plastic DIP	N74S138N • N74LS138N	
Ceramic DIP	N74S138F • N74LS138F	S54S138F • S54LS138F
Flatpak		S54S138W • S54LS138W

INPUT AND OUTPUT LOADING AND FAN-OUT TABLE [a]

PINS	DESCRIPTION		54/74	54S/74S	54LS/74LS
A_0-A_2	Address inputs	I_{IH} (μA) I_{IL} (mA)		50 -2.0	20 -0.36
$\bar{E}_1$, $\bar{E}_2$	Enable (Active LOW) inputs	I_{IH} (μA) I_{IL} (mA)		50 -2.0	20 -0.36
E_3	Enable (Active HIGH) input	I_{IH} (μA) I_{IL} (mA)		50 -2.0	20 -0.36
$\bar{0}$-$\bar{7}$	Decoder ouputs	I_{OH} (μA) I_{OL} (mA)		-1000 20	-400 4/8 [a]

NOTE

a. The slashed numbers indicate different parametric values for Military/Commercial temperature ranges respectively.

2. 74xxx Family Data Sheets

FUNCTIONAL DESCRIPTION

The "138" decoder accepts three binary weighted inputs (A_0,A_1,A_2) and when enabled provides eight mutually exclusive active LOW outputs ($\overline{0}$-$\overline{7}$). The device features three Enable inputs: two active LOW ($\overline{E}_1$,$\overline{E}_2$) and one active HIGH (E_3). Every output will be HIGH unless $\overline{E}_1$ and $\overline{E}_2$ are LOW and E_3 is HIGH. This multiple Enable function allows easy parallel expansion of the device to a 1-of-32 (5 lines to 32 lines) decoder with just four "138's" and one inverter.

The device can be used as an eight output demultiplexer by using one of the active LOW Enable inputs as the data input and the remaining Enable inputs as strobes. Enable inputs not used must be permanently tied to their appropriate active HIGH or active LOW state.

TRUTH TABLE

INPUTS						OUTPUTS							
$\overline{E}_1$	$\overline{E}_2$	E_3	A_0	A_1	A_2	$\overline{0}$	$\overline{1}$	$\overline{2}$	$\overline{3}$	$\overline{4}$	$\overline{5}$	$\overline{6}$	$\overline{7}$
H	X	X	X	X	X	H	H	H	H	H	H	H	H
X	H	X	X	X	X	H	H	H	H	H	H	H	H
X	X	L	X	X	X	H	H	H	H	H	H	H	H
L	L	H	L	L	L	L	H	H	H	H	H	H	H
L	L	H	H	L	L	H	L	H	H	H	H	H	H
L	L	H	L	H	L	H	H	L	H	H	H	H	H
L	L	H	H	H	L	H	H	H	L	H	H	H	H
L	L	H	L	L	H	H	H	H	H	L	H	H	H
L	L	H	H	L	H	H	H	H	H	H	L	H	H
L	L	H	L	H	H	H	H	H	H	H	H	L	H
L	L	H	H	H	H	H	H	H	H	H	H	H	L

NOTES

H = HIGH voltage level
L = LOW voltage level
X = Don't care

LOGIC DIAGRAM

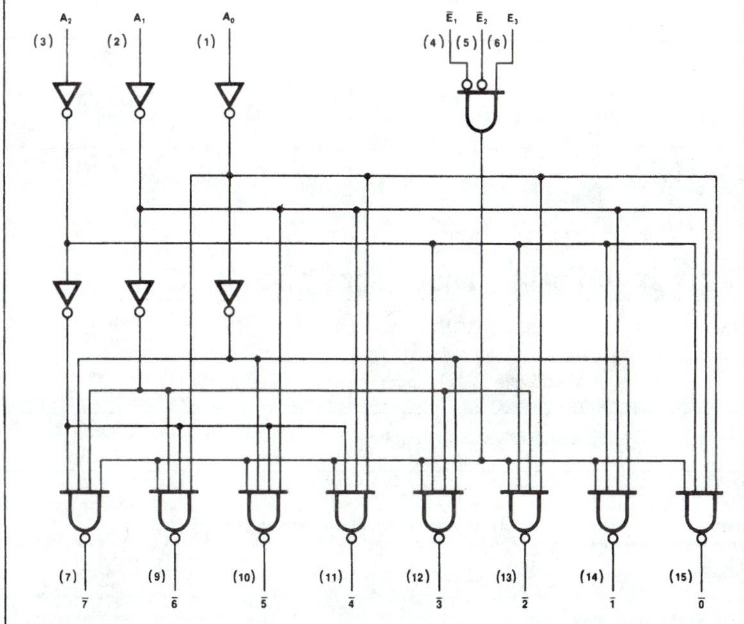

( ) = Pin Number
V_{CC} = Pin 16
GND = Pin 8

DC CHARACTERISTICS OVER OPERATING TEMPERATURE RANGE

PARAMETER		TEST CONDITIONS	54/74		54S/74S		54LS/74LS		UNIT
			Min	Max	Min	Max	Min	Max	
I_{CC}	Supply current	V_{CC} = Max				74		10	mA

2. 74xxx Family Data Sheets

54/74147

DESCRIPTION

The "147" 10-Line-to-4-Line Priority Encoder decodes the nine inputs to ensure that only the highest-order data line is encoded to four-line (8421) BCD outputs. Only nine data inputs are provided since the implied decimal "zero" condition requires no data input. The "zero" is encoded when all nine data lines are at a HIGH logic level. All of the "147" inputs and outputs are active LOW which simplifies mechanical switch encoding.

FEATURES

- Encodes 10-line decimal to 4-line BCD
- Useful for 10-position switch encoding
- Used in code converters and generators

LOGIC SYMBOL

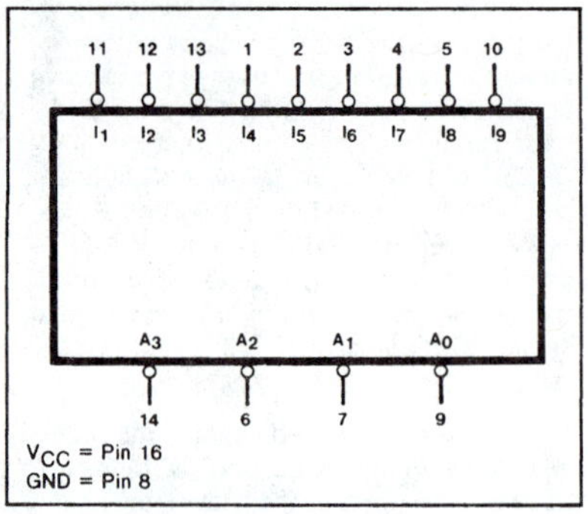

PIN CONFIGURATION

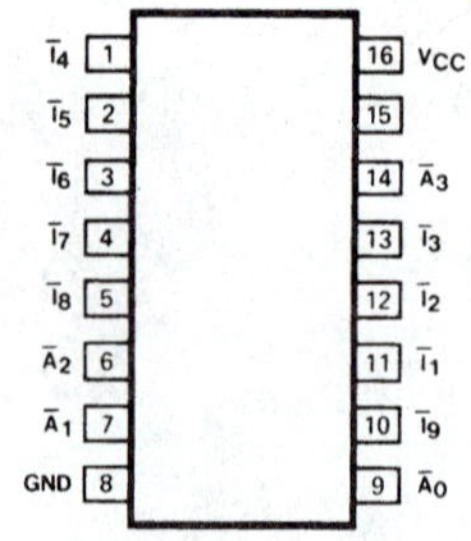

ORDERING CODE (See Section 9 for further Package and Ordering Information)

PACKAGES	COMMERCIAL RANGES $V_{CC}=5V\pm5\%$; $T_A=0°C$ to $+70°C$	MILITARY RANGES $V_{CC}=5V\pm10\%$; $T_A=-55°C$ to $+125°C$
Plastic DIP	N74147N	
Ceramic DIP	N74147F	S54147F
Flatpak		S54147W

INPUT AND OUTPUT LOADING AND FAN-OUT TABLE[(a)]

PINS	DESCRIPTION		54/74	54S/74S	54LS/74LS
$\bar{I}_1-\bar{I}_9$	Priority (active LOW) inputs	I_{IH} (μA) I_{IL} (mA)	40 −1.6		
$\bar{A}_0-\bar{A}_3$	Address (active LOW) outputs	I_{OH} (μA) I_{OL} (mA)	−800 16		

NOTE

a. The slashed numbers indicate different parametric values for Military/Commercial temperature ranges respectively.

2. 74xxx Family Data Sheets

FUNCTIONAL DESCRIPTION

The "147" 9-input priority encoder accepts data from nine active LOW inputs ($\bar{I}_1$-$\bar{I}_9$) and provides a binary representation on the four active LOW outputs (A_0-A_3). A priority is assigned to each input so that when two or more inputs are simultaneously active, the input with the highest priority is represented on the output, with input line $\bar{I}_9$ having the highest priority.

The device provides the 10-line-to-4-line priority encoding function by use of the implied decimal "zero." The "zero" is encoded when all nine data inputs are HIGH forcing all four outputs HIGH.

TRUTH TABLE

INPUTS									OUTPUTS			
$\bar{I}_1$	$\bar{I}_2$	$\bar{I}_3$	$\bar{I}_4$	$\bar{I}_5$	$\bar{I}_6$	$\bar{I}_7$	$\bar{I}_8$	$\bar{I}_9$	$\bar{A}_3$	$\bar{A}_2$	$\bar{A}_1$	$\bar{A}_0$
H	H	H	H	H	H	H	H	H	H	H	H	H
X	X	X	X	X	X	X	X	L	L	H	H	L
X	X	X	X	X	X	X	L	H	L	H	H	H
X	X	X	X	X	X	L	H	H	H	L	L	L
X	X	X	X	X	L	H	H	H	H	L	L	H
X	X	X	X	L	H	H	H	H	H	L	H	L
X	X	X	L	H	H	H	H	H	H	L	H	H
X	X	L	H	H	H	H	H	H	H	H	L	L
X	L	H	H	H	H	H	H	H	H	H	L	H
L	H	H	H	H	H	H	H	H	H	H	H	L

H = HIGH voltage level
L = LOW voltage level
X = Don't care

LOGIC DIAGRAM

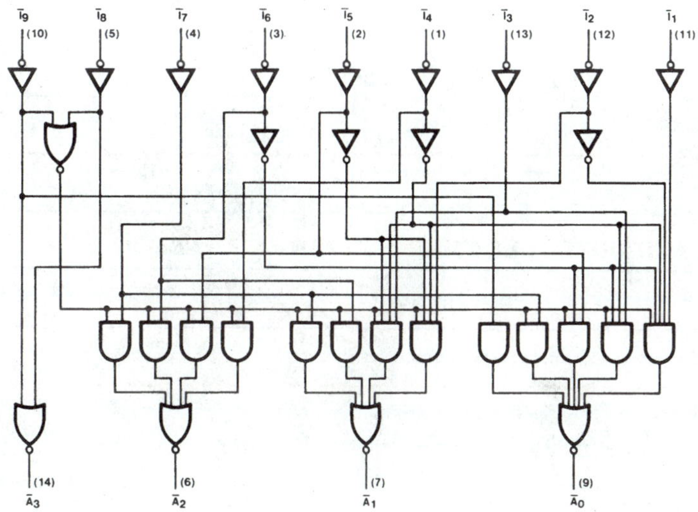

V_{CC} = Pin 16
GND = Pin 8
( ) = Pin numbers

DC CHARACTERISTICS OVER OPERATING TEMPERATURE RANGE

PARAMETER		TEST CONDITIONS	54/74		54S/74S		54LS/74LS		UNIT
			Min	Max	Min	Max	Min	Max	
I_{OS}	Output short circuit current	V_{CC} = Max, V_{OUT} = OV	−35	−85					mA
I_{CC}	Supply current	V_{CC} = Max, $V_{\bar{7}}$ = OV		70					mA
		V_{CC} = Max, inputs open		62					mA

2. 74xxx Family Data Sheets

54/74153
54S/74S153
54LS/74LS153

LOGIC SYMBOL

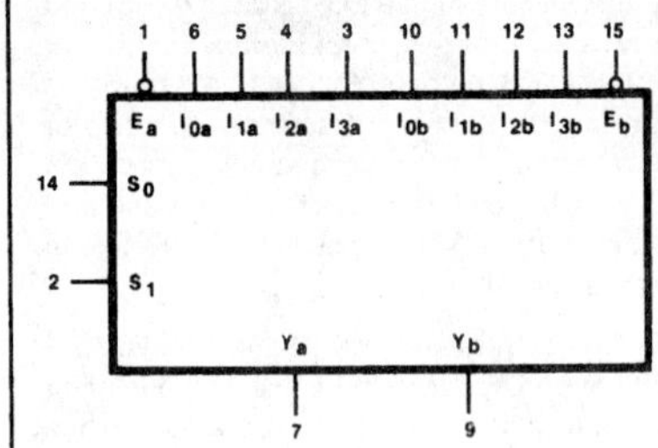

DESCRIPTION

The "153" is a high speed Dual 4-Input Multiplexer with common Select inputs and individual Enable inputs for each section. The device can select two bits of data from four sources. The two buffered outputs present data in the non-inverted (true) form. The "153" can generate any two functions of three variables in addition to multiplexer operation.

FEATURES

- **Non-inverting outputs**
- **Separate Enable for each section**
- **Common Select inputs**
- **See "253" for 3-State version**

PIN CONFIGURATION

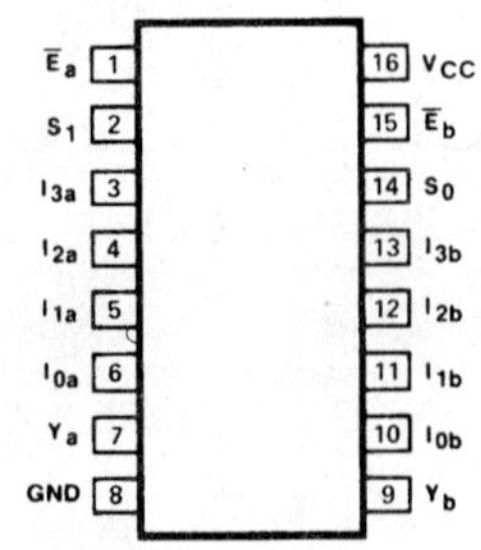

ORDERING CODE (See Section 9 for further Package and Ordering Information)

PACKAGES	COMMERCIAL RANGES V_{CC} = 5V ± 5%; T_A = 0°C to 70°C	MILITARY RANGES V_{CC} = 5V ± 10%; T_A = -55°C to +125°C
Plastic DIP	N74153N N74S153N • N74LS153N	
Ceramic DIP	N74153F N74S153F • N74LS153F	S54153F S54S153F • S54LS153F
Flatpak		S54153W S54S153W • S54LS153W

INPUT AND OUTPUT LOADING AND FAN-OUT TABLE (a)

PINS	DESCRIPTION		54/74	54S/74S	54LS/74LS
S_0, S_1	Common Select inputs	I_{IH} (μA) I_{IL} (mA)	40 -1.6	50 -2.0	20 -0.4
$\bar{E}_a$, $\bar{E}_b$	Enable (Active LOW) inputs	I_{IH} (μA) I_{IL} (mA)	40 -1.6	50 -2.0	20 -0.4
I_{0a} - I_{3a}	Data inputs from multiplexer "a"	I_{IH} (μA) I_{IL} (mA)	40 -1.6	50 -2.0	20 -0.4
I_{0b} - I_{3b}	Data inputs from multiplexer "b"	I_{IH} (μA) I_{IL} (mA)	40 -1.6	50 -2.0	20 -0.4
Y	Multiplexer outputs	I_{OH} (μA) I_{OL} (mA)	-800 16	-1000 20	-400 4/8 (a)

NOTE

a. The slashed numbers indicate different parametric values for Military/Commercial temperature ranges respectively.

2. 74xxx Family Data Sheets

FUNCTIONAL DESCRIPTION

The "153" is a Dual 4-input Multiplexer that can select two bits of data from up to four sources under control of the common Select inputs (S_0, S_1). The two 4-input multiplexer circuits have individual active LOW Enables ($\bar{E}_a$, $\bar{E}_b$) which can be used to strobe the outputs independently. Outputs (Y_a, Y_b) are forced LOW when the corresponding Enables ($\bar{E}_a$, $\bar{E}_b$) are HIGH.

The device is the logical implementation of a 2-pole, 4-position switch, where the position of the switch is determined by the logic levels supplied to the two Select inputs. The logic equations for the outputs are shown below.

$$Y_a = \bar{E}_a \bullet (I_{0a} \bullet \bar{S}_1 \bullet \bar{S}_0 + I_{1a} \bullet \bar{S}_1 \bullet S_0 + I_{2a} \bullet S_1 \bullet \bar{S}_0 + I_{3a} \bullet S_1 \bullet S_0)$$

$$Y_b = \bar{E}_b \bullet (I_{0b} \bullet \bar{S}_1 \bullet \bar{S}_0 + I_{1b} \bullet \bar{S}_1 \bullet S_0 + I_{2b} \bullet S_1 \bullet \bar{S}_0 + I_{3b} \bullet S_1 \bullet S_0)$$

The "153" can be used to move data to a common output bus from a group of registers. The state of the Select inputs would determine the particular register from which the data came. An alternative application is as a function generator. The device can generate two functions of three variables. This is useful for implementing highly irregular random logic.

TRUTH TABLE

SELECT INPUTS		INPUTS (a or b)					OUTPUT
S_0	S_1	$\bar{E}$	I_0	I_1	I_2	I_3	Y
X	X	H	X	X	X	X	L
L	L	L	L	X	X	X	L
L	L	L	H	X	X	X	H
H	L	L	X	L	X	X	L
H	L	L	X	H	X	X	H
L	H	L	X	X	L	X	L
L	H	L	X	X	H	X	H
H	H	L	X	X	X	L	L
H	H	L	X	X	X	H	H

H = HIGH voltage level
L = LOW voltage level
X = Don't care

LOGIC DIAGRAM

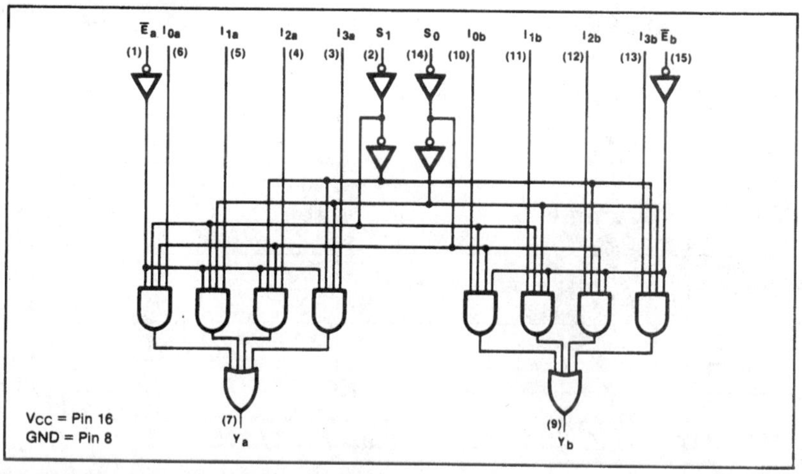

DC CHARACTERISTICS OVER OPERATING TEMPERATURE RANGE [b]

PARAMETER		TEST CONDITIONS		54/74 Min	54/74 Max	54S/74S Min	54S/74S Max	54LS/74LS Min	54LS/74LS Max	UNIT
I_{CC}	Supply current	V_{CC} = Max	Mil		52		70		10	mA
			Com		60		70		10	mA

NOTE

b. For family dc characteristics, see 54/74, 54H/74H, 54S/74S, and 54LS/74LS specifications.

2. 74xxx Family Data Sheets

54/74173
54LS/74LS173

DESCRIPTION

The "173" is a Quad D-Type Flip-Flop with 3-State buffered outputs, a gated input enable and a gated output enable. The device is used as a 4-Bit Register capable of driving a 3-State bus directly. Data is loaded into the register during the LOW-to-HIGH clock transition. The Master Reset asynchronously clears all flip-flops.

FEATURES

- **Edge triggered D-Type register**
- **Gated input enable for "hold" do nothing mode**
- **3-State output buffers**
- **Gated output enable control**
- **Pin compatible with the 8T10 and DM8551**

LOGIC SYMBOL

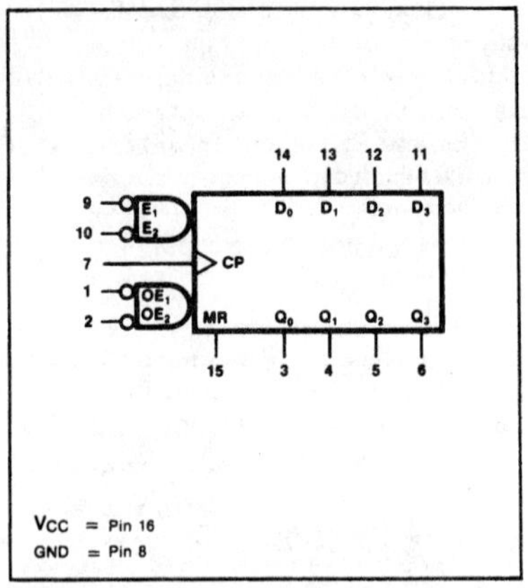

PIN CONFIGURATION

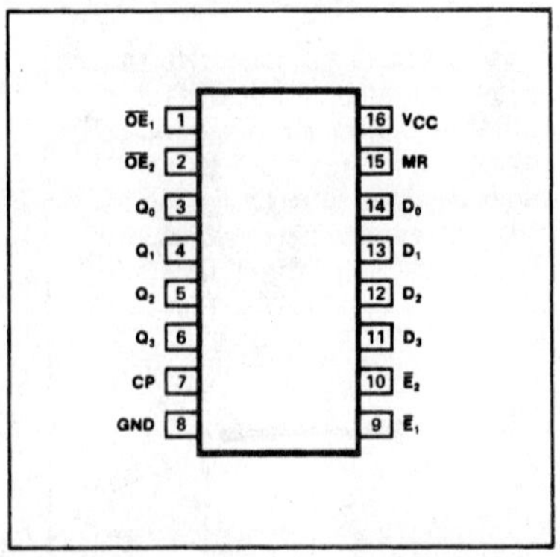

ORDERING CODE (See Section 9 for further Package and Ordering Information)

PACKAGES	COMMERCIAL RANGES V_{CC} = 5V ± 5%; T_A = 0°C to +70°C	MILITARY RANGES V_{CC} = 5V ± 10%; T_A = -55°C to +125°C
Plastic DIP	N74173N • N74LS173N	
Ceramic DIP	N74173F • N74LS173F	S54LS173F
Flatpak		S54LS173W

INPUT AND OUTPUT LOADING AND FAN-OUT TABLE (a)

PINS	DESCRIPTION		54/74	54S/74S	54LS/74LS
D_0-D_3	Data inputs	I_{IH} (μA) I_{IL} (mA)	40 -1.6		20 -0.4
$\overline{E}_1$,$\overline{E}_2$	Clock Enable inputs	I_{IH} (μA) I_{IL} (mA)	40 -1.6		20 -0.4
CP	Clock (Active HIGH going edge) input	I_{IH} (μA) I_{IL} (mA)	40 -1.6		20 -0.4
$\overline{OE}_1$,$\overline{OE}_2$	3-State Output Enable inputs	I_{IH} (μA) I_{IL} (mA)	40 -1.6		20 -0.4
MR	Asynchronous Master Reset	I_{IH} (μA) I_{IL} (mA)	40 -1.6		20 -0.4
Q_0-Q_3	3-State outputs	I_{OH} (μA) I_{OL} (mA)	-5.2 16		-1/-2.6 (a) 12/24 (a)

NOTE

a. The slashed numbers indicate different parametric values for Military/Commercial temperature ranges respectively.

2. 74xxx Family Data Sheets

FUNCTIONAL DESCRIPTION

The "173" is a 4-Bit Parallel Load Register with clock enable control, 3-State buffered outputs and master reset. When the two Clock Enable ($\overline{E}_1$ and $\overline{E}_2$) inputs are LOW, the data on the D inputs is loaded into the register synchronously with the LOW-to-HIGH Clock (CP) transition. When one or both $\overline{E}$ inputs are HIGH one setup time before the LOW-to-HIGH clock transition, the register will retain the previous data. The Data inputs and Clock Enable inputs are fully edge-triggered and must be stable only one setup time before the LOW-to-HIGH clock transition.

The Master Reset (MR) is an active HIGh asynchronous input. When the Mr is HIGH all four flip-flops are reset (cleared) independently of any other input condition.

The 3-State output buffers are controlled by a 2-input NOR gate. When both Output Enable ($\overline{OE}_1$ and $\overline{OE}_2$) inputs are LOW, the data in the register is presented at the Q outputs. When one or both $\overline{OE}$ inputs is HIGH the outputs are forced to a high impedance "off" state. The 3-State output buffers are completely independent of the register operation; the $\overline{OE}$ transition do not affect the clock and reset operations.

MODE SELECT—FUNCTION TABLE

REGISTER OPERATING MODES	INPUTS					OUTPUTS
	MR	CP	$\overline{E}_1$	$\overline{E}_2$	D_n	Q_n (Register)
Reset (clear)	H	X	X	X	X	L
Parallel Load	L	↑	l	l	l	L
	L	↑	l	l	h	H
Hold (No change)	L	X	h	X	X	q_n
	L	X	X	h	X	q_n

3-STATE BUFFER OPERATING MODES	INPUTS			OUTPUTS
	Q_n (Register)	$\overline{OE}_1$	$\overline{OE}_2$	Q_0, Q_1, Q_2, Q_3
Read	L	L	L	L
	H	L	L	H
Disabled	X	H	X	(Z)
	X	X	H	(Z)

NOTES

H = HIGH voltage level
h = HIGH voltage level one setup time prior to the LOW-to-HIGH clock transition
L = LOW voltage level
l = LOW voltage level one setup time prior to the LOW-to-HIGH clock transition
q_n = Lower case letters indicate the state of the referenced input (or output) one setup time prior to the LOW-to-HIGH clock transition
X = Don't care
(Z) = High impedance "off" state
↑ = LOW-to-HIGH transition

LOGIC DIAGRAM

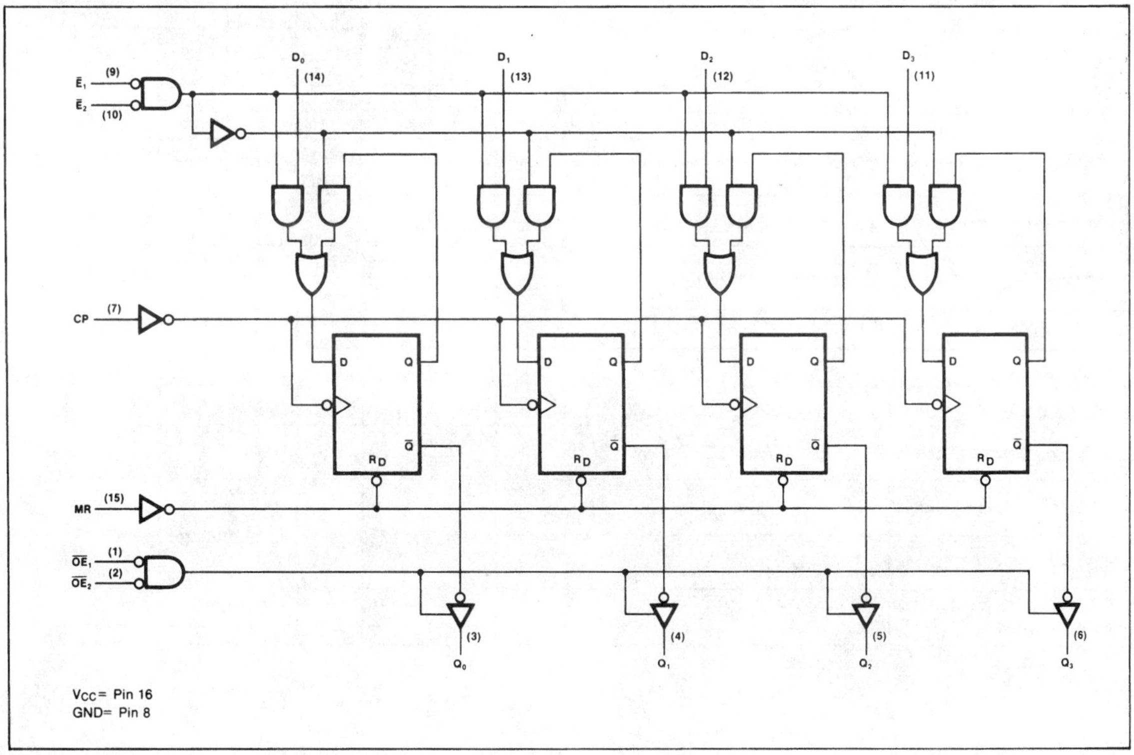

2. 74xxx Family Data Sheets

DC CHARACTERISTICS OVER OPERATING TEMPERATURE RANGE[b]

PARAMETER		TEST CONDITIONS		54/74		54S/74S		54LS/74LS		UNIT
				Min	Max	Min	Max	Min	Max	
V_{OL}	Output LOW voltage	V_{CC} = Min	I_{OL} = 16mA		0.4					V
			I_{OL} = 12mA						0.4	V
			I_{OL} = 24mA						0.5[c]	V
V_{OH}	Output HIGH voltage	V_{CC} = Min I_{OH} = See Fan Out Table		2.4				2.4		V
I_{OS}	Output short circuit current	V_{CC} = Max, V_{OUT} = 0V		-30	-70			-30	-100	mA
I_{CC}	Supply current	V_{CC} = Max			72				30	mA

AC CHARACTERISTICS T_A = 25°C (See Section 4 for Test Circuits and Conditions)

PARAMETER		TEST CONDITIONS	54/74 C_L = 50pF R_L = 400Ω		54S/74S		54LS/74LS C_L = 45pF R_L = 667Ω		UNIT
			Min	Max	Min	Max	Min	Max	
f_{MAX}	Maximum clock frequency	Figure 1	25				30		MHz
t_{PLH} t_{PHL}	Propagation delay clock to output	Figure 1		43 31				18 26	ns ns
t_{PHL}	Propagation delay MR to output	Figure 4		27				36	ns
t_{PZH}	Output enable to HIGH level	Figure 2		30				15	ns
t_{PZL}	Output enable to LOW level	Figure 3		30				18	ns
t_{PHZ}	Output disable from HIGH level	Figure 2		14				13	ns
t_{PLZ}	Output disable from LOW level	Figure 3		20				20	ns

AC SETUP REQUIREMENTS T_A = 25° C (See Section 4 for test Circuits and Conditions)

PARAMETER		TEST CONDITIONS	54/74		54S/74S		54LS/74LS		UNIT
			Min	Max	Min	Max	Min	Max	
t_W(CP)	Clock pulse width	Figure 1	20				18		ns
t_W(MR)	MR pulse width	Figure 4	20				20		ns
t_s(D)	Set up time Data to Clock	Figure 5	10				13		ns
t_h(D)	Hold time Data to Clock	Figure 5	10				0		ns
$t_s(\bar{E})$	Set up time Data to Clock	Figure 5	17				17		ns
$t_h(\bar{E})$	Hold time Enable to Clock	Figure 5	2				0		ns
t_{rec}(MR)	Recovery time Master Reset to Clock	Figure 4	10				17		ns

NOTES

(c) This parameter for Commercial Range only.

2. 74xxx Family Data Sheets

AC WAVEFORMS[d]

CLOCK TO OUTPUT DELAYS
AND CLOCK PULSE WIDTH

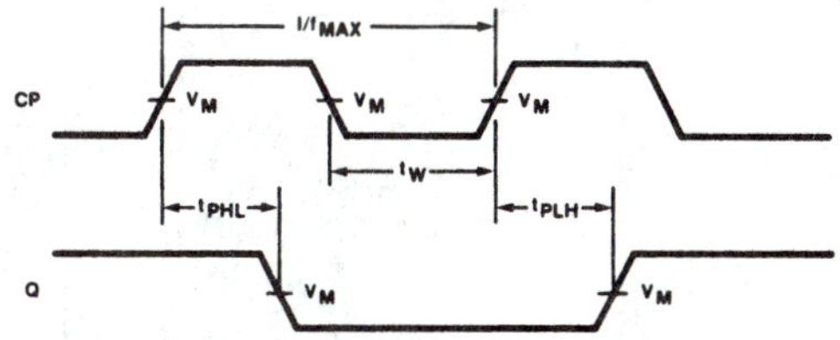

Figure 1

3-STATE ENABLE TIME TO HIGH LEVEL
AND DISABLE TIME FROM HIGH LEVEL

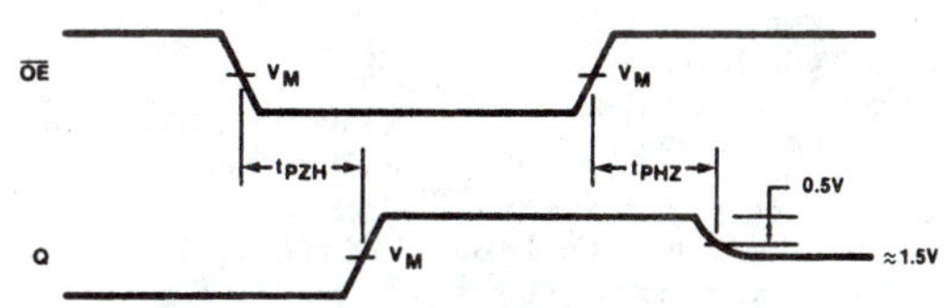

Figure 2

3-STATE ENABLE TIME TO LOW LEVEL
AND DISABLE TIME FROM LOW LEVEL

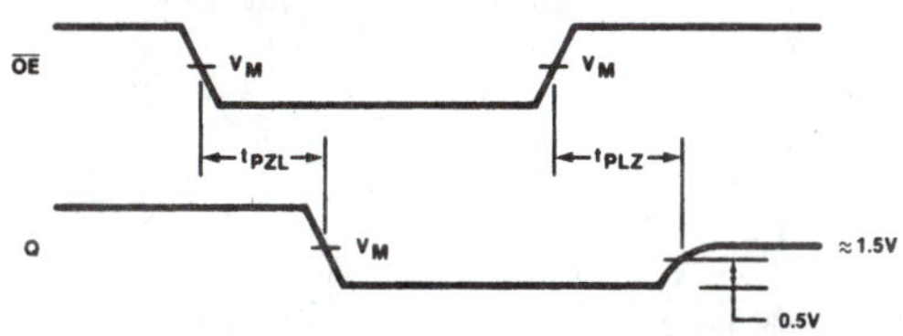

Figure 3

MASTER RESET PULSE WIDTH,
MASTER RESET TO OUTPUT DELAY AND
MASTER RESET TO CLOCK RECOVERY TIME

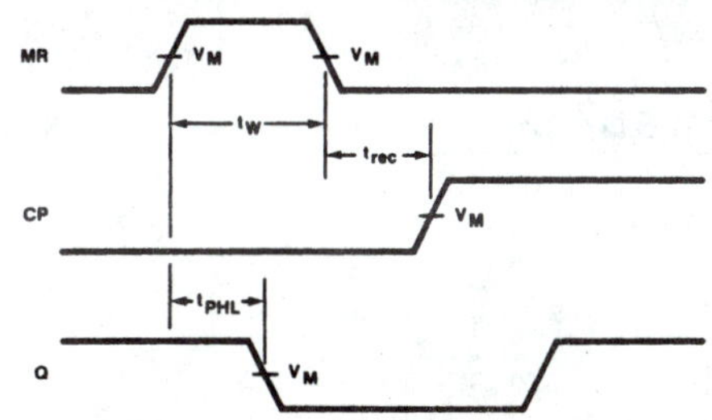

Figure 4

SETUP (ts) AND HOLD (th) TIMES
FOR DATA (D) AND ENABLE (Ē) INPUTS

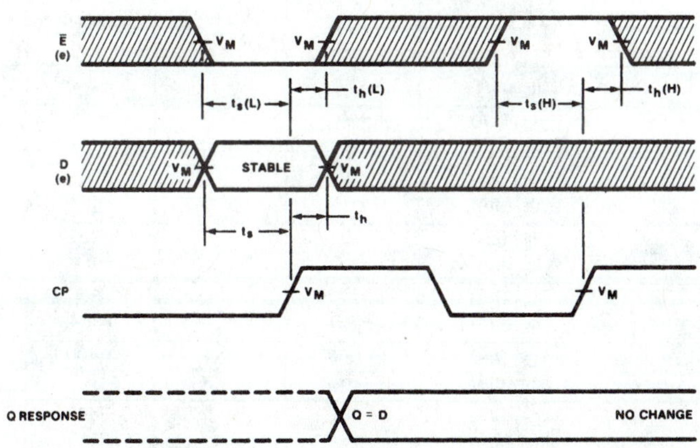

Figure 5

NOTES

(d) V_M = 1.5V for 54/74 & 54S/74S; V_M = 1.3V for 54LS/74LS.

(e) The shaded areas indicate when the input is permitted to change for predictable

2. 74xxx Family Data Sheets

54/74191
54LS/74LS191

DESCRIPTION

The "191" is a presettable 4-Bit Binary Up/Down Counter with state changes of the counter synchronous with the LOW-to-HIGH transition of the Clock Pulse input.

The circuit features an asynchronous Parallel Load ($\overline{PL}$) input which overrides counting and loads the data present on the D_n inputs into the flip-flops. Synchronous expansion in a multistage counter is made possible by a Count Enable ($\overline{CE}$) input. The count up or count down mode is determined by an Up/Down ($\overline{U}$/D) input. A variety of methods for generating carry/borrow signals in multistage counter application is made possible by Terminal Count (TC) and Ripple Clock ($\overline{RC}$) outputs.

FEATURES

- **Synchronous, reversible 4-bit binary counting**
- **Asynchronous parallel load capability**
- **Count Enable control for synchronous expansion**
- **Single Up/Down control input**

LOGIC SYMBOL

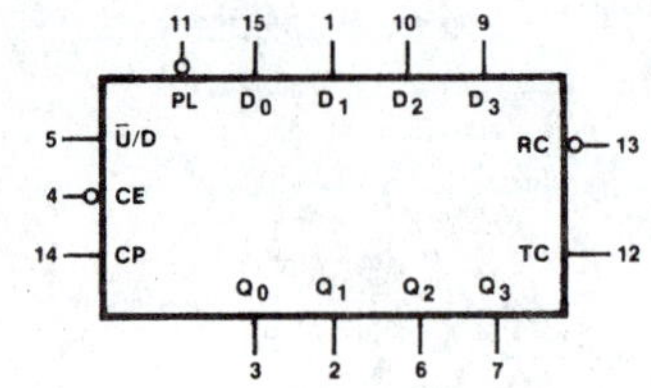

V_{CC} = Pin 16
GND = Pin 8

PIN CONFIGURATION

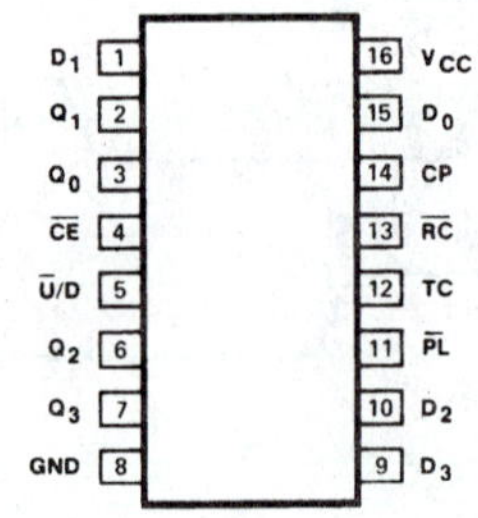

ORDERING CODE (See Section 9 for further Package and Ordering Information)

PACKAGES	COMMERCIAL RANGES V_{CC}=5V ± 5%; T_A=0°C to +70°C	MILITARY RANGES V_{CC}=5V ± 10%; T_A=−55°C to +125°C
Plastic DIP	N74191N • N74LS191N	
Ceramic DIP	N74191F • N74LS191F	S54191F • S54LS191F
Flatpak		S54191W • S54LS191W

INPUT AND OUTPUT LOADING AND FAN-OUT TABLE(a)

PINS	DESCRIPTION		54/74	54S/74S	54LS/74LS
CP	Clock Pulse (active HIGH going edge) input	I_{IH} (μA) I_{IL} (mA)	40 −1.6		20 −0.4
$\overline{CE}$	Clock Enable (active LOW) input	I_{IH} (μA) I_{IL} (mA)	120 −4.8		60 −1.2
$\overline{U}$/D	Up/Down count control input	I_{IH} (μA) I_{IL} (mA)	40 −1.6		20 −0.4
$\overline{PL}$	Parallel Load (active LOW) input	I_{IH} (μA) I_{IL} (mA)	40 −1.6		20 −0.4
D_n	Parallel Data inputs	I_{IH} (μA) I_{IL} (mA)	40 −1.6		20 −0.4
Q_n	Counter outputs	I_{OH} (μA) I_{OL} (mA)	−800 16		−400 4/8(a)
TC	Terminal Count output	I_{OH} (μA) I_{OL} (mA)	−800 16		−400 4/8(a)
$\overline{RC}$	Ripple Clock (active LOW pulse) output	I_{OH} (μA) I_{OL} (mA)	−800 16		−400 4/8(a)

NOTE

a. The slashed numbers indicate different parametric values for Military/Commercial temperature ranges respectively.

2. 74xxx Family Data Sheets

FUNCTIONAL DESCRIPTION

The "191" is an asynchronously presettable Up/Down 4-Bit Binary Counter. It contains four master/slave flip-flops with internal gating and steering logic to provide asynchronous preset and synchronous count-up and count-down operation.

Asynchronous parallel load capability permits the counter to be preset to any desired number. Information present on the parallel Data inputs (D_0-D_3) is loaded into the counter and appears on the outputs when the Parallel Load ($\overline{PL}$) input is LOW. As indicated in the Mode Select Table, this operation overrides the counting function.

Counting is inhibited by a HIGH level on the Count Enable ($\overline{CE}$) input. When $\overline{CE}$ is LOW, internal state changes are initiated synchronously by the LOW-to-HIGH transition of the Clock input. The Up/Down ($\overline{U}$/D) input signal determines the direction of counting as indicated in the Mode Select Table. The $\overline{CE}$ input may go LOW when the clock is in either state, however, the LOW-to-HIGH $\overline{CE}$ transition must occur only while the Clock is HIGH. Also, the $\overline{U}$/D input should be changed only when either $\overline{CE}$ or CP is HIGH.

MODE SELECT-FUNCTION TABLE

OPERATING MODE	INPUTS					OUTPUTS
	$\overline{PL}$	$\overline{U}$/D	$\overline{CE}$	CP	D_n	Q_n
Parallel load	L	X	X	X	L	L
	L	X	X	X	H	H
Count up	H	L	I	↑	X	count up
Count down	H	H	I	↑	X	count down
Hold "do nothing"	H	X	H	X	X	no change

TC AND $\overline{RC}$ TRUTH TABLE

INPUTS			TERMINAL COUNT STATE				OUTPUTS	
$\overline{U}$/D	$\overline{CE}$	CP	Q_0	Q_1	Q_2	Q_3	TC	$\overline{RC}$
H	X	X	H	H	H	H	L	H
L	H	X	H	H	H	H	H	H
L	L	⊔	H	H	H	H	H	⊔
L	X	X	L	L	L	L	L	H
H	H	X	L	L	L	L	H	H
H	L	⊔	L	L	L	L	H	⊔

H = HIGH voltage level steady state.
L = LOW voltage level steady state.
I = LOW voltage level one setup time prior to the LOW-to-HIGH clock transition.
X = Don't care.
↑ = LOW-to-HIGH clock transition.
⊔ = LOW pulse.

LOGIC DIAGRAM

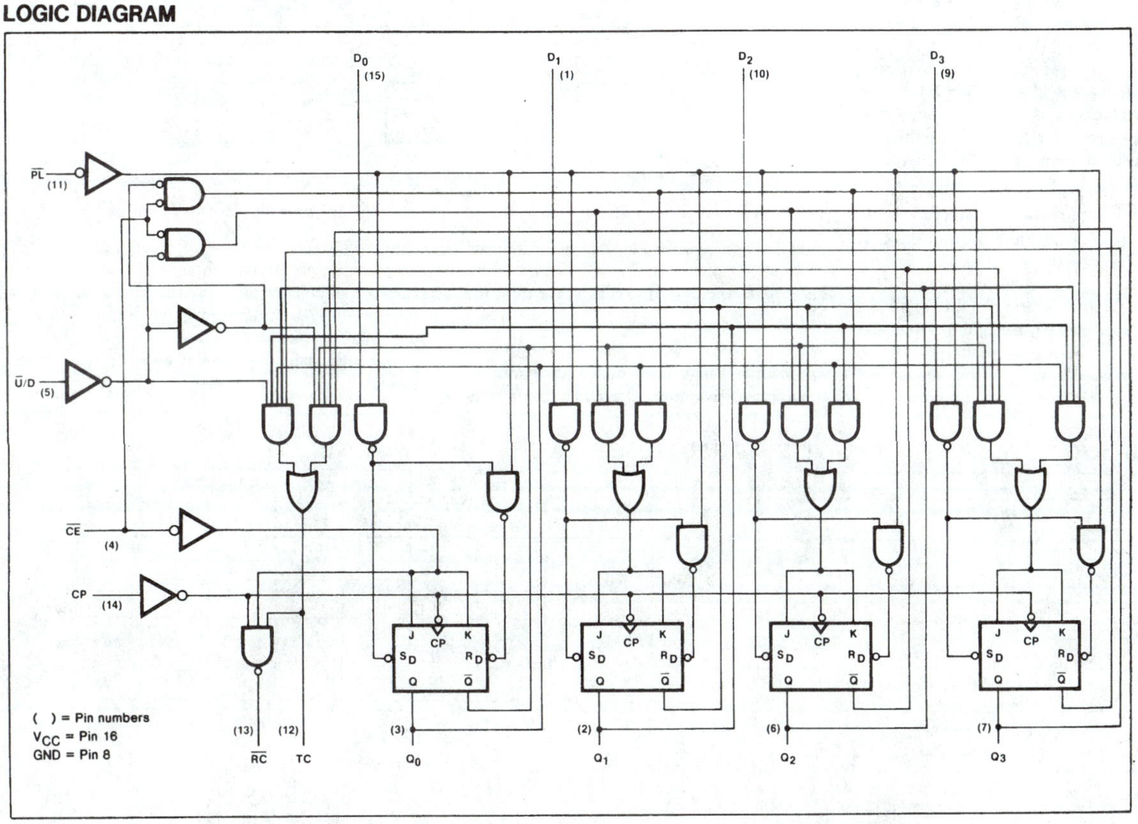

2. 74xxx Family Data Sheets

Overflow/underflow indications are produced by two types of outputs, the Terminal Count (TC) and Ripple Clock ($\overline{RC}$). The Terminal Count (TC) output is normally LOW and goes HIGH when a circuit reaches zero in the count-down mode or reaches "15" in the count-up mode. The TC output will remain HIGH until a state change occurs, either by counting or presetting, or until $\overline{U}/D$ is changed. Do not use the TC output as a clock signal because it is subject to decoding spikes.

The TC signal is used internally to enable the Ripple Clock ($\overline{RC}$) output. When TC is HIGH and $\overline{CE}$ is LOW, the Ripple Clock follows the Clock Pulse (CP) delayed by two gate delays. The $\overline{RC}$ output essentially duplicates the LOW clock pulse width, although delayed in time by two gate delays. This feature simplifies the design of multistage counters as indicated in Figures A and B. In Figure A, each $\overline{RC}$ output is used as the clock input for the next higher stage. When the clock source has a limited drive capability this configuration is particularly advantageous, since the clock source drives only the first stage. It is only necessary to inhibit the first stage to prevent counting in all stages, since a HIGH signal on $\overline{CE}$ inhibits the $\overline{RC}$ output pulse as indicated in the Mode Select Table. The timing skew between state changes in the first and last stages is represented by the cumulative delay of the clock as it ripples through the preceding stages. This is a disadvantage of the configuration in some applications.

Figure B shows a method of causing state changes to occur simultaneously in all stages. The $\overline{RC}$ outputs propagate the carry/borrow signals in ripple fashion and all clock inputs are driven in parallel. The LOW state duration of the clock in this configuration must be long enough to allow the negative-going edge of the carry/borrow signal to ripple through to the last stage before the clock goes HIGH. Since the $\overline{RC}$ output of any package goes HIGH shortly after its CP input goes HIGH, there is no such restriction on the HIGH state duration of the clock.

In Figure C the configuration shown avoids ripple delays and their associated restrictions. Combining the TC signals from all the preceding stages forms the $\overline{CE}$ input signal for a given stage. An enable signal must be included in each carry gate in order to inhibit counting. The TC output of a given stage is not affected by its own $\overline{CE}$ therefore the simple inhibit scheme of Figure A and B does not apply.

N-STAGE COUNTER USING RIPPLE CLOCK

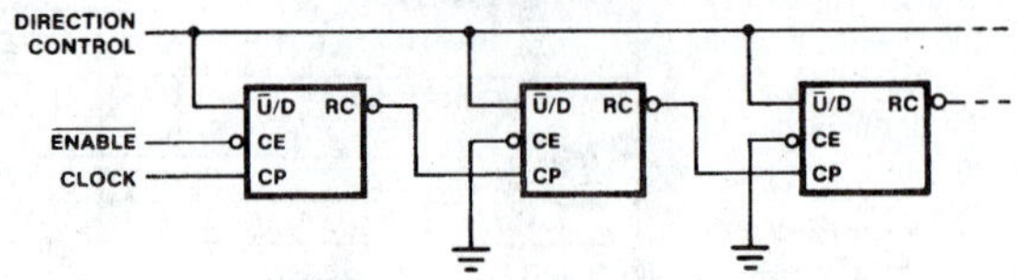

Figure A

SYNCHRONOUS N-STAGE COUNTER USING RIPPLE CARRY/BORROW

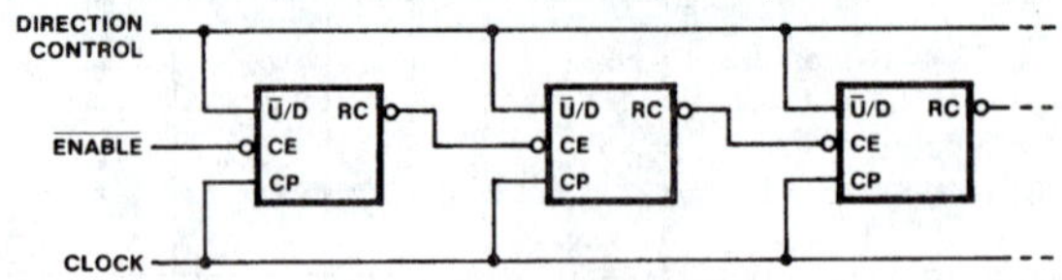

Figure B

SYNCHRONOUS N-STAGE COUNTER WITH PARALLEL GATED CARRY/BORROW

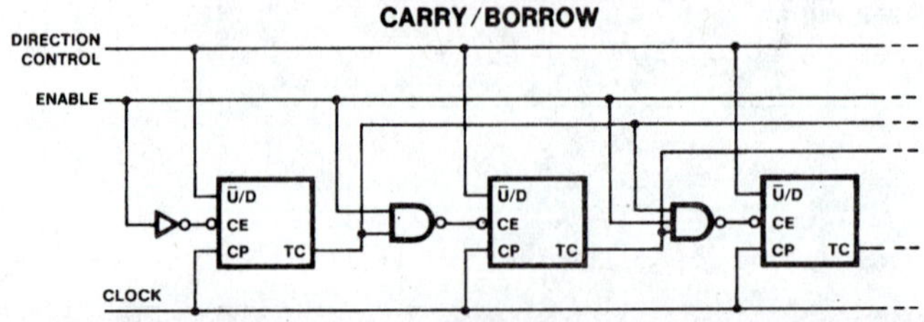

Figure C

DC CHARACTERISTICS OVER OPERATING TEMPERATURE RANGE[b]

PARAMETER		TEST CONDITIONS		54/74		54S/74S		54LS/74LS		UNIT
				Min	Max	Min	Max	Min	Max	
I_{OS}	Output short circuit current	V_{CC} = Max	Mil	−20	−65			−15	−100	mA
			Com	−18	−65			−15	−100	mA
I_{CC}	Supply current	V_{CC} = Max	Mil		99				35	mA
			Com		105				35	mA

NOTE
b. For family dc characteristics, see 54/74, 54H/74H, 54S/74S, and 54LS/74LS specifications.

2. 74xxx Family Data Sheets

AC CHARACTERISTICS: T_A=25°C (See Section 4 for Test Circuits and Conditions)

PARAMETER		TEST CONDITIONS	54/74		54S/74S		54LS/74LS		UNIT
			C_L = 15pF R_L = 400Ω				C_L = 15pF R_L = 2kΩ		
			Min	Max	Min	Max	Min	Max	
f_{MAX}	Maximum input count frequency	Figure 1	20				20		MHz
t_{PLH} t_{PHL}	Propagation delay, Clock to Q outputs	Figure 1		24 36				24 36	ns ns
t_{PLH} t_{PHL}	Propagation delay, Clock to $\overline{RC}$ output	Figure 2		20 24				20 24	ns ns
t_{PLH} t_{PHL}	Propagation delay, Clock to TC output	Figure 1		42 52				42 52	ns ns
t_{PLH} t_{PHL}	Propagation delay $\overline{U}$/D to $\overline{RC}$ output	Figure 7		45 45				45 45	ns ns
t_{PLH} t_{PHL}	Propagation delay $\overline{U}$/D to TC output	Figure 7		33 33				33 33	ns ns
t_{PLH} t_{PHL}	Propagation delay Data to Q outputs	Figure 3		22 50				32 40	ns ns
t_{PLH} t_{PHL}	Propagation delay $\overline{PL}$ to any output	Figure 4		33 50				33 50	ns ns
t_{PLH} t_{PHL}	Propagation delay $\overline{CE}$ to $\overline{RC}$ output	Figure 2		33 33				33 33	ns ns

AC SETUP REQUIREMENTS: T_A = 25°C (See Section 4 for Test Circuits and Conditions)

PARAMETER		TEST CONDITIONS	54/74		54S/74S		54LS/74LS		UNIT
			Min	Max	Min	Max	Min	Max	
t_W	CP pulse width	Figure 1	25				25		ns
t_W	$\overline{PL}$ pulse width	Figure 5	35				35		ns
t_s	Setup time Data to $\overline{PL}$	Figure 6	20				20		ns
t_h	Hold time Data to $\overline{PL}$	Figure 6	0				0		ns
t_{rec}	Recovery time $\overline{PL}$ to CP	Figure 5	40				40		ns
$t_s(L)$	Setup time LOW $\overline{CE}$ to clock	Figure 8	40				40		ns
$t_h(L)$	Hold time LOW $\overline{CE}$ to Clock	Figure 8	0				0		ns

2. 74xxx Family Data Sheets

AC WAVEFORMS

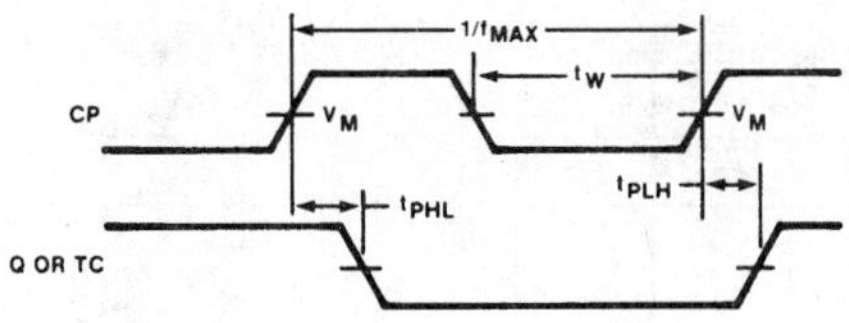

V_M = 1.5V for 54/74 and 54S/74S; V_M = 1.3V for 54LS/74LS.

Figure 1

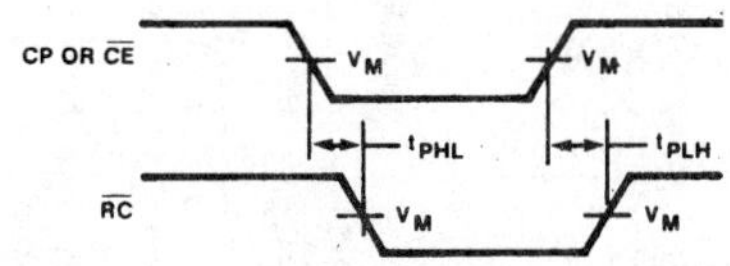

V_M = 1.5V for 54/74 and 54S/74S; V_M = 1.3V for 54LS/74LS.

Figure 2

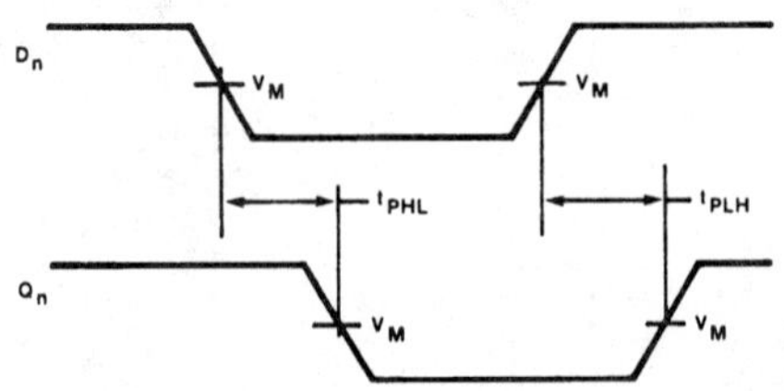

V_M = 1.5V for 54/74 and 54S/74S; V_M = 1.3V for 54LS/74LS.

Figure 3

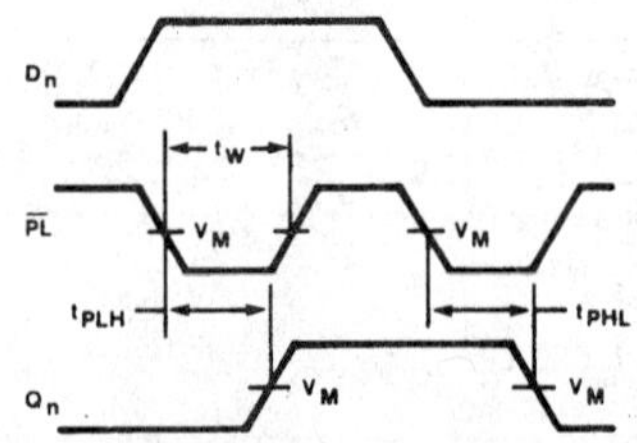

V_M = 1.5V for 54/74 and 54S/74S; V_M = 1.3V for 54LS/74LS.

Figure 4

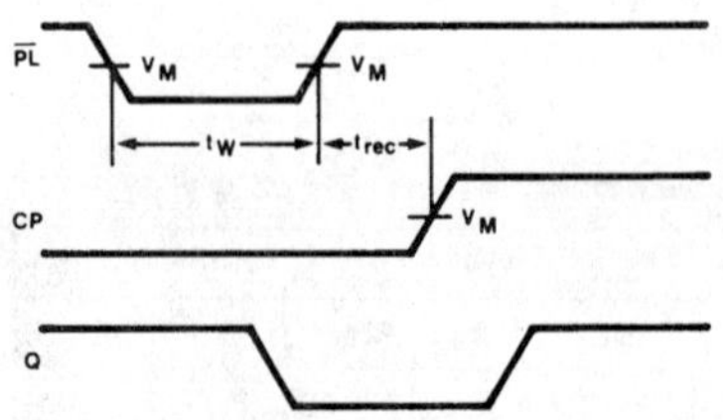

V_M = 1.5V for 54/74 and 54S/74S; V_M = 1.3V for 54LS/74LS.

Figure 5

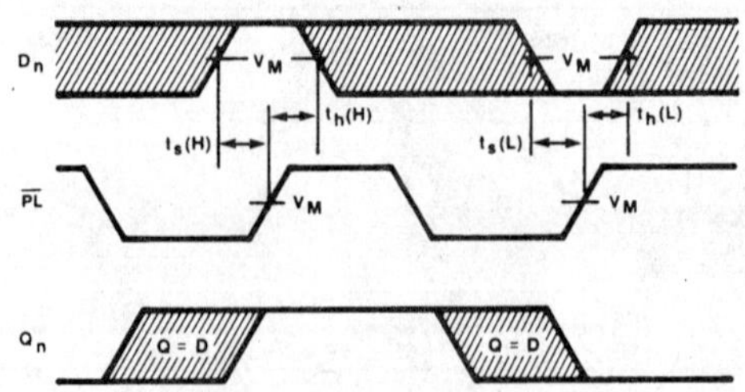

V_M = 1.5V for 54/74 and 54S/74S; V_M = 1.3V for 54LS/74LS.

The shaded areas indicate when the input is permitted to change for predictable output performance.

Figure 6

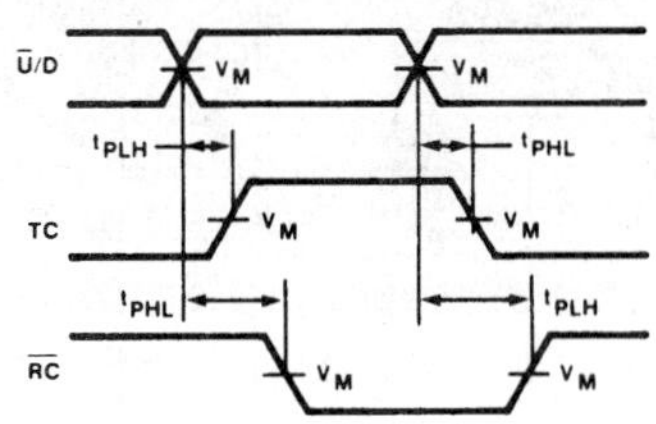

V_M = 1.5V for 54/74 and 54S/74S; V_M = 1.3V for 54LS/74LS.

Figure 7

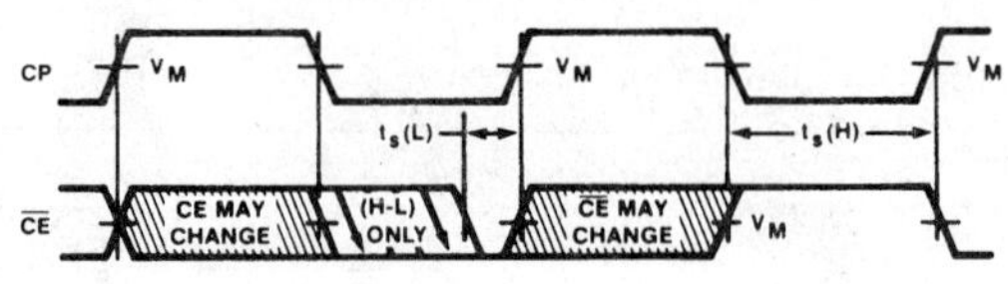

V_M = 1.5V for 54/74 and 54S/74S; V_M = 1.3V for 54LS/74LS.

The shaded areas indicate when the input is permitted to change for predictable output performance.

Figure 8

3. GAL16V8

National Semiconductor

GAL16V8 Generic Array Logic

General Description

The NSC E2CMOS™ GAL® device combines a high performance CMOS process with electrically erasable floating gate technology. This programmable memory technology applied to array logic provides designers with reconfigurable logic and bipolar performance at significantly reduced power levels.

The 20-pin GAL16V8 features 8 programmable Output Logic Macrocells (OLMCs) allowing each TRI-STATE® output to be configured by the user. Additionally, the GAL16V8 is capable of emulating, in a functional/fuse map/parametric compatible device, all common 20-pin PAL® device architectures.

Programming is accomplished using readily available hardware and software tools. NSC guarantees a minimum 100 erase/write cycles.

Unique test circuitry and reprogrammable cells allow complete AC, DC, cell and functionality testing during manufacture. Therefore, NSC guarantees 100% field programmability and functionality of the GAL devices. In addition, electronic signature is available to provide positive device ID. A security circuit is built-in, providing proprietary designs with copy protection.

Features

- Electrically erasable cell technology
  - Reconfigurable logic
  - Reprogrammable cells
  - Guaranteed 100% yields
- High performance E2CMOS technology
  - Low power: 45 mA/90 mA max active
  - High Speed: 20 ns–35 ns max access
- Eight output logic macrocells
  - Maximum flexibility for complex logic designs
  - Also emulates 20-pin PAL devices with full function/fuse map/parametric compatibility
- Preload and power-on reset of all registers
  - 100% functional testability
- Fully supported by National PLAN™ development software
- High speed programming algorithm
- Security cell prevents copying logic

PAL Replacement by Device Type

"Small-PAL" Mode				"Registered-PAL" Mode			"Medium-PAL" Mode
10L8	12L6	14L4	16L2	16R8	16R6	16R4	16L8
10H8	12H6	14H4	16H2	16RP8	16RP6	16RP4	16H8
10P8	12P6	14P4	16P2				16P8

PAL Replacement by Speed/Power

PAL			GAL
Rev	Speed	Power*	Rev
B-2	25 ns	90 mA	25L or 20L or 25Q (45 mA)
A	25 ns	180 mA	25L or 20L (90 mA)
B-2 MIL	30 ns	90 mA	30L or 25L (110 mA) or 30Q (55 mA)
A MIL	30 ns	180 mA	30L or 25L (110 mA)
B-4	35 ns	45 mA	30Q
A-2	35 ns	90 mA	30Q (45 mA)
STD	35 ns	180 mA	25L (90 mA)

*Shown for Medium PAL products

Block Diagram—GAL16V8

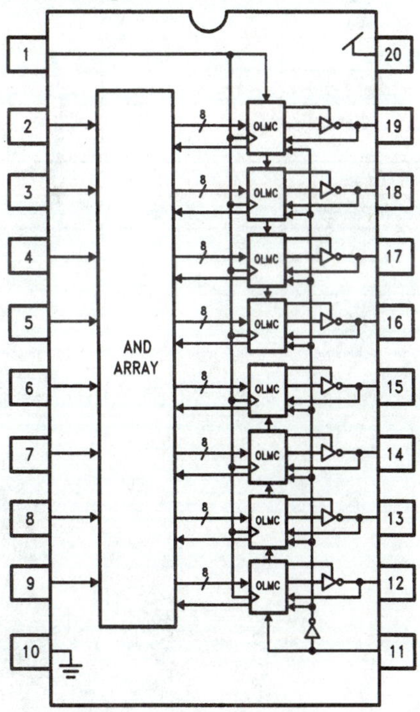

3. GAL16V8

Absolute Maximum Ratings (Note 1)

If Military/Aerospace specified devices are required, please contact the National Semiconductor Sales Office/Distributors for availability and specifications.

Supply Voltage (V_{CC}) (Note 2)	−0.5V to +7.0V
Input Voltage (Note 2)	−2.5V to V_{CC} +1.0V
Off-State Output Voltage (Note 2)	−2.5V to V_{CC} +1.0V
Output Current	+100 mA
Storage Temperature	−65°C to +150°C
Ambient Temperature with Power Applied	−65°C to +125°C
Junction Temperature	−65°C to +150°C
Lead Temperature (Soldering, 10 seconds)	260°C
ESD Tolerance	500V

C_{ZAP} = 100 pF
R_{ZAP} = 150Ω
Test Method: Human Body Model
Test Specification: NSC SOP-5-028

Recommended Operating Conditions

SUPPLY VOLTAGE AND TEMPERATURE

Symbol	Parameter	Commercial			Industrial			Military			Units
		Min	Nom	Max	Min	Nom	Max	Min	Nom	Max	
V_{CC}	Supply Voltage	4.75	5	5.25	4.5	5	5.5	4.5	5	5.5	V
T_A	Operating Free-Air Temperature	0	25	75	−40	25	85	−55	25		°C
T_C	Operating Case Temperature									125	°C

AC TIMING REQUIREMENTS

Symbol	Parameter		GAL16V8-20L		GAL16V8-25Q GAL16V8-25L		GAL16V8-30Q GAL16V8-30L		Units
			COM/IND		COM/IND MIL*		COM/IND* MIL		
			Min	Max	Min	Max	Min	Max	
t_{SU}	Set-Up Time (Input or Feedback before Clock)		15		20		25		ns
t_H	Hold Time (Input after Clock)		0		0		0		ns
t_W	Clock Pulse Width (High/Low)		12		15		15		ns
t_{CYCLE}	Clock Cycle Period (with Feedback) (Note 3)		30		35		45		ns
f_{CLK}	Clock Frequency (Note 4)	With Feedback		33.3		28.5		22.2	MHz
		Without Feedback		41.6		33.3		33.3	MHz
f_I	Input Frequency (Note 5)			50.0		40.0		33.3	MHz
t_{PR}	Clock Valid after Power-Up			100		100		100	ns

* -25Q available only in COM/IND operating ranges;
-30L available only in MIL operating range (for COM/IND, use -25L or -30Q).

Note 1: Absolute maximum ratings are those values beyond which the device may be permanently damaged. Proper operation is not guaranteed outside the specified recommended operating conditions.

Note 2: Some device pins may be raised above these limits during programming and preload operations according to the applicable specification.

Note 3: $t_{CYCLE} = t_{SU} + t_{CLK}$

Note 4: f_{CLK} (with feedback) $= (t_{CYCLE})^{-1}$
f_{CLK} (without feedback) $= (2\ t_w)^{-1}$

Note 5: $f_I = (t_{PD})^{-1}$

3. GAL16V8

Electrical Characteristics Over Recommended Operating Conditions

Symbol	Parameter		Conditions		Temperature Range	Min	Typ	Max	Units
V_{IH}	High Level Input Voltage					2.0		$V_{CC}+1$	V
V_{IL}	Low Level Input Voltage					−1.0		0.8	V
V_{OH}	High Level Output Voltage		V_{CC} = Min	$I_{OH} = -3.2$ mA	COM/IND	2.4			V
				$I_{OH} = -2.0$ mA	MIL	2.4			V
V_{OL}	Low Level Output Voltage		V_{CC} = Min	I_{OL} = 24 mA	COM/IND			0.5	V
				I_{OL} = 12 mA	MIL			0.5	V
I_{OZH}	High Level Off State Output Current		V_{CC} = Max, $V_O = V_{CC}$ (Max)					10	μA
I_{OZL}	Low Level Off State Output Current		V_{CC} = Max, V_O = GND					−10	μA
I_I	Maximum Input Current		V_{CC} = Max, $V_I = V_{CC}$ (Max)					10	μA
I_{IH}	High Level Input Current		V_{CC} = Max, $V_I = V_{CC}$ (Max)					10	μA
I_{IL}	Low Level Input Current		V_{CC} = Max, V_I = GND					−10	μA
I_{OS}*	Output Short Circuit Current		V_{CC} = 5.0V, V_O = GND			−30		−130	mA
I_{CC}	Supply Current	Quarter Power (GAL16V8-25Q, -30Q)	f = 15 MHz, V_{CC} = Max		COM			45	mA
					MIL/IND			55	mA
		Half Power (GAL16V8-20L, -25L, -30L)	f = 15 MHz, V_{CC} = Max		COM			90	mA
					MIL/IND			110	mA
C_I	Input Capacitance		V_{CC} = 5.0V, V_I = 2.0V					12	pF
$C_{I/O}$	I/O Capacitance		V_{CC} = 5.0V, $V_{I/O}$ = 2.0V					15	pF

*One output at a time for a maximum duration of one second @ 25°C.

Switching Characteristics Over Recommended Operating Conditions

Symbol	Parameter	Conditions	GAL16V8-20L COM/IND Min	GAL16V8-20L COM/IND Max	GAL16V8-25Q GAL16V8-25L COM/IND MIL* Min	GAL16V8-25Q GAL16V8-25L COM/IND MIL* Max	GAL16V8-30Q GAL16V8-30L COM/IND* MIL Min	GAL16V8-30Q GAL16V8-30L COM/IND* MIL Max	Units
t_{PD}	Input or Feedback to Combinatorial Output	S1 Closed, C_L = 50 pF		20		25		30	ns
t_{CLK}	Clock to Registered Output or Feedback	S1 Closed, C_L = 50 pF		15		15		20	ns
t_{PZXG}	$\overline{G}$ ↓ to Registered Output Enabled	Active High: S1 Open, C_L = 50 pF Active Low: S1 Closed, C_L = 50 pF		18		20		25	ns
t_{PXZG}	$\overline{G}$ ↑ to Registered Output Disabled	From V_{OH}: S1 Open, C_L = 5 pF From V_{OL}: S1 Closed, C_L = 5 pF		18		20		25	ns
t_{PZXI}	Input to Combinatorial Output Enabled via Product Term	Active High: S1 Open, C_L = 50 pF Active Low: S1 Closed, C_L = 50 pF		20		25		30	ns
t_{PXZI}	Input to Combinatorial Output Disabled via Product Term	From V_{OH}: S1 Open, C_L = 5 pF From V_{OL}: S1 Closed, C_L = 5 pF		20		25		30	ns
t_{RESET}	Power-Up to Registered Output High	S1 Closed, C_L = 50 pF		45		45		45	μs

* -25Q available only in COM/IND operating ranges;
-30L available only in MIL operating range (for COM/IND, use -25L or -30Q).

3. GAL16V8

Functional Description

The GAL logic array consists of a programmable AND array with fixed OR-gate connections, similar to the bipolar PAL architecture. The logic array is organized as 16 complementary input lines crossing 64 "product term" lines with a programmable E^2PROM cell at each intersection (2048 cells). Each programmable cell may establish a connection between an input line (true or complement phase of an array input signal) and a product term. A product term is satisfied (logically true) while all of the input lines "connected" to it are in the high logic state.

The 64 product terms are organized into eight output groups with eight terms each. Seven or eight of the product terms in each output group feed into an OR-gate to produce each output logic function; one of the product terms may instead be used to control the associated TRI-STATE device output. The fundamental transfer function of each GAL output is the familiar Boolean sum-of-products. Design development software is available which accepts Boolean equations and converts them automatically into GAL programming patterns.

As shown in the GAL16V8 Block Diagram *(Figure 1)*, a total of eight output logic functions are available. Each of the AND/OR logic functions feeds into an "output logic macrocell" (OLMC). The eight OLMCs control the flow of input and output signals between the logic array and the device's I/O pins.

Under control of an OLMC, each output may be designated either registered or combinatorial (non-registered). In the registered output configuration, the logic function output passes through a D-type flip-flop triggered by the rising edge of the clock input. Additionally, the logic function's output polarity may be designated active-low or active-high (adjusted before the register, if present). OLMC options such as these are selected using a set of programmable architecture control cells. These architecture cells are normally configured automatically by the development software or programming hardware.

All of the possible I/O configurations of the GAL16V8 are classified into three basic modes: "Small-PAL" mode, "Registered-PAL" mode and "Medium-PAL" mode. These modes correspond to the architectures of the PAL families which the GAL16V8 can emulate. The modes determine the mixture of OLMC configurations which can be selected for the device. The OLMC Selection table (Table I) lists which functions can be selected on the device pin* 1 and pins* 11 through 19 for each of the three modes. The logic diagrams in *Figure 3* illustrate these OLMC functions.

"OUTPUT" represents the always-active combinatorial output configuration available in the "Small-PAL" mode. "REGISTER" is the registered output with register feedback available in the "Registered-PAL" mode. "I/O" is the combinatorial bidirectional I/O available in "Registered-PAL" and "Medium-PAL" modes. "TRI-STATE" is the TRI-STATE combinatorial output function appearing on pins* 12 and 19 in the "Medium-PAL" mode. "INPUT" in Table I denotes an OLMC used as a dedicated input only.

*Applies to both 20-pin DIP and 20-lead PCC packages for GAL16V8.

20-Lead PCC Connection Diagram

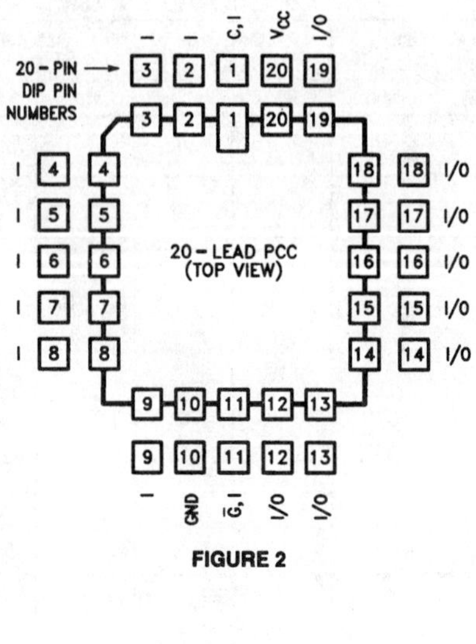

FIGURE 2

3. GAL16V8

OLMC Selection Table

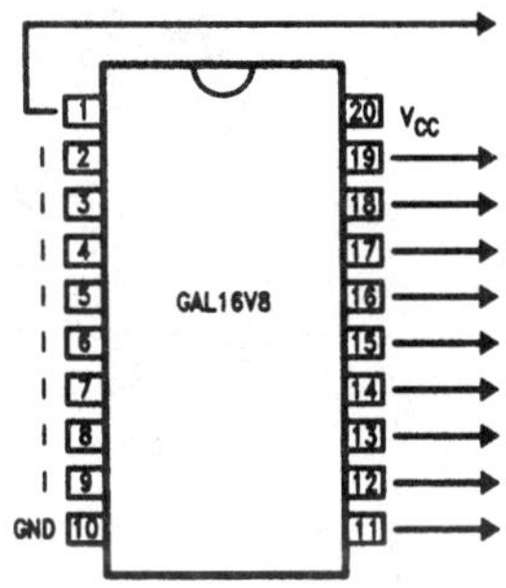

TABLE I

"Small-PAL" Mode	"Registered-PAL" Mode	"Medium-PAL" Mode
INPUT	CLOCK	INPUT
INPUT or OUTPUT*	REGISTER or I/O	TRI-STATE**
INPUT or OUTPUT*	REGISTER or I/O	I/O
INPUT or OUTPUT*	REGISTER or I/O	I/O
OUTPUT*	REGISTER or I/O	I/O
OUTPUT*	REGISTER or I/O	I/O
INPUT or OUTPUT*	REGISTER or I/O	I/O
INPUT or OUTPUT*	REGISTER or I/O	I/O
INPUT or OUTPUT*	REGISTER or I/O	TRI-STATE**
INPUT	OUPUT ENABLE ($\overline{G}$)	INPUT

*Active combinatorial output

**TRI-STATE combinatorial output

PAL Replacement Configurations

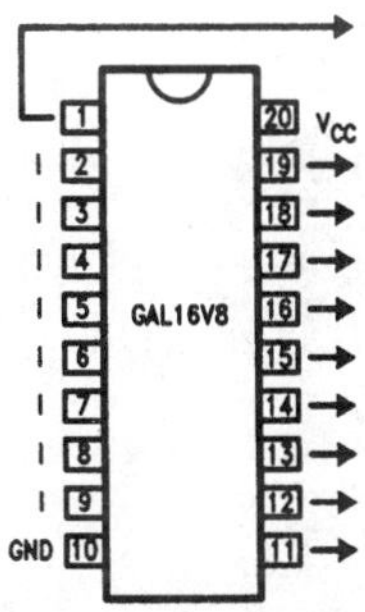

TABLE II

	"Small PAL" Mode				"Registered-PAL"			"Medium-PAL"
	INPUT	INPUT	INPUT	INPUT	CLOCK	CLOCK	CLOCK	INPUT
	OUTPUT*	INPUT	INPUT	INPUT	REGISTER	I/O	I/O	TRI-STATE**
	OUTPUT*	OUTPUT*	INPUT	INPUT	REGISTER	REGISTER	I/O	I/O
	OUTPUT*	OUTPUT*	OUTPUT*	INPUT	REGISTER	REGISTER	REGISTER	I/O
	OUTPUT*	OUTPUT*	OUTPUT*	OUTPUT*	REGISTER	REGISTER	REGISTER	I/O
	OUTPUT*	OUTPUT*	OUTPUT*	OUTPUT*	REGISTER	REGISTER	REGISTER	I/O
	OUTPUT*	OUTPUT*	OUTPUT*	INPUT	REGISTER	REGISTER	REGISTER	I/O
	OUTPUT*	OUTPUT*	INPUT	INPUT	REGISTER	REGISTER	I/O	I/O
	OUTPUT*	INPUT	INPUT	INPUT	REGISTER	I/O	I/O	TRI-STATE**
	INPUT	INPUT	INPUT	INPUT	$\overline{G}$	$\overline{G}$	$\overline{G}$	INPUT
EMULATED PAL PRODUCTS	10L8 10H8 10P8	12L6 12H6 12P6	14L4 14H4 14P4	16L2 16H2 16P2	16R8 16RP8	16R6 16RP6	16R4 16RP4	16L8 16H8 16P8

*Active combinatorial output.

**TRI-STATE combinatorial output.

3. GAL16V8

OLMC Configurations

OUTPUT (Active Combinatorial Output)

REGISTER (Registered Output)

I/O (Combinatorial Input/Output)

TRI-STATE (TRI-STATE Combinatorial Output)

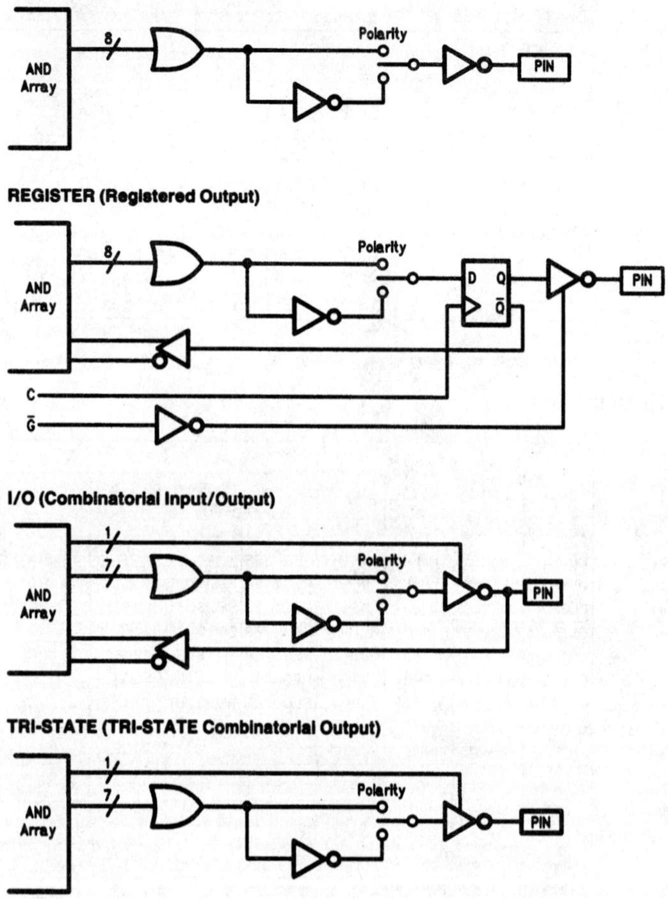

FIGURE 3

3. GAL16V8

Functional Description (Continued)

In the "Small-PAL" and "Medium-PAL" modes (Table I), pins* 1 and 11 are always dedicated inputs. In the "Registered-PAL" mode, however, pin* 1 becomes the clock input controlling all OLMC registers, and pin* 11 becomes the output enable ($\overline{G}$) input controlling the TRI-STATE outputs of all registered OLMCs. Within the "Small-PAL" and "Registered-PAL" modes in Table I, the functions of pins* 12 through 19 can be selected individually from either of the two functions listed. For example, in "Registered-PAL" mode, pins* 12 through 19 can each be designated as either a registered output or a combinatorial I/O. The "Medium-PAL" mode represents a single fixed configuration used to emulate combinatorial medium PAL devices (16L8, 16H8, 16P8).

Table II lists the bipolar PAL products which the GAL16V8 can emulate, and the specific input/output configurations used. This is just a subset, however, of all the configurations provided in Table I.

All registers in a GAL device are reset to the low state upon power-up. The active-low outputs, in turn, assume high logic levels (if enabled) regardless of the selected output polarity. This may simplify sequential circuit design and test. To ensure successful power-up reset, V_{CC} must rise monotonically until the specified operating voltage is attained. During power-up, the clock input should assume a valid, stable logic state as early as possible (within the specified time, t_{PR}) to avoid interfering with the reset operation. The clock input should also remain stable until after the power-up reset operation is completed to allow the registers to capture the proper next state on the first high-going clock transition.

It should be noted that the switching of any input not logically connected to a product term or logic function has no effect on the associated output logic state. To minimize power consumption, however, unused inputs should be connected to a stable logic level such as ground or V_{CC} (CMOS GAL inputs may be tied directly to the supply voltage without causing excessive loading conditions).

*Applies to both 20-pin DIP and 20-lead PCC packages for GAL16V8.

Clock/Input Frequency Specifications

The clock frequency (f_{CLK}) parameter listed in the Recommended Operating Conditions table specifies the maximum speed at which the GAL registers are guaranteed to operate. Clock frequency is defined differently for the two cases in which register feedback is used versus when it is not. In a data-path type application, when the logic functions fed into the registers are not dependent on register feedback from the previous cycle (i.e. based only on external inputs), the minimum required cycle period (f_{CLK}^{-1} without feedback) is defined as the greater of the minimum clock period (t_w high + t_w low) and the minimum "data window" period (t_{SU} + t_H). This assumes optimal alignment between data inputs and the clock input. In sequential logic applications such as state machines, the minimum required cycle period (t_{CYCLE} = f_{CLK}^{-1} with feedback) is defined as t_{CLK} + t_{SU}. This provides sufficient time for outputs from the registers to feed back through the logic array and set up on the inputs to the registers before the end of each cycle.

The input frequency (f_I) parameter specifies the maximum rate at which each GAL input can be toggled and still produce valid logic transitions on each combinatorial output. The f_I specification is derived as the inverse of the combinatorial propagation delay (t_{PD}).

Design Development Support

A variety of software tools and programming equipment is available to support the development of designs using GAL products. Typical software packages accept Boolean logic equations to define desired functions. Most are available to run on personal computers and generate a JEDEC-compatible "cell-map" (analogous to a PAL "fuse-map"). The industry-standard JEDEC format ensures that the resulting cell-map file can be down-loaded into a variety of programming equipment. Many software packages and programming units support a large variety of programmable logic products as well. The PLAN™ software package from National Semiconductor supports all programmable logic products available from National and is fully JEDEC-compatible. PLAN software also provides automatic device selection based on the designer's Boolean logic equations.

National strongly recommends using only approved programming hardware and software for developing GAL designs. Programming using unapproved equipment generally voids all guarantees. Approved programmers incorporate specialized programming algorithms that program the array and automatically configure the architecture cells. To ensure data retention and reliability, the programming algorithm also tracks the number of programming cycles to which each GAL device has been subjected since shipment, and stores this information automatically in the device.

The special GAL programming algorithm can also program a GAL device using a standard fuse-map developed for any of the emulated PAL products. PAL fuse-maps can be created by any JEDEC-compatible PAL development software or by loading the fuse pattern from an existing programmed PAL device into the programming unit (provided the PAL device has not been secured). However, to utilize the full flexibility of the GAL architecture, true GAL development software (such as PLAN software) is recommended.

Detailed logic diagrams showing all JEDEC cell-map addresses in the GAL logic array and OLMC are provided for direct map editing and diagnostic purposes (see "Programming Details"). For a list of current software and programming support tools available for these devices, please contact your local National sales representative or distributor. If detailed specifications of the GAL programming algorithm are needed, please contact the National Semiconductor Programmable Device Support department.

3. GAL16V8

Security Cell

A security cell is provided on all GAL16V8 devices as a deterrent to unauthorized copying of the array configuration patterns. Once programmed, the circuitry enabling array access is disabled, preventing further programming or verification of the array. The security cell can be erased only in conjunction with the array during a bulk erase cycle, so the original configuration can never be examined once this cell is programmed.

Electronic Signature

Each GAL device contains an electronic signature word consisting of 64 bits of reprogrammable memory. The electronic signature word can be programmed to contain any identification information desired by the user. Some uses include pattern identification labels, revision numbers, dates, inventory control information, etc. The data stored in the electronic signature word has no effect on the functionality of the device. The information is read out of the device using the normal program verification procedure provided by the programming equipment. The information may be accessed at any time independent of the state of the security cell. National's PLAN development software allows electronic signature data to be entered by the user and downloaded to the programming equipment.

Bulk Erase

The programming equipment automatically performs a bulk erase operation prior to each programming operation. No special erase operation need be performed by the user. Bulk erase clears the logic array, architecture cells, security cell, and electronic signature information. The GAL device is thereby reverted back to its virgin state.

Latch-Up Protection

GAL devices are designed with an on-chip charge pump to negatively bias the substrate. The negative bias is of sufficient magnitude to prevent input undershoots from causing the circuitry to latch. Additionally, outputs are designed with n-channel pullups instead of the traditional p-channel pullups to eliminate any possibility of SCR induced latching.

Manufacturer Testing

Because of E²CMOS technology, GAL devices can be reprogrammed in milliseconds. This allows each device to be completely tested by the manufacturer using numerous logic array and architecture patterns prior to shipping. Every programmable cell and every logic path through every device is fully tested for programmability, functionality and performance to all AC and DC parameters. The customer can therefore expect 100% programming and functional yield and 100% compliance of all GAL products to datasheet specifications.

The testing procedure performed on all GAL devices by the manufacturer tests all aspects of device operation. Extensive testing of all programmable cells in the device include margin testing, internal verify, and program retention during high-temperature bake. All DC and AC parameters are tested at hot and cold temperatures using a variety of worst case logic and signal patterns. Functional tests include reprogramming each OLMC to all valid architectural configurations.

Register Preload

The register preload feature allows OLMC registers to be directly loaded with any desired data pattern. It also allows the present state of OLMC registers to be examined regardless of TRI-STATE control conditions. This simplifies testing of devices after programming. A device may be put into any desired register state at any point during the functional test sequence. The test sequence may then be resumed to verify proper next-state transitions. This allows complete verification of sequential logic circuits, including states that are normally impossible or difficult to reach. It may also shorten the overall test time significantly.

Register preload is not an operational mode and is not intended for board-level testing because elevated voltage levels must be applied to the device. The programming equipment normally provides the register preload capability as part of its functional test facility. Note that the testing of GAL devices after programming by the user may be considered unnecessary because all E²CMOS GAL products are completely tested by the manufacturer, guaranteeing 100% post-programming functional yield.

The register preload algorithm is described for those users who wish to test programmed GAL devices using test equipment other than approved GAL programming equipment. As shown in the Register Preload Waveform in *Figure 5*, the preload sequence must not begin until the normal power-up reset operation has completed (after time t_{RESET}). The device is placed into preload mode by raising the "PRLD" input (pin* 11) to voltage V_{IES}, as specified in the Register Preload Specifications (Table III).

To preload the OLMC registers, a series of data bits are shifted into the device on the "S_{DIN}" input (pin* 9), one bit for each OLMC in which registered output has been selected. (Non-registered OLMCs are bypassed.) The shift sequence is clocked by the rising edge of the "D_{CLK}" input (pin* 1). The data stream is shifted in through the registered OLMC with the lowest corresponding pin number, and then "upward" through all remaining registered OLMCs in pin-number ascending order. Therefore, the first data bit in the series is ultimately loaded into the registered OLMC with the highest corresponding pin number, as shown in *Figure 4*.

As the data series is shifted into the S_{DIN} input, the contents of all registers (in registered OLMCs) are shifted "upward" and out onto the "S_{DOUT}" output (pin* 12). Complete present-state information can be examined in this manner. Test fixtures can be devised to test several GAL devices in which the S_{DOUT} pin of each chip is connected to the S_{DIN} pin of the next, and all preload and present-state data can be shifted around a single serial loop.

Note that when shifting register data into S_{DIN} or out of S_{DOUT}, V_{IL}/V_{OL} = register reset (0), and V_{IH}/V_{OH} = register set (1). These 0 and 1 register states are always inverted (active-low) on the normal output pins regardless of the selected output polarity (polarity affects logic function values before register inputs).

*Applies to both 20-pin DIP and 20-lead PCC Packages for GAL16V8.

3. GAL16V8

Register Preload (Continued)

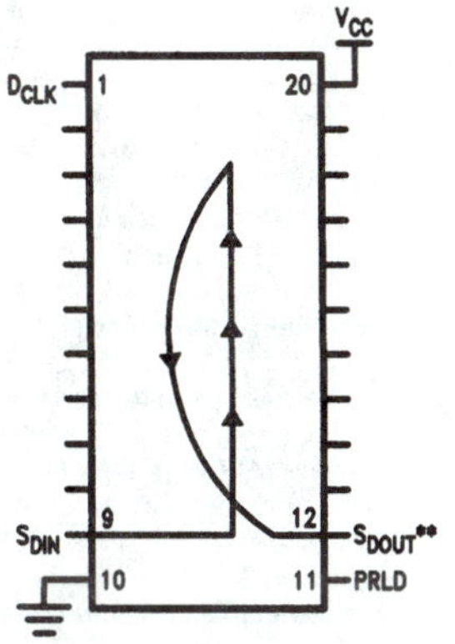

**The S_{DOUT} output buffer is an open drain output during preload. This pin should be terminated to V_{CC} with a 10 kΩ resistor.

FIGURE 4. Output Register Preload Pinout

Register Preload Specifications

TABLE III

Symbol	Parameter	Conditions	Min	Typ	Max	Units
V_{IH}	Input Voltage (High)		2.40		V_{CC}	V
V_{IL}	Input Voltage (Low)		0.00		0.50	V
V_{IES}	Registered Preload Input Voltage		14.5	15	15.5	V
V_{OH}	Output Voltage (High) (Note 1)				V_{CC}	V
V_{OL}	Output Voltage (Low) (Note 1)	$I_{OL} \leq 12$ mA	0.00		0.50	V
I_{IH}, I_{IL}	Input Current (Programming)			±1	±10	μA
I_{OH}	High Level Output Current (Note 1)	$V_{OH} \leq V_{CC}$			10	μA
t_{PWV}	Verify Pulse Width		1	5	10	μs
t_D	Pulse Sequence Delay		1	5	10	μs
t_{RESET}	Register Reset Time from Valid V_{CC}				45	μs

Note 1: The S_{DOUT} output buffer is an open drain output. This pin should be terminated to V_{CC} with a 10k resistor.

Register Preload Waveforms

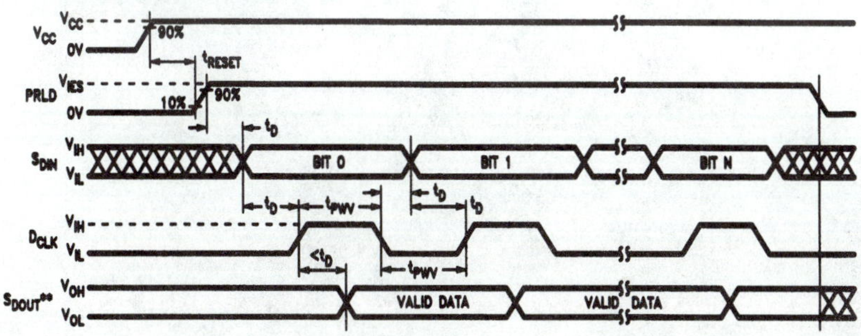

**The S_{DOUT} output buffer is an open drain output during preload. This pin should be terminated to V_{CC} with a 10 kΩ resistor.

FIGURE 5

3. GAL16V8

Programming Details

Understanding the information in this section is not essential when using approved programming equipment and software for developing GAL designs. This is a more thorough disclosure of the GAL architecture provided for direct JEDEC cell-map editing and diagnostic purposes. This section alone, however, does not contain sufficient information to implement the GAL programming algorithm. If detailed specifications of the GAL programming algorithm are needed, please contact the National Semiconductor Programmable Device Support department.

As mentioned in the Functional Description, the OLMC is responsible for selecting input and/or output paths, registered vs. combinatorial outputs, active-high or low polarity, and common vs. locally-controlled TRI-STATE control. Additionally, the OLMCs select between alternate logic array input paths to maintain JEDEC cell-map compatibility with either "small-PAL" or "medium-PAL" logic arrays.

The various configurations of the OLMCs are controlled by a set of programmable "architecture" cells, separate from the logic-defining array cells. Each GAL device contains two "global" architecture cells, "SYN" and "AC0", which affect all OLMCs. Each of the device's eight OLMCs also contains two "local" cells, "AC1" and "XOR". The OLMC Logic Diagram in *Figure 6* shows how the architecture cells select the different paths through the OLMC.

The SYN bit controls whether a device will have any registered outputs (SYN = 0) or will be purely combinatorial (SYN = 1). The SYN bit determines whether device pins* 1 and 11 are used as the clock and global TRI-STATE control inputs (SYN = 0) or whether they are ordinary inputs (SYN = 1). The AC0 bit selects between the "Small-PAL" mode and the "Medium/Registered-PAL" modes. The function of the AC1 bits depend on the state of the AC0 bit. In "Small-PAL" mode (AC0 = 0), the AC1 bit in each OLMC determines whether the associated device pin is an output (AC1 = 0) or an input (AC1 = 1). In "Registered-PAL" mode (AC0 = 1), the AC1 bit determines whether each OLMC is registered (AC1 = 0) or combinatorial (AC1 = 1). In "Medium-PAL" mode (AC0 = 1), the AC1 bits in all OLMCs must be set to 1 (combinatorial). All of the valid architecture bit configurations are shown in the OLMC Architecture table (Table IV), which has the same familiar format used in the OLMC Selection table (Table I).

Independent of SYN, AC0 and the AC1 bits, the XOR bit in each OLMC selects between active-low (XOR = 0) or active-high (XOR = 1) output polarity.

*Applies to both 20-pin DIP and 20-lead PCC packages for GAL16V8.

3. GAL16V8

OLMC Logic Diagram

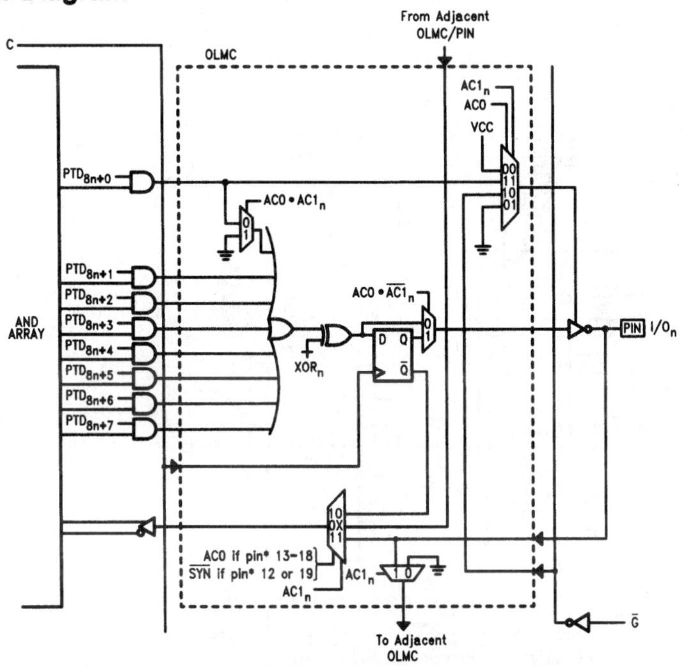

*Applies to both 20-pin DIP and 20-lead PCC packages for GAL16V8.

FIGURE 6

OLMC Architecture Programming

TABLE IV

	"Small-PAL" Mode			"Registered-PAL" Mode			"Medium-PAL" Mode	
	Function		JEDEC Input Line #s (Note 1)	Function		JEDEC Input Line #s (Note 1)	Function	JEDEC Input Lines #s (Note 1)
Pin 1	INPUT	INPUT	2,3	CLOCK	CLOCK		INPUT	2,3
*** Pin 19	OUTPUT*	INPUT	6,7	REGISTER	I/O	2,3	TRI-STATE**	
*** Pin 18	OUTPUT*	INPUT	10,11	REGISTER	I/O	6,7	I/O	6,7
*** Pin 17	OUTPUT*	INPUT	14,15	REGISTER	I/O	10,11	I/O	10,11
*** Pin 16	OUTPUT*	NC		REGISTER	I/O	14,15	I/O	14,15
*** Pin 15	OUTPUT*	NC		REGISTER	I/O	18,19	I/O	18,19
*** Pin 14	OUTPUT*	INPUT	18,19	REGISTER	I/O	22,23	I/O	22,23
*** Pin 13	OUTPUT*	INPUT	22,23	REGISTER	I/O	26,27	I/O	26,27
*** Pin 12	OUTPUT*	INPUT	26,27	REGISTER	I/O	30,13	TRI-STATE**	
Pin 11	INPUT	INPUT	30,31	$\overline{G}$	$\overline{G}$		INPUT	30,31
	$AC1_n = 0$	$AC1_n = 1$		$AC1_n = 0$	$AC1_n = 1$		$AC1_n = 1$	
	SYN = 1, AC0 = 0			SYN = 0, AC0 = 1			SYN = 1, AC0 = 1	
	All outputs are combinatorial and always active.			At least one output is registered.			All I/O pins are combinatorial.	

Note: Pin numbers above apply to both 20-pin DIP and 20-lead PCC packages for GAL16V8.

Note 1: All even and odd numbered JEDEC input line numbers correspond to true and complement array inputs, respectively.

*Active combinatorial output.

**TRI-STATE combinatorial output.

*** $AC1_n$ applies to these I/O pins only.

3. GAL16V8

GAL16V8 Logic Diagram

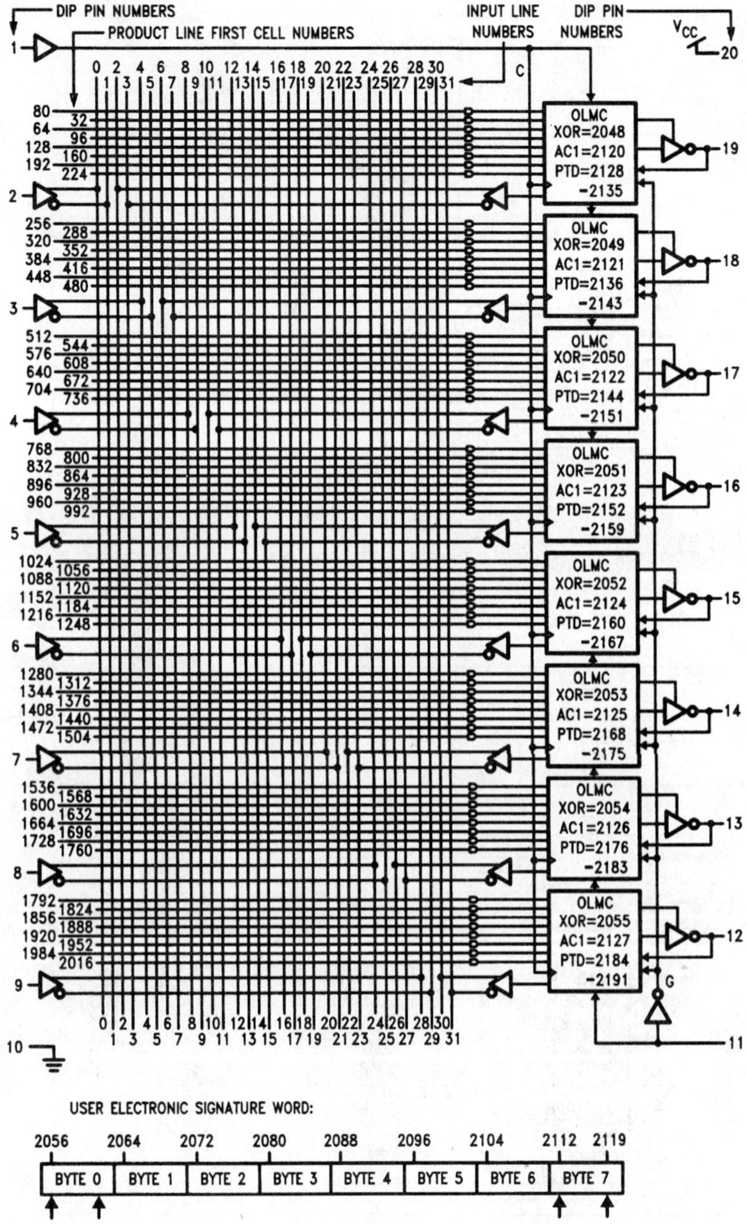

SYN=2192
ACO=2193

JEDEC Logic Array Cell Number = Product Line First Cell Number + Input Line Number

FIGURE 7

4. ADC0808

National Semiconductor

ADC0808/ADC0809 8-Bit μP Compatible A/D Converters with 8-Channel Multiplexer

General Description

The ADC0808, ADC0809 data acquisition component is a monolithic CMOS device with an 8-bit analog-to-digital converter, 8-channel multiplexer and microprocessor compatible control logic. The 8-bit A/D converter uses successive approximation as the conversion technique. The converter features a high impedance chopper stabilized comparator, a 256R voltage divider with analog switch tree and a successive approximation register. The 8-channel multiplexer can directly access any of 8-single-ended analog signals.

The device eliminates the need for external zero and full-scale adjustments. Easy interfacing to microprocessors is provided by the latched and decoded multiplexer address inputs and latched TTL TRI-STATE® outputs.

The design of the ADC0808, ADC0809 has been optimized by incorporating the most desirable aspects of several A/D conversion techniques. The ADC0808, ADC0809 offers high speed, high accuracy, minimal temperature dependence, excellent long-term accuracy and repeatability, and consumes minimal power. These features make this device ideally suited to applications from process and machine control to consumer and automotive applications. For 16-channel multiplexer with common output (sample/hold port) see ADC0816 data sheet. (See AN-247 for more information.)

Features

- Easy interface to all microprocessors
- Operates ratiometrically or with 5 V_{DC} or analog span adjusted voltage reference
- No zero or full-scale adjust required
- 8-channel multiplexer with address logic
- 0V to 5V input range with single 5V power supply
- Outputs meet TTL voltage level specifications
- Standard hermetic or molded 28-pin DIP package
- 28-pin molded chip carrier package
- ADC0808 equivalent to MM74C949
- ADC0809 equivalent to MM74C949-1

Key Specifications

- Resolution: 8 Bits
- Total Unadjusted Error: ±½ LSB and ±1 LSB
- Single Supply: 5 V_{DC}
- Low Power: 15 mW
- Conversion Time: 100 μs

Block Diagram

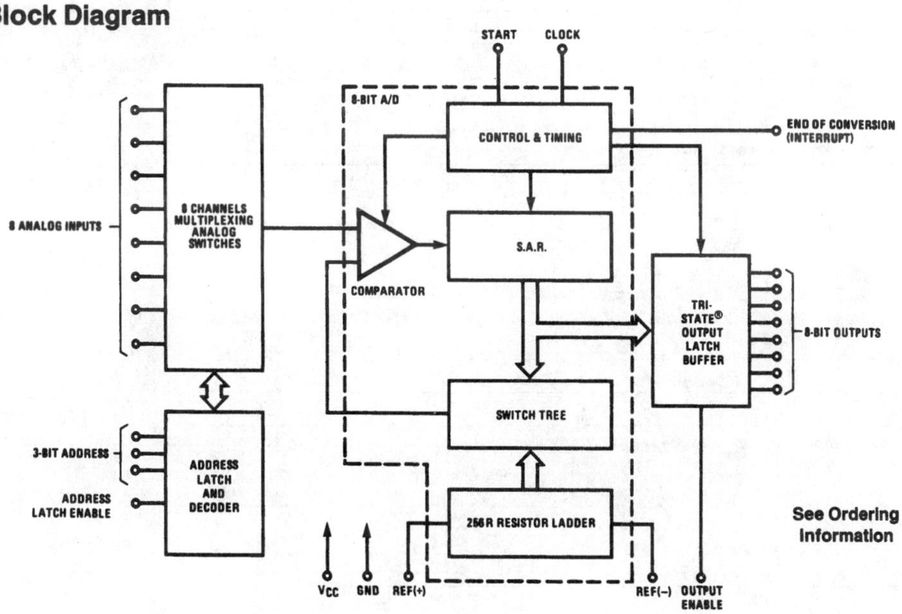

See Ordering Information

4. ADC0808

Absolute Maximum Ratings (Notes 1 & 2)

If Military/Aerospace specified devices are required, please contact the National Semiconductor Sales Office/Distributors for availability and specifications.

Supply Voltage (V_{CC}) (Note 3)	6.5V
Voltage at Any Pin	−0.3V to (V_{CC}+0.3V)
Except Control Inputs	
Voltage at Control Inputs	−0.3V to +15V
(START, OE, CLOCK, ALE, ADD A, ADD B, ADD C)	
Storage Temperature Range	−65°C to +150°C
Package Dissipation at T_A=25°C	875 mW
Lead Temp. (Soldering, 10 seconds)	
Dual-In-Line Package (plastic)	260°C
Dual-In-Line Package (ceramic)	300°C
Molded Chip Carrier Package	
Vapor Phase (60 seconds)	215°C
Infrared (15 seconds)	220°C
ESD Susceptibility (Note 11)	400V

Operating Conditions (Notes 1 & 2)

Temperature Range (Note 1)	$T_{MIN} \le T_A \le T_{MAX}$
ADC0808CJ	$-55°C \le T_A \le +125°C$
ADC0808CCJ, ADC0808CCN, ADC0809CCN	$-40°C \le T_A \le +85°C$
ADC0808CCV, ADC0809CCV	$-40°C \le T_A \le +85°C$
Range of V_{CC} (Note 1)	4.5 V_{DC} to 6.0 V_{DC}

Electrical Characteristics

Converter Specifications: V_{CC}=5 V_{DC}=V_{REF+}, $V_{REF(-)}$=GND, $T_{MIN} \le T_A \le T_{MAX}$ and f_{CLK}=640 kHz unless otherwise stated.

Symbol	Parameter	Conditions	Min	Typ	Max	Units
	ADC0808 Total Unadjusted Error (Note 5)	25°C T_{MIN} to T_{MAX}			±½ ±¾	LSB LSB
	ADC0809 Total Unadjusted Error (Note 5)	0°C to 70°C T_{MIN} to T_{MAX}			±1 ±1¼	LSB LSB
	Input Resistance	From Ref(+) to Ref(−)	1.0	2.5		kΩ
	Analog Input Voltage Range	(Note 4) V(+) or V(−)	GND−0.10		V_{CC}+0.10	V_{DC}
$V_{REF(+)}$	Voltage, Top of Ladder	Measured at Ref(+)		V_{CC}	V_{CC}+0.1	V
$\frac{V_{REF(+)}+V_{REF(-)}}{2}$	Voltage, Center of Ladder		V_{CC}/2-0.1	V_{CC}/2	V_{CC}/2+0.1	V
$V_{REF(-)}$	Voltage, Bottom of Ladder	Measured at Ref(−)	−0.1	0		V
I_{IN}	Comparator Input Current	f_c=640 kHz, (Note 6)	−2	±0.5	2	μA

Electrical Characteristics

Digital Levels and DC Specifications: ADC0808CJ $4.5V \le V_{CC} \le 5.5V$, $-55°C \le T_A \le +125°C$ unless otherwise noted ADC0808CCJ, ADC0808CCN, ADC0808CCV, ADC0809CCN and ADC0809CCV, $4.75 \le V_{CC} \le 5.25V$, $-40°C \le T_A \le +85°C$ unless otherwise noted

Symbol	Parameter	Conditions	Min	Typ	Max	Units
ANALOG MULTIPLEXER						
$I_{OFF(+)}$	OFF Channel Leakage Current	V_{CC}=5V, V_{IN}=5V, T_A=25°C T_{MIN} to T_{MAX}		10	200 1.0	nA μA
$I_{OFF(-)}$	OFF Channel Leakage Current	V_{CC}=5V, V_{IN}=0, T_A=25°C T_{MIN} to T_{MAX}	−200 −1.0	−10		nA μA

4. ADC0808

Electrical Characteristics (Continued)

Digital Levels and DC Specifications: ADC0808CJ $4.5V \le V_{CC} \le 5.5V$, $-55°C \le T_A \le +125°C$ unless otherwise noted ADC0808CCJ, ADC0808CCN, ADC0808CCV, ADC0809CCN and ADC0809CCV, $4.75 \le V_{CC} \le 5.25V$, $-40°C \le T_A \le +85°C$ unless otherwise noted

Symbol	Parameter	Conditions	Min	Typ	Max	Units
CONTROL INPUTS						
$V_{IN(1)}$	Logical "1" Input Voltage		$V_{CC}-1.5$			V
$V_{IN(0)}$	Logical "0" Input Voltage				1.5	V
$I_{IN(1)}$	Logical "1" Input Current (The Control Inputs)	$V_{IN}=15V$			1.0	μA
$I_{IN(0)}$	Logical "0" Input Current (The Control Inputs)	$V_{IN}=0$	−1.0			μA
I_{CC}	Supply Current	$f_{CLK}=640$ kHz		0.3	3.0	mA
DATA OUTPUTS AND EOC (INTERRUPT)						
$V_{OUT(1)}$	Logical "1" Output Voltage	$I_O=-360$ μA	$V_{CC}-0.4$			V
$V_{OUT(0)}$	Logical "0" Output Voltage	$I_O=1.6$ mA			0.45	V
$V_{OUT(0)}$	Logical "0" Output Voltage EOC	$I_O=1.2$ mA			0.45	V
I_{OUT}	TRI-STATE Output Current	$V_O=5V$ $V_O=0$	 −3		3	μA μA

Electrical Characteristics

Timing Specifications $V_{CC}=V_{REF(+)}=5V$, $V_{REF(-)}=GND$, $t_r=t_f=20$ ns and $T_A=25°C$ unless otherwise noted.

Symbol	Parameter	Conditions	Min	Typ	Max	Units
t_{WS}	Minimum Start Pulse Width	*(Figure 5)*		100	200	ns
t_{WALE}	Minimum ALE Pulse Width	*(Figure 5)*		100	200	ns
t_s	Minimum Address Set-Up Time	*(Figure 5)*		25	50	ns
t_H	Minimum Address Hold Time	*(Figure 5)*		25	50	ns
t_D	Analog MUX Delay Time From ALE	$R_S=0\Omega$ *(Figure 5)*		1	2.5	μS
t_{H1}, t_{H0}	OE Control to Q Logic State	$C_L=50$ pF, $R_L=10k$ *(Figure 8)*		125	250	ns
t_{1H}, t_{0H}	OE Control to Hi-Z	$C_L=10$ pF, $R_L=10k$ *(Figure 8)*		125	250	ns
t_c	Conversion Time	$f_c=640$ kHz, *(Figure 5)* (Note 7)	90	100	116	μS
f_c	Clock Frequency		10	640	1280	kHz
t_{EOC}	EOC Delay Time	*(Figure 5)*	0		8+2 μS	Clock Periods
C_{IN}	Input Capacitance	At Control Inputs		10	15	pF
C_{OUT}	TRI-STATE Output Capacitance	At TRI-STATE Outputs, (Note 12)		10	15	pF

Note 1: Absolute Maximum Ratings indicate limits beyond which damage to the device may occur. DC and AC electrical specifications do not apply when operating the device beyond its specified operating conditions.

Note 2: All voltages are measured with respect to GND, unless othewise specified.

Note 3: A zener diode exists, internally, from V_{CC} to GND and has a typical breakdown voltage of 7 V_{DC}.

Note 4: Two on-chip diodes are tied to each analog input which will forward conduct for analog input voltages one diode drop below ground or one diode drop greater than the V_{CC}n supply. The spec allows 100 mV forward bias of either diode. This means that as long as the analog V_{IN} does not exceed the supply voltage by more than 100 mV, the output code will be correct. To achieve an absolute $0V_{DC}$ to $5V_{DC}$ input voltage range will therefore require a minimum supply voltage of 4.900 V_{DC} over temperature variations, initial tolerance and loading.

Note 5: Total unadjusted error includes offset, full-scale, linearity, and multiplexer errors. See *Figure 3.* None of these A/Ds requires a zero or full-scale adjust. However, if an all zero code is desired for an analog input other than 0.0V, or if a narrow full-scale span exists (for example: 0.5V to 4.5V full-scale) the reference voltages can be adjusted to achieve this. See *Figure 13.*

Note 6: Comparator input current is a bias current into or out of the chopper stabilized comparator. The bias current varies directly with clock frequency and has little temperature dependence *(Figure 6).* See paragraph 4.0.

Note 7: The outputs of the data register are updated one clock cycle before the rising edge of EOC.

Note 8: Human body model, 100 pF discharged through a 1.5 kΩ resistor.

4. ADC0808

Functional Description

Multiplexer. The device contains an 8-channel single-ended analog signal multiplexer. A particular input channel is selected by using the address decoder. Table I shows the input states for the address lines to select any channel. The address is latched into the decoder on the low-to-high transition of the address latch enable signal.

TABLE I

SELECTED ANALOG CHANNEL	ADDRESS LINE C	B	A
IN0	L	L	L
IN1	L	L	H
IN2	L	H	L
IN3	L	H	H
IN4	H	L	L
IN5	H	L	H
IN6	H	H	L
IN7	H	H	H

CONVERTER CHARACTERISTICS

The Converter

The heart of this single chip data acquisition system is its 8-bit analog-to-digital converter. The converter is designed to give fast, accurate, and repeatable conversions over a wide range of temperatures. The converter is partitioned into 3 major sections: the 256R ladder network, the successive approximation register, and the comparator. The converter's digital outputs are positive true.

The 256R ladder network approach *(Figure 1)* was chosen over the conventional R/2R ladder because of its inherent monotonicity, which guarantees no missing digital codes. Monotonicity is particularly important in closed loop feedback control systems. A non-monotonic relationship can cause oscillations that will be catastrophic for the system. Additionally, the 256R network does not cause load variations on the reference voltage.

The bottom resistor and the top resistor of the ladder network in *Figure 1* are not the same value as the remainder of the network. The difference in these resistors causes the output characteristic to be symmetrical with the zero and full-scale points of the transfer curve. The first output transition occurs when the analog signal has reached $+\frac{1}{2}$ LSB and succeeding output transitions occur every 1 LSB later up to full-scale.

The successive approximation register (SAR) performs 8 iterations to approximate the input voltage. For any SAR type converter, n-iterations are required for an n-bit converter. *Figure 2* shows a typical example of a 3-bit converter. In the ADC0808, ADC0809, the approximation technique is extended to 8 bits using the 256R network.

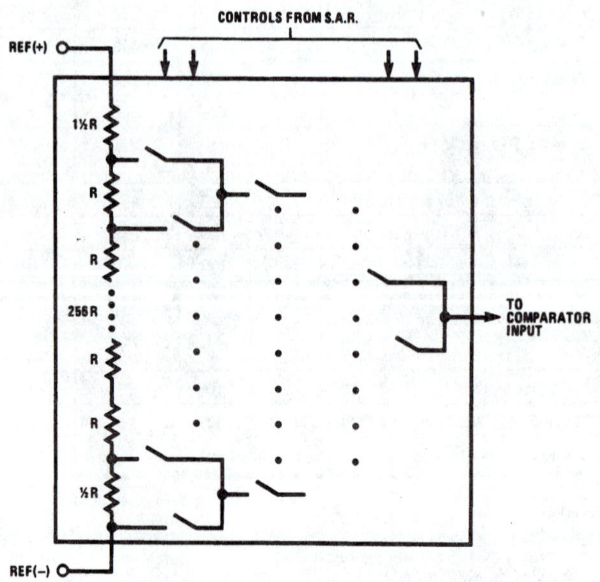

FIGURE 1. Resistor Ladder and Switch Tree

4. ADC0808

Functional Description (Continued)

The A/D converter's successive approximation register (SAR) is reset on the positive edge of the start conversion (SC) pulse. The conversion is begun on the falling edge of the start conversion pulse. A conversion in process will be interrupted by receipt of a new start conversion pulse. Continuous conversion may be accomplished by tying the end-of-conversion (EOC) output to the SC input. If used in this mode, an external start conversion pulse should be applied after power up. End-of-conversion will go low between 0 and 8 clock pulses after the rising edge of start conversion.

The most important section of the A/D converter is the comparator. It is this section which is responsible for the ultimate accuracy of the entire converter. It is also the comparator drift which has the greatest influence on the repeatability of the device. A chopper-stabilized comparator provides the most effective method of satisfying all the converter requirements.

The chopper-stabilized comparator converts the DC input signal into an AC signal. This signal is then fed throught a high gain AC amplifier and has the DC level restored. This technique limits the drift component of the amplifier since the drift is a DC component which is not passed by the AC amplifier. This makes the entire A/D converter extremely insensitive to temperature, long term drift and input offset errors.

Figure 4 shows a typical error curve for the ADC0808 as measured using the procedures outlined in AN-179.

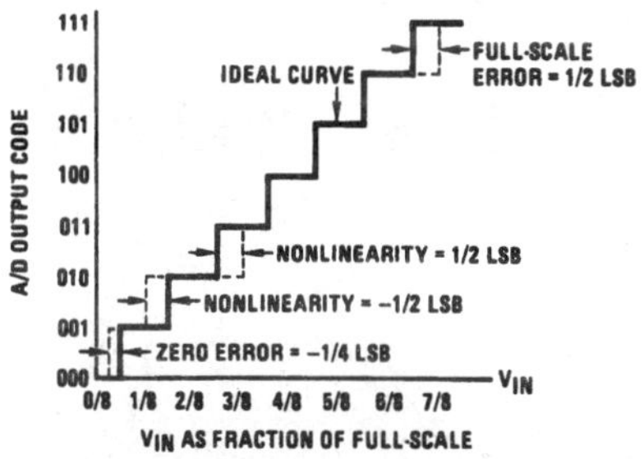

FIGURE 2. 3-Bit A/D Transfer Curve

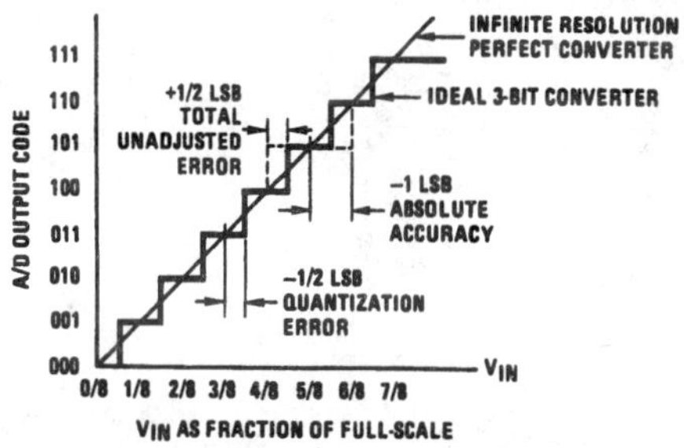

FIGURE 3. 3-Bit A/D Absolute Accuracy Curve

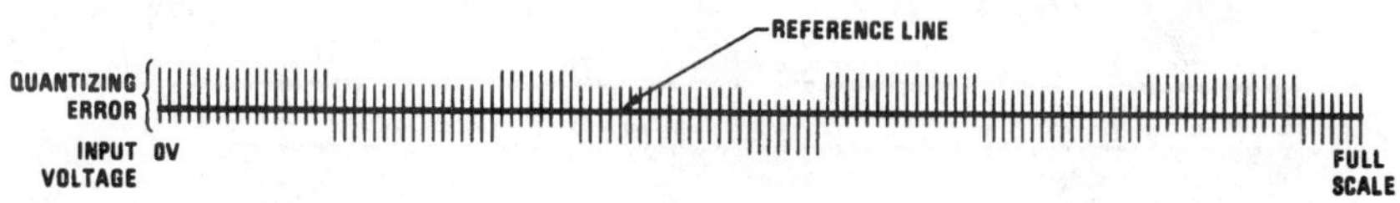

FIGURE 4. Typical Error Curve

4. ADC0808

Connection Diagrams

Dual-In-Line Package

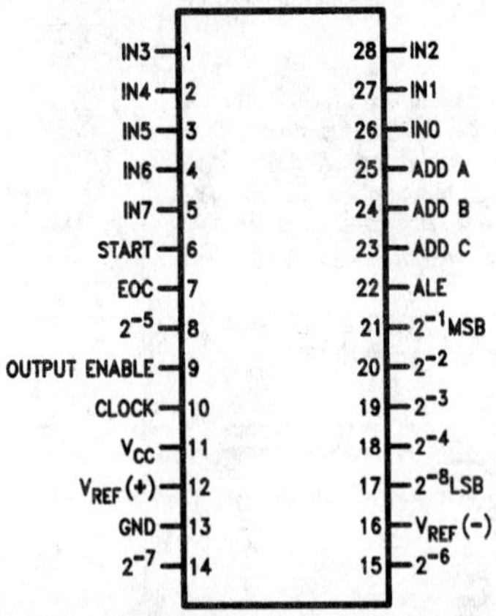

Order Number ADC0808CCN, ADC0809CCN, ADC0808CCJ or ADC0808CJ
See NS Package J28A or N28A

Molded Chip Carrier Package

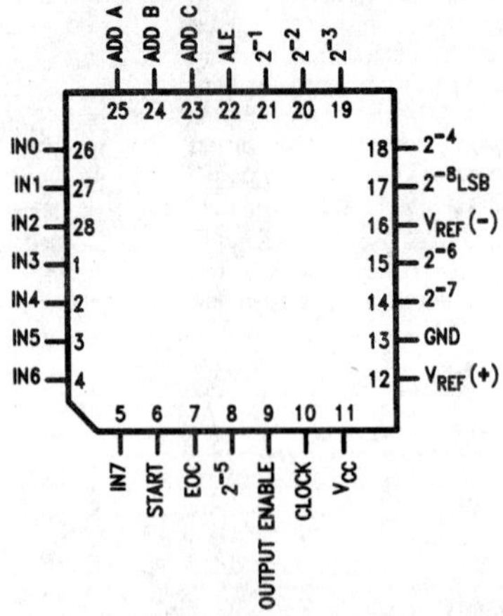

Order Number ADC0808CCV or ADC0809CCV
See NS Package V28A

Timing Diagram

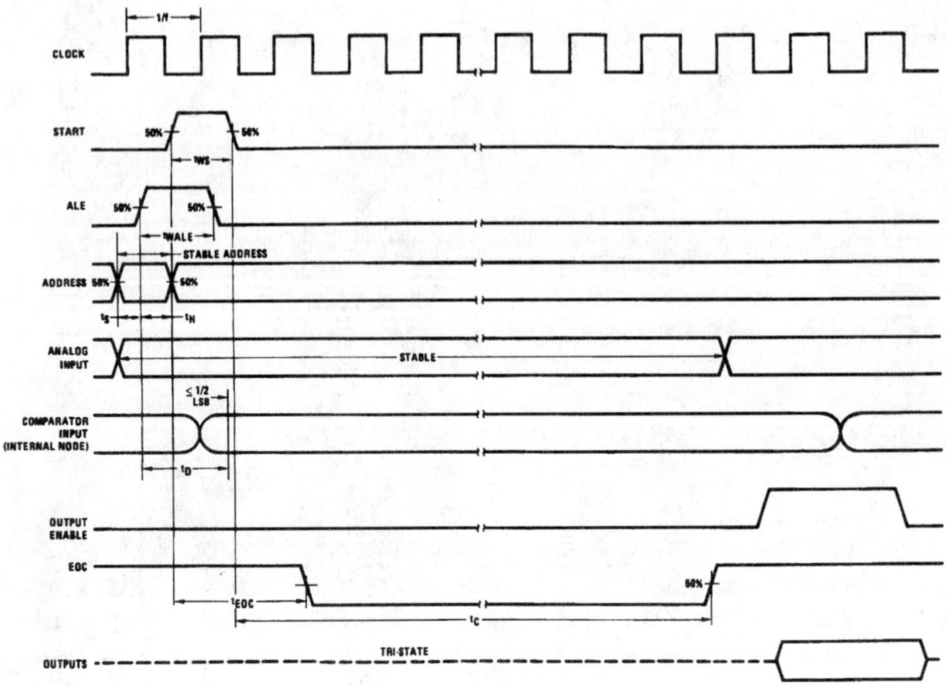

FIGURE 5

5. DAC0830

National Semiconductor

DAC0830/DAC0831/DAC0832 8-Bit μP Compatible, Double-Buffered D to A Converters

General Description

The DAC0830 is an advanced CMOS/Si-Cr 8-bit multiplying DAC designed to interface directly with the 8080, 8048, 8085, Z80®, and other popular microprocessors. A deposited silicon-chromium R-2R resistor ladder network divides the reference current and provides the circuit with excellent temperature tracking characteristics (0.05% of Full Scale Range maximum linearity error over temperature). The circuit uses CMOS current switches and control logic to achieve low power consumption and low output leakage current errors. Special circuitry provides TTL logic input voltage level compatibility.

Double buffering allows these DACs to output a voltage corresponding to one digital word while holding the next digital word. This permits the simultaneous updating of any number of DACs.

The DAC0830 series are the 8-bit members of a family of microprocessor-compatible DACs (MICRO-DAC™). For applications demanding higher resolution, the DAC1000 series (10-bits) and the DAC1208 and DAC1230 (12-bits) are available alternatives.

Features

- Double-buffered, single-buffered or flow-through digital data inputs
- Easy interchange and pin-compatible with 12-bit DAC1230 series
- Direct interface to all popular microprocessors
- Linearity specified with zero and full scale adjust only—NOT BEST STRAIGHT LINE FIT.
- Works with ±10V reference-full 4-quadrant multiplication
- Can be used in the voltage switching mode
- Logic inputs which meet TTL voltage level specs (1.4V logic threshold)
- Operates "STAND ALONE" (without μP) if desired
- Available in 20-pin small-outline or molded chip carrier package

Key Specifications

■ Current settling time	1 μs
■ Resolution	8 bits
■ Linearity (guaranteed over temp.)	8, 9, or 10 bits
■ Gain Tempco	0.0002% FS/°C
■ Low power dissipation	20 mW
■ Single power supply	5 to 15 V_{DC}

Typical Application

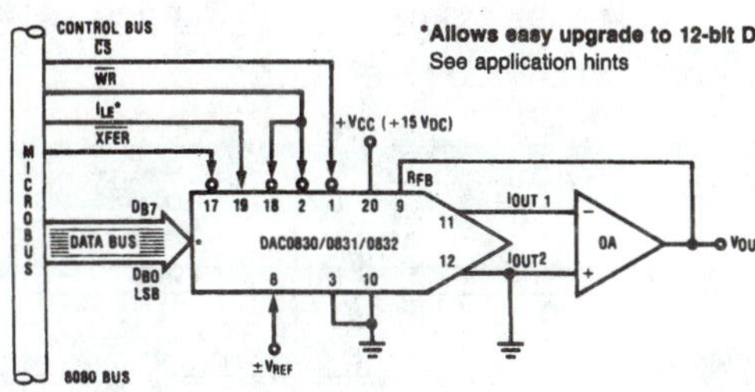

*Allows easy upgrade to 12-bit DAC1230, See application hints

TL/H/5608–1

Connection Diagrams (Top Views)

Dual-In-Line and Small-Outline Packages

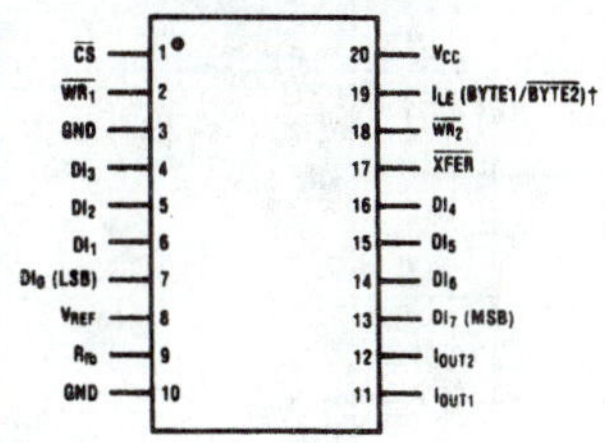

†This is necessary for the 12-bit DAC1230 series to permit interchanging from an 8-bit to a 12-bit DAC with **No PC board changes** and no software changes, See applications section.

Molded Chip Carrier Package

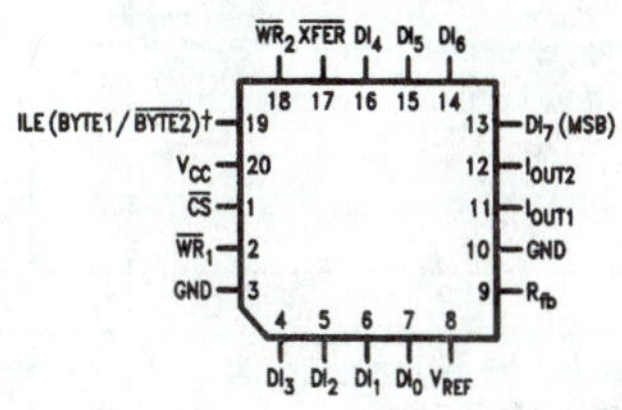

5. DAC0830

Absolute Maximum Ratings (Notes 1 & 2)

If Military/Aerospace specified devices are required, please contact the National Semiconductor Sales Office/Distributors for availability and specifications.

Supply Voltage (V_{CC})	17 V_{DC}
Voltage at Any Digital Input	V_{CC} to GND
Voltage at V_{REF} Input	±25V
Storage Temperature Range	−65°C to +150°C
Package Dissipation at $T_A = 25°C$ (Note 3)	500 mW
DC Voltage Applied to I_{OUT1} or I_{OUT2} (Note 4)	−100 mV to V_{CC}
ESD Susceptability (Note 14)	800V
Lead Temperature (soldering, 10 sec.)	
Dual-In-Line Package (plastic)	260°C
Dual-In-Line Package (ceramic)	300°C
Surface Mount Package	
Vapor Phase (60 sec.)	215°C
Infrared (15 sec.)	220°C

Operating Conditions

Temperature Range	$T_{MIN} \le T_A \le T_{MAX}$
Part numbers with 'LCN' suffix	0°C to +70°C
Part numbers with 'LCWM' suffix	0°C to +70°C
Part numbers with 'LCV' suffix	0°C to +70°C
Part numbers with 'LCJ' suffix	−40°C to +85°C
Part numbers with 'LJ' suffix	−55°C to +125°C
Voltage at Any Digital Input	V_{CC} to GND

Electrical Characteristics

$V_{REF} = 10.000\ V_{DC}$ unless otherwise noted. **Boldface limits apply over temperature, $T_{MIN} \le T_A \le T_{MAX}$.** For all other limits $T_A = 25°C$.

Parameter	Conditions	See Note	$V_{CC} = 4.75\ V_{DC}$, $V_{CC} = 15.75\ V_{DC}$: Typ (Note 12)	$V_{CC} = 4.75\ V_{DC}$, $V_{CC} = 15.75\ V_{DC}$: Tested Limit (Note 5)	$V_{CC} = 5\ V_{DC} \pm 5\%$, $V_{CC} = 12\ V_{DC} \pm 5\%$ to $15\ V_{DC} \pm 5\%$: Design Limit (Note 6)	Limit Units
CONVERTER CHARACTERISTICS						
Resolution			8	8	**8**	bits
Linearity Error Max	Zero and full scale adjusted $-10V \le V_{REF} \le +10V$	4, 8				
DAC0830LJ & LCJ				**0.05**	**0.05**	% FSR
DAC0832LJ & LCJ				**0.2**	**0.2**	% FSR
DAC0830LCN, LCWM & LCV				0.05	**0.05**	% FSR
DAC0831LCN				0.1	**0.1**	% FSR
DAC0832LCN, LCWM & LCV				0.2	**0.2**	% FSR
Differential Nonlinearity Max	Zero and full scale adjusted $-10V \le V_{REF} \le +10V$	4, 8				
DAC0830LJ & LCJ				**0.1**	**0.1**	% FSR
DAC0832LJ & LCJ				**0.4**	**0.4**	% FSR
DAC0830LCN, LCWM & LCV				0.1	**0.1**	% FSR
DAC0831LCN				0.2	**0.2**	% FSR
DAC0832LCN, LCWM & LCV				0.4	**0.4**	% FSR
Monotonicity	$-10V \le V_{REF} \le +10V$ LJ & LCJ	4		**8**	**8**	bits
	$-10V \le V_{REF} \le +10V$ LCN, LCWM & LCV			8	**8**	bits
Gain Error Max	Using Internal R_{fb} $-10V \le V_{REF} \le +10V$	7	±0.2	±1	**±1**	% FS
Gain Error Tempco Max	Using internal R_{fb}		**0.0002**		**0.0006**	%FS/°C
Power Supply Rejection	All digital inputs latched high $V_{CC} = 14.5V$ to 15.5V		0.0002	0.0025		%FSR/V
	11.5V to 12.5V		0.0006			
	4.5V to 5.5V		0.013	0.015		
Reference Input Max			**15**	**20**	**20**	kΩ
Reference Input Min			**15**	**10**	**10**	kΩ
Output Feedthrough Error	$V_{REF} = 20$ Vp-p, f = 100 kHz All data inputs latched low		3			mVp-p

5. DAC0830

Electrical Characteristics $V_{REF} = 10.000\ V_{DC}$ unless otherwise noted. **Boldface limits apply over temperature, $T_{MIN} \le T_A \le T_{MAX}$.** For all other limits $T_A = 25°C$. (Continued)

Parameter		Conditions		See Note	$V_{CC} = 4.75\ V_{DC}$, $V_{CC} = 15.75\ V_{DC}$ Typ (Note 12)	$V_{CC} = 4.75\ V_{DC}$, $V_{CC} = 15.75\ V_{DC}$ Tested Limit (Note 5)	$V_{CC} = 5\ V_{DC} \pm 5\%$, $V_{CC} = 12\ V_{DC} \pm 5\%$ to $15\ V_{DC} \pm 5\%$ Design Limit (Note 6)	Limit Units
CONVERTER CHARACTERISTICS (Continued)								
Output Leakage Current Max	I_{OUT1}	All data inputs latched low	LJ & LCJ	10		**100**	**100**	nA
			LCN, LCWM & LCV			50	**100**	
	I_{OUT2}	All data inputs latched high	LJ & LCJ			**100**	**100**	nA
			LCN, LCWM & LCV			50	**100**	
Output Capacitance	I_{OUT1}	All data inputs latched low			45			pF
	I_{OUT2}				115			
	I_{OUT1}	All data inputs latched high			130			pF
	I_{OUT2}				30			
DIGITAL AND DC CHARACTERISTICS								
Digital Input Voltages	Max	Logic Low	LJ 4.75V			**0.6**		V_{DC}
			LJ 15.75V			**0.8**		
			LCJ 4.75V			**0.7**		
			LCJ 15.75V			**0.8**		
			LCN, LCWM, LCV			0.95	**0.8**	
	Min	Logic High	LJ & LCJ			**2.0**	**2.0**	V_{DC}
			LCN, LCWM, LCV			1.9	**2.0**	
Digital Input Currents	Max	Digital inputs <0.8V	LJ & LCJ		−50	**−200**	**−200**	μA
			LCN, LCWM, LCV			−160	**−200**	μA
		Digital inputs >2.0V	LJ & LCJ		0.1	**+10**	**+10**	μA
			LCN, LCWM, LCV			+8	**+10**	
Supply Current Drain	Max		LJ & LCJ		1.2	**3.5**	**3.5**	mA
			LCN, LCWM, LCV			1.7	**2.0**	

5. DAC0830

Electrical Characteristics V_{REF} = 10.000 V_{DC} unless otherwise noted. **Boldface limits apply over temperature, $T_{MIN} \le T_A \le T_{MAX}$.** For all other limits T_A = 25°C. (Continued)

Symbol	Parameter	Conditions	See Note	V_{CC} = 15.75 V_{DC} Typ (Note 12)	V_{CC} = 15.75 V_{DC} Tested Limit (Note 5)	V_{CC} = 12 V_{DC} ±5% to 15 V_{DC} ±5% Design Limit (Note 6)	V_{CC} = 4.75 V_{DC} Typ (Note 12)	V_{CC} = 4.75 V_{DC} Tested Limit (Note 5)	V_{CC} = 5 V_{DC} ±5% Design Limit (Note 6)	Limit Units
AC CHARACTERISTICS										
t_s	Current Setting Time	V_{IL} = 0V, V_{IH} = 5V		1.0			1.0			µs
t_W	Write and XFER Pulse Width Min	V_{IL} = 0V, V_{IH} = 5V	11 9	100	250 **320**	**320**	375	600 **900**	**900**	
t_{DS}	Data Setup Time Min	V_{IL} = 0V, V_{IH} = 5V	9	100	250 **320**	**320**	375	600 **900**	**900**	
t_{DH}	Data Hold Time Min	V_{IL} = 0V, V_{IH} = 5V	9		30 **30**			50 **50**		ns
t_{CS}	Control Setup Time Min	V_{IL} = 0V, V_{IH} = 5V	9	110	250 **320**	**320**	600	900 **1100**	**1100**	
t_{CH}	Control Hold Time Min	V_{IL} = 0V, V_{IH} = 5V	9	0	0 **0**	**10**	0	0 **0**		

Note 1: Absolute Maximum Ratings indicate limits beyond which damage to the device may occur. DC and AC electrical specifications do not apply when operating the device beyond its specified operating conditions.

Note 2: All voltages are measured with respect to GND, unless otherwise specified.

Note 3: The maximum power dissipation must be derated at elevated temperatures and is dictated by T_{JMAX}, θ_{JA}, and the ambient temperature, T_A. The maximum allowable power dissipation at any temperature is $P_D = (T_{JMAX} - T_A)/\theta_{JA}$ or the number given in the Absolute Maximum Ratings, whichever is lower. For this device, T_{JMAX} = 125°C (plastic) or 150°C (ceramic), and the typical junction-to-ambient thermal resistance of the J package when board mounted is 80°C/W. For the N package, this number increases to 100°C/W and for the V package this number is 120°C/W.

Note 4: For current switching applications, both I_{OUT1} and I_{OUT2} must go to ground or the "Virtual Ground" of an operational amplifier. The linearity error is degraded by approximately $V_{OS} \div V_{REF}$. For example, if V_{REF} = 10V then a 1 mV offset, V_{OS}, on I_{OUT1} or I_{OUT2} will introduce an additional 0.01% linearity error.

Note 5: Tested limits are guaranteed to National's AOQL (Average Outgoing Quality Level).

Note 6: Guaranteed, but not 100% production tested. These limits are not used to calculate outgoing quality levels.

Note 7: Guaranteed at V_{REF} = ±10 V_{DC} and V_{REF} = ±1 V_{DC}.

Note 8: The unit "FSR" stands for "Full Scale Range." "Linearity Error" and "Power Supply Rejection" specs are based on this unit to eliminate dependence on a particular V_{REF} value and to indicate the true performance of the part. The "Linearity Error" specification of the DAC0830 is "0.05% of FSR (MAX)". This guarantees that after performing a zero and full scale adjustment (see Sections 2.5 and 2.6), the plot of the 256 analog voltage outputs will each be within 0.05% × V_{REF} of a straight line which passes through zero and full scale.

Note 9: Boldface tested limits apply to the LJ and LCJ suffix parts only.

Note 10: A 100nA leakage current with R_{fb} = 20k and V_{REF} = 10V corresponds to a zero error of $(100 \times 10^{-9} \times 20 \times 10^3) \times 100/10$ which is 0.02% of FS.

Note 11: The entire write pulse must occur within the valid data interval for the specified t_W, t_{DS}, t_{DH}, and t_S to apply.

Note 12: Typicals are at 25°C and represent most likely parametric norm.

Note 13: Human body model, 100 pF discharged through a 1.5 kΩ resistor.

5. DAC0830

Switching Waveform

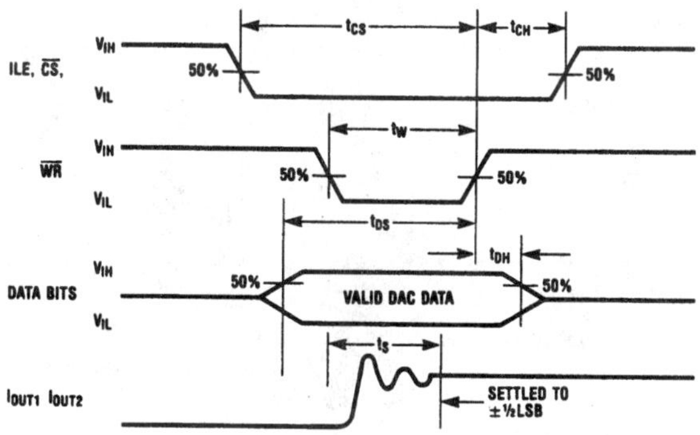

TL/H/5608-2

Definition of Package Pinouts

Control Signals (All control signals level actuated)

$\overline{CS}$: **Chip Select** (active low). The $\overline{CS}$ in combination with ILE will enable $\overline{WR_1}$.

ILE: **Input Latch Enable** (active high). The ILE in combination with $\overline{CS}$ enables $\overline{WR_1}$.

$\overline{WR_1}$: **Write 1.** The active low $\overline{WR_1}$ is used to load the digital input data bits (DI) into the input latch. The data in the input latch is latched when $\overline{WR_1}$ is high. To update the input latch—$\overline{CS}$ and $\overline{WR_1}$ must be low while ILE is high.

$\overline{WR_2}$: **Write 2** (active low). This signal, in combination with $\overline{XFER}$, causes the 8-bit data which is available in the input latch to transfer to the DAC register.

$\overline{XFER}$: **Transfer control signal** (active low). The $\overline{XFER}$ will enable $\overline{WR_2}$.

Other Pin Functions

DI_0-DI_7: **Digital Inputs.** DI_0 is the least significant bit (LSB) and DI_7 is the most significant bit (MSB).

I_{OUT1}: **DAC Current Output 1.** I_{OUT1} is a maximum for a digital code of all 1's in the DAC register, and is zero for all 0's in DAC register.

I_{OUT2}: **DAC Current Output 2.** I_{OUT2} is a constant minus I_{OUT1}, or $I_{OUT1} + I_{OUT2} = \text{constant}$ (I full scale for a fixed reference voltage).

R_{fb}: **Feedback Resistor.** The feedback resistor is provided on the IC chip for use as the shunt feedback resistor for the external op amp which is used to provide an output voltage for the DAC. This on-chip resistor should always be used (not an external resistor) since it matches the resistors which are used in the on-chip R-2R ladder and tracks these resistors over temperature.

V_{REF}: **Reference Voltage Input.** This input connects an external precision voltage source to the internal R-2R ladder. V_{REF} can be selected over the range of +10 to −10V. This is also the analog voltage input for a 4-quadrant multiplying DAC application.

V_{CC}: **Digital Supply Voltage.** This is the power supply pin for the part. V_{CC} can be from +5 to +15V_{DC}. Operation is optimum for +15V_{DC}.

GND: The pin 10 voltage must be at the same ground potential as I_{OUT1} and I_{OUT2} for current switching applications. Any difference of potential (V_{OS} pin 10) will result in a linearity change of

$$\frac{V_{OS}\text{ pin 10}}{3V_{REF}}$$

For example, if V_{REF} = 10V and pin 10 is 9mV offset from I_{OUT1} and I_{OUT2} the linearity change will be 0.03%.

Pin 3 can be offset ±100mV with no linearity change, but the logic input threshold will shift.

5. DAC0830

Linearity Error

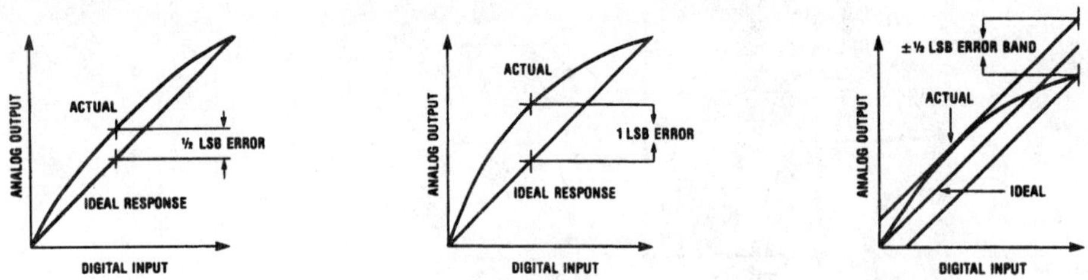

a) End point test after zero and fs adj. **b) Best straight line** **c) Shifting fs adj. to pass best straight line test**

Definition of Terms

Resolution: Resolution is directly related to the number of switches or bits within the DAC. For example, the DAC0830 has 2^8 or 256 steps and therefore has 8-bit resolution.

Linearity Error: Linearity Error is the maximum deviation from a *straight line passing through the endpoints of the DAC transfer characteristic.* It is measured after adjusting for zero and full-scale. Linearity error is a parameter intrinsic to the device and cannot be externally adjusted.

National's linearity "end point test" (a) and the "best straight line" test (b,c) used by other suppliers are illustrated above. The "end point test" greatly simplifies the adjustment procedure by eliminating the need for multiple iterations of checking the linearity and then adjusting full scale until the linearity is met. The "end point test" guarantees that linearity is met after a single full scale adjust. (One adjustment vs. multiple iterations of the adjustment.) The "end point test" uses a standard zero and F.S. adjustment procedure and is a much more stringent test for DAC linearity.

Power Supply Sensitivity: Power supply sensitivity is a measure of the effect of power supply changes on the DAC full-scale output.

Settling Time: Settling time is the time required from a code transition until the DAC output reaches within ±½LSB of the final output value. Full-scale settling time requires a zero to full-scale or full-scale to zero output change.

Full-Scale Error: Full scale error is a measure of the output error between an ideal DAC and the actual device output. Ideally, for the DAC0830 series, full-scale is V_{REF} −1LSB. For V_{REF}=10V and unipolar operation, $V_{FULL\text{-}SCALE}$= 10.0000V−39mV=9.961V. Full-scale error is adjustable to zero.

Differential Nonlinearity: The difference between any two consecutive codes in the transfer curve from the theoretical 1 LSB is differential nonlinearity.

Monotonic: If the output of a DAC increases for increasing digital input code, then the DAC is monotonic. An 8-bit DAC which is monotonic to 8 bits simply means that increasing digital input codes will produce an increasing analog output.

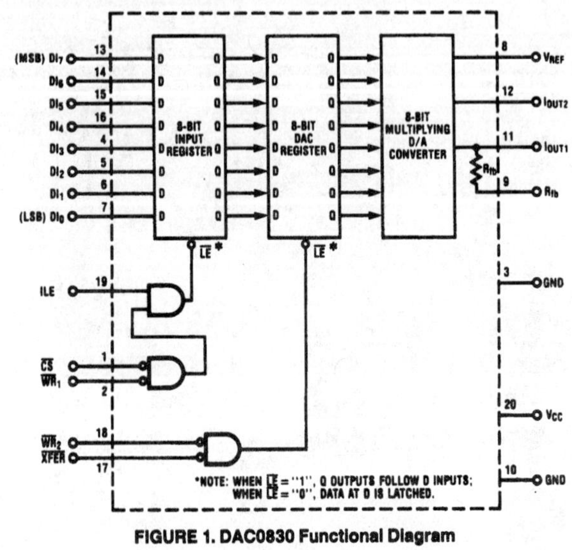

FIGURE 1. DAC0830 Functional Diagram

5. DAC0830

Typical Performance Characteristics

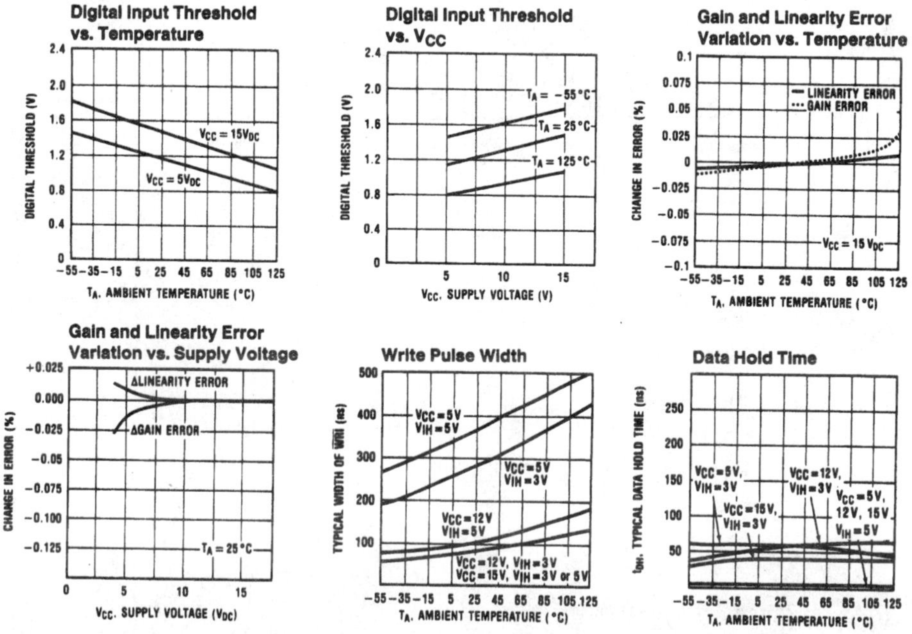

DAC0830 Series Application Hints

These DAC's are the industry's first microprocessor compatible, double-buffered 8-bit multiplying D to A converters. Double-buffering allows the utmost application flexibility from a digital control point of view. This 20-pin device is also pin for pin compatible (with one exception) with the DAC1230, a 12-bit MICRO-DAC. In the event that a system's analog output resolution and accuracy must be upgraded, substituting the DAC1230 can be easily accomplished. By tying address bit A_0 to the ILE pin, a two-byte μP write instruction (double precision) which automatically increments the address for the second byte write (starting with A_0 = "1") can be used. This allows either an 8-bit or the 12-bit part to be used with no hardware or software changes. For the simplest 8-bit application, this pin should be tied to V_{CC} (also see other uses in section 1.1).

Analog signal control versatility is provided by a precision R-2R ladder network which allows full 4-quadrant multiplication of a wide range bipolar reference voltage by an applied digital word.

1.0 DIGITAL CONSIDERATIONS

A most unique characteristic of these DAC's is that the 8-bit digital input byte is double-buffered. This means that the data must transfer through two independently controlled 8-bit latching registers before being applied to the R-2R ladder network to change the analog output. The addition of a second register allows two useful control features. First, any DAC in a system can simultaneously hold the current DAC data in one register (DAC register) and the next data word in the second register (input register) to allow fast updating of the DAC output on demand. Second, and probably more important, double-buffering allows any number of DAC's in a system to be updated to their new analog output levels simultaneously via a common strobe signal.

The timing requirements and logic level convention of the register control signals have been designed to minimize or eliminate external interfacing logic when applied to most popular microprocessors and development systems. It is easy to think of these converters as 8-bit "write-only" memory locations that provide an analog output quantity. All inputs to these DAC's meet TTL voltage level specs and can also be driven directly with high voltage CMOS logic in non-microprocessor based systems. To prevent damage to the chip from static discharge, all unused digital inputs should be tied to V_{CC} or ground. If any of the digital inputs are inadvertantly left floating, the DAC interprets the pin as a logic "1".

1.1 Double-Buffered Operation

Updating the analog output of these DAC's in a double-buffered manner is basically a two step or double write operation. In a microprocessor system two unique system addresses must be decoded, one for the input latch controlled by the $\overline{CS}$ pin and a second for the DAC latch which is controlled by the $\overline{XFER}$ line. If more than one DAC is being driven, *Figure 2*, the $\overline{CS}$ line of each DAC would typically be decoded individually, but all of the converters could share a common $\overline{XFER}$ address to allow simultaneous updating of any number of DAC's. The timing for this operation is shown, *Figure 3*.

It is important to note that the analog outputs that will change after a simultaneous transfer are those from the DAC's whose input register had been modified prior to the $\overline{XFER}$ command.

5. DAC0830

DAC0830 Series Application Hints (Continued)

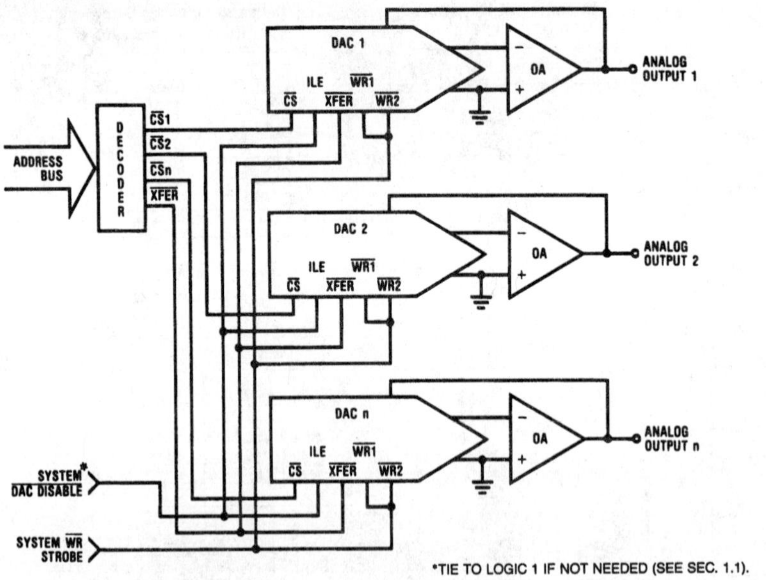

*TIE TO LOGIC 1 IF NOT NEEDED (SEE SEC. 1.1).

FIGURE 2. Controlling Mutiple DACs

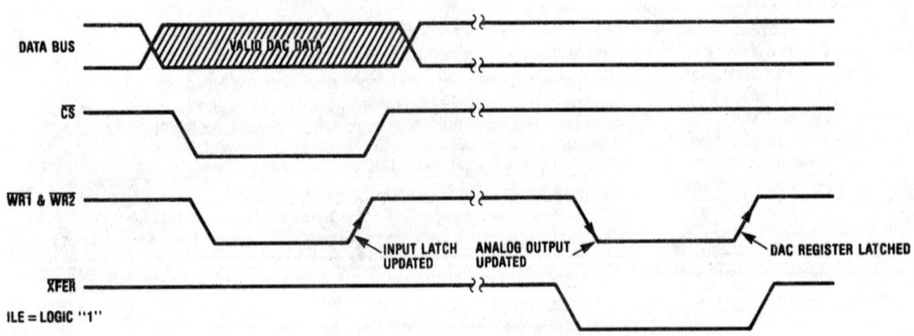

FIGURE 3

The ILE pin is an active high chip select which can be decoded from the address bus as a qualifier for the normal $\overline{CS}$ signal generated during a write operation. This can be used to provide a higher degree of decoding unique control signals for a particular DAC, and thereby create a more efficient addressing scheme.

Another useful application of the ILE pin of each DAC in a multiple DAC system is to tie these inputs together and use this as a control line that can effectively "freeze" the outputs of all the DAC's at their present value. Pulling this line low latches the input register and prevents new data from being written to the DAC. This can be particularly useful in multiprocessing systems to allow a processor other than the one controlling the DAC's to take over control of the data bus and control lines. If this second system were to use the same addresses as those decoded for DAC control (but for a different purpose) the ILE function would prevent the DAC's from being erroneously altered.

In a "Stand-Alone" system the control signals are generated by discrete logic. In this case double-buffering can be controlled by simply taking $\overline{CS}$ and $\overline{XFER}$ to a logic "0", ILE to a logic "1" and pulling $\overline{WR_1}$ low to load data to the input latch. Pulling $\overline{WR_2}$ low will then update the analog output. A logic "1" on either of these lines will prevent the changing of the analog output.

5. DAC0830

DAC0830 Series Application Hints (Continued)

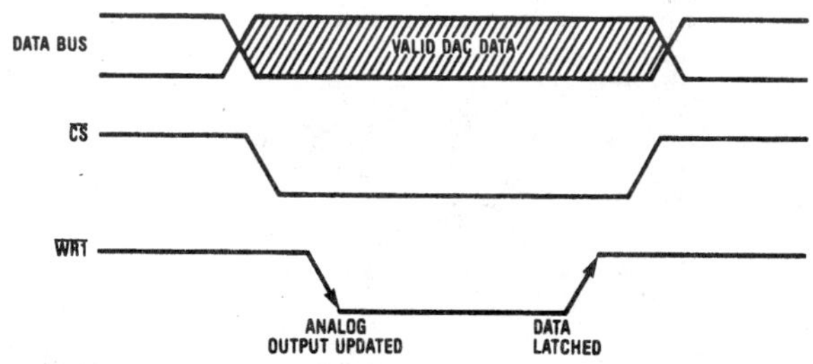

ILE = LOGIC "1"; WR2 and XFER GROUNDED

FIGURE 4

1.2 Single-Buffered Operation

In a microprocessor controlled system where maximum data throughput to the DAC is of primary concern, or when only one DAC of several needs to be updated at a time, a single-buffered configuration can be used. One of the two internal registers allows the data to flow through and the other register will serve as the data latch.

Digital signal feedthrough (see Section 1.5) is minimized if the input register is used as the data latch. Timing for this mode is shown in *Figure 4*.

Single-buffering in a "stand-alone" system is achieved by strobing $\overline{WR_1}$ low to update the DAC with $\overline{CS}$, $\overline{WR_2}$ and $\overline{XFER}$ grounded and ILE tied high.

1.3 Flow-Through Operation

Though primarily designed to provide microprocessor interface compatibility, the MICRO-DAC's can easily be configured to allow the analog output to continuously reflect the state of an applied digital input. This is most useful in applications where the DAC is used in a continuous feedback control loop and is driven by a binary up-down counter, or in function generation circuits where a ROM is continuously providing DAC data.

Simply grounding $\overline{CS}$, $\overline{WR_1}$, $\overline{WR_2}$, and $\overline{XFER}$ and tying ILE high allows both internal registers to follow the applied digital inputs (flow-through) and directly affect the DAC analog output.

1.4 Control Signal Timing

When interfacing these MICRO-DAC to any microprocessor, there are two important time relationships that must be considered to insure proper operation. The first is the minimum $\overline{WR}$ strobe pulse width which is specified as 900 ns for all valid operating conditions of supply voltage and ambient temperature, but typically a pulse width of only 180ns is adequate if $V_{CC} = 15V_{DC}$. A second consideration is that the guaranteed minimum data hold time of 50ns should be met or erroneous data can be latched. This hold time is defined as the length of time data must be held valid on the digital inputs *after* a qualified (via $\overline{CS}$) $\overline{WR}$ strobe makes a low to high transition to latch the applied data.

If the controlling device or system does not inherently meet these timing specs the DAC can be treated as a slow memory or peripheral and utilize a technique to extend the write strobe. A simple extension of the write time, by adding a wait state, can simultaneously hold the write strobe active and data valid on the bus to satisfy the minimum $\overline{WR}$ pulsewidth. If this does not provide a sufficient data hold time at the end of the write cycle, a negative edge triggered one-shot can be included between the system write strobe and the $\overline{WR}$ pin of the DAC. This is illustrated in *Figure 5* for an exemplary system which provides a 250ns $\overline{WR}$ strobe time with a data hold time of less than 10ns.

The proper data set-up time prior to the latching edge (LO to HI transition) of the $\overline{WR}$ strobe, is insured if the $\overline{WR}$ pulsewidth is within spec and the data is valid on the bus for the duration of the DAC $\overline{WR}$ strobe.

1.5 Digital Signal Feedthrough

When data is latched in the internal registers, but the digital inputs are changing state, a narrow spike of current may flow out of the current output terminals. This spike is caused by the rapid switching of internal logic gates that are responding to the input changes.

There are several recommendations to minimize this effect. When latching data in the DAC, always use the input register as the latch. Second, reducing the V_{CC} supply for the DAC from +15V to +5V offers a factor of 5 improvement in the magnitude of the feedthrough, but at the expense of internal logic switching speed. Finally, increasing C_C (*Figure 8*) to a value consistent with the actual circuit bandwidth requirements can provide a substantial damping effect on any output spikes.

5. DAC0830

DAC0830 Series Application Hints (Continued)

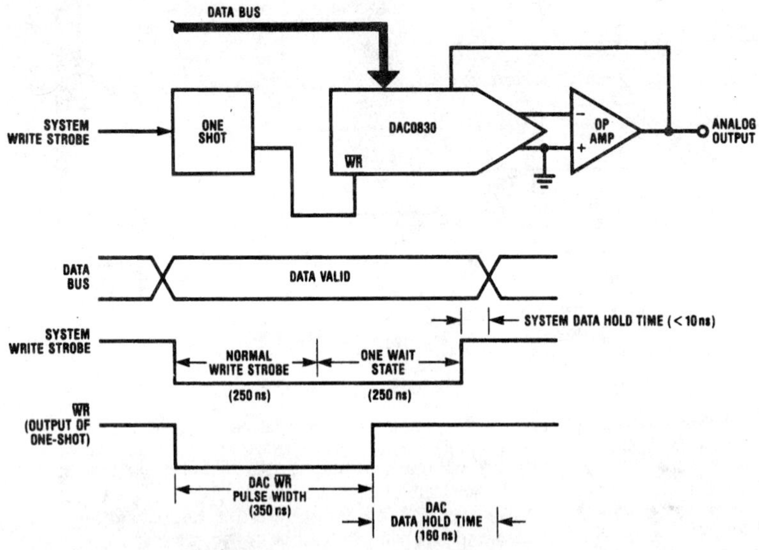

FIGURE 5. Accommodating a High Speed System

2.0 ANALOG CONSIDERATIONS

The fundamental purpose of any D to A converter is to provide an accurate analog output quantity which is representative of the applied digital word. In the case of the DAC0830, the output, I_{OUT1}, is a current directly proportional to the product of the applied reference voltage and the digital input word. For application versatility, a second output, I_{OUT2}, is provided as a current directly proportional to the complement of the digital input. Basically:

$$I_{OUT1} = \frac{V_{REF}}{15\ k\Omega} \times \frac{\text{Digital Input}}{256};$$

$$I_{OUT2} = \frac{V_{REF}}{15\ k\Omega} \times \frac{255 - \text{Digital Input}}{256}$$

where the digital input is the decimal (base 10) equivalent of the applied 8-bit binary word (0 to 255), V_{REF} is the voltage at pin 8 and 15 kΩ is the nominal value of the internal resistance, R, of the R-2R ladder network (discussed in Section 2.1).

Several factors external to the DAC itself must be considered to maintain analog accuracy and are covered in subsequent sections.

2.1 The Current Switching R-2R Ladder

The analog circuitry, *Figure 6*, consists of a silicon-chromium (SiCr or Si-chrome) thin film R-2R ladder which is deposited on the surface oxide of the monolithic chip. As a result, there are no parasitic diode problems with the ladder (as there may be with diffused resistors) so the reference voltage, V_{REF}, can range −10V to +10V even if V_{CC} for the device is $5V_{DC}$.

The digital input code to the DAC simply controls the position of the SPDT current switches and steers the available ladder current to either I_{OUT1} or I_{OUT2} as determined by the logic input level ("1" or "0") respectively, as shown in *Figure 6*. The MOS switches operate in the current mode with a small voltage drop across them and can therefore switch currents of either polarity. This is the basis for the 4-quadrant multiplying feature of this DAC.

2.2 Basic Unipolar Output Voltage

To maintain linearity of output current with changes in the applied digital code, it is important that the voltages at both of the current output pins be as near ground potential ($0V_{DC}$) as possible. With V_{REF} = +10V every millivolt appearing at either I_{OUT1} or I_{OUT2} will cause a 0.01% linearity error. In most applications this output current is converted to a voltage by using an op amp as shown in *Figure 7*.

The inverting input of the op amp is a "virtual ground" created by the feedback from its output through the internal 15 kΩ resistor, R_{fb}. All of the output current (determined by the digital input and the reference voltage) will flow through R_{fb} to the output of the amplifier. Two-quadrant operation can be obtained by reversing the polarity of V_{REF} thus causing I_{OUT1} to flow into the DAC and be sourced from the output of the amplifier. The output voltage, in either case, is always equal to $I_{OUT1} \times R_{fb}$ and is the opposite polarity of the reference voltage.

The reference can be either a stable DC voltage source or an AC signal anywhere in the range from −10V to +10V. The DAC can be thought of as a digitally controlled attenuator: the output voltage is always less than or equal to the applied reference voltage. The V_{REF} terminal of the device presents a nominal impedance of 15 kΩ to ground to external circuitry.

Always use the internal R_{fb} resistor to create an output voltage since this resistor matches (and tracks with temperature) the value of the resistors used to generate the output current (I_{OUT1}).

5. DAC0830

DAC0830 Series Application Hints (Continued)

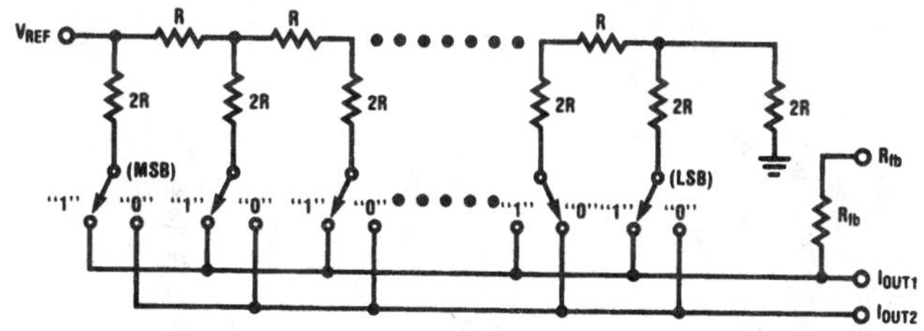

FIGURE 6

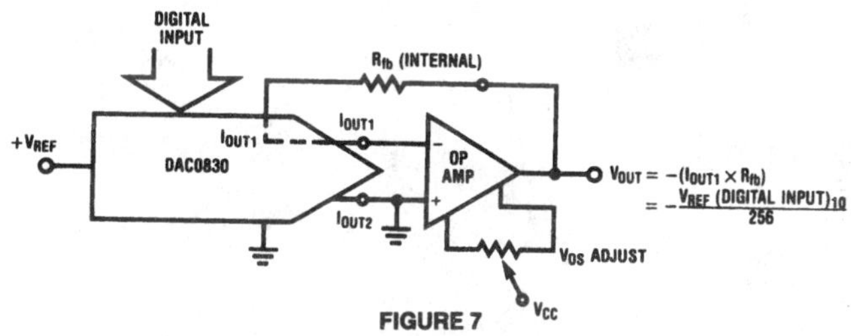

FIGURE 7

2.3 Op Amp Considerations

The op amp used in *Figure 7* should have offset voltage nulling capability (See Section 2.5).

The selected op amp should have as low a value of input bias current as possible. The product of the bias current times the feedback resistance creates an output voltage error which can be significant in low reference voltage applications. BI-FET op amps are highly recommended for use with these DACs because of their very low input current.

Transient response and settling time of the op amp are important in fast data throughput applications. The largest stability problem is the feedback pole created by the feedback resistance, R_{fb}, and the output capacitance of the DAC. This appears from the op amp output to the (−) input and includes the stray capacitance at this node. Addition of a lead capacitance, C_C in *Figure 8*, greatly reduces overshoot and ringing at the output for a step change in DAC output current.

Finally, the output voltage swing of the amplifier must be greater than V_{REF} to allow reaching the full scale output voltage. Depending on the loading on the output of the amplifier and the available op amp supply voltages (only ±12 volts in many development systems), a reference voltage less than 10 volts may be necessary to obtain the full analog output voltage range.

2.4 Bipolar Output Voltage with a Fixed Reference

The addition of a second op amp to the previous circuitry can be used to generate a bipolar output voltage from a fixed reference voltage. This, in effect, gives sign significance to the MSB of the digital input word and allows two-quadrant multiplication of the reference voltage. The polarity of the reference can also be reversed to realize full 4-quadrant multiplication: $\pm V_{REF} \times \pm$Digital Code$= \pm V_{OUT}$. This circuit is shown in *Figure 9*.

This configuration features several improvements over existing circuits for bipolar outputs with other multiplying DACs. Only the offset voltage of amplifier 1 has to be nulled to preserve linearity of the DAC. The offset voltage error of the second op amp (although a constant output voltage error) has no effect on linearity. It should be nulled only if absolute output accuracy is required. Finally, the values of the resistors around the second amplifier do not have to match the internal DAC resistors, they need only to match and temperature track each other. A thin film 4-resistor network available from Beckman Instruments, Inc. (part no. 694-3-R10K-D) is ideally suited for this application. These resistors are matched to 0.1% and exhibit only 5 ppm/°C resistance tracking temperature coefficient. Two of the four available 10 kΩ resistors can be paralleled to form R in *Figure 9* and the other two can be used independently as the resistances labeled 2R.

2.5 Zero Adjustment

For accurate conversions, the input offset voltage of the output amplifier must always be nulled. Amplifier offset errors create an overall degradation of DAC linearity.

The fundamental purpose of zeroing is to make the voltage appearing at the DAC outputs as near $0V_{DC}$ as possible. This is accomplished for the typical DAC — op amp connection (*Figure 7*) by shorting out R_{fb}, the amplifier feedback resistor, and adjusting the V_{OS} nulling potentiometer of the op amp until the output reads zero volts. This is done, of course, with an applied digital code of all zeros if I_{OUT1} is driving the op amp (all one's for I_{OUT2}). The short around R_{fb} is then removed and the converter is zero adjusted.

Appendix B

CUPL Reference Tables

Table **B.1**

CUPL Reserve Words

APPEND	ASSEMBLY	ASSY	COMPANY	CONDITION
DATE	DEFAULT	DESIGNER	DEVICE	ELSE
FIELD	FLD	FORMAT	FUSE	IF
JUMP	LOC	LOCATION	MACRO	MIN
NAME	NODE	OUT	PARTNO	PIN
PINNODE	PRESENT	REV	REVISION	SEQUENCE
SEQUENCED	SEQUENCEJK	SEQUENCERS	SEQUENCET	TABLE
&	*	:	;	@
#	+	.	'	$
( )	[ ]	{ }	..	!
^	–	/	/* */	=
=>	'			

Table **B.2**

Commonly Used CUPL Variable (V) Name Extensions

Used with a PLD whose output is designated with a V, like GAL16V8, as part of the dot extension when writing a CUPL logic expression.

Example: Q3.D = A & B # C; signifies that the input to a D flip-flop whose output is Q_3 is the expression AB + C.

EXTENSION	MEANING
D	Input to D flip-flop
L	Input to a latch
J	J input to J–K flip-flop
K	K input to J–K flip-flop
S	S input to R–S flip-flop
R	R input to R–S flip-flop
T	Toggle input to toggle flip-flop
DQ	Q output of a D flip-flop
LQ	Q output of an input latch
AP	Asynchronous PRESET (set)
AR	Asynchronous RESET (clear)
SP	Synchronous PRESET
SR	Synchronous RESET
OE	Output enable—three-state control

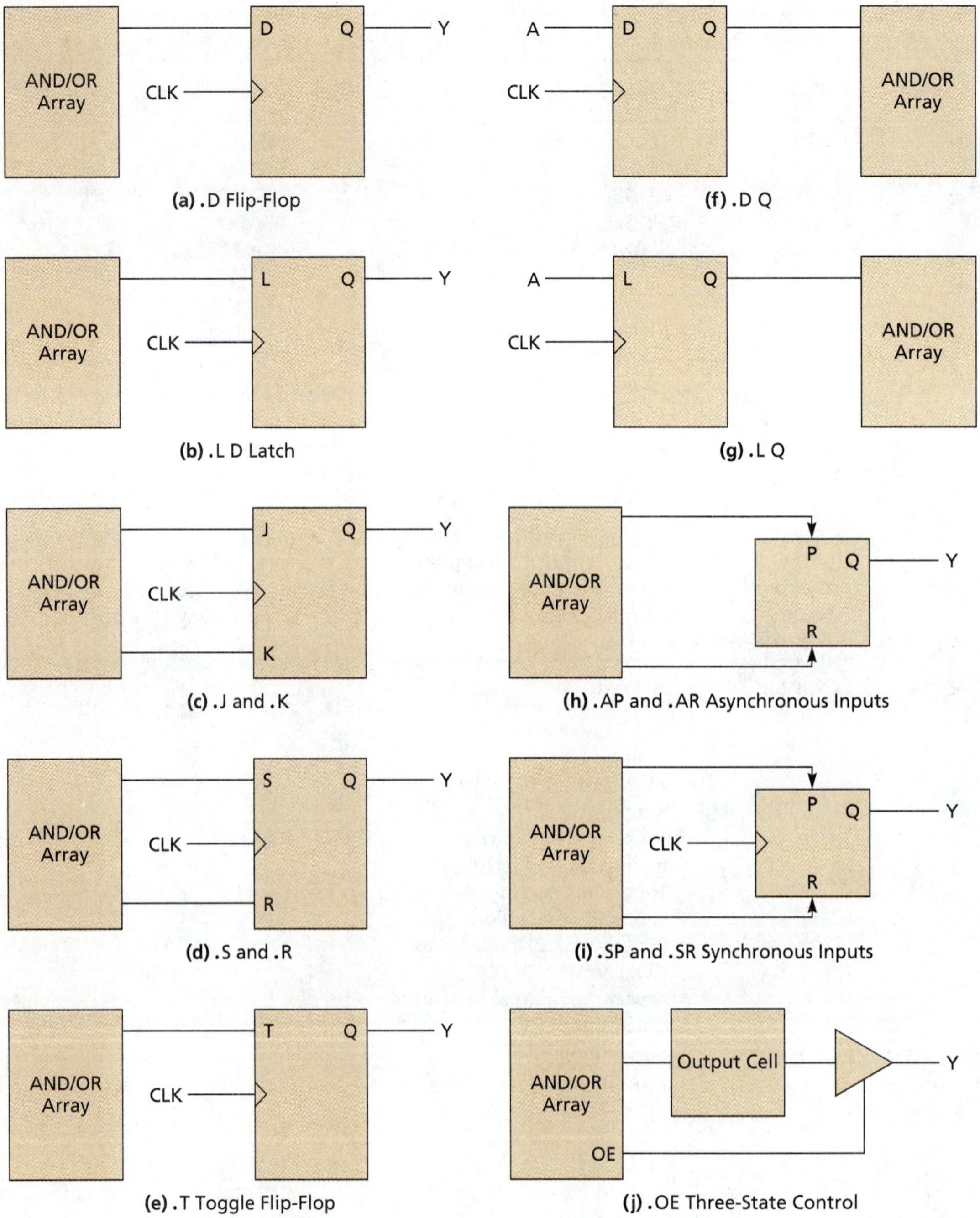

Figure B.1 Extension Diagrams

```
/*******************************************************************************/
/***********--MAIN MENU--***************************Message Center**********/
/*                                         /*                                 */
/*    <E>dit Design File                   /*                                 */
/*    <C>ompile CUPL File                  /*                                 */
/*    <L>ook at DOC File                   /*                                 */
/*    View Expanded <M>acro (MX) File      /*        Create or Edit           */
/*    <R>eview Error LST File              /*      a design (.PLD) file       */
/*    <J>EDEC File Editor                  /*                                 */
/*    Ed<I>t Simulation Input File         /*                                 */
/*    <S>imulate CUPL File                 /*                                 */
/*    <V>iew Simulation Results            /*                                 */
/*    <D>evice Selection                   /**********************************/
/*    <B>uild/View Device Library          /**********Operator Input**********/
/*    <P>ALASM to CUPL                     /*                                 */
/*    <H>elp (CUPL Quick Reference)        /*                                 */
/*    <T>utorial for PLD's                 /*                                 */
/*    Design E<X>amples                    /*                                 */
/*    <U>ser's Editor                      /*                                 */
/*    <A>dditional Product Information     /*                                 */
/*    S<Y>stem Information                 /*                                 */
/*    <Q>uit                               /*                                 */
/*                                         /*                                 */
/************Device = none*****************************************************/
/************************CUPL/Starter Kit v4.2a - 10************************/
```

Figure B.2 CUPL Main Menu

```
/*********************************************************************************/
/*********Device Selection Guide*********************Message Center**********/
/*                                        /*                                   */
/*     <1> P16L8                          /*                                   */
/*     <2> P16R4                          /*                                   */
/*     <3> P16R6                          /*                                   */
/*     <4> P16R8                          /*         Press F1 for a list       */
/*     <5> P20L8                          /*       of device information       */
/*     <6> P20L10                         /*                                   */
/*     <7> P20R4                          /*                                   */
/*     <8> P20R6                          /*                                   */
/*     <9> P20R8                          /*                                   */
/*     <A> P22V10                         /*************************************/
/*     <B> G16V8                          /***********Operator Input***********/
/*     <C> G16V8A                         /*                                   */
/*     <D> G20V8                          /*                                   */
/*     <E> G20V8A                         /*                                   */
/*     <U>nselect Device                  /*                                   */
/*                                        /*                                   */
/*     <R>eturn to main menu              /*                                   */
/*                                        /*                                   */
/*************Device = none***************************************************/
/**************************CUPL/Starter Kit v4.2a - 10/93**********************/
```

Figure B.3 CUPL Device Selection Menu

```
/******************************************************************************/
/************CUPL Compiler****************************Message Center**********/
/*                                        /*                                */
/*     <J> EDEC download format           /*                                */
/*     <H>EX download format              /*                                */
/*     <I> HL download format             /*                                */
/*     <N>ame of JEDEC file = PLD file    /*    Produces a JEDEC download   */
/*     <A>bsolute file                    /*          format file           */
/*     <L>isting file                     /*                                */
/*     <E>xpanded macro definition file   /*                                */
/*     <X>panded product terms            /*                                */
/*     <F>use plot/chip diagram           /*                                */
/*     <B>erkeley PLA file                /*******Press F5 when finished*******/
/*     <C>reate PALASM file               /***********Operator Input***********/
/*     <U>se a specified library          /*                                */
/*     <S>imulate after compilation       /*                                */
/*     <W>aveform simulation output       /*                                */
/*     <O>ne-hot-bit state machine        /*                                */
/*     <K> Optimization methods           /*                                */
/*     <Z> Output file formats            /*                                */
/*     <M>inimization Levels              /*>                               */
/*     <R> More Options                   /*                                */
/**************Flags = m1******************************************************/
/*************************CUPL/Starter Kit v4.2a - 10/93**********************/
```

Figure B.4 CUPL Compiler Options

```
Name         XXXXX;
Partno       XXXXX;
Date         XX/XX/XX;
Revision     XX;
Designer     XXXXX;
Company      XXXXX;
Assembly     XXXXX;
Location     XXXXX;

/*****************************************************************************/
/*                                                                           */
/*                                                                           */
/*                                                                           */
/*****************************************************************************/
/*  Allowable Target Device Types:
/*****************************************************************************/

/**  Inputs  **/

Pin         =         ;       /*                                             */
Pin         =         ;       /*                                             */
Pin         =         ;       /*                                             */
Pin         =         ;       /*                                             */
Pin         =         ;       /*                                             */
Pin         =         ;       /*                                             */
Pin         =         ;       /*                                             */
Pin         =         ;       /*                                             */
Pin         =         ;       /*                                             */
Pin         =         ;       /*                                             */
Pin         =         ;       /*                                             */
Pin         =         ;       /*                                             */

/**  Outputs  **/

Pin         =         ;       /*                                             */
Pin         =         ;       /*                                             */
Pin         =         ;       /*                                             */
Pin         =         ;       /*                                             */
Pin         =         ;       /*                                             */
Pin         =         ;       /*                                             */
Pin         =         ;       /*                                             */
Pin         =         ;       /*                                             */

/**  Declarations and Intermediate Variable Definitions  **/

/**  Logic Equations  **/
```

Figure B.6 TMPL.PLD File

```
0                                  EZ Edit 2.00
Name     960sampl;
Partno   XXXXX;
Date*****************************************Help**********************************/
Revision *                                                                       */
Designer * Up Arrow       : Cursor up          /* Home      : Beginning of line  */
Company  * Down Arrow     : Cursor down        /* End       : End of line        */
Assembly * Left Arrow     : Cursor left        /* PgUp      : Page up            */
Location * Right Arrow    : Cursor right       /* PgDn      : Page down          */
Device***************************************************************************/
         * Ctrl-Home      : Top of screen      /* F7        : Border color       */
/********* Ctrl-End       : Bottom of screen   /* Shift-F7  : Border bkgnd       */
/*  A sa * Ctrl-PgUp      : Top of file        /* F8        : Text color         */
/*  syst * Ctrl-PgDn      : Bottom of file     /* Shift-F8  : Text bkgnd         */
/*  two  * Ctrl-Bkspace   : Deletes a line     /*                                */
/********* Backspace      : Deletes a char     /*                                */
/*  Allo * F2             : Switch Window      /*                                */
/********* Alt-X          : Save/Quit          /*                                */

/**************************Press any key to continue*************************/

Pin [2..9] = [i7..0];   /* inputs */
Pin   25   = W_R;       /* Write/Read */

  <D:\CUPL\960SAMPL.PLD>                               Press F1 for HELP < 1 1 >
```

Figure B.5 CUPL EZ Edit

CUPL for WINDOWS

A new version of the CUPL starter kit is included with the text. This version runs the CUPL compiler under WINDOWS. To install the WINDOWS version, insert the disk in drive A or B and from windows, run the SETUP program from that disk. Once CUPL is installed, to use it, double click on the CUPL icon. This brings up the working window of CUPL.

Follow this sequence to compile the CUPL .PLD file:

1) Select OPEN from the FILE drop down menu and load the appropriate .PLD file using drive and path information as well as filename and extension.
2) Click on OPTIONS, DEVICE SELECTION, GAL, g16V8, OK
3) Click on OPTIONS, COMPILE OPTIONS, JEDEC Name = filename, OUTPUT file, list, expanded macro, JEDEC, equations, OK, OK
4) Click on RUN, device specific compile

You will get the appropriate message at the bottom of the screen when CUPL is finished compiling or runs into an error. A detailed description of the CUPL WINDOWS version is included in the accompanying lab manual to the text.

Appendix C

Answers to Odd-Numbered Problems

Chapter 1

1.1 Less heat dissipated, smaller size, uses less power.

1.3 less power used, less heat, greater packing density.

1.5 Pack a lot more circuits into smaller area.

1.7 Microwave oven controllers, CD players, TV remote controls, VCRs, watches, car stereos, etc.

1.9 Analog has many levels (of voltage, current), digital signals have only two levels.

1.11 State is the current logic level of a signal.

1.13

QUARTERS	DOLLAR	DRINK	CHANGE
Inserted	Not	Yes	No
Not	Inserted	Yes	Yes
Not	Not	No	No

1.15 A buffer output state is the same logic level as its input state. In comparison, an inverter's output state is the opposite of its input state.

1.17

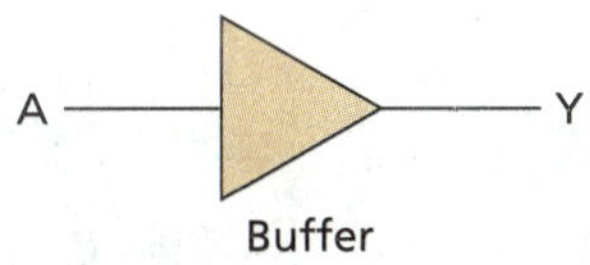

A	Y
0	0
1	1

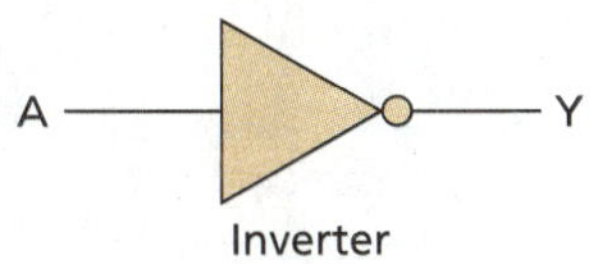

A	Y
0	1
1	0

Figure 1.K1 Problem 1.17

1.19

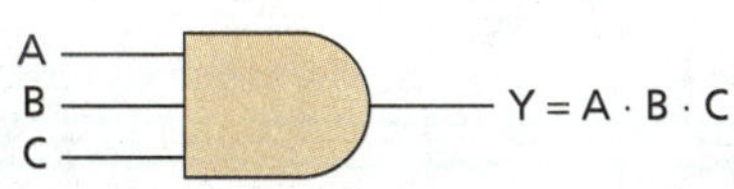

A	B	C	Y
0	0	0	0
0	0	1	0
0	1	0	0
0	1	1	0
1	0	0	0
1	0	1	0
1	1	0	0
1	1	1	1

Figure 1.K3 Problem 1.19

1.21 An AND gate is defined as a gate whose output is a logic 1 when all its inputs are logic 1's.

1.23

WHEEL3	WHEEL2	WHEEL1	WHEEL0	UNLOCK
0	0	0	0	0
0	0	0	1	0
0	0	1	0	0
0	0	1	1	0
0	1	0	0	0
0	1	0	1	0
0	1	1	0	0
0	1	1	1	0
1	0	0	0	0
1	0	0	1	0
1	0	1	0	0
1	0	1	1	0
1	1	0	0	0
1	1	0	1	0
1	1	1	0	0
1	1	1	1	1

UNLOCK = WHEEL3 · WHEEL2 · WHEEL1 · WHEEL0

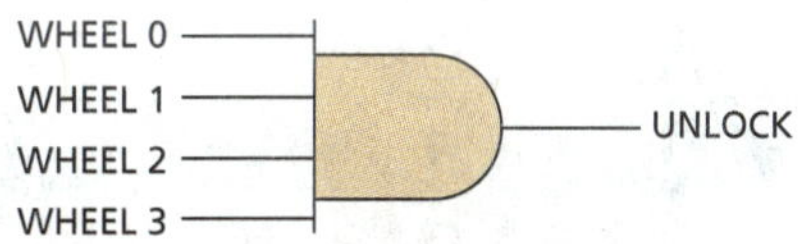

Figure 1.K5 Problem 1.23

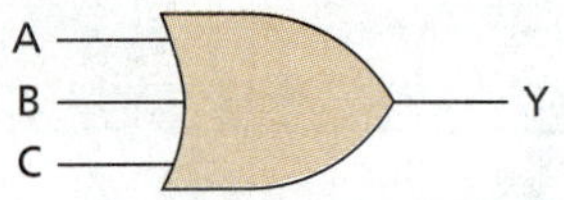

Figure 1.K7 Problem 1.25

1.25

A	B	C	Y
0	0	0	0
0	0	1	1
0	1	0	1
0	1	1	1
1	0	0	1
1	0	1	1
1	1	0	1
1	1	1	1

$Y = A + B + C$

1.27 An OR gate is defined as a gate whose output is a logic 1 if any of its inputs is a logic 1.

1.29 AND gate with this expression:

$$\overline{DOOR} = \overline{CARD1} \cdot \overline{CARD2} \cdot \overline{CARD3} \cdot \overline{CARD4}$$

and an OR gate with this expression:

$$ALARM = CARD1 + CARD2 + CARD3 + CARD4$$

1.31

NAND

A	B	C	Y
0	0	0	1
0	0	1	1
0	1	0	1
0	1	1	1
1	0	0	1
1	0	1	1
1	1	0	1
1	1	1	0

NOR

A	B	C	Y
0	0	0	1
0	0	1	0
0	1	0	0
0	1	1	0
1	0	0	0
1	0	1	0
1	1	0	0
1	1	1	0

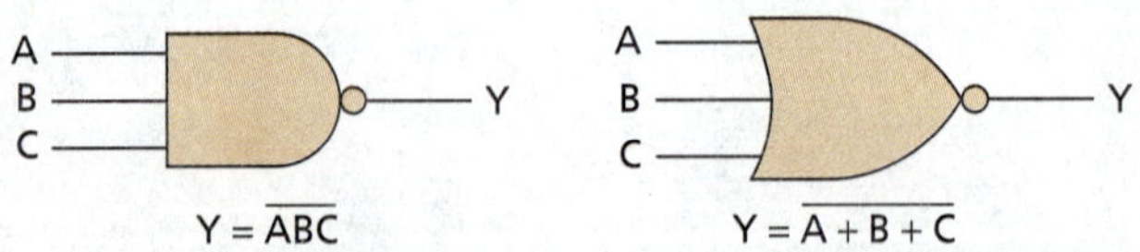

Figure 1.K9 Problem 1.31

1.33 $RESET = B3 \cdot B2 \cdot B1 \cdot B0$

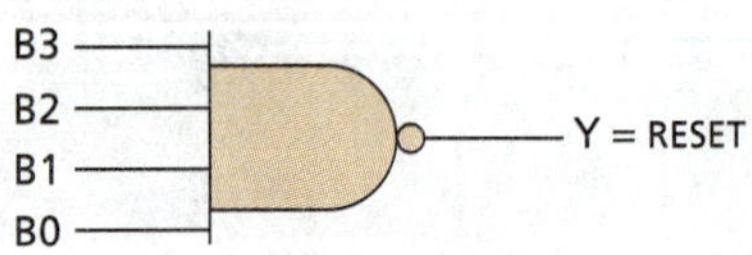

Figure 1.K11 Problem 1.33

B3	B2	B1	B0	RESET
0	0	0	0	1
0	0	0	1	1
0	0	1	0	1
0	0	1	1	1
0	1	0	0	1
0	1	0	1	1
0	1	1	0	1
0	1	1	1	1
1	0	0	0	1
1	0	0	1	1
1	0	1	0	1
1	0	1	1	1
1	1	0	0	1
1	1	0	1	1
1	1	1	0	1
1	1	1	1	0

1.35 $Y = \overline{\overline{A} \cdot \overline{B}}$

A	B	$\overline{A}$	$\overline{B}$	$\overline{A} \cdot \overline{B}$	$\overline{\overline{A} \cdot \overline{B}}$
0	0	1	1	1	0
0	1	1	0	0	1
1	0	0	1	0	1
1	1	0	0	0	1

The same truth table as for an OR gate, so $\overline{\overline{A} \cdot \overline{B}} = A + B$

1.37 $Y = \overline{A\overline{B} + \overline{BC}}$

A	B	C	$\overline{B}$	$A\overline{B}$	BC	$\overline{BC}$	$A\overline{B} + \overline{BC}$	$D = \overline{A\overline{B} + \overline{BC}}$
0	0	0	1	0	0	1	1	0
0	0	1	1	0	0	1	1	0
0	1	0	0	0	0	1	1	0
0	1	1	0	0	1	0	0	1
1	0	0	1	1	0	1	1	0
1	0	1	1	1	0	1	1	0
1	1	0	0	0	0	1	1	0
1	1	1	0	0	1	0	0	1

1.39 $Y = \overline{(A + \overline{B})\overline{C} + \overline{BC}}$

1.41 $Y = A \oplus B \oplus C$

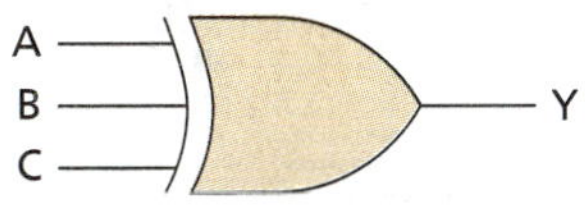

Figure 1.K13 Problem 1.41(a)

A	B	C	Y
0	0	0	0
0	0	1	1
0	1	0	1
0	1	1	0
1	0	0	1
1	0	1	0
1	1	0	0
1	1	1	1

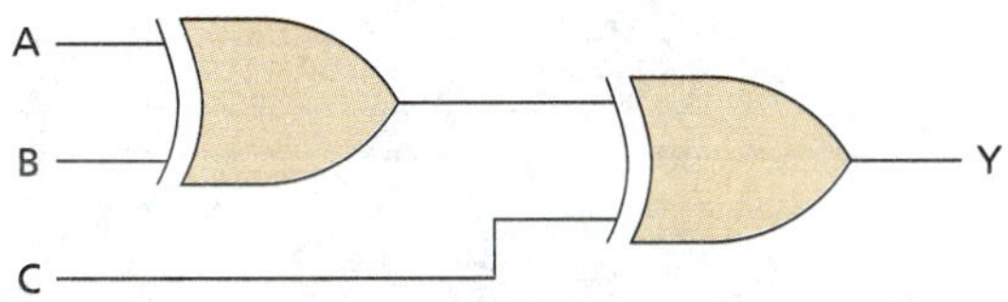

Figure 1.K14 Problem 1.41(b)

1.43

a.

B3	B2	B1	B0	B3 ⊕ B2	B2 ⊕ B1	B1 ⊕ B0	ENABLE
0	0	0	0	0	0	0	0
0	0	0	1	0	0	1	1
0	0	1	0	0	1	1	1
0	0	1	1	0	1	0	1
0	1	0	0	1	1	0	1
0	1	0	1	1	1	1	1
0	1	1	0	1	0	1	1
0	1	1	1	1	0	0	1
1	0	0	0	1	0	0	1
1	0	0	1	1	0	1	1
1	0	1	0	1	1	1	1
1	0	1	1	1	1	0	1
1	1	0	0	0	1	0	1
1	1	0	1	0	1	1	1
1	1	1	0	0	0	1	1
1	1	1	1	0	0	0	0

b. ENABLE = B3 ⊕ B2 + B2 ⊕ B1 + B1 ⊕ B0

c. Gate 2 is bad—should be a high (logic 1) out.

1.45 A logic probe gives a quick check of a logic circuit's function and is usable for tracing logic states through a circuit. Its limitations include occasional incorrect state information for open inputs: It may not show the exact voltage level of a logic state or the frequency at which it changes.

1.47 Output stage may be open, neither showing a logic 1 or logic 0, but some voltage level in between.

1.49 AND gate 3 is giving the wrong state out.

1.51 ENABLE = A + B + C + D ⊕ ABCD

A	B	C	D	A + B + C + D	ABCD	ENABLE
0	0	0	0	0	0	0
0	0	0	1	1	0	1
0	0	1	0	1	0	1
0	0	1	1	1	0	1
0	1	0	0	1	0	1
0	1	0	1	1	0	1
0	1	1	0	1	0	1
0	1	1	1	1	0	1
1	0	0	0	1	0	1
1	0	0	1	1	0	1
1	0	1	0	1	0	1
1	0	1	1	1	0	1
1	1	0	0	1	0	1
1	1	0	1	1	0	1
1	1	1	0	1	0	1
1	1	1	1	1	1	0

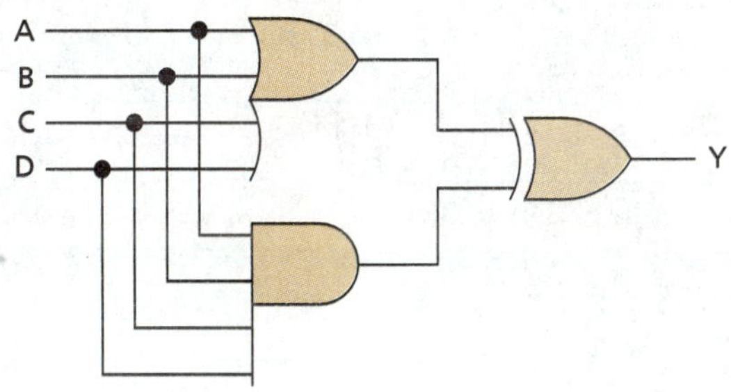

Figure 1.K17 Problem 1.51

1.53 $\text{PLAYS} = \text{SONG} \cdot \text{CENTS} \cdot \overline{\text{DOLLAR}} + \text{SONG} \cdot \overline{\text{CENTS}} \cdot \text{DOLLAR} = \text{SONG} \cdot (\text{CENTS} \oplus \text{DOLLAR})$

$\text{CHANGE} = \text{SONG} \cdot \text{DOLLAR} \cdot \overline{\text{CENTS}}$

$\text{ERROR} = \text{CENTS} \cdot \text{DOLLAR}$

SONG	CENTS	DOLLAR	PLAYS	CHANGE	ERROR
0	X	X	0	0	0
0	1	1	0	0	1
1	0	0	0	0	0
1	0	1	1	1	0
1	1	0	1	0	0
1	1	1	0	0	1

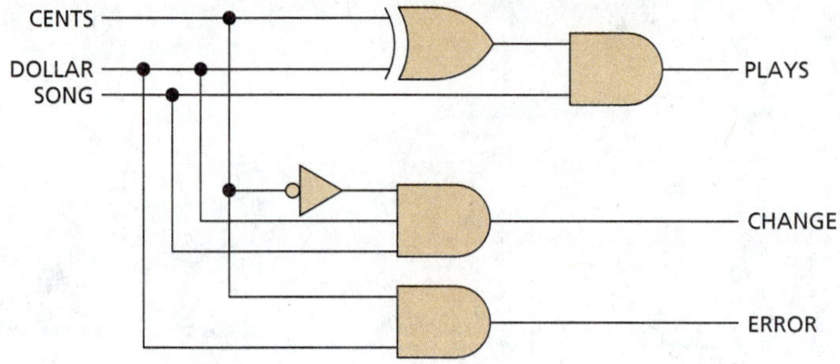

Figure 1.K19 Problem 1.53

Chapter 2

2.1

a. $\overline{A}AC + \overline{A}AD + \overline{A}BC + \overline{A}BD + ACD + ADD + BCD + BDD$

b. $AA\overline{B}CD + A\overline{B}B + ACCD + BC$

c. $\overline{A}A + \overline{A}B + \overline{A}C + A\overline{B} + B\overline{B} + \overline{B}C + A\overline{C} + B\overline{C} + C\overline{C}$

d. $AA + AC + \overline{A} \cdot \overline{B} \cdot \overline{C} + \overline{B} \cdot \overline{C} \cdot \overline{C} + AACD + ACCD$

e. $\overline{A}BB + BB\overline{B} + BB\overline{C} + \overline{A} \cdot B \cdot \overline{C} + B \cdot \overline{B} \cdot \overline{C} + B \cdot \overline{C} \cdot \overline{C}$

f. $AAA\overline{B}C\overline{C} + AAB\overline{B}C\overline{C} + AAB\overline{B}C\overline{C} + ABB\overline{B}C\overline{C} + AAB\overline{C} + ABB\overline{C} + ABB\overline{C} + BBB\overline{C}$

2.3

a. $A[C(B + \overline{B} + D) + \overline{B}D]$

b. $AB + D[\overline{C}(A + \overline{B}) + A]$ or $A[B + D(\overline{C} + 1)] + \overline{B} \cdot \overline{C} \cdot D$

c. $A[D(BC + 1) + \overline{D}(\overline{B}C + 1)] + BCD$

d. $E\{\overline{B}D + C[A(1 + B + \overline{D}) + B]\}$

e. $S\{T[R + U(1 + B)] + BR\}$

f. $I\{T[H(S + 1) + S] + S\}$

2.5

a. $AD + BD + AB\overline{C}$

b. $ACD + BC$

c. $\overline{A}B + \overline{A}C + A\overline{B} + A\overline{C}$

d. A

e. $\overline{A}B + B\overline{C}$

f. $B\overline{C}$

2.7

a. $AC + A\overline{B}D$

b. $AB + AD + \overline{B} \cdot \overline{C} \cdot D$

c. $A + BCD$

d. $E(\overline{B}D + AC + BC) = E\overline{B}D + EAC + EBC$

e. $S(RT + BR + TU) = SRT + SBR + STU$

f. $I(S + HT) = IS + IHT$

2.9

a. $Y = \overline{A} \cdot \overline{B} + B \cdot \overline{C}$

b. $Y = A \cdot C$

c. $Y = A \cdot \overline{B} + A \cdot C$

d. $T = Z + X \cdot \overline{W}$

e. $V = S + T + \overline{U}$

f. $Y = B + \overline{A} \cdot \overline{D} + C \cdot \overline{D}$

2.11

a. $\overline{A} \cdot \overline{C} + B \cdot C + \overline{B} \cdot \overline{C}$ or $\overline{B} \cdot \overline{C} + BC + \overline{A}B$

b. $\overline{N} \cdot P + N \cdot \overline{P} + N \cdot \overline{T}$ or $\overline{N}P + N\overline{P} + P\overline{T}$

c. $A \cdot B + \overline{A} \cdot \overline{B} \cdot \overline{C} + B \cdot \overline{C} \cdot D + B \cdot C \cdot \overline{D}$

Alternate answer: $\overline{A} \cdot \overline{C} \cdot D$ in place of $B \cdot \overline{C} \cdot D$

d. $\overline{A} \cdot B + B \cdot C + \overline{A} \cdot C \cdot D + A \cdot \overline{B} \cdot \overline{C} \cdot D + A \cdot C \cdot \overline{D}$

2.13

a. $A \cdot B + C$

b. $W \cdot \overline{X} + W \cdot \overline{Y} + \overline{W} \cdot X \cdot T$

c. $A \cdot B \cdot C \cdot D$

d. $\overline{R} \cdot S \cdot \overline{U} + R \cdot \overline{T} + S \cdot \overline{T}$

e. $A \cdot B \cdot C$

2.15

a.

	$\overline{B}$	B
$\overline{A}$		1
A	1	1

Figure 2.K1 Problem 2.15(a)

b.

	$\overline{C}$	C
$\overline{A}\,\overline{B}$		1
$\overline{A}\,B$	1	
$A\,B$		1
$A\,\overline{B}$	1	

Figure 2.K2 Problem 2.15(b)

c.

	$\overline{C}\,\overline{D}$	$\overline{C}D$	CD	$C\overline{D}$
$\overline{A}\,\overline{B}$		1		
$\overline{A}B$				1
AB	1			
$A\overline{B}$		1		

Figure 2.K3 Problem 2.15(c)

d.

	$\overline{C}$	C
$\overline{A}\,\overline{B}$		1
$\overline{A}\,B$		1
$A\,B$	1	1
$A\,\overline{B}$	1	1

Figure 2.K4 Problem 2.15(d)

e.

	$\overline{C}\,\overline{D}$	$\overline{C}D$	CD	$C\overline{D}$
$\overline{A}\,\overline{B}$	1	1	1	1
$\overline{A}B$	1	1	1	1
AB	1		1	1
$A\overline{B}$	1	1		

Figure 2.K5 Problem 2.15(e)

2.17

a. $A + B$

b. $B \cdot \overline{C} + A \cdot \overline{C} + \overline{A} \cdot \overline{B} \cdot C$

c. $\overline{B} \cdot \overline{C} \cdot D + A \cdot B \cdot \overline{C} \cdot \overline{D} + \overline{A} \cdot B \cdot C \cdot \overline{D}$

d. $C + A$

e. $\overline{A} + \overline{B} \cdot \overline{C} + B \cdot C + B \cdot \overline{D}$. Last term could be $\overline{C} \cdot \overline{D}$.

2.19

a. $\overline{A} \cdot B + \overline{B} \cdot \overline{C}$

b. $\overline{Y} + W \cdot X$

c. $\overline{A} \cdot B + B \cdot \overline{D} + \overline{A} \cdot \overline{C} \cdot D$

d. $\overline{N} \cdot \overline{V} + \overline{N} \cdot G + R \cdot \overline{V} + \overline{G} \cdot \overline{V}$

2.21

a.

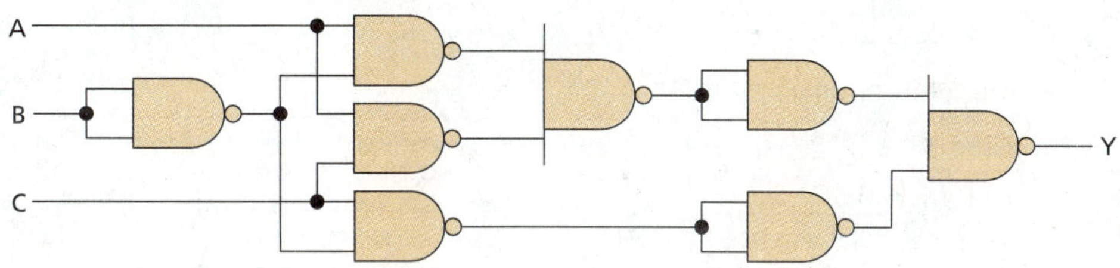

Figure 2.K11 Problem 2.21(a)

b.

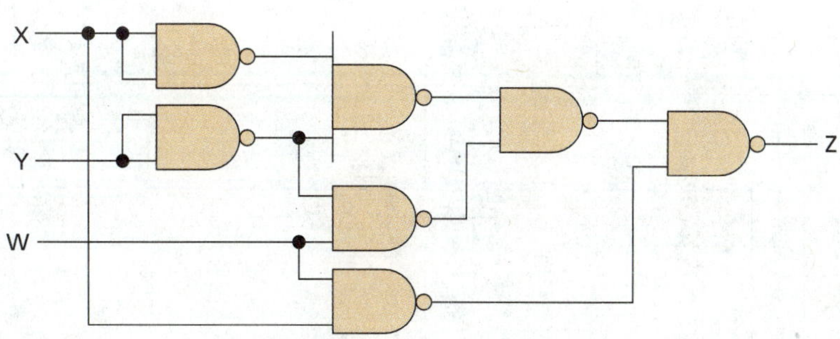

Figure 2.K12 Problem 2.21(b)

2.23

a.

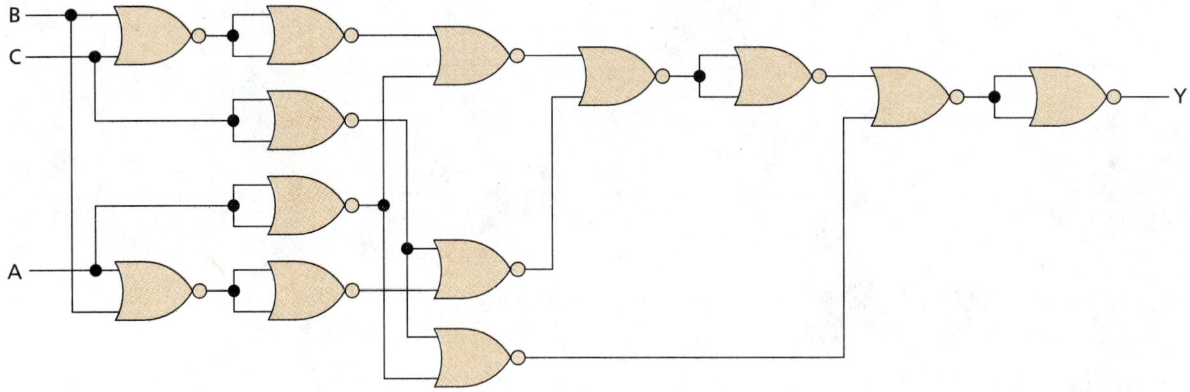

Figure 2.K15 Problem 2.23(a)

b.

AB \ C	$\overline{C}$	C
$\overline{A}\,\overline{B}$	1	
$\overline{A}\,B$		1
A B	1	
$A\,\overline{B}$		1

Figure 2.K16 Problem 2.23(b)

2.25

a. $A \cdot B \cdot C + B \cdot C \cdot D + A \cdot B$

b. $\overline{C} \cdot \overline{D} + \overline{A} \cdot C + \overline{B} \cdot C$

c. $\overline{A} + B + \overline{C} + \overline{D}$

2.27

a. $\overline{(\overline{A \cdot B} + \overline{B \cdot C})(B \cdot C + \overline{A} \cdot \overline{C} + A \cdot C)}$

b. $\overline{A} \cdot \overline{B} \cdot C + A \cdot \overline{C} + A \cdot B$

c.

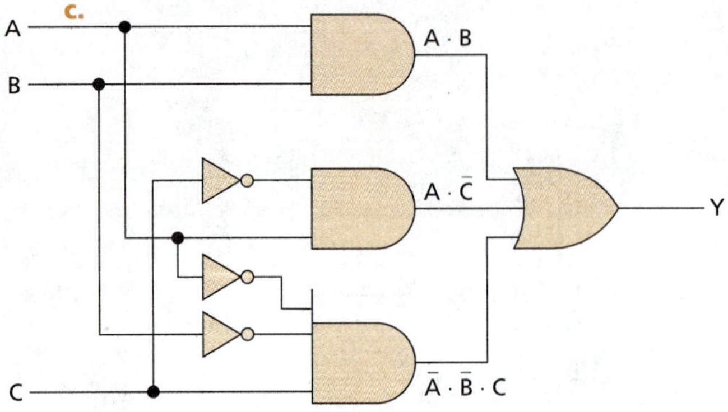

Figure 2.K19 Problem 2.27 (solution)

2.29

	$\overline{C}\,\overline{D}$	$\overline{C}D$	CD	$C\overline{D}$
$\overline{A}\,\overline{B}$	1	1		
$\overline{A}B$	X	1		1
AB	1			
$A\overline{B}$	X	X	1	1

Figure 2.K21 Problem 2.29 (solution)

$Y = \overline{A} \cdot \overline{C} + \overline{C} \cdot D + A \cdot \overline{B} + \overline{A} \cdot B \cdot \overline{D}$

2.31

AB \ C	$\overline{C}$	C
$\overline{A}\,\overline{B}$	1	
$\overline{A}\,B$		1
A B	1	
$A\,\overline{B}$		1

Figure 2.K23 Problem 2.31 (solution)

$Y = \overline{A \oplus B \oplus C}$

Chapter 3

3.1

a. $3 \times 1{,}000 = 3{,}000$ $4 \times 100 = 400$ $9 \times 10 = 90$ $7 \times 1 = 7$ $3{,}000 + 400 + 90 + 7 = 3{,}497$

b. $2 \times 100 = 200$ $2 \times 10 = 20$ $5 \times 1 = 5$ $3 \times 0.1 = 0.3$ $9 \times 0.01 = 0.09$ $6 \times 0.001 = 0.006$ $200 + 20 + 5 + 0.3 + 0.09 + 0.006 = 225.396$

c. $6 \times 1{,}000 = 6{,}000$ $3 \times 1 = 3$ $2 \times 0.1 = 0.2$ $5 \times 0.001 = 0.005$ $6{,}000 + 3 + 0.2 + 0.005 = 6{,}003.205$

d. $3 \times 10 = 30$ $4 \times 1 = 4$ $6 \times 0.001 = 0.006$ $7 \times 0.0001 = 0.0007$ $8 \times 0.00001 = 0.0008$ $30 + 4 + 0.006 + 0.0007 + 0.00008 = 34.00678$

e. $9 \times 10{,}000 = 90{,}000$ $8 \times 1{,}000 = 8{,}000$ $2 \times 100 = 200$ $7 \times 1 = 7$ $5 \times 0.1 = 0.5$ $1 \times 0.01 = 0.01$ $2 \times 0.0001 = 0.0002$ $90{,}000 + 8{,}000 + 200 + 7 + 0.5 + 0.01 + 0.0002 = 98{,}207.5102$

3.3

a. $1 \times 512 + 2 \times 64 + 3 \times 8 + 4 \times 1 = 512 + 128 + 24 + 4 = 668$

b. $8 \times 4{,}096 + 3 \times 256 + 5 \times 16 + 5 \times 1 = 32{,}768 + 768 + 80 + 5 = 33{,}621$

c. $16 + 4 + 2 + 1 = 23$

d. $6 \times 4{,}096 + 5 \times 256 + 12 \times 16 + 13 \times 1 = 24{,}576 + 1{,}280 + 192 + 13 = 26{,}061$

e. $128 + 32 + 8 + 4 + 2 = 174$

3.5

a. 10011 = 19* 10001 = 17

b. 11101101 = 237* 11100101 = 229

c. 100001100001 = 2,145 100011000001 = 2,241*

* = larger values

3.7

a. 5.921875

b. 1,234.567

c. 11.011

d. 59.75

e. 9.5625

3.9

a. 10100010010

b. 1001110100.01011001

c. 111011.111

d. 111101001000010

e. 10000000000.11

3.11

a. 10000000

b. 1110110

c. 10011000

d. 11100

e. 101000.1011

3.13

a. $93 + 35 = 128$

b. $59 + 59 = 118$

c. $39 + 113 = 152$

d. $13.8125 + 14.1875 = 28$

e. $35.5 + 5.1875 = 40.6875$

3.15

a. 111010

b. 0

c. −1001010

d. −.0110

e. 11110.0101

3.17

a. $93 - 35 = 58$

b. $59 - 59 = 0$

c. $39 - 113 = -74$

d. $13.8125 - 14.1875 = -0.375$

e. $35.5 - 5.1875 = 30.3125$

3.19

a. 274

b. 8,968

c. 46.22

d. 40,877

e. 187,431

3.21

a. 0011100

b. 011101

c. 01100.101

d. 100000

e. 010010.011

3.23

a.
$$\begin{array}{r} 1011101 \\ +1011101 \\ \hline 10111010 \end{array}$$

b.
$$\begin{array}{r} 111011 \\ +000101 \\ \hline 1000000 \end{array}$$

c.
$$\begin{array}{r} 0100111 \\ +0001111 \\ \hline 0110110 \\ -1001010 \end{array}$$

d.
$$\begin{array}{r} 1101.1101 \\ +0001.1010 \\ \hline 1111.1010 \\ -0000.0110 \end{array}$$

e.
$$\begin{array}{r} 100011.1000 \\ +111010.1101 \\ \hline 1011110.0101 \end{array}$$

3.25

a. +27

b. −45

c. −27

d. +243

e. −127

3.27 *Unsigned:* 0 to 65,535
Positive: +0 to +32,767
Negative: −1 to −32,768

3.29

a. 1 1011 1110 1110B = 7,150
1BEEH = 4,096 + 11 × 256 + 14 × 16 + 14 = 7,150

b. 1 0011 1100 1101B = 5,069
13CDH = 4,096 + 3 × 256 + 12 × 16 + 13 = 5,069

c. 11 1111 1011 0001B = 16,305
3FB1H = 3 × 4,096 + 15 × 256 + 11 × 16 + 1 = 16,305

d. 1 1000 0111 1110 1101B = 100,333
187EDH = 65,536 + 8 × 4,096 + 7 × 256 + 14 × 16 + 13 = 100,333

e. 10 1011 0011 1000B = 11,064
2B38H = 2 × 4,096 + 11 × 256 + 3 × 16 + 8 = 11,064

f. 1110 1111 0111 0011B = 61,299
EF73H = 14 × 4,096 + 15 × 256 + 7 × 16 + 3 = 61,299

3.31

a. 3C0EH = 15,374
+A37EH = +41,854
DF8CH = 57,228

b. D20C7H = 860,359
+1FBE3H = +130,019
F1CAAH = 990,378

c. 5B0H = 1,456
+AAAH = +2,730
105AH = 4,186

3.33

a. D1F3AH
D1F3BH

b. 5CDFFFH
5CE000H

c. 3EEA44H
3EEA45H

3.35

a. AC9H AC9H
−25BH +DA5H
86EH

b. C005H C005H
−2E71H +D18FH
9194H

c. 350H 350H
−25AH +DA6H
F6H

d. C50H C50H
−86BH +795H
3E5H

e. 9000H 9000H
−7A28H +85D8H
15D8H

f. 13F2H 13F2H
−C00H +F400H
7F2H

3.37

a. 1001 0011 0111 0110BCD

b. 0101 0000 0000 0111 1000 0101BCD

c. 0111 0000 0110 0110 0011BCD

d. 1000 0100 0000 1001 0011BCD

e. 0111 1001 0010 0001BCD

f. 0111 1000 0000 0010 1001BCD

3.39 Sum greater than 9
carry over

3.41

a. 33 37 20 2B 20 35 32 20 3D 20 38 39

b. 1B 20 54 48 49 53 20 1A

3.43 The SUM includes three inputs, A-word, B-word, and carry in for the full adder, while the half adder lacks the carry in input. The carry out expression for the half adder is merely the ANDing of inputs A and B, while for a full adder the expression is the sum of products of all three inputs. The half adder only does half the required addition, since it does not include adding carryover from preceding stages.

3.45 A high state on the carry out of a two's complementer adder/subtracter indicates an overflow from addition (SUB = 0) or the need to borrow when subtracting (SUB = 1) through the use of an XOR gate. The inputs to the gate are SUB and C_{OUT} of the four-bit adder. Carry from addition is a result of C_{OUT} = 1 and SUB = 0. Need for a borrow occurs when SUB = 1 and C_{OUT} = 0. For all other combinations of C_{OUT} and SUB, there is no indication of carry or borrow. A truth table of these conditions shows that the XOR function is used to produce a CARRY FLAG, which indicates both the carry and borrow conditions:

C_{OUT}	SUB	CARRY FLAG
0	0	0
0	1	1 = borrow
1	0	1 = carry
1	1	0

3.47 The carry out from the least significant bit to the next place adder is always low.

3.49 Additional two's complement of the output when subtracting and the initial results are negative (borrow required).

3.51 Output of AND gate 2 should be a 1 instead of 0.

3.53 Use a four-input NOR gate. If any bit is a 1, the zero indicator is low. When all four bits are 0, the NOR gate is high, indicating a zero condition.

3.55

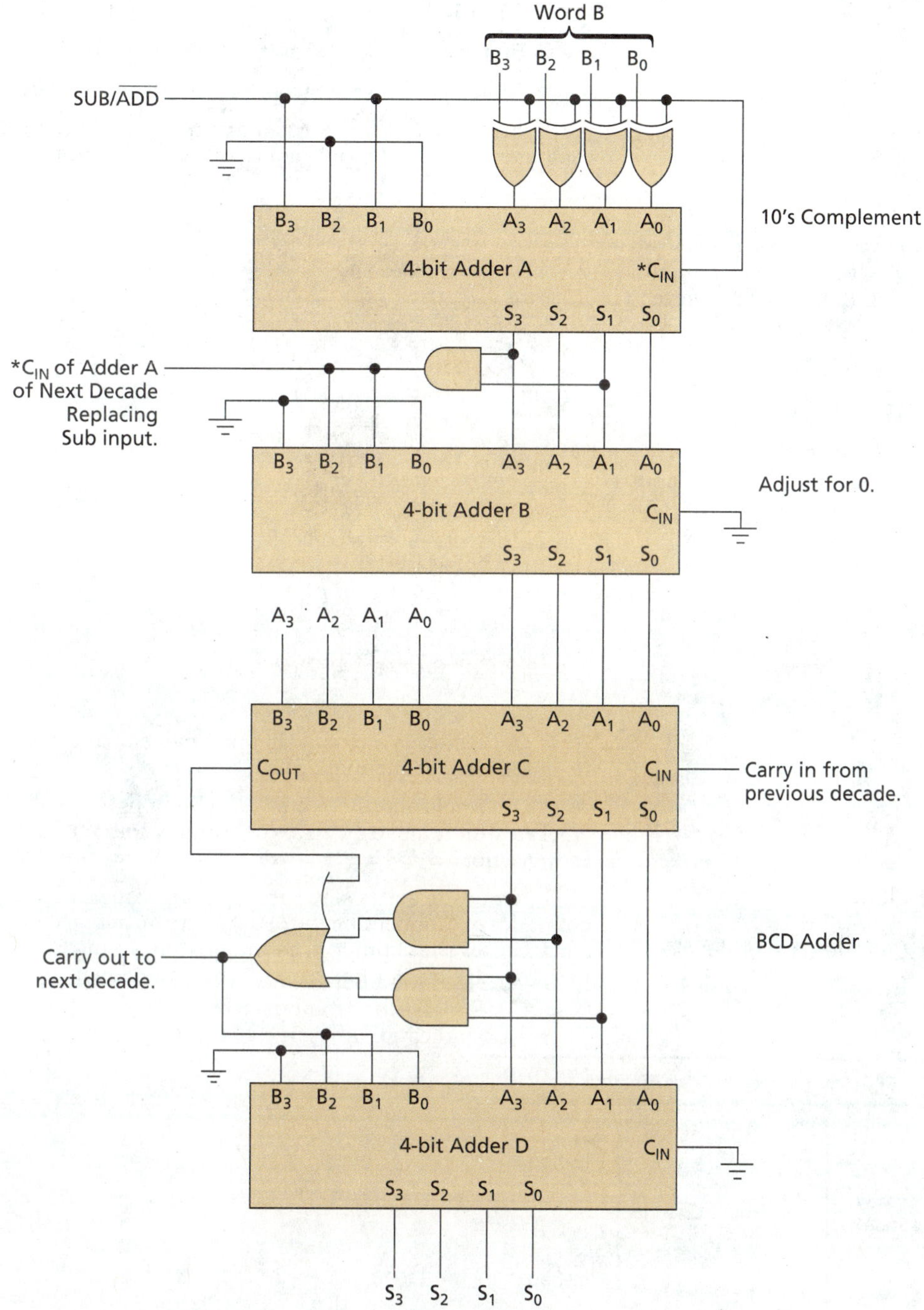

Figure 3.K1 Ten's Complement BCD Adder/Subtracter, Problem 3.55 (solution)

Chapter 4

4.1 Low

4.3

S	R	Q	$\overline{Q}$
0	0	No change	
0	1	0	1
1	0	1	0
1	1	Undefined	

4.5 Changes in the output of an edge-triggered flip-flop only occur at the leading (positive edge trigger) or trailing (negative edge trigger) edge of the clock input. Changes for a level-triggered flip-flop occur whenever the level of the clock input is in the active state.

4.7 Applying zeros to both $\overline{R}$ and $\overline{S}$ input forces both Q and $\overline{Q}$ to a high state. This is an undefined condition for a flip-flop by definition that the outputs are always in opposite state.

4.9 Internally, the D input is connected to the S input and an inverter driving the R input of a R–S latch. This causes the D flip-flop to be set when D = 1 (S = 1, R = 0) and reset when D = 0 (S = 0, R = 1).

4.11

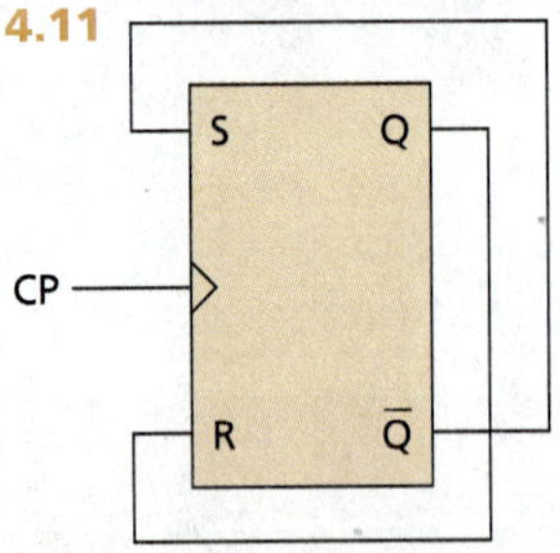

Figure 4.K4 Problem 4.11 (solution)

4.13 The frequency of the Q output is ½ of the clock input frequency.

4.15

$\overline{S}$	$\overline{R}$	J	K	CP	Q	Function	
1	0	X	X	X	0	Reset	R–S Flip-flop
0	1	X	X	X	1	Set	
1	1	X	X	Steady State	No	Change	Clocked R–S Flip-flop
1	1	0	1	↓	0	Reset	
1	1	1	0	↓	1	Set	
1	1	0	0	↓	No	Change	Toggle Flip-flop
1	1	1	1	↓	Toggle		

X = don't care state

Figure 4.K7 Problem 4.15 (solution)

4.17 $R_A = 578\ \Omega$ $R_B = 5.78\ K\Omega$

4.19 $10.91\ K\Omega$

4.21 States of the asynchronous inputs and outputs.

4.23 Indicates transition from present to next state.

4.25 State tables show present and next states on each line. Truth tables only contain next state results.

4.27

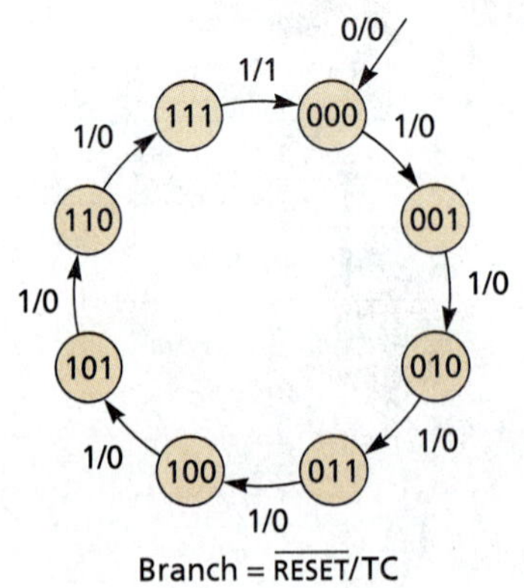

Figure 4.K10 Problem 4.27 (solution)

4.29 A counter is a circuit that is capable of altering its output in a sequential order.

4.31 In synchronous counters, all flip-flops are clocked simultaneously. In asynchronous counters, they are not.

4.33 Opposite active states to count up, same states to count down. Connecting the Q output to an active low clock causes the counter to count up. Connecting Q to a positive-edge clock input causes it to count down.

4.35 256

4.37

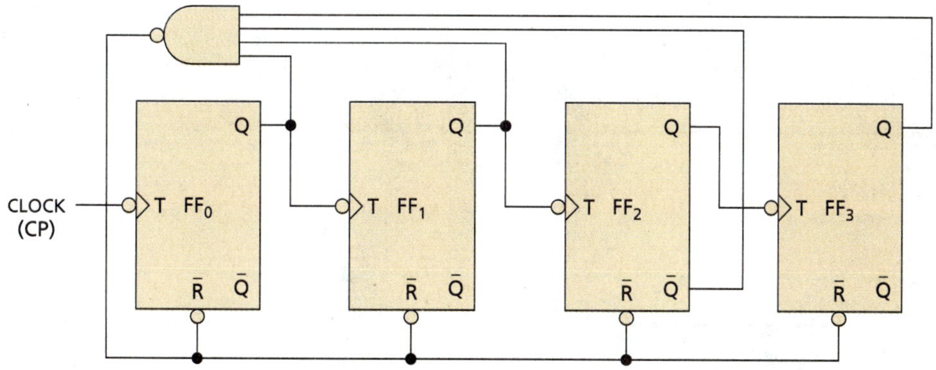

Figure 4.K13 Problem 4.37 (solution)

4.39

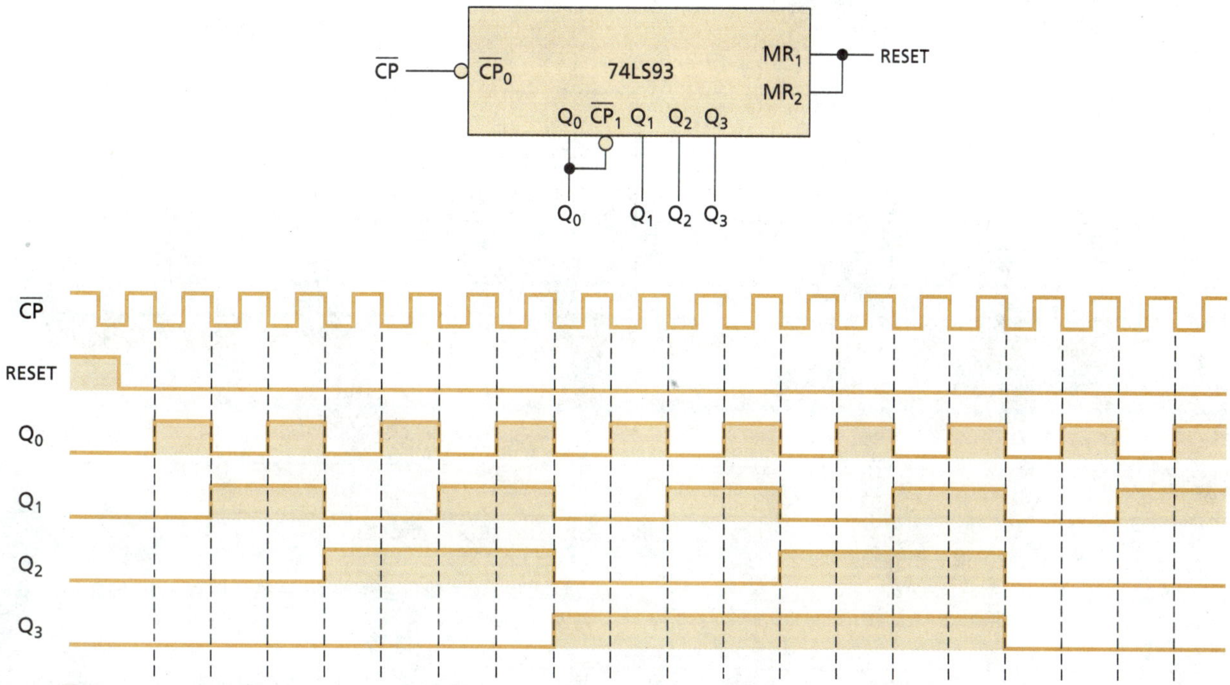

Figure 4.K15 Problem 4.39 (solution)

4.41

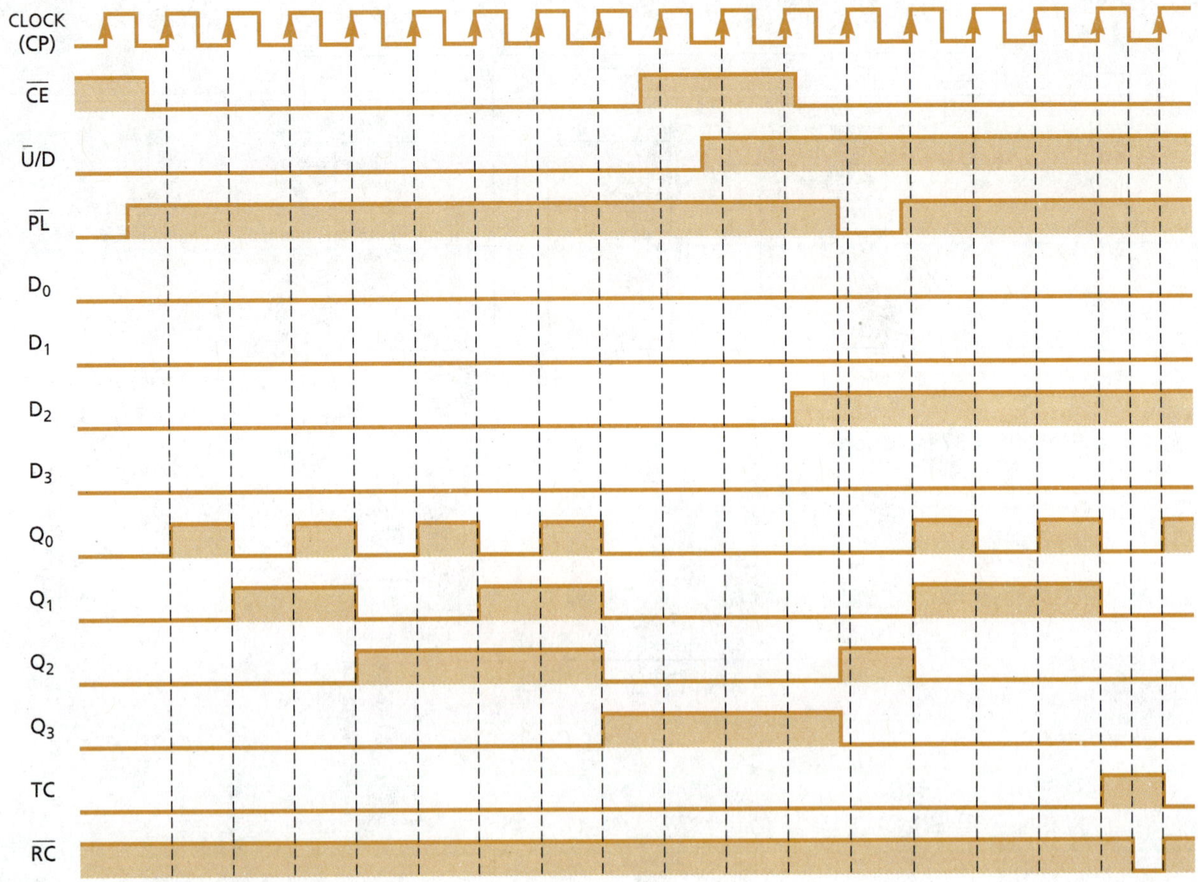

Figure 4.K17 Problem 4.41 (solution)

4.43

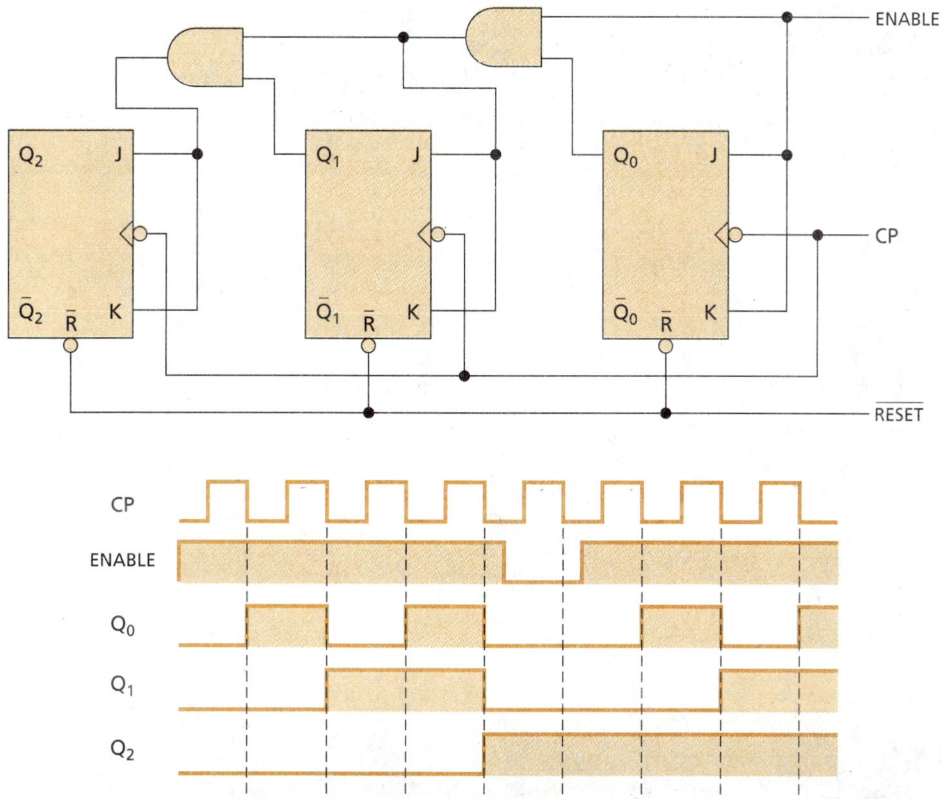

Figure 4.K19 Problem 4.43 (solution)

4.45 An eight-stage ring counter has its D inputs connected to the Q outputs of the previous stage. The first flip-flop is set and the remaining seven reset when the circuit is initialized. An eight-bit Johnson counter is connected similarly to a ring counter except that D_0 is connected to $\overline{Q_7}$ instead of Q_7 and the first flip-flop is reset along with the others when the circuit is initialized. The ring counter requires no output decoding, whereas the mod number for a Johnson counter is twice that of the ring counter.

4.47 Flip-flop 4, because Q_4 is always low.

4.49

Q_n								Q_{n+1}			
Q_3	Q_2	Q_1	Q_0	JK_3	JK_2	JK_1	JK_0	Q_3	Q_2	Q_1	Q_0
0	0	0	0	0	0	0	1	0	0	0	1
0	0	0	1	0	0	1	1	0	0	1	0
0	0	1	0	0	0	0	1	0	0	1	1
0	0	1	1	0	1	1	1	0	1	0	0
0	1	0	0	0	0	0	1	0	1	0	1
0	1	0	1	0	0	1	1	0	1	1	0
0	1	1	0	0	0	0	1	0	1	1	1
0	1	1	1	1	1	1	1	1	0	0	0
1	0	0	0	0	0	0	1	1	0	0	1
1	0	0	1	0	0	1	1	1	0	1	0
1	0	1	0	0	0	0	1	1	0	1	1
1	0	1	1	0	1	1	1	1	1	0	0
1	1	0	0	1	1	0	0	0	0	0	0

(a) State Table Solution

$\bar{J}_0 = \bar{K}_0 = Q_3 \cdot Q_2$ or $J_0 = K_0 = \bar{Q}_3 + \bar{Q}_2$

$J_1 = K_1 = Q_0$

$J_2 = K_2 = Q_1 \cdot Q_0 + Q_3 \cdot Q_2$

$J_3 = K_3 = Q_2 \cdot Q_1 \cdot Q_0 + Q_3 \cdot Q_2$

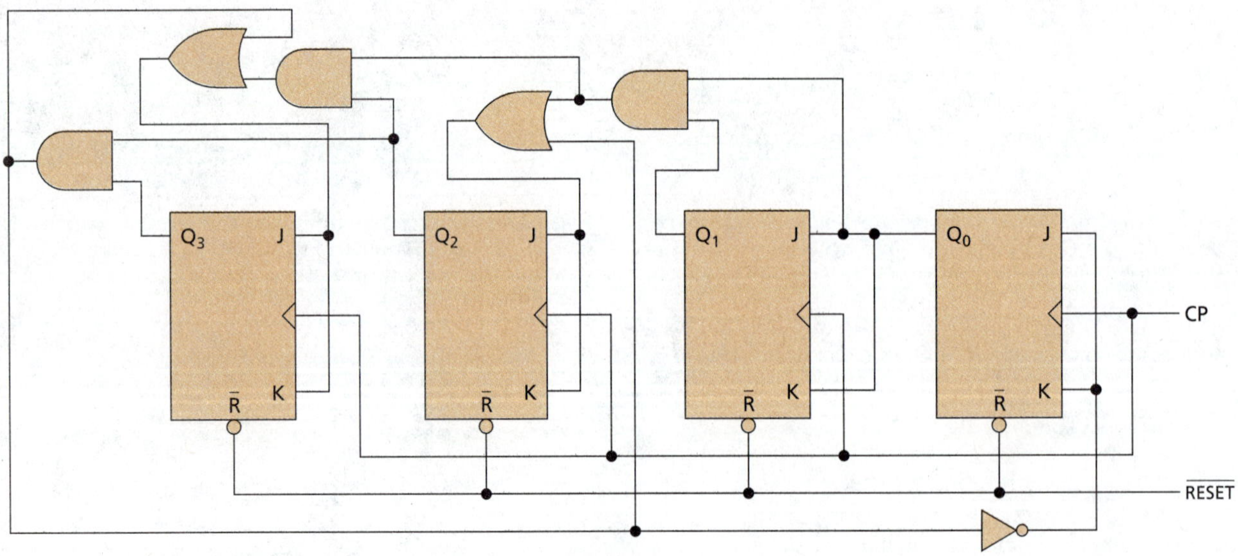

(b) Counter

Figure **4.K22** Problem 4.49 (solution)

4.51

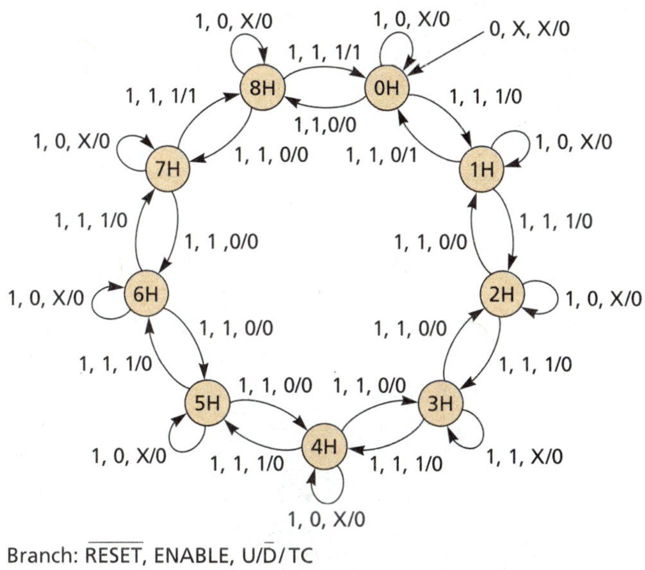

Figure 4.K24 Problem 4.51 (solution)

4.53

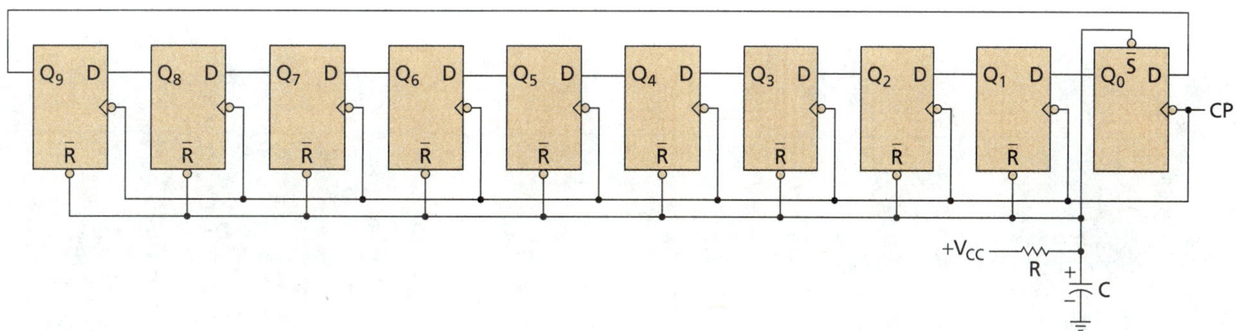

Figure 4.K26 Problem 4.53 (solution)

4.55

State Table

Q_n				Q_{n+1}				Data Inputs			
Q_3	Q_2	Q_1	Q_0	Q_3	Q_2	Q_1	Q_0	D_3	D_2	D_1	D_0
0	0	0	0	0	0	0	1	0	0	0	1
0	0	0	1	0	0	1	0	0	0	1	0
0	0	1	0	0	0	1	1	0	0	1	1
0	0	1	1	0	1	0	0	0	1	0	0
0	1	0	0	0	1	0	1	0	1	0	1
0	1	0	1	0	1	1	0	0	1	1	0
0	1	1	0	0	1	1	1	0	1	1	1
0	1	1	1	1	0	0	0	1	0	0	0
1	0	0	0	1	0	0	1	1	0	0	1
1	0	0	1	0	0	0	0	0	0	0	0
A	B	C	D								

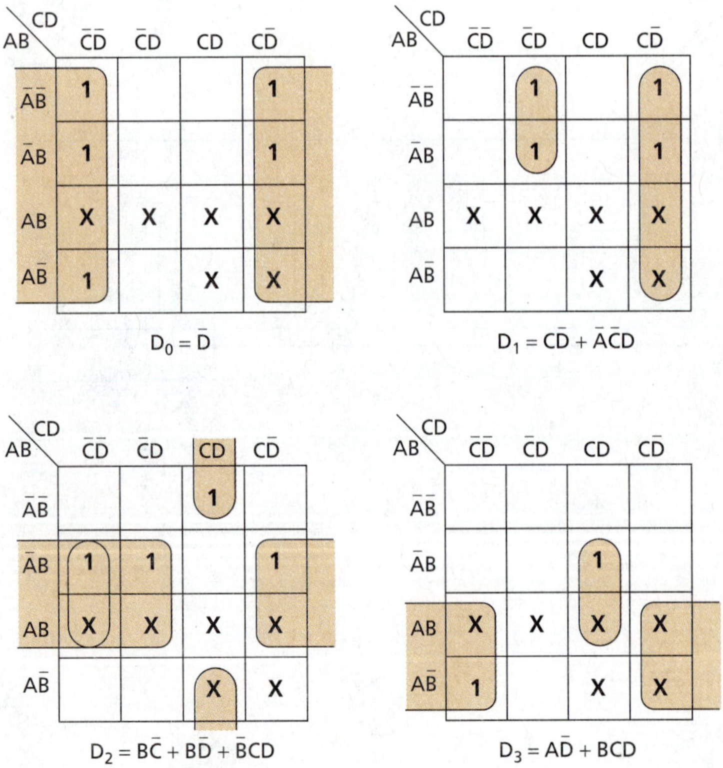

Figure 4.K28 Problem 4.55 (solution)

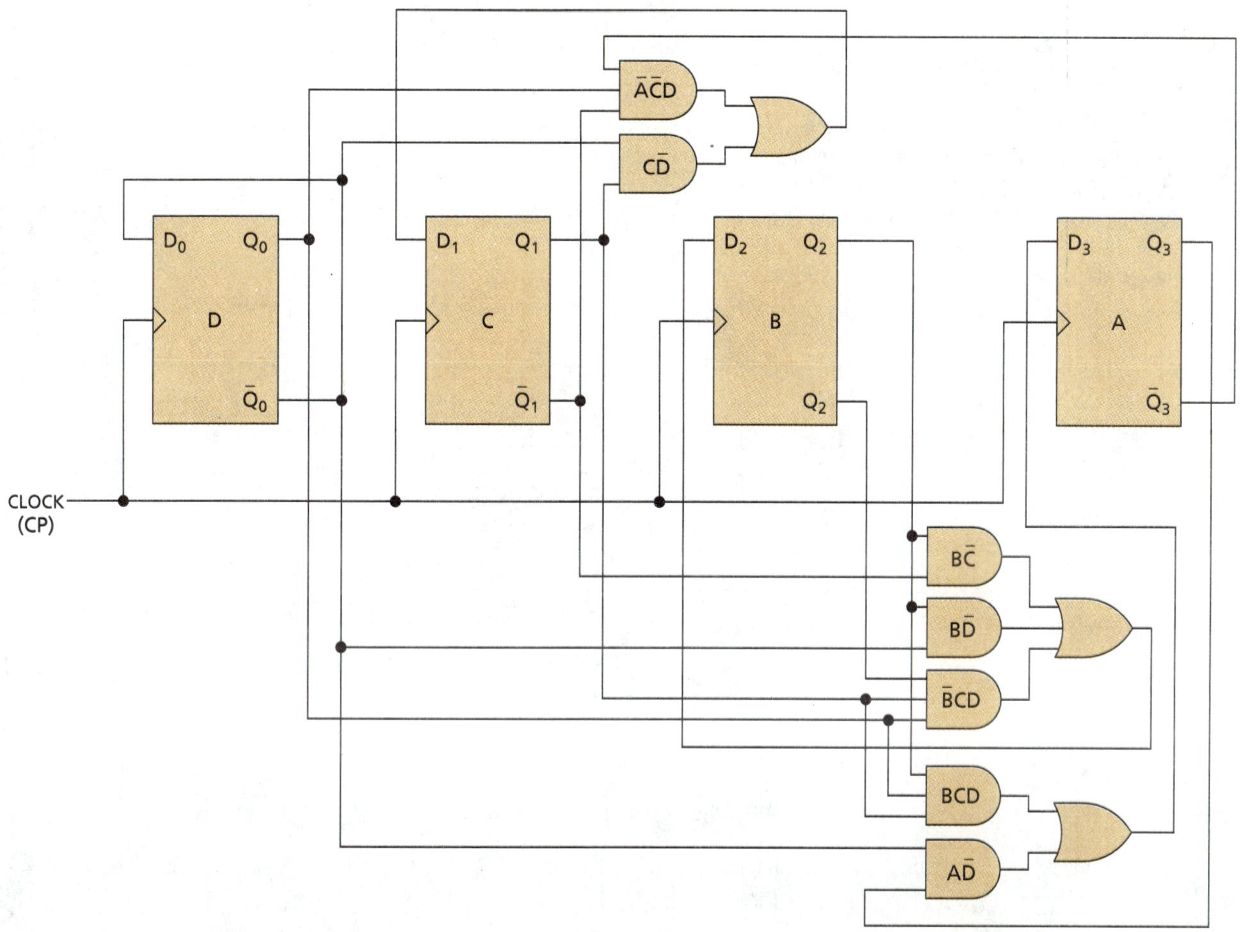

Figure **4.K28** Problem 4.55 (solution)

Chapter 5

5.1

a. Register—temporary storage device

b. Parallel data—all bits of a data word transferred at one time on multiple data paths

c. Serial data—data words move one bit at a time on a single data path

5.3

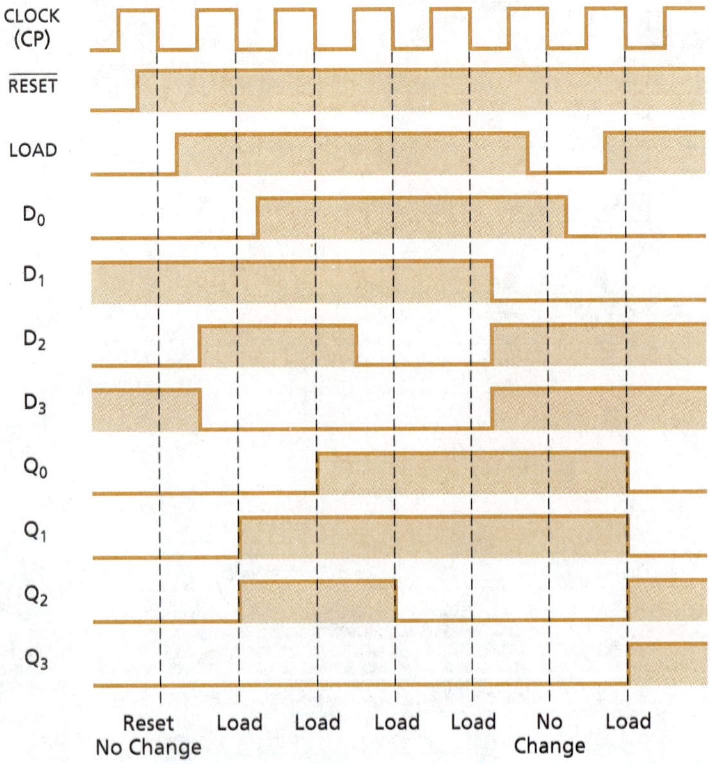

Figure 5.K1 Problem 5.3 (solution)

5.5 Two output enables.

5.7 Decoder and multiplexer (AND/OR).

5.9 Shift left moves data from the least significant bit position toward the most significant bit. Shift right moves data in the opposite direction.

5.11 Receive parallel data from a computer and transfer them serially to another computer that is quite a distance away. Transfer data from computer memory (RAM) to a floppy or hard disk.

5.13

a. Volatile—data are lost when power is removed

b. Access time—time between addressing memory and receiving a data word out

c. Memory cell—circuit or element that holds one binary bit of data

d. Address—binary code that defines a memory location

e. Chip enable (CE) or chip select (CS)—signal input that allows access to a memory chip

5.15 Data are lost when the memory is read. Data must be rewritten (refreshed) after they are read.

5.17 No, bipolar memories are not destructive read memories. When they are addressed during a read cycle, their individual states are sensed through the output lines. However, like any flip-flop, they retain their current state until they are written into.

5.19 Apply A_0 to A_4 to the address inputs and assert $\overline{CS}$ and $\overline{RAS}$ inputs. MUX signal switches address inputs from A_0 to A_4 to lines A_5 to A_9 and $\overline{CAS}$ is asserted. A shift register is used to generate RAS, MUX, and CAS signals in the correct sequence.

5.21 Dynamic memories are refreshed because the charge on C_{gs} leaks off through the high impedance of an IGFET's gate to source.

5.23 MOS memories have high packing density because they have very little power consumption, resulting in low heat dissipation, small cell size, and a high packing density.

5.25 Permanent, nonvolatile memory.

5.27

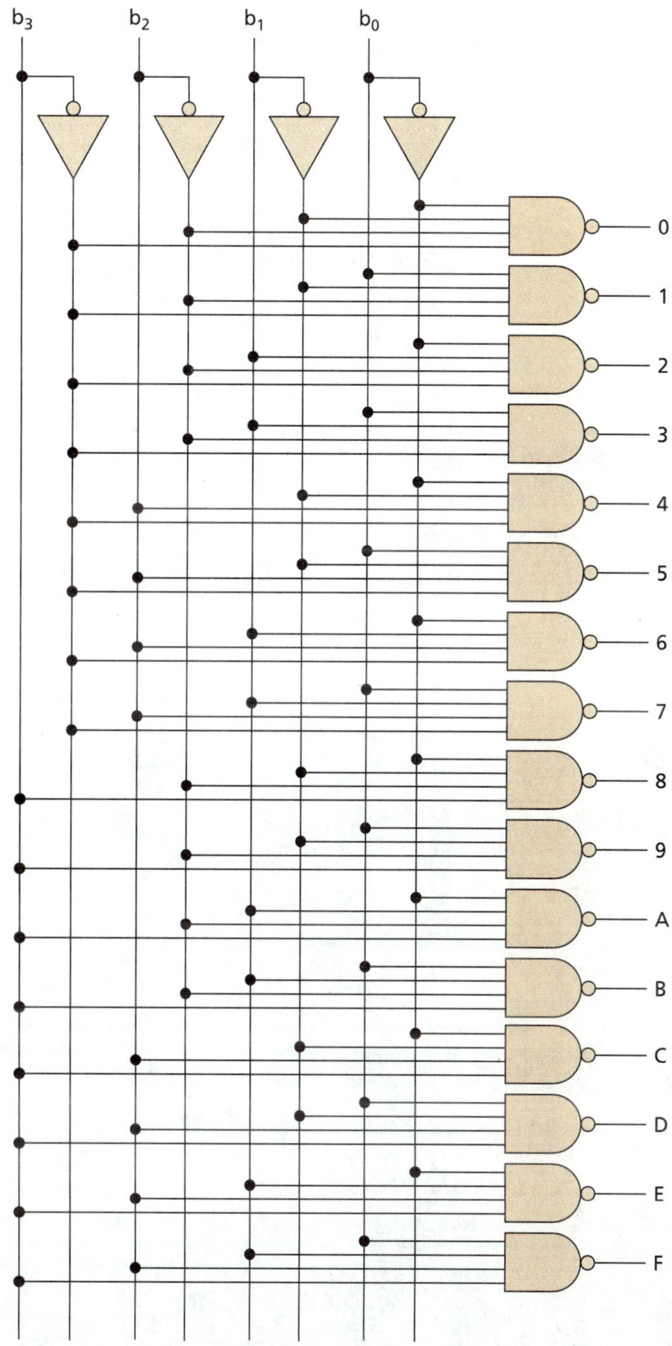

Figure 5.K5 Problem 5.27 (solution)

5.29 Output $\overline{6}$ is asserted low.

5.31 A multiplexer channels one of many inputs to a single output. A demultiplexer channels one input to one of many outputs.

5.33

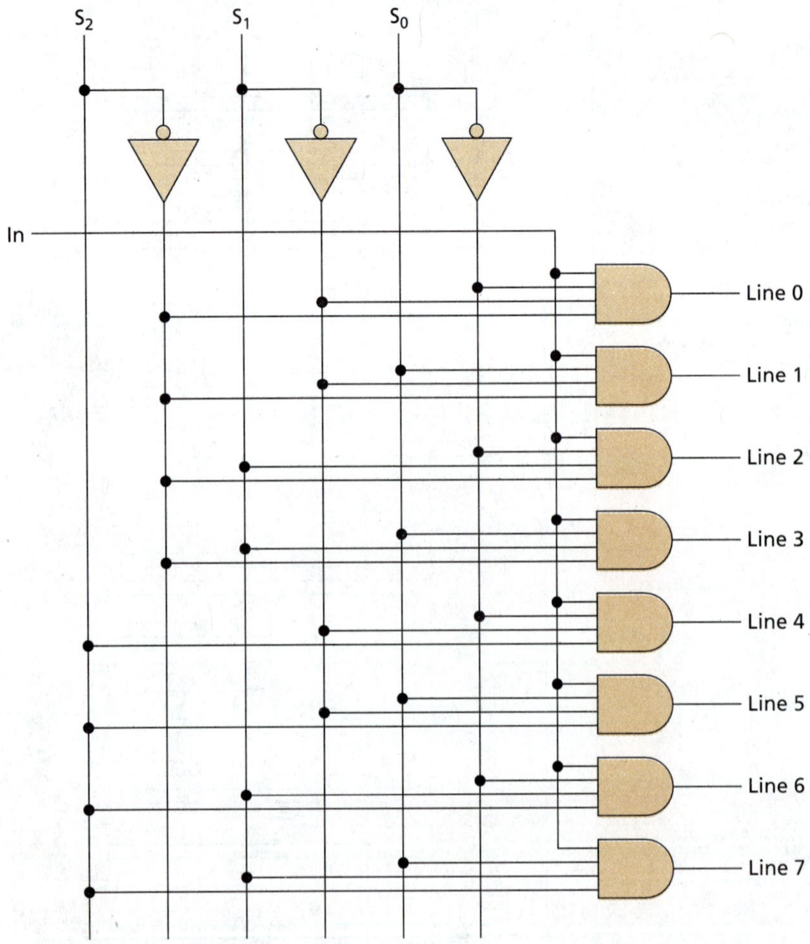

Figure 5.K9 Problem 5.33 (solution)

Chapter 6

6.1 AND/OR array or multiplexer

6.3 PAL—fixed OR array
PROM—fixed AND array
PLA—programmable AND and OR arrays
GAL—generic array logic

6.5

a.

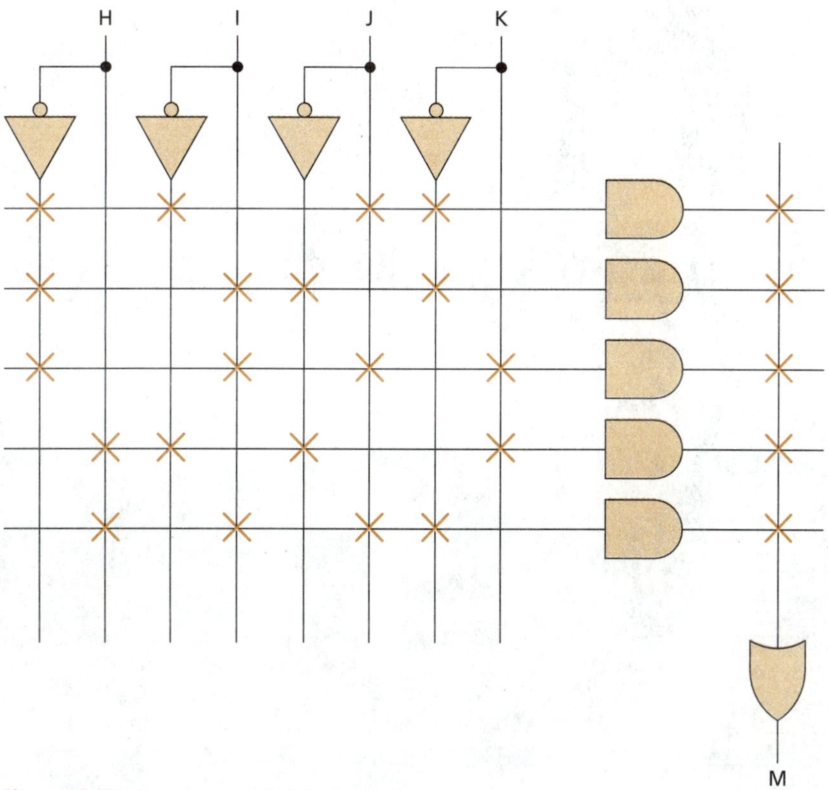

Figure 6.K1 Problem 6.5(a) (solution)

b.

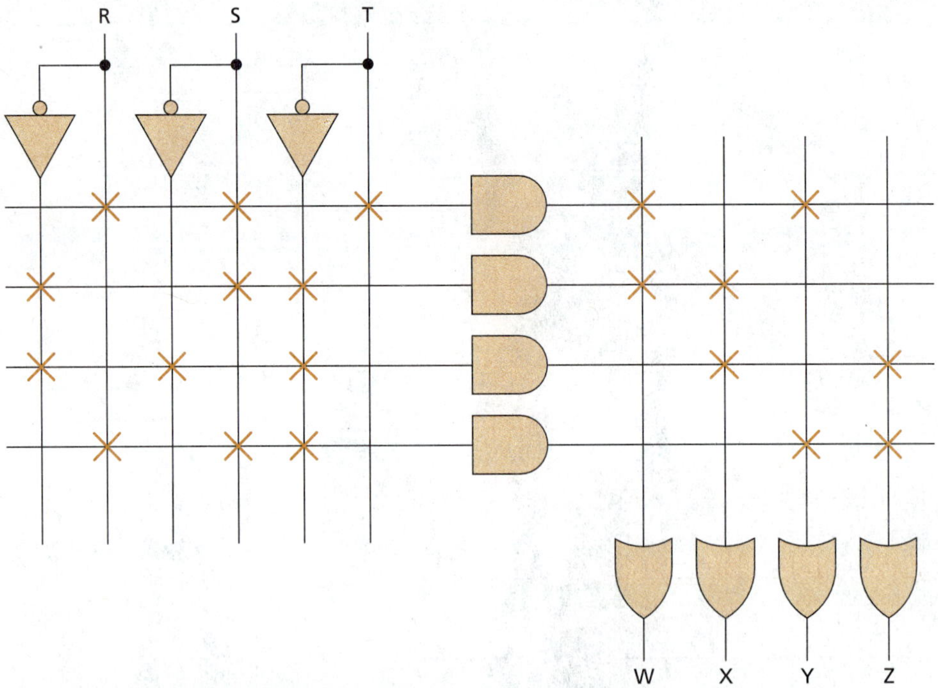

Figure 6.K2 Problem 6.5(b) (solution)

c.

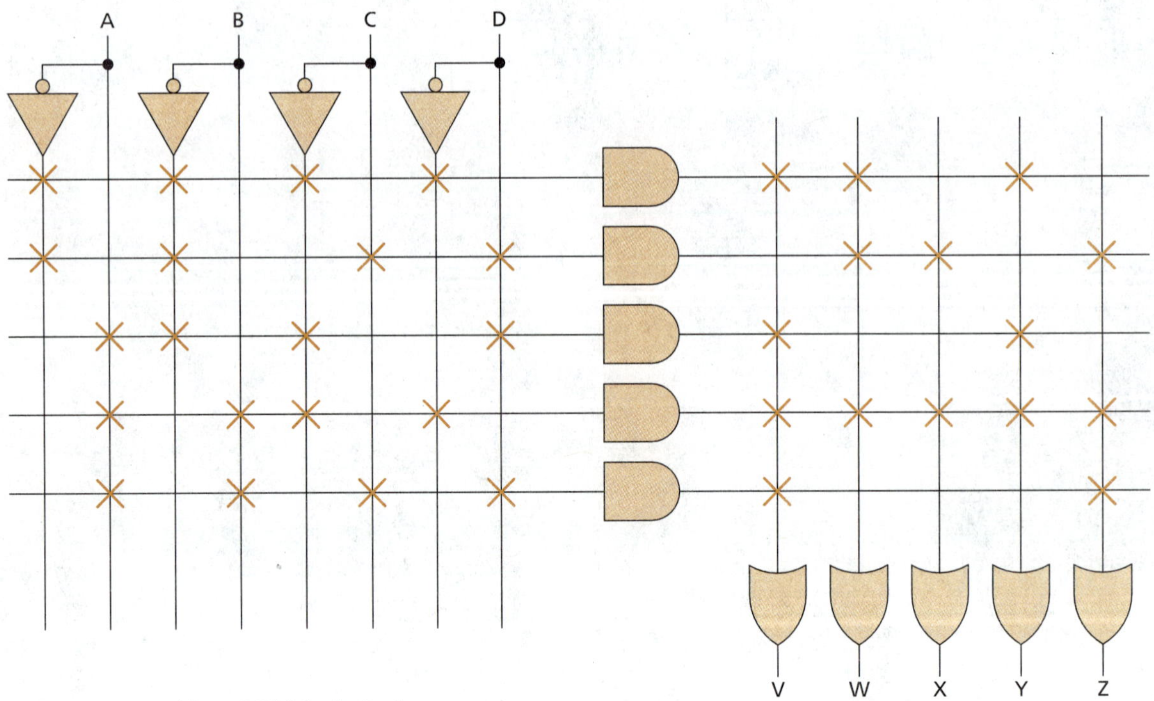

Figure 6.K3 Problem 6.5(c) (solution)

6.7

a.

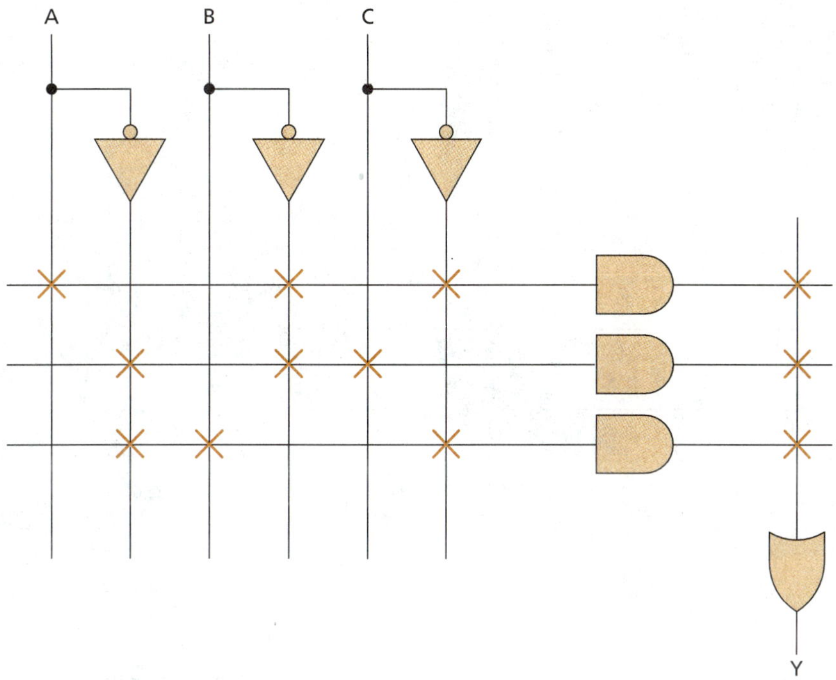

Figure 6.K7 Problem 6.7(a)

b.

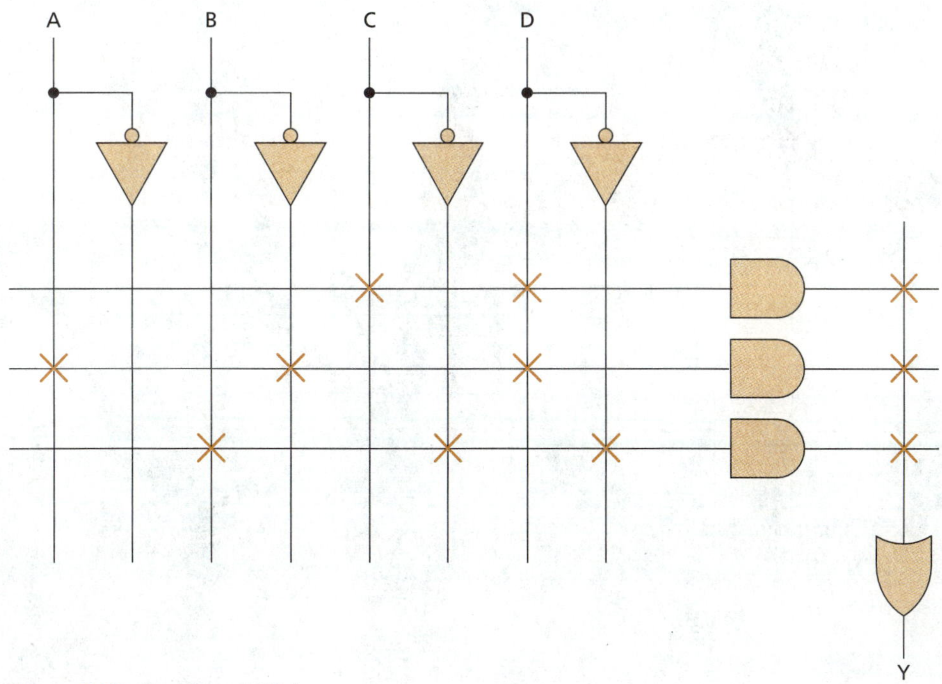

Figure 6.K8 Problem 6.7(b)

c.

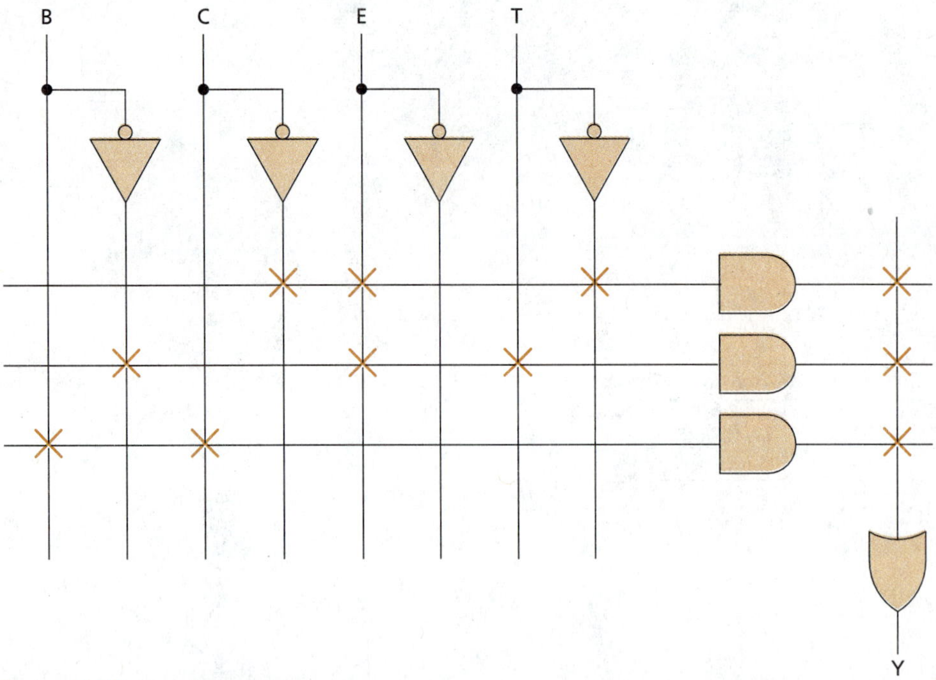

Figure 6.K9 Problem 6.7(c)

d.

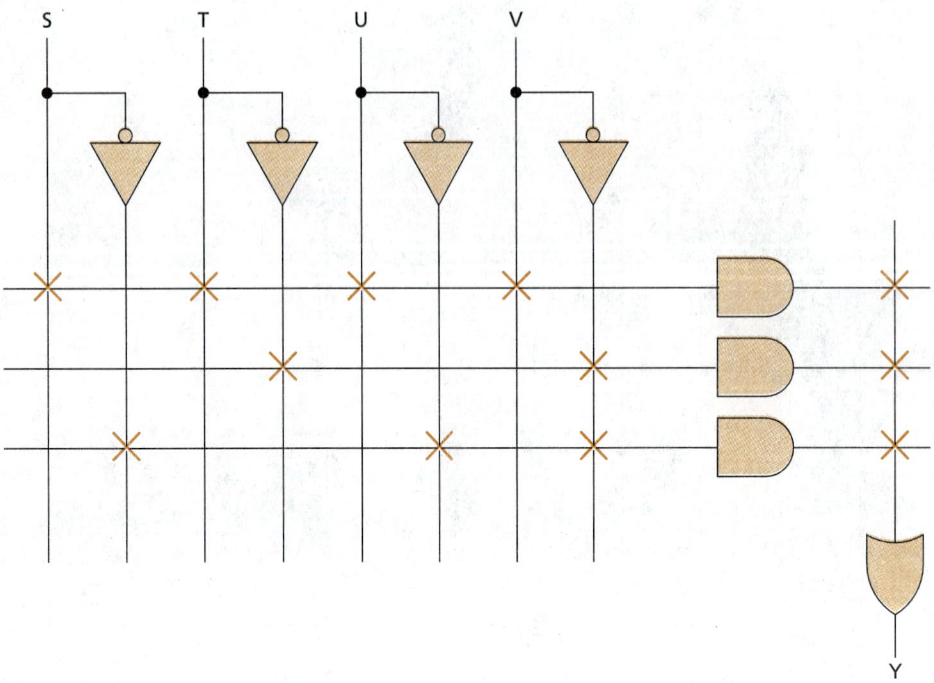

Figure 6.K10 Problem 6.7(d)

6.9

a. 24 inputs, 10 active high outputs
b. 8 inputs, 12 variable outputs
c. 16 inputs, 8 registered outputs
d. 22 inputs, 10 active low outputs
e. 24 inputs, 8 variable outputs

6.11 .PLD—ASCII text file with PLD design
.JED—JEDEC fuse plot file
.DOC—documentation text file
.LST—listing file which will show syntax errors

6.13

a.
```
FIELD INPUT = [A1, A2, A3];
FIELD OUTPUT = [B5, B3, B1, B0];
TABLE INPUT => OUTPUT {
   'b'000 => 0; 'b'001 => 5; 'b'010 => A;
   'b'011 => C; 'b'100 => 3; 'b'101 => 9;
   'b'110 => 6; 'b'111 => F; }
```

b.
```
FIELD INPUT = [DO, IT, TO, DAY];
FIELD OUTPUT = [AND, BE, DONE];
TABLE INPUT => OUTPUT {
   0 => 'b'010; 1 => 'b'011; 2 => 'b'001;
   3 => 'b'101; 4 => 'b'100; 5 => 'b'100;
   6 => 'b'011; 7 => 'b'010; 8 => 'b'001;
   9 => 'b'011; A => 'b'001; B => 'b'010;
   C => 'b'111; D => 'b'111; E => 'b'101;
   F => 'b'100; }
```

6.15

```
            Listing of File: D:\DIGTXT\PLD_FILE\CGO_STOP.PLD Page: 1

Name          GO_STOP.PLD;
Partno        GAL16V8;
Date          08/14/94;
Revision      XX;
Designer      MILLER;
Company       XXXXX;
Assembly      XXXXX;
Location      PHOENIX;

/*************************************************************************/
/*                                                                       */
/*  Figure 6.K15 Problem 6.15 (solution)                                 */
/*                                                                       */
/*************************************************************************/
/*  Allowable Target Device Types: GAL16V8                               */
/*************************************************************************/

/**  Inputs  **/

Pin 2 = GO;
Pin 3 = STOP;
Pin 4 = OOPS;
Pin 5 = HELP;

/**  Outputs  **/

Pin 18 = RUN;
Pin 17 = DONE;

/**  Logic Equations  **/

RUN = GO # !( STOP & OOPS ) # HELP;
DONE = GO & !HELP # ( !STOP $ !OOPS );
```

Figure 6.K15 Problem 6.15 (solution)

6.17

a. Valid

b. Valid

c. Invalid—has a space character

d. Valid

e. Invalid—uses a reserve word

f. Invalid—does not have at least one alpha character

g. Valid

h. Invalid—contains illegal ? character

6.19 Select the PLD device.

6.21

```
                Listing of File: D:\DIGTXT\PLD_FILE\CDECODE.PLD Page: 1

Name            DECODE.PLD;
Partno          16V8;
Date            06/06/95;
Revision        XX;
Designer        Miller;
Company         XXXXX;
Assembly        XXXXX;
Location        Phoenix;

/*******************************************************************************/
/*                                                                             */
/*  Figure 6.K17 Problem 6.21 (solution) Template 1                            */
/*                  Outputs 0 - 7. For outputs 8 - 15,                         */
/*                  repeat template using 8 - F => OUT8 - OUT15                */
/*                                                                             */
/*******************************************************************************/
/*  Allowable Target Device Types: GAL16V8                                     */
/*******************************************************************************/

/**  Inputs  **/

Pin [ 2..5 ] = [ IN0..IN3 ];  /*  Input Array  */

/**  Outputs  **/

Pin [ 12..19 ] = OUT0..OUT7 ]; /* Lower 8 output array */

/** Declarations and Intermediate Variable Definitions **/

FIELD INPUT = [ IN 3..IN0 ];
FIELD OUTPUT = [ OUT7..OUT0 ];

/**  Logic Equations  **/

TABLE INPUT => OUTPUT {
        0 => 01;        1 => 02;        2 => 04;
        3 => 08;        4 => 10;        5 => 20;
        6 => 40;        7 => 80;
        'b'1XXX => 00;
        }

/** Repeat same type of template for the second half of the decoder **/
```

Figure 6.K17 Problem 6.21 (solution) Template 1

Chapter 7

7.1

a. Resolution—what part of the whole is one step

b. Accuracy—how close to the actual value are the results

c. Precision—repeatability

7.3 $V_A = 3.619$ V

7.5 $R_f = 533\ \Omega$

7.7 0.7874%

7.9 29.53 mV

7.11 The divisor used calculating V_A for a weighted resistor ladder is $2^n - 1$ and R/2R is 2^n. R_{TH} for a weighted resistor ladder is found by taking the parallel resistances of all the resistors in a weighted resistor ladder. For an R/2R network, R_{TH} is the value of R.

R/2R ladders are preferable because they use only two different resistor values and have a better resolution as a result.

7.13

a. 5.90625 V

b. 1.304 KΩ

7.15 A monotonicity curve is a step voltage waveform for a DAC. It is generated by applying a free-running up counter to the binary inputs and observing V_{OUT} using an oscilloscope. It shows the performance of the DAC and can be used to detect problems or inaccurate results.

7.17 $\overline{CS}$, $\overline{WR_1}$, $\overline{WR_2}$ uses ILE and $\overline{XFR}$ as strobes for double buffering.

7.19 0.3922%

7.21 R/2R

7.23

a. Maximum conversion time—the longest time it takes an ADC to convert an analog voltage to an equivalent binary value.

b. Average conversion time is ½ maximum conversion time.

7.25 The advantage of a flash converter is speed. The disadvantages are a large amount of circuitry, which results in poor resolution and a need for an encoder.

7.27 99.3 mS

7.29 Tracking up and down analog voltages and longer conversion times compared to flash type.

7.31 A successive approximations converter sets each flip-flop in turn and then compares V_{REF} to V_{IN}. The circuit then resets the flip-flop if V_{REF} is higher or leaves it set if V_{REF} is lower or the same. This process is repeated for each flip-flop in the converter until the last one is set and tested.

7.33 24.35 μS

7.35 Maximum conversion time for a ramp converter is the total number of steps times the time period of a clock cycle while maximum conversion time for a successive approximations converter is the number of bits times the time of a clock cycle.

7.37 The channel number in binary is placed on A_2–A_0 and strobed in using ALE. The analog input on the corresponding IN line is sent to the successive approximation converter to be converted to digital on the application of a SC pulse.

7.39 Ramp—resolution and no encoder
Up/down ramp or continuous—tracking increasing and decreasing analog inputs
Successive approximation—conversion speed

7.41 Reduced noise and increased accuracy

7.43 A dual-slope ADC has a maximum conversion time that is twice as long as a counter ramp ADC and is $[2(2^n - 1)]/(n)$ times longer than a successive approximations ADC.

Chapter 8

8.1 IP, temporary register, memory data buffer, A-register, B-register, ALU, I/O register

8.3 Keep track of program fetches

8.5 COUNT

8.7 LIP

8.9 During the T_3 phase of the instruction cycle

8.11 An accumulator is a register that retains results of arithmetic operations

8.13 MOV A[<mem>], MOV A,<immed>, ADD [<mem>], ADD <immed>, ADD B, same set of SUB instructions, INCA, DECA, IN

8.15 Accumulator (A-register) and buffer (B-register).

8.17 An opcode is the binary code for a computer instruction operation.

8.19 T_0: IP —> address buffer; the contents of the instruction pointer are copied into the address buffer.
T_1: memory data —> instruction register; the contents of the memory location accessed by the address buffer are copied into the instruction register.
T_2: increment IP

8.21 An opcode fetch cycle loads retrieved data into the instruction register, whereas an operand fetch loads the temporary register.

8.23 MOV A, [<mem>]; MOV B, [<mem>]; MOV [<mem>],A; MOV [<mem>],B; ADD [<mem>]; SUB [<mem>]

8.25

T_0: IP —> address buffer	EIP, LADD
T_1: memory —> instruction register	CE.LINST
T_2: increment IP	COUNT
T_3: increment A-register	SEL = 100, EALU, LACC

8.27 The memory address register receives the address of the data. Memory is enabled onto the data bus and loaded into the buffer register.

8.29 Hexadecimal or 1-of-16 decoder

8.31 Mod 6

8.33 The IN and OUT interface is used to interface peripherals such as keyboards and display units to the BABE computer.

8.35

a. 11 μS
b. 4 μS
c. 4 μS
d. 3 μS
e. 10 μmS
f. 8 μS

8.37 The BABE computer starts by fetching the JMP opcode and then the JMP operand (5C06H). The operand value is placed in the temporary register. Lastly, the contents of the temporary register are transferred to the IP so that the next program fetch will be from location 5C06H.

8.39

	CF	SF	ZF
a.	0	1	0
b.	1	0	1
c.	1	1	0
d.	1	0	0

8.41 Figure k1a and k1b

8.43

```
          MOV A, [NUMB1]
          ADD [NUMB2]
          JC SHIFT_IT
          SHR
   SHOW   OUT
          HALT
SHIFT_IT  SHL
          JMP SHOW
```

8.45

```
1400 01 <NUMB1>
1403 11 <NUMB2>
1406 D1 0C 14
1409 29
140A 5F
140B 6F
140C 28
140D 81 0A 14
```

8.47

```
          IN
          SUB [BOTTOM]
          JC SMALL
          ADD [BOTTOM]
          SUB [TOP]
          JC GOOD
          JZ GOOD
          MOV A, [MUCH]
SHOW_IT   OUT
          HALT
SMALL     MOV A, [LESS]
          JMP SHOW_IT
GOOD      MOV A,[OKEH]
          JMP SHOW_IT
```

8.49

```
        IN
        MOV [COUNT],A
        SUB OH
        JZ MOD_0
        MOV B,[COUNT]
        MOV A,01
REPEAT  SHL
        DECB
        JNZ REPEAT
MOD_0   OUT
        HALT
```

```
                Listing of File: D:\DIGTXT\PLD_FILE\CINST_D1.PLD Page: 1

Name            INST_D1.PLD;
Partno          16V8;
Date            07/16/94;
Revision        XX;
Designer        MILLER;
Company         XXXXX;
Assembly        XXXXX;
Location        PHOENIX;

/******************************************************************************/
/*                                                                            */
/*  Figure 8.K1a Instruction Decoder Lower Nibble                             */
/*                                                                            */
/******************************************************************************/
/*  Allowable Target Device Types: GAL16V8                                    */
/******************************************************************************/

/**  Inputs  **/

Pin 2 = OPH_A;  /*  Lower Nibble LSB  */
Pin 3 = OPH_B;
Pin 4 = OPH_C;
Pin 5 = OPH_D;

/**  Outputs  **/

Pin 18 = MEMA;
Pin 17 = MEMB;
Pin 16 = IMMEDA;
Pin 15 = IMMEDB;
Pin 14 = STORA;
Pin 13 = STORB;
Pin 12 = HALT;

/** Declarations and Intermediate Variable Definitions **/

FIELD UPNIB = [ OPH_D, OPH_C, OPH_B, OPH_A ];
FIELD DECODE = [ MEMA, MEMB, IMMEDA, IMMEDB, STORA, STORB, HALT ];

/* EQUATIONS */

TABLE UPNIB => DECODE {
        1 => 40;   2 => 20;   3 => 10;   4 => 08;   5 => 04;
        6 => 02;   E => 01;
        }
```

Figure 6.K1 Problem 8.41 (solution)

```
              Listing of File: D:\DIGTXT\PLD_FILE\CINST_D2.PLD Page: 1

Name           INST_D3.PLD;
Partno         16V8;
Date           07/16/94;
Revision       XX;
Designer       MILLER;
Company        XXXXX;
Assembly       XXXXX;
Location       PHOENIX;

/******************************************************************************/
/*                                                                            */
/*  Figure 8.K1b Instruction Decoder Upper Nibble                             */
/*                                                                            */
/******************************************************************************/
/*  Allowable Target Device Types: GAL16V8                                    */
/******************************************************************************/

/**  Inputs  **/

Pin 2 = OPL_A;  /* Opcode Upper Nibble LSB */
Pin 3 = OPL_B;
Pin 4 = OPL_C;
Pin 5 = OPL_D;

/**  Outputs  **/

Pin 18 = MOV;
Pin 17 = ADD_SUB;
Pin 16 = LOGIC;
Pin 15 = I_O;
Pin 14 = JUMP;
Pin 13 = HALT;

/** Declarations and Intermediate Variable Definitions **/

FIELD LONIB = [ OPL_D, OPL_C, OPL_B, OPL_A ];
FIELD DECODE = [ MOV, ADD_SUB, LOGIC, I_O, JUMP, HALT ];

/**  Logic Equations  **/

TABLE LONIB => DECODE {
0 => 20;  1 => 10;  2 => 08;  3 => 04;  A => 02;  E => 01;
}
```

Figure **8.K1b** Instruction Decoder Upper Nibble

Chapter 9

9.1 DTL or diode-transistor logic uses a diodes to develop the logic function for the device. An inverting common emitter amplifier drives output loads. This family generally operated at a V_{CC} of 5 V and ground. TTL also operates in these ranges, but the logic function is developed using multiple base/emitter junctions. Outputs are driven by a totem pole-type arrangement for source and sink currents. TTL also has the advantage of speed and a larger fanout.

9.3 Noise margin is the degree to which a logic level can degrade and still represent its specified logic level.

9.5

a. NAND

b. $NM_H = 3.6\text{ V} - 3.2\text{ V} = 0.4\text{ V}$
$NM_L = 0.98 - 0.85\text{ V} = 0.13\text{ V}$

c. Q_1 – AND function, Q_2 – phase splitter that drives output transistors, Q_3 (source) and Q_4 (sink)

d. $FO_L = 8.1/0.6 = 13$ $FO_H = 33\text{ m}/25\ \mu = 1{,}320$
FO = 13

9.7 Open collector—has a single common emitter amplifier without a collector resistor or current path, as an output stage. This configuration is used for wire-ANDing several gates or for providing logic outputs greater than 5 V.

Totem pole—standard TTL output with a sink and source current path. Used for basic logic circuits and functions.

Three-state buffer—output can either pass logic state out or appear to be open. Used to connect numerous outputs to a single signal path.

Darlington pair—uses a Darlington configuration to drive the source current. Used where high-speed applications are required.

9.9 The wired AND can be considered to be a logic OR function if you consider all inputs and outputs as complements. Applying DeMorgan's theorem, $C = \overline{AB}$ is the same as $\overline{C} = \overline{A} + \overline{B}$.

9.11 Fastest (lowest propagation delay): 74F00
Least power dissipation: 74LS00
Best fanout: 74LS00 with 20

9.13 Absolute maximum ratings refer to specifications that should never be exceeded. Exceeding these values, such as a V_{CC} of 7.5 V, most likely will cause damage to the chip.

9.15 ECL has a high power dissipation since its circuits are continually on. ECL also has a poor fanout, since loads added to ECL outputs force output voltages to drop quickly. Logic levels for ECL, because the transistors are neither fully on or off, are nonstandard.

9.17 MOS devices dissipate less power and heat. Therefore, they have a higher packing density than TTL. Fanout is higher for MOS devices.

9.19 MOS devices are preferred over TTL in notebook computers because they consume less power, thereby giving longer battery life. They have a high packing density, which means that more logic circuits can take less room.

9.21 An open input to a TTL NAND gate prevents the input transistor, Q_1, from conducting similar to applying a logic 1 (+5 V) to the same input. An open input to a CMOS device leaves the input FET to react to any electrical noise. The high input impedance of the FET along with C_{gs} makes the input appear to "float" between sensing a logic 1 and logic 0. The output is indeterminable because of this changing input.

9.23

a. Fanout = 2.4 ma/1 μa = 2,400

b. Power dissipation = 30 μa × 15 V = 450 μa

c. $NM_H = 15 - 11 = 4\text{ V}$ $NM_L = 4 - 0.05 = 3.95\text{ V}$

9.25 Today's high-speed CMOS is rated as being TTL compatible. This simply means that it uses the same logic levels and can drive one TTL input dependably.

Index

Credits

We are grateful for permission to reprint the following:

Materials for Appendix B and elsewhere in the text, from the *CUPL User's Guide* (91-10128-5) REV 3.0. Permission is granted by Logical Devices, Inc., 692 South Military Trail, Deerfield Beach, FL 33442.

Materials for Appendix A and elsewhere in the text, graphic representations and descriptions of the GAL16V8, DAC0830, and ADC0808 devices, from *Programmable Logic Devices and Data Acquisition Linear Devices Data Books* (copyright 1989). Permission granted by National Semiconductor, International Communications, 1090 Kifer Road, Mail Stop 16-300, Sunnyvale, CA 94086-3737.

In Appendix A, the *Signetics Logic TTL Data Manual,* revised edition copyright 1992. Permission granted by Phillips Semiconductors.

CUPL Software
Logical Devices, Inc.
130 Capital Drive # A
Golden, CO 80401

Phone: 303-279-6868
Fax: 303-279-6869

WEB SITE: http://www.logicaldevices.com
E-Mail: logdev@henge.com

Call for upgrades to:

CUPL Total Designer
VHDL
PLDmaster
PALExpert